BASIC THEORY OF

FRACTIONAL DIFFERENTIAL EQUATIONS

Third Edition

BASIC THEORY OF FRACTIONAL DIFFERENTIAL EQUATIONS

Third Edition

Yong Zhou
Macau University of Science and Technology, China
Xiangtan University, China

World Scientific

NEW JERSEY • LONDON • SINGAPORE • BEIJING • SHANGHAI • HONG KONG • TAIPEI • CHENNAI • TOKYO

Published by

World Scientific Publishing Co. Pte. Ltd.
5 Toh Tuck Link, Singapore 596224
USA office: 27 Warren Street, Suite 401-402, Hackensack, NJ 07601
UK office: 57 Shelton Street, Covent Garden, London WC2H 9HE

British Library Cataloguing-in-Publication Data
A catalogue record for this book is available from the British Library.

BASIC THEORY OF FRACTIONAL DIFFERENTIAL EQUATIONS
Third Edition

ISBN 978-981-127-168-7 (hardcover)
ISBN 978-981-127-169-4 (ebook for institutions)
ISBN 978-981-127-170-0 (ebook for individuals)

For any available supplementary material, please visit
https://www.worldscientific.com/worldscibooks/10.1142/13289#t=suppl

Preface to the Third Edition

The concept of fractional derivative dates back to a famous correspondence between G.A. de L'Hospital and G.W. Leibniz, in 1695. Many mathematicians contributed to the development of this branch of mathematical analysis with the pioneer work owing to L. Euler, J.L. Lagrange, P.S. Laplace, J.B.J. Fourier, N.H. Abel, J. Liouville, B. Riemann, H.L. Greer, H. Holmgren, A.K. Grünwald, A.V. Letnikov, N.Ya. Sonin, H. Laurent, P.A. Nekrassov, A. Krug, J. Hadamard, O. Heaviside, S. Pincherle, G.H. Hardy, J.E. Littlewood, H. Weyl, P. Lévy, A. Marchaud, H.T. Davis, A. Zygmund, E.R. Love, A. Erdélyi, H. Kober, D.V. Widder, M. Riesz and W. Feller. In the past sixty years, fractional calculus had played a very important role in various fields such as physics, chemistry, mechanics, electricity, biology, economics, control theory, signal and image processing, biophysics, blood flow phenomena, aerodynamics and fitting of experimental data.

In the last decade, fractional calculus has been recognized as one of the best tools to describe long-memory processes. Such models are interesting for engineers and physicists but also for pure mathematicians. The most important among such models are those described by differential equations containing fractional-order derivatives. Their evolutions behave in a much more complex way than in the classical integer-order case and the study of the corresponding theory is a hugely demanding task. Although some results of qualitative analysis for fractional differential equations can be similarly obtained, many classical methods are hardly applicable directly to fractional differential equations. New theories and methods are thus required to be specifically developed, whose investigation becomes more challenging. Comparing with classical theory of differential equations, the researches on the theory of fractional differential equations are only on their initial stage of development.

This monograph is devoted to a rapidly developing area of the research for the qualitative theory of fractional differential equations. In particular, we are interested in the basic theory of fractional differential equations. The development of such basic theory should be the starting point for further research concerning the dynamics, control, numerical analysis and applications of fractional differential equations.

The third edition of this book is divided into eight chapters. Chapter 1 introduces preliminary facts from fractional calculus, nonlinear analysis and semigroup theory. In Chapter 2, we present a unified framework to investigate the basic existence theory for discontinuous fractional functional differential equations with bounded delay, unbounded delay and infinite delay, respectively. Chapter 3 is devoted to the study of fractional differential equations in Banach spaces via measure of noncompactness method, topological degree method and Picard operator technique. In Chapter 4, we discuss fractional evolution equations with Riemann-Liouville fractional derivative, Caputo fractional derivative and Hilfer fractional derivative, respectively. Chapter 5 deals with initial boundary value problems of fractional impulsive differential equations including Langevin equations and evolution equations. In Chapter 6, by using critical point theory, we study existence and multiplicity of solutions for boundary value problems to fractional differential equations. In Chapter 7, we investigate the existence and multiplicity of homoclinic solutions for fractional Hamiltonian systems via variational methods. And in the last Chapter, we introduce the recent works on fractional partial differential equations including fractional Navier-Stokes equations, fractional Rayleigh-Stokes equations, fractional Euler-Lagrange equations, fractional diffusion equations and wave equations.

The book is self-contained and unified in presentation, and it provides the necessary background material required to go further into the subject and explore the rich research literature. Each chapter concludes with a section devoted to notes and bibliographical remarks and all abstract results are illustrated by examples. The tools used include many classical and modern nonlinear analysis methods. This book is useful for researchers working in the areas of pure and applied mathematics, and related disciplines. It may also be used as a valuable source for graduate level advanced courses on fractional differential equations.

I wish to express my appreciation to Professors B. Ahmad, D. Baleanu, M. Benchohra, L. Bourdin, M. Fečkan, V. Kiryakova, J.J. Nieto, H.R. Sun, J.J. Trujillo, J.R. Wang, M. Yamamoto and X.F. Zhou for their support. I also thank the editorial assistance of World Scientific Publishing Co., especially Ms. L.F. Kwong and Dr. S.C. Lim.

This monograph was first published in 2014. I acknowledge with gratitude the support of National Natural Science Foundation of China (12071396, 11671339, 11271309, 10971173, 10371103) and the Macau Science and Technology Development Fund (Grant No. 0092/2022/A).

Yong Zhou
Macau University of Science and Technology
Xiangtan University

About the Author

Yong Zhou is a Full Professor at School of Mathematics and Computational Science, Xiangtan University since 2000. He is also a Distinguished Guest Professor at Macau University of Science and Technology since 2018. His research fields include fractional differential equations, functional differential equations, evolution equations and inclusions, control theory. Zhou has published six monographs with Springer, Elsevier, De Gruyter, World Scientific and Science Press, respectively, and more than three hundred research papers in international journals including Mathematische Annalen, Journal of Functional Analysis, Inverse Problems, Nonlinearity, Proceedings of the Royal Society of Edinburgh A: Mathematics, Zeitschrift für Angewandte Mathematik und Physik, Discrete and Continuous Dynamical System, International Journal of Bifurcation and Chaos, Bulletin des Sciences Mathématiques, Comptes rendus Mathematique, and so on. He was included in the Highly Cited Researchers list from Thompson Reuters (2014) and Clarivate

Analytics (2015–2021). Zhou has undertaken five projects from National Natural Science Foundation of China, and two projects from the Macau Science and Technology Development Fund. He won the second prize of Chinese University Natural Science Award in 2000, and the second prize of Natural Science Award of Hunan Province, China in 2017 and 2021. He was the Editor-in-Chief of International Journal of Dynamical Systems and Differential Equations from 2007 to 2011. In addition, he had worked as an Associate Editor for IEEE Transactions on Fuzzy Systems, Journal of Applied Mathematics & Computing, Mathematical Inequalities & Applications, and an Editorial Board Member of Fractional Calculus and Applied Analysis.

Contents

Preface to the Third Edition v

About the Author vii

1. Preliminaries 1

1.1 Introduction 1
1.2 Some Notations, Concepts and Lemmas 1
1.3 Fractional Calculus 4
1.3.1 Definitions 5
1.3.2 Properties 9
1.3.3 Mittag-Leffler Functions 12
1.4 Some Results from Nonlinear Analysis 13
1.4.1 Laplace and Fourier Transforms 13
1.4.2 Sobolev Spaces 14
1.4.3 Measure of Noncompactness 15
1.4.4 Topological Degree 17
1.4.5 Picard Operator 18
1.4.6 Fixed Point Theorems 19
1.4.7 Critical Point Theorems 21
1.5 Semigroups 24
1.5.1 C_0-Semigroup 24
1.5.2 Almost Sectorial Operators 25

2. Fractional Functional Differential Equations 29

2.1 Introduction 29
2.2 Neutral Equations with Bounded Delay 30
2.2.1 Introduction 30
2.2.2 Existence and Uniqueness 30
2.2.3 Extremal Solutions 35
2.3 p-Type Neutral Equations 44
2.3.1 Introduction 44

2.3.2 Existence and Uniqueness 46
2.3.3 Continuous Dependence 57
2.4 Neutral Equations with Infinite Delay 60
2.4.1 Introduction 60
2.4.2 Existence and Uniqueness 62
2.4.3 Continuation of Solutions 69
2.5 Iterative Functional Differential Equations 73
2.5.1 Introduction 73
2.5.2 Existence 74
2.5.3 Data Dependence 80
2.5.4 Examples and General Cases 81
2.6 Oscillations and Nonoscillations 88
2.6.1 Introduction 88
2.6.2 Preliminaries 88
2.6.3 Oscillation of Neutral Differential Systems 90
2.6.4 Existence of Nonoscillatory Solutions 98
2.6.5 Fractional Partial Functional Differential Equations 108
2.7 Notes and Remarks 119

3. Fractional Ordinary Differential Equations in Banach Spaces 121

3.1 Introduction 121
3.2 Cauchy Problems via Measure of Noncompactness Method 123
3.2.1 Introduction 123
3.2.2 Existence 123
3.3 Cauchy Problems via Topological Degree Method 132
3.3.1 Introduction 132
3.3.2 Qualitative Analysis 132
3.4 Cauchy Problems via Picard Operators Technique 136
3.4.1 Introduction 136
3.4.2 Results via Picard Operators 136
3.4.3 Results via Weakly Picard Operators 143
3.5 Notes and Remarks 147

4. Fractional Abstract Evolution Equations 149

4.1 Introduction 149
4.2 Evolution Equations with Riemann-Liouville Derivative 150
4.2.1 Introduction 150
4.2.2 Definition of Mild Solutions 151
4.2.3 Preliminary Lemmas 154
4.2.4 Compact Semigroup Case 160
4.2.5 Noncompact Semigroup Case 165
4.3 Evolution Equations with Caputo Derivative 168

4.3.1 Introduction 168
4.3.2 Definition of Mild Solutions 168
4.3.3 Preliminary Lemmas 170
4.3.4 Compact Semigroup Case 174
4.3.5 Noncompact Semigroup Case 177
4.4 Nonlocal Problems for Evolution Equations 179
4.4.1 Introduction 179
4.4.2 Definition of Mild Solutions 179
4.4.3 Existence 181
4.5 Optimal Controls of Fractional Evolution Equations 187
4.5.1 Introduction 187
4.5.2 Preliminaries 187
4.5.3 Existence of α-Mild Solutions 190
4.5.4 Existence of Fractional Optimal Controls 196
4.6 Abstract Cauchy Problems with Almost Sectorial Operators 200
4.6.1 Introduction 200
4.6.2 Properties of Operators 205
4.6.3 Linear Problems 211
4.6.4 Nonlinear Problems 216
4.6.5 Applications 224
4.7 Evolution Equations with Hilfer Derivative 226
4.7.1 Introduction 226
4.7.2 Preliminaries 227
4.7.3 Some Lemmas 229
4.7.4 Existence Results 234
4.8 Infinite Interval Problems with Hilfer Derivative 237
4.8.1 Introduction 237
4.8.2 Preliminaries 238
4.8.3 Lemmas 242
4.8.4 Existence on Infinite Interval 248
4.9 Notes and Remarks 252

5. Fractional Impulsive Differential Equations 253

5.1 Introduction 253
5.2 Impulsive Initial Value Problems 254
5.2.1 Introduction 254
5.2.2 Formula of Solutions 254
5.2.3 Existence 257
5.3 Impulsive Boundary Value Problems 262
5.3.1 Introduction 262
5.3.2 Formula of Solutions 262
5.3.3 Existence 265

5.4 Impulsive Langevin Equations 269
5.4.1 Introduction 269
5.4.2 Formula of Solutions 270
5.4.3 Existence 278
5.5 Impulsive Evolution Equations 285
5.5.1 Introduction 285
5.5.2 Cauchy Problems 286
5.5.3 Nonlocal Problems 288
5.6 Notes and Remarks 294

6. Fractional Boundary Value Problems 295

6.1 Introduction 295
6.2 Solutions for BVP with Left and Right Fractional Integrals 295
6.2.1 Introduction 295
6.2.2 Fractional Derivative Space 298
6.2.3 Variational Structure 303
6.2.4 Existence under Ambrosetti-Rabinowitz Condition 310
6.2.5 Superquadratic Case 315
6.2.6 Asymptotically Quadratic Case 319
6.3 Multiple Solutions for BVP with Parameters 322
6.3.1 Introduction 322
6.3.2 Existence 323
6.4 Infinite Solutions for BVP with Left and Right Fractional Integrals 333
6.4.1 Introduction 333
6.4.2 Existence 334
6.5 Solutions for BVP with Left and Right Fractional Derivatives . . . 343
6.5.1 Introduction 343
6.5.2 Variational Structure 344
6.5.3 Existence of Weak Solutions 347
6.5.4 Existence of Solutions 351
6.6 Notes and Remarks 355

7. Fractional Hamiltonian Systems 357

7.1 Introduction 357
7.2 Existence and Multiplicity of Homoclinic Solutions (I) 357
7.2.1 Fractional Derivative Space 357
7.2.2 Some Lemmas 358
7.2.3 Existence of Homoclinic Solutions 365
7.3 Existence and Multiplicity of Homoclinic Solutions (II) 379
7.3.1 Introduction 379
7.3.2 Some Lemmas 379
7.3.3 Existence and Multiplicity 382

7.4 Notes and Remarks 385

8. Fractional Partial Differential Equations 387

8.1 Introduction 387
8.2 Fractional Navier-Stokes Equations 388
8.2.1 Introduction 388
8.2.2 Preliminaries 389
8.2.3 Global Existence 392
8.2.4 Local Existence 399
8.2.5 Regularity 403
8.3 Fractional Rayleigh-Stokes Equations 411
8.3.1 Introduction 411
8.3.2 Preliminaries 412
8.3.3 Globally Lispchitz Source Term 415
8.3.4 Locally Lispchitz Source Term 423
8.4 Fractional Euler-Lagrange Equations 431
8.4.1 Introduction 431
8.4.2 Functional Spaces 433
8.4.3 Variational Structure 437
8.4.4 Existence of Weak Solution 439
8.5 Fractional Diffusion Equations 444
8.5.1 Introduction 444
8.5.2 Regularity and Unique Existence 445
8.6 Fractional Wave Equations 451
8.6.1 Introduction 451
8.6.2 Preliminaries 453
8.6.3 Approximation Solution 456
8.6.4 Energy Estimates 462
8.6.5 Well-Posedness and Regularity 467
8.7 Notes and Remarks 472

Bibliography 473

Index 497

Chapter 1

Preliminaries

1.1 Introduction

In this chapter, we introduce some notations and basic facts on fractional calculus, special functions, semigroups, Laplace and Fourier transforms, measure of non-compactness, fixed point theorems and critical point theorems which are needed throughout this book.

1.2 Some Notations, Concepts and Lemmas

As usual $\mathbb{N}$ denotes the set of positive integer numbers and $\mathbb{N}_0$ the set of nonnegative integer numbers. $\mathbb{R}$ denotes the real numbers, $\mathbb{R}_+$ denotes the set of nonnegative reals and $\mathbb{R}^+$ the set of positive reals. Let $\mathbb{C}$ be the set of complex numbers.

We recall that a vector space X equipped with a norm $|\cdot|$ is called a normed vector space. A subset E of a normed vector space X is said to be bounded if there exists a number K such that $|x| \leq K$ for all $x \in E$. A subset E of a normed vector space X is called convex if for any $x, y \in E$, $ax + (1-a)y \in E$ for all $a \in [0,1]$.

A sequence $\{x_n\}$ in a normed vector space X is said to converge to the vector x in X if and only if the sequence $\{|x_n - x|\}$ converges to zero as $n \to \infty$. A sequence $\{x_n\}$ in a normed vector space X is called a Cauchy sequence if for every $\varepsilon > 0$ there exists an $N = N(\varepsilon)$ such that for all $n, m \geq N(\varepsilon)$, $|x_n - x_m| < \varepsilon$. Clearly a convergent sequence is also a Cauchy sequence, but the converse may not be true. A space X where every Cauchy sequence of elements of X converges to an element of X is called a complete space. A complete normed vector space is said to be a Banach space.

Let E be a subset of a Banach space X. A point $x \in X$ is said to be a limit point of E if there exists a sequence of vectors in E which converges to x. We say a subset E is closed if E contains all of its limit points. The union of E and its limit points is called the closure of E and will be denoted by $\bar{E}$. Let E, F be normed vector spaces, and E be a subset of X. An operator $\mathscr{T} : E \to F$ is continuous at a point $x \in E$ if and only if for any $\varepsilon > 0$ there is a $\delta > 0$ such that $|\mathscr{T}x - \mathscr{T}y| < \varepsilon$ for all $y \in E$ with $|x - y| < \delta$. Further, $\mathscr{T}$ is continuous on E, or simply continuous, if it is continuous at all points of E.

We say that a subset E of a Banach space X is compact if every sequence of vectors in E contains a subsequence which converges to a vector in E. We say that E is relatively compact in X if every sequence of vectors in E contains a subsequence which converges to a vector in X, i.e., E is relatively compact in X if $\bar{E}$ is compact.

Let $J = [a, b]$ $(-\infty < a < b < \infty)$ be a finite interval of $\mathbb{R}$. We assume that X is a Banach space with the norm $|\cdot|$. Denote $C(J, X)$ be the Banach space of all continuous functions from J into X with the norm

$$\|x\| = \sup_{t\in J} |x(t)|,$$

where $x \in C(J, X)$. $C^n(J, X)$ $(n \in \mathbb{N}_0)$ denotes the set of mappings, which have continuous derivatives up to order n on J, $AC(J, X)$ is the space of functions which are absolutely continuous on J and $AC^n(J, X)$ $(n \in \mathbb{N})$ is the space of functions f such that $f \in C^{n-1}(J, X)$ and $f^{(n-1)} \in AC(J, X)$. In particular, $AC^1(J, X) = AC(J, X)$. We also introduce the set of functions $PC(J, X) = \big\{x : J \to X \mid x$ is continuous at $t \in J\backslash\{t_1, t_2, ..., t_\delta\}$, and x is continuous from left and has right hand limits at $t \in \{t_1, t_2, ..., t_\delta\}\big\}$ endowed with the norm

$$\|x\|_{PC} = \max\left\{\sup_{t\in J} |x(t+0)|,\ \sup_{t\in J} |x(t-0)|\right\},$$

it is easy to see $(PC(J, X), \|\cdot\|_{PC})$ is a Banach space. Denote $PC^1(J, \mathbb{R}) \equiv \{x \in PC(J, \mathbb{R}) \mid x' \in PC(J, \mathbb{R})\}$. Set $\|x\|_{PC^1} = \|x\|_{PC} + \|x'\|_{PC}$. It can be seen that endowed with the norm $\|\cdot\|_{PC^1}$, $PC^1(J, \mathbb{R})$ is also a Banach space.

Let $1 \le p \le \infty$. $L^p(J, X)$ denotes the Banach space of all measurable functions $f : J \to X$. $L^p(J, X)$ is normed by

$$\|f\|_{L^pJ} = \begin{cases} \left(\displaystyle\int_J |f(t)|^p dt\right)^{\frac{1}{p}}, & 1 \le p < \infty, \\ \displaystyle\inf_{\mu(\bar{J})=0} \left\{\sup_{t\in J\backslash\bar{J}} |f(t)|\right\}, & p = \infty. \end{cases}$$

In particular, $L^1(J, X)$ is the Banach space of measurable functions $f : J \to X$ with the norm

$$\|f\|_{LJ} = \int_J |f(t)|dt,$$

and $L^\infty(J, X)$ is the Banach space of measurable functions $f : J \to X$ which are bounded, equipped with the norm

$$\|f\|_{L^\infty J} = \inf\{c > 0 \mid |f(t)| \le c, \text{ a.e. } t \in J\}.$$

Lemma 1.1. *(Hölder inequality) Assume that* $p, q \ge 1$, *and* $\frac{1}{p} + \frac{1}{q} = 1$. *If* $f \in L^p(J, X), g \in L^q(J, X)$, *then for* $1 \le p \le \infty$, $fg \in L^1(J, X)$ *and*

$$\|fg\|_{LJ} \le \|f\|_{L^pJ}\|g\|_{L^qJ}.$$

A family F in $C(J,X)$ is called uniformly bounded if there exists a positive constant K such that $|f(t)| \leq K$ for all $t \in J$ and all $f \in F$. Further, F is called equicontinuous, if for every $\varepsilon > 0$ there exists a $\delta = \delta(\varepsilon) > 0$ such that $|f(t_1) - f(t_2)| < \varepsilon$ for all $t_1, t_2 \in J$ with $|t_1 - t_2| < \delta$ and all $f \in F$.

Lemma 1.2. *(Arzela-Ascoli theorem) If a family $F = \{f(t)\}$ in $C(J,\mathbb{R})$ is uniformly bounded and equicontinuous on J, then F has a uniformly convergent subsequence $\{f_n(t)\}_{n=1}^{\infty}$. If a family $F = \{f(t)\}$ in $C(J,X)$ is uniformly bounded and equicontinuous on J, and for any $t^* \in J$, $\{f(t^*)\}$ is relatively compact, then F has a uniformly convergent subsequence $\{f_n(t)\}_{n=1}^{\infty}$.*

Arzela-Ascoli theorem is the key to the following result: a subset F in $C(J,\mathbb{R})$ is relatively compact if and only if it is uniformly bounded and equicontinuous on J.

Lemma 1.3. *(PC-type Arzela-Ascoli theorem) Let X be a Banach space and $\mathcal{W} \subset PC(J,X)$. If the following conditions are satisfied:*

(i) *$\mathcal{W}$ is an equicontinuous function subset of $PC(J,X)$;*
(ii) *$\mathcal{W}$ is equicontinuous in (t_k, t_{k+1}), $k = 0, 1, 2, \ldots, m$, where $t_0 = 0$, $t_{m+1} = T$;*
(iii) *$\mathcal{W}(t) = \{u(t) \mid u \in \mathcal{W}, t \in J\backslash\{t_1, \ldots, t_m\}\}$, $\mathcal{W}(t_k^+) = \{u(t_k^+) \mid u \in \mathcal{W}\}$ and $\mathcal{W}(t_k^-) \equiv \{u(t_k^-) \mid u \in \mathcal{W}\}$ is a relatively compact subsets of X.*

Then $\mathcal{W}$ is a relatively compact subset of $PC(J,X)$.

Lemma 1.4. *(The generalized Arzela-Ascoli theorem) The set $\Lambda \subset C^1([0,\infty),X)$ is relatively compact if and only if the following conditions hold:*

(a) *for any $h > 0$, the set $V = \{v : \ v(t) = x(t)/(1+t), x \in \Lambda\}$ is equicontinuous on $[0,h]$;*
(b) $\lim_{t\to\infty} |x(t)|/(1+t) = 0$ *uniformly for $x \in \Lambda$;*
(c) *for any $t \in [0,\infty)$, $V(t) = \{v(t) : v(t) = x(t)/(1+t), x \in \Lambda\}$ is relatively compact in X.*

Lemma 1.5. *(Lebesgue dominated convergence theorem) Let E be a measurable set and let $\{f_n\}$ be a sequence of measurable functions such that $\lim_{n\to\infty} f_n(x) = f(x)$ a.e. in E, and for every $n \in \mathbb{N}$, $|f_n(x)| \leq g(x)$ a.e. in E, where g is integrable on E. Then*

$$\lim_{n\to\infty} \int_E f_n(x)dx = \int_E f(x)dx.$$

Finally, we state Bochner theorem.

Lemma 1.6. *(Bochner theorem) A measurable function $f : (a,b) \to X$ is Bochner integrable if $|f|$ is Lebesgue integrable.*

1.3 Fractional Calculus

The gamma function $\Gamma(z)$ is defined by

$$\Gamma(z) = \int_0^\infty t^{z-1}e^{-t}dt \quad (\mathrm{Re}(z) > 0),$$

where $t^{z-1} = e^{(z-1)\log(t)}$. This integral is convergent for all complex $z \in \mathbb{C}$ $(\mathrm{Re}(z) > 0)$.

For this function the reduction formula

$$\Gamma(z+1) = z\Gamma(z) \qquad (\mathrm{Re}(z) > 0)$$

holds. In particular, if $z = n \in \mathbb{N}_0$, then

$$\Gamma(n+1) = n! \qquad (n \in \mathbb{N}_0)$$

with (as usual) $0! = 1$.

Let us consider some of the starting points for a discussion of fractional calculus. One development begins with a generalization of repeated integration. Thus if f is locally integrable on (c, ∞), then the n-fold iterated integral is given by

$$\begin{aligned} {}_cD_t^{-n}f(t) &= \int_c^t ds_1 \int_c^{s_1} ds_2 \cdots \int_c^{s_{n-1}} f(s_n)ds_n \\ &= \frac{1}{(n-1)!}\int_c^t (t-s)^{n-1}f(s)ds \end{aligned}$$

for almost all t with $-\infty \le c < t < \infty$ and $n \in \mathbb{N}$. Writing $(n-1)! = \Gamma(n)$, an immediate generalization is the integral of f of fractional order $\alpha > 0$,

$${}_cD_t^{-\alpha}f(t) = \frac{1}{\Gamma(\alpha)}\int_c^t (t-s)^{\alpha-1}f(s)ds \quad \text{(left hand)}$$

and similarly for $-\infty < t < d \le \infty$

$${}_tD_d^{-\alpha}f(t) = \frac{1}{\Gamma(\alpha)}\int_t^d (s-t)^{\alpha-1}f(s)ds \quad \text{(right hand)}$$

both being defined for suitable f.

A number of definitions for the fractional derivative have emerged over the years, we refer the reader to Diethelm, 2010; Hilfer, 2006; Kilbas, Srivastava and Trujillo, 2006; Miller and Ross, 1993; Podlubny, 1999. In this book, we restrict our attention to the use of the Riemann-Liouville, Caputo and Hilfer fractional derivatives. In this section, we introduce some basic definitions and properties of the fractional integrals and fractional derivatives which are used further in this book. The materials in this section are taken from Kilbas, Srivastava and Trujillo, 2006.

1.3.1 *Definitions*

Definition 1.1. (Left and right Riemann-Liouville fractional integrals) Let $J = [a,b]$ $(-\infty < a < b < \infty)$ be a finite interval of $\mathbb{R}$. The left and right Riemann-Liouville fractional integrals ${}_aD_t^{-\alpha}f(t)$ and ${}_tD_b^{-\alpha}f(t)$ of order $\alpha \in \mathbb{R}^+$, are defined by

$$ {}_aD_t^{-\alpha}f(t) = \frac{1}{\Gamma(\alpha)}\int_a^t (t-s)^{\alpha-1}f(s)ds, \quad t > a, \quad \alpha > 0 \tag{1.1} $$

and

$$ {}_tD_b^{-\alpha}f(t) = \frac{1}{\Gamma(\alpha)}\int_t^b (s-t)^{\alpha-1}f(s)ds, \quad t < b, \quad \alpha > 0, \tag{1.2} $$

respectively, provided the right-hand sides are pointwise defined on $[a,b]$. When $\alpha = n \in \mathbb{N}$, the definitions (1.1) and (1.2) coincide with the n-th integrals of the form

$$ {}_aD_t^{-n}f(t) = \frac{1}{(n-1)!}\int_a^t (t-s)^{n-1}f(s)ds $$

and

$$ {}_tD_b^{-n}f(t) = \frac{1}{(n-1)!}\int_t^b (s-t)^{n-1}f(s)ds. $$

Definition 1.2. (Left and right Riemann-Liouville fractional derivatives) The left and right Riemann-Liouville fractional derivatives ${}_aD_t^{\alpha}f(t)$ and ${}_tD_b^{\alpha}f(t)$ of order $\alpha \in \mathbb{R}_+$, are defined by

$$ \begin{aligned} {}_aD_t^{\alpha}f(t) &= \frac{d^n}{dt^n}{}_aD_t^{-(n-\alpha)}f(t) \\ &= \frac{1}{\Gamma(n-\alpha)}\frac{d^n}{dt^n}\left(\int_a^t (t-s)^{n-\alpha-1}f(s)ds\right), \quad t > a \end{aligned} $$

and

$$ \begin{aligned} {}_tD_b^{\alpha}f(t) &= (-1)^n\frac{d^n}{dt^n}{}_tD_b^{-(n-\alpha)}f(t) \\ &= \frac{1}{\Gamma(n-\alpha)}(-1)^n\frac{d^n}{dt^n}\left(\int_t^b (s-t)^{n-\alpha-1}f(s)ds\right), \quad t < b, \end{aligned} $$

respectively, where $n = [\alpha]+1$, $[\alpha]$ means the integer part of α. In particular, when $\alpha = n \in \mathbb{N}_0$, then

$$ {}_aD_t^{0}f(t) = {}_tD_b^{0}f(t) = f(t), $$

$$ {}_aD_t^{n}f(t) = f^{(n)}(t) \quad \text{and} \quad {}_tD_b^{n}f(t) = (-1)^n f^{(n)}(t), $$

where $f^{(n)}(t)$ is the usual derivative of $f(t)$ of order n. If $0 < \alpha < 1$, then

$$ {}_aD_t^{\alpha}f(t) = \frac{1}{\Gamma(1-\alpha)}\frac{d}{dt}\left(\int_a^t (t-s)^{-\alpha}f(s)ds\right), \quad t > a $$

and

$$ {}_tD_b^{\alpha}f(t) = -\frac{1}{\Gamma(1-\alpha)}\frac{d}{dt}\left(\int_t^b (s-t)^{-\alpha}f(s)ds\right), \quad t < b. $$

Remark 1.1. If $f \in C([a,b],\mathbb{R}^N)$, it is obvious that Riemann-Liouville fractional integral of order $\alpha > 0$ exists on $[a,b]$. On the other hand, following Lemma 2.2 in Kilbas, Srivastava and Trujillo, 2006, we know that the Riemann-Liouville fractional derivative of order $\alpha \in [n-1,n)$ exists almost everywhere on $[a,b]$ if $f \in AC^n([a,b],\mathbb{R}^N)$.

The left and right Caputo fractional derivatives are defined via above Riemann-Liouville fractional derivatives.

Definition 1.3. (Left and right Caputo fractional derivatives) The left and right Caputo fractional derivatives ${}_a^C D_t^\alpha f(t)$ and ${}_t^C D_b^\alpha f(t)$ of order $\alpha \in \mathbb{R}_+$ are defined by

$$ {}_a^C D_t^\alpha f(t) = {}_a D_t^\alpha \left(f(t) - \sum_{k=0}^{n-1} \frac{f^{(k)}(a)}{k!}(t-a)^k \right) $$

and

$$ {}_t^C D_b^\alpha f(t) = {}_t D_b^\alpha \left(f(t) - \sum_{k=0}^{n-1} \frac{f^{(k)}(b)}{k!}(b-t)^k \right), $$

respectively, where

$$ n = [\alpha] + 1 \text{ for } \alpha \notin \mathbb{N}_0; \; n = \alpha \text{ for } \alpha \in \mathbb{N}_0. \tag{1.3} $$

In particular, when $0 < \alpha < 1$, then

$$ {}_a^C D_t^\alpha f(t) = {}_a D_t^\alpha (f(t) - f(a)) $$

and

$$ {}_t^C D_b^\alpha f(t) = {}_t D_b^\alpha (f(t) - f(b)). $$

The Riemann-Liouville fractional derivative and the Caputo fractional derivative are connected with each other by the following relations.

Proposition 1.1.

(i) *If $\alpha \notin \mathbb{N}_0$ and $f(t)$ is a function for which the Caputo fractional derivatives ${}_a^C D_t^\alpha f(t)$ and ${}_t^C D_b^\alpha f(t)$ of order $\alpha \in \mathbb{R}^+$ exist together with the Riemann-Liouville fractional derivatives ${}_a D_t^\alpha f(t)$ and ${}_t D_b^\alpha f(t)$, then*

$$ {}_a^C D_t^\alpha f(t) = {}_a D_t^\alpha f(t) - \sum_{k=0}^{n-1} \frac{f^{(k)}(a)}{\Gamma(k-\alpha+1)}(t-a)^{k-\alpha} $$

and

$$ {}_t^C D_b^\alpha f(t) = {}_t D_b^\alpha f(t) - \sum_{k=0}^{n-1} \frac{f^{(k)}(b)}{\Gamma(k-\alpha+1)}(b-t)^{k-\alpha}, $$

where $n = [\alpha] + 1$. In particular, when $0 < \alpha < 1$, we have

$$ {}_a^C D_t^\alpha f(t) = {}_a D_t^\alpha f(t) - \frac{f(a)}{\Gamma(1-\alpha)}(t-a)^{-\alpha} $$

and

$$ {}_t^C D_b^\alpha f(t) = {}_t D_b^\alpha f(t) - \frac{f(b)}{\Gamma(1-\alpha)}(b-t)^{-\alpha}. $$

(ii) *If $\alpha = n \in \mathbb{N}_0$ and the usual derivative $f^{(n)}(t)$ of order n exists, then ${}_a^CD_t^n f(t)$ and ${}_t^CD_b^n f(t)$ are represented by*

$$ {}_a^CD_t^n f(t) = f^{(n)}(t) \text{ and } {}_t^CD_b^n f(t) = (-1)^n f^{(n)}(t). \quad (1.4)$$

Proposition 1.2. *Let $\alpha \in \mathbb{R}_+$ and let n be given by (1.3). If $f \in AC^n([a,b],\mathbb{R}^N)$, then the Caputo fractional derivatives ${}_a^CD_t^\alpha f(t)$ and ${}_t^CD_b^\alpha f(t)$ exist almost everywhere on $[a,b]$.*

(i) *If $\alpha \notin \mathbb{N}_0$, ${}_a^CD_t^\alpha f(t)$ and ${}_t^CD_b^\alpha f(t)$ are represented by*

$$ {}_a^CD_t^\alpha f(t) = \frac{1}{\Gamma(n-\alpha)}\left(\int_a^t (t-s)^{n-\alpha-1} f^{(n)}(s)ds\right)$$

and

$$ {}_t^CD_b^\alpha f(t) = \frac{(-1)^n}{\Gamma(n-\alpha)}\left(\int_t^b (s-t)^{n-\alpha-1} f^{(n)}(s)ds\right),$$

respectively, where $n = [\alpha]+1$. In particular, when $0<\alpha<1$ and $f \in AC([a,b],\mathbb{R}^N)$,

$$ {}_a^CD_t^\alpha f(t) = \frac{1}{\Gamma(1-\alpha)}\left(\int_a^t (t-s)^{-\alpha} f'(s)ds\right) \quad (1.5)$$

and

$$ {}_t^CD_b^\alpha f(t) = -\frac{1}{\Gamma(1-\alpha)}\left(\int_t^b (s-t)^{-\alpha} f'(s)ds\right). \quad (1.6)$$

(ii) *If $\alpha = n \in \mathbb{N}_0$ then ${}_a^CD_t^\alpha f(t)$ and ${}_t^CD_b^\alpha f(t)$ are represented by (1.4). In particular,*

$$ {}_a^CD_t^0 f(t) = {}_t^CD_b^0 f(t) = f(t).$$

Remark 1.2. If f is an abstract function with values in Banach space X, then integrals which appear in above definitions are taken in Bochner's sense.

Definition 1.4. (Hilfer fractional derivative) The Hilfer fractional derivative ${}_a^HD_t^{\mu,\nu} f(t)$ of order $n-1<\mu<n$ and $0 \le \nu \le 1$ is defined by

$$ {}_a^HD_t^{\mu,\nu} f(t) = {}_aD_t^{-\nu(n-\mu)} \frac{d^n}{dt^n} {}_aD_t^{-(1-\nu)(n-\mu)} f(t),$$

provided the right-hand side is pointwise defined on $[a,b]$.

Remark 1.3.

(i) When $\nu = 0$ and $n-1<\mu<n$, the Hilfer fractional derivative corresponds to the classical Riemann-Liouville fractional derivative:

$$ {}_a^HD_t^{\mu,0} f(t) = \frac{d^n}{dt^n} {}_aD_t^{-(1-\mu)} f(t) = {}_aD_t^\mu f(t).$$

(ii) When $\nu = 1$, $n-1 < \mu < n$, the Hilfer fractional derivative corresponds to the classical Caputo fractional derivative:

$$ {}_a^H D_t^{\mu,1} f(t) = {}_a D_t^{-(n-\mu)} \frac{d^n}{dt^n} f(t) = {}_a^C D_t^{\mu} f(t). $$

The fractional integrals and derivatives, defined on a finite interval $[a, b]$ of $\mathbb{R}$, are naturally extended to whole axis $\mathbb{R}$.

Definition 1.5. (Left and right Liouville-Weyl fractional integrals on the real axis) The left and right Liouville-Weyl fractional integrals ${}_{-\infty}D_t^{-\alpha} f(t)$ and ${}_t D_{+\infty}^{-\alpha} f(t)$ of order $\alpha > 0$ on the whole axis $\mathbb{R}$ are defined by

$$ {}_{-\infty}D_t^{-\alpha} f(t) = \frac{1}{\Gamma(\alpha)} \int_{-\infty}^{t} (t-s)^{\alpha-1} f(s) ds \tag{1.7} $$

and

$$ {}_t D_{+\infty}^{-\alpha} f(t) = \frac{1}{\Gamma(\alpha)} \int_{t}^{\infty} (s-t)^{\alpha-1} f(s) ds, $$

respectively, where $t \in \mathbb{R}$ and $\alpha > 0$.

Definition 1.6. (Left and right Liouville-Weyl fractional derivatives on the real axis) The left and right Liouville-Weyl fractional derivatives ${}_{-\infty}D_t^{\alpha} f(t)$ and ${}_t D_{+\infty}^{\alpha} f(t)$ of order α on the whole axis $\mathbb{R}$ are defined by

$$ \begin{aligned} {}_{-\infty}D_t^{\alpha} f(t) &= \frac{d^n}{dt^n} ({}_{-\infty}D_t^{-(n-\alpha)} f(t)) \\ &= \frac{1}{\Gamma(n-\alpha)} \frac{d^n}{dt^n} \left(\int_{-\infty}^{t} (t-s)^{n-\alpha-1} f(s) ds \right) \end{aligned} $$

and

$$ \begin{aligned} {}_t D_{+\infty}^{\alpha} f(t) &= (-1)^n \frac{d^n}{dt^n} ({}_t D_{+\infty}^{-(n-\alpha)} f(t)) \\ &= \frac{1}{\Gamma(n-\alpha)} (-1)^n \frac{d^n}{dt^n} \left(\int_{t}^{\infty} (s-t)^{n-\alpha-1} f(s) ds \right), \end{aligned} $$

respectively, where $n = [\alpha] + 1$, $\alpha \geq 0$ and $t \in \mathbb{R}$.

In particular, when $\alpha = n \in \mathbb{N}_0$, then

$$ {}_{-\infty}D_t^{0} f(t) = {}_t D_{+\infty}^{0} f(t) = f(t), $$

$$ {}_{-\infty}D_t^{n} f(t) = f^{(n)}(t) \quad \text{and} \quad {}_t D_{+\infty}^{n} f(t) = (-1)^n f^{(n)}(t), $$

where $f^{(n)}(t)$ is the usual derivative of $f(t)$ of order n. If $0 < \alpha < 1$ and $t \in \mathbb{R}$, then

$$ \begin{aligned} {}_{-\infty}D_t^{\alpha} f(t) &= \frac{1}{\Gamma(1-\alpha)} \frac{d}{dt} \left(\int_{-\infty}^{t} (t-s)^{-\alpha} f(s) ds \right) \\ &= \frac{\alpha}{\Gamma(1-\alpha)} \int_{0}^{\infty} \frac{f(t) - f(t-s)}{s^{\alpha+1}} ds \end{aligned} $$

and

$$ \begin{aligned} {}_t D_{+\infty}^{\alpha} f(t) &= -\frac{1}{\Gamma(1-\alpha)} \frac{d}{dt} \left(\int_{t}^{\infty} (s-t)^{-\alpha} f(s) ds \right) \\ &= \frac{\alpha}{\Gamma(1-\alpha)} \int_{0}^{\infty} \frac{f(t) - f(t+s)}{s^{\alpha+1}} ds. \end{aligned} $$

Formulas (1.5) and (1.6) can be used for the definition of the Caputo fractional derivatives on the whole axis $\mathbb{R}$.

Definition 1.7. (Left and right Caputo fractional derivatives on the real axis) The left and right Caputo fractional derivatives ${}_{-\infty}^{\ C}D_t^\alpha f(t)$ and ${}_t^C D_{+\infty}^\alpha f(t)$ of order α (with $\alpha > 0$ and $\alpha \notin \mathbb{N}$) on the whole axis $\mathbb{R}$ are defined by

$$ {}_{-\infty}^{\ C}D_t^\alpha f(t) = \frac{1}{\Gamma(n-\alpha)} \int_{-\infty}^{t} (t-s)^{n-\alpha-1} f^{(n)}(s)ds \tag{1.8} $$

and

$$ {}_t^C D_{+\infty}^\alpha f(t) = \frac{(-1)^n}{\Gamma(n-\alpha)} \int_{t}^{\infty} (s-t)^{n-\alpha-1} f^{(n)}(s)ds, \tag{1.9} $$

respectively.

When $0 < \alpha < 1$, the relations (1.8) and (1.9) take the following forms

$$ {}_{-\infty}^{\ C}D_t^\alpha f(t) = \frac{1}{\Gamma(1-\alpha)} \int_{-\infty}^{t} (t-s)^{-\alpha} f'(s)ds $$

and

$$ {}_t^C D_{+\infty}^\alpha f(t) = -\frac{1}{\Gamma(1-\alpha)} \int_{t}^{\infty} (s-t)^{-\alpha} f'(s)ds. $$

1.3.2 *Properties*

We present here some properties of the fractional integral and fractional derivative operators that will be useful throughout this book.

Proposition 1.3. *If $\alpha \geq 0$ and $\beta > 0$, then*

$$ {}_aD_t^{-\alpha}(t-a)^{\beta-1} = \frac{\Gamma(\beta)}{\Gamma(\beta+\alpha)}(t-a)^{\beta+\alpha-1} \quad (\alpha > 0), $$
$$ {}_aD_t^{\alpha}(t-a)^{\beta-1} = \frac{\Gamma(\beta)}{\Gamma(\beta-\alpha)}(t-a)^{\beta-\alpha-1} \quad (\alpha \geq 0) $$

and

$$ {}_tD_b^{-\alpha}(b-t)^{\beta-1} = \frac{\Gamma(\beta)}{\Gamma(\beta+\alpha)}(b-t)^{\beta+\alpha-1} \quad (\alpha > 0), $$
$$ {}_tD_b^{\alpha}(b-t)^{\beta-1} = \frac{\Gamma(\beta)}{\Gamma(\beta-\alpha)}(b-t)^{\beta-\alpha-1} \quad (\alpha \geq 0). $$

In particular, if $\beta = 1$ and $\alpha \geq 0$, then the Riemann-Liouville fractional derivatives of a constant are, in general, not equal to zero:

$$ {}_aD_t^\alpha 1 = \frac{(t-a)^{-\alpha}}{\Gamma(1-\alpha)}, \quad {}_tD_b^\alpha 1 = \frac{(b-t)^{-\alpha}}{\Gamma(1-\alpha)}. $$

On the other hand, for $j = 1, 2, ..., [\alpha] + 1$,

$$ {}_aD_t^\alpha (t-a)^{\alpha-j} = 0, \quad {}_tD_b^\alpha (b-t)^{\alpha-j} = 0. $$

The semigroup properties of the fractional integral operators ${}_aD_t^{-\alpha}$ and ${}_tD_b^{-\alpha}$ are given by the following result.

Proposition 1.4. *If $\alpha > 0$ and $\beta > 0$, then the equations*

$$ {}_aD_t^{-\alpha}\left({}_aD_t^{-\beta} f(t)\right) = {}_aD_t^{-\alpha-\beta} f(t) \ \ \text{and} \ \ {}_tD_b^{-\alpha}\left({}_tD_b^{-\beta} f(t)\right) = {}_tD_b^{-\alpha-\beta} f(t) \quad (1.10) $$

are satisfied at almost every point $t \in [a,b]$ for $f \in L^p([a,b],\mathbb{R}^N)$ $(1 \le p < \infty)$. If $\alpha + \beta > 1$, then the relations in (1.10) hold at any point of $[a,b]$.

Proposition 1.5.

(i) *If $\alpha > 0$ and $f \in L^p([a,b],\mathbb{R}^N)$ $(1 \le p \le \infty)$, then the following equalities*

$$ {}_aD_t^{\alpha}\left({}_aD_t^{-\alpha} f(t)\right) = f(t) \ \ \text{and} \ \ {}_tD_b^{\alpha}\left({}_tD_b^{-\alpha} f(t)\right) = f(t) \quad (\alpha > 0) $$

hold almost everywhere on $[a,b]$.

(ii) *If $\alpha > \beta > 0$, then, for $f \in L^p([a,b],\mathbb{R}^N)$ $(1 \le p \le \infty)$, the relations*

$$ {}_aD_t^{\beta}\left({}_aD_t^{-\alpha} f(t)\right) = {}_aD_t^{-\alpha+\beta} f(t) \ \ \text{and} \ \ {}_tD_b^{\beta}\left({}_tD_b^{-\alpha} f(t)\right) = {}_tD_b^{-\alpha+\beta} f(t) $$

hold almost everywhere on $[a,b]$.

In particular, when $\beta = k \in \mathbb{N}$ and $\alpha > k$, then

$$ {}_aD_t^{k}\left({}_aD_t^{-\alpha} f(t)\right) = {}_aD_t^{-\alpha+k} f(t) \ \ \text{and} \ \ {}_tD_b^{k}\left({}_tD_b^{-\alpha} f(t)\right) = (-1)^k {}_tD_b^{-\alpha+k} f(t). $$

To present the next property, we use the spaces of functions ${}_aD_t^{-\alpha}(L^p)$ and ${}_tD_b^{-\alpha}(L^p)$ defined for $\alpha > 0$ and $1 \le p \le \infty$ by

$$ {}_aD_t^{-\alpha}(L^p) = \{f :\ f = {}_aD_t^{-\alpha}\varphi,\ \varphi \in L^p([a,b],\mathbb{R}^N)\} $$

and

$$ {}_tD_b^{-\alpha}(L^p) = \{f :\ f = {}_tD_b^{-\alpha}\phi,\ \phi \in L^p([a,b],\mathbb{R}^N)\}, $$

respectively. The composition of the fractional integral operator ${}_aD_t^{-\alpha}$ with the fractional derivative operator ${}_aD_t^{\alpha}$ is given by the following result.

Proposition 1.6. *Let $\alpha > 0$, $n = [\alpha] + 1$ and let $f_{n-\alpha}(t) = {}_aD_t^{-(n-\alpha)} f(t)$ be the fractional integral (1.1) of order $n - \alpha$.*

(i) *If $1 \le p \le \infty$ and $f \in {}_aD_t^{-\alpha}(L^p)$, then*

$$ {}_aD_t^{-\alpha}\left({}_aD_t^{\alpha} f(t)\right) = f(t). $$

(ii) *If $f \in L^1([a,b],\mathbb{R}^N)$ and $f_{n-\alpha} \in AC^n([a,b],\mathbb{R}^N)$, then the equality*

$$ {}_aD_t^{-\alpha}\left({}_aD_t^{\alpha} f(t)\right) = f(t) - \sum_{j=1}^{n} \frac{f_{n-\alpha}^{(n-j)}(a)}{\Gamma(\alpha - j + 1)}(t-a)^{\alpha-j} $$

holds almost everywhere on $[a,b]$.

Proposition 1.7. *Let $\alpha > 0$ and $n = [\alpha] + 1$. Also let $g_{n-\alpha}(t) = {}_tD_b^{-(n-\alpha)}g(t)$ be the fractional integral (1.2) of order $n - \alpha$.*

(i) *If $1 \leq p \leq \infty$ and $g \in {}_tD_b^{-\alpha}(L^p)$, then*

$${}_tD_b^{-\alpha}\Big({}_tD_b^{\alpha}g(t)\Big) = g(t).$$

(ii) *If $g \in L^1([a,b],\mathbb{R}^N)$ and $g_{n-\alpha} \in AC^n([a,b],\mathbb{R}^N)$, then the equality*

$${}_tD_b^{-\alpha}\Big({}_tD_b^{\alpha}g(t)\Big) = g(t) - \sum_{j=1}^{n} \frac{(-1)^{n-j}g_{n-\alpha}^{(n-j)}(a)}{\Gamma(\alpha-j+1)}(b-t)^{\alpha-j}$$

holds almost everywhere on $[a,b]$.

In particular, if $0 < \alpha < 1$, then

$${}_tD_b^{-\alpha}\Big({}_tD_b^{\alpha}g(t)\Big) = g(t) - \frac{g_{1-\alpha}(a)}{\Gamma(\alpha)}(b-t)^{\alpha-1},$$

where $g_{1-\alpha}(t) = {}_tD_b^{\alpha-1}g(t)$ while for $\alpha = n \in \mathbb{N}$, the following equality holds:

$${}_tD_b^{-n}\Big({}_tD_b^{n}g(t)\Big) = g(t) - \sum_{k=0}^{n-1} \frac{(-1)^k g^{(k)}(a)}{k!}(b-t)^k.$$

Proposition 1.8. *Let $\alpha > 0$ and let $y \in L^\infty([a,b],\mathbb{R}^N)$ or $y \in C([a,b],\mathbb{R}^N)$. Then*

$${}_a^C D_t^{\alpha}\Big({}_aD_t^{-\alpha}y(t)\Big) = y(t) \quad \text{and} \quad {}_t^C D_b^{\alpha}\Big({}_tD_b^{-\alpha}y(t)\Big) = y(t).$$

Proposition 1.9. *Let $\alpha > 0$ and let n be given by (1.3). If $y \in AC^n([a,b],\mathbb{R}^N)$ or $y \in C^n([a,b],\mathbb{R}^N)$, then*

$${}_aD_t^{-\alpha}\Big({}_a^C D_t^{\alpha}y(t)\Big) = y(t) - \sum_{k=0}^{n-1} \frac{y^{(k)}(a)}{k!}(t-a)^k$$

and

$${}_tD_b^{-\alpha}\Big({}_t^C D_b^{\alpha}y(t)\Big) = y(t) - \sum_{k=0}^{n-1} \frac{(-1)^k y^{(k)}(b)}{k!}(b-t)^k.$$

In particular, if $0 < \alpha \leq 1$ and $y \in AC([a,b],\mathbb{R}^N)$ or $y \in C([a,b],\mathbb{R}^N)$, then

$${}_aD_t^{-\alpha}\Big({}_a^C D_t^{\alpha}y(t)\Big) = y(t) - y(a) \quad \text{and} \quad {}_tD_b^{-\alpha}\Big({}_t^C D_b^{\alpha}y(t)\Big) = y(t) - y(b). \tag{1.11}$$

On the other hand, we have the following properties of fractional integration.

Proposition 1.10. *Let $\alpha > 0$, $p \geq 1$, $q \geq 1$, and $\frac{1}{p} + \frac{1}{q} \leq 1 + \alpha$ ($p \neq 1$ and $q \neq 1$ in the case when $\frac{1}{p} + \frac{1}{q} = 1 + \alpha$).*

(i) *If $\varphi \in L^p([a,b],\mathbb{R}^N)$ and $\psi \in L^q([a,b],\mathbb{R}^N)$, then*

$$\int_a^b \varphi(t)\ {}_aD_t^{-\alpha}\psi(t)dt = \int_a^b \psi(t)\ {}_tD_b^{-\alpha}\varphi(t)dt. \tag{1.12}$$

(ii) *If* $f \in {}_tD_b^{-\alpha}(L^p)$ *and* $g \in {}_aD_t^{-\alpha}(L^q)$, *then*

$$\int_a^b f(t)\ {}_aD_t^\alpha g(t)dt = \int_a^b g(t)\ {}_tD_b^\alpha f(t)dt. \tag{1.13}$$

Then applying Proposition 1.1, we can derive the integration by parts formula for the left and right Riemann-Liouville fractional derivatives looks as follows.

Proposition 1.11.

$$\int_a^b {}_aD_t^\alpha f(t)\cdot g(t)dt = \int_a^b {}_tD_b^\alpha g(t)\cdot f(t)dt, \quad 0 < \alpha \le 1,$$

provided the boundary conditions

$$f(a) = f(b) = 0, \quad f' \in L^\infty([a,b],\mathbb{R}^N),\ g \in L^1([a,b],\mathbb{R}^N),$$

or

$$g(a) = g(b) = 0, \quad g' \in L^\infty([a,b],\mathbb{R}^N),\ f \in L^1([a,b],\mathbb{R}^N)$$

are fulfilled.

Remark 1.4. If f, g are abstract functions with values in Banach space X, then integrals which appear in above properties are taken in Bochner's sense.

1.3.3 *Mittag-Leffler Functions*

Definition 1.8. (Miller and Ross, 1993; Podlubny, 1999) The generalized Mittag-Leffler function $E_{\alpha,\beta}$ is defined by

$$E_{\alpha,\beta}(z) := \sum_{k=0}^{\infty} \frac{z^k}{\Gamma(\alpha k+\beta)} = \frac{1}{2\pi i}\int_\Upsilon \frac{\lambda^{\alpha-\beta}e^\lambda}{\lambda^\alpha - z}d\lambda, \quad \alpha,\ \beta > 0, z \in \mathbb{C},$$

where Υ is a contour which starts and ends as $-\infty$ and encircles the disc $|\lambda| \le |z|^{1/\alpha}$ counter-clockwise.

If $0 < \alpha < 1,\ \beta > 0$, then the asymptotic expansion of $E_{\alpha,\beta}$ as $z \to \infty$ is given by

$$E_{\alpha,\beta}(z) = \begin{cases} \dfrac{1}{\alpha} z^{(1-\beta)/\alpha}\exp(z^{1/\alpha}) + \varepsilon_{\alpha,\beta}(z), & |\arg z| \le \dfrac{1}{2}\alpha\pi, \\ \varepsilon_{\alpha,\beta}(z), & |\arg(-z)| < (1-\dfrac{1}{2}\alpha)\pi, \end{cases} \tag{1.14}$$

where

$$\varepsilon_{\alpha,\beta}(z) = -\sum_{n=1}^{N-1} \frac{z^{-n}}{\Gamma(\beta-\alpha n)} + O\left(|z|^{-N}\right), \quad \text{as} \quad z \to \infty.$$

For short, set

$$E_\alpha(z) := E_{\alpha,1}(z), \quad e_\alpha(z) := E_{\alpha,\alpha}(z).$$

Then Mittag-Leffler functions have the following properties.

Proposition 1.12. *For* $\alpha \in (0,1)$ *and* $t \in \mathbb{R}$,

(i) $E_\alpha(t),\ e_\alpha(t) > 0$;
(ii) $(E_\alpha(t))' = \frac{1}{\alpha}e_\alpha(t)$;
(iii) $\lim\limits_{t\to-\infty} E_\alpha(t) = \lim\limits_{t\to-\infty} e_\alpha(t) = 0$;
(iv) ${}_0^C D_t^\alpha E_\alpha(\omega t^\alpha) = \omega E_\alpha(\omega t^\alpha),\ {}_0D_t^{\alpha-1}(t^{\alpha-1}e_\alpha(\omega t^\alpha)) = E_\alpha(\omega t^\alpha),\ \omega \in \mathbb{C}$.

Definition 1.9. (Mainardi, Paraddisi and Forenflo, 2000) The Wright function M_α is defined by

$$\begin{aligned} M_\alpha(z) &:= \sum_{n=0}^{\infty} \frac{(-z)^n}{n!\Gamma(-\alpha n + 1 - \alpha)} \\ &= \frac{1}{\pi}\sum_{n=1}^{\infty}\frac{(-z)^n}{(n-1)!}\Gamma(n\alpha)\sin(n\pi\alpha), \quad z \in \mathbb{C} \end{aligned}$$

with $0 < \alpha < 1$.

For $-1 < r < \infty, \lambda > 0$, the following results hold.

Proposition 1.13.

(W1) $M_\alpha(t) \geq 0,\ t > 0$;

(W2) $\displaystyle\int_0^\infty \frac{\alpha}{t^{\alpha+1}} M_\alpha(\frac{1}{t^\alpha})e^{-\lambda t}dt = e^{-\lambda^\alpha}$;

(W3) $\displaystyle\int_0^\infty M_\alpha(t)t^r dt = \frac{\Gamma(1+r)}{\Gamma(1+\alpha r)}$;

(W4) $\displaystyle\int_0^\infty M_\alpha(t)e^{-zt}dt = E_\alpha(-z), \quad z \in \mathbb{C}$;

(W5) $\displaystyle\int_0^\infty \alpha t M_\alpha(t)e^{-zt}dt = e_\alpha(-z), \quad z \in \mathbb{C}$.

1.4 Some Results from Nonlinear Analysis

1.4.1 *Laplace and Fourier Transforms*

In this subsection we present definitions and some properties of Laplace and Fourier transforms.

Definition 1.10. The Laplace transform of a function $f(t)$ of a real variable $t \in \mathbb{R}^+$ is defined by

$$(\mathcal{L}f)(s) = \mathcal{L}[f(t)](s) = \bar{f}(s) := \int_0^\infty e^{-st}f(t)dt \quad (s \in \mathbb{C}). \tag{1.15}$$

The inverse Laplace transform is given for $x \in \mathbb{R}^+$ by the formula

$$(\mathcal{L}^{-1}g)(x) = \mathcal{L}^{-1}[g(s)](x) := \frac{1}{\pi i}\int_{\gamma-i\infty}^{\gamma+i\infty} e^{sx}g(s)ds \quad (\gamma = \mathrm{Re}(s)). \tag{1.16}$$

Proposition 1.14. *Let $f(t)$ be defined on $(0,\infty)$ and $0<\alpha<1$. Then Laplace transform of fractional integral and fractional differential operator satisfies*

(i) $\overline{{}_0D_t^{-\alpha}f}(s) = s^{-\alpha}\bar{f}(s)$;

(ii) $\overline{{}_0D_t^{\alpha}f}(s) = s^{\alpha}\bar{f}(s) - ({}_0D_t^{\alpha-1}f)(0)$;

(iii) $\overline{{}_0^CD_t^{-\alpha}f}(s) = s^{\alpha}\bar{f}(s) - s^{\alpha-1}f(0)$.

Definition 1.11. The Fourier transform of a function $f(t)$ of a real variable $t\in\mathbb{R}$ is defined by

$$(\mathcal{F}f)(w) = \mathcal{F}[f(t)](w) = \hat{f}(w) := \int_{-\infty}^{\infty} e^{-it\cdot w}f(t)dt \quad (w\in\mathbb{R}). \tag{1.17}$$

The inverse Fourier transform is given by the formula

$$(\mathcal{F}^{-1}g)(t) = \mathcal{F}^{-1}[g(w)](t) = \frac{1}{2\pi}\hat{g}(-s) := \frac{1}{2\pi}\int_{-\infty}^{\infty} e^{it\cdot w}g(w)dw \quad (t\in\mathbb{R}). \tag{1.18}$$

The integrals in (1.17) and (1.18) converge absolutely for functions $f,g\in L^1(\mathbb{R})$ and in the norm of the space $L^2(\mathbb{R})$ for $f,g\in L^2(\mathbb{R})$.

Proposition 1.15. *Let $f(t)$ be defined on $(-\infty,\infty)$ and $0<\alpha<1$. Then Fourier transform of Liouville-Weyl fractional integral and fractional differential operator satisfies*

(i) $\widehat{{}_{-\infty}D_t^{-\alpha}f}(w) = (iw)^{-\alpha}\hat{f}(w)$;

(ii) $\widehat{{}_tD_{\infty}^{-\alpha}f}(w) = (-iw)^{-\alpha}\hat{f}(w)$;

(iii) $\widehat{{}_{-\infty}D_t^{\alpha}f}(w) = (iw)^{\alpha}\hat{f}(w)$;

(iv) $\widehat{{}_tD_{\infty}^{\alpha}f}(w) = (-iw)^{\alpha}\hat{f}(w)$.

1.4.2 *Sobolev Spaces*

We refer to Cazenave and Haraux, 1998, for the definitions and results given below.

Consider an open subset Ω of $\mathbb{R}^N$. $\mathcal{D}(\Omega)$ is the space of C^{∞}(real-valued or complex valued) functions with compact support in Ω and $\mathcal{D}'(\Omega)$ is the space of distributions on Ω. A distribution $T\in\mathcal{D}'(\Omega)$ is said to belong to $L^p(\Omega)(1\le p\le\infty)$ if there exists a function $f\in L^p(\Omega)$ such that

$$\langle T,\varphi\rangle = \int_{\Omega} f(x)\varphi(x)dx,$$

for all $\varphi\in\mathcal{D}(\Omega)$. In that case, it is well known that f is unique. Let $m\in\mathbb{N}$ and let $p\in[1,\infty]$. Define

$$W^{m,p}(\Omega) = \{f\in L^p(\Omega)|\ D^{\alpha}f\in L^p(\Omega) \ \text{ for all } \alpha\in\mathbb{N}^N \ \text{ such that } |\alpha|\le m\}.$$

$W^{m,p}(\Omega)$ is a Banach space which equipped with the norm

$$\|f\|_{W^{m,p}(\Omega)} = \sum_{|\alpha|\leq m} \|D^\alpha f\|_{L^p},$$

for all $f \in W^{m,p}(\Omega)$. For all m, p as above, we denote by $W_0^{m,p}(\Omega)$ the closure of $\mathcal{D}(\Omega)$ in $W^{m,p}(\Omega)$. If $p = 2$, one sets $W^{m,2}(\Omega) = H^m(\Omega)$, $W_0^{m,2}(\Omega) = H_0^m(\Omega)$ and one equips $H^m(\Omega)$ with the following equivalent norm:

$$\|f\|_{H^m} = \left(\sum_{|\alpha|\leq m} \|D^\alpha u\|_{L^2}^2 \right)^{\frac{1}{2}}.$$

Then $H^m(\Omega)$ is a Hilbert space with the scalar product

$$\langle u, v\rangle_{H^m} = \sum_{|\alpha|\leq m} \int_\Omega D^\alpha u \cdot D^\alpha v dx.$$

If Ω is bounded, there exists a constant $C(\Omega)$ such that

$$\|u\|_{L^2} \leq C(\Omega)\|\nabla u\|_{L^2},$$

for all $u \in H_0^1(\Omega)$ (this is Poincaré inequality). It may be more convenient to equip $H_0^1(\Omega)$ with the following scalar product

$$\langle u, v\rangle = \int_\Omega \nabla u \cdot \nabla v dx,$$

which defines an equivalent norm to $\|\cdot\|_{H^1}$ on the closed space $H_0^1(\Omega)$.

1.4.3 *Measure of Noncompactness*

We recall here some definitions and properties of measure of noncompactness.

Assume that X is a Banach space with the norm $|\cdot|$. The measure of noncompactness α is said to be:

(i) *Monotone* if for all bounded subsets B_1, B_2 of X, $B_1 \subseteq B_2$ implies $\alpha(B_1) \leq \alpha(B_2)$.
(ii) *Nonsingular* if $\alpha(\{x\} \cup B) = \alpha(B)$ for every $x \in X$ and every nonempty subset $B \subseteq X$.
(iii) *Regular* $\alpha(B) = 0$ if and only if B is relatively compact in X.

One of the most important examples of measure of noncompactness is the Hausdorff measure of noncompactness α defined on each bounded subset B of X by

$$\alpha(B) = \inf\{\varepsilon > 0 : \ B \subset \bigcup_{j=1}^{m} B_\varepsilon(x_j) \text{ where } x_j \in X\},$$

where $B_\varepsilon(x_j)$ is a ball of radius $\leq \varepsilon$ centered at x_j, $j = 1, 2, ..., m$, m is a positive integer number. Without confusion, Kuratowski measure of noncompactness α_1 defined on each bounded subset B of X by

$$\alpha_1(B) = \inf\{\varepsilon > 0 : \ B \subset \bigcup_{j=1}^{m} M_j \text{ and } \operatorname{diam}(M_j) \leq \varepsilon\},$$

where the diameter of M_j is defined by $\text{diam}(M_j) = \sup\{|x - y| : x, y \in M_j\}$, $j = 1, 2, ..., m$.

It is well known that Hausdorff measure of noncompactness α and Kuratowski measure of noncompactness α_1 enjoy the above properties (i)-(iii) and other properties. We refer the reader to Banaš and Goebel, 1980; Deimling, 1985; Heinz, 1983; Lakshmikantham and Leela, 1969.

(iv) $\alpha(B_1 + B_2) \leq \alpha(B_1) + \alpha(B_2)$, where $B_1 + B_2 = \{x + y : x \in B_1,\ y \in B_2\}$;

(v) $\alpha(B_1 \cup B_2) \leq \max\{\alpha(B_1), \alpha(B_2)\}$;

(vi) $\alpha(\lambda B) \leq |\lambda|\alpha(B)$ for any $\lambda \in \mathbb{R}$.

In particular, the relationship of Hausdorff measure of noncompactness α and Kuratowski measure of noncompactness α_1 is given by

(vii) $\alpha(B) \leq \alpha_1(B) \leq 2\alpha(B)$.

Let $J = [0, a], a \in \mathbb{R}^+$. For any $W \subset C(J, X)$, we define

$$\int_0^t W(s)ds = \left\{ \int_0^t u(s)ds : u \in W \right\}, \ \text{for } t \in [0, a],$$

where $W(s) = \{u(s) \in X : u \in W\}$.

We present here some useful properties.

Proposition 1.16. *If $W \subset C(J, X)$ is bounded and equicontinuous, then $\overline{co}W \subset C(J, X)$ is also bounded and equicontinuous.*

Proposition 1.17. *(Guo, Lakshmikantham and Liu, 1996) If $W \subset C(J, X)$ is bounded and equicontinuous, then $t \to \alpha(W(t))$ is continuous on J, and*

$$\alpha(W) = \max_{t \in J} \alpha(W(t)), \quad \alpha\left(\int_0^t W(s)ds\right) \leq \int_0^t \alpha(W(s))ds, \ \textit{for } t \in [0, a].$$

Proposition 1.18. *(Mönch, 1980) Let $\{u_n\}_{n=1}^{\infty}$ be a sequence of Bochner integrable functions from J into X with $|u_n(t)| \leq \tilde{m}(t)$ for almost all $t \in J$ and every $n \geq 1$, where $\tilde{m} \in L(J, \mathbb{R}^+)$, then the function $\psi(t) = \alpha(\{u_n(t)\}_{n=1}^{\infty})$ belongs to $L(J, \mathbb{R}^+)$ and satisfies*

$$\alpha\left(\left\{ \int_0^t u_n(s)ds : n \geq 1 \right\}\right) \leq 2 \int_0^t \psi(s)ds.$$

Proposition 1.19. *(Liu, 2008) Let $\{u_n(t)\}_{n=1}^{\infty} : [0, \infty) \to X$ be a continuous function family. If there exists $\rho \in L^1[0, \infty)$ such that*

$$|u_n(t)| \leq \rho(t), \quad t \in [0, \infty), \ n = 1, 2, \ldots.$$

Then $\alpha_1(\{u_n(t)\}_{n=1}^{\infty})$ is integrable on $[0, \infty)$, and

$$\alpha_1\left(\left\{ \int_0^t u_n(s)ds : \ n = 1, 2, \ldots \right\}\right) \leq 2 \int_0^t \alpha_1(\{u_n(s) : \ n = 1, 2, \ldots\})ds.$$

Proposition 1.20. *(Bothe, 1998) If W is bounded, then for each $\varepsilon > 0$, there is a sequence $\{u_n\}_{n=1}^{\infty} \subset W$, such that*

$$\alpha(W) \leq 2\alpha(\{u_n\}_{n=1}^{\infty}) + \varepsilon.$$

1.4.4 *Topological Degree*

For a minute description of the following notions we refer the reader to Banaš and Goebel, 1980; Deimling, 1985; Heinz, 1983; Lakshmikantham and Leela, 1969.

Definition 1.12. Consider $\Omega \subset X$ and $\mathscr{F} : \Omega \to X$ a continuous bounded mapping. We say that $\mathscr{F}$ is α-Lipschitz if there exists $k \geq 0$ such that

$$\alpha(\mathscr{F}(B)) \leq k\alpha(B), \quad \forall\ B \subset \Omega \text{ bounded.}$$

If, in addition, $k < 1$, then we say that $\mathscr{F}$ is a strict α-contraction.

We say that $\mathscr{F}$ is α-condensing if

$$\alpha(\mathscr{F}(B)) < \alpha(B), \quad \forall\ B \subset \Omega \text{ bounded with } \alpha(B) > 0.$$

In other words, $\alpha(\mathscr{F}(B)) \geq \alpha(B)$ implies $\alpha(B) = 0$. The class of all strict α-contractions $\mathscr{F} : \Omega \to X$ is denoted by $\mathcal{SC}_{\alpha}(\Omega)$ and the class of all α-condensing mappings $\mathscr{F} : \Omega \to X$ is denoted by $C_{\alpha}(\Omega)$.

We remark that $\mathcal{SC}_{\alpha}(\Omega) \subset C_{\alpha}(\Omega)$ and every $\mathscr{F} \in C_{\alpha}(\Omega)$ is α-Lipschitz with constant $k = 1$. We also recall that $\mathscr{F} : \Omega \to X$ is Lipschitz if there exists $k > 0$ such that

$$|\mathscr{F}x - \mathscr{F}y| \leq k|x - y|, \quad \forall\ x, y \in \Omega$$

and that $\mathscr{F}$ is a strict contraction if $k < 1$.

Next, we collect some properties of the applications defined above.

Proposition 1.21. *If $\mathscr{F}, \mathscr{G} : \Omega \to X$ are α-Lipschitz mappings with constants k, k', respectively, then $\mathscr{F} + \mathscr{G} : \Omega \to X$ is α-Lipschitz with constant $k + k'$.*

Proposition 1.22. *If $\mathscr{F} : \Omega \to X$ is compact, then $\mathscr{F}$ is α-Lipschitz with constant $k = 0$.*

Proposition 1.23. *If $\mathscr{F} : \Omega \to X$ is Lipschitz with constant k, then $\mathscr{F}$ is α-Lipschitz with the same constant k.*

The theorem below asserts the existence and the basic properties of the topological degree for α-condensing perturbations of the identity. For more details, see Isaia, 2006.

Let

$$\mathcal{T} = \left\{(I - \mathscr{F}, \Omega, y) : \Omega \subset X \text{ open and bounded}, \mathscr{F} \in C_{\alpha}(\overline{\Omega}), y \in X \setminus (I - \mathscr{F})(\partial\Omega)\right\}$$

be the family of the admissible triplets.

Theorem 1.1. *There exists one degree function $D : \mathcal{T} \to \mathbb{N}_0$ which satisfies the properties:*

(i) *Normalization* $D(I, \Omega, y) = 1$ *for every* $y \in \Omega$.

(ii) *Additivity on domain For every disjoint, open sets* $\Omega_1, \Omega_2 \subset \Omega$ *and every* y *does not belong to* $(I - \mathscr{F})(\overline{\Omega} \backslash (\Omega_1 \cup \Omega_2))$, *we have*

$$D(I - \mathscr{F}, \Omega, y) = D(I - \mathscr{F}, \Omega_1, y) + D(I - \mathscr{F}, \Omega_2, y).$$

(iii) *Invariance under homotopy* $D(I - H(t, \cdot), \Omega, y(t))$ *is independent of* $t \in [0, 1]$ *for every continuous, bounded mapping* $H : [0, 1] \times \overline{\Omega} \to X$ *which satisfies*

$$\alpha(H([0, 1] \times B)) < \alpha(B), \quad \forall\ B \subset \overline{\Omega} \text{ with } \alpha(B) > 0$$

and every continuous function $y : [0, 1] \to x$ *which satisfies*

$$y(t) \neq x - H(t, x), \quad \forall\ t \in [0, 1],\ \forall\ x \in \partial\Omega.$$

(iv) *Existence* $D(I - \mathscr{F}, \Omega, y) \neq 0$ *implies* $y \in (I - \mathscr{F})(\Omega)$.

(v) *Excision* $D(I - \mathscr{F}, \Omega, y) = D(I - \mathscr{F}, \Omega_1, y)$ *for every open set* $\Omega_1 \subset \Omega$ *and every* y *does not belong to* $(I - \mathscr{F})(\overline{\Omega} \backslash \Omega_1)$.

Having in hand a degree function defined on $\mathcal{T}$, we collect the usability of the a priori estimate method by means of this degree.

Theorem 1.2. *Let* $\mathscr{F} : X \to X$ *be* α*-condensing and*

$$\mathcal{S} = \{x \in X :\ \exists\ \lambda \in [0, 1] \quad \text{such that} \quad x = \lambda \mathscr{F} x\}.$$

If $\mathcal{S}$ *is a bounded set in* X, *so there exists* $r > 0$ *such that* $\mathcal{S} \subset B_r(0)$, *then*

$$D(I - \lambda \mathscr{F}, B_r(0), 0) = 1, \quad \forall\ \lambda \in [0, 1].$$

Consequently, $\mathscr{F}$ *has at least one fixed point and the set of the fixed points of* $\mathscr{F}$ *lies in* $B_r(0)$.

1.4.5 *Picard Operator*

Let (X, d) be a metric space and $A : X \to X$ an operator. We shall use the following notations:

$P(X) = \{Y \subseteq X \mid Y \neq \varnothing\}$; $F_A = \{x \in X \mid A(x) = x\}$–the fixed point set of A;
$I(A) = \{Y \in P(X) \mid A(Y) \subseteq Y\}$;
$O_A(x) = \{x, A(x), A^2(x), ..., A^n(x), ...\}$–the A-orbit of $x \in X$;
$H : P(X) \times P(X) \to \mathbb{R}_+ \cup \{+\infty\}$;
$H(Y, Z) = \max\{\sup_{a \in Y} \inf_{b \in Z} d(a, b),\ \sup_{b \in Z} \inf_{a \in Y} d(a, b)\}$
–the Pompeiu-Hausdorff functional on $P(X)$.

Definition 1.13. (Rus, 1987) Let (X, d) be a metric space. An operator $A : X \to X$ is a Picard operator if there exists $x^* \in X$ such that $F_A = \{x^*\}$ and the sequence $\{A^n(x_0)\}_{n \in \mathbb{N}}$ converges to x^* for all $x_0 \in X$.

Definition 1.14. (Rus, 1993) Let (X, d) be a metric space. An operator $A : X \to X$ is a weak Picard operator if the sequence $\{A^n(x_0)\}_{n \in \mathbb{N}}$ converges for all $x_0 \in X$ and its limit (which may depend on x_0) is a fixed point of A.

If A is a weak Picard operator, then we consider the operator

$$A^\infty : X \to X, \quad A^\infty(x) = \lim_{n\to\infty} A^n(x).$$

The following results are useful in what follows.

Definition 1.15. (Rus, 1979) Let (Y, d) be a complete metric space and $A, B : Y \to Y$ two operators. Suppose that:

(i) A is a contraction with contraction constant ρ and $F_A = \{x_A^*\}$;
(ii) B has fixed points and $x_B^* \in F_B$;
(iii) there exists $\eta > 0$ such that $d(A(x), B(x)) \leq \eta$, for all $x \in Y$.

Then $d(x_A^*, x_B^*) \leq \frac{\eta}{1-\rho}$.

Definition 1.16. (Rus and Mureşan, 2000) Let (X, d) be a complete metric space and $A, B : X \to X$ two orbitally continuous operators. Assume that:

(i) there exists $\rho \in [0, 1)$ such that
$$d(A^2(x), A(x)) \leq \rho d(x, A(x)),\ d(B^2(x), B(x)) \leq \rho d(x, B(x)), \text{ for all } x \in X;$$
(ii) there exists $\eta > 0$ such that $d(A(x), B(x)) \leq \eta$ for all $x \in X$.

Then $H(F_A, F_B) \leq \frac{\eta}{1-\rho}$, where H denotes the Pompeiu-Hausdorff functional.

Theorem 1.3. *(Rus, 1993) Let (X, d) be a metric space. Then $A : X \to X$ is a weak Picard operator if and only if there exists a partition $X = \bigcup_{\lambda\in\Lambda} X_\lambda$ of X such that*

(i) *$X_\lambda \in I(A)$;*
(ii) *$A|_{X_\lambda} : X_\lambda \to X_\lambda$ is a Picard operator, for all $\lambda \in \Lambda$.*

1.4.6 *Fixed Point Theorems*

In this subsection, we present some fixed point theorems which will be used in the following chapters.

Theorem 1.4. *(Banach contraction mapping principle) Let (X, d) be a complete metric space, and let $\mathscr{T} : \Omega \to \Omega$ be a contraction mapping:*

$$d(\mathscr{T}x, \mathscr{T}y) \leq kd(x, y),$$

where $0 < k < 1$, for each $x, y \in \Omega$. Then, there exists a unique fixed point x of $\mathscr{T}$ in Ω, i.e., $\mathscr{T}x = x$.

Theorem 1.5. *(Schauder fixed point theorem) Let X be a Banach space and $\Omega \subset X$ a convex and closed set. If $\mathscr{T} : \Omega \to \Omega$ is a continuous operator such that $\mathscr{T}\Omega \subset X$, $\mathscr{T}\Omega$ is relatively compact, then $\mathscr{T}$ has at least one fixed point in Ω.*

Theorem 1.6. *(Schaefer fixed point theorem) Let X be a Banach space and let $F : X \to X$ be a completely continuous mapping. Then either*

(i) *the equation* $x = \lambda Fx$ *has a solution for* $\lambda = 1$, *or*
(ii) *the set* $\{x \in X : x = \lambda Fx$ *for some* $\lambda \in (0, 1)\}$ *is unbounded.*

Theorem 1.7. *(Darbo-Sadovskii fixed point theorem) If* Ω *is bounded closed and convex subset of Banach space* X, *the continuous mapping* $\mathscr{T} : \Omega \to \Omega$ *is an* α-*contraction, then the mapping* $\mathscr{T}$ *has at least one fixed point in* Ω.

Theorem 1.8. *(Krasnoselskii fixed point theorem) Let* X *be a Banach space, let* Ω *be a bounded closed convex subset of* X *and let* $\mathscr{S}$, $\mathscr{T}$ *be mappings of* Ω *into* X *such that* $\mathscr{S}z + \mathscr{T}w \in \Omega$ *for every pair* $z, w \in \Omega$. *If* $\mathscr{S}$ *is a contraction and* $\mathscr{T}$ *is completely continuous, then the equation* $\mathscr{S}z + \mathscr{T}z = z$ *has a solution on* Ω.

Theorem 1.9. *(Nonlinear alternative of Leray-Schauder type) Let* $\mathcal{C}$ *be a nonempty convex subset of* X. *Let* U *be a nonempty open subset of* $\mathcal{C}$ *with* $0 \in U$ *and* $F : \overline{U} \to \mathcal{C}$ *be a compact and continuous operator. Then either*

(i) F *has fixed points, or*
(ii) *there exist* $y \in \partial U$ *and* $\lambda^* \in [0, 1]$ *with* $y = \lambda^* F(y)$.

Theorem 1.10. *(O'Regan fixed point theorem) Let* U *be an open set in a closed, convex set* $\mathcal{C}$ *of* X. *Assume* $0 \in U$, $T(\overline{U})$ *is bounded and* $T : \overline{U} \to \mathcal{C}$ *is given by* $T = T_1 + T_2$, *where* $T_1 : \overline{U} \to X$ *is completely continuous, and* $T_2 : \overline{U} \to X$ *is a nonlinear contraction. Then either*

(i) T *has a fixed point in* $\overline{U}$, *or*
(ii) *there is a point* $x \in \partial U$ *and* $\lambda \in (0, 1)$ *with* $x = \lambda T(x)$.

A non-empty closed set K in a Banach space X is called a cone if:

(i) $K + K \subseteq K$;
(ii) $\lambda K \subseteq K$ for $\lambda \in \mathbb{R}, \lambda \geq 0$;
(iii) $\{-K\} \cap K = \{0\}$, where 0 is the zero element of X.

We introduce an order relation " $\leq$ " in X as follows. Let $z, y \in X$. Then $z \leq y$ if and only if $y - z \in K$. A cone K is called normal if the norm $\|\cdot\|_X$ is semi-monotone increasing on K, that is, there is a constant $N > 0$ such that $\|z\|_X \leq N\|y\|_X$ for all $z, y \in K$ with $z \leq y$. It is known that if the cone K is normal in X, then every order-bounded set in X is norm-bounded. Similarly, the cone K in X is called regular if every monotone increasing (resp. decreasing) order bounded sequence in X converges in norm.

For any $a, b \in X, a \leq b$, the order interval $[a, b]$ is a set in X given by

$$[a, b] = \{z \in X : \ a \leq z \leq b\}.$$

Let X and Y be two ordered Banach spaces. A mapping $\mathscr{T} : X \to Y$ is said to be nondecreasing or monotone increasing if $z \leq y$ implies $\mathscr{T}z \leq \mathscr{T}y$ for all $z, y \in [a, b]$.

Theorem 1.11. *(Hybrid fixed point theorem) (Dhage, 2006) Let X be a Banach space and $A, B, C : X \to X$ be three monotone increasing operators such that*

(i) *A is a contraction with contraction constant $k < 1$;*
(ii) *B is completely continuous;*
(iii) *C is totally bounded;*
(iv) *there exist elements a and b in X such that $a \leq Aa + Ba + Ca$ and $b \geq Ab + Bb + Cb$ with $a \leq b$.*

Further if the cone K in X is normal, then the operator equation $Az+Bz+Cz = z$ has a least and a greatest solution in $[a, b]$.

1.4.7 *Critical Point Theorems*

Let H be a real Banach space and $C^1(H, \mathbb{R}^N)$ denotes the set of functionals that are Fréchet differentiable and their Fréchet derivatives are continuous on H.

We need to use the critical point theorems to consider the fractional boundary value problems. For the reader's convenience, we state some necessary definitions and theorems and skip the proofs.

Definition 1.17. (Rabinowitz, 1986) Let $\psi \in C^1(H, \mathbb{R}^N)$. If any sequence $\{u_k\} \subset H$ for which $\{\psi(u_k)\}$ is bounded and $\psi'(u_k) \to 0$ as $k \to \infty$ possesses a convergent subsequence, then we say ψ satisfies Palais-Smale condition (denoted by (PS) condition for short).

Definition 1.18. (Mawhin and Willem, 1989) Let H be a real Banach space, $\psi : H \to \mathbb{R}$ is differentiable and $c \in \mathbb{R}$. We say that ψ satisfies the $(\mathrm{PS})_c$ condition if the existence of a sequence $\{u_k\}$ in H such that

$$\psi(u_k) \to c, \quad \psi'(u_k) \to 0,$$

as $k \to \infty$, implies that c is a critical value of ψ.

Theorem 1.12. *(Mawhin and Willem, 1989) Let H be a real Banach space and $I \in C^1(H, \mathbb{R})$ satisfies Palais-Smale condition. If I is bounded from below, then $c = \inf_H I$ is a critical value of I.*

Theorem 1.13. *(Rabinowitz, 1986) Let H be a real Banach space and $I \in C^1(H, \mathbb{R})$ with I even, bounded from below, and satisfy Palais-Smale condition. Suppose that $I(0) = 0$, there is a set $K \subset H$ such that K is homeomorphic to S^{d-1} (unit sphere) by an odd map, and $\sup_K I < 0$. Then I possesses at least d distinct pairs of critical points.*

Theorem 1.14. *(Mawhin and Willem, 1989) Let H be a real reflexive Banach space. If the functional $\psi : H \to \mathbb{R}^N$ is weakly lower semi-continuous and coercive, i.e., $\lim_{|z|\to\infty} \psi(z) = +\infty$, then there exists $z_0 \in H$ such that $\psi(z_0) = \inf_{z \in H} \psi(z)$. Moreover, if ψ is also Fréchet differentiable on H, then $\psi'(z_0) = 0$.*

Let B_r be the open ball in H with the radius r and centered at 0 and ∂B_r denote its boundary.

Theorem 1.15. *(Mountain pass theorem) (Rabinowitz, 1986) Let H be a real Banach space and $I \in C^1(H, \mathbb{R})$ satisfies (PS) condition. Suppose that I satisfies the following conditions:*

(i) $I(0) = 0$;
(ii) *there exist constants $\rho, \beta > 0$ such that $I|_{\partial B_\rho(0)} \geq \beta$;*
(iii) *there exists $e \in H \setminus \overline{B}_\rho(0)$ such that $I(e) \leq 0$.*

Then I possesses a critical value $c \geq \beta$ given by

$$c = \inf_{g \in \Gamma} \max_{s \in [0,1]} I(g(s)),$$

where $B_\rho(0)$ is an open ball in H of radius ρ centered at 0, and

$$\Gamma = \{g \in C([0,1], H) : \ g(0) = 0, \ g(1) = e\}.$$

Theorem 1.16. *(Rabinowitz, 1986) Let H be a real Banach space and $I \in C^1(H, \mathbb{R})$ with I even. Suppose that I satisfies Palais-Smale condition,* (i), (ii) *of Theorem 1.15 and the following condition:*

(iv) *for each finite-dimensional subspace $H' \subset H$, there is $r = r(H') > 0$ such that $I(u) \leq 0$ for $u \in H' \setminus B_r(0)$, where $B_r(0)$ is an open ball in H of radius r centered at 0.*

Then I possesses an unbounded sequence of critical values.

Let X be a reflexive and separable Banach space, then there are $e_j \in X$ and $e_j^* \in X^*$ such that

$$X = \overline{\text{span}\{e_j : j = 1, 2, ...\}} \quad \text{and} \quad X^* = \overline{\text{span}\{e_j^* : j = 1, 2, ...\}},$$

and

$$\langle e_i^*, e_j \rangle = \begin{cases} 1, & i = j, \\ 0, & i \neq j. \end{cases}$$

For convenience, we write

$$X_j := \text{span}\{e_j\}, \quad Y_k := \bigoplus_{j=1}^{k} X_j, \quad Z_k := \overline{\bigoplus_{j=k}^{\infty} X_j}. \tag{1.19}$$

And let

$$B_k := \{u \in Y_k : \ |u| \leq \rho_k\}, \quad N_k := \{u \in Z_k : \ |u| = \gamma_k\}.$$

Theorem 1.17. *(Fountain theorem) (Bartsch, 1993) Suppose:*

(H1) *X is a Banach space, $\varphi \in C^1(X,\mathbb{R})$ is an even functional, the subspace X_k, Y_k and Z_k are defined by (1.19).*

If for every $k \in \mathbb{N}$, there exist $\rho_k > r_k > 0$ such that

(H2) $a_k := \max_{\substack{u\in Y_k \\ |u|=\rho_k}} \varphi(u) \le 0$;

(H3) $b_k := \inf_{\substack{u\in Z_k \\ |u|=r_k}} \varphi(u) \to \infty$, *as* $k \to \infty$;

(H4) *φ satisfies the $(PS)_c$ condition for every $c > 0$.*

Then φ has an unbounded sequence of critical values.

Theorem 1.18. *(Dual Fountain theorem) (Bartsch, 1993) Assume (H1) is satisfied, and there is a $k_0 > 0$ so as to for each $k \ge k_0$, there exist $\rho_k > r_k > 0$ such that*

(H5) $d_k := \inf_{\substack{u\in Z_k \\ |u|\le\rho_k}} \varphi(u) \to 0$, *as* $k \to \infty$;

(H6) $i_k := \max_{\substack{u\in Y_k \\ |u|=r_k}} \varphi(u) < 0$;

(H7) $\inf_{\substack{u\in Z_k \\ |u|=\rho_k}} \varphi(u) \ge 0$;

(H8) *φ satisfies the $(\mathrm{PS})_c^*$ condition for every $c \in [d_{k_0}, 0)$.*

Then φ has a sequence of negative critical values converging to 0.

Remark 1.5. φ satisfies the $(\mathrm{PS})_c^*$ condition means that: if any sequence $\{u_{n_j}\} \subset X$ such that $n_j \to \infty, u_{n_j} \in Y_{n_j}, \varphi(u_{n_j}) \to c$ and $(\varphi|_{Y_{n_j}})'(u_{n_j}) \to 0$, then $\{u_{n_j}\}$ contains a subsequence converging to a critical point of φ. It is obvious that if φ satisfies the $(\mathrm{PS})_c^*$ condition, then φ satisfies the $(\mathrm{PS})_c$ condition.

Let X be a nonempty set and $\Phi, \tilde{\Psi} : X \to \mathbb{R}$ be two functionals. For $r, r_1, r_2, r_3 \in \mathbb{R}$ with $r_1 < \sup_X \Phi$, $r_2 > \inf_X \Phi$, $r_2 > r_1$, and $r_3 > 0$, we define

$$\varphi(r) := \inf_{u\in\Phi^{-1}(-\infty,r)} \frac{\sup_{u\in\Phi^{-1}(-\infty,r)} \tilde{\Psi}(u) - \tilde{\Psi}(u)}{r - \Phi(u)}, \tag{1.20}$$

$$\beta(r_1, r_2) := \inf_{u\in\Phi^{-1}(-\infty,r_1)} \sup_{v\in\Phi^{-1}[r_1,r_2)} \frac{\tilde{\Psi}(v) - \tilde{\Psi}(u)}{\Phi(v) - \Phi(u)}, \tag{1.21}$$

$$\gamma(r_2, r_3) := \frac{\sup_{u\in\Phi^{-1}(-\infty,r_2+r_3)} \tilde{\Psi}(u)}{r_3}, \tag{1.22}$$

$$\alpha(r_1, r_2, r_3) := \max\{\varphi(r_1), \varphi(r_2), \gamma(r_2, r_3)\}. \tag{1.23}$$

Theorem 1.19. *(Averna and Bonanno, 2009; Bonanno and Candito, 2008) Let X be a reflexive real Banach space, $\Phi : X \to \mathbb{R}$ be a convex, coercive, and continuously*

Gâteaux differentiable functional whose Gâteaux derivative admits a continuous inverse on X^*, $\tilde{\Psi} : X \to \mathbb{R}$ *be a continuously Gâteaux differentiable functional whose Gâteaux derivative is compact, such that*

(i) $\inf_X \Phi = \Phi(0) = \tilde{\Psi}(0) = 0$;
(ii) *for every* u_1, u_2 *satisfying* $\tilde{\Psi}(u_1) \geq 0$ *and* $\tilde{\Psi}(u_2) \geq 0$, *one has*
$$\inf_{t\in[0,1]} \tilde{\Psi}\,(tu_1 + (1-t)u_2) \geq 0.$$

Assume further that there exist three positive constants r_1, r_2 *and* r_3, *with* $r_1 < r_2$, *such that*

(iii) $\alpha(r_1, r_2, r_3)) < \beta(r_1, r_2)$.

Then, for each $\lambda \in \big(1/\beta(r_1,r_2), 1/\alpha(r_1,r_2,r_3)\big)$, *the functional* $\Phi - \lambda\tilde{\Psi}$ *has three distinct critical points* u_1, u_2 *and* u_3 *such that* $u_1 \in \Phi^{-1}(-\infty, r_1)$, $u_2 \in \Phi^{-1}[r_1, r_2)$ *and* $u_3 \in \Phi^{-1}(-\infty, r_2 + r_3)$.

1.5 Semigroups

1.5.1 C_0-*Semigroup*

Let X be a Banach space and $B(X)$ be the Banach space of linear bounded operators.

Definition 1.19. A semigroup is a one parameter family $\{T(t)\}_{t\geq 0} \subset B(X)$ satisfying the conditions:

(i) $T(t)T(s) = T(t+s)$, for $t, s \geq 0$;
(ii) $T(0) = I$.

Here I denotes the identity operator in X.

Definition 1.20. A semigroup $\{T(t)\}_{t\geq 0}$ is uniformly continuous if
$$\lim_{t\to 0+} \|T(t) - T(0)\|_{B(X)} = 0,$$
that is if
$$\lim_{|t-s|\to 0} \|T(t) - T(s)\|_{B(X)} = 0.$$

Definition 1.21. We say that the semigroup $\{T(t)\}_{t\geq 0}$ is a C_0-semigroup if the map $t \to T(t)x$ is strongly continuous, for each $x \in X$, i.e.
$$\lim_{t\to 0+} T(t)x = x, \quad \forall\, x \in X.$$

Definition 1.22. Let $T(t)$ be a C_0-semigroup defined on X. The infinitesimal generator A of $T(t)$ is the linear operator defined by
$$A(x) = \lim_{t\to 0+} \frac{T(t)x - x}{t}, \quad \text{for } x \in D(A),$$
where $D(A) = \{x \in X : \lim_{t\to 0+} \frac{T(t)x-x}{t} \text{ exists in } X\}$.

1.5.2 *Almost Sectorial Operators*

We firstly introduce some special functions and classes of functions which will be used in the following, for more details, we refer to Markus, 2006; Periago and Straub, 2002.

Let S_μ^0 with $0 < \mu < \pi$ be the open sector

$$\{z \in \mathbb{C}\backslash\{0\} : |\arg z| < \mu\}$$

and S_μ be its closure, that is

$$S_\mu = \{z \in \mathbb{C}\backslash\{0\} : \ |\arg z| \le \mu\} \cup \{0\}.$$

Denote by $D(A)$ the domain of A, by $\sigma(A)$ its spectrum, while $\rho(A) := \mathbb{C} - \sigma(A)$ is the resolvent set of A. We state the concept of almost sectorial operators as follows.

Definition 1.23. (Periago and Straub, 2002) Let $-1 < \gamma < 0$ and $0 < \omega < \pi/2$. By $\Theta_\omega^\gamma(X)$ we denote the family of all linear closed operators $A : D(A) \subset X \to X$ which satisfy:

(i) $\sigma(A) \subset S_\omega$;
(ii) for every $\omega < \mu < \pi$ there exists a constant C_μ such that

$$\|R(z; A)\|_{B(X)} \le C_\mu |z|^\gamma, \quad \text{for all } z \in \mathbb{C} \setminus S_\mu,$$

where $R(z; A) = (zI - A)^{-1}$, $z \in \rho(A)$, which are bounded linear operators the resolvent of A. A linear operator A will be called an almost sectorial operator on X if $A \in \Theta_\omega^\gamma(X)$.

Remark 1.6. Let $A \in \Theta_\omega^\gamma(X)$. Then the definition implies that $0 \in \rho(A)$.

Set

$$\mathcal{F}_0^\gamma(S_\mu^0) = \bigcup_{s<0} \Psi_s^\gamma(S_\mu^0) \cup \Psi_0(S_\mu^0),$$

$$\mathcal{F}(S_\mu^0) = \{f \in \mathcal{H}(S_\mu^0)| \quad \text{there exist} \quad k, n \in \mathbb{N} \quad \text{such that} \quad f\psi_n^k \in \mathcal{F}_0^\gamma(S_\mu^0)\},$$

where

$$\mathcal{H}(S_\mu^0) = \{f : S_\mu^0 \mapsto \mathbb{C}| \ f \text{ is holomorphic}\},$$

$$\mathcal{H}^\infty(S_\mu^0) = \{f \in \mathcal{H}(S_\mu^0)| \ f \text{ is bounded}\},$$

$$\varphi_0(z) = \frac{1}{1+z}, \quad \psi_n(z) := \frac{z}{(1+z)^n}, \quad z \in \mathbb{C}\backslash\{-1\}, n \in \mathbb{N} \cup \{0\},$$

$$\Psi_0(S_\mu^0) = \left\{ f \in \mathcal{H}(S_\mu^0)| \ \sup_{z \in S_\mu^0} \left| \frac{f(z)}{\varphi_0(z)} \right| < \infty \right\},$$

and for each $s < 0$,

$$\Psi_s^\gamma(S_\mu^0) := \left\{ f \in \mathcal{H}(S_\mu^0)| \ \sup_{z \in S_\mu^0} |\psi_n^s(z) f(z)| < \infty \right\},$$

where n is the smallest integer such that $n \ge 2$ and $\gamma + 1 < -(n-1)s$.

Observe that the classes of functions introduced above satisfy the inclusions

$$\mathcal{F}_0^\gamma(S_\mu^0) \subset \mathcal{H}^\infty(S_\mu^0) \subset \mathcal{F}(S_\mu^0) \subset \mathcal{H}(S_\mu^0).$$

Moreover, taking $k, n \in \mathbb{N} \cup \{0\}$ with $n > k$, one easily sees that $\psi_n^k \in \mathcal{F}_0^\gamma(S_\mu^0)$.

Assume that $A \in \Theta_\omega^\gamma(X)$ with $-1 < \gamma < 0$ and $0 < \omega < \pi/2$. Following Periago and Straub, 2002 (see also McIntosh, 1986; Cowling, Doust, McIntosh *et al.*, 1996), a closed linear operator $f \to f(A)$ can be constructed for every $f \in \mathcal{F}(S_\mu^0)$ via a extended functional calculus. In the following we give a short overview to this construction.

For $f \in \mathcal{F}_0^\gamma(S_\mu^0)$, via Dunford-Riesz integral, the operator $f(A)$ is defined by

$$f(A) = \frac{1}{2\pi i} \int_{\Gamma_\theta} f(z) R(z; A) dz, \tag{1.24}$$

where the integral contour $\Gamma_\theta := \{\mathbb{R}_+ e^{i\theta}\} \cup \{\mathbb{R}_+ e^{-i\theta}\}$, is oriented counter-clockwise and $\omega < \theta < \mu < \pi$. It follows that the integral is absolutely convergent and defines a bounded linear operator on X, and its value does not depend on the choice of θ.

Notice in particular that for $k, n \in \mathbb{N} \cup \{0\}$ with $n > k$,

$$\psi_n^k(A) = A^k (A + 1)^{-n}$$

and the operator $\psi_n^k(A)$ is injective. Notice also that if $f \in \mathcal{F}(S_\mu^0)$, then there exist $k, n \in \mathbb{N}$ such that $f\psi_n^k \in \mathcal{F}_0^\gamma(S_\mu^0)$. Hence, for $f \in \mathcal{F}(S_\mu^0)$, one can define a closed linear operator, still denoted by $f(A)$,

$$D(f(A)) = \{x \in X |\ (f\psi_n^k)(A)x \in D(A^{(n-1)k})\},$$

$$f(A) = (\psi_n^k(A))^{-1} (f\psi_n^k)(A),$$

and the definition of $f(A)$ does not depend on the choice of k and n. We emphasize that $f(A)$ is indeed an extension of the original and the triple $(\mathcal{F}_0^\gamma(S_\mu^0), \mathcal{F}(S_\mu^0), f(A))$ is called an *abstract functional calculus* on X (see Markus, 2006).

With respect to this construction we collect some basic properties. For more details, we refer to Periago and Straub, 2002.

Proposition 1.24. *The following assertions hold.*

(i) $\alpha f(A) + \beta g(A) = (\alpha f + \beta g)(A)$, $(fg)(A) = f(A)g(A)$ *for* $\forall\ f, g \in \mathcal{F}_0^\gamma(S_\mu^0)$, $\alpha, \beta \in \mathbb{C}$;

(ii) $f(A)g(A) \subset (fg)(A)$ *for* $\forall\ f, g \in \mathcal{F}(S_\mu^0)$, *and*

(iii) $f(A)g(A) = (fg)(A)$, *provided that* $g(A)$ *is bounded or* $D((fg)(A)) \subset D(g(A))$.

Since for each $\beta \in \mathbb{C}, z^\beta \in \mathcal{F}(S_\mu^0)$ $(z \in \mathbb{C} \setminus (-\infty, 0]$, $0 < \mu < \pi)$, one can define, via the triple $(\mathcal{F}_0^\gamma(S_\mu^0), \mathcal{F}(S_\mu^0), f(A))$, the complex powers of A which are closed by

$$A^\beta = z^\beta(A), \quad \beta \in \mathbb{C}.$$

However, in difference to the case of sectorial operators, having $0 \in \rho(A)$ does not imply that the complex powers $A^{-\beta}$ with $Re(\beta) > 0$, are bounded. The operator $A^{-\beta}$ belongs to $\mathscr{L}(X)$ whenever $Re(\beta) > 1+\gamma$. So, in this situation, the linear space $X^\beta := D(A^\beta)$, $\beta > 1+\gamma$, endowed with the graph norm $|x|_\beta = |A^\beta x|$, $x \in X^\beta$, is a Banach space.

Next, we turn our attention to the semigroup associated with A. Since given $t \in S^0_{\frac{\pi}{2}-\omega}$, $e^{-tz} \in \mathcal{H}^\infty(S^0_\mu)$ satisfies the conditions (a) and (b) of Lemma 2.13 of Periago and Straub, 2002, the family

$$T(t) = e^{-tz}(A) = \frac{1}{2\pi i}\int_{\Gamma_\theta} e^{-tz}R(z;A)dz, \quad t \in S^0_{\frac{\pi}{2}-\omega}, \tag{1.25}$$

here $\omega < \theta < \mu < \frac{\pi}{2} - |\arg t|$, forms an analytic semigroup of growth order $1+\gamma$.

Remark 1.7. From Periago and Straub, 2002, note that if $A \in \Theta^\gamma_\omega(X)$, then A generates a semigroup $T(t)$ with a singular behavior at $t = 0$ in a sense, called semigroup of growth $1+\gamma$. Moreover, the semigroup $T(t)$ is analytic in an open sector of the complex plane $\mathbb{C}$, but the strong continuity fails at $t = 0$ for data which are not sufficiently smooth.

For more properties on $T(t)$, please see the following proposition.

Proposition 1.25. *(Periago and Straub, 2002) Let $A \in \Theta^\gamma_\omega(X)$ with $-1 < \gamma < 0$ and $0 < \omega < \pi/2$. Then the following properties remain true.*

(i) *$T(t)$ is analytic in $S^0_{\frac{\pi}{2}-\omega}$ and*

$$\frac{d^n}{dt^n}T(t) = (-A)^n T(t), \quad \text{for all } t \in S^0_{\frac{\pi}{2}-\omega}.$$

(ii) *The functional equation $T(s+t) = T(s)T(t)$ for all s, $t \in S^0_{\frac{\pi}{2}-\omega}$ holds.*

(iii) *There exists a constant $C_0 = C_0(\gamma) > 0$ such that*

$$\|T(t)\|_{B(X)} \le C_0 t^{-\gamma-1}, \quad \text{for all } t > 0.$$

(iv) *The range $R(T(t))$ of $T(t)$, $t \in S^0_{\frac{\pi}{2}-\omega}$ is contained in $D(A^\infty)$. Particularly, $R(T(t)) \subset D(A^\beta)$ for all $\beta \in \mathbb{C}$ with $Re(\beta) > 0$,*

$$A^\beta T(t)x = \frac{1}{2\pi i}\int_{\Gamma_\theta} z^\beta e^{-tz}R(z;A)x dz, \quad \text{for all } x \in X,$$

and hence there exists a constant $C' = C'(\gamma,\beta) > 0$ such that

$$\|A^\beta T(t)\|_{B(X)} \le C' t^{-\gamma-Re(\beta)-1}, \quad \text{for all } t > 0.$$

(v) *If $\beta > 1+\gamma$, then $D(A^\beta) \subset \Sigma_T$, where Σ_T is the continuity set of the semigroup $\{T(t)\}_{t\ge 0}$, that is,*

$$\Sigma_T = \left\{x \in X \,\Big|\, \lim_{t\to 0+} T(t)x = x\right\}.$$

Remark 1.8. We note that the condition (ii) of the proposition does not satisfy for $t = 0$ or $s = 0$.

Recall that semigroups of growth $1+\gamma$ were investigated earlier in deLaubenfels, 1994 and Toropova, 2003.

The relation between the resolvent operators of A and the semigroup $T(t)$ is characterized by

Proposition 1.26. *(Periago and Straub, 2002) Let $A \in \Theta_\omega^\gamma(X)$ with $-1 < \gamma < 0$ and $0 < \omega < \pi/2$. Then for every $\lambda \in \mathbb{C}$ with $\mathrm{Re}(\lambda) > 0$, one has*

$$R(\lambda; -A) = \int_0^\infty e^{-\lambda t} T(t) dt.$$

Chapter 2

Fractional Functional Differential Equations

2.1 Introduction

The main objective of this chapter is to present a unified framework to investigate the basic existence theory for a variety of fractional functional differential equations with applications. As far as we know, many complex processes in nature and technology are described by functional differential equations which are dominant nowadays because the functional components in equations allow one to consider prehistory or after-effect influence. Various classes of functional differential equations are of fundamental importance in many problems arising in bionomics, epidemiology, electronics, theory of neural networks, automatic control, etc. Quite long ago delay differential equations had shown their efficiency in the study of the behavior of real populations. One can show that even though the delay terms occurring in the equations are unbounded, the domain of the initial data (past history or memory) may be finite or infinite. Consequently, those two cases need to be discussed independently. Moreover, one can consider functional differential equations so that the delay terms also occur in the derivative of the unknown solution. Since the general formulation of such a problem is difficult to state, a special kind of equations called neutral functional differential equations has been introduced.

On the other hand, fractional calculus is one of the best tools to characterize long-memory processes and materials, anomalous diffusion, long-range interactions, long-term behaviors, power laws, allometric scaling laws, and so on. So the corresponding mathematical models are fractional differential equations. Their evolutions behave in a much more complicated way so to study the corresponding dynamics is much more difficult. Although the existence theorems for the fractional differential equations can be similarly obtained, not all the classical theory of differential equation can be directly applied to the fractional differential equations. Hence, a somewhat theoretical frame needs to be established.

In Section 2.2, we discuss the existence and uniqueness of solutions and the existence of extremal solutions of initial value problem for the fractional neutral differential equations with bounded delay. Section 2.3 is devoted to study of the basic existence theory for fractional p-type neutral differential equations with unbounded delay but finite memory. In Section 2.4, we present a unified treatment of fundamental existence theory of fractional neutral differential equations with infinite memory. In Section 2.5, we consider a fractional iterative functional differential equation with parameter. Some theorems to prove the existence of the iterative series solutions are presented under some natural conditions. In Section 2.6, we discuss oscillation and existence of nonoscillatory solutions of fractional functional differential equations.

2.2 Neutral Equations with Bounded Delay

2.2.1 *Introduction*

Let $I_0 = [-\tau, 0]$, $\tau > 0$, $t_0 \geq 0$ and $I = [t_0, t_0+\sigma], \sigma > 0$ be two closed and bounded intervals in $\mathbb{R}$. Denote $J = [t_0 - \tau, t_0 + \sigma]$.

Let $\mathcal{C} = C(I_0, \mathbb{R}^n)$ be the space of continuous functions on I_0. For any element $\varphi \in \mathcal{C}$, define the norm

$$\|\varphi\|_* = \sup_{\theta \in I_0} |\varphi(\theta)|.$$

If $z \in C(J, \mathbb{R}^n)$, then for any $t \in I$ define $z_t \in \mathcal{C}$ by

$$z_t(\theta) = z(t + \theta), \quad \theta \in [-\tau, 0].$$

Consider the initial value problems (fractional IVP for short) of fractional neutral functional differential equations with bounded delay of the form

$$\begin{cases} {}_{t_0}^{C}D_t^{\alpha}(x(t) - k(t, x_t)) = F(t, x_t), & \text{a.e. } t \in (t_0, t_0 + \sigma], \\ x_{t_0} = \varphi, \end{cases} \tag{2.1}$$

where ${}_{t_0}^{C}D_t^{\alpha}$ is Caputo fractional derivative of order $0 < \alpha < 1$, $F : I \times \mathcal{C} \to \mathbb{R}^n$ is a given function satisfying some assumptions that will be specified later, and $\varphi \in \mathcal{C}$.

In Subsection 2.2.2, we establish the existence and uniqueness theorems of fractional IVP (2.1). In Subsection 2.2.3, we discuss the existence of extremal solutions for fractional IVP (2.1). We firstly give the definitions of $L^{\frac{1}{\beta}}$-Carathéodory, $L^{\frac{1}{\gamma}}$-Chandrabhan and $L^{\frac{1}{\delta}}$-Lipschitz, where β, γ, δ are some given numbers. Next, we apply Hybrid fixed point theorem to prove the existence results of extremal solutions for fractional IVP (2.1) under $L^{\frac{1}{\beta}}$-Carathéodory, $L^{\frac{1}{\gamma}}$-Chandrabhan and $L^{\frac{1}{\delta}}$-Lipschitz conditions. We do not require the continuity of the nonlinearity involved in the equation (2.1). In the end, we will present an example to illustrate our main results.

2.2.2 *Existence and Uniqueness*

Let $A(\sigma, \gamma) = \{x \in C([t_0 - \tau, t_0 + \sigma], \mathbb{R}^n) : x_{t_0} = \varphi, \ \sup_{t_0 \leq t \leq t_0 + \sigma} |x(t) - \varphi(0)| \leq \gamma\}$, where σ, γ are positive constants.

Before stating and proving the main results, we introduce the following hypotheses:

(H1) $F(t, \varphi)$ is measurable with respect to t on I;
(H2) $F(t, \varphi)$ is continuous with respect to φ on $C(I_0, \mathbb{R}^n)$;
(H3) there exist $\alpha_1 \in (0, \alpha)$ and a real-valued function $m(t) \in L^{\frac{1}{\alpha_1}} I$ such that for any $x \in A(\sigma, \gamma)$, $|F(t, x_t)| \leq m(t)$, for $t \in I$;
(H4) for any $x \in A(\sigma, \gamma)$, $k(t, x_t) = k_1(t, x_t) + k_2(t, x_t)$;
(H5) k_1 is continuous and for any $x', x'' \in A(\sigma, \gamma)$, $t \in I$

$$|k_1(t, x'_t) - k_1(t, x''_t)| \leq l\|x' - x''\|, \quad \text{where} \quad l \in (0, 1);$$

(H6) k_2 is completely continuous and for any bounded set Λ in $A(\sigma, \gamma)$, the set $\{t \to k_2(t, x_t) : x \in \Lambda\}$ is equicontinuous in $C(I, \mathbb{R}^n)$.

Lemma 2.1. *If there exist $\sigma \in (0, a)$ and $\gamma \in (0, \infty)$ such that (H1)-(H3) are satisfied, then for $t \in (t_0, t_0 + \sigma]$, fractional IVP* (2.1) *is equivalent to the following equation*

$$\begin{cases} x(t) = \varphi(0) - k(t_0, \varphi) + k(t, x_t) + \dfrac{1}{\Gamma(\alpha)} \displaystyle\int_{t_0}^{t} (t - s)^{\alpha - 1} F(s, x_s) ds, \quad t \in (t_0, t_0 + \sigma], \\ x_{t_0} = \varphi. \end{cases} \tag{2.2}$$

Proof. First, it is easy to obtain that $F(t, x_t)$ is Lebesgue measurable on I according to conditions (H1) and (H2). A direct calculation gives that $(t - s)^{\alpha - 1} \in L^{\frac{1}{1-\alpha_1}}([t_0, t], \mathbb{R})$, for $t \in I$. In the light of Hölder inequality and (H3), we obtain that $(t - s)^{\alpha - 1} F(s, x_s)$ is Lebesgue integrable with respect to $s \in [t_0, t]$ for all $t \in I$ and $x \in A(\sigma, \gamma)$, and

$$\int_{t_0}^{t} |(t - s)^{\alpha - 1} F(s, x_s)| ds \leq \|(t - s)^{\alpha - 1}\|_{L^{\frac{1}{1-\alpha_1}}[t_0, t]} \|m\|_{L^{\frac{1}{\alpha_1}} I}. \tag{2.3}$$

According to Definitions 1.1 and 1.3, it is easy to see that if x is a solution of the fractional IVP (2.1), then x is a solution of the equation (2.2).

On the other hand, if (2.2) is satisfied, then for every $t \in (t_0, t_0 + \sigma]$, we have

$$\begin{aligned} {}_{t_0}^{C}D_t^{\alpha}(x(t) - k(t, x_t)) &= {}_{t_0}^{C}D_t^{\alpha}\left(\varphi(0) - k(t_0, \varphi) + \frac{1}{\Gamma(\alpha)} \int_{t_0}^{t} (t - s)^{\alpha - 1} F(s, x_s) ds\right) \\ &= {}_{t_0}^{C}D_t^{\alpha}\left(\frac{1}{\Gamma(\alpha)} \int_{t_0}^{t} (t - s)^{\alpha - 1} F(s, x_s) ds\right) \\ &= {}_{t_0}^{C}D_t^{\alpha}\left({}_{t_0}D_t^{-\alpha} F(t, x_t)\right) \\ &= {}_{t_0}D_t^{\alpha}\left({}_{t_0}D_t^{-\alpha} F(t, x_t)\right) - {}_{t_0}D_t^{-\alpha} F(t, x_t)\big|_{t=t_0} \frac{(t - t_0)^{-\alpha}}{\Gamma(1 - \alpha)} \\ &= F(t, x_t) - {}_{t_0}D_t^{-\alpha} F(t, x_t)\big|_{t=t_0} \frac{(t - t_0)^{-\alpha}}{\Gamma(1 - \alpha)}. \end{aligned}$$

According to (2.3), we know that ${}_{t_0}D_t^{-\alpha} F(t, x_t)\big|_{t=t_0} = 0$, which means that ${}_{t_0}^{C}D_t^{\alpha}(x(t) - k(t, x_t)) = F(t, x_t)$, $t \in (t_0, t_0 + \sigma]$, and this completes the proof. □

Theorem 2.1. *Assume that there exist $\sigma \in (0,a)$ and $\gamma \in (0,\infty)$ such that (H1)-(H6) are satisfied. Then the fractional IVP (2.1) has at least one solution on $[t_0, t_0+\eta]$ for some positive number η.*

Proof. According to (H4), the equation (2.2) is equivalent to the following equation

$$\begin{cases} x(t) = \varphi(0) - k_1(t_0,\varphi) - k_2(t_0,\varphi) + k_1(t,x_t) + k_2(t,x_t) \\ \qquad + \dfrac{1}{\Gamma(\alpha)}\displaystyle\int_{t_0}^{t}(t-s)^{\alpha-1}F(s,x_s)ds, \qquad t \in I, \\ x_{t_0} = \varphi. \end{cases}$$

Let $\widetilde{\varphi} \in A(\sigma,\gamma)$ be defined as $\widetilde{\varphi}_{t_0} = \varphi$, $\widetilde{\varphi}(t_0+t) = \varphi(0)$ for all $t \in [0,\sigma]$. If x is a solution of the fractional IVP (2.1), let $x(t_0+t) = \widetilde{\varphi}(t_0+t) + y(t)$, $t \in [-\tau,\sigma]$, then we have $x_{t_0+t} = \widetilde{\varphi}_{t_0+t} + y_t$, $t \in [0,\sigma]$. Thus y satisfies the equation

$$\begin{aligned} y(t) = &- k_1(t_0,\varphi) - k_2(t_0,\varphi) + k_1(t_0+t, y_t + \widetilde{\varphi}_{t_0+t}) + k_2(t_0+t, y_t + \widetilde{\varphi}_{t_0+t}) \\ &+ \frac{1}{\Gamma(\alpha)}\int_0^t (t-s)^{\alpha-1}F(t_0+s, y_s + \widetilde{\varphi}_{t_0+s})ds, \quad t \in [0,\sigma]. \end{aligned} \tag{2.4}$$

Since k_1, k_2 are continuous and x_t is continuous in t, there exists $\sigma' > 0$, when $0 < t < \sigma'$,

$$|k_1(t_0+t, y_t + \widetilde{\varphi}_{t_0+t}) - k_1(t_0,\varphi)| < \frac{\gamma}{3}, \tag{2.5}$$

and

$$|k_2(t_0+t, y_t + \widetilde{\varphi}_{t_0+t}) - k_2(t_0,\varphi)| < \frac{\gamma}{3}. \tag{2.6}$$

Choose

$$\eta = \min\left\{\sigma, \sigma', \left(\frac{\gamma\Gamma(\alpha)(1+\beta)^{1-\alpha_1}}{3M}\right)^{\frac{1}{(1+\beta)(1-\alpha_1)}}\right\} \tag{2.7}$$

where $\beta = \frac{\alpha-1}{1-\alpha_1} \in (-1,0)$ and $M = \|m\|_{L^{\frac{1}{\alpha_1}}I}$.

Define $E(\eta,\gamma)$ as follows

$$E(\eta,\gamma) = \{y \in C([-\tau,\eta],\mathbb{R}^n) : \ y(s) = 0 \ \text{ for } s \in [-\tau,0] \ \text{ and } \|y\| \le \gamma\}.$$

Then $E(\eta,\gamma)$ is a closed bounded and convex subset of $C([-\tau,\sigma],\mathbb{R}^n)$. On $E(\eta,\gamma)$ we define the operators S and U as follows

$$(Sy)(t) = \begin{cases} 0, & t \in [-\tau,0], \\ -k_1(t_0,\varphi) + k_1(t_0+t, y_t + \widetilde{\varphi}_{t_0+t}), & t \in [0,\eta], \end{cases}$$

$$(Uy)(t) = \begin{cases} 0, & t \in [-\tau,0], \\ -k_2(t_0,\varphi) + k_2(t_0+t, y_t + \widetilde{\varphi}_{t_0+t}) & \\ \quad + \dfrac{1}{\Gamma(\alpha)}\displaystyle\int_0^t (t-s)^{\alpha-1}F(t_0+s, y_s + \widetilde{\varphi}_{t_0+s})ds, & t \in [0,\eta]. \end{cases}$$

It is easy to see that the operator equation

$$y = Sy + Uy \tag{2.8}$$

has a solution $y \in E(\eta, \gamma)$ if and only if y is a solution of the equation (2.4). Thus $x(t_0 + t) = y(t) + \widetilde{\varphi}(t_0 + t)$ is a solution of the equation (2.1) on $[0, \eta]$. Therefore, the existence of a solution of the fractional IVP (2.1) is equivalent that (2.8) has a fixed point in $E(\eta, \gamma)$.

Now we show that $S + U$ has a fixed point in $E(\eta, \gamma)$. The proof is divided into three steps.

Claim I. $Sz + Uy \in E(\eta, \gamma)$ for every pair $z, y \in E(\eta, \gamma)$.

In fact, for every pair $z, y \in E(\eta, \gamma)$, $Sz + Uy \in C([-\tau, \eta], \mathbb{R}^n)$. Also, it is obvious that $(Sz + Uy)(t) = 0$, $t \in [-\tau, 0]$.

Moreover, for $t \in [0, \eta]$, by (2.5)-(2.7) and the condition (H3), we have

$$\begin{aligned}
&|(Sz)(t) + (Uy)(t)| \\
&\le |-k_1(t_0, \varphi) + k_1(t_0 + t, z_t + \widetilde{\varphi}_{t_0+t})| + |-k_2(t_0, \varphi) + k_2(t_0 + t, y_t + \widetilde{\varphi}_{t_0+t})| \\
&\quad + \frac{1}{\Gamma(\alpha)} \int_0^t |(t-s)^{\alpha-1} F(t_0 + s, y_s + \widetilde{\varphi}_{t_0+s})| ds \\
&\le \frac{2\gamma}{3} + \frac{1}{\Gamma(\alpha)} \left(\int_0^t (t-s)^{\frac{\alpha-1}{1-\alpha_1}} ds \right)^{1-\alpha_1} \left(\int_{t_0}^{t_0+t} (m(s))^{\frac{1}{\alpha_1}} ds \right)^{\alpha_1} \\
&\le \frac{2\gamma}{3} + \frac{1}{\Gamma(\alpha)} \left(\int_0^t (t-s)^{\frac{\alpha-1}{1-\alpha_1}} ds \right)^{1-\alpha_1} \left(\int_{t_0}^{t_0+\sigma} (m(s))^{\frac{1}{\alpha_1}} ds \right)^{\alpha_1} \\
&\le \frac{2\gamma}{3} + \frac{M\eta^{(1+\beta)(1-\alpha_1)}}{\Gamma(\alpha)(1+\beta)^{1-\alpha_1}} \\
&\le \gamma.
\end{aligned}$$

Therefore,

$$\|Sz + Uy\| = \sup_{t \in [0,\eta]} |(Sz)(t) + (Uy)(t)| \le \gamma,$$

which means that $Sz + Uy \in E(\eta, \gamma)$ for any $z, y \in E(\eta, \gamma)$.

Claim II. S is a contraction on $E(\eta, \gamma)$.

For any $y', y'' \in E(\eta, \gamma)$, $y'_t + \widetilde{\varphi}_{t_0+t}, y''_t + \widetilde{\varphi}_{t_0+t} \in A(\delta, \gamma)$. So by (H5), we get that

$$\begin{aligned}
|(Sy')(t) - (Sy'')(t)| &= |k_1(t_0 + t, y'_t + \widetilde{\varphi}_{t_0+t}) - k_1(t_0 + t, y''_t + \widetilde{\varphi}_{t_0+t})| \\
&\le l\|y' - y''\|,
\end{aligned}$$

which implies that

$$\|Sy' - Sy''\| \le l\|y' - y''\|.$$

In view of $0 < l < 1$, S is a contraction on $E(\eta, \gamma)$.

Claim III. U is a completely continuous operator.

Let

$$(U_1 y)(t) = \begin{cases} 0, & t \in [-\tau, 0], \\ -k_2(t_0, \varphi) + k_2(t_0 + t, y_t + \widetilde{\varphi}_{t_0+t}), & t \in [0, \eta] \end{cases}$$

and

$$(U_2y)(t) = \begin{cases} 0, & t \in [-\tau, 0], \\ \dfrac{1}{\Gamma(\alpha)} \displaystyle\int_0^t (t-s)^{\alpha-1} F(t_0+s, y_s + \widetilde{\varphi}_{t_0+s}) ds, & t \in [0, \eta]. \end{cases}$$

Clearly, $U = U_1 + U_2$.

Since k_2 is completely continuous, U_1 is continuous and $\{U_1y : y \in E(\eta, \gamma)\}$ is uniformly bounded. From the condition that the set $\{t \to k_2(t, x_t) : x \in \Lambda\}$ is equicontinuous for any bounded set Λ in $A(\sigma, \gamma)$, we can conclude that U_1 is a completely continuous operator.

On the other hand, for any $t \in [0, \eta]$, we have

$$\begin{aligned} |(U_2y)(t)| &\leq \frac{1}{\Gamma(\alpha)} \int_0^t (t-s)^{\alpha-1} |F(t_0+s, y_s + \widetilde{\varphi}_{t_0+s})| ds \\ &\leq \frac{1}{\Gamma(\alpha)} \left(\int_0^t (t-s)^{\frac{\alpha-1}{1-\alpha_1}} ds \right)^{1-\alpha_1} \left(\int_{t_0}^{t_0+t} (m(s))^{\frac{1}{\alpha_1}} ds \right)^{\alpha_1} \\ &\leq \frac{M\eta^{(1+\beta)(1-\alpha_1)}}{\Gamma(\alpha)(1+\beta)^{1-\alpha_1}}. \end{aligned}$$

Hence, $\{U_2y : y \in E(\eta, \gamma)\}$ is uniformly bounded.

Now, we will prove that $\{U_2y : y \in E(\eta, \gamma)\}$ is equicontinuous. For any $0 \leq t_1 < t_2 \leq \eta$ and $y \in E(\eta, \gamma)$, we get that

$$\begin{aligned} &|(U_2y)(t_2) - (U_2y)(t_1)| \\ &= \left| \frac{1}{\Gamma(\alpha)} \int_0^{t_1} \Big((t_2-s)^{\alpha-1} - (t_1-s)^{\alpha-1} \Big) F(t_0+s, y_s + \widetilde{\varphi}_{t_0+s}) ds \right. \\ &\quad \left. + \frac{1}{\Gamma(\alpha)} \int_{t_1}^{t_2} (t_2-s)^{\alpha-1} F(t_0+s, y_s + \widetilde{\varphi}_{t_0+s}) ds \right| \\ &\leq \frac{1}{\Gamma(\alpha)} \int_0^{t_1} \Big((t_1-s)^{\alpha-1} - (t_2-s)^{\alpha-1} \Big) |F(t_0+s, y_s + \widetilde{\varphi}_{t_0+s})| ds \\ &\quad + \frac{1}{\Gamma(\alpha)} \int_{t_1}^{t_2} (t_2-s)^{\alpha-1} |F(t_0+s, y_s + \widetilde{\varphi}_{t_0+s})| ds \\ &\leq \frac{M}{\Gamma(\alpha)} \left(\int_0^{t_1} \Big((t_1-s)^{\alpha-1} - (t_2-s)^{\alpha-1} \Big)^{\frac{1}{1-\alpha_1}} ds \right)^{1-\alpha_1} \\ &\quad + \frac{M}{\Gamma(\alpha)} \left(\int_{t_1}^{t_2} \Big((t_2-s)^{\alpha-1} \Big)^{\frac{1}{1-\alpha_1}} ds \right)^{1-\alpha_1} \\ &\leq \frac{M}{\Gamma(\alpha)} \left(\int_0^{t_1} (t_1-s)^{\beta} - (t_2-s)^{\beta} ds \right)^{1-\alpha_1} + \frac{M}{\Gamma(\alpha)} \left(\int_{t_1}^{t_2} (t_2-s)^{\beta} ds \right)^{1-\alpha_1} \\ &\leq \frac{M}{\Gamma(\alpha)(1+\beta)^{1-\alpha_1}} \Big(t_1^{1+\beta} - t_2^{1+\beta} + (t_2-t_1)^{1+\beta} \Big)^{1-\alpha_1} \\ &\quad + \frac{M}{\Gamma(\alpha)(1+\beta)^{1-\alpha_1}} (t_2-t_1)^{(1+\beta)(1-\alpha_1)} \end{aligned}$$

$$\leq \frac{2M}{\Gamma(\alpha)(1+\beta)^{1-\alpha_1}}(t_2 - t_1)^{(1+\beta)(1-\alpha_1)},$$

which means that $\{U_2 y : y \in E(\eta, \gamma)\}$ is equicontinuous. Moreover, it is clear that U_2 is continuous. So U_2 is a completely continuous operator. Then $U = U_1 + U_2$ is a completely continuous operator.

Therefore, Krasnoselskii fixed point theorem shows that $S + U$ has a fixed point on $E(\eta, \gamma)$, and hence the fractional IVP (2.1) has a solution $x(t) = \varphi(0) + y(t - t_0)$ for all $t \in [t_0, t_0 + \eta]$. □

In the case where $k_1 \equiv 0$, we get the following result.

Corollary 2.1. *Assume that there exist $\sigma \in (0, a)$ and $\gamma \in (0, \infty)$ such that (H1)-(H3) hold and*

(H5)$'$ *k is continuous and for any $x', x'' \in A(\sigma, \gamma)$, $t \in I$*

$$|k(t, x'_t) - k(t, x''_t)| \leq l\|x' - x''\|, \quad \text{where} \quad l \in (0, 1).$$

Then fractional IVP (2.1) has at least one solution on $[t_0, t_0 + \eta]$ for some positive number η.

In the case where $k_2 \equiv 0$, we have the following result.

Corollary 2.2. *Assume that there exist $\sigma \in (0, a)$ and $\gamma \in (0, \infty)$ such that (H1)-(H3) hold and*

(H6)$'$ *k is completely continuous and for any bounded set Λ in $A(\sigma, \gamma)$, the set $\{t \to k(t, x_t) : x \in \Lambda\}$ is equicontinuous on $C(I, \mathbb{R}^n)$.*

Then fractional IVP (2.1) has at least one solution on $[t_0, t_0 + \eta]$ for some positive number η.

2.2.3 *Extremal Solutions*

Define the order relation " $\leq$ " by the cone K in $C(J, \mathbb{R}^n)$, given by

$$K = \{z \in C(J, \mathbb{R}^n) \mid z(t) \geq 0 \ \text{ for all } \ t \in J\}.$$

Clearly, the cone K is normal in $C(J, \mathbb{R}^n)$. Note that the order relation " $\leq$ " in $C(J, \mathbb{R}^n)$ also induces the order relation in the space $\mathcal{C}$ which we also denote by " $\leq$ " itself when there is no confusion.

We give the following definitions in the sequel.

Definition 2.1. A mapping $f : I \times \mathcal{C} \to \mathbb{R}^n$ is called $L^{\frac{1}{\delta}}$-Lipschitz if

(i) $t \mapsto f(t, z)$ is Lebesgue measurable for each $z \in \mathcal{C}$;
(ii) there exist a constant $\delta \in [0, \alpha)$ and a function $l \in L^{\frac{1}{\delta}}(I, \mathbb{R}_+)$ such that

$$|f(t, z) - f(t, y)| \leq l(t)\|z - y\|_*, \quad \text{a.e.} \ \ t \in I$$

for all $z, y \in \mathcal{C}$.

Definition 2.2. A mapping $g : I \times \mathcal{C} \to \mathbb{R}^n$ is said to be Carathéodory if
(i) $t \mapsto g(t, z)$ is Lebesgue measurable for each $z \in \mathcal{C}$;
(ii) $z \mapsto g(t, z)$ is continuous almost everywhere for $t \in I$.

Furthermore, a Carathéodory function $g(t, z)$ is called $L^{\frac{1}{\beta}}$-Carathéodory if
(iii) for each real number $r > 0$, there exist a constant $\beta \in [0, \alpha)$ and a function $m_r \in L^{\frac{1}{\beta}}(I, \mathbb{R}_+)$ such that

$$|g(t, z)| \leq m_r(t), \quad \text{a.e. } t \in I$$

for all $z \in \mathcal{C}$ with $\|z\|_* \leq r$.

Definition 2.3. A mapping $h : I \times \mathcal{C} \to \mathbb{R}^n$ is said to be Chandrabhan if
(i) $t \mapsto h(t, z)$ is Lebesgue measurable for each $z \in \mathcal{C}$;
(ii) $z \mapsto h(t, z)$ is nondecreasing almost everywhere for $t \in I$.

Furthermore, a Chandrabhan function $h(t, z)$ is called $L^{\frac{1}{\gamma}}$-Chandrabhan if
(iii) for each real number $r > 0$, there exist a constant $\gamma \in [0, \alpha)$ and a function $w_r \in L^{\frac{1}{\gamma}}(I, \mathbb{R}_+)$ such that

$$|h(t, z)| \leq w_r(t), \quad \text{a.e. } t \in I$$

for all $z \in \mathcal{C}$ with $\|z\|_* \leq r$.

Definition 2.4. A function $x \in C(J, \mathbb{R}^n)$ is called a solution of fractional IVP (2.1) on J if
(i) the function $[x(t) - k(t, x_t)]$ is absolutely continuous on I;
(ii) $x_{t_0} = \varphi$, and
(iii) x satisfies the equation in (2.1).

Definition 2.5. A function $a \in C(J, \mathbb{R}^n)$ is called a lower solution of fractional IVP (2.1) on J if the function $[a(t) - k(t, a_t)]$ is absolutely continuous on I, and

$$\begin{cases} {}^C_{t_0}D^\alpha_t(a(t) - k(t, a_t)) \leq F(t, a_t), & \text{a.e. } t \in (t_0, t_0 + \sigma] \\ a_{t_0} \leq \varphi. \end{cases}$$

Again, a function $b \in C(J, \mathbb{R}^n)$ is called an upper solution of fractional IVP (2.1) on J if the function $[b(t) - k(t, b_t)]$ is absolutely continuous on I, and

$$\begin{cases} {}^C_{t_0}D^\alpha_t(b(t) - k(t, b_t)) \geq F(t, b_t), & \text{a.e. } t \in (t_0, t_0 + \sigma] \\ b_{t_0} \geq \varphi. \end{cases}$$

Finally, a function $x \in C(J, \mathbb{R}^n)$ is a solution of fractional IVP (2.1) on J if it is a lower as well as a upper solution of fractional IVP (2.1) on J.

Definition 2.6. A solution x_M of fractional IVP (2.1) is said to be maximal if for any other solution x to fractional IVP (2.1), one has $x(t) \leq x_M(t)$ for all $t \in J$. Again, a solution x_m of fractional IVP (2.1) is said to be minimal if $x_m(t) \leq x(t)$ for all $t \in J$, where x is any solution for fractional IVP (2.1) on J.

We need the following hypotheses in the sequel.

(F1) $F(t, z_t) = f(t, z_t) + g(t, z_t) + h(t, z_t)$, where $f, g, h : I \times \mathcal{C} \to \mathbb{R}^n$;
(F2) fractional IVP (2.1) has a lower solution a and an upper solution b with $a \leq b$;
(k0) $k(t, z)$ is continuous with respect to t on I for any $z \in \mathcal{C}$;
(k1) $|k(t, z) - k(t, y)| \leq k_0\|z - y\|_*$, for $z, y \in \mathcal{C}$, $t \in I$, where $k_0 > 0$;
(k2) $k(t, z)$ is nondecreasing with respect to z for any $z \in \mathcal{C}$ and almost all $t \in I$;
(f1) f is $L^{\frac{1}{\delta}}$-Lipschitz, and there exists $\eta \in [0, \alpha)$ such that $|f(t, 0)| \in L^{\frac{1}{\eta}}(I, \mathbb{R}_+)$;
(f2) $f(t, z)$ is nondecreasing with respect to z for any $z \in \mathcal{C}$ and almost all $t \in I$;
(g1) g is $L^{\frac{1}{\beta}}$-Carathéodory;
(g2) $g(t, z)$ is nondecreasing with respect to z for any $z \in \mathcal{C}$ and almost all $t \in I$;
(h1) h is $L^{\frac{1}{\gamma}}$-Chandrabhan.

For any positive constant r, let $B_r = \{z \in C(J, \mathbb{R}^n) : \|z\| \leq r\}$. Set

$$q_1 = \frac{\alpha - 1}{1 - \delta} \in (-1, 0), \qquad L = \|l\|_{L^{\frac{1}{\delta}} I}$$

and

$$q_2 = \frac{\alpha - 1}{1 - \beta} \in (-1, 0), \qquad M_r = \|m_r\|_{L^{\frac{1}{\beta}} I}.$$

In order to prove our main results, we need the following lemma.

Lemma 2.2. *Assume that the hypotheses (F1), (f1), (g1) and (h1) hold.* $x \in C(J, \mathbb{R}^n)$ *is a solution for fractional IVP* (2.1) *on* J *if and only if* x *satisfies the following relation*

$$\begin{cases} x(t) = \varphi(0) + k(t, x_t) - k(t_0, \varphi) + \dfrac{1}{\Gamma(\alpha)} \displaystyle\int_{t_0}^{t} (t - s)^{\alpha - 1} F(s, x_s) ds, & \text{for } t \in I, \\ x(t_0 + \theta) = \varphi(\theta), & \text{for } \theta \in I_0. \end{cases} \tag{2.9}$$

Proof. For any positive constant r and $x \in B_r$, since x_t is continuous in t, according to (g_1) and Definition 2.2(i)-(ii), $g(t, x_t)$ is a measurable function on I. Direct calculation gives that $(t - s)^{\alpha - 1} \in L^{\frac{1}{1-\beta}}[t_0, t]$, for $t \in I$ and $\beta \in [0, \alpha)$. By using Lemma 1.1 (Hölder inequality) and Definition 2.2(iii), for $t \in I$, we obtain that

$$\begin{aligned} \int_{t_0}^{t} |(t - s)^{\alpha - 1} g(s, x_s)| ds &\leq \left(\int_{t_0}^{t} (t - s)^{\frac{\alpha - 1}{1 - \beta}} ds \right)^{1 - \beta} \|m_r\|_{L^{\frac{1}{\beta}}[t_0, t]} \\ &= \left(\int_{t_0}^{t} (t - s)^{q_2} ds \right)^{1 - \beta} \|m_r\|_{L^{\frac{1}{\beta}}[t_0, t]} \\ &\leq \frac{M_r}{(1 + q_2)^{1 - \beta}} \, \sigma^{(1 + q_2)(1 - \beta)}, \end{aligned}$$

which means that $(t - s)^{\alpha - 1} g(s, x_s)$ is Lebesgue integrable with respect to $s \in [t_0, t]$ for all $t \in I$ and $x \in B_r$.

According to (f1), for $t \in I$ and $x \in B_r$, we get that

$$|f(t,x_t)| \le l(t)\|x_t\|_* + |f(t,0)| \le l(t)r + |f(t,0)|.$$

Using the similar argument and noting that (f1) and (h1), we can get that $(t-s)^{\alpha-1}f(s,x_s)$ and $(t-s)^{\alpha-1}h(s,x_s)$ are Lebesgue integrable with respect to $s \in [t_0,t]$ for all $t \in I$ and $x \in B_r$. Thus, according to (F1), we get that $(t-s)^{\alpha-1}F(s,x_s)$ is Lebesgue integrable with respect to $s \in [t_0,t]$ for all $t \in I$ and $x \in B_r$.

Let $G(\theta,s) = (t-\theta)^{-\alpha}|\theta - s|^{\alpha-1}m_r(s)$. Since $G(\theta,s)$ is a nonnegative, measurable function on $D = [t_0,t] \times [t_0,t]$ for $t \in I$, we have

$$\int_{t_0}^{t}\left(\int_{t_0}^{t} G(\theta,s)ds\right)d\theta = \int_D G(\theta,s)dsd\theta = \int_{t_0}^{t}\left(\int_{t_0}^{t} G(\theta,s)d\theta\right)ds$$

and

$$\begin{aligned}
\int_D G(\theta,s)dsd\theta &= \int_{t_0}^{t}\left(\int_{t_0}^{t} G(\theta,s)ds\right)d\theta \\
&= \int_{t_0}^{t}(t-\theta)^{-\alpha}\left(\int_{t_0}^{t}|\theta-s|^{\alpha-1}m_r(s)ds\right)d\theta \\
&= \int_{t_0}^{t}(t-\theta)^{-\alpha}\left(\int_{t_0}^{\theta}(\theta-s)^{\alpha-1}m_r(s)ds\right)d\theta \\
&\quad + \int_{t_0}^{t}(t-\theta)^{-\alpha}\left(\int_{\theta}^{t}(s-\theta)^{\alpha-1}m_r(s)ds\right)d\theta \\
&\le \frac{2M_r}{(1+q_2)^{1-\beta}}\,\sigma^{(1+q_2)(1-\beta)}\int_{t_0}^{t}(t-\theta)^{-\alpha}d\theta \\
&\le \frac{2M_r}{(1-\alpha)(1+q_2)^{1-\beta}}\,\sigma^{(1+q_2)(1-\beta)+1-\alpha}.
\end{aligned}$$

Therefore, $G_1(\theta,s) = (t-\theta)^{-\alpha}(\theta-s)^{\alpha-1}g(s,x_s)$ is a Lebesgue integrable function on $D = (t_0,t) \times (t_0,t)$, then we have

$$\int_{t_0}^{t} d\theta \int_{t_0}^{\theta} G_1(\theta,s)ds = \int_{t_0}^{t} ds \int_{s}^{t} G_1(\theta,s)d\theta.$$

We now prove that

$${}_{t_0}D_t^{\alpha}\left({}_{t_0}D_t^{-\alpha}F(t,x_t)\right) = F(t,x_t), \quad \text{for } t \in (t_0, t_0+\sigma].$$

Indeed, we have

$$\begin{aligned}
{}_{t_0}D_t^{\alpha}\left({}_{t_0}D_t^{-\alpha}g(t,x_t)\right) &= \frac{1}{\Gamma(1-\alpha)\Gamma(\alpha)}\frac{d}{dt}\int_{t_0}^{t}(t-\theta)^{-\alpha}\left(\int_{t_0}^{\theta}(\theta-s)^{\alpha-1}g(s,x_s)ds\right)d\theta \\
&= \frac{1}{\Gamma(1-\alpha)\Gamma(\alpha)}\frac{d}{dt}\int_{t_0}^{t}d\theta\int_{t_0}^{\theta}G_1(\theta,s)ds \\
&= \frac{1}{\Gamma(1-\alpha)\Gamma(\alpha)}\frac{d}{dt}\int_{t_0}^{t}ds\int_{s}^{t}G_1(\theta,s)d\theta
\end{aligned}$$

$$
\begin{aligned}
&= \frac{1}{\Gamma(1-\alpha)\Gamma(\alpha)} \frac{d}{dt} \int_{t_0}^{t} g(s, x_s) ds \int_{s}^{t} (t-\theta)^{-\alpha}(\theta - s)^{\alpha-1} d\theta \\
&= \frac{d}{dt} \int_{t_0}^{t} g(s, x_s) ds \\
&= g(t, x_t), \quad \text{for } t \in (t_0, t_0 + \sigma].
\end{aligned}
$$

Similarly, we can get

$$
{}_{t_0}D_t^{\alpha}\Big({}_{t_0}D_t^{-\alpha} f(t, x_t)\Big) = f(t, x_t),\ {}_{t_0}D_t^{\alpha}\Big({}_{t_0}D_t^{-\alpha} h(t, x_t)\Big) = h(t, x_t), \quad \text{for } t \in (t_0, t_0+\sigma],
$$

which implies

$$
{}_{t_0}D_t^{\alpha}\Big({}_{t_0}D_t^{-\alpha} F(t, x_t)\Big) = F(t, x_t), \quad \text{for } t \in (t_0, t_0 + \sigma].
$$

If x satisfies the relation (2.9), then we get that $x(t) - k(t, x_t)$ is absolutely continuous on I. In fact, for any disjoint family of open intervals $\{(a_i, b_i)\}_{1 \le i \le n}$ on I with $\sum_{i=1}^{n}(b_i - a_i) \to 0$, we have

$$
\begin{aligned}
&\sum_{i=1}^{n} \frac{1}{\Gamma(\alpha)} \left| \int_{t_0}^{b_i} (b_i - s)^{\alpha-1} g(s, x_s) ds - \int_{t_0}^{a_i} (a_i - s)^{\alpha-1} g(s, x_s) ds \right| \\
&\le \sum_{i=1}^{n} \frac{1}{\Gamma(\alpha)} \left| \int_{a_i}^{b_i} (b_i - s)^{\alpha-1} g(s, x_s) ds \right| \\
&\quad + \sum_{i=1}^{n} \frac{1}{\Gamma(\alpha)} \left| \int_{t_0}^{a_i} (b_i - s)^{\alpha-1} g(s, x_s) ds - \int_{t_0}^{a_i} (a_i - s)^{\alpha-1} g(s, x_s) ds \right| \\
&\le \sum_{i=1}^{n} \frac{1}{\Gamma(\alpha)} \int_{a_i}^{b_i} (b_i - s)^{\alpha-1} m_r(s) ds \\
&\quad + \sum_{i=1}^{n} \frac{1}{\Gamma(\alpha)} \int_{t_0}^{a_i} ((a_i - s)^{\alpha-1} - (b_i - s)^{\alpha-1}) m_r(s) ds \\
&\le \sum_{i=1}^{n} \frac{1}{\Gamma(\alpha)} \left(\int_{a_i}^{b_i} (b_i - s)^{\frac{\alpha-1}{1-\beta}} ds \right)^{1-\beta} \|m_r\|_{L^{\frac{1}{\beta}} I} \\
&\quad + \sum_{i=1}^{n} \frac{1}{\Gamma(\alpha)} \left(\int_{t_0}^{a_i} \left((a_i - s)^{\frac{\alpha-1}{1-\beta}} - (b_i - s)^{\frac{\alpha-1}{1-\beta}} \right) ds \right)^{1-\beta} \|m_r\|_{L^{\frac{1}{\beta}} I} \\
&\le \sum_{i=1}^{n} \frac{(b_i - a_i)^{(1+q_2)(1-\beta)}}{\Gamma(\alpha)(1+q_2)^{1-\beta}} \|m_r\|_{L^{\frac{1}{\beta}} I} \\
&\quad + \sum_{i=1}^{n} \frac{(a_i^{1+q_2} - b_i^{1+q_2} + (b_i - a_i)^{1+q_2})^{1-\beta}}{\Gamma(\alpha)(1+q_2)^{1-\beta}} \|m_r\|_{L^{\frac{1}{\beta}} I} \\
&\le 2 \sum_{i=1}^{n} \frac{(b_i - a_i)^{(1+q_2)(1-\beta)}}{\Gamma(\alpha)(1+q_2)^{1-\beta}} M_r \\
&\to 0.
\end{aligned}
$$

Using the similar method, as $\sum_{i=1}^{n}(b_i - a_i) \to 0$, we can get that

$$\sum_{i=1}^{n} \frac{1}{\Gamma(\alpha)} \left| \int_{t_0}^{b_i} (b_i - s)^{\alpha-1} f(s, x_s) ds - \int_{t_0}^{a_i} (a_i - s)^{\alpha-1} f(s, x_s) ds \right| \to 0$$

and

$$\sum_{i=1}^{n} \frac{1}{\Gamma(\alpha)} \left| \int_{t_0}^{b_i} (b_i - s)^{\alpha-1} h(s, x_s) ds - \int_{t_0}^{a_i} (a_i - s)^{\alpha-1} h(s, x_s) ds \right| \to 0.$$

Hence, $\sum_{i=1}^{n} \|x(b_i) - k(b_i, x_{b_i}) - x(a_i) + k(a_i, x_{a_i})\| \to 0$, as $\sum_{i=1}^{n}(b_i - a_i) \to 0$. Therefore, $x(t) - k(t, x_t)$ is absolutely continuous on I which implies that $x(t) - k(t, x_t)$ is differentiable for almost all $t \in I$. According to the argument above, for almost all $t \in (t_0, t_0 + \sigma]$, we have

$$\begin{aligned}
{}_{t_0}^{C}D_t^{\alpha}(x(t) - k(t, x_t)) &= {}_{t_0}^{C}D_t^{\alpha}\left(\varphi(0) - k(t_0, \varphi) + \frac{1}{\Gamma(\alpha)} \int_{t_0}^{t} (t - s)^{\alpha-1} F(s, x_s) ds\right) \\
&= {}_{t_0}^{C}D_t^{\alpha}\left(\frac{1}{\Gamma(\alpha)} \int_{t_0}^{t} (t - s)^{\alpha-1} F(s, x_s) ds\right) \\
&= {}_{t_0}^{C}D_t^{\alpha}\left({}_{t_0}D_t^{-\alpha} F(t, x_t)\right) \\
&= {}_{t_0}D_t^{\alpha}\left({}_{t_0}D_t^{-\alpha} F(t, x_t)\right) - {}_{t_0}D_t^{-\alpha} F(t, x_t)\big|_{t=t_0} \frac{(t - t_0)^{-\alpha}}{\Gamma(1 - \alpha)} \\
&= F(t, x_t) - {}_{t_0}D_t^{-\alpha} F(t, x_t)\big|_{t=t_0} \frac{(t - t_0)^{-\alpha}}{\Gamma(1 - \alpha)}.
\end{aligned}$$

Since $(t-s)^{\alpha-1}F(s, x_s)$ is Lebesgue integrable with respect to $s \in [t_0, t]$ for all $t \in I$, we know that ${}_{t_0}D_t^{-\alpha} F(t, x_t)\big|_{t=t_0} = 0$, which means that ${}_{t_0}^{C}D_t^{q} x(t) = F(t, x_t)$, a.e. $t \in (t_0, t_0 + \sigma]$. Hence, $x \in C(J, \mathbb{R}^n)$ is a solution of fractional IVP (2.1). On the other hand, it is obvious that if $x \in C(J, \mathbb{R}^n)$ is a solution of fractional IVP (2.1), then x satisfies the relation (2.9), and this completes the proof. □

Theorem 2.2. *Assume that the hypotheses (F1), (F2), (k0)-(k2), (f1), (f2), (g1), (g2) and (h1) hold. Then fractional IVP* (2.1) *has a minimal and a maximal solution in* $[a, b]$ *defined on* J *provided that*

$$k_0 + \frac{L\sigma^{(1+q_1)(1-\delta)}}{\Gamma(\alpha)(1 + q_1)^{1-\delta}} < 1. \tag{2.10}$$

Proof. Define three operators A, B and C on $C(J, \mathbb{R}^n)$ as follows

$$\begin{cases}
(Ax)(t) = k(t, x_t) - k(t_0, \varphi) + \dfrac{1}{\Gamma(\alpha)} \displaystyle\int_{t_0}^{t} (t - s)^{\alpha-1} f(s, x_s) ds, & \text{for } t \in I, \\
(Ax)(t_0 + \theta) = 0, & \text{for } \theta \in I_0,
\end{cases}$$

$$\begin{cases}
(Bx)(t) = \varphi(0) + \dfrac{1}{\Gamma(\alpha)} \displaystyle\int_{t_0}^{t} (t - s)^{\alpha-1} g(s, x_s) ds, & \text{for } t \in I, \\
(Bx)(t_0 + \theta) = \varphi(\theta), & \text{for } \theta \in I_0,
\end{cases}$$

and

$$\begin{cases} (Cx)(t) = \dfrac{1}{\Gamma(\alpha)} \displaystyle\int_{t_0}^{t} (t-s)^{\alpha-1} h(s, x_s) ds, & \text{for } t \in I, \\ (Cx)(t_0 + \theta) = 0, & \text{for } \theta \in I_0, \end{cases}$$

where $x \in C(J, \mathbb{R}^n)$.

Obviously, $Ax + Bx + Cx \in C(J, \mathbb{R}^n)$ for every $x \in C(J, \mathbb{R}^n)$. From Lemma 2.2, we get that fractional IVP (2.1) is equivalent to the operator equation $(Ax)(t) + (Bx)(t) + (Cx)(t) = x(t)$ for $t \in J$. Now we show that the operator equation $Ax + Bx + Cx = x$ has a least and a greatest solution in $[a, b]$. The proof is divided into three steps.

Claim I. A is a contraction in $C(J, \mathbb{R}^n)$.

For any $x, y \in C(J, \mathbb{R}^n)$ and $t \in I$, according to (k1) and (f1), we have

$$\begin{aligned} |(Ax)(t) - (Ay)(t)| &\leq |k(t, x_t) - k(t, y_t)| + \frac{1}{\Gamma(\alpha)} \int_{t_0}^{t} (t-s)^{\alpha-1} |f(s, x_s) - f(s, y_s)| ds \\ &\leq k_0 \|x_t - y_t\|_* + \frac{1}{\Gamma(\alpha)} \int_{t_0}^{t} (t-s)^{\alpha-1} l(s) \|x_s - y_s\|_* \, ds \\ &\leq k_0 \|x - y\| + \frac{1}{\Gamma(\alpha)} \left(\int_{t_0}^{t} (t-s)^{\frac{\alpha-1}{1-\delta}} ds \right)^{1-\delta} \|l\|_{L^{\frac{1}{\delta}}[t_0, t]} \|x - y\| \\ &\leq k_0 \|x - y\| + \frac{L\sigma^{(1+q_1)(1-\delta)}}{\Gamma(\alpha)(1+q_1)^{1-\delta}} \|x - y\| \\ &= \left(k_0 + \frac{L\sigma^{(1+q_1)(1-\delta)}}{\Gamma(\alpha)(1+q_1)^{1-\delta}} \right) \|x - y\|, \end{aligned}$$

which implies that $\|Ax - Ay\| \leq \left(k_0 + \frac{L\sigma^{(1+q_1)(1-\delta)}}{\Gamma(\alpha)(1+q_1)^{1-\delta}} \right) \|x - y\|$. Therefore, A is a contraction in $C(J, \mathbb{R}^n)$ according to (2.10).

Claim II. B is a completely continuous operator and C is a totally bounded operator.

For any $x \in C(J, \mathbb{R}^n)$, we can choose a positive constant r such that $\|x\| \leq r$. Firstly, we will prove that B is continuous on B_r. For $x^n, x \in B_r, n = 1, 2, ...$ with $\lim_{n\to\infty} \|x^n - x\| = 0$, we get

$$\lim_{n\to\infty} x_s^n = x_s, \quad \text{for} \quad s \in I.$$

Thus, by (g1) and Definition 2.2(ii), and noting that x_s is continuous with respect to s on I, we have

$$\lim_{n\to\infty} g(s, x_s^n) = g(s, x_s), \quad \text{a.e } s \in I.$$

On the other hand, noting that $|g(s, x_s^n) - g(s, x_s)| \leq 2m_r(s)$, by Lebesgue dominated convergence theorem, we have

$$|(Bx^n)(t) - (Bx)(t)| \leq \frac{1}{\Gamma(\alpha)} \int_{t_0}^{t} (t-s)^{\alpha-1} |g(s, x_s^n) - g(s, x_s)| ds \to 0, \quad \text{as } n \to \infty,$$

which implies

$$\|Bx^n - Bx\| \to 0 \quad \text{as} \quad n \to \infty.$$

This means that B is continuous.

Next, we will show that for any positive constant r, $\{Bx : x \in B_r\}$ is relatively compact. It suffices to show that the family of functions $\{Bx : x \in B_r\}$ is uniformly bounded and equicontinuous.

For any $x \in B_r$ and $t \in I$, we have

$$\begin{aligned}
|(Bx)(t)| &\le |\varphi(0)| + \frac{1}{\Gamma(\alpha)} \int_{t_0}^{t} (t-s)^{\alpha-1} |g(s, x_s)| ds \\
&\le |\varphi(0)| + \frac{1}{\Gamma(\alpha)} \left(\int_0^t (t-s)^{\frac{\alpha-1}{1-\beta}} ds \right)^{1-\beta} \|m_r\|_{L^{\frac{1}{\beta}}[t_0,t]} \\
&\le |\varphi(0)| + \frac{M_r}{\Gamma(\alpha)} \left(\int_0^t (t-s)^{q_2} ds \right)^{1-\beta} \\
&\le |\varphi(0)| + \frac{M_r \sigma^{(1+q_2)(1-\beta)}}{\Gamma(\alpha)(1+q_2)^{1-\beta}}.
\end{aligned}$$

For $\theta \in I_0$, we have $|(Bx)(t_0 + \theta)| = |\varphi(\theta)|$. Thus $\{Bx : x \in B_r\}$ is uniformly bounded. In the following, we will show that $\{Bx : x \in B_r\}$ is a family of equicontinuous functions.

For any $x \in B_r$ and $t_0 \le t_1 < t_2 \le t_0 + \sigma$, we get

$$\begin{aligned}
&|(Bx)(t_2) - (Bx)(t_1)| \\
&= \frac{1}{\Gamma(\alpha)} \left| \int_{t_0}^{t_1} \left((t_2-s)^{\alpha-1} - (t_1-s)^{\alpha-1} \right) g(s, x_s) ds + \int_{t_1}^{t_2} (t_2-s)^{\alpha-1} g(s, x_s) ds \right| \\
&\le \frac{1}{\Gamma(\alpha)} \int_{t_0}^{t_1} \left| \left((t_2-s)^{\alpha-1} - (t_1-s)^{\alpha-1} \right) g(s, x_s) \right| ds \\
&\quad + \frac{1}{\Gamma(\alpha)} \int_{t_1}^{t_2} |(t_2-s)^{\alpha-1} g(s, x_s)| ds \\
&\le \frac{1}{\Gamma(\alpha)} \int_{t_0}^{t_1} \left((t_1-s)^{\alpha-1} - (t_2-s)^{\alpha-1} \right) m_r(s) ds + \frac{1}{\Gamma(\alpha)} \int_{t_1}^{t_2} (t_2-s)^{\alpha-1} m_r(s) ds \\
&\le \frac{1}{\Gamma(\alpha)} \left(\int_{t_0}^{t_1} \left((t_1-s)^{\alpha-1} - (t_2-s)^{\alpha-1} \right)^{\frac{1}{1-\beta}} ds \right)^{1-\beta} \|m_r\|_{L^{\frac{1}{\beta}}[t_0,t_1]} \\
&\quad + \frac{1}{\Gamma(\alpha)} \left(\int_{t_1}^{t_2} \left((t_2-s)^{\alpha-1} \right)^{\frac{1}{1-\beta}} ds \right)^{1-\beta} \|m_r\|_{L^{\frac{1}{\beta}}[t_1,t_2]} \\
&\le \frac{M_r}{\Gamma(\alpha)} \left(\int_{t_0}^{t_1} (t_1-s)^{q_2} - (t_2-s)^{q_2} ds \right)^{1-\beta} + \frac{M_r}{\Gamma(\alpha)} \left(\int_{t_1}^{t_2} (t_2-s)^{q_2} ds \right)^{1-\beta} \\
&\le \frac{M_r}{\Gamma(\alpha)(1+q_2)^{1-\beta}} \left((t_1-t_0)^{1+q_2} - (t_2-t_0)^{1+q_2} + (t_2-t_1)^{1+q_2} \right)^{1-\beta} \\
&\quad + \frac{M_r}{\Gamma(\alpha)(1+q_2)^{1-\beta}} (t_2-t_1)^{(1+q_2)(1-\beta)}
\end{aligned}$$

$$\le \frac{2M_r}{\Gamma(\alpha)(1+q_2)^{1-\beta}}(t_2-t_1)^{(1+q_2)(1-\beta)}.$$

As $t_2 - t_1 \to 0$, the right-hand side of the above inequality tends to zero independently of $x \in B_r$. In view of the continuity of φ, we can get that $\{Bx : x \in B_r\}$ is a family of equicontinuous functions. Therefore, $\{Bx : x \in B_r\}$ is relatively compact by Arzela-Ascoli theorem.

Using the similar argument, we can get that $\{Cx : x \in B_r\}$ is also relatively compact, which means that C is totally bounded.

Claim III. A, B and C are three monotone increasing operators.

Since $x, y \in C(J, \mathbb{R}^n)$ with $x \le y$ implies that $x_t \le y_t$ for $t \in I$, according to (k2) and (f2), we have

$$\begin{aligned}(Ax)(t) &= k(t,x_t) - k(t_0,\varphi) + \frac{1}{\Gamma(\alpha)}\int_{t_0}^{t}(t-s)^{\alpha-1}f(s,x_s)ds\\ &\le k(t,y_t) - k(t_0,\varphi) + \frac{1}{\Gamma(\alpha)}\int_{t_0}^{t}(t-s)^{\alpha-1}f(s,y_s)ds\\ &= (Ay)(t).\end{aligned}$$

Hence A is a monotone increasing operator. Similarly, we can conclude that B and C are also monotone increasing operators according to (g2), (h1) and Definition 2.3(ii).

Clearly, K is a normal cone. From (F2) and Definition 2.5, we have that $a \le Aa + Ba + Ca$ and $b \ge Ab + Bb + Cb$ with $a \le b$. Thus the operators A, B and C satisfy all the conditions of Theorem 1.11 and hence the operator equation $Ax + Bx + Cx = x$ has a least and a greatest solution in $[a, b]$. Therefore, fractional IVP (2.1) has a minimal and a maximal solution on J. □

Example 2.1. Consider the following IVP of scalar discontinuous fractional functional differential equation

$$\begin{cases} {}_0^C D_t^{\frac{1}{2}} x(t) &= F(t,x_t)\\ &= f(t) + \zeta(t)x(t) + \frac{1}{t^{1/3}}x(t-1) + \zeta(t)h(x(t)), \quad \text{a.e. } t \in (0,\sigma],\\ x(\theta) = 0, & \qquad \theta \in [-1,0], \end{cases} \tag{2.11}$$

where $0 < \sigma \le (\frac{1}{2\Gamma(3/2)})^2 = \frac{1}{\pi}$ and we take functions $f(t)$, $\zeta(t)$ and $h(x(t))$ as follows

$$f(t) = \begin{cases} t, & 0 \le t \le \frac{\sigma}{2},\\ 0, & \frac{\sigma}{2} < t \le \sigma, \end{cases} \qquad \zeta(t) = \begin{cases} 0, & 0 \le t \le \frac{\sigma}{2},\\ 1, & \frac{\sigma}{2} < t \le \sigma, \end{cases}$$

and

$$h(x(t)) = \begin{cases} x(t), & x(t) \ge 0,\\ x(t) - 1, & x(t) < 0. \end{cases}$$

Evidently, the function

$$F(t,\varphi)=f(t,\varphi)+g(t,\varphi)+h(t,\varphi),\quad \varphi\in C([-1,0],\mathbb{R}),$$

where

$$f(t,\varphi)=f(t)+\zeta(t)\varphi(0),\quad g(t,\varphi)=\frac{1}{t^{1/3}}\varphi(-1)\quad\text{and}\quad h(t,\varphi)=\zeta(t)h(\varphi(0)).$$

One can easily check that $a(t)=0$ is a lower solution of fractional IVP (2.11). On the other hand, let

$$b(t)=\begin{cases}t, & t\in[0,\sigma],\\ 0, & t\in[-1,0].\end{cases}$$

Then, $b\in C([-1,\sigma],\mathbb{R})$ is a upper solution of fractional IVP (2.11). In fact, direct calculation gives that

$${}_0^CD_t^{\frac{1}{2}}b(t)=\frac{t^{\frac{1}{2}}}{\Gamma(\frac{3}{2})}\geq 2t\geq F(t,b_t)=\begin{cases}t, & 0<t\leq\dfrac{\sigma}{2},\\ 2t, & \dfrac{\sigma}{2}<t\leq\sigma,\end{cases}\quad\text{for } t\in(0,\sigma].$$

Moreover, noting that $\Gamma(\frac{3}{2})=\frac{\sqrt{\pi}}{2}$, it is easy to verify that conditions (k0)-(k2), (f1)-(f2), (g1)-(g2), (h1) and (2.10) are satisfied. Therefore, Theorem 2.2 allows us to conclude that fractional IVP (2.11) has a minimal and a maximal solution in $[0,b]$ defined on $[-1,\sigma]$.

2.3 *p*-Type Neutral Equations

2.3.1 *Introduction*

Let $\mathcal{C}=C([-1,0],\mathbb{R}^n)$ denote the space of continuous functions on $[-1,0]$. For any element $\varphi\in\mathcal{C}$, define the norm $\|\varphi\|_*=\sup_{\theta\in[-1,0]}|\varphi(\theta)|$.

Consider the IVP of fractional p-type neutral functional differential equations of the form

$${}_{t_0}^CD_t^q g(t,x_t)=f(t,x_t), \tag{2.12}$$

$$x_{t_0}=\varphi,\qquad (t_0,\varphi)\in\Omega, \tag{2.13}$$

where ${}_{t_0}^CD_t^q$ is Caputo fractional derivative of order $0<q<1$, Ω is an open subset of $[0,\infty)\times\mathcal{C}$ and $g,f:\Omega\to\mathbb{R}^n$ are given functionals satisfying some assumptions that will be specified later. $x_t\in\mathcal{C}$ is defined by $x_t(\theta)=x(p(t,\theta))$, where $-1\leq\theta\leq 0$, $p(t,\theta)$ is a p-function.

Definition 2.7. (Lakshmikantham, Wen and Zhang, 1994) A function $p\in C(J\times[-1,0],\mathbb{R})$ is called a p-function if it has the following properties:

(i) $p(t,0)=t$;
(ii) $p(t,-1)$ is a nondecreasing function of t;
(iii) there exists a $\sigma\geq-\infty$ such that $p(t,\theta)$ is an increasing function for θ for each $t\in(\sigma,\infty)$;

(iv) $p(t, 0) - p(t, -1) > 0$ for $t \in (\sigma, \infty)$.

In the following, we suppose $t \in (\sigma, \infty)$.

Definition 2.8. (Lakshmikantham, Wen and Zhang, 1994) Let $t_0 \geq 0, A > 0$ and $x \in C([p(t_0, -1), t_0 + A], \mathbb{R}^n)$. For any $t \in [t_0, t_0 + A]$, we define x_t by

$$x_t(\theta) = x(p(t, \theta)), \quad -1 \leq \theta \leq 0,$$

so that $x_t \in \mathcal{C} = C([-1, 0], \mathbb{R}^n)$.

Note that the frequently used symbol "x_t" (in Hale, 1977; Lakshmikantham, 2008; Lakshmikantham, Wen and Zhang, 1994, $x_t(\theta) = x(t + \theta)$, where $-r \leq \theta \leq 0, r > 0, r = \text{const}$) in the theory of functional differential equations with bounded delay is a partial case of the above definition. Indeed, in this case we can put $p(t, \theta) = t + r\theta, \ \theta \in [-1, 0]$.

Definition 2.9. A function x is said to be a solution of fractional IVP (2.12)-(2.13) on $[p(t_0, -1), t_0 + \alpha]$, if there are $t_0 \geq 0, \alpha > 0$, such that

(i) $x \in C([p(t_0, -1), t_0 + \alpha], \mathbb{R}^n)$ and $(t, x_t) \in \Omega$, for $t \in [t_0, t_0 + \alpha]$;
(ii) $x_{t_0} = \varphi$;
(iii) $g(t, x_t)$ is differentiable and (2.12) holds almost everywhere on $[t_0, t_0 + \alpha]$.

We need the following lemma relative to p-function before we proceed further, which is taken from Lakshmikantham, Wen and Zhang, 1994.

Lemma 2.3. *(Lakshmikantham, Wen and Zhang, 1994) Suppose that $p(t, \theta)$ is a p-function. For $A > 0$, $\tau \in (\sigma, \infty)$ (τ may be σ if $\sigma > -\infty$), let $x \in C([p(\tau, -1), \tau + A], \mathbb{R}^n)$ and $\varphi \in C([-1, 0], \mathbb{R}^n)$. Then we have*

(i) *x_t is continuous in t on $[\tau, \tau + A]$ and $\tilde{p}(t, \theta) = p(\tau + t, \theta) - \tau$ is also a p-function;*
(ii) *if $p(\tau + t, -1) < \tau$ for $t > 0$, then there exists $-1 < s(\tau, t) < 0$ such that $p(\tau + t, s(\tau, t)) = \tau$ and*

$$\begin{cases} p(\tau + t, -1) \leq p(\tau + t, \theta) \leq \tau, & \text{for } -1 \leq \theta \leq s(\tau, t), \\ \tau \leq p(\tau + t, \theta) \leq \tau + t, & \text{for } s(\tau, t) \leq \theta \leq 0. \end{cases}$$

Moreover, $s \to 0$ uniformly in τ as $t \to 0$;
(iii) *there exists a function $\eta \in C([p(\tau, -1), \tau], \mathbb{R}^n)$ such that*

$$\eta(p(\tau, \theta)) = \varphi(\theta) \ \text{ for } -1 \leq \theta \leq 0.$$

It is well known that a neutral functional differential equation (NFDE for short) is one in which the derivatives of the past history or derivatives of functionals of the past history are involved as well as the present state of the system. In other words, in order to guarantee that the equation (2.12) is NFDE, the coefficient of $x(t)$ that is contained in $g(t, x_t)$ can not be equal to zero. Then we need introduce the concept of atomic.

Let $g \in C(\mathbb{R}^+ \times \mathcal{C}, \mathbb{R}^n)$ and $g(t, \varphi)$ be linear in φ. Then Riesz representation theorem shows that there exists a $n \times n$ matrix function $\eta(t, \theta)$ of bounded variation such that

$$g(t, \varphi) = \int_{-\gamma}^{0} [d_\theta \eta(t, \theta)]\varphi(\theta).$$

For $t_0 \geq 0$ and $\theta_0 \in (-\gamma, 0)$, if

$$\det[\eta(t_0, \theta_0^+) - \eta(t_0, \theta_0^-)] \neq 0,$$

then we say that $g(t, \varphi)$ is atomic at θ_0 for t_0. Similarly, one can define $g(t, \varphi)$ to be atomic at the endpoints $-r$ and 0 for t_0. If for every $t \geq 0$, $g(t, \varphi)$ is atomic at θ_0 for t, then we say that $g(t, \varphi)$ is atomic at θ_0 for $\mathbb{R}^+$. If $g(t, \varphi)$ is not linear in φ, suppose that $g(t, \varphi)$ has a Fréchet derivative with respect to φ, then $g'_\varphi(t, \varphi)\psi \in \mathbb{R}^n$ for $(t, \varphi) \in \mathbb{R}^+ \times \mathcal{C}$ and $\psi \in \mathcal{C}$, where g'_φ denote the Fréchet derivative of g with respect to φ. Then $g'_\varphi(t, \varphi)$ is a linear mapping from $\mathcal{C}$ into $\mathbb{R}^n$ and therefore

$$g'_\varphi(t, \varphi)\psi = \int_{-\gamma}^{0} [d_\theta \mu(t, \varphi, \theta)]\psi(\theta),$$

where $\mu(t, \varphi, \theta)$ is a matrix function of bounded variation. As before, if $\det[\mu(t_0, \varphi_0, \theta_0^+) - \mu(t_0, \varphi_0, \theta_0^-)] \neq 0$, for $t_0 \geq 0$, then we say that, the nonlinear $g(t, \varphi)$ is atomic at θ_0 for (t_0, φ_0). If $g(t, \varphi)$ is atomic at θ_0, for every (t, φ), then we say that $g(t, \varphi)$ is atomic at θ_0 for $\mathbb{R}^+ \times \mathcal{C}$.

For a detailed discussion on atomic concept we refer the reader to the books Hale, 1977; Lakshmikantham, Wen and Zhang, 1994.

Lemma 2.4. *(Hale, 1977; Lakshmikantham, Wen and Zhang, 1994) Suppose that $g(t, \varphi)$ is atomic at zero on Ω. Then there are a continuous $n \times n$ matrix function $A(t, \varphi)$ with $\det A(t, \varphi) \neq 0$ on Ω and a functional $L(t, \varphi, \psi)$ which is linear in ψ such that*

$$g'_\varphi(t, \varphi)\psi = A(t, \varphi)\psi(0) + L(t, \varphi, \psi).$$

Moreover, there exists a continuous function $\gamma : \Omega \times [0, 1] \to \mathbb{R}^+$ with $\gamma(t, \varphi, 0) = 0$ such that for every $s \in [0, 1]$ and ψ with $(t, \psi) \in \Omega$, $\psi(\theta) = 0$ for $-1 \leq \theta \leq -s$,

$$|L(t, \varphi, \psi)| \leq \gamma(t, \varphi, s)\|\psi\|_*.$$

In Subsection 2.3.2, we discuss various criteria on existence and uniqueness of solutions for fractional IVP (2.12)-(2.12). Subsection 2.3.3 is devoted to the continuous dependence on data for solutions.

2.3.2 *Existence and Uniqueness*

Assume that the functional $f : \Omega \to \mathbb{R}^n$ satisfies the following conditions.

(H1) $f(t, \varphi)$ is Lebesgue measurable with respect to t for any $(t, \varphi) \in \Omega$;
(H2) $f(t, \varphi)$ is continuous with respect to φ for any $(t, \varphi) \in \Omega$;

(H3) there exist a constant $q_1 \in (0, q)$ and a $L^{\frac{1}{q_1}}$-integrable function m such that $|f(t, \varphi)| \le m(t)$ for any $(t, \varphi) \in \Omega$.

For each $(t_0, \varphi) \in \Omega$, let $\tilde{p}(t, \theta) = p(t_0 + t, \theta) - t_0$. Define the function $\eta \in C([\tilde{p}(0, -1), \infty), \mathbb{R}^n)$ by

$$\begin{cases} \eta(\tilde{p}(0, \theta)) = \varphi(\theta), & \text{for } \theta \in [-1, 0], \\ \eta(t) = \varphi(0), & \text{for } t \in [0, \infty). \end{cases}$$

Let $x \in C([p(t_0, -1), t_0 + \alpha], \mathbb{R}^n)$, $\alpha < A$ and let

$$x(t_0 + t) = \eta(t) + z(t) \quad \text{for } \tilde{p}(0, -1) \le t \le \alpha. \tag{2.14}$$

Lemma 2.5. *$x(t)$ is a solution of fractional IVP (2.12)-(2.13) on $[p(t_0, -1), t_0 + \alpha]$ if and only if $z(t)$ satisfies the relation*

$$\begin{cases} g(t_0 + t, \tilde{\eta}_t + \tilde{z}_t) - g(t_0, \varphi) = \dfrac{1}{\Gamma(q)} \displaystyle\int_0^t (t - s)^{q-1} f(t_0 + s, \tilde{\eta}_s + \tilde{z}_s) ds, & t \in [0, \alpha], \\ \tilde{z}_0 = 0, \end{cases} \tag{2.15}$$

where $\tilde{\eta}_t(\theta) = \eta(\tilde{p}(t, \theta))$, $\tilde{z}_t(\theta) = z(\tilde{p}(t, \theta))$, for $-1 \le \theta \le 0$.

Proof. Since x_t is continuous in t, x_t is a measurable function, therefore according to conditions (H1) and (H2), $f(t, x_t)$ is Lebesgue measurable on $[t_0, t_0 + \alpha]$. Direct calculation gives that $(t - s)^{q-1} \in L^{\frac{1}{1-q_1}}[t_0, t]$, for $t \in [t_0, t_0 + \alpha]$ and $q_1 \in (0, q)$. In light of Hölder inequality, we obtain that $(t - s)^{q-1} f(s, x_s)$ is Lebesgue integrable with respect to $s \in [t_0, t]$ for all $t \in [t_0, t_0 + \alpha]$, and

$$\int_{t_0}^t |(t - s)^{q-1} f(s, x_s)| ds \le \|(t - s)^{q-1}\|_{L^{\frac{1}{1-q_1}}[t_0, t]} \|m\|_{L^{\frac{1}{q_1}}[t_0, t_0 + \alpha]}.$$

Hence $x(t)$ is the solution of fractional IVP (2.12)-(2.13) if and only if it satisfies the relation

$$\begin{cases} g(t, x_t) - g(t_0, x_{t_0}) = \dfrac{1}{\Gamma(q)} \displaystyle\int_{t_0}^t (t - u)^{q-1} f(u, x_u) du, & \text{for } t \in [t_0, t_0 + \alpha], \\ x_{t_0} = \varphi, \end{cases}$$

or setting $u = t_0 + s$,

$$\begin{cases} g(t_0 + t, x_{t_0 + t}) - g(t_0, x_{t_0}) = \dfrac{1}{\Gamma(q)} \displaystyle\int_0^t (t - s)^{q-1} f(t_0 + s, x_{t_0 + s}) ds, & \text{for } t \in [0, \alpha], \\ x_{t_0} = \varphi. \end{cases} \tag{2.16}$$

In view of (2.14), we have

$$\begin{aligned} x_{t_0 + t}(\theta) = x(p(t_0 + t, \theta)) &= x(\tilde{p}(t, \theta) + t_0) \\ &= \eta(\tilde{p}(t, \theta)) + z(\tilde{p}(t, \theta)) \\ &= \tilde{\eta}_t(\theta) + \tilde{z}_t(\theta), \quad \text{for } t \in [0, \alpha]. \end{aligned}$$

In particular $x_{t_0}(\theta) = \tilde{\eta}_0(\theta) + \tilde{z}_0(\theta)$. Hence $x_{t_0} = \varphi$ if and only if $\tilde{z}_0 = 0$ according to $\tilde{\eta} = \varphi$. It is clear that $x(t)$ satisfies (2.16) if and only if $z(t)$ satisfies (2.15). $\square$

For any $\sigma, \xi > 0$, let

$$E(\sigma,\xi) = \{z \in C([\tilde{p}(0,-1),\sigma],\mathbb{R}^n): \ \tilde{z}_0 = 0, \ \|\tilde{z}_t\|_* \le \xi \ \text{ for } t \in [0,\sigma]\},$$

which is a bounded closed convex subset of the Banach space $C([\tilde{p}(0,-1),\sigma],\mathbb{R}^n)$ endowed with supremum norm $\|\cdot\|$.

Lemma 2.6. *Suppose $\Omega \subseteq R \times \mathcal{C}$ is open, $W \subset \Omega$ is compact. For any a neighborhood $V' \subset \Omega$ of W, there is a neighborhood $V'' \subset V'$ of W and there exist positive numbers σ and ξ such that $(t_0 + t, \tilde{\eta}_t + \lambda\tilde{z}_t) \in V'$ with $0 \le \lambda \le 1$ for any $(t_0, \varphi) \in V'', t \in [0,\sigma]$ and $z \in E(\sigma,\xi)$.*

The proof of Lemma 2.6 is similar to that of (iii) of Lemma 2.1.8 in Lakshmikantham, Wen and Zhang, 1994, thus it is omitted.

Suppose g is atomic at 0 on Ω. Define two operators S and T on $E(\alpha,\beta)$ as follows

$$\begin{cases} (Sz)(t) = 0, & \text{for } t \in [\tilde{p}(0,-1),0], \\ A(t_0+t,\tilde{\eta}_t)(Sz)(t) = g(t_0,\varphi) - g(t_0+t,\tilde{\eta}_t+\tilde{z}_t) & \\ \qquad + g'_\varphi(t_0+t,\tilde{\eta}_t)\tilde{z}_t - L(t_0+t,\tilde{\eta}_t,\tilde{z}_t), & \text{for } t \in [0,\alpha] \end{cases} \tag{2.17}$$

and

$$\begin{cases} (Tz)(t) = 0, & \text{for } t \in [\tilde{p}(0,-1),0], \\ A(t_0+t,\tilde{\eta}_t)(Tz)(t) & \\ \qquad = \dfrac{1}{\Gamma(q)}\displaystyle\int_0^t (t-s)^{q-1} f(t_0+s,\tilde{\eta}_s+\tilde{z}_s)ds, & \text{for } t \in [0,\alpha], \end{cases} \tag{2.18}$$

where $A(t_0+t,\tilde{\eta}_t)$, $L(t_0+t,\tilde{\eta}_t,\tilde{z}_t)$ are functions described in Lemma 2.4.

It is clear that the operator equation

$$z = Sz + Tz \tag{2.19}$$

has a solution $z \in E(\alpha,\beta)$ if and only if z is a solution of (2.15). Therefore the existence of a solution of fractional IVP (2.12)-(2.13) is equivalent to determining $\alpha, \beta > 0$ such that $S+T$ has a fixed point on $E(\alpha,\beta)$.

We are now in a position to prove the following existence results, and the proof is based on Krasnoselskii fixed point theorem.

Theorem 2.3. *Suppose $g : \Omega \to \mathbb{R}^n$ is continuous together with its first Fréchet derivative with respect to the second argument, and g is atomic at 0 on Ω. $f : \Omega \to \mathbb{R}^n$ satisfies conditions (H1)-(H3). $W \subset \Omega$ is a compact set. Then there exist a neighborhood $V \subset \Omega$ of W and a constant $\alpha > 0$ such that for any $(t_0,\varphi) \in V$, fractional IVP* (2.12)-(2.13) *has a solution which exists on $[p(t_0,-1), t_0+\alpha]$.*

Proof. As we have mentioned above, we only need to discuss operator equation (2.19). For any $(t,\varphi) \in \Omega$, the property of the matrix function $A(t,\varphi)$ which is nonsingular and continuous on Ω implies that its inverse matrix $A^{-1}(t,\varphi)$ exists and is continuous on Ω. Let $V_0 \subset \Omega$ be the neighborhood of W, suppose that there is an $M > 0$ such that

$$|A^{-1}(t^0, \varphi)| \leq M, \quad \text{for every } (t^0, \varphi) \in V_0. \tag{2.20}$$

Note the complete continuity of the function $(m(t))^{\frac{1}{q_1}}$, hence, for a given positive number N, there must exist a number $\alpha_0 > 0$ satisfying

$$\left(\int_{t_0}^{t_0+\alpha_0} (m(s))^{\frac{1}{q_1}} ds\right)^{q_1} \leq N. \tag{2.21}$$

Due to the continuity of functions γ and g'_φ described in Lemma 2.4, there exist a neighborhood $V_1 \subset \Omega$ of W and constants $h_1 > 0$, $h_2 \in (0, 1]$ such that

$$|\gamma(t_0 + t, \tilde{\eta}_t, -s)| = |\gamma(t_0 + t, \tilde{\eta}_t, -s) - \gamma(t_0 + t, \tilde{\eta}_t, 0)| < \frac{1}{4M}, \tag{2.22}$$

$$|g'_\varphi(t_0 + t, \tilde{\eta}_t + \psi) - g'_\varphi(t_0 + t, \tilde{\eta}_t)| < \frac{1}{8M}, \tag{2.23}$$

whenever $(t_0 + t, \tilde{\eta}_t)$, $(t_0 + t, \tilde{\eta}_t + \psi) \in V_1$ and $\|\psi\|_* < h_1$, $-s \in [0, h_2]$.

Let $V_2 = V_0 \cap V_1$. According to Lemma 2.6, we can find a neighborhood $V \subset V_2$ of W and positive numbers α_1 and β with $\alpha_1 < \alpha_0$ and $\beta \leq h_1$ such that $(t_0+t, \tilde{\eta}_t+\lambda\tilde{z}_t) \in V_2$ with $0 \leq \lambda \leq 1$ for any $(t_0, \varphi) \in V$, $t \in [0, \alpha_1]$ and $z \in E(\alpha_1, \beta)$. Let

$$h(t_0 + t, \tilde{\eta}_t, \tilde{z}_t) = g(t_0 + t, \tilde{\eta}_t + \tilde{z}_t) - g(t_0 + t, \tilde{\eta}_t) - g'_\varphi(t_0 + t, \tilde{\eta}_t)\tilde{z}_t.$$

Then we have

$$\begin{aligned} |h(t_0 + t, \tilde{\eta}_t, \tilde{z}_t)| &= \left|\left(\int_0^1 g'_\varphi(t_0 + t, \tilde{\eta}_t + \lambda\tilde{z}_t)d\lambda - g'_\varphi(t_0 + t, \tilde{\eta}_t)\right)\tilde{z}_t\right| \\ &\leq \left|\int_0^1 g'_\varphi(t_0 + t, \tilde{\eta}_t + \lambda\tilde{z}_t) - g'_\varphi(t_0 + t, \tilde{\eta}_t)d\lambda\right| \|\tilde{z}_t\|_*. \end{aligned} \tag{2.24}$$

According to (2.20), (2.23) and (2.24), for any $(t_0, \varphi) \in V$, we have

$$|A^{-1}(t_0 + t, \tilde{\eta}_t)\ h(t_0 + t, \tilde{\eta}_t, \tilde{z}_t)| \leq \frac{\beta}{8}. \tag{2.25}$$

On the other hand, for any $z, w \in E(\alpha_1, \beta)$ and $t \in [0, \alpha_1]$

$$\|\lambda\tilde{z}_t + (1 - \lambda)\tilde{w}_t\|_* \leq \|\lambda\tilde{z}_t\|_* + \|(1 - \lambda)\tilde{w}_t\|_* \leq \lambda\beta + (1 - \lambda)\beta = \beta,$$

thus, $(t_0 + t, \tilde{\eta}_t + \lambda\tilde{z}_t + (1 - \lambda)\tilde{w}_t) \in V_2$, and

$$\begin{aligned} &|h(t_0 + t, \tilde{\eta}_t, \tilde{z}_t) - h(t_0 + t, \tilde{\eta}_t, \tilde{w}_t)| \\ &= |g(t_0 + t, \tilde{\eta}_t + \tilde{z}_t) - g(t_0 + t, \tilde{\eta}_t + \tilde{w}_t) - g'_\varphi(t_0 + t, \tilde{\eta}_t)(\tilde{z}_t - \tilde{w}_t)| \\ &= \left|\left(\int_0^1 g'_\varphi(t_0 + t, \tilde{\eta}_t + \tilde{w}_t + \lambda(\tilde{z}_t - \tilde{w}_t))d\lambda - g'_\varphi(t_0 + t, \tilde{\eta}_t)\right)(\tilde{z}_t - \tilde{w}_t)\right| \\ &\leq \left|\int_0^1 (g'_\varphi(t_0 + t, \tilde{\eta}_t + \lambda\tilde{z}_t + (1 - \lambda)\tilde{w}_t) - g'_\varphi(t_0 + t, \tilde{\eta}_t))\, d\lambda\right|\ \|\tilde{z}_t - \tilde{w}_t\|_*. \end{aligned} \tag{2.26}$$

From (2.20), (2.23) and (2.26), we have

$$|A^{-1}(t_0+t,\tilde{\eta}_t)[h(t_0+t,\tilde{\eta}_t,\tilde{z}_t)-h(t_0+t,\tilde{\eta}_t,\tilde{w}_t)]|\le\frac{1}{8}\|\tilde{z}_t-\tilde{w}_t\|_*. \tag{2.27}$$

By (ii) of Lemma 2.3, we can also choose $\alpha_2<\alpha_1$ such that for $t\in[0,\alpha_2]$, $-s(0,t)\in[0,h_2]$. From (2.20) and (2.22), we have

$$\begin{aligned}&|A^{-1}(t_0+t,\tilde{\eta}_t)||L(t_0+t,\tilde{\eta}_t,\tilde{z}_t)|\\&\le|A^{-1}(t_0+t,\tilde{\eta}_t)|\gamma(t_0+t,\tilde{\eta}_t,-s(0,t))\|\tilde{z}_t\|_*\\&\le\frac{1}{4}\|\tilde{z}_t\|_*,\end{aligned} \tag{2.28}$$

whenever $t\in[0,\alpha_2]$ and $z\in E(\alpha_2,\beta)$.

Now consider the expression $g(t_0,\varphi)-g(t_0+t,\tilde{\eta}_t)$. Since g is continuous in Ω and noting the facts that $\tilde{\eta}_t$ is continuous in t and $\tilde{\eta}_0=\varphi$, there exists a constant $\alpha_3<\alpha_2$ such that

$$|g(t_0,\varphi)-g(t_0+t,\tilde{\eta}_t)|<\frac{\beta}{8M}, \tag{2.29}$$

whenever $t\in[0,\alpha_3]$.

Set

$$\alpha=\min\left\{\alpha_3,(1+b)^{\frac{1}{1+b}}\left(\frac{\Gamma(q)\beta}{2MN}\right)^{\frac{1}{(1-q_1)(1+b)}}\right\}, \tag{2.30}$$

where $b=\frac{q-1}{1-q_1}\in(-1,0)$.

Now we show that for any $(t_0,\varphi)\in V$, $S+T$ has a fixed point on $E(\alpha,\beta)$, where S and T are defined as in (2.17) and (2.18) respectively. The proof is divided into three steps.

Claim I. $Sz+Tw\in E(\alpha,\beta)$ whenever $z,w\in E(\alpha,\beta)$.

Obviously, for every pair $z,w\in E(\alpha,\beta)$, $(Sz)(t)$ and $(Tw)(t)$ are continuous in $t\in[0,\alpha]$. From (2.25), (2.28) and (2.29), for $t\in[0,\alpha]$, we have

$$\begin{aligned}|(Sz)(t)|&\le|A^{-1}(t_0+t,\tilde{\eta}_t)|\,\{|g(t_0,\varphi)-g(t_0+t,\tilde{\eta}_t)|+|L(t_0+t,\tilde{\eta}_t,\tilde{z}_t)|\\&\quad+|g(t_0+t,\tilde{\eta}_t)-g(t_0+t,\tilde{\eta}_t+\tilde{z}_t)+g'_\varphi(t_0+t,\tilde{\eta}_t)\tilde{z}_t|\}\\&\le\frac{\beta}{2}.\end{aligned}$$

For $t\in[0,\alpha]$, by using (2.20), (2.21), (2.30) and Hölder inequality, we have

$$\begin{aligned}|(Tw)(t)|&\le|A^{-1}(t_0+t,\tilde{\eta}_t)|\,\frac{1}{\Gamma(q)}\left|\int_0^t(t-s)^{q-1}f(t_0+s,\tilde{\eta}_s+\tilde{w}_s)ds\right|\\&\le\frac{M}{\Gamma(q)}\left(\int_0^t\left((t-s)^{q-1}\right)^{\frac{1}{1-q_1}}ds\right)^{1-q_1}\left(\int_{t_0}^{t_0+\alpha}(m(s))^{\frac{1}{q_1}}ds\right)^{q_1}\\&\le\frac{MN}{\Gamma(q)}\left(\frac{1}{1+b}\alpha^{1+b}\right)^{1-q_1}\\&\le\frac{\beta}{2}.\end{aligned} \tag{2.31}$$

Thus $|(Sz)(t) + (Tw)(t)| \leq \beta$ i.e. $Sz + Tw \in E(\alpha, \beta)$, whenever $z, w \in E(\alpha, \beta)$.

Claim II. S is a contraction mapping from $E(\alpha, \beta)$ into itself whose contraction constant is independent of $(t_0, \varphi) \in V$.

For any $z, w \in E(\alpha, \beta)$, $\tilde{w}_0 - \tilde{z}_0 = 0$. Hence (ii) of Lemma 2.3 and Lemma 2.4 are applicable to $\tilde{w}_t - \tilde{z}_t$. For every pair $z, w \in E(\alpha, \beta)$, from (2.27), (2.28) and noting the fact that

$$\begin{aligned}
\sup_{0\leq t\leq\alpha} \|\tilde{z}_t - \tilde{w}_t\|_* &= \sup_{0\leq t\leq\alpha} \sup_{-1\leq\theta\leq 0} |z(\tilde{p}(t,\theta) - w(\tilde{p}(t,\theta))| \\
&= \sup_{0\leq t\leq\alpha} \sup_{\tilde{p}(t,-1)\leq s\leq t} |z(s) - w(s)| \\
&= \sup_{\tilde{p}(0,-1)\leq s\leq\alpha} |z(s) - w(s)| \\
&= \|z - w\|,
\end{aligned}$$

we have

$$\begin{aligned}
\|Sz - Sw\| &= \sup_{\tilde{p}(0,-1)\leq t\leq\alpha} |(Sz)(t) - (Sw)(t)| \\
&= \sup_{0\leq t\leq\alpha} |(Sz)(t) - (Sw)(t)| \\
&\leq \sup_{0\leq t\leq\alpha} \big\{|A^{-1}(t_0 + t, \tilde{\eta}_t)|\big(|L(t_0 + t, \tilde{\eta}_t, \tilde{w}_t - \tilde{z}_t)| \\
&\quad + |h(t_0 + t, \tilde{\eta}_t, \tilde{z}_t) - h(t_0 + t, \tilde{\eta}_t, \tilde{w}_t)|\big)\big\} \\
&\leq \left(\frac{1}{8} + \frac{1}{4}\right) \sup_{0\leq t\leq\alpha} \|\tilde{z}_t - \tilde{w}_t\|_* \\
&\leq \frac{3}{8}\|z - w\|.
\end{aligned}$$

Therefore S is a contraction mapping from $E(\alpha, \beta)$ into itself whose contraction constant is independent of $(t_0, \varphi) \in V$.

Claim III. Now we show that T is a completely continuous operator.

For any $z \in E(\alpha, \beta)$ and $0 \leq t_1 < t_2 \leq \alpha$, we get

$$\begin{aligned}
&|(Tz)(t_2) - (Tz)(t_1)| \\
&= \Bigg|A^{-1}(t_0 + t_2, \tilde{\eta}_{t_2})\frac{1}{\Gamma(q)}\int_0^{t_2} (t_2 - s)^{q-1} f(t_0 + s, \tilde{\eta}_s + \tilde{z}_s)ds \\
&\quad - A^{-1}(t_0 + t_1, \tilde{\eta}_{t_1})\frac{1}{\Gamma(q)}\int_0^{t_1} (t_1 - s)^{q-1} f(t_0 + s, \tilde{\eta}_s + \tilde{z}_s)ds\Bigg| \\
&= \Bigg|A^{-1}(t_0 + t_2, \tilde{\eta}_{t_2})\frac{1}{\Gamma(q)}\int_{t_1}^{t_2} (t_2 - s)^{q-1} f(t_0 + s, \tilde{\eta}_s + \tilde{z}_s)ds \\
&\quad + A^{-1}(t_0 + t_2, \tilde{\eta}_{t_2})\frac{1}{\Gamma(q)}\int_0^{t_1} (t_2 - s)^{q-1} f(t_0 + s, \tilde{\eta}_s + \tilde{z}_s)ds \\
&\quad - A^{-1}(t_0 + t_2, \tilde{\eta}_{t_2})\frac{1}{\Gamma(q)}\int_0^{t_1} (t_1 - s)^{q-1} f(t_0 + s, \tilde{\eta}_s + \tilde{z}_s)ds
\end{aligned}$$

$$
\begin{aligned}
&+ A^{-1}(t_0+t_2,\tilde{\eta}_{t_2})\frac{1}{\Gamma(q)}\int_0^{t_1}(t_1-s)^{q-1}f(t_0+s,\tilde{\eta}_s+\tilde{z}_s)ds \\
&- A^{-1}(t_0+t_1,\tilde{\eta}_{t_1})\frac{1}{\Gamma(q)}\int_0^{t_1}(t_1-s)^{q-1}f(t_0+s,\tilde{\eta}_s+\tilde{z}_s)ds\Big| \\
\le\ & \frac{|A^{-1}(t_0+t_2,\tilde{\eta}_{t_2})|}{\Gamma(q)}\left|\int_{t_1}^{t_2}(t_2-s)^{q-1}f(t_0+s,\tilde{\eta}_s+\tilde{z}_s)ds\right| \\
&+ \frac{|A^{-1}(t_0+t_2,\tilde{\eta}_{t_2})|}{\Gamma(q)}\left|\int_0^{t_1}[(t_2-s)^{q-1}-(t_1-s)^{q-1}]f(t_0+s,\tilde{\eta}_s+\tilde{z}_s)ds\right| \\
&+ \frac{|A^{-1}(t_0+t_2,\tilde{\eta}_{t_2})-A^{-1}(t_0+t_1,\tilde{\eta}_{t_1})|}{\Gamma(q)}\left|\int_0^{t_1}(t_1-s)^{q-1}f(t_0+s,\tilde{\eta}_s+\tilde{z}_s)ds\right| \\
=\ & \frac{|A^{-1}(t_0+t_2,\tilde{\eta}_{t_2})|}{\Gamma(q)}(I_1+I_2)+\frac{|A^{-1}(t_0+t_2,\tilde{\eta}_{t_2})-A^{-1}(t_0+t_1,\tilde{\eta}_{t_1})|}{\Gamma(q)}I_3,
\end{aligned}
$$

where

$$
\begin{aligned}
I_1 &= \left|\int_{t_1}^{t_2}(t_2-s)^{q-1}f(t_0+s,\tilde{\eta}_s+\tilde{z}_s)ds\right|, \\
I_2 &= \left|\int_0^{t_1}\left((t_2-s)^{q-1}-(t_1-s)^{q-1}\right)f(t_0+s,\tilde{\eta}_s+\tilde{z}_s)ds\right|, \\
I_3 &= \left|\int_0^{t_1}(t_1-s)^{q-1}f(t_0+s,\tilde{\eta}_s+\tilde{z}_s)ds\right|.
\end{aligned}
$$

By using analogous argument performed in (2.31), we can conclude that

$$
\begin{aligned}
I_1 &\le \frac{N}{(1+b)^{1-q_1}}(t_2-t_1)^{(1+b)(1-q_1)}, \\
I_3 &\le \frac{N}{(1+b)^{1-q_1}}\left({t_1}^{1+b}\right)^{1-q_1},
\end{aligned}
$$

and

$$
\begin{aligned}
I_2 &\le \left(\int_0^{t_1}\left|(t_2-s)^{q-1}-(t_1-s)^{q-1}\right|^{\frac{1}{1-q_1}}ds\right)^{1-q_1}\left(\int_{t_0}^{t_0+t_1}|f(s,x_s)|^{\frac{1}{q_1}}ds\right)^{q_1} \\
&\le N\left(\int_0^{t_1}(t_1-s)^b-(t_2-s)^b ds\right)^{1-q_1} \\
&= \frac{N}{(1+b)^{1-q_1}}\left({t_1}^{1+b}-{t_2}^{1+b}+(t_2-t_1)^{1+b}\right)^{1-q_1} \\
&\le \frac{N}{(1+b)^{1-q_1}}(t_2-t_1)^{(1+b)(1-q_1)},
\end{aligned}
$$

where $b=\frac{q-1}{1-q_1}\in(-1,0)$. Therefore

$$
\begin{aligned}
&|(Tz)(t_2)-(Tz)(t_1)| \\
&\le \frac{|A^{-1}(t_0+t_2,\tilde{\eta}_{t_2})|}{\Gamma(q)}\frac{2N}{(1+b)^{1-q_1}}(t_2-t_1)^{(1+b)(1-q_1)}
\end{aligned}
$$

$$+\frac{|A^{-1}(t_0+t_2,\tilde{\eta}_{t_2})-A^{-1}(t_0+t_1,\tilde{\eta}_{t_1})|}{\Gamma(q)}\frac{N}{(1+b)^{1-q_1}}\left({t_1}^{1+b}\right)^{1-q_1}.$$

Since $A^{-1}(t_0+t,\tilde{\eta}_t)$ is continuous in $t\in[0,\alpha]$, then $\{Tz; z\in E(\alpha,\beta)\}$ is equicontinuous. In addition, T is continuous from the condition (H2) and $\{Tz; z\in E(\alpha,\beta)\}$ is uniformly bounded from (2.31), thus T is a completely continuous operator by Arzela-Ascoli theorem.

Therefore, by Theorem 1.7, for every $(t_0,\varphi)\in V$, $S+T$ has a fixed point on $E(\alpha,\beta)$. Hence, fractional IVP (2.12)-(2.13) has a solution defined on $[p(t_0,-1),t_0+\alpha]$. □

Corollary 2.3. *Suppose that* $(t_0,\varphi)\in\Omega$ *is given,* g,f *are defined as in Theorem 2.3. Then there exists a solution of fractional IVP* (2.12)-(2.13).

Corollary 2.4. *Suppose that* Ω,f *are defined as in Theorem 2.3. If* $(t_0,\varphi)\in\Omega$ *is given, then the fractional IVP relative to fractional p-type retarded differential equations of the form*

$$\begin{cases} {}_{t_0}^{C}D_t^q x(t)=f(t,x_t),\\ x_{t_0}=\varphi \end{cases}$$

has a solution.

The following existence and uniqueness result for fractional IVP (2.12)-(2.13) is based on Banach contraction mapping principle.

Theorem 2.4. *Suppose* $(t_0,\varphi)\in\Omega$ *is given,* g *is defined as in Theorem 2.3.* $f:\Omega\to\mathbb{R}^n$ *satisfies the condition (H3) and*

(H4) $f(t,x_t)$ *is measurable for every* $(t,x_t)\in\Omega$;

(H5) *let* $A>0$, *there exists a nonnegative function* $\ell:[0,A]\to[0,\infty)$ *continuous at* $t=0$ *and* $\ell(0)=0$ *such that for any* (t,x_t), $(t,y_t)\in\Omega$, $t\in[t_0,t_0+A]$, *we have*

$$\left|\int_{t_0}^{t}(t-s)^{q-1}(f(s,x_s)-f(s,y_s))ds\right|\le\ell(t-t_0)\sup_{t_0\le s\le t}\|x_s-y_s\|_*.$$

Then fractional IVP (2.12)-(2.13) *has a unique solution.*

Proof. According to the argument of Theorem 2.3, it suffices to prove that $S+T$ has a unique fixed point on $E(\alpha,\beta)$, where $\alpha,\beta>0$ sufficiently small. Now, choose $\alpha\in(0,A]$ such that (2.30) holds and

$$c=\frac{3}{8}+\sup_{0\le s\le\alpha}\frac{|A^{-1}(t_0+s,\tilde{\eta}_s)|\,|\ell(s)|}{\Gamma(q)}<1.$$

Obviously, $S+T$ is a mapping from $E(\alpha,\beta)$ into itself. Using the same argument as that of Theorem 2.3, for any $z,w\in E(\alpha,\beta), t\in[0,\alpha]$, we get

$$|(Sz)(t)-(Sw)(t)|\le\frac{3}{8}\|z-w\|,$$

and

$$\begin{aligned}
|(Tz)(t)-(Tw)(t)| &\le \frac{|A^{-1}(t_0+t,\tilde{\eta}_t)|}{\Gamma(q)}\left|\int_0^t (t-s)^{q-1}f(t_0+s,\tilde{\eta}_s+\tilde{z}_s)ds\right.\\
&\quad \left.-\int_0^t (t-s)^{q-1}f(t_0+s,\tilde{\eta}_s+\tilde{w}_s)ds\right|\\
&\le \frac{|A^{-1}(t_0+t,\tilde{\eta}_t)|}{\Gamma(q)}\,|\ell(t)|\sup_{0\le s\le t}\|\tilde{z}_s-\tilde{w}_s\|_*\\
&\le \frac{\sup_{0\le s\le \alpha}|A^{-1}(t_0+s,\tilde{\eta}_s)|\;|\ell(s)|}{\Gamma(q)}\|z-w\|.
\end{aligned}$$

Therefore

$$\begin{aligned}
|(S+T)z(t)-(S+T)w(t)| &\le \left(\frac{3}{8}+\sup_{0\le s\le\alpha}\frac{|A^{-1}(t_0+s,\tilde{\eta}_s)|\;|\ell(s)|}{\Gamma(q)}\right)\|z-w\|\\
&= c\|z-w\|.
\end{aligned}$$

Hence, we have

$$\|(S+T)z-(S+T)w\|\le c\|z-w\|,$$

where $c<1$. By applying Theorem 1.4, we know that $S+T$ has a unique fixed point on $E(\alpha,\beta)$. The proof is completed. □

Corollary 2.5. *Suppose the condition (H5) of Theorem 2.4 is replaced by the following condition:*

(H5)$'$ *let $A>0$, there exist $q_2\in(0,q)$ and a real-valued function $\ell_1\in L^{\frac{1}{q_2}}[t_0,t_0+A]$ such that for any (t,x_t), $(t,y_t)\in\Omega$, $t\in[t_0,t_0+A]$, we have*

$$|f(t,x_t)-f(t,y_t)|\le \ell_1(t)\sup_{t_0\le s\le t}\|x_s-y_s\|_*.$$

Then the result of Theorem 2.4 holds.

Proof. It suffices to prove that the condition (H5) of Theorem 2.4 holds. Note that $\ell_1\in L^{\frac{1}{q_2}}[t_0,t_0+A]$, let $K=\|\ell_1\|_{L^{\frac{1}{q_2}}[t_0,\,t_0+A]}$. Then for any (t,x_t), $(t,y_t)\in\Omega$ we have

$$\begin{aligned}
&\left|\int_{t_0}^t (t-s)^{q-1}(f(s,x_s)-f(s,y_s))ds\right|\\
&\le \int_{t_0}^t (t-s)^{q-1}|f(s,x_s)-f(s,y_s)|ds\\
&\le \int_{t_0}^t (t-s)^{q-1}\ell_1(s)\,ds\sup_{t_0\le s\le t}\|x_s-y_s\|_*\\
&\le \frac{K}{(1+b_1)^{1-q_2}}\,(t-t_0)^{(1+b_1)(1-q_2)}\sup_{t_0\le s\le t}\|x_s-y_s\|_*,
\end{aligned}$$

where $b_1 = \frac{q-1}{1-q_2} \in (-1, 0)$. Let

$$\ell(t - t_0) = \frac{K}{(1 + b_1)^{1-q_2}} (t - t_0)^{(1+b_1)(1-q_2)}.$$

Obviously, $\ell : [0, A] \to [0, \infty)$ continuous at $t = 0$ and $\ell(0) = 0$. Then the condition (H5) of Theorem 2.4 holds. □

The next result is concerned with the uniqueness of solutions.

Theorem 2.5. *Suppose that g is defined as in Theorem 2.3 and the condition (H5)′ of Corollary 2.5 holds. If x is a solution of fractional IVP* (2.12)-(2.13), *then x is unique.*

Proof. Suppose (for contradiction) x and y are the solutions of fractional IVP (2.12)-(2.13) on $[p(t_0, -1), t_0 + A]$ with $x \neq y$, let

$$t_1 = \inf\{t \in [t_0, t_0 + A] : x(t) \neq y(t)\}.$$

Then $t_0 \leq t_1 < t_0 + A$ and

$$x(t) = y(t) \quad \text{for } p(t_0, -1) \leq t < t_1,$$

which implies that

$$x_t(\theta) = x(p(t, \theta)) = y(p(t, \theta)) = y_t(\theta), \ t_0 \leq t < t_1, \ -1 \leq \theta \leq 0. \tag{2.32}$$

Choose $\alpha > 0$ such that $t_1 + \alpha < t_0 + A$. According to (i) of Definition 2.9, we have

$$\{(t, x_t), \ t_1 \leq t \leq t_1 + \alpha\} \bigcup \{(t, y_t), \ t_1 \leq t \leq t_1 + \alpha\} \subset \Omega.$$

On the one hand, x and y satisfy (2.12)-(2.13) on $[t_0, t_0 + A]$, thus from (2.32) and the condition (H5)′, for $t \in [t_0, t_1 + \alpha]$, we have

$$\begin{aligned} |g(t, x_t) - g(t, y_t)| &\leq \frac{1}{\Gamma(q)} \left| \int_{t_0}^{t} (t - s)^{q-1} (f(s, x_s) - f(s, y_s)) ds \right| \\ &= \frac{1}{\Gamma(q)} \left| \int_{t_1}^{t} (t - s)^{q-1} (f(s, x_s) - f(s, y_s)) ds \right| \\ &\leq \frac{1}{\Gamma(q)} \int_{t_1}^{t} (t - s)^{q-1} \ell_1(s) \, ds \sup_{t_0 \leq s \leq t} \|x_s - y_s\|_* \\ &\leq \frac{K}{\Gamma(q)(1 + b_1)^{1-q_2}} \alpha^{(1+b_1)(1-q_2)} \sup_{t_1 \leq s \leq t_1 + \alpha} \|x_s - y_s\|_*, \end{aligned} \tag{2.33}$$

where $b_1 = \frac{q-1}{1-q_2} \in (-1, 0), K = \|\ell_1\|_{L^{\frac{1}{q_2}}[t_0, \, t_0 + A]}$.

On the other hand, since $g(t, \varphi)$ is continuously differentiable in φ, we have

$$g(t, x_t) - g(t, y_t) = g'_\varphi(t, y_t)(x_t - y_t) + k\|x_t - y_t\|_* \tag{2.34}$$

with $k \to 0$ as $\|x_t - y_t\|_* \to 0$.

By the hypothesis that $g(t,\varphi)$ is atomic at 0 on Ω, there exist a nonsingular continuous matrix function $A(t,y_t)$ and a function $L(t,y_t,\psi)$ which is linear in ψ such that

$$g'_\varphi(t,y_t)\psi = A(t,y_t)\psi(0) + L(t,y_t,\psi). \tag{2.35}$$

Moreover, there is a positive real-valued continuous function $\gamma(t,y_t,-s)$ such that for every $s\in[-1,0]$,

$$|L(t,y_t,\psi)| \le \gamma(t,y_t,-s)\|\psi\|_* \tag{2.36}$$

if $\psi(\theta)=0$ for $-1\le\theta\le s$.

Hence for every $t\in[t_1,t_1+\alpha]$, by (ii) of Lemma 2.3, there is $s(t_1,t-t_1)\in[-1,0]$ with $s(t_1,t-t_1)\to 0$ as $t\to t_1$ such that

$$|L(t,y_t,x_t-y_t)| \le \gamma(t,y_t,-s(t_1,t-t_1))\|x_t-y_t\|_*.$$

From (2.34)-(2.36), it follows that

$$g(t,x_t)-g(t,y_t) = A(t,y_t)(x(t)-y(t)) + L(t,y_t,x_t-y_t) + k\|x_t-y_t\|_*,$$

therefore

$$\begin{aligned}|x(t)-y(t)| \le |A^{-1}(t,y_t)|[|g(t,x_t)-g(t,y_t)|\\ +\gamma(t,y_t,-s(t_1,t-t_1))\|x_t-y_t\|_* + k\|x_t-y_t\|_*].\end{aligned}$$

Let $M_1 = \max\{|A^{-1}(t,y_t)| : t_1\le t\le t_1+\alpha\}$. Then by relation (2.33), for $t\in[t_1,t_1+\alpha]$, we have

$$|x(t)-y(t)| \le c_1 \sup_{t_1\le s\le t_1+\alpha}\|x_s-y_s\|_*,$$

where $c_1 = M_1\left(\frac{K}{\Gamma(q)(1+b_1)^{1-q_2}}\alpha^{(1+b_1)(1-q_2)} + \gamma(t,y_t,-s(t_1,t-t_1)) + k\right)$.
Noting that

$$\begin{aligned}\sup_{t_1\le s\le t_1+\alpha}\|x_s-y_s\|_* &= \sup_{t_1\le s\le t_1+\alpha}\ \sup_{-1\le\theta\le 0}|x(p(s,\theta))-y(p(s,\theta))|\\ &= \sup_{t_1\le s\le t_1+\alpha}\ \sup_{p(s,-1)\le\rho\le s}|x(\rho)-y(\rho)|\\ &= \sup_{p(t_1,-1)\le s\le t_1+\alpha}|x(s)-y(s)|,\end{aligned}$$

we have

$$\sup_{p(t_1,-1)\le s\le t_1+\alpha}|x(s)-y(s)| \le c_1 \sup_{p(t_1,-1)\le s\le t_1+\alpha}|x(s)-y(s)|.$$

Choose α so small that $c_1<1$. Thus

$$\sup_{p(t_1,-1)\le s\le t_1+\alpha}|x(s)-y(s)| = 0, \quad \text{i.e. } x(t)\equiv y(t), \quad \text{for } t_1\le t\le t_1+\alpha,$$

contradicting the definition of t_1. □

2.3.3 *Continuous Dependence*

The following lemma is introduced in Lakshmikantham, Wen and Zhang, 1994. However, for the sake of completeness, we outline its proof here.

Lemma 2.7. *Assume $x \in C([p(0,-1), A], \mathbb{R}^n)$. Then for every $t \in [0, A]$*

$$\|x_t\| \le \sup_{0 \le s \le t} |x(s)| + \|x_0\|.$$

Proof. By definition, $\|x_0\| = \sup_{-1\le\theta\le 0} |x(p(0,\theta))|$. If $p(t,-1) \ge 0$, then

$$0 \le p(t,\theta) \le t, \quad \text{for} \quad -1 \le \theta \le 0.$$

Thus,

$$\sup_{-1\le\theta\le 0} |x(p(t,\theta))| \le \sup_{0\le s\le t} |x(s)| \le \sup_{0\le s\le t} |x(s)| + \|x_0\|.$$

If $p(t,-1) < 0$, then by Lemma 2.3, there exists an $s \in [-1, 0]$ such that

$$p(t,-1) \le p(t,\theta) \le p(0,\theta), \quad \text{for} \quad -1 \le \theta \le s,$$

while

$$0 \le p(t,\theta) \le t, \quad \text{for} \;\; s \le \theta \le 0.$$

Hence

$$\begin{aligned}
\sup_{-1\le\theta\le 0} |x(p(t,\theta))| &\le \sup_{-1\le\theta\le s} |x(p(t,\theta))| + \sup_{s\le\theta\le 0} |x(p(t,\theta))| \\
&\le \sup_{-1\le\theta\le 0} |x(p(0,\theta))| + \sup_{s\le\theta\le 0} |x(p(t,\theta))| \\
&= \|x_0\| + \sup_{0\le s\le t} |x(s)|,
\end{aligned}$$

completing the proof. □

We can now prove the following result on continuous dependence.

Theorem 2.6. *Let $(t_0, \varphi) \in \Omega$ be given. Suppose that the solution $x = x(t_0, \varphi)$ of (2.12) through (t_0, φ) defined on $[t_0, A]$ is unique. Then for every $\epsilon > 0$, there exists a $\delta(\epsilon) > 0$ such that $(s, \psi) \in \Omega$, $|s - t_0| < \delta$ and $\|\psi - \varphi\| < \varphi$ imply*

$$\|x_t(s,\psi) - x_t(t_0,\varphi)\| < \epsilon, \quad \textit{for all} \;\; t \in [\sigma, A],$$

where $x(s, \psi)$ is the solution of (2.12) through (s, ψ) and $\sigma = \max\{s, t_0\}$.

Proof. In order to prove the theorem, it is enough to show that if $\{(t_k, \varphi^k)\} \subset \Omega$, with $t_k \to t_0$ and $\varphi^k \to \varphi$ as $k \to \infty$, then there is a natural number N such that each solution $x^k = x(t^k, \varphi^k)$ with $k \ge N$ of (2.12) through (t^k, φ^k) exists on $[p(t_k, -1), A]$ and $x^k(t) \to x(t)$ uniformly on $[p(\sigma, -1), A]$, where $\sigma = \sup\{t_0, t^k : k \ge N\}$.

Since $x_t(t_0, \varphi)$ is continuous in $t \in [t_0, A]$, the set $W = \{(t, x_t(t_0, \varphi)) : t \in [t_0, A]\}$ is compact in Ω. By Theorem 2.3, there exist a neighborhood V of W and number

$\alpha > 0$ such that for any $(s, \psi) \in V$, there is a solution $x(s, \psi)$ of (2.12) through (s, ψ) which exists at least on $[s, s + \alpha]$. Without loss of generality, we let $V = V(W, r)$, choose N so large that $|t_k - t_0| < \frac{r}{2}$ and $\|\varphi^k - \varphi\| < \frac{r}{2}$, so that $(t_k, \varphi^k) \in V$ for $k \geq N$. Thus $x^k = x(t_k, \varphi^k)$ exists at least on $[t_k, t_k + \alpha]$. For convenience, we shall denote $\varphi = \varphi^0$, $x = x^0$ and $x^k = x(t_k, \varphi^k)$, $k = 0, 1, \ldots$.

Let $p_k(t, \theta) = p(t_k + t, \theta) - t_k$. Define η^k, y^k the same way as in Lemma 2.5. Recalling the proof of Lemma 2.5, we see that y^k satisfies:

$$g(t_k+t, \eta_t^k+y_t^k)-g(t_k, \varphi^k) = \frac{1}{\Gamma(q)} \int_0^t (t-s)^{q-1} f(t_k+s, \eta_s^k+y_s^k)ds, \ \ t \in [0, \alpha] \quad (2.37)$$

if and only if x^k is the solution of (2.12) on $[p(t_k, -1), t_k+\alpha]$, where $\eta_t^k = \eta^k(p_k(t, \theta))$, $y_t^k = y^k(p_k(t, \theta))$.

Set $\bar{y}^k = y^k|_{[0,\alpha]}$, the restriction of y^k to $[0, \alpha]$. Let $\Lambda = \{\bar{y}^k : k = 0, 1, 2, \ldots\}$. For every $z_k = (t_k, \varphi^k)$, define operators $S(z_k) : \Lambda \to C([0, \alpha], \mathbb{R}^n)$ and $T(z_k) : \Lambda \to C([0, \alpha], \mathbb{R}^n)$ as follows:

$$\begin{aligned} S(z_k)z(t) = {} & A^{-1}(t_k + t, \eta_t^k)[g(t_k, \varphi^k) - g(t_k + t, \eta_t^k + z_t) \\ & + g'_\varphi(t_k + t, \eta_t^k)z_t - L(t_k + t, \eta_t^k + z_t)], \quad 0 \leq t \leq \alpha, \end{aligned}$$

and

$$T(z_k)z(t) = A^{-1}(t_k + t, \eta_t^k) \frac{1}{\Gamma(q)} \int_0^t (t - s)^{q-1} f(t_k + s, \eta_s^k + z_s)ds, \quad 0 \leq t \leq \alpha,$$

where $z_t(\theta) = z(\bar{p}_k(t, \theta))$ with $z_0 = 0$.

It is easy to see that $\{T(z_k)\bar{y}^k\}$ is compact in $C([0, \alpha], \mathbb{R}^n)$. Recalling the Theorem 2.3, we see that there exists a constant $\gamma \in [0, 1)$ which is independent to z_k such that

$$\|Sz - Sy\| \leq \gamma \|z - y\| \quad \text{for any} \quad z, y \in \Lambda. \quad (2.38)$$

Let $\{z_k : k = 0, 1, 2, \ldots\} = \bar{\Lambda}$. Denote the Kuratowskii noncompact measure of $A \subset C([0, \alpha], \mathbb{R}^n)$ by $\alpha(A)$. Then (2.38) implies that

$$\alpha\Bigg(\bigcup_{z_k \in \bar{\Lambda}} S(z_k)(\Lambda) \Bigg) \leq \gamma \alpha(\Lambda).$$

Let $R = S + T$. Thus $\bar{y}^k = R(z_k)\bar{y}_k$. By the well-known properties of Kuratowskii noncompact measure α, we immediately obtain that

$$\begin{aligned} \alpha(\Lambda) &= \alpha(\{R(z_k)\bar{y}_k\}) \leq \alpha(\{S(z_k)\bar{y}^k\}) + \alpha(\{T(z_k)\bar{y}^k\}) \\ &= \alpha(\{S(z_k)\bar{y}^k\}) \leq \alpha(\bigcup_{z_k \in \bar{\Lambda}} S(z_k)(\Lambda)) \leq \gamma \alpha(\Lambda). \end{aligned}$$

This means that $\alpha(\Lambda) = 0$ which implies Λ is relatively compact in $C([0, \alpha], \mathbb{R}^n)$. Hence there exists a subsequence of Λ, say $\{\bar{y}^{k_i}\}$, which converges uniformly on $[0, \alpha]$. Assume that

$$\bar{y}^{k_i}(t) \to \bar{y}^*(t) \quad \text{uniformly on} \quad [0, \alpha].$$

Define a function $y^* : [p_0(0,-1), \alpha] \to \mathbb{R}^n$ by

$$\begin{cases} y^*(t) = \bar{y}^*(t), & \text{for } 0 \le t \le \alpha, \\ y_0^* = 0, \end{cases}$$

where p_0 is such a p-function that $p_0(t,\theta) = p(t+t_0,\theta) - t_0$. Let $\delta = \inf\{p_k(0,-1) : k = 0,1,2,...\}$ and $\hat{y}^k$ denote the extension of y^k to $[\delta, \alpha]$ which is defined by

$$\begin{cases} \hat{y}^{k_i}(t) = y^{k_i}(t), & \text{for } 0 \le t \le \alpha, \\ \hat{y}^{k_i}(t) = 0, & \text{for } \delta \le t \le 0. \end{cases}$$

Obviously, $\{\hat{y}^{k_i}(t)\}$ converges uniformly on $[\delta, \alpha]$ as $k_i \to \infty$. Consequently, $\{\hat{y}^{k_i}\}$ is a relatively compact set.

We claim that $y_t^{k_i} \to y_t^*$ uniformly in $t \in [0, \alpha]$. In fact,

$$\begin{aligned} |p_{k_i}(t,\theta) - p_0(t,\theta)| &= |(p(t_{k_i} + t, \theta) - t_{k_i}) - (p(t_0 + t, \theta) - t_0)| \\ &\le |p(t_{k_i} + t, \theta) - p(t_0 + t, \theta)| + |t_0 - t_{k_i}|. \end{aligned}$$

Hence, for every $\mu > 0$ there exists a number L such that

$$|p_{k_i}(t,\theta) - p_0(t,\theta)| < \mu, \quad \text{whenever} \quad k_i \ge L.$$

We have the inequality

$$\begin{aligned} \left\| y_t^{k_i} - y_t^* \right\| &= \sup_{-1 \le \theta \le 0} \left| \hat{y}^{k_i}(p_{k_i}(t,\theta)) - y^*(p_0(t,\theta)) \right| \\ &= \sup_{-1 \le \theta \le 0} \left| \hat{y}^{k_i}(p_{k_i}(t,\theta)) - \hat{y}^{k_i}(p_0(t,\theta)) + \hat{y}^{k_i}(p_0(t,\theta)) - y^*(p_0(t,\theta)) \right| \\ &\le \sup_{-1 \le \theta \le 0} \left| \hat{y}^{k_i}(p_{k_i}(t,\theta)) - \hat{y}^{k_i}(p_0(t,\theta)) \right| \\ &\quad + \sup_{-1 \le \theta \le 0} \left| \hat{y}^{k_i}(p_0(t,\theta)) - y^*(p_0(t,\theta)) \right|. \end{aligned}$$

By Lemma 2.7, we get

$$\begin{aligned} \sup_{-1 \le \theta \le 0} \left| \hat{y}^{k_i}(p_0(t,\theta)) - y^*(p_0(t,\theta)) \right| &\le \sup_{0 \le \theta \le \alpha} \left| y^{k_i}(t) - y^*(t) \right| + \left\| \hat{y}_0^{k_i} - y_0^* \right\| \\ &= \sup_{0 \le \theta \le \alpha} \left| y^{k_i}(t) - y^*(t) \right|. \end{aligned}$$

For every $\epsilon > 0$ there exists a number L_1 such that

$$\sup_{0 \le \theta \le \alpha} \left| y^{k_i}(t) - y^*(t) \right| < \frac{\epsilon}{2}, \quad \text{for} \quad k_i \ge L_1,$$

by the definition of y^*. On the other hand, since $\{\hat{y}^{k_i}\}$ is an equi-continuous set, for the given ϵ, there exists a $\mu > 0$ such that

$$\left| \hat{y}^{k_i}(t) - \hat{y}^{k_i}(\tau) \right| < \frac{\epsilon}{2}, \quad \text{for} \quad |t - \tau| < \mu. \tag{2.39}$$

We can choose $L \ge L_1$ so that $|p_{k_i}(t,\theta) - p_0(t,\theta)| < \mu$. Thus (2.39) holds as long as $k_i \ge L$. Furthermore,

$$\left\| y_t^{k_i} - y_t \right\| < \epsilon, \quad \text{whenever} \quad k_i \ge L,$$

which is just our claim. A similar argument shows that $\eta_t^{k_i} \to \eta_t$ uniformly in $t \in [0, \alpha]$. The limiting process upon (2.37) yields

$$\begin{cases} g(t_0 + t, \eta_t + y_t^*) - g(t_0, \varphi) = \dfrac{1}{\Gamma(q)} \displaystyle\int_0^t (t-s)^{q-1} f(t_0 + s, \eta_s + y_s^*) ds, & 0 \le t \le \alpha, \\ y_0^* = 0, \end{cases}$$

which demonstrates that y^* as well as y^0 is a solution of the fractional IVP

$$\begin{cases} g(t_0 + t, \eta_t + y_t^*) - g(t_0, \varphi) = \dfrac{1}{\Gamma(q)} \displaystyle\int_0^t (t-s)^{q-1} f(t_0 + s, \eta_s + y_s) ds, & 0 \le t \le \alpha, \\ y_0 = 0. \end{cases}$$

The hypothesis that $x(t_0, \varphi)$ is unique, that is, y^0 is unique implies that $y^* = y^0$. Thus $y^{k_i}(t) \to y^0(t)$ uniformly on $[0, \alpha]$ as $k_i \to \infty$. The verified fact that every subsequence of sequence $\{y^k\}$ has a convergent subsequence with a same limit y^0 implies that the entire sequence $\{y^k\}$ converges to y^0. Translating these remarks back into x^k, we have indeed obtained the result stated in this theorem for the interval $[p(\sigma, -1), \sigma + \alpha]$.

Let $b = \sigma + \alpha$. If $b < A$, $(b, x_b) \in W$, we can choose $N_1 \ge N$ such that $(b, x_b^k) \in V$ as long as $k \ge N_1$. By Theorem 2.3, for every point (b, x_b^k) the solution $x^k(b, x_b^k)$ exists at least on $[p(b, -1), b + \alpha]$. The above argument can be adapted to this interval which yields the assertion that $x^k(b, x_b^k)(t) \to x^0(t_0, \varphi)(t)$ uniformly on the same interval. The conclusion stated in theorem can be verified by successive steps of finite intervals of length α. Hence the proof is completed. □

2.4 Neutral Equations with Infinite Delay

2.4.1 *Introduction*

In Section 2.4, we consider the initial value problem of fractional neutral functional differential equations with infinite delay of the form

$$ {}_{t_0}^{C}D_t^q g(t, x_t) = f(t, x_t), \qquad t \in [t_0, \infty), \tag{2.40}$$

$$ x_{t_0} = \varphi, \qquad (t_0, \varphi) \in [0, \infty) \times \Omega, \tag{2.41}$$

where ${}_{t_0}^{C}D_t^q$ is Caputo fractional derivative of order $0 < q < 1$, Ω is an open subset of B and $g, f : [t_0, \infty) \times \Omega \to \mathbb{R}^n$ are given functionals satisfying some assumptions that will be specified later. B is called a phase space that will be defined later.

If $x : (-\infty, A) \to \mathbb{R}^n$, $A \in (0, \infty)$, then for any $t \in [0, A)$ define x_t by $x_t(\theta) = x(t + \theta)$, for $\theta \in (-\infty, 0]$.

Denote by $BC(J, \mathbb{R}^n)$ the Banach space of all continuous and bounded functions from J into $\mathbb{R}^n$ with the norm $\| \cdot \|$.

To describe fractional neutral functional differential equations with infinite delay, we need to discuss a phase space B in a convenient way. We shall provide a general description of phase spaces of neutral differential equations with infinite delay which is taken from Lakshmikantham, Wen and Zhang, 1994.

Let B be a real vector space either

(i) of continuous functions that map $(-\infty, 0]$ to $\mathbb{R}^n$ with $\varphi = \psi$ if $\varphi(s) = \psi(s)$ on $(-\infty, 0]$ or

(ii) of measurable functions that map $(-\infty, 0]$ to $\mathbb{R}^n$ with $\varphi = \psi$ (or φ is equivalent to ψ) in B if $\varphi(s) = \psi(s)$ almost everywhere on $(-\infty, 0]$, and $\varphi(0) = \psi(0)$.

Let B be endowed with a norm $\|\cdot\|_B$ such that B is complete with respect to $\|\cdot\|_B$. Thus B equipped with norm $\|\cdot\|_B$ is a Banach space. We denote this space by $(B, \|\cdot\|_B)$ or simply by B, whenever no confusion arises.

Let $0 \leq a < A$. If $x : (-\infty, A) \to \mathbb{R}^n$ is given such that $x_a \in B$ and $x \in [a, A) \to \mathbb{R}^n$ is continuous, then $x_t \in B$ for all $t \in [a, A)$.

This is a very weak condition that the common admissible phase spaces and BC satisfy. For more details of the phase spaces, we refer the reader to Hino, Murakami and Naito, 1991; Lakshmikantham, Wen and Zhang, 1994.

Definition 2.10. A function $x : (-\infty, t_0 + \sigma) \to \mathbb{R}^n (t_0 \in [0, \infty), \sigma > 0)$ is said to be a solution of fractional IVP (2.40)-(2.41) through (t_0, φ) on $[t_0, t_0 + \sigma)$, if

(i) $x_{t_0} = \varphi$;

(ii) x is continuous on $[t_0, t_0 + \sigma)$;

(iii) $g(t, x_t)$ is absolutely continuous on $[t_0, t_0 + \sigma)$;

(iv) (2.40) holds almost everywhere on $[t_0, t_0 + \sigma)$.

Let $\Omega \subseteq B$ be an open set such that for any $(t_0, \varphi) \in [0, \infty) \times \Omega$, there exist constants $\sigma_1, \gamma_1 > 0$ so that $x_t \in \Omega$ provided that $x \in A(t_0, \varphi, \sigma_1, \gamma_1)$ and $t \in [t_0, t_0 + \sigma_1]$, where $A(t_0, \varphi, \sigma_1, \gamma_1)$ is defined as

$$A(t_0, \varphi, \sigma_1, \gamma_1) = \left\{ x : (-\infty, t_0 + \sigma_1] \to \mathbb{R}^n, x_{t_0} = \varphi, \sup_{t_0 \leq t \leq t_0 + \sigma_1} |x(t) - \varphi(0)| \leq \gamma_1 \right\}.$$

In order to guarantee that equation (2.40) is NFDE, the coefficient of $x(t)$ that is contained in $g(t, x_t)$ cannot be equal to zero. Then we need to introduce the generalized atomic concept.

Definition 2.11. (Lakshmikantham, Wen and Zhang, 1994) The functional $g : [0, \infty) \times \Omega \to \mathbb{R}^n$ is said to be generalized atomic on Ω, if

$$g(t, \varphi) - g(t, \psi) = K(t, \varphi, \psi)(\varphi(0) - \psi(0)) + L(t, \varphi, \psi)$$

where $(t, \varphi, \psi) \in [0, \infty) \times \Omega \times \Omega$, $K : [0, \infty) \times \Omega \times \Omega \to \mathbb{R}^{n \times n}$ and $L : [0, \infty) \times \Omega \times \Omega \to \mathbb{R}^n$ satisfy

(i) $\det K(t, \varphi, \varphi) \neq 0$ for all $(t, \varphi) \in [0, \infty) \times \Omega$;

(ii) for any $(t_0, \varphi) \in [0, \infty) \times \Omega$, there exist constants $\delta_1, \gamma_1 > 0$, and $k_1, k_2 > 0$, with $2k_2 + k_1 < 1$ such that for all $x, y \in A(t_0, \varphi, \sigma_1, \gamma_1)$, $g(t, x_t), K(t, x_t, y_t)$ and $L(t, x_t, y_t)$ are continuous in $t \in [t_0, t_0 + \sigma_1]$, and

$$|K^{-1}(t_0, \varphi, \varphi) L(t, x_t, y_t)| \leq k_1 \sup_{t_0 \leq s \leq t} |x(s) - y(s)|,$$

$$|K^{-1}(t_0, \varphi, \varphi) K(t, x_t, y_t) - I| \leq k_2,$$

where I is the $n \times n$ unit matrix.

For a detailed discussion on the atomic concept we refer the reader to the books Hale, 1977; Lakshmikantham, Wen and Zhang, 1994.

In Subsection 2.4.2, we shall discuss existence and uniqueness of solutions for fractional IVP (2.40)-(2.41) on a class of comparatively comprehensive phase spaces. We establish various criteria on existence and uniqueness of solutions for fractional IVP (2.40)-(2.41). In Subsection 2.4.3, we proceed to consider the continuation of solutions.

2.4.2 *Existence and Uniqueness*

The following existence result for fractional IVP (2.40)-(2.41) is based on Krasnoselskii fixed point theorem.

Theorem 2.7. *Assume that g is generalized atomic on Ω, and that for any $(t_0, \varphi) \in [0, \infty) \times \Omega$, there exist constants $\sigma_1, \gamma_1 \in (0, \infty)$, $q_1 \in (0, q)$ and a real-valued function $m(t) \in L^{\frac{1}{q_1}}[t_0, t_0 + \sigma_1]$ such that*

(H1) *for any $x \in A(t_0, \varphi, \sigma_1, \gamma_1)$, $f(t, x_t)$ is measurable;*
(H2) *for any $x \in A(t_0, \varphi, \sigma_1, \gamma_1)$, $|f(t, x_t)| \le m(t)$, for $t \in [t_0, t_0 + \sigma_1]$;*
(H3) *$f(t, \phi)$ is continuous with respect to ϕ on Ω.*

Then fractional IVP (2.40)-(2.41) has a solution.

Proof. We know that $f(t, x_t)$ is Lebesgue measurable in $[t_0, t_0 + \sigma_1]$ according to condition (H1). Direct calculation gives that $(t - s)^{q-1} \in L^{\frac{1}{1-q_1}}[t_0, t]$, for $t \in [t_0, t_0 + \sigma_1]$. In light of Hölder inequality and the condition (H2), we obtain that $(t-s)^{q-1} f(s, x_s)$ is Lebesgue integrable with respect to $s \in [t_0, t]$ for all $t \in [t_0, t_0 + \sigma_1]$, and

$$\int_{t_0}^{t} |(t-s)^{q-1} f(s, x_s)| ds \le \|(t-s)^{q-1}\|_{L^{\frac{1}{1-q_1}}[t_0,t]} \|m\|_{L^{\frac{1}{q_1}}[t_0,t_0+\sigma_1]}. \tag{2.42}$$

According to Definition 2.10, fractional IVP (2.40)-(2.41) is equivalent to the following equation

$$g(t, x_t) = g(t_0, \varphi) + \frac{1}{\Gamma(q)} \int_{t_0}^{t} (t-s)^{q-1} f(s, x_s) ds \quad \text{for } t \in [t_0, t_0 + \sigma_1]. \tag{2.43}$$

Let $\hat{\varphi} \in A(t_0, \varphi, \sigma_1, \gamma_1)$ be defined as $\hat{\varphi}_{t_0} = \varphi$, $\hat{\varphi}(t_0 + t) = \varphi(0)$ for all $t \in [0, \sigma_1]$. If x is a solution of fractional IVP (2.40)-(2.41), let $x(t_0 + t) = \hat{\varphi}(t_0 + t) + z(t)$, $t \in (-\infty, \sigma_1]$, then we have $x_{t_0+t} = \hat{\varphi}_{t_0+t} + z_t$, $t \in [0, \sigma_1]$. Thus (2.43) implies that z satisfies the equation

$$g(t_0 + t, \hat{\varphi}_{t_0+t} + z_t) = g(t_0, \varphi) + \frac{1}{\Gamma(q)} \int_{0}^{t} (t-s)^{q-1} f(t_0 + s, \hat{\varphi}_{t_0+s} + z_s) ds, \tag{2.44}$$

for $0 \le t \le \sigma_1$.

Since g is generalized atomic on Ω, there exist positive constant $\alpha > 1$ and a positive function $\sigma_2(\gamma)$ defined in $(0, \gamma_1]$, such that for any $\gamma \in (0, \gamma_1]$, when $0 \le t \le \sigma_2(\gamma)$, we have

$$\alpha(2k_2 + k_1) < 1, \tag{2.45}$$

$$|K^{-1}(t_0, \varphi, \varphi)K(t_0 + t, x_{t_0+t}, y_{t_0+t}) - I| \le k_2, \tag{2.46}$$

$$|I - K^{-1}(t_0 + t, \hat{\varphi}_{t_0+t}, \hat{\varphi}_{t_0+t})K(t_0, \varphi, \varphi)| \le \min\{\alpha k_2, \alpha - 1\}, \tag{2.47}$$

$$|K^{-1}(t_0 + t, \hat{\varphi}_{t_0+t}, \hat{\varphi}_{t_0+t})||g(t_0 + t, \hat{\varphi}_{t_0+t}) - g(t_0, \varphi)| \le \frac{1 - \alpha(2k_2 + k_1)}{2}\gamma. \tag{2.48}$$

Note the completely continuity of the function $(m(t))^{\frac{1}{q_1}}$. Hence, for a given positive number M, there must exist a number $h > 0$, satisfying

$$\int_{t_0}^{t_0+h} (m(s))^{\frac{1}{q_1}} ds \le M.$$

For a given $\gamma \in (0, \gamma_1]$, choose

$$\sigma = \min\left\{\sigma_1, \sigma_2(\gamma), h, (1+\beta)^{\frac{1}{1+\beta}} \left(\frac{(1 - \alpha(2k_2 + k_1))\Gamma(q)\gamma}{2\alpha|K^{-1}(t_0, \varphi, \varphi)|M^{q_1}}\right)^{\frac{1}{(1-q_1)(1+\beta)}}\right\}, \tag{2.49}$$

where $\beta = \frac{q-1}{1-q_1} \in (-1, 0)$.

For any $(t_0, \varphi) \in [0, \infty) \times \Omega$, define $E(\sigma, \gamma)$ as follows:

$$E(\sigma, \gamma) = \{z : (-\infty, \sigma) \to \mathbb{R}^n \text{ is continuous; } z(s) = 0 \ \text{ for } s \in (-\infty, 0] \text{ and } \|z\| \le \gamma\}$$

where $\|z\| = \sup_{0 \le s \le \sigma} |z(t)|$. Then $E(\sigma, \gamma)$ is a closed bounded and convex subset of Banach space $BC((-\infty, \sigma_1], \mathbb{R}^n)$.

Now, on $E(\sigma, \gamma)$ define two operators S and U as follows:

$$(Sz)(t) = \begin{cases} 0, & t \in (-\infty, 0], \\ K^{-1}(t_0 + t, \hat{\varphi}_{t_0+t}, \hat{\varphi}_{t_0+t})[-g(t_0 + t, \hat{\varphi}_{t_0+t} + z_t) & \\ \quad + g(t_0, \varphi) + K(t_0 + t, \hat{\varphi}_{t_0+t}, \hat{\varphi}_{t_0+t})z(t)], & t \in [0, \sigma], \end{cases}$$

and

$$(Uz)(t) = \begin{cases} 0, & t \in (-\infty, 0], \\ K^{-1}(t_0 + t, \hat{\varphi}_{t_0+t}, \hat{\varphi}_{t_0+t}) & \\ \quad \times \dfrac{1}{\Gamma(q)} \displaystyle\int_0^t (t - s)^{q-1} f(t_0 + s, \hat{\varphi}_{t_0+s} + z_s) ds, & t \in [0, \sigma], \end{cases}$$

where $z \in E(\sigma, \gamma)$.

It is easy to see that the operator equation

$$z = Sz + Uz \tag{2.50}$$

has a solution $z \in E(\sigma, \gamma)$ if and only if z is a solution of the equation (2.44). Thus, $x_{t+t_0} = \hat{\varphi}_{t_0+t} + z_t$ is a solution of the equation (2.40) on $[0, \sigma]$. Therefore, the existence of a solution of fractional IVP (2.40)-(2.41) is equivalent to determining $\sigma, \gamma > 0$ such that (2.50) has a fixed point in $E(\sigma, \gamma)$.

Now we show that $S+U$ has a fixed point in $E(\sigma,\gamma)$. The proof is divided into three steps.

Claim I. $Sz+Uw \in E(\sigma,\gamma)$ for every pair $z,w \in E(\sigma,\gamma)$.

Obviously, for every pair $z,w \in E(\sigma,\gamma)$, $(Sz)(t)$ and $(Uw)(t)$ are continuous in $t\in[0,\sigma]$, and for $t\in[0,\sigma]$, by using the Hölder inequality and (2.47), we have

$$\begin{aligned}
|(Uw)(t)| &\le |K^{-1}(t_0+t,\hat{\varphi}_{t_0+t},\hat{\varphi}_{t_0+t})\ K(t_0,\varphi,\varphi)\ K^{-1}(t_0,\varphi,\varphi)| \\
&\quad\times \frac{1}{\Gamma(q)}\left|\int_0^t (t-s)^{q-1} f(t_0+s,\hat{\varphi}_{t_0+s}+w_s)ds\right| \\
&\le \alpha|K^{-1}(t_0,\varphi,\varphi)|\ \frac{1}{\Gamma(q)}\ \left(\int_0^t \left((t-s)^{q-1}\right)^{\frac{1}{1-q_1}}ds\right)^{1-q_1} \\
&\quad\times \left(\int_{t_0}^{t_0+\sigma}(m(s))^{\frac{1}{q_1}}ds\right)]^{q_1} \\
&\le \alpha|K^{-1}(t_0,\varphi,\varphi)|\ \frac{M^{q_1}}{\Gamma(q)}\ \left(\frac{1}{1+\beta}\sigma^{1+\beta}\right)^{1-q_1} \\
&\le \frac{1-\alpha(2k_2+k_1)}{2}\gamma,
\end{aligned} \tag{2.51}$$

where $\beta=\frac{q-1}{1-q_1}\in(-1,0)$, and

$$\begin{aligned}
|(Sz)(t)| &= \big|K^{-1}(t_0+t,\hat{\varphi}_{t_0+t},\hat{\varphi}_{t_0+t})[-g(t_0+t,\hat{\varphi}_{t_0+t}+z_t)+g(t_0+t,\hat{\varphi}_{t_0+t}) \\
&\quad -g(t_0+t,\hat{\varphi}_{t_0+t})+g(t_0,\varphi)+K(t_0+t,\hat{\varphi}_{t_0+t},\hat{\varphi}_{t_0+t})z(t)]\big| \\
&= \big|K^{-1}(t_0+t,\hat{\varphi}_{t_0+t},\hat{\varphi}_{t_0+t})[-K(t_0+t,\hat{\varphi}_{t_0+t}+z_t,\hat{\varphi}_{t_0+t})z(t) \\
&\quad -L(t_0+t,\hat{\varphi}_{t_0+t}+z_t,\hat{\varphi}_{t_0+t})-g(t_0+t,\hat{\varphi}_{t_0+t})+g(t_0,\varphi) \\
&\quad +K(t_0+t,\hat{\varphi}_{t_0+t},\hat{\varphi}_{t_0+t})z(t)]\big| \\
&= \big|K^{-1}(t_0+t,\hat{\varphi}_{t_0+t},\hat{\varphi}_{t_0+t})[K(t_0+t,\hat{\varphi}_{t_0+t},\hat{\varphi}_{t_0+t}) \\
&\quad -K(t_0+t,\hat{\varphi}_{t_0+t}+z_t,\hat{\varphi}_{t_0+t})]z(t)+K^{-1}(t_0+t,\hat{\varphi}_{t_0+t},\hat{\varphi}_{t_0+t}) \\
&\quad\times[-L(t_0+t,\hat{\varphi}_{t_0+t}+z_t,\hat{\varphi}_{t_0+t})-g(t_0+t,\hat{\varphi}_{t_0+t})+g(t_0,\varphi)]\big| \\
&= \big|K^{-1}(t_0+t,\hat{\varphi}_{t_0+t},\hat{\varphi}_{t_0+t})K(t_0,\varphi,\varphi) \\
&\quad\times[K^{-1}(t_0,\varphi,\varphi)K(t_0+t,\hat{\varphi}_{t_0+t},\hat{\varphi}_{t_0+t})-I]z(t) \\
&\quad -K^{-1}(t_0+t,\hat{\varphi}_{t_0+t},\hat{\varphi}_{t_0+t})K(t_0,\varphi,\varphi) \\
&\quad\times[K^{-1}(t_0,\varphi,\varphi)K(t_0+t,\hat{\varphi}_{t_0+t}+z_t,\hat{\varphi}_{t_0+t})-I]z(t) \\
&\quad +K^{-1}(t_0+t,\hat{\varphi}_{t_0+t},\hat{\varphi}_{t_0+t})[-L(t_0+t,\hat{\varphi}_{t_0+t}+z_t,\hat{\varphi}_{t_0+t}) \\
&\quad -g(t_0+t,\hat{\varphi}_{t_0+t})+g(t_0,\varphi)]\big| \\
&\le |K^{-1}(t_0+t,\hat{\varphi}_{t_0+t},\hat{\varphi}_{t_0+t})K(t_0,\varphi,\varphi)| \\
&\quad\times[(|K^{-1}(t_0,\varphi,\varphi)K(t_0+t,\hat{\varphi}_{t_0+t},\hat{\varphi}_{t_0+t})-I| \\
&\quad +|K^{-1}(t_0,\varphi,\varphi)K(t_0+t,\hat{\varphi}_{t_0+t}+z_t,\hat{\varphi}_{t_0+t})-I|)|z(t)| \\
&\quad +|K^{-1}(t_0,\varphi,\varphi)L(t_0+t,\hat{\varphi}_{t_0+t}+z_t,\hat{\varphi}_{t_0+t})|\ ] \\
&\quad +|K^{-1}(t_0+t,\hat{\varphi}_{t_0+t},\hat{\varphi}_{t_0+t})||g(t_0+t,\hat{\varphi}_{t_0+t})-g(t_0,\varphi)|.
\end{aligned}$$

According to (2.45)-(2.48), we have

$$|(Sz)(t)| \le \alpha(2k_2+k_1)\gamma + \frac{1-\alpha(2k_2+k_1)}{2}\gamma = \frac{1+\alpha(2k_2+k_1)}{2}\gamma.$$

Therefore, $|(Sz)(t)+(Uw)(t)| \le \gamma$ for $t \in [0,\sigma]$. This means that $Sz+Uw \in E(\sigma,\gamma)$ whenever $z, w \in E(\sigma,\gamma)$.

Claim II. S is a contraction mapping on $E(\sigma,\gamma)$.

For any $z, w \in E(\sigma,\gamma)$, we obtain

$$\begin{aligned}
&|(Sz)(t)-(Sw)(t)| \\
&\le |K^{-1}(t_0+t,\hat{\varphi}_{t_0+t},\hat{\varphi}_{t_0+t})||K(t_0+t,\hat{\varphi}_{t_0+t},\hat{\varphi}_{t_0+t}) \\
&\quad - K(t_0+t,\hat{\varphi}_{t_0+t}+z_t,\hat{\varphi}_{t_0+t}+w_t)||z(t)-w(t)| \\
&\quad + |K^{-1}(t_0+t,\hat{\varphi}_{t_0+t},\hat{\varphi}_{t_0+t})L(t_0+t,\hat{\varphi}_{t_0+t}+z_t,\hat{\varphi}_{t_0+t}+w_t)| \\
&\le \big|[I-K^{-1}(t_0+t,\hat{\varphi}_{t_0+t},\hat{\varphi}_{t_0+t})K(t_0,\varphi,\varphi)] \\
&\quad - [K^{-1}(t_0+t,\hat{\varphi}_{t_0+t},\hat{\varphi}_{t_0+t})K(t_0,\varphi,\varphi)] \\
&\quad \times [K^{-1}(t_0,\varphi,\varphi)K(t_0+t,\hat{\varphi}_{t_0+t}+z_t,\hat{\varphi}_{t_0+t}+w_t)-I]\big|\,|z(t)-w(t)| \\
&\quad + |K^{-1}(t_0+t,\hat{\varphi}_{t_0+t},\hat{\varphi}_{t_0+t})K(t_0,\varphi,\varphi)K^{-1}(t_0,\varphi,\varphi) \\
&\quad \times L(t_0+t,\hat{\varphi}_{t_0+t}+z_t,\hat{\varphi}_{t_0+t}+w_t)| \\
&\le (\alpha k_2+\alpha k_2)|z(t)-w(t)| + \alpha k_1 \sup_{0\le s\le t}|z(s)-w(s)| \\
&\le \alpha(2k_2+k_1)\sup_{0\le s\le t}|z(s)-w(s)|,
\end{aligned}$$

where $\alpha(2k_2+k_1)<1$, and therefore S is a contraction mapping on $E(\sigma,\gamma)$.

Claim III. Now we show that U is a completely continuous operator.

For any $z \in E(\sigma,\gamma), 0 \le \tau < t \le \sigma$, we get

$$\begin{aligned}
&|(Uz)(t)-(Uz)(\tau)| \\
&= \Big|K^{-1}(t_0+t,\hat{\varphi}_{t_0+t},\hat{\varphi}_{t_0+t})\frac{1}{\Gamma(q)}\int_0^t (t-s)^{q-1}f(t_0+s,\hat{\varphi}_{t_0+s}+z_s)ds \\
&\quad - K^{-1}(t_0+\tau,\hat{\varphi}_{t_0+\tau},\hat{\varphi}_{t_0+\tau})\frac{1}{\Gamma(q)}\int_0^\tau (\tau-s)^{q-1}f(t_0+s,\hat{\varphi}_{t_0+s}+z_s)ds\Big| \\
&= \Big|K^{-1}(t_0+t,\hat{\varphi}_{t_0+t},\hat{\varphi}_{t_0+t})\frac{1}{\Gamma(q)}\int_\tau^t (t-s)^{q-1}f(t_0+s,\hat{\varphi}_{t_0+s}+z_s)ds \\
&\quad + K^{-1}(t_0+t,\hat{\varphi}_{t_0+t},\hat{\varphi}_{t_0+t})\frac{1}{\Gamma(q)}\int_0^\tau (t-s)^{q-1}f(t_0+s,\hat{\varphi}_{t_0+s}+z_s)ds \\
&\quad - K^{-1}(t_0+t,\hat{\varphi}_{t_0+t},\hat{\varphi}_{t_0+t})\frac{1}{\Gamma(q)}\int_0^\tau (\tau-s)^{q-1}f(t_0+s,\hat{\varphi}_{t_0+s}+z_s)ds \\
&\quad + K^{-1}(t_0+t,\hat{\varphi}_{t_0+t},\hat{\varphi}_{t_0+t})\frac{1}{\Gamma(q)}\int_0^\tau (\tau-s)^{q-1}f(t_0+s,\hat{\varphi}_{t_0+s}+z_s)ds \\
&\quad - K^{-1}(t_0+\tau,\hat{\varphi}_{t_0+\tau},\hat{\varphi}_{t_0+\tau})\frac{1}{\Gamma(q)}\int_0^\tau (\tau-s)^{q-1}f(t_0+s,\hat{\varphi}_{t_0+s}+z_s)ds\Big|
\end{aligned}$$

$$
\begin{aligned}
&\leq \frac{|K^{-1}(t_0+t,\hat{\varphi}_{t_0+t},\hat{\varphi}_{t_0+t})|}{\Gamma(q)}\left|\int_\tau^t (t-s)^{q-1} f(t_0+s,\hat{\varphi}_{t_0+s}+z_s)ds\right| \\
&\quad + \frac{|K^{-1}(t_0+t,\hat{\varphi}_{t_0+t},\hat{\varphi}_{t_0+t})|}{\Gamma(q)} \\
&\quad \times \left|\int_0^\tau [(t-s)^{q-1}-(\tau-s)^{q-1}] f(t_0+s,\hat{\varphi}_{t_0+s}+z_s)ds\right| \\
&\quad + \frac{|K^{-1}(t_0+t,\hat{\varphi}_{t_0+t},\hat{\varphi}_{t_0+t}) - K^{-1}(t_0+\tau,\hat{\varphi}_{t_0+\tau},\hat{\varphi}_{t_0+\tau})|}{\Gamma(q)} \\
&\quad \times \left|\int_0^\tau (\tau-s)^{q-1} f(t_0+s,\hat{\varphi}_{t_0+s}+z_s)ds\right| \\
&= \frac{|K^{-1}(t_0+t,\hat{\varphi}_{t_0+t},\hat{\varphi}_{t_0+t})|}{\Gamma(q)}(I_1+I_2) \\
&\quad + \frac{|K^{-1}(t_0+t,\hat{\varphi}_{t_0+t},\hat{\varphi}_{t_0+t}) - K^{-1}(t_0+\tau,\hat{\varphi}_{t_0+\tau},\hat{\varphi}_{t_0+\tau})|}{\Gamma(q)} I_3,
\end{aligned}
$$

where

$$
\begin{aligned}
I_1 &= \left|\int_\tau^t (t-s)^{q-1} f(t_0+s,\hat{\varphi}_{t_0+s}+z_s)ds\right|, \\
I_2 &= \left|\int_0^\tau \left((t-s)^{q-1}-(\tau-s)^{q-1}\right) f(t_0+s,\hat{\varphi}_{t_0+s}+z_s)ds\right|, \\
I_3 &= \left|\int_0^\tau (\tau-s)^{q-1} f(t_0+s,\hat{\varphi}_{t_0+s}+z_s)ds\right|.
\end{aligned}
$$

By using an analogous argument presented in (2.51), we can conclude that

$$
\begin{aligned}
I_1 &\leq \frac{M^{q_1}}{(1+\beta)^{1-q_1}}\left((t-\tau)^{1+\beta}\right)^{1-q_1}, \\
I_3 &\leq \frac{M^{q_1}}{(1+\beta)^{1-q_1}}\left(\tau^{1+\beta}\right)^{1-q_1},
\end{aligned}
$$

and

$$
\begin{aligned}
I_2 &\leq \left(\int_0^\tau \left|(t-s)^{q-1}-(\tau-s)^{q-1}\right|^{\frac{1}{1-q_1}} ds\right)^{1-q_1} \left(\int_{t_0}^{t_0+\tau} |f(s,x_s)|^{\frac{1}{q_1}} ds\right)^{q_1} \\
&\leq M^{q_1}\left(\int_0^\tau (\tau-s)^\beta-(t-s)^\beta ds\right)^{1-q_1} \\
&\leq \frac{M^{q_1}}{(1+\beta)^{1-q_1}}\left(\tau^{1+\beta}-t^{1+\beta}+(t-\tau)^{1+\beta}\right)^{1-q_1} \\
&\leq \frac{M^{q_1}}{(1+\beta)^{1-q_1}}\left((t-\tau)^{1+\beta}\right)^{1-q_1},
\end{aligned}
$$

where $\beta = \frac{q-1}{1-q_1} \in (-1,0)$. Therefore

$$
|(Uz)(t)-(Uz)(\tau)| \leq \frac{|K^{-1}(t_0+t,\hat{\varphi}_{t_0+t},\hat{\varphi}_{t_0+t})|}{\Gamma(q)} \frac{2M^{q_1}}{(1+\beta)^{1-q_1}}\left((t-\tau)^{1+\beta}\right)^{1-q_1}
$$

$$+\frac{|K^{-1}(t_0+t,\hat{\varphi}_{t_0+t},\hat{\varphi}_{t_0+t})-K^{-1}(t_0+\tau,\hat{\varphi}_{t_0+\tau},\hat{\varphi}_{t_0+\tau})|}{\Gamma(q)}\frac{M^{q_1}}{(1+\beta)^{1-q_1}}\left(\tau^{1+\beta}\right)^{1-q_1}.$$

Since the property of the matrix function $K(t_0+t,\hat{\varphi}_{t_0+t},\hat{\varphi}_{t_0+t})$ which is nonsingular and continuous in $t\in[0,\sigma]$ implies that its inverse matrix $K^{-1}(t_0+t,\hat{\varphi}_{t_0+t},\hat{\varphi}_{t_0+t})$ exists and is continuous in $t\in[0,\sigma]$, then $\{Uz: z\in E(\sigma,\gamma)\}$ is equicontinuous. On the other hand, U is continuous from condition (H3) and $\{Uz: z\in E(\sigma,\gamma)\}$ is uniformly bounded from (2.51), thus U is a completely continuous operator by Arzela-Ascoli theorem.

Therefore, Krasnoselskii fixed point theorem shows that $S+U$ has a fixed point on $E(\sigma,\gamma)$, and hence fractional IVP (2.40)-(2.41) has a solution $x(t)=\varphi(0)+z(t-t_0)$ for all $t\in[t_0,t_0+\sigma]$. □

Remark 2.1. If we replace condition (H1) by

(H1)′ $f(t,\phi)$ is measurable with respect to t on $[t_0,t_0+\sigma_1]$.

Then we can also conclude that the result of Theorem 2.7 holds. In fact, for any $x\in A(t_0,\varphi,\sigma_1,\gamma_1)$, suppose $x_{t_0+t}=\hat{\varphi}_{t_0+t}+z_t$, $t\in[0,\sigma_1]$, then, according to the definition of $\hat{\varphi}_{t_0+t}$ and z_t, we know that x_{t_0+t} is a measurable function. It follows that from (H1)′ and (H3), $f(t,x_t)$ is measurable in t, where $x\in A(t_0,\varphi,\sigma_1,\gamma_1)$ and satisfies $x_{t_0+t}=\hat{\varphi}_{t_0+t}+z_t$, $t\in[0,\sigma_1]$.

Remark 2.2. If we replace condition (H3) by a weaker condition:

(H3)′ for any $x,y\in A(t_0,\varphi,\sigma,\gamma)$ with $\sup_{t_0\le s\le t_0+\sigma}|x(s)-y(s)|\to 0$,

$$\left|\int_{t_0}^{t}(t-s)^{q-1}(f(s,x_s)-f(s,y_s))ds\right|\to 0,\quad t\in[t_0,t_0+\sigma],$$

where σ satisfy (2.48), then we can also conclude that the result of Theorem 2.7 holds.

The following existence and uniqueness result for fractional IVP (2.40)-(2.41) is based on Banach contraction mapping principle.

Theorem 2.8. *Assume that g is generalized atomic on Ω, and that for any $(t_0,\varphi)\in[0,\infty)\times\Omega$, there exist constants $\sigma_1,\gamma_1\in(0,\infty)$, $q_1\in(0,q)$ and a real-valued function $m(t)\in L^{\frac{1}{q_1}}[t_0,t_0+\sigma_1]$ such that conditions (H1)-(H2) of Theorem 2.7 hold. Further assume that:*

(H4) *there exists a nonnegative function $\ell:[0,\sigma_1]\to[0,\infty)$ continuous at $t=0$ and $\ell(0)=0$ such that for any $x,y\in A(t_0,\varphi,\sigma_1,\gamma_1)$, we have*

$$\left|\int_{t_0}^{t}(t-s)^{q-1}[f(s,x_s)-f(s,y_s)]ds\right|\le\ell(t-t_0)\sup_{t_0\le s\le t}|x(s)-y(s)|,\quad t\in[t_0,t_0+\sigma_1],$$

then fractional IVP (2.40)-(2.41) *has a unique solution.*

Proof. According to the argument of Theorem 2.7, it suffices to prove that $S+U$ has a unique fixed point on $E(\sigma,\gamma)$, where $\sigma,\gamma>0$ are sufficiently small. Now, choose $\sigma\in(0,\sigma_1)$, $\gamma\in(0,\gamma_1]$, such that (2.49) holds and that

$$c=\alpha(2k_2+k_1)+\sup_{0\le s\le\sigma}\frac{|K^{-1}(t_0+s,\hat{\varphi}_{t_0+s},\hat{\varphi}_{t_0+s})||\ell(s)|}{\Gamma(q)}<1.$$

Obviously, $S+U$ is a mapping from $E(\sigma,\gamma)$ into itself. Using the same argument as that of Theorem 2.7, for any $z,w\in E(\sigma,\gamma)$, we get

$$|(Sz)(t)-(Sw)(t)|\le\alpha(2k_2+k_1)\sup_{0\le s\le\sigma}|z(s)-w(s)|,$$

and

$$\begin{aligned}
&|(Uz)(t)-(Uw)(t)|\\
&\le\frac{|K^{-1}(t_0+t,\hat{\varphi}_{t_0+t},\hat{\varphi}_{t_0+t})|}{\Gamma(q)}\left|\int_0^t(t-s)^{q-1}f(t_0+s,\hat{\varphi}_{t_0+s}+z_s)ds\right.\\
&\quad\left.-\int_0^t(t-s)^{q-1}f(t_0+s,\hat{\varphi}_{t_0+s}+w_s)ds\right|\\
&\le\frac{|K^{-1}(t_0+t,\hat{\varphi}_{t_0+t},\hat{\varphi}_{t_0+t})|}{\Gamma(q)}|\ell(t)|\sup_{0\le s\le t}|z(s)-w(s)|\\
&\le\frac{\sup_{0\le s\le\sigma}|K^{-1}(t_0+s,\hat{\varphi}_{t_0+s},\hat{\varphi}_{t_0+s})||\ell(s)|}{\Gamma(q)}\sup_{0\le s\le\sigma}|z(s)-w(s)|.
\end{aligned}$$

Therefore

$$\begin{aligned}
&|(S+U)z(t)-(S+U)w(t)|\\
&\le\left[\alpha(2k_2+k_1)+\sup_{0\le s\le\sigma}\frac{|K^{-1}(t_0+s,\hat{\varphi}_{t_0+s},\hat{\varphi}_{t_0+s})|\,|\ell(s)|}{\Gamma(q)}\right]\sup_{0\le s\le\sigma}|z(s)-w(s)|\\
&=c\sup_{0\le s\le\sigma}|z(s)-w(s)|.
\end{aligned}$$

Hence, we have

$$\|(S+U)z-(S+U)w\|\le c\|z-w\|,$$

where $c<1$. By applying Banach contraction mapping principle, we know that $S+U$ has a unique fixed point on $E(\sigma,\gamma)$. □

Corollary 2.6. *If the condition (H4) of Theorem 2.8 is replaced by the following condition:*

(H4)′ *there exist* $q_2\in(0,q)$ *and a function* $\ell_1\in L^{\frac{1}{q_2}}[t_0,t_0+\sigma_1]$*, such that for any* $x,y\in A(t_0,\varphi,\sigma_1,\gamma_1)$ *we have*

$$|f(t,x_t)-f(t,y_t)|\le\ell_1(t)\sup_{t_0\le s\le t}|x(s)-y(s)|,\quad t\in[t_0,t_0+\sigma_1],$$

then the result of Theorem 2.8 holds.

Proof. It suffices to prove that the condition (H4) of Theorem 2.8 holds. Note that $\ell_1 \in L^{\frac{1}{q_2}}[t_0, t_0+\sigma_1]$, hence, there must exist a positive number N, such that $N = \|\ell_1\|_{L^{\frac{1}{q_2}}[t_0,\, t_0+\sigma_1]}$. Then for any $x, y \in A(t_0, \varphi, \sigma_1, \gamma_1)$ we have

$$\begin{aligned}
&\left|\int_{t_0}^{t}(t-s)^{q-1}(f(s,x_s)-f(s,y_s))ds\right| \\
&\le \int_{t_0}^{t}(t-s)^{q-1}|f(s,x_s)-f(s,y_s)|ds \\
&\le \int_{t_0}^{t}(t-s)^{q-1}\ell_1(s)\,ds \sup_{t_0\le s\le t}|x(s)-y(s)| \\
&\le \frac{N}{(1+\beta')^{1-q_2}}\ (t-t_0)^{(1+\beta')(1-q_2)} \sup_{t_0\le s\le t}|x(s)-y(s)|,
\end{aligned}$$

where $\beta' = \frac{q-1}{1-q_2} \in (-1, 0)$. Let

$$\ell(t-t_0) = \frac{N}{(1+\beta')^{1-q_2}}\ (t-t_0)^{(1+\beta')(1-q_2)},\ t \in [t_0, t_0+\sigma_1].$$

Obviously, $\ell : [0, \sigma_1] \to [0, \infty)$ continuous at $t = 0$ and $\ell(0) = 0$. Then the condition (H4) of Theorem 2.8 holds. □

2.4.3 *Continuation of Solutions*

For any $t_0, \varphi \in [0,\infty) \times \Omega$, $\omega \subset \Omega$ and positive constants $\sigma, \gamma > 0$, define $B_\omega(t_0, \varphi, \sigma, \gamma)$ as the set of all maps $x : (-\infty, t_0+\sigma) \to \mathbb{R}^n$ such that $x_{t_0} = \varphi$, $x : [t_0, t_0+\sigma) \to \mathbb{R}^n$ is continuous with $|x(t) - \varphi(0)| \le \gamma$ and $x_t \in \omega$ for all $t \in [t_0, t_0+\sigma)$. In the following theorem, W is a set of all subsets of Ω such that for any $(t_0, \varphi) \in [0,\infty) \times \Omega$, constants $\sigma, \gamma > 0$ and a set $\omega \in W$, if $x \in B_\omega(t_0, \varphi, \sigma, \gamma)$ and if $x(t_0+\sigma) = \lim_{t\to(t_0+\sigma)^-} x(t)$ exists, then $x_{t_0+\sigma} \in \Omega$.

Theorem 2.9. *Let all conditions of Theorem 2.7 hold. Besides, suppose that $\sigma \in (0, \sigma_1], \gamma \in (0, \gamma_1]$ and for any $x \in B_\omega(t_0, \varphi, \sigma, \gamma)$,*

(H5) *there exist constants $q_\omega \in (0, q)$ and a real-valued function $m_\omega(t) \in L^{\frac{1}{q_\omega}}[t_0, t_0+\sigma]$ such that $f(t, x_t)$ is measurable and $|f(t, x_t)| \le m_\omega(t)$ for $t \in [t_0, t_0+\sigma)$;*

(H6) $\lim_{\tau\to0^+}[g(t, x_{t-\tau}) - g(t-\tau, x_{t-\tau})] = 0$ *uniformly for $t \in [t_0+\tau, t_0+\sigma)$;*

(H7) $K(t, x_t, x_t) - K(t, x_t, x_{t-\tau}) \to 0$ *uniformly for $t \in [t_0+\tau, t_0+\sigma)$ as $\tau \to 0^+$ and as* $\sup_{t_0+\tau\le s\le t}|x(s) - x(s-\tau)| \to 0$*;*

(H8) *there exists a constant H such that $|K^{-1}(t, x_t, x_t)| \le H$ for all $t \in [t_0, t_0+\sigma)$;*

(H9) *there exists a continuous function $\ell_\omega : [0,\infty) \to [0,\infty)$ with $\ell_\omega(0) = 0$ such that*

$$|L(t, x_t, x_{t-\tau}) - L_b^*(t, x_t, x_{t-\tau})| \le \ell_\omega(b) \sup_{-b\le\theta\le0}|x(t+\theta) - x(t-\tau+\theta)|$$

where for a given $b > 0$, $\lim_{\tau\to0^+} L_b^*(t, x_t, x_{t-\tau}) = 0$ *uniformly for $t \in [t_0 + \tau, t_0+\sigma)$.*

Then for any $\omega \in W$ and any $\gamma > 0$, if $x(t)$ is a noncontinuable solution of fractional IVP (2.40)-(2.41) defined on $[t_0, t_0+\sigma)$, there exists a $t^* \in [t_0, t_0+\sigma)$ such that $|x(t^*) - \varphi(0)| > \gamma$ or $x_{t^*} \notin \omega$.

Proof. By way of contradiction, if there exists a noncontinuable solution $x(t)$ of fractional IVP (2.40)-(2.41) on $[t_0, t_0+\sigma)$ such that $|x(t) - \varphi(0)| \leq \gamma$ and $x_t \in \omega$ for all $t \in [t_0, t_0+\sigma)$, that is, $x \in B_\omega(t_0, \varphi, \sigma, \gamma)$, then first, $x(t)$ is not uniformly continuous on $[t_0, t_0+\sigma)$. Otherwise, $x(t_0+\sigma) = \lim_{t \to (t_0+\sigma)^-} x(t)$ exists and thus $x_{t_0+\sigma} \in \Omega$. By Theorem 2.7, $x(t)$ can be continued beyond $t_0+\sigma$.

Therefore, there exist a sufficiently small constant $\varepsilon > 0$, and sequences $\{t_k\} \subseteq [t_0, t_0+\sigma)$, $\{\Delta_k\}$ with $\Delta_k \to 0^+$ as $k \to \infty$, such that

$$|x(t_k) - x(t_k - \Delta_k)| \geq \varepsilon, \quad \text{for all } k = 1, 2, \dots .$$

Now choose a constant $H > 0$ so that

$$|K^{-1}(t, x_t, x_t)| \leq H, \quad \text{for all } t \in [t_0, t_0+\sigma).$$

For given H and $\varepsilon > 0$, by (H5)-(H7) and (H9), we can find positive constants b and σ_0 so that

$$\frac{2HM_\omega}{\Gamma(q)(1+\beta_\omega)^{1-q_\omega}} \left(\sigma_0^{1+\beta_\omega}\right)^{1-q_\omega} < \frac{\varepsilon}{5},$$

where $\beta_\omega = \frac{q-1}{1-q_\omega} \in (-1, 0)$, $M_\omega = \left(\int_{t_0}^{t_0+\sigma} (m_\omega(s))^{\frac{1}{q_\omega}} ds\right)^{q_\omega}$,

$$H|K(t, x_t, x_t) - K(t, x_t, x_{t-\tau})| < \frac{1}{5}, \quad \text{as} \quad \sup_{t_0+\tau \leq s \leq t} |x(s) - x(s-\tau)| \leq \varepsilon,$$

and

$$H|g(t, x_{t-\tau}) - g(t-\tau, x_{t-\tau})| < \frac{\varepsilon}{5},$$

$$H\ell_\omega(b) < \frac{1}{5}, \; b < \frac{\delta - \sigma_0}{2},$$

$$H|L_b^*(t, x_t, x_{t-\tau})| < \frac{\varepsilon}{5},$$

for all $t \in [t_0+\tau, t_0+\sigma)$ and $0 < \tau < \sigma_0$.

Since $x(t)$ is uniformly continuous on $[t_0, t_0+\sigma-b]$, we can find a constant $H_1 > 0$ so that for all $k \geq H_1$, we have $\Delta_k < \sigma_0$ and $|x(t) - x(t-\Delta_k)| < \varepsilon$ for all $t \in [t_0+\sigma_k, t_0+\sigma-b]$. Now for all $k \geq H_1$, define a sequence $\{s_k\}$ in the following pattern

$$s_k = \inf\{t \in (t_0+\sigma-b, t_0+\sigma) : |x(t) - x(t-\Delta_k)| \geq \varepsilon\}.$$

Then

$$|x(s_k) - x(s_k - \Delta_k)| = \varepsilon.$$

Thus we get

$$\frac{2HM_\omega}{\Gamma(q)(1+\beta_\omega)^{1-q_\omega}} \left(\Delta_k^{1+\beta_\omega}\right)^{1-q_\omega} < \frac{\varepsilon}{5},$$

$$H|K(s_k, x_{s_k}, x_{s_k}) - K(s_k, x_{s_k}, x_{s_k-\Delta_k})| < \frac{1}{5},$$
$$H|g(s_k, x_{s_k-\Delta_k}) - g(s_k - \Delta_k, x_{s_k-\Delta_k})| < \frac{\varepsilon}{5}$$

and

$$\begin{aligned}&H|L(s_k, x_{s_k}, x_{s_k-\Delta_k}) - L_b^*(s_k, x_{s_k}, x_{s_k-\Delta_k})| \\ \leq\ & \frac{1}{5} \sup_{-b\leq\theta\leq 0} |x(s_k+\theta) - x(s_k - \Delta_k + \theta)| \leq \frac{\varepsilon}{5}.\end{aligned}$$

On the other hand, we see that

$$\begin{aligned}&g(s_k, x_{s_k}) - g(s_k - \Delta_k, x_{s_k-\Delta_k}) \\ &= g(s_k, x_{s_k}) - g(s_k, x_{s_k-\Delta_k}) + g(s_k, x_{s_k-\Delta_k}) - g(s_k - \Delta_k, x_{s_k-\Delta_k}) \\ &= (K(s_k, x_{s_k}, x_{s_k-\Delta_k}) - K(s_k, x_{s_k}, x_{s_k}))\ \ (x(s_k) - x(s_k - \Delta_k)) \\ &\quad + K(s_k, x_{s_k}, x_{s_k})\,(x(s_k) - x(s_k - \Delta_k)) + L(s_k, x_{s_k}, x_{s_k-\Delta_k}) \\ &\quad - L_b^*(s_k, x_{s_k}, x_{s_k-\Delta_k}) + L_b^*(s_k, x_{s_k}, x_{s_k-\Delta_k}) \\ &\quad + g(s_k, x_{s_k-\Delta_k}) - g(s_k - \Delta_k, x_{s_k-\Delta_k}).\end{aligned}$$

By using the same argument as that of Claim III in Theorem 2.7, we have

$$\begin{aligned}&|g(s_k, x_{s_k}) - g(s_k - \Delta_k, x_{s_k-\Delta_k})| \\ &\leq \frac{1}{\Gamma(q)} \left| \int_{s_k-\Delta_k}^{s_k} (s_k - s)^{q-1} f(s, x_s) ds \right| \\ &\quad + \frac{1}{\Gamma(q)} \left| \int_{t_0}^{s_k-\Delta_k} [(s_k - s)^{q-1} - (s_k - \Delta_k - s)^{q-1}] f(s, x_s) ds \right| \\ &\leq \frac{2M_\omega}{\Gamma(q)(1+\beta_\omega)^{1-q_\omega}} \left(\Delta_k^{1+\beta_\omega}\right)^{1-q_\omega}.\end{aligned}$$

Therefore

$$\begin{aligned}&|x(s_k) - x(s_k - \Delta_k)| \\ &\leq |K^{-1}(s_k, x_{s_k}, x_{s_k})|[|g(s_k, x_{s_k}) - g(s_k - \Delta_k, x_{s_k-\Delta_k})| \\ &\quad + |K(s_k, x_{s_k}, x_{s_k-\Delta_k}) - K(s_k, x_{s_k}, x_{s_k})||x(s_k) - x(s_k - \Delta_k)| \\ &\quad + |L(s_k, x_{s_k}, x_{s_k-\Delta_k}) - L_b^*(s_k, x_{s_k}, x_{s_k-\Delta_k})| \\ &\quad + |L_b^*(s_k, x_{s_k}, x_{s_k-\Delta_k})| + |g(s_k, x_{s_k-\Delta_k}) - g(s_k - \Delta_k, x_{s_k-\Delta_k})|] \\ &< \varepsilon.\end{aligned}$$

This is contrary to $|x(s_k) - x(s_k - \Delta_k)| = \varepsilon$. The proof is completed. □

Remark 2.3. If we replace conditions of Theorem 2.7 by conditions of Remark 2.1, the result of Theorem 2.9 holds.

Remark 2.4. If we replace the condition (H3) of Theorem 2.7 by a weaker condition:

(H3)″ for any $x, y \in A(t_0, \varphi, \sigma, \gamma)$ with $\sup_{t_0 \le s \le t_0+\sigma} |x(s) - y(s)| \to 0$,

$$\left| \int_{t_0}^{t} (t-s)^{q-1}[f(s, x_s) - f(s, y_s)]ds \right| \to 0, \quad t \in [t_0, t_0 + \sigma],$$

where $\sigma \in (0, \sigma_1], \gamma \in (0, \gamma_1]$.

Then we can also conclude that the result of Theorem 2.9 holds.

In the following, for any $(t_0, \varphi) \in [0, \infty) \times \Omega$ and any constants $\varepsilon, \sigma, \gamma > 0$, $C_\varepsilon(t_0, \varphi, \delta, \gamma)$ denotes the set of all functions $x : (-\infty, t_0 + \sigma] \to \mathbb{R}^n$ so that $\|x_{t_0} - \varphi\|_B < \varepsilon$, $x : [t_0, t_0 + \sigma] \to \mathbb{R}^n$ is continuous and $|x(t) - \varphi(0)| \le \gamma$.

Theorem 2.10. *Suppose that for any $(t_0, \varphi) \in [0, \infty) \times \Omega$, the solution of fractional IVP (2.40)-(2.41) is unique. Besides, suppose that $\sigma \in (0, \sigma_1], \gamma \in (0, \gamma_1]$ and for any $x \in C_\varepsilon(t_0, \varphi, \sigma, \gamma)$, (H5)-(H9) hold and*

(H10) *for any $x, y \in C_\varepsilon(t_0, \varphi, \sigma, \gamma)$, if $\|x_{t_0} - y_{t_0}\|_B \to 0$ and $\sup_{t_0 \le s \le t_0+\sigma} |x(s) - y(s)| \to 0$, then $g(t, x_t) \to g(t, y_t)$ and $|\int_{t_0}^{t} (t-s)^{q-1}[f(s, x_s) - f(s, y_s)]ds| \to 0$ for $t \in [t_0, t_0 + \sigma]$.*

If x is a noncontinuable solution of fractional IVP(2.40)-(2.41) defined on $[t_0, t_0 + \sigma_1)$, then for any $\varepsilon > 0$ and $\sigma \in (0, \sigma_1)$, we can find a $\sigma > 0$ so that if $\|\varphi - \psi\|_B < \sigma$, then $|x(t) - y(t)| < \varepsilon$ for $t \in [t_0, t_0 + \sigma]$, where $y(t)$ is a solution of (2.40) through (t_0, ψ).

Proof. By way of contradiction, if the conclusion above is not true, then there exist $\varepsilon > 0$, sequences $\{t_k\} \subseteq [t_0, t_0 + \sigma]$ and $\{\varphi^k\} \subseteq \Omega$ such that

$$\|\varphi^k - \varphi\|_B < \frac{1}{k},$$
$$|y^k(t_k) - x(t_k)| = \varepsilon$$

and

$$|y^k(t) - x(t)| < \varepsilon, \quad \text{for } t \in [t_0, t_k),$$

where $y^k(t)$ is a solution of following fractional IVP

$$ {}_{t_0}^{C}D_t^\alpha g(t, y_t) = f(t, y_t), \; y_{t_0} = \varphi^k. \tag{2.52}$$

Without loss of generality, we may assume $t_k \to \bar{t} \in [t_0, t_0 + \sigma]$ as $k \to \infty$. Now define a sequence of functions $\{z^k\}$ as follows:

$$z^k(t) = \begin{cases} y^k(t), & \text{for } t \in [t_0, t_k], \\ y^k(t_k), & \text{for } t \in [t_k, \bar{t}\,], \quad \text{if } t_k < \bar{t}. \end{cases}$$

Using the same argument as that of Theorem 2.7, we can assume that $\{z^k\}$ is equicontinuous in $t \in [t_0, \bar{t}\,]$. By Arzela-Ascoli theorem, without loss of generality, we can find a function $y : (-\infty, \bar{t}\,] \to \mathbb{R}^n$ such that $\lim_{k\to\infty} \sup_{t_0 \le s \le \bar{t}} |z^k(s) - y(s)| = 0$ and $y(s) = \varphi(s)$ for $s \le t_0$.

Now considering the equation (2.52), we get

$$g(t, y_t^k) - g(t_0, \varphi^k) = \frac{1}{\Gamma(q)} \int_{t_0}^{t} (t-s)^{q-1} f(s, y_s^k) ds, \quad \text{for } t \in [t_0, \bar{t}\,].$$

By (H10) and Lebesgue dominated convergence theorem and let $k \to \infty$, we obtain

$$g(t, y_t) - g(t_0, \varphi) = \frac{1}{\Gamma(q)} \int_{t_0}^{t} (t-s)^{q-1} f(s, y_s) ds, \quad \text{for } t \in [t_0, \bar{t}\,].$$

This means that $y(t) = x(t)$ for $t \in [t_0, \bar{t}\,]$ by the uniqueness assumption of the solutions of fractional IVP (2.40)-(2.41). This is contrary to

$$|y^k(t_k) - x(t_k)| = \varepsilon$$

and

$$\lim_{k\to\infty} \sup_{t_0 \le s \le \bar{t}} |z^k(s) - y(s)| = 0.$$

The proof is completed. □

2.5 Iterative Functional Differential Equations

2.5.1 *Introduction*

In Section 2.5, we consider the following fractional iterative functional differential equations with parameter

$$\begin{cases} {}_a^C D_t^q x(t) = f(t, x(t), x(x^v(t))) + \lambda, & t \in [a,b],\ v \in \mathbb{R} \setminus \{0\},\ q \in (0,1),\ \lambda \in \mathbb{R}, \\ x(t) = \varphi(t), & t \in [a_1, a], \\ x(t) = \psi(t), & t \in [b, b_1], \end{cases} \tag{2.53}$$

where ${}_a^C D_t^q$ is Caputo fractional derivative of order q and

(C1) $a_1 \le a < b \le b_1$, $a_1 \le a_1^v$ and $b_1^v \le b_1$;
(C2) $f \in C([a,b] \times [a_1, b_1]^2, \mathbb{R})$;
(C3) $\varphi \in C([a_1, a], [a_1, b_1])$ and $\psi \in C([b, b_1], [a_1, b_1])$.

Definition 2.12. A function $x \in C([a_1, b_1], [a_1, b_1])$ is said to be a solution of the problem (2.53) if x satisfies the equation ${}_a^C D_t^q x(t) = f(t, x(t), x(x^v(t))) + \lambda$ on $[a,b]$, and the conditions $x(t) = \varphi(t), t \in [a_1, a]$, $x(t) = \psi(t), t \in [b, b_1]$.

The purpose of this section is to determine the pair (x, λ), $x \in C([a_1, b_1], [a_1, b_1])$ (or $C_L^q([a_1, b_1], [a_1, b_1])$), $\lambda \in \mathbb{R}$, which satisfies the problem (2.53). In Subsection 2.5.2, by using Schauder fixed point theorem, we establish existence theorems in $C([a_1, b_1], [a_1, b_1])$ and $C_L^q([a_1, b_1], [a_1, b_1])$ respectively. Unfortunately, uniqueness results can not be obtained since the solution operator is not Lipschitz continuous but only Hölder continuous. Meanwhile, data dependence results of solutions and parameters provide possible way to describe the error estimates between explicit and approximative solutions for such problems. In Subsection 2.5.4, We make some examples to illustrate our results and conclude some possible extensions to general parametrized fractional iterative functional differential equations.

2.5.2 *Existence*

We first give existence result in $C([a_1, b_1], [a_1, b_1])$. Let (x, λ) be a solution of the problem (2.53). Then this problem is equivalent to the following fixed point equation

$$x(t) = \begin{cases} \varphi(t), & \text{for } t \in [a_1, a], \\ \varphi(a) + \dfrac{1}{\Gamma(q)} \displaystyle\int_a^t (t-s)^{q-1} f(s, x(s), x(x^v(s)))ds & \\ \qquad + \dfrac{\lambda}{\Gamma(q+1)}(t-a)^q, & \text{for } t \in [a, b], \\ \psi(t), & \text{for } t \in [b, b_1]. \end{cases} \tag{2.54}$$

From the condition of continuity of x in $t = b$, we have that

$$\lambda = \frac{\Gamma(q+1)(\psi(b) - \varphi(a))}{(b-a)^q} - \frac{q}{(b-a)^q} \int_a^b (b-s)^{q-1} f(s, x(s), x(x^v(s)))ds.$$

Now we consider the operator

$$A : C([a_1, b_1], [a_1, b_1]) \to C([a_1, b_1], \mathbb{R}),$$

where

$$(Ax)(t) := \begin{cases} \varphi(t), & \text{for } t \in [a_1, a], \\ \varphi(a) + \dfrac{(t-a)^q}{(b-a)^q}(\psi(b) - \varphi(a)) - \dfrac{(t-a)^q}{\Gamma(q)(b-a)^q} & \\ \qquad \times \displaystyle\int_a^b (b-s)^{q-1} f(s, x(s), x(x^v(s)))ds & \\ \qquad + \dfrac{1}{\Gamma(q)} \displaystyle\int_a^t (t-s)^{q-1} f(s, x(s), x(x^v(s)))ds, & \text{for } t \in [a, b], \\ \psi(t), & \text{for } t \in [b, b_1]. \end{cases} \tag{2.55}$$

It is clear that (x, λ) is a solution of the problem (2.53) if and only if x is a fixed point of the operator A and λ is given by (2.54). So, the problem is to study the fixed point equation

$$x = A(x).$$

Now, we are ready to state our first result in this section.

Theorem 2.11. *We suppose that*

(i) *conditions (C1)-(C3) are satisfied;*

(ii) *there are* m_f, $M_f \in \mathbb{R}$ *such that*

$$m_f \leq f(t,u,w) \leq M_f,\ \forall\, t \in [a,b], u,w \in [a_1,b_1],$$

along with

$$a_1 \leq \min\{\varphi(a),\psi(b)\} - \max\left\{0, \frac{M_f(b-a)^q}{\Gamma(q+1)}\right\} + \min\left\{0, \frac{m_f(b-a)^q}{\Gamma(q+1)}\right\},$$

and

$$\max\{\varphi(a),\psi(b)\} - \min\left\{0, \frac{m_f(b-a)^q}{\Gamma(q+1)}\right\} + \max\left\{0, \frac{M_f(b-a)^q}{\Gamma(q+1)}\right\} \leq b_1.$$

Then problem (2.53) *has a solution in* $C([a_1,b_1],[a_1,b_1])$.

Proof. In what follow we consider on $C([a_1,b_1],\mathbb{R})$ with the Chebyshev norm $\|\cdot\|_C$.

Condition (ii) assures that the set $C([a_1,b_1],[a_1,b_1])$ is an invariant subset for the operator A, that is, we have

$$A(C([a_1,b_1],[a_1,b_1])) \subset C([a_1,b_1],[a_1,b_1]).$$

Indeed, for $t \in [a_1,a] \cup [b,b_1]$, we have $A(x)(t) \in [a_1,b_1]$. Furthermore, we obtain

$$a_1 \leq A(x)(t) \leq b_1,\ \forall\, t \in [a,b],$$

if and only if

$$a_1 \leq \min_{t\in[a,b]} A(x)(t) \tag{2.56}$$

and

$$\max_{t\in[a,b]} A(x)(t) \leq b_1 \tag{2.57}$$

hold.

Since

$$\min_{t\in[a,b]} A(x)(t) \geq \min\{\varphi(a),\psi(b)\} - \max\left\{0, \frac{M_f(b-a)^q}{\Gamma(q+1)}\right\} + \min\left\{0, \frac{m_f(b-a)^q}{\Gamma(q+1)}\right\},$$

and

$$\max_{t\in[a,b]} A(x)(t) \leq \max\{\varphi(a),\psi(b)\} - \min\left\{0, \frac{m_f(b-a)^q}{\Gamma(q+1)}\right\} + \max\left\{0, \frac{M_f(b-a)^q}{\Gamma(q+1)}\right\},$$

respectively, the requirements (2.56) and (2.57) are satisfied with the conditions appearing in (ii).

So, in the above conditions we have a self-mapping operator

$$A : C([a_1,b_1],[a_1,b_1]) \to C([a_1,b_1],[a_1,b_1]).$$

Further, we check A is a completely continuous operator.

Let $\{x_n\}$ be a sequence such that $x_n \to x$ in $C([a_1,b_1],[a_1,b_1])$. Then for each $t \in [a_1,b_1]$, we have that

$$|(Ax_n)(t) - (Ax)(t)| \leq \begin{cases} 0, & \text{for } t \in [a_1,a], \\ \dfrac{2(b-a)^q}{\Gamma(q+1)} \|f(\cdot, x_n(x_n^v(\cdot))) - f(\cdot, x(x^v(\cdot)))\|_C, & \text{for } t \in [a,b], \\ 0, & \text{for } t \in [b,b_1]. \end{cases}$$

Since $f \in C([a,b] \times [a_1,b_1]^2, \mathbb{R})$, we have that

$$\|Ax_n - Ax\|_C \to 0, \quad \text{as } n \to \infty.$$

Now, consider $a_1 \le t_1 < t_2 \le a$. Then,

$$|(Ax)(t_2) - (Ax)(t_1)| = |\varphi(t_2) - \varphi(t_1)|.$$

Similarly, for $b \le t_1 < t_2 \le b_1$,

$$|(Ax)(t_2) - (Ax)(t_1)| = |\psi(t_2) - \psi(t_1)|.$$

On the other hand, for $a \le t_1 < t_2 \le b$,

$$\begin{aligned} |(Ax)(t_2) - (Ax)(t_1)| \le{}& \frac{(t_2-t_1)^q}{(b-a)^q}|\psi(b) - \varphi(a)| \\ &+ \frac{4(t_2-t_1)^q \max\{|m_f|, |M_f|\}}{\Gamma(q+1)}. \end{aligned} \tag{2.58}$$

Together with Arzela-Ascoli theorem and A is a continuous operator, we can conclude that A is a completely continuous operator.

It is obvious that the set $C([a_1,b_1],[a_1,b_1]) \subseteq C([a_1,b_1],\mathbb{R})$ is a bounded convex closed subset of the Banach space $C([a_1,b_1],\mathbb{R})$. Thus, the operator A has a fixed point due to Schauder fixed point theorem. This completes the proof. □

In the following, we present the existence and estimate results in $C^q_L([a_1,b_1],[a_1,b_1])$. Let $L > 0$ and $I \subset \mathbb{R}$ be a compact interval, and introduce the following notation:

$$C^q_L(I,\mathbb{R}) = \{x \in C(I,\mathbb{R}) |\ |x(t_1) - x(t_2)| \le L|t_1 - t_2|^q\}$$

for all $t_1, t_2 \in I$. Remark that $C^q_L(I,\mathbb{R}) \subseteq C(I,\mathbb{R})$ is a complete metric space. Then (2.58) implies that under assumptions of Theorem 2.11 any solution of problem (2.53) belongs to $C^q_{L_*}([a,b],\mathbb{R})$ for

$$L_* = \frac{|\psi(b) - \varphi(a)|}{(b-a)^q} + \frac{4\max\{|m_f|,|M_f|\}}{\Gamma(q+1)}. \tag{2.59}$$

Now we present our second result in this section.

Theorem 2.12. *We suppose that*

(i) *conditions of Theorem 2.11 hold and* $\varphi \in C^q_{L_\varphi}([a_1,a],[a_1,b_1])$, $\psi \in C^q_{L_\psi}([b,b_1],[a_1,b_1])$ *for some* $L_\varphi, L_\psi \ge 0$.

Then problem (2.53) *has a solution in* $X = C^q_L([a_1,b_1],[a_1,b_1])$ *and all its solution belongs to* X *for*

$$L = \left(\sqrt[1-q]{L_\varphi} + \sqrt[1-q]{L_\psi} + \sqrt[1-q]{L_*}\right)^{1-q},$$

where L_* *is defined by* (2.59).

Assume in addition

(ii) *there exist* $L_u > 0$ *and* $L_w > 0$ *such that*

$$|f(t,u_1,w_1) - f(t,u_2,w_2)| \le L_u|u_1 - u_2| + L_w|w_1 - w_2|,$$

for $\forall\ t \in [a,b],\ u_i, w_i \in [a_1,b_1],\ i = 1,2.$

Then two solutions x_1 *and* x_2 *of problem* (2.53) *satisfy*

$$\|x_1 - x_2\|_C \le L_A^{\frac{1}{1-q\min\{1,v\}}} \tag{2.60}$$

for

$$L_A := \frac{2(b-a)^q}{\Gamma(q+1)}\left((L_u + L_w)b_1^{1-q\min\{1,v\}} + \max\{1,v^q\}b_1^{q\max\{v-1,0\}}L_wL\right). \tag{2.61}$$

If in addition

$$\Gamma(q+1) > 2(b-a)^q L_u, \tag{2.62}$$

then

$$\|x_1 - x_2\|_C \le \left(\frac{2(b-a)^q L_w\left(b_1^{1-q\min\{1,v\}} + \max\{1,v^q\}b_1^{q\max\{v-1,0\}}L\right)}{\Gamma(q+1) - 2(b-a)^q L_u}\right)^{\frac{1}{1-q\min\{1,v\}}}. \tag{2.63}$$

Proof. Consider the operator A given by (2.55). From Theorem 2.11, we have

$$A : C([a_1,b_1],[a_1,b_1]) \to C([a_1,b_1],[a_1,b_1])$$

and A has a fixed point in $C([a_1,b_1],[a_1,b_1])$.

Now, consider $a_1 \le t_1 < t_2 \le a$. Then,

$$|(Ax)(t_2) - (Ax)(t_1)| = |\varphi(t_2) - \varphi(t_1)| \le L_\varphi|t_1 - t_2|^q \le L_*|t_1 - t_2|^q$$

as $\varphi \in C^q_{L_\varphi}([a_1,a],[a_1,b_1])$, due to (i).

Similarly, for $b \le t_1 < t_2 \le b_1$,

$$|(Ax)(t_2) - (Ax)(t_1)| = |\psi(t_2) - \psi(t_1)| \le L_\psi|t_1 - t_2|^q \le L_*|t_1 - t_2|^q$$

that follows from (i), too.

On the other hand, for $a \le t_1 < t_2 \le b$, we already know (see (2.58))

$$|(Ax)(t_2) - (Ax)(t_1)| \le L_*|t_1 - t_2|^q.$$

Next, if $a_1 \le t_1 \le a \le t_2 \le b$, then by Hölder inequality with $q' = \frac{1}{q}$ and $p' = \frac{1}{1-q}$ (note $q', p' > 1$),

$$\begin{aligned}
|(Ax)(t_2) - (Ax)(t_1)| &\le |(Ax)(a) - (Ax)(t_1)| + |(Ax)(t_2) - (Ax)(a)| \\
&\le L_\varphi(a-t_1)^q + L_*(t_2-a)^q \\
&\le \sqrt[p']{L_\varphi^{p'} + L_*^{p'}}\ \sqrt[q']{(a-t_1)^{qq'} + (t_2-a)^{qq'}} \\
&\le L|t_1 - t_2|^q.
\end{aligned}$$

Furthermore, if $a_1 \leq t_1 \leq a < b \leq t_2 \leq b_1$, then again by the Hölder inequality with $q' = \frac{1}{q}$ and $p' = \frac{1}{1-q}$,

$$
\begin{aligned}
&|(Ax)(t_2) - (Ax)(t_1)| \\
&\leq |(Ax)(a) - (Ax)(t_1)| + |(Ax)(b) - (Ax)(a)| + |(Ax)(t_2) - (Ax)(b)| \\
&\leq L_\varphi (a - t_1)^q + L_*(b-a)^q + L_\psi (t_2 - b)^q \\
&\leq \sqrt[p']{L_\varphi^{p'} + L_*^{p'} + L_\psi^{p'}} \sqrt[q']{(a-t_1)^{qq'} + (b-a)^{qq'} + (t_2 - b)^{qq'}} \\
&= L|t_1 - t_2|^q.
\end{aligned}
$$

Therefore, the function $A(x)(t)$ belongs to X. This proves the first statement.

Take $x_1, x_2 \in X$. Then for all $t \in [a_1, a] \cup [b, b_1]$, we have

$$|A(x_1)(t) - A(x_2)(t)| = 0.$$

Moreover, for $t \in [a, b]$, from our conditions, we get

$$
\begin{aligned}
&|A(x_1)(t) - A(x_2)(t)| \\
&\leq \frac{(t-a)^q}{\Gamma(q)(b-a)^q} \int_a^b (b-s)^{q-1} \left| f(s, x_1(s), x_1(x_1^v(s))) - f(s, x_2(s), x_2(x_2^v(s))) \right| ds \\
&\quad + \frac{1}{\Gamma(q)} \int_a^t (t-s)^{q-1} \left| f(s, x_1(s), x_1(x_1^v(s))) - f(s, x_2(s), x_2(x_2^v(s))) \right| ds \\
&\leq \frac{1}{\Gamma(q)} \int_a^b (b-s)^{q-1} \left(L_u |x_1(s) - x_2(s)| + L_w \left| x_1(x_1^v(s)) - x_2(x_2^v(s)) \right| \right) ds \\
&\quad + \frac{1}{\Gamma(q)} \int_a^t (t-s)^{q-1} \left(L_u |x_1(s) - x_2(s)| + L_w \left| x_1(x_1^v(s)) - x_2(x_2^v(s)) \right| \right) ds \\
&\leq \frac{1}{\Gamma(q)} \int_a^b (b-s)^{q-1} \Big(L_u |x_1(s) - x_2(s)| + L_w \left| x_1(x_1^v(s)) - x_1(x_2^v(s)) \right| \\
&\quad + L_w \left| x_1(x_2^v(s)) - x_2(x_2^v(s)) \right| \Big) ds \\
&\quad + \frac{1}{\Gamma(q)} \int_a^t (t-s)^{q-1} \Big(L_u |x_1(s) - x_2(s)| + L_w \left| x_1(x_1^v(s)) - x_1(x_2^v(s)) \right| \\
&\quad + L_w \left| x_1(x_2^v(s)) - x_2(x_2^v(s)) \right| \Big) ds \\
&\leq \frac{1}{\Gamma(q)} \int_a^b (b-s)^{q-1} \Big((L_u + L_w) \|x_1 - x_2\|_C + L_w L \left| x_1^v(s) - x_2^v(s) \right|^q \Big) ds \\
&\quad + \frac{1}{\Gamma(q)} \int_a^t (t-s)^{q-1} \Big((L_u + L_w) \|x_1 - x_2\|_C + L_w L \left| x_1^v(s) - x_2^v(s) \right|^q \Big) ds \\
&\leq \frac{1}{\Gamma(q)} \int_a^b (b-s)^{q-1} \Big((L_u + L_w) \|x_1 - x_2\|_C \\
&\quad + \max\{1, v^q\} b_1^{q \max\{v-1, 0\}} L_w L \|x_1 - x_2\|_C^{q \min\{1, v\}} \Big) ds \\
&\quad + \frac{1}{\Gamma(q)} \int_a^t (t-s)^{q-1} \Big((L_u + L_w) \|x_1 - x_2\|_C
\end{aligned}
$$

$$
\begin{aligned}
&+\max\{1,v^q\}b_1^{q\max\{v-1,0\}}L_wL\|x_1-x_2\|_C^{q\min\{1,v\}}\Big)ds\\
&\le \frac{2(b-a)^q}{\Gamma(q+1)}\Big((L_u+L_w)\|x_1-x_2\|_C\\
&+\max\{1,v^q\}b_1^{q\max\{v-1,0\}}L_wL\|x_1-x_2\|_C^{q\min\{1,v\}}\Big),
\end{aligned}
$$

where we use the inequality

$$r^v-s^v\le\max\{1,v\}r^{\max\{v-1,0\}}(r-s)^{\min\{1,v\}}$$

for any $r\ge s\ge 0$ and $v>0$. From $\|x_1-x_2\|_C\le b_1$ we get

$$
\begin{aligned}
\|A(x_1)-A(x_2)\|_C &\le \frac{2(b-a)^q}{\Gamma(q+1)}\Big((L_u+L_w)b_1^{1-q\min\{1,v\}}\\
&\quad+\max\{1,v^q\}b_1^{q\max\{v-1,0\}}L_wL\Big)\|x_1-x_2\|_C^{q\min\{1,v\}} \qquad (2.64)\\
&=L_A\|x_1-x_2\|_C^{q\min\{1,v\}}.
\end{aligned}
$$

So A is Hölder continuous but not Lipschitz continuous, since $q\min\{1,v\}\le q<1$. If x_1 and x_2 are fixed points of A then

$$\|x_1-x_2\|_C=\|A(x_1)-A(x_2)\|_C\le L_A\|x_1-x_2\|_C^{q\min\{1,v\}}$$

which implies (2.60). In general, we have

$$
\begin{aligned}
\|x_1-x_2\|_C &\le \frac{2(b-a)^q}{\Gamma(q+1)}L_u\|x_1-x_2\|_C\\
&\quad+\frac{2(b-a)^qL_w}{\Gamma(q+1)}\Big(b_1^{1-q\min\{1,v\}}\\
&\quad+\max\{1,v^q\}b_1^{q\max\{v-1,0\}}L\Big)\|x_1-x_2\|_C^{q\min\{1,v\}},
\end{aligned}
$$

which implies (2.63) under (2.62). The proof is completed. □

We do not know about uniqueness. But this is not so surprising, since A is not Lipschitzian in general. So we cannot apply metric fixed point theorems, only topological one. This can be simply illustrated on a simpler problem

$$ {}_0^CD_t^{\frac{1}{2}}x(t)=x(\sqrt{x(t)}),\quad x(0)=0,\quad t\in[0,1]. \qquad (2.65)$$

Rewriting (2.65) as

$$x(t)=B(x)(t)=\frac{1}{\Gamma(\frac{1}{2})}\int_0^t\frac{x(\sqrt{x(s)})}{\sqrt{t-s}}ds,$$

it follows that $B: C_{\frac{1}{2}}([0,1],[0,1])\to C_{\frac{1}{2}}([0,1],\mathbb{R})$ satisfies

$$\|B(x_1)-B(x_2)\|_C\le\frac{2}{\sqrt{\pi}}\left(\|x_1-x_2\|_C+\frac{1}{2}\sqrt{\|x_1-x_2\|_C}\right),$$

so it is not Lipschitzian. Hence (2.65) should have a nonzero solution, and it does have $x(t)=\frac{4}{\pi}t$.

2.5.3 *Data Dependence*

Consider the following two problems

$$\begin{cases} {}_a^C D_t^q x(t) = f_i(t, x(t), x(x^v(t))) + \lambda_i, & t \in [a,b],\ v \in (0,1],\ q \in (0,1), \\ x(t) = \varphi_i(t), & t \in [a_1, a], \\ x(t) = \psi_i(t), & t \in [b, b_1], \end{cases} \tag{2.66}$$

where f_i, λ_i, φ_i and ψ_i, $i = 1, 2$ be as in the Theorem 2.12.

Consider the operators

$$A_i : C([a_1, b_1], [a_1, b_1]) \to C([a_1, b_1], [a_1, b_1])$$

given by (2.55) when φ, ψ, f and λ are replaced by φ_i, ψ_i, f_i and λ_i, respectively.

We are ready to state the third result in this section.

Theorem 2.13. *Suppose the conditions of the Theorem 2.12 hold, and, moreover*

(i) *there exists* $\eta_1 > 0$ *such that*

$$|\varphi_1(t) - \varphi_2(t)| \le \eta_1, \ \forall\, t \in [a_1, a],$$

and

$$|\psi_1(t) - \psi_2(t)| \le \eta_1, \ \forall\, t \in [b, b_1];$$

(ii) *there exists* $\eta_2 > 0$ *such that*

$$|f_1(t, u, w) - f_2(t, u, w)| \le \eta_2, \ \forall\, t \in [a, b],\ u, w \in [a_1, b_1].$$

Let r_* *be a positive root of equation*

$$r_* = L^* r_*^{q\min\{1,v\}} + 3\eta_1 + \frac{2(b-a)^q}{\Gamma(q+1)}\eta_2, \tag{2.67}$$

where $L^* = \min\{L_{A_1}, L_{A_2}\}$ *(see* (2.61)*). Then*

$$\|x_1^* - x_2^*\|_C \le r_*, \tag{2.68}$$

and

$$|\lambda_1^* - \lambda_2^*| \le \frac{\Gamma(q+1)}{(b-a)^q}\left(2\eta_1 + \frac{L^*}{2} r_*^{q\min\{1,v\}}\right) + \eta_2, \tag{2.69}$$

where (x_i^*, λ_i^*), $i = 1, 2$ *are solutions of the corresponding problems* (2.66)*. Note* r_* *is uniquely defined.*

Proof. Using the condition (i), it is easy to see that for $x \in C([a_1, b_1], [a_1, b_1])$ and $t \in [a_1, a] \cup [b, b_1]$, we have

$$\|A_1(x) - A_2(x)\|_C \le \eta_1.$$

On the other hand, for $t \in [a, b]$, using the condition (ii), we obtain

$$|A_1(x)(t) - A_2(x)(t)|$$

$$
\begin{aligned}
&\le |\varphi_1(a) - \varphi_2(a)| + \frac{(t-a)^q}{(b-a)^q}(|\psi_1(b) - \psi_2(b)| + |\varphi_1(a) - \varphi_2(a)|) \\
&\quad + \frac{(t-a)^q}{\Gamma(q)(b-a)^q}\int_a^b (b-s)^{q-1}\,|f_1(s, x(s), x(x^v(s))) - f_2(s, x(s), x(x^v(s)))|\,ds \\
&\quad + \frac{1}{\Gamma(q)}\int_a^t (t-s)^{q-1}\,|f_1(s, x(s), x(x^v(s))) - f_2(s, x(s), x(x^v(s)))|\,ds \\
&\le 3\eta_1 + \frac{2(b-a)^q}{\Gamma(q+1)}\eta_2.
\end{aligned}
$$

So, we have

$$
\|A_1(x) - A_2(x)\|_C \le 3\eta_1 + \frac{2(b-a)^q}{\Gamma(q+1)}\eta_2.
$$

Next, (2.64) holds for both A_i with L_{f_i}. Without loss of generality, we may suppose that $L^* = L_{A_1} = \min\{L_{A_1}, L_{A_2}\}$. Consequently, we obtain

$$
\begin{aligned}
\|x_1^* - x_2^*\|_C &= \|A_1(x_1^*) - A_2(x_2^*)\|_C \\
&\le \|A_1(x_1^*) - A_1(x_2^*)\|_C + \|A_1(x_2^*) - A_2(x_2^*)\|_C \\
&\le L^*\|x_1^* - x_2^*\|_C^{q\min\{1,v\}} + 3\eta_1 + \frac{2(b-a)^q}{\Gamma(q+1)}\eta_2,
\end{aligned}
$$

which implies (2.68). Moreover, we get

$$
\begin{aligned}
&|\lambda_1^* - \lambda_2^*| \\
&\le \frac{\Gamma(q+1)(|\psi_1(b) - \psi_2(b)| + |\varphi_1(a) - \varphi_2(a)|)}{(b-a)^q} \\
&\quad + \frac{q}{(b-a)^q}\int_a^b (b-s)^{q-1}\,|f_1(s, x_1^*(s), x_1^*(x_1^{*v}(s))) - f_1(s, x_2^*(s), x_2^*(x_2^{*v}(s)))|\,ds \\
&\quad + \frac{q}{(b-a)^q}\int_a^b (b-s)^{q-1}\,|f_1(s, x_2^*(s), x_2^*(x_2^{*v}(s))) - f_2(s, x_2^*(s), x_2^*(x_2^{*v}(s)))|\,ds \\
&\le \frac{\Gamma(q+1)}{(b-a)^q}\left(2\eta_1 + \frac{L^*}{2} r_*^{q\min\{1,v\}}\right) + \eta_2.
\end{aligned}
$$

The proof is completed. □

2.5.4 *Examples and General Cases*

Example 2.2. Consider the following problem:

$$
\begin{cases}
{}_0^C D_t^{\frac{1}{2}} x(t) = \mu x(x(t)) + \lambda, & t \in [0,1],\ \mu > 0,\ \lambda \in \mathbb{R}, \\
x(t) = 0, & t \in [-h, 0],\ h > 0, \\
x(t) = 1, & t \in [1, 1+h],
\end{cases}
\tag{2.70}
$$

where $x \in C([-h, 1+h], [-h, 1+h])$.

Proposition 2.1. *Suppose that*

$$\mu \leq \frac{\Gamma(\frac{3}{2})h}{1+2h}.$$

Then the problem (2.70) *has a solution in* $C([-h, 1+h], [-h, 1+h])$.

Proof. First of all notice that accordingly to the Theorem 2.11 we have $\upsilon = 1$, $q = \frac{1}{2}$, $a = 0$, $b = 1$, $\psi(b) = 1$, $\varphi(a) = 0$ and $f(t, u_1, u_2) = \mu u_2$. Moreover, $a_1 = -h$ and $b_1 = 1 + h$ can be taken. Therefore, from the relation

$$m_f \leq f(t, u_1, u_2) \leq M_f, \ \forall\, t \in [0, 1], \ u_1, u_2 \in [-h, 1+h],$$

we can choose $m_f = -h\mu$ and $M_f = (1+h)\mu$. For these data it can be easily verified that the condition (ii) from the Theorem 2.11 is equivalent to the relation

$$\mu \leq \frac{\Gamma(\frac{3}{2})h}{1+2h},$$

consequently we complete the proof. □

Example 2.3. Consider the following problem:

$$\begin{cases} {}_{2h}^{C}D_t^{\frac{1}{2}} x(t) = \mu x^2(x(t)) + \lambda, & t \in [2h, 3h], \ \mu > 0, \ \lambda \in \mathbb{R}, \\ x(t) = \dfrac{1}{2}, & t \in [h, 2h], \\ x(t) = \dfrac{1}{2}, & t \in [3h, 4h], \ h \in \left[\dfrac{1}{8}, \dfrac{1}{2}\right], \end{cases} \tag{2.71}$$

where $x \in C([h, 4h], [h, 4h])$. Note $\frac{1}{2} \in [h, 4h]$ for $h \in [\frac{1}{8}, \frac{1}{2}]$.

Proposition 2.2. *We suppose that*

$$0 < \mu \leq \frac{(-1+8h)\sqrt{\pi}}{64h^{5/2}}, \quad \text{for } h \in \left(\frac{1}{8}, \frac{1}{5}\right],$$

$$0 < \mu \leq \frac{(1-2h)\sqrt{\pi}}{64h^{5/2}}, \quad \text{for } h \in \left[\frac{1}{5}, \frac{1}{2}\right).$$

Then the problem (2.71) *has a solution in* $C_L^{\frac{1}{2}}([h, 4h], [h, 4h])$ *with* $L = \frac{128\mu h^2}{\sqrt{\pi}}$. *Note* $0 < \mu \leq \frac{15\sqrt{5\pi}}{64} \doteq 0.928905$. *Furthermore, any two solutions* $x_1, x_2 \in C_L^{\frac{1}{2}}([h, 4h], [h, 4h])$ *of* (2.71) *satisfy*

$$\|x_1 - x_2\|_C \leq \frac{4096h^4\mu^2\left(64h^{\frac{3}{2}}\mu + \sqrt{\pi}\right)^2}{\pi^2}. \tag{2.72}$$

Proof. First of all notice that accordingly to the Theorem 2.12 we have $\upsilon = 1$, $q = \frac{1}{2}$, $a = 2h$, $b = 3h$, $\psi(b) = \frac{1}{2}$, $\varphi(a) = \frac{1}{2}$, $a_1 = h$, $b_1 = 4h$. Observe that $|f(t, u_1, u_2) - f(t, w_1, w_2)| = \mu|u_2 + w_2||u_2 - w_2| \leq 8h\mu|u_2 - w_2|$, $u_2, w_2 \in [h, 4h]$. So $L_u = 0$ and $L_w = 8h\mu$. Next, we choose $m_f = \mu h^2$ and $M_f = 16\mu h^2$. By a common check in the conditions of Theorem 2.12 we can make sure that

$$a_1 \leq \min\{\varphi(a), \psi(b)\} - \max\left\{0, \frac{M_f(b-a)^q}{\Gamma(q+1)}\right\} + \min\left\{0, \frac{m_f(b-a)^q}{\Gamma(q+1)}\right\}$$

$$\Longleftrightarrow h + \frac{16\mu h^{\frac{5}{2}}}{\Gamma(\frac{3}{2})} \leq \frac{1}{2},$$

$$\max\{\varphi(a), \psi(b)\} - \min\left\{0, \frac{m_f(b-a)^q}{\Gamma(q+1)}\right\} + \max\left\{0, \frac{M_f(b-a)^q}{\Gamma(q+1)}\right\} \leq b_1$$

$$\Longleftrightarrow \frac{1}{2} \leq 4h - \frac{16\mu h^{\frac{5}{2}}}{\Gamma(\frac{3}{2})}.$$

These inequalities are equivalent to

$$0 < \mu \leq \min\left\{\frac{(1-2h)\sqrt{\pi}}{64h^{\frac{5}{2}}}, \frac{(-1+8h)\sqrt{\pi}}{64h^{\frac{5}{2}}}\right\}.$$

The function $\kappa(h) = \min\left\{-\frac{(-1+2h)\sqrt{\pi}}{64h^{\frac{5}{2}}}, \frac{(-1+8h)\sqrt{\pi}}{64h^{\frac{5}{2}}}\right\}$ is increasing from 0 to $\frac{15\sqrt{5\pi}}{64} \doteq 0.928905$ on $[\frac{1}{8}, \frac{1}{5}]$ and then it is decreasing to 0 on $[\frac{1}{5}, \frac{1}{2}]$. Next, we derive $L_\varphi = L_\psi = 0$ and

$$L_* = \frac{|\psi(b) - \varphi(a)|}{(b-a)^q} + \frac{4\max\{|m_f|, |M_f|\}}{\Gamma(q+1)} = \frac{64\mu h^2}{\Gamma(\frac{3}{2})} = \frac{128\mu h^2}{\sqrt{\pi}},$$

so $L = L_*$. By (2.61) we derive

$$L_A = \frac{2\sqrt{h}}{\Gamma(\frac{3}{2})}\left(8h\mu\sqrt{4h} + 8h\mu\frac{128\mu h^2}{\sqrt{\pi}}\right) = \frac{64h^2\mu\left(64h^{\frac{3}{2}}\mu + \sqrt{\pi}\right)}{\pi}.$$

This gives (2.72) by (2.60). Therefore, by Theorem 2.12 the proof is completed. □

Example 2.4. Now take the following problems

$$\begin{cases} {}^{C}_{2h}D^{\frac{1}{2}}_{t} x(t) = \mu x^2(x(t)) + \lambda_i, & t \in [2h, 3h],\ \mu_i = \mu,\ \lambda_i \in \mathbb{R}, \\ x(t) = \varphi_i, & t \in [h, 2h],\ h > 0, \\ x(t) = \psi_i, & t \in [3h, 4h] \end{cases} \tag{2.73}$$

for $i = 1, 2$. Suppose the following assumptions.

(H1) $\varphi_i \in C^{\frac{1}{2}}_{L_*}([h, 2h], [h, 4h])$, $\psi_i \in C^{\frac{1}{2}}_{L_*}([3h, 4h], [h, 4h])$ such that $\varphi_i(2h) = \frac{1}{2}$, $\psi_i(3h) = \frac{1}{2}$, $i = 1, 2$ and $L_* = \frac{128\mu h^2}{\sqrt{\pi}}$;
(H2) we are in the conditions of Proposition 2.2 for both of the problems (2.73).

Let (x_i^*, λ_i^*) be solutions of the problems (2.73). We are looking for an estimation for $\|x_1^* - x_2^*\|_C$ and $|\lambda_1^* - \lambda_2^*|$.

Then, build upon Theorem 2.13, by a common substitution one can make sure that we have

Proposition 2.3. *Consider the problems* (2.73) *and suppose the requirements (H1)-(H2) hold. Additionally, there exists* $\eta_1 > 0$ *such that*

$$|\varphi_1(t) - \varphi_2(t)| \leq \eta_1,\ \forall\, t \in [h, 2h],$$

and

$$|\psi_1(t) - \psi_2(t)| \le \eta_1, \ \forall\, t \in [3h, 4h].$$

Then

$$\|x_1^* - x_2^*\|_C \le r_*,$$

and

$$|\lambda_1^* - \lambda_2^*| \le \frac{2}{\sqrt{\pi h}}\left(2\eta_1 + \frac{L^*}{2}\sqrt{r_*}\right),$$

where L^* *and* r_* *are given by* (2.74) *and* (2.75), *respectively.*

Proof. Results follow from Theorem 2.13 as follows. By Proposition 2.2, we have $L = \sqrt{3}L_* = \frac{128\mu h^2\sqrt{3}}{\sqrt{\pi}}$ and then (see (2.61))

$$L^* = L_{A_1} = L_{A_2} = \frac{64h^2\mu(64\sqrt{3}h^{3/2}\mu + \sqrt{\pi})}{\pi}. \tag{2.74}$$

Realizing that now $\eta_2 = 0$, equation (2.67) has the form

$$r_* = L^*\sqrt{r_*} + 3\eta_1,$$

which has the positive solution

$$r_* = \frac{1}{\pi^2}\Bigg(25165824h^7\mu^4 + 64h^2\mu\pi^{\frac{3}{2}} + \eta_1\pi^2 + 8\sqrt{2}\sqrt{4947802324992h^{14}\mu^8 + 25165824h^9\mu^5\pi^{\frac{3}{2}} + 393216h^7\mu^4\eta_1\pi^2}\Bigg). \tag{2.75}$$

The estimate for $|\lambda_1^* - \lambda_2^*|$ follows directly from (2.69). The proof is finished. □

We conclude this section by considering a general fractional order iterative functional differential equations with parameter given by

$$\begin{cases} {}_a^C D_t^q x(t) = f(t, x(t), x(x^v(t)), \lambda), & t \in [a,b],\ v \in (0,1],\ q \in (0,1),\ \lambda \in J, \\ x(t) = \varphi(t), & t \in [a_1, a], \\ x(t) = \psi(t), & t \in [b, b_1], \end{cases} \tag{2.76}$$

when $J \subset \mathbb{R}$ is an open interval, conditions (C1), (C3) are supposed and (C2) is extended to

(C4) $f \in C([a,b] \times [a_1, b_1]^2 \times J, \mathbb{R})$.

Then by (2.55) we have an operator $A(\lambda, x)$. It is easy to see that $A(\lambda, x) = A(x)$ for the problem (2.53). Supposing the assumptions of Theorem 2.11 for the problem (2.76) uniformly with respect to $\lambda \in J$, we can find its fixed point $x^*(\lambda, \cdot) \in C([a_1, b_1], [a_1, b_1])$. In order to get a solution of the problem (2.76), we need to solve

$$\Upsilon(\lambda) = \Gamma(q)(\psi(b) - \varphi(a)) - \int_a^b (b-s)^{q-1} f(s, x^*(\lambda, s), x^*(\lambda, x^*(\lambda, s)^v))ds = 0. \tag{2.77}$$

If there is an $\lambda_0 \in J$ solving (2.77) then $x^*(\lambda_0, t)$ is a solution of the problem (2.76). Since $x^*(\lambda, \cdot)$ is not unique in general, function $\Upsilon(\lambda)$ is multivalued. Consequently this way is not very useful. We propose another approach. The problem (2.76) is equivalent to the following fixed point equation

$$x(t) = \begin{cases} \varphi(t), & \text{for } t \in [a_1, a], \\ \varphi(a) + \dfrac{1}{\Gamma(q)} \displaystyle\int_a^t (t-s)^{q-1} f(s, x(s), x(x^v(s)), \lambda) ds, & \text{for } t \in [a, b], \\ \psi(t), & \text{for } t \in [b, b_1]. \end{cases}$$

From the condition of continuity of x in $t = b$, we have that

$$\psi(b) = \varphi(a) + \frac{1}{\Gamma(q)} \int_a^b (b-s)^{q-1} f(s, x(s), x(x^v(s)), \lambda) ds.$$

Now we consider the operator

$$A : C^b([a_1, b_1], [a_1, b_1]) \times J \to C^b([a_1, b_1], \mathbb{R})$$

where

$$C^b([a_1, b_1], [a_1, b_1]) = \{x \in C([a_1, b], [a_1, b_1]) \cap C^b((b, b_1], [a_1, b_1]) : \exists \lim_{s \to b_+} x(s)\},$$

$$C^b([a_1, b_1], \mathbb{R}) = \{x \in C([a_1, b], \mathbb{R}) \cap C^b((b, b_1], \mathbb{R}) : \exists \lim_{s \to b_+} x(s)\}$$

and

$$A(x, \lambda)(t) := \begin{cases} \varphi(t), & \text{for } t \in [a_1, a], \\ \varphi(a) + \dfrac{1}{\Gamma(q)} \displaystyle\int_a^t (t-s)^{q-1} f(s, x(s), x(x^v(s)), \lambda) ds, & \text{for } t \in [a, b], \\ \psi(t), & \text{for } t \in (b, b_1]. \end{cases}$$

Now, we are ready to state the following result.

Theorem 2.14. *Suppose that*

(i) *conditions (C1), (C3) and (C4) are satisfied;*
(ii) *there are m_f, $M_f \in \mathbb{R}$ such that*

$$m_f \le f(t, u, w, \lambda) \le M_f, \ \forall \ t \in [a, b], u, w \in [a_1, b_1], \lambda \in J$$

along with

$$a_1 \le \varphi(a) + \min\left\{0, \frac{m_f(b-a)^q}{\Gamma(q+1)}\right\},$$

and

$$\varphi(a) + \max\left\{0, \frac{M_f(b-a)^q}{\Gamma(q+1)}\right\} \le b_1.$$

Then operator $A(x, \lambda)$ has a fixed point in $C^b([a_1, b_1], [a_1, b_1])$ for any $\lambda \in J$.

Proof. Like in the proof of Theorem 2.11, condition (ii) assures that the set $C^b([a_1,b_1],[a_1,b_1])$ is an invariant subset for the operator A, that is, we have

$$A(C^b([a_1,b_1],[a_1,b_1]) \times J) \subset C^b([a_1,b_1],[a_1,b_1]). \tag{2.78}$$

Similarly, A is a completely continuous operator. It is obvious that the set $C^b([a_1,b_1],[a_1,b_1]) \subseteq C^b([a_1,b_1],\mathbb{R})$ is a bounded convex closed subset of the Banach space $C^b([a_1,b_1],\mathbb{R})$. Thus, the operator $A(x,\lambda)$ has a fixed point due to Schauder fixed point theorem. This completes the proof. □

We still do not have uniqueness result. For this purpose, we suppose

(C5) f is nonnegative and nondecreasing, i.e. $m_f \geq 0$ and $0 \leq f(s_1,u_1,v_1,\lambda) \leq f(s_2,u_2,v_2,\lambda)$ for any $s_1 \leq s_2 \in [a,b]$, $u_1 \leq u_2, v_1 \leq v_2 \in [a_1,b_1]$ and $\lambda \in J$.

We introduce the Banach space

$$C^b_m([a_1,b_1],[a_1,b_1]) = \left\{x \in C^b([a_1,b_1],[a_1,b_1]) \mid x \text{ is nondecreasing on } [a_1,b_1]\right\}.$$

Now, we have the next result.

Theorem 2.15. *We suppose conditions (i), (ii) of Theorem 2.14, (C5) as well $\varphi(t)$, $\psi(t)$ are nondecreasing with $\varphi(a) \leq \psi(b)$. Then operator $A(x,\lambda)$ is monotone nondecreasing in x on $C^b_m([a_1,b_1],[a_1,b_1])$ for any $\lambda \in J$. Consequently it has a unique smallest and largest fixed points $x_m(\lambda), x_M(\lambda)$ in $C^b_m([a_1,b_1],[a_1,b_1])$. Moreover, a nondecreasing sequence $\{A^k(a_1,\lambda)(t)\}_{k\geq 1}$ and a nonincreasing sequence $\{A^k(b_1,\lambda)(t)\}_{k\geq 1}$ satisfy*

$$a_1 \leq A^k(a_1,\lambda)(t) \leq x_m(\lambda)(t) \leq x_M(\lambda)(t) \leq A^k(b_1,\lambda)(t) \leq b_1, \quad t \in J$$

for any $k \geq 1$ and $\lim_{k\to\infty} A^k(a_1,\lambda)(t) = x_m(\lambda)(t)$ and $\lim_{k\to\infty} A^k(b_1,\lambda)(t) = x_M(\lambda)(t)$ uniformly on $[a_1,b_1]$.

Proof. We already know (2.78). Let $x \in C^b_m([a_1,b_1],[a_1,b_1])$ then clearly $A(x,\lambda)(t_1) \leq A(x,\lambda)(t_2)$ for $t_1 \leq t_2 \in [a_1,a]$ and $t_1 \leq t_2 \in (b,b_1]$. Next for $s_1 \leq s_2 \in [a,b]$, we have $x(s_1) \leq x(s_2)$, $x^v(s_1) \leq x^v(s_2)$ and $x(x^v(s_1)) \leq x(x^v(s_2))$, which imply

$$f(s_1,x(s_1),x(x^v(s_1)),\lambda) \leq f(s_2,x(s_2),x(x^v(s_2)),\lambda). \tag{2.79}$$

Furthermore, for $t_1 \leq t_2 \in [a,b]$, following El-Sayed, 1995 and Darwish, 2008, and using (2.79), we derive

$$\begin{aligned}
&A(x,\lambda)(t_2) - A(x,\lambda)(t_1)\\
&= \frac{1}{\Gamma(q)}\int_a^{t_2}(t_2-s)^{q-1}f(s,x(s),x(x^v(s)),\lambda)ds\\
&\quad - \frac{1}{\Gamma(q)}\int_a^{t_1}(t_1-s)^{q-1}f(s,x(s),x(x^v(s)),\lambda)ds
\end{aligned}$$

$$= \frac{1}{\Gamma(q)} \int_a^{t_1} \left((t_2 - s)^{q-1} - (t_1 - s)^{q-1}\right) f(s, x(s), x(x^v(s)), \lambda) ds$$
$$+ \frac{1}{\Gamma(q)} \int_{t_1}^{t_2} (t_2 - s)^{q-1} f(s, x(s), x(x^v(s)), \lambda) ds$$
$$\geq \frac{1}{\Gamma(q)} f(t_1, x(t_1), x(x^v(t_1)), \lambda)$$
$$\times \left(\int_a^{t_1} (t_2 - s)^{q-1} - (t_1 - s)^{q-1} ds + \int_{t_1}^{t_2} (t_2 - s)^{q-1} ds \right)$$
$$= \frac{1}{\Gamma(q+1)} f(t_1, x(t_1), x(x^v(t_1)), \lambda) \left((t_2 - a)^q - (t_1 - a)^q\right)$$
$$\geq 0.$$

Consequently, we obtain

$$A(C_m^b([a_1, b_1], [a_1, b_1]), \lambda) \subset C_m^b([a_1, b_1], [a_1, b_1])$$

for any $\lambda \in J$.

Next, if $x_1, x_2 \in C_m^b([a_1, b_1], [a_1, b_1])$ with $x_1(t) \leq x_2(t)$, $t \in [a_1, b_1]$ then clearly we have $A(x_1, \lambda)(t) \leq A(x_2, \lambda)(t)$ for $t \in [a_1, a] \cup (b, b_1]$. For $s \in [a, b]$, we have $x_1(s) \leq x_2(s)$, $x_1^v(s) \leq x_2^v(s)$ and $x_1(x_1^v(s)) \leq x_2(x_2^v(s))$, which imply

$$f(s, x_1(s), x_1(x_1^v(s)), \lambda) \leq f(s, x_2(s), x_2(x_2^v(s)), \lambda).$$

Then for $t \in [a, b]$, we have

$$A(x_2, \lambda)(t) - A(x_1, \lambda)(t) = \frac{1}{\Gamma(q)} \int_a^t (t - s)^{q-1} \Big(f(s, x_2(s), x_2(x_2^v(s)), \lambda)$$
$$- f(s, x_1(s), x_1(x_1^v(s)), \lambda) \Big) ds \geq 0.$$

This means that operator $A(x, \lambda)$ is monotone nondecreasing in x on $C_m^b([a_1, b_1], [a_1, b_1])$ for any $\lambda \in J$. We also know that A is a completely continuous operator. Then results follow from the general theory of nondecreasing compact operators in Banach spaces (see, e.g., Deimling, 1985). The proof is completed. □

To get continuous solution, we need to solve either

$$\Upsilon_m(\lambda) = \psi(b) - \varphi(a) - \frac{1}{\Gamma(q)} \int_a^b (b - s)^{q-1} f(s, x_m(\lambda)(s), x_m(\lambda)(x_m^v(\lambda)(s)), \lambda) ds = 0$$

or

$$\Upsilon_M(\lambda) = \psi(b) - \varphi(a) - \frac{1}{\Gamma(q)} \int_a^b (b - s)^{q-1} f(s, x_M(\lambda)(s), x_M(\lambda)(x_M^v(\lambda)(s)), \lambda) ds = 0.$$

We can use to handle these equations also an analytical-numerical method like in Ronto, 2009. This means that first successive approximation is used

$$x_{n+1}(\lambda, t) = A(x_n(\lambda, t), \lambda)$$

for up to some order j with either $x_0(\lambda, t) = a_1$ or $x_0(\lambda, t) = b_1$. Then approximations

$$\Upsilon_j(\lambda) = \psi(b) - \varphi(a) - \frac{1}{\Gamma(q)} \int_a^b (b - s)^{q-1} f(s, x_j(\lambda, s), x_j(\lambda, x_j^v(\lambda, s)), \lambda) ds$$

of Υ_m and Υ_M are numerically drawn to check if they change the sign over J.

2.6 Oscillations and Nonoscillations

2.6.1 *Introduction*

The objective of oscillation theory is to acquire as much information as possible about the qualitative properties of solutions of differential equations. Oscillation theory of functional differential equations with integer derivative has been developed in the past thirty years (see, Györi and Ladas, 1991). However, to the best of our knowledge, there are few results on oscillation for fractional differential equations. In this section, we discuss oscillation and existence of nonoscillatory solutions of fractional functional differential equations.

2.6.2 *Preliminaries*

In this subsection, we introduce preliminary facts which are used throughout this section.

Definition 2.13. (Kilbas, Srivastava and Trujillo, 2006) Let $[a,b](-\infty < a < b < \infty)$ be a finite interval and let $AC[a,b]$ be the space of functions f which are absolutely continuous on $[a,b]$. It is known that $AC[a,b]$ coincides with the space of primitives of Lebesgue summable functions:

$$f(x) \in AC[a,b] \ \Rightarrow f(x) = c + \int_a^x \psi(t)dt \ (\psi(t) \in L(a,b)).$$

Firstly, we consider the fractional delay differential systems

$${}_0D_t^{\alpha}x(t) + \sum_{i=1}^{n} P_i x(t-\tau_i) = 0, \quad t \geq 0, \tag{2.80}$$

where $x(t) = [x_1(t), x_2(t), ..., x_m(t)]^T$, ${}_0D_t^{\alpha}x(t) = [{}_0D_t^{\alpha_1}x_1(t), {}_0D_t^{\alpha_2}x_2(t), ..., {}_0D_t^{\alpha_m}x_m(t)]^T$ is Riemann-Liouville fractional derivative of order $0 < \alpha, \alpha_j < 1$, $\alpha_j = p_j/q_j$, p_j, q_j are odd numbers, for $j = 1,2,...,m$, and $P_i \in \mathbb{R}^{m\times m}$, $\tau_i \in [0,\infty)$ for $i = 1,2,...,n$.

Without loss of generality, we will assume the coefficients P_i of (2.80) are all nonzero and that $\tau_1 = \max\{\tau_1, ..., \tau_n\}$.

Definition 2.14. By a solution of (2.80) in $[0,\infty)$ with initial function $\varphi \in AC[-\tau_1, 0]$, we mean a function $x \in AC[-\tau_1,\infty)$ such that $x(t) = \varphi(t)$, $t \in [-\tau_1, 0]$, ${}_0D_t^{\alpha}x(t)$ exists and $x(t)$ satisfies (2.80) in $[0,\infty)$. A solution $x(t) = [x_1(t), ..., x_m(t)]^T$ of system (2.80) is said to oscillate if every component $x_i(t)$ of the solution has arbitrarily large zeros. Otherwise the solution is called non-oscillatory.

We recall some facts about Laplace transforms. If $X(s)$ is the Laplace transform of $x(t)$,

$$X(s) = (\mathcal{L}x)(s) = \int_0^{\infty} e^{-st}x(t)dt,$$

then the abscissa of convergence of $X(s)$ is defined by

$$b = \inf\{\delta \in \mathbb{R} : X(\delta) \text{ exists}\}.$$

Then $X(s)$ is analytic for $\mathrm{Re}(s) > b$.

We call a function $x(t)$ to be eventually positive if there exists a $c \geq 0$ such that $x_c(t) > 0$ for all $t > 0$, where $x_c(t) = x(t+c)$.

For any m-dimensional vector $x = (x_1, x_2, ..., x_m)^T \in \mathbb{R}^m$, $\|x\|$ denotes its norm. For any $m \times m$ real matrix A, the associated matrix norm is then defined by $\|A\| = \max_{\|x\|=1} \|Ax\|$. Denote $\mu(P_i)$ is the logarithmic norm with $\mu(P_i) = \max_{\|u\|=1}(P_i u, u)$.

Lemma 2.8. *(Kilbas, Srivastava and Trujillo, 2006) Let $(\mathcal{L}_0 D_t^\alpha x)(s)$ is the Laplace transform of the Riemann-Liouville fractional derivative of order α with the lower limit 0 for a function x, and $X(s)$ is the Laplace transform of $x(t) \in AC[0,b]$, for any $b > 0$, and the following estimate*

$$|x(t)| \leq Ae^{p_0 t} \quad (t > b > 0)$$

holds for constants $A > 0$ and $p_0 > 0$. Then the relation

$$(\mathcal{L}_0 D_t^\alpha x)(s) = s^\alpha BX(s) - {}_0D_t^{-(1-\alpha)} x(0), \ 0 < \alpha < 1$$

is valid for $\mathrm{Re}(s) > p_0$, where

$$X(s) = [X_1(s), X_2(s), ..., X_m(s)]^T, \quad s^\alpha = [s^{\alpha_1}, s^{\alpha_2}, ..., s^{\alpha_m}],$$

$${}_0D_t^{-(1-\alpha)} x(0) = \left({}_0D_t^{-(1-\alpha_1)} x_1(0), {}_0D_t^{-(1-\alpha_2)} x_2(0), ..., {}_0D_t^{-(1-\alpha_m)} x_m(0)\right)^T,$$

$$B = [B_1, B_2, ..., B_m]^T, \ B_i = (b_{ij})_{m \times m}, \ b_{ij} = \begin{cases} 1, & i = j, \\ 0, & i \neq j. \end{cases}$$

Lemma 2.9. *(Vladimirov, 1981) If $X(s)$ is the Laplace transform of a non-negative function $x(t)$ and has abscissa of convergence $b > -\infty$, then $X(s)$ has a singularity at the point $s = b$.*

Lemma 2.10. *(Henry, 1981) Let $v, w : [0,\infty) \to [0,\infty)$ be continuous functions. If $w(\cdot)$ is nondecreasing and there are constants $a > 0$ and $0 < \beta < 1$ such that*

$$v(t) \leq w(t) + a \int_0^t \frac{v(s)}{(t-s)^\beta} ds,$$

then there exists a constant $k = k(\beta)$ such that

$$v(t) \leq w(t) + ka \int_0^t \frac{w(s)}{(t-s)^\beta} ds$$

for every $t \in [0,\infty)$.

2.6.3 *Oscillation of Neutral Differential Systems*

In this subsection, we discuss the linear autonomous system of neutral delay differential equations with Riemann-Liouville fractional derivative

$$ {}_0D_t^\alpha\left[x(t)+\sum_{j=1}^{l}P_jx(t-\tau_j)\right]+\sum_{i=1}^{n}Q_ix(t-\delta_i)=0 $$

where ${}_0D_t^\alpha x(t) = [{}_0D_t^{\alpha_1}x_1(t), {}_0D_t^{\alpha_2}x_2(t), ..., {}_0D_t^{\alpha_m}x_m(t)]^T$ is Riemann-Liouville fractional derivative, the coefficients $P_j(j = 1, 2, ..., l)$ and $Q_i(i = 1, 2, ..., n)$ are real $m \times m$ matrices and the delays $\tau_j(j = 1, 2, ..., l)$ and $\delta_i(i = 1, 2, ..., n)$ are non-negative real numbers. Sufficient conditions for all solutions of the given equation to be oscillatory are obtained by using fractional calculus and Laplace transform.

Lemma 2.11. *For any $c \in \mathbb{R}$, the Laplace transform $X_c(s)$ of $x_c(t)$ exists and has the same abscissa of convergence as $X(s)$.*

Proof. Given that

$$ \begin{aligned} X_c(s) &= \int_0^\infty e^{-st}x_c(t)dt = \int_0^\infty e^{-st}x(t+c)dt = e^{sc}\int_c^\infty e^{-st}x(t)dt \\ &= e^{sc}\left[X(s)-\int_0^c e^{-st}x(t)dt\right]. \end{aligned} $$

Since the last integral defines an entire function of the complex variable s, therefore $X(s)$ and $X_c(s)$ converge or diverge for the same values of s, and have their singularities at the same points. This completes the proof. □

Lemma 2.12. *The solution of equation* (2.80) *has an exponent estimate*

$$ x(t) = o(e^{q_0t}) \qquad (t > b > 0) $$

for constant $q_0 > 0$.

Proof. Taking Riemann-Liouville integral of equation (2.80), we get

$$ \begin{aligned} x(t) &= \frac{t^{\alpha-1}}{\Gamma(\alpha)}Bx_0 - \sum_{i=1}^{n}P_i\int_0^t\frac{1}{\Gamma(\alpha)}(t-s)^{\alpha-1}Bx(s-\tau_i)ds \\ &= \frac{t^{\alpha-1}}{\Gamma(\alpha)}Bx_0 - \sum_{i=1}^{n}P_iF_i(t), \end{aligned} \tag{2.81} $$

where

$$ x_0 = {}_0D_t^{-(1-\alpha)}x(0), $$

$$ F_i(t)=\int_0^t\frac{1}{\Gamma(\alpha)}(t-s)^{\alpha-1}Bx(s-\tau_i)ds, \quad \frac{t^{\alpha-1}}{\Gamma(\alpha)}=\left[\frac{t^{\alpha_1-1}}{\Gamma(\alpha_1)},\frac{t^{\alpha_2-1}}{\Gamma(\alpha_2)},...,\frac{t^{\alpha_m-1}}{\Gamma(\alpha_m)}\right]. $$

As $AC[-\tau_1, 0]$ is the Banach space with the norm $\|\varphi\|_{AC} = [\|\varphi_1\|_{AC}, \|\varphi_2\|_{AC}, ..., \|\varphi_m\|_{AC}]^T$.
Then we have

$$\begin{aligned}\|F_i(t)\| &\leq \int_0^t \frac{1}{\Gamma(\alpha)}(t-s)^{\alpha-1} B\|x(s-\tau_i)\|ds \\ &\leq \frac{1}{\Gamma(\alpha)} \int_0^t (t-s)^{\alpha-1} B \max_{s-\tau_i \leq \eta \leq s} \|x(\eta)\|ds \\ &\leq \frac{1}{\Gamma(\alpha)} \int_0^t (t-s)^{\alpha-1} B \max_{s-\tau_1 \leq \eta \leq s} \|x(\eta)\|ds,\end{aligned}$$

or

$$\|F_i(t)\| \leq \frac{1}{\Gamma(\alpha)} \int_0^t (t-s)^{\alpha-1} B \left(\max_{s-\tau_1 \leq \eta \leq s} \|x(\eta)\| + \|\varphi\|_{AC} \right) ds. \tag{2.82}$$

From (2.82), it follows that

$$\begin{aligned}\|x(t)\| &\leq \frac{b^{\alpha-1}}{\Gamma(\alpha)} B|x_0| + \sum_{i=1}^{n} \frac{\|P_i\|}{\Gamma(\alpha)} \left[\int_0^t (t-s)^{\alpha-1} B \left(\max_{s-\tau_1 \leq \eta \leq s} \|x(\eta)\| + \|\varphi\|_{AC} \right) ds \right] \\ &\leq \frac{b^{\alpha-1}}{\Gamma(\alpha)} B|x_0| + \frac{np}{\Gamma(\alpha)} \frac{t^\alpha}{\alpha} B\|\varphi\|_{AC} + \frac{np}{\Gamma(\alpha)} \int_0^t (t-s)^{\alpha-1} B \max_{s-\tau_1 \leq \eta \leq s} \|x(\eta)\|ds,\end{aligned}$$

where $p = \max\{\|P_i\|\}$, for $i = 1, 2, ..., n$.

Next we introduce a nondecreasing function $m(t)$ as

$$m(t) = \frac{b^{\alpha-1}}{\Gamma(\alpha)} B|x_0| + \frac{np}{\Gamma(\alpha)} \frac{t^\alpha}{\alpha} B\|\|\varphi\|_{AC}\|.$$

By Lemma 2.10, there exists a number α_0 in $\{\alpha_i\}$ such that

$$\begin{aligned}\|x(t)\| &\leq \max_{t-\tau_1 \leq s \leq t} \|x(s)\| \\ &\leq m(t) + \frac{knp}{\Gamma(\alpha_0)} \int_0^t (t-s)^{\alpha_0-1} m(s)ds \\ &\leq m(t)\left(1 + \frac{knp}{\alpha_0 \Gamma(\alpha_0)} t^{\alpha_0}\right).\end{aligned} \tag{2.83}$$

Obviously, from (2.83) we infer that $x(t)$ has an exponent estimate. The proof is completed. □

Theorem 2.16. *If the characteristic equation*

$$\det\left(\lambda^\alpha B + \sum_{i=1}^{n} P_i e^{-\lambda \tau_i}\right) = 0 \tag{2.84}$$

has no real roots, then every solution of (2.80) *is oscillatory, where* $\lambda^\alpha = [\lambda^{\alpha_1}, \lambda^{\alpha_2}, ..., \lambda^{\alpha_m}]$.

Proof. For the sake of contradiction, let us assume that (2.84) has no real roots and that (2.80) has a non-oscillatory solution $x(t) = [x_1(t), ..., x_m(t)]^T$. This means that one of the components of $x(t)$ is non-oscillatory. Without loss of generality we assume that the component $x_1(t)$ is eventually positive, such that for some $c \geq 0$, $x_c(t) > 0$ for $t \geq 0$. As (2.80) is autonomous, it follows by Lemma 2.11 that $X_1(s)$ and $X_c(s)$ have the same convergence. Then we assume that $x_1(t) > 0$ for $t \geq -\tau_1$. Taking Laplace transform of both sides of (2.80), we obtain

$$s^\alpha BX(s) - {}_0D_t^{-(1-\alpha)}x(0) + \sum_{i=1}^{n} P_i \int_0^\infty e^{-st}x(t-\tau_i)dt = 0,$$

i.e.

$$s^\alpha BX(s) - {}_0D_t^{-(1-\alpha)}x(0) + \sum_{i=1}^{n} P_i e^{-s\tau_i} X(s) + \sum_{i=1}^{n} P_i e^{-s\tau_i} \int_{-\tau_i}^{0} e^{-st}x(t)dt = 0.$$

Hence

$$\left(s^\alpha B + \sum_{i=1}^{n} P_i e^{-s\tau_i}\right) X(s) = {}_0D_t^{-(1-\alpha)}x(0) - \sum_{i=1}^{n} P_i e^{-s\tau_i} \int_{-\tau_i}^{0} e^{-st}x(t)dt. \quad (2.85)$$

Let

$$F(s) = s^\alpha B + \sum_{i=1}^{n} P_i e^{-s\tau_i}, \; x_0 = {}_0D_t^{-(1-\alpha)}x(0),$$

$$\Phi(s) = x_0 - \sum_{i=1}^{n} P_i e^{-s\tau_i} \int_{-\tau_i}^{0} e^{-st}x(t)dt.$$

Then, from (2.85) we get

$$F(s)X(s) = \Phi(s), \; \mathrm{Re}(s) > b. \quad (2.86)$$

Since $\det[F(s)] = 0$ has no real roots, $\det[F(s)] > 0, \; s \in \mathbb{R}$. By Cramer's rule, we have

$$X_1(s) = \frac{\det[D(s)]}{\det[F(s)]}, \; \mathrm{Re}(s) > b, \quad (2.87)$$

where

$$D(s) = \begin{pmatrix} \Phi_1(s) & F_{12}(s) & \cdots & F_{1m}(s) \\ \vdots & \vdots & & \vdots \\ \Phi_m(s) & F_{m2}(s) & \cdots & F_{mm}(s) \end{pmatrix},$$

$\Phi_i(s)$ is the ith component of the vector $\Phi(s)$ and $F_{ij}(s)$ is the (i, j)th component of the matrix $F(s)$. Clearly, for all $i, j = 1, 2, ..., m$ the functions $\Phi_i(s)$ and $F_{ij}(s)$ are entire and hence $\det[D(s)]$ and $\det[F(s)]$ are also entire functions.

Since $\det[F(s)] > 0$ for $s \in \mathbb{R}$, so $\det[D(s)]/\det[F(s)]$ holds for $s \in \mathbb{R}$ and thus (2.87) becomes

$$X_1(s) = \frac{\det[D(s)]}{\det[F(s)]}, \; s \in \mathbb{R}. \quad (2.88)$$

As $x_1(t) > 0$, it follows that $X_1(s) > 0$ for all $s \in \mathbb{R}$ and, by $\det[F(s)] > 0\ s \in \mathbb{R}$ and (2.88), $\det[D(s)] > 0,\ s \in \mathbb{R}$. Now one can see from the definitions of $D(s)$, $F(s)$ and $\Phi(s)$ that there exist positive constants M, β, and s_0 such that

$$\det[D(s)] \leq Me^{-\beta s}, \quad \text{for } s \leq -s_0. \tag{2.89}$$

Since $\det[F(s)]$ is a continuous function in the variables $s, e^{-s\tau_1}, ..., e^{-s\tau_n}$, and $\det[F(s)] > 0,\ s \in \mathbb{R}$, it follows that there exists a positive number m_0 such that

$$\det[F(s)] \geq m_0, \quad \text{for } s \in \mathbb{R}. \tag{2.90}$$

From (2.88), (2.89) and (2.90), it follows that

$$X_1(s) = \int_0^\infty e^{-st}x_1(t)dt \geq \int_T^\infty e^{-st}x_1(t)dt \geq e^{-sT}\int_T^\infty x_1(t)dt > 0$$

and so

$$0 < \int_T^\infty x_1(t)dt \leq \frac{M}{m_0}e^{s(T-\beta)} \to 0, \quad \text{as } s \to -\infty.$$

This implies that $x_1(t) \equiv 0$ for $t \geq T$, which is a contradiction. The proof is completed. □

In Theorem 2.16, the characteristic equation (2.84) plays an important role in the investigation of the oscillation of equation (2.80). However, to determine whether (2.84) has a real root, is quite an issue in itself. In the following we derive some sufficient conditions for the oscillation of equation (2.80) which can easily be applied.

Before proceeding for it, we need the following lemma which is interesting in its own right.

Lemma 2.13. *Assume that $P_i \in \mathbb{R}^{m\times m}$, $\tau_i \geq 0$ for $i = 1, 2, ..., n$, and $\bar{\alpha} = \min\{\alpha_j\}$, for $j = 1, 2, ..., m$ with*

$$\sum_{i=1}^n \mu(-P_i)e^{-\lambda\tau_i} < 0, \quad \textit{for } \lambda \in \mathbb{R} \tag{2.91}$$

and

$$\inf_{\lambda<0}\left[\frac{1}{\lambda^{\bar{\alpha}}}\sum_{i=1}^n \mu(-P_i)e^{-\lambda\tau_i}\right] > 1. \tag{2.92}$$

Then every solution of (2.80) oscillates.

Proof. Assume, for the sake of contraction, that (2.80) has a non-oscillatory solution. Then, by Theorem 2.16, the characteristic equation (2.84) has a real root λ_0. In consequence, there exists a vector $u \in \mathbb{R}^n$ with $\|u\| = 1$ such that

$$\left(\lambda_0^\alpha B + \sum_{i=1}^n P_i e^{-\lambda_0\tau_i}\right)u = 0,$$

i.e.

$$\lambda_0^{\alpha} Bu = -\sum_{i=1}^{n} P_i e^{-\lambda_0 \tau_i} u.$$

Hence

$$\lambda_0^{\bar{\alpha}} = (\lambda_0^{\bar{\alpha}} u, u) \leq (\lambda_0^{\alpha} Bu, u) = (-\sum_{i=1}^{n} P_i e^{-\lambda_0 \tau_i} u, u)$$
$$= (-\sum_{i=1}^{n} P_i u, u) e^{-\lambda_0 \tau_i} \leq \sum_{i=1}^{n} \mu(-P_i) e^{-\lambda_0 \tau_i}.$$

Then by (2.91), $\lambda_0 < 0$ such that

$$\left[\frac{1}{\lambda_0^{\bar{\alpha}}} \sum_{i=1}^{n} \mu(-P_i) e^{-\lambda_0 \tau_i}\right] \leq 1 \quad \text{or} \quad -\lambda_0 \geq \left[\sum_{i=1}^{n} -\mu(-P_i)\right]^{\frac{1}{\bar{\alpha}}}. \tag{2.93}$$

This contradicts (2.92) and completes the proof. □

Theorem 2.17. *Assume that for each $i = 1, 2, ..., n$,*

$$P_i \in \mathbb{R}^{m \times m}, \quad \tau_i \geq 0 \quad \text{and} \quad \mu(-P_i) \leq 0.$$

Then each of the following two conditions is sufficient for the oscillation of all solutions of (2.80):

(i) $\displaystyle\sum_{i=1}^{n} -\mu(-P_i)\tau_i \left[\sum_{i=1}^{n} -\mu(-P_i)\right]^{\frac{1-\bar{\alpha}}{\bar{\alpha}}} > \frac{1}{e}$;

(ii) $\displaystyle\left[\prod_{i=1}^{n} (-\mu(-P_i))\right]^{\frac{1}{n}} \sum_{i=1}^{n} \tau_i \left[\sum_{i=1}^{n} -\mu(-P_i)\right]^{\frac{1-\bar{\alpha}}{\bar{\alpha}}} > \frac{1}{e}$.

Proof. We employ Lemma 2.13. As $\mu(-P_i) \leq 0$, (2.91) is satisfied and so it suffices to establish (2.92). First, assume that (i) holds. Then, by using the inequality $e^x \geq ex$, we see that for all $\lambda < 0$,

$$\frac{1}{\lambda^{\bar{\alpha}}} \sum_{i=1}^{n} \mu(-P_i) e^{-\lambda \tau_i} \geq \frac{1}{\lambda^{\bar{\alpha}}} \sum_{i=1}^{n} \mu(-P_i) e(-\lambda \tau_i)$$
$$= -e \frac{1}{\lambda^{\bar{\alpha}}} \sum_{i=1}^{n} \mu(-P_i) \tau_i \lambda^{\bar{\alpha}} (-\lambda)^{1-\bar{\alpha}}$$
$$\geq e \sum_{i=1}^{n} -\mu(-P_i)\tau_i \left[\sum_{i=1}^{n} -\mu(-P_i)\right]^{\frac{1-\bar{\alpha}}{\bar{\alpha}}},$$

which, together with (i), implies that (2.92) holds. Next, assume that (ii) holds. Then, by using the arithmetic mean–geometric mean inequality we find that for all $\lambda < 0$,

$$\frac{1}{\lambda^{\bar{\alpha}}} \sum_{i=1}^{n} \mu(-P_i) e^{-\lambda \tau_i} = -\frac{1}{\lambda^{\bar{\alpha}}} \sum_{i=1}^{n} -\mu(-P_i) e^{-\lambda \tau_i}$$

$$\geq -\frac{1}{\lambda^{\bar{\alpha}}}n\left[\prod_{i=1}^{n} -\mu(-P_i)e^{-\lambda\tau_i}\right]^{\frac{1}{n}}$$

$$= -\frac{1}{\lambda^{\bar{\alpha}}}n\left[\prod_{i=1}^{n} -\mu(-P_i)\right]^{\frac{1}{n}} \exp\left(-\frac{1}{n}\lambda\sum_{i=1}^{n}\tau_i\right)$$

$$\geq -\frac{1}{\lambda^{\bar{\alpha}}}\left[\prod_{i=1}^{n} -\mu(-P_i)\right]^{\frac{1}{n}} e\left(-\lambda\sum_{i=1}^{n}\tau_i\right)$$

$$= \left[\prod_{i=1}^{n} -\mu(-P_i)\right]^{\frac{1}{n}} e(-\lambda)^{1-\bar{\alpha}}\sum_{i=1}^{n}\tau_i$$

$$\geq e\left[\prod_{i=1}^{n}(-\mu(-P_i))\right]^{\frac{1}{n}}\sum_{i=1}^{n}\tau_i\left[\sum_{i=1}^{n} -\mu(-P_i)\right]^{\frac{1-\bar{\alpha}}{\bar{\alpha}}}.$$

From this and (ii), it follows that (2.92) holds. The proof is completed. □

As a special case of the delay differential system with one delay,

$$_0D_t^{\alpha}x(t) + Px(t-\tau) = 0, \tag{2.94}$$

where

$$P \in \mathbb{R}^{m\times m} \quad \text{and} \quad \tau \geq 0,$$

the conditions (i) and (ii) coincide and each reduces to

$$[-\mu(-P)]^{\frac{1}{\bar{\alpha}}}\tau > \frac{1}{e}. \tag{2.95}$$

Note that (2.95) is sharp in the sense that the lower bound $1/e$ cannot be improved. Moreover, when P is a scalar, (2.95) is a sufficient condition for the oscillation of all solutions to equation (2.94).

For the delay differential system (2.94), we also have the following explicit sufficient condition for the oscillation of all solutions.

Theorem 2.18. *Assume that*

$$P \in \mathbb{R}^{m\times m} \quad \textit{and} \quad \tau \geq 0.$$

If P has no real eigenvalues in the interval $(-\infty, 1/(e\tau)^{\bar{\alpha}}]$ (when $\tau = 0$, replace $1/(e\tau)^{\bar{\alpha}}$ by $+\infty$), then every solution of (2.94) oscillates.

Proof. For $\tau = 0$, this result follows immediately from Theorem 2.16. So assume $\tau > 0$. Note that the characteristic equation $\det(\lambda^{\alpha}B + Pe^{-\lambda\tau}) = 0$ has a real root λ_0, that is, $\det(\lambda_0^{\alpha}e^{\lambda_0\tau}B + P) = 0$ if and only if $\mu_0^{\alpha} = -\lambda_0^{\alpha}e^{\lambda_0\tau}$ is a real eigenvalue of P. For convenience, we take one element $\lambda_0^{\alpha_i}$ of λ_0^{α}:

$$\mu_0^{\alpha_i} = -\lambda_0^{\alpha_i}e^{\lambda_0\tau}. \tag{2.96}$$

Observe that (2.96) holds if $\lambda_0^{\alpha_i} + \mu_0^{\alpha_i}e^{-\lambda_0\tau} = 0$, that is, the equation $\lambda^{\alpha_i} + \mu_0^{\alpha_i}e^{-\lambda\tau} = 0$ has a real root. If $\mu_0 \leq 1/e\tau$, then the eigenvalue μ_0^{α} of P should lie in the interval $(-\infty, 1/(e\tau)^{\bar{\alpha}}]$. The proof is completed. □

Definition 2.15. We say that (2.80) is oscillatory, globally in the delays, if for all $\tau_i \geq 0$ for $i = 1, 2, ..., n$, every solution of (2.80) oscillates.

The following corollary is an immediate consequence of Theorem 2.18.

Corollary 2.7. *Equation* (2.94) *is oscillatory globally in the delay* τ *if* P *has no real eigenvalues.*

Next, we consider the linear autonomous system of neutral delay differential equations

$$ {}_0D_t^\alpha\left[x(t) + \sum_{j=1}^{l} P_j x(t-\tau_j)\right] + \sum_{i=1}^{n} Q_i x(t-\delta_i) = 0, \tag{2.97} $$

where the coefficients P_j and Q_i are real $m \times m$ matrices and the delays τ_j and δ_i are non-negative real numbers. Associated with (2.97), the characteristic equation is

$$ det\left(\lambda^\alpha B + \lambda^\alpha \sum_{j=1}^{l} P_j e^{-\lambda\tau_j} + \sum_{i=1}^{n} Q_i e^{-\lambda\delta_i}\right) = 0. \tag{2.98} $$

Lemma 2.14. *If* $lp < 1$, *then the solution of equation* (2.97) *has an exponent estimate*

$$ \|x(t)\| \leq A_0 e^{b_0 t} \qquad (t > b > 0) $$

for constants $A_0 > 0$ *and* $b_0 > 0$.

Proof. Let $\delta = \max_{1\leq i\leq n}\{\delta_i\}$, $\bar{\delta} = \max_{1\leq j\leq l}\{\tau_j,\ \delta\}$, $q = \max_{1\leq i\leq n}\{\|Q_i\|\}$, and take $x_0 = {}_0D_t^{\alpha-1}x(0) + \sum_{j=1}^{l} P_{j0}D_t^{\alpha-1}x(-\tau_j)$ with $x(t) \in AC[0,b]$. Then there exists a constant M such that $\|x(t)\| \leq M$. Then, for $t > b$,

$$ \begin{aligned} \|x(t)\| &\leq c\|x_0\| + lp\|x(t-\tau_j)\| + \frac{nq}{\Gamma(\alpha_0)}\int_0^t (t-s)^{\alpha_0-1}\|x(s-\delta_i)\|ds \\ &\leq c\|x_0\| + lp\max_{t-\tau_i\leq s\leq t}\|x(s)\| + \frac{nq}{\Gamma(\alpha_0)}\int_0^t (t-s)^{\alpha_0-1}\max_{s-\delta_i\leq\eta\leq s}\|x(\eta)\|ds \\ &\leq c\|x_0\| + lp(M + \|\varphi\|_{AC} + \max_{b\leq s\leq t}\|x(s)\|) \\ &\quad + \frac{nq}{\Gamma(\alpha_0)}\int_0^b (t-s)^{\alpha_0-1}\max_{s-\bar{\delta}\leq\eta\leq b}\|x(\eta)\|ds \\ &\quad + \frac{nq}{\Gamma(\alpha_0)}\int_b^t (t-s)^{\alpha_0-1}\max_{b\leq\eta\leq s}\|x(\eta)\|ds \\ &\leq c\|x_0\| + (M + \|\varphi\|_{AC})\left(lp + \frac{nq}{\Gamma(\alpha_0)}\frac{t^{\alpha_0}}{\alpha_0}\right) + lp\max_{b\leq s\leq t}\|x(s)\| \\ &\quad + \frac{nq}{\Gamma(\alpha_0)}\int_b^t (t-s)^{\alpha_0-1}\max_{b\leq\eta\leq s}\|x(\eta)\|ds. \end{aligned} $$

Set

$$a = c\|x_0\| + (M + \|\varphi\|_{AC})\left(lp + \frac{nq}{\Gamma(\alpha_0)}\frac{t^{\alpha_0}}{\alpha_0}\right),$$

which yields

$$\max_{b\le s\le t}\|x(s)\| \le \frac{1}{1-lp}\left(a + \frac{nq}{\Gamma(\alpha_0)}\int_b^t (t-s)^{\alpha_0-1}\max_{b\le\eta\le s}\|x(\eta)\|ds\right).$$

Consequently, by Lemma 2.12, we obtain

$$\begin{aligned}
\|x(t)\| &\le \max_{b\le s\le t}\|x(s)\| \\
&\le \frac{a(t)}{1-lp} + \frac{1}{(1-lp)^2}\frac{knq}{\Gamma(\alpha_0)}\int_0^t (t-s)^{\alpha_0-1}a(s)ds \\
&\le \frac{a(t)}{1-lp}\left(1 + \frac{1}{1-lp}\frac{knq}{\alpha_0\Gamma(\alpha_0)}t^{\alpha_0}\right).
\end{aligned}$$

The proof is completed. □

A slight modification in the proof of Theorem 2.16 shows that the following result is also true.

Theorem 2.19. *Assume that for $j = 1, 2, ..., l$ and $i = 1, 2, ..., n$,*

$$P_j, Q_i \in \mathbb{R}^{m\times m}, \quad \tau_j \in (0,\infty) \quad \textit{and} \quad \delta_i \in [0,\infty).$$

If the characteristic equation (2.98) *has no real roots, then every solution of* (2.97) *oscillates.*

Proof. If we modify the functions $F(s)$ and $\Phi(s)$, defined in Theorem 2.16, as

$$F(s) = s^\alpha B + s^\alpha B\sum_{j=1}^{l} P_j e^{-s\tau_j} + \sum_{i=1}^{n} Q_i e^{-s\delta_i},$$

$$\begin{aligned}
\Phi(s) = x_0 &- \sum_{i=1}^{n} Q_i e^{-s\delta_i}\int_{-\delta_i}^{0} e^{-st}x(t)dt \\
&- s^\alpha B\sum_{j=1}^{l} P_j e^{-s\tau_j}\int_{\tau_j}^{0} e^{-st}x(t)dt + \sum_{j=1}^{l} P_j({}_0D_t^{-(1-\alpha)}x)(-\tau_j).
\end{aligned}$$

Then, following the method of proof for Theorem 2.16, one can complete the proof. □

2.6.4 *Existence of Nonoscillatory Solutions*

In this subsection, we discuss the nonoscillatory characteristics of solutions for the following fractional neutral functional differential equation

$$ {}_tD^{\alpha}_{+\infty}[x(t)+cx(t-\tau)]' + \sum_{i=1}^{m} P_i(t)F_i(x(t-\sigma_i)) = 0, \quad t \geq t_0, \tag{2.99}$$

where ${}_tD^{\alpha}_{+\infty}$ is Liouville-Weyl fractional derivatives of order $\alpha \geq 0$ on the half-axis, $c \in \mathbb{R}$, τ, $\sigma_i \in \mathbb{R}^+$, $P_i \in C([t_0,\infty),\mathbb{R})$, $F_i \in C(\mathbb{R},\mathbb{R})$, $i = 1,2,...,m$, $m \geq 1$ is an integer.

Let $r = \max_{1\leq i\leq m}\{\tau,\sigma_i\}$. By a solution of equation (2.99), we mean a function $x \in C([t_1-r,\infty),\mathbb{R})$ for some $t_1 \geq t_0$ such that ${}_tD^{\alpha}_{+\infty}[x(t)+cx(t-\tau)]$ exists on $[t_1,\infty)$ and that equation (2.99) is satisfied for $t \geq t_1$.

A nontrivial solution x of equation (2.99) is said to be oscillatory if it has an arbitrarily large number of zeros. Otherwise, x is said to be nonoscillatory, that is, x is nonoscillatory if there exists a $T > t_1$ such that $x(t) \neq 0$ for $t \geq T$. In other words, a nonoscillatory solution must be eventually positive or eventually negative.

We will consider the two cases: $c \neq \pm 1$ and $c = -1$. Our main results are the following theorems.

Theorem 2.20. *Assume that* $c \neq \pm 1$ *and that*

$$\int_{t_0}^{\infty} t^{\alpha}|P_i(t)|dt < \infty, \ i = 1,2,...,m. \tag{2.100}$$

Then (2.99) *has a bounded nonoscillatory solution.*

Proof. Case I. $-1 < c \leq 0$. By (2.100), we choose a $T > t_0$ sufficiently large so that

$$\frac{1}{\Gamma(\alpha+1)}\int_T^{\infty} s^{\alpha}\Big(\sum_{i=1}^{m}|P_i(s)|M_1\Big)ds \leq \frac{1+c}{3},$$

where

$$M_1 = \max_{2(1+c)/3\leq x\leq 4/3}\{|F_i(x)| : 1 \leq i \leq m\}.$$

Let $C([t_0,\infty),\mathbb{R})$ be the set of all continuous functions with the norm $||x|| = \sup_{t\geq t_0}|x(t)| < \infty$. Then $C([t_0,\infty),\mathbb{R})$ is a Banach space. We define a closed, bounded and convex subset Ω of $C([t_0,\infty),\mathbb{R})$ by

$$\Omega = \{x = x(t) \in C([t_0,\infty),\mathbb{R}) : \frac{2(1+c)}{3} \leq x(t) \leq \frac{4}{3}, t \geq t_0\}.$$

Define two maps $\mathcal{A}_1$ and $\mathcal{A}_2 : \Omega \to C([t_0,\infty),\mathbb{R})$ as follows:

$$(\mathcal{A}_1x)(t) = \begin{cases} 1+c-cx(t-\tau), & t \geq T, \\ (\mathcal{A}_1x)(T), & t_0 \leq t \leq T. \end{cases}$$

$$(\mathcal{A}_2x)(t) = \begin{cases} \dfrac{1}{\Gamma(\alpha+1)}\displaystyle\int_t^\infty (s-t)^\alpha\Big(\sum_{i=1}^m P_i(s)F_i(x(s-\sigma_i))\Big)ds, & t \geq T, \\ (\mathcal{A}_2x)(T), & t_0 \leq t \leq T. \end{cases}$$

(i) We shall show that $\mathcal{A}_1x + \mathcal{A}_2y \in \Omega$ for any $x, y \in \Omega$.

Indeed, for every $x, y \in \Omega$ and $t \geq T$, we get

$$\begin{aligned}
&(\mathcal{A}_1x)(t) + (\mathcal{A}_2y)(t) \\
&\leq 1 + c - cx(t-\tau) + \frac{1}{\Gamma(\alpha+1)}\int_t^\infty (s-t)^\alpha\Big(\sum_{i=1}^m |P_i(s)||F_i(y(s-\sigma_i))|\Big)ds \\
&\leq 1 + c - \frac{4}{3}c + \frac{1}{\Gamma(\alpha+1)}\int_T^\infty s^\alpha\Big(\sum_{i=1}^m |P_i(s)|M_1\Big)ds \\
&\leq 1 + c - \frac{4}{3}c + \frac{1+c}{3} = \frac{4}{3}.
\end{aligned}$$

Furthermore, we have

$$\begin{aligned}
&(\mathcal{A}_1x)(t) + (\mathcal{A}_2y)(t) \\
&\geq 1 + c - cx(t-\tau) - \frac{1}{\Gamma(\alpha+1)}\int_t^\infty (s-t)^\alpha\Big(\sum_{i=1}^m |P_i(s)||F_i(y(s-\sigma_i))|\Big)ds \\
&\geq 1 + c - \frac{1}{\Gamma(\alpha+1)}\int_T^\infty s^\alpha\Big(\sum_{i=1}^m |P_i(s)|M_1\Big)ds \\
&\geq 1 + c - \frac{1+c}{3} = \frac{2(1+c)}{3}.
\end{aligned}$$

From the above two inequalities, it follows that

$$\frac{2(1+c)}{3} \leq (\mathcal{A}_1x)(t) + (\mathcal{A}_2y)(t) \leq \frac{4}{3}, \quad \text{for} \quad t \geq t_0.$$

Thus $\mathcal{A}_1x + \mathcal{A}_2y \in \Omega$ for any $x, y \in \Omega$.

(ii) We show that $\mathcal{A}_1$ is a contraction mapping on Ω.

In fact, for $x, y \in \Omega$ and $t \geq T$, we have

$$|(\mathcal{A}_1x)(t) - (\mathcal{A}_1y)(t)| \leq -c|x(t-\tau) - y(t-\tau)| \leq -c||x - y||,$$

which implies that

$$||\mathcal{A}_1x - \mathcal{A}_1y|| \leq -c||x - y||.$$

Since $0 < -c < 1$, we conclude that $\mathcal{A}_1$ is a contraction mapping on Ω.

(iii) Here we show that $\mathcal{A}_2$ is completely continuous.

First, we will show that $\mathcal{A}_2$ is continuous. Let $x_k = x_k(t) \in \Omega$ be such that $x_k(t) \to x(t)$ as $k \to \infty$. Because Ω is closed, $x = x(t) \in \Omega$. For $t \geq T$, we have

$$\begin{aligned}
&|(\mathcal{A}_2x_k)(t) - (\mathcal{A}_2x)(t)| \\
&\leq \frac{1}{\Gamma(\alpha+1)}\int_t^\infty s^\alpha\Big(\sum_{i=1}^m |P_i(s)||F_i(x_k(s-\sigma_i)) - F_i(x(s-\sigma_i))|\Big)ds
\end{aligned}$$

$$\leq \frac{1}{\Gamma(\alpha+1)}\int_T^\infty s^\alpha\Big(\sum_{i=1}^m |P_i(s)||F_i(x_k(s-\sigma_i))-F_i(x(s-\sigma_i))|\Big)ds.$$

Since $|F_i(x_k(t-\sigma_i))-F_i(x(t-\sigma_i))|\to 0$ as $k\to\infty$ for $i=1,2,...,m$, by applying the Lebesgue dominated convergence theorem, we deduce that $\lim_{k\to\infty}||(\mathcal{A}_2x_k)(t)-(\mathcal{A}_2x)(t)||=0$. This means that $\mathcal{A}_2$ is continuous.

Next, we show $\mathcal{A}_2\Omega$ is relatively compact. It suffices to show that the family of functions $\{\mathcal{A}_2x : x\in\Omega\}$ is uniformly bounded and equicontinuous on $[t_0,\infty)$. The uniform boundedness is obvious. For the equicontinuity, according to Levitan's result, we only need to show that, for any given $\varepsilon>0$, $[T,\infty)$ can be decomposed into finite subintervals in such a way that on each subinterval all functions of the family have change of amplitude less than ε. By (2.100), for any $\varepsilon>0$, take $T^*\geq T$ large enough so that

$$\frac{1}{\Gamma(\alpha+1)}\int_{T^*}^\infty s^\alpha\Big(M_1\sum_{i=1}^m|P_i(s)|\Big)ds<\frac{\varepsilon}{2}.$$

Then, for $x\in\Omega,\ t_2>t_1\geq T^*$, we have

$$\begin{aligned}
&|(\mathcal{A}_2x)(t_2)-(\mathcal{A}_2x)(t_1)|\\
&\leq \frac{1}{\Gamma(\alpha+1)}\int_{t_2}^\infty s^\alpha\Big(\sum_{i=1}^m|P_i(s)||F_i(x(s-\sigma_i))|\Big)ds\\
&\quad+\frac{1}{\Gamma(\alpha+1)}\int_{t_1}^\infty s^\alpha\Big(\sum_{i=1}^m|P_i(s)||F_i(x(s-\sigma_i))|\Big)ds\\
&\leq \frac{1}{\Gamma(\alpha+1)}\int_{t_2}^\infty s^\alpha\Big(M_1\sum_{i=1}^m|P_i(s)|\Big)ds\\
&\quad+\frac{1}{\Gamma(\alpha+1)}\int_{t_1}^\infty s^\alpha\Big(M_1\sum_{i=1}^m|P_i(s)|\Big)ds\\
&<\frac{\varepsilon}{2}+\frac{\varepsilon}{2}=\varepsilon.
\end{aligned}$$

For $x\in\Omega$ and $T\leq t_1<t_2\leq T^*$, we obtain

$$\begin{aligned}
&|(\mathcal{A}_2x)(t_2)-(\mathcal{A}_2x)(t_1)|\\
&\leq \frac{1}{\Gamma(\alpha+1)}\int_{t_1}^{t_2} s^\alpha\Big(\sum_{i=1}^m|P_i(s)||F_i(x(s-\sigma_i))|\Big)ds\\
&\leq \frac{1}{\Gamma(\alpha+1)}\int_{t_1}^{t_2} s^\alpha\Big(M_1\sum_{i=1}^m|P_i(s)|\Big)ds\\
&\leq \frac{1}{\Gamma(\alpha+1)}\max_{T\leq s\leq T^*}\Big\{s^\alpha\Big(M_1\sum_{i=1}^m|P_i(s)|\Big)\Big\}(t_2-t_1).
\end{aligned}$$

Thus there exists a $\delta>0$ such that

$$|(\mathcal{A}_2x)(t_2)-(\mathcal{A}_2x)(t_1)|<\varepsilon,\ \text{ if }\ 0<t_2-t_1<\delta.$$

For any $x \in \Omega$, $t_0 \le t_1 < t_2 \le T$, it is easy to see that

$$|(\mathcal{A}_2x)(t_2) - (\mathcal{A}_2x)(t_1)| = 0 < \varepsilon.$$

Therefore $\{\mathcal{A}_2x : x \in \Omega\}$ is uniformly bounded and equicontinuous on $[t_0, \infty)$, and hence $\mathcal{A}_2\Omega$ is relatively compact. Hence, the conclusion of Theorem 1.8 (Krasnoselskii fixed point theorem) applies and there exits $x_0 \in \Omega$ such that $\mathcal{A}_1x_0+\mathcal{A}_2x_0 = x_0$, that is,

$$x_0(t) = 1 + c - cx_0(t-\tau) + \frac{1}{\Gamma(\alpha+1)}\int_t^{\infty}(s-t)^{\alpha}\Big(\sum_{i=1}^{m}P_i(s)F_i(x_0(s-\sigma_i))\Big)ds,$$

which implies that

$$x_0(t) = 1+c-cx_0(t-\tau)+\frac{1}{\Gamma(\alpha)}\int_t^{\infty}ds\int_t^{s}(s-u)^{\alpha-1}\Big(\sum_{i=1}^{m}P_i(s)F_i(x_0(s-\sigma_i))\Big)du.$$

Hence

$$[x_0(t) + cx_0(t-\tau)]' = \frac{1}{\Gamma(\alpha)}\int_t^{\infty}(s-t)^{\alpha-1}\Big(\sum_{i=1}^{m}P_i(s)F_i(x_0(s-\sigma_i))\Big)ds.$$

It is easy to see that $x_0(t)$ is a nonoscillatory solution of equation (2.99).

Case II. $-\infty < c < -1$. By (2.100), we choose a $T > t_0$ sufficiently large such that

$$-\frac{1}{c\Gamma(\alpha+1)}\int_{T+\tau}^{\infty}s^{\alpha}\Big(\sum_{i=1}^{m}|P_i(s)|M_2\Big)ds \le -\frac{c+1}{2},$$

where

$$M_2 = \max_{-(c+1)/2\le x\le -2c}\{|F_i(x)| : 1 \le i \le m\}.$$

Let $C([t_0,\infty),\mathbb{R})$ be the set as in the proof of Theorem 2.20. We define a closed, bounded and convex subset Ω of $C([t_0,\infty),\mathbb{R})$ as follows:

$$\Omega = \{x = x(t) \in C([t_0,\infty),\mathbb{R}) : -\frac{c+1}{2} \le x(t) \le -2c, t \ge t_0\}.$$

Define two maps $\mathcal{A}_1$ and $\mathcal{A}_2 : \Omega \to C([t_0,\infty),\mathbb{R})$ by

$$(\mathcal{A}_1x)(t) = \begin{cases} -c-1-\dfrac{1}{c}x(t+\tau), t \ge T, \\ (\mathcal{A}_1x)(T), \qquad t_0 \le t \le T. \end{cases}$$

$$(\mathcal{A}_2x)(t) = \begin{cases} \dfrac{1}{c\Gamma(\alpha+1)}\displaystyle\int_{t+\tau}^{\infty}(s-t-\tau)^{\alpha}\Big(\sum_{i=1}^{m}P_i(s)F_i(x(s-\sigma_i))\Big)ds, t \ge T, \\ (\mathcal{A}_2x)(T), \qquad t_0 \le t \le T. \end{cases}$$

In the first step, let us show that $\mathcal{A}_1x + \mathcal{A}_2y \in \Omega$ for any $x, y \in \Omega$.

In fact, for every $x, y \in \Omega$ and $t \ge T$, we get

$$(\mathcal{A}_1x)(t) + (\mathcal{A}_2y)(t)$$

$$
\begin{aligned}
&\le -c - 1 - \frac{1}{c}x(t+\tau) \\
&\quad - \frac{1}{c}\frac{1}{\Gamma(\alpha+1)}\int_{t+\tau}^{\infty}(s-t-\tau)^{\alpha}\Big(\sum_{i=1}^{m}|P_i(s)||F_i(y(s-\sigma_i))|\Big)ds \\
&\le -c - 1 + 2 - \frac{1}{c}\frac{1}{\Gamma(\alpha+1)}\int_{T+\tau}^{\infty}s^{\alpha}\Big(\sum_{i=1}^{m}|P_i(s)|M_2\Big)ds \\
&\le -c + 1 - \frac{c+1}{2} \le -2c
\end{aligned}
$$

and

$$
\begin{aligned}
&(\mathcal{A}_1x)(t) + (\mathcal{A}_2y)(t) \\
&\ge -c - 1 - \frac{1}{c}x(t+\tau) \\
&\quad + \frac{1}{c}\frac{1}{\Gamma(\alpha+1)}\int_{t+\tau}^{\infty}(s-t)^{\alpha}\Big(\sum_{i=1}^{m}|P_i(s)||F_i(y(s-\sigma_i))|\Big)ds \\
&\ge -c - 1 + \frac{1}{c}\frac{1}{\Gamma(\alpha+1)}\int_{T}^{\infty}s^{\alpha}\Big(\sum_{i=1}^{m}|P_i(s)|M_2\Big)ds \\
&\ge -c - 1 + \frac{c+1}{2} = -\frac{c+1}{2},
\end{aligned}
$$

which imply that

$$
-\frac{c+1}{2} \le (\mathcal{A}_1x)(t) + (\mathcal{A}_2y)(t) \le -2c, \quad \text{for} \quad t \ge t_0.
$$

Thus $\mathcal{A}_1x + \mathcal{A}_2y \in \Omega$ for any $x, y \in \Omega$.
Next we show that $\mathcal{A}_1$ is a contraction mapping on Ω.

For $x, y \in \Omega$ and $t \ge T$, we have

$$
|(\mathcal{A}_1x)(t) - (\mathcal{A}_1y)(t)| \le -\frac{1}{c}|x(t+\tau) - y(t+\tau)| \le -\frac{1}{c}||x-y||,
$$

which implies that

$$
||\mathcal{A}_1x - \mathcal{A}_1y|| \le -\frac{1}{c}||x-y||.
$$

In view of the condition $0 < -1/c < 1$, it follows that $\mathcal{A}_1$ is a contraction mapping on Ω.

As in the proof of Case I, we can obtain that the mapping $\mathcal{A}_2$ is completely continuous. Therefore, all the conditions of Theorem 1.8 are satisfied. Hence there exists $x_0 \in \Omega$ such that $\mathcal{A}_1x_0 + \mathcal{A}_2x_0 = x_0$. Clearly, $x_0 = x_0(t)$ is a bounded positive solution of equation (2.99).

Case III. $0 \le c < 1$. By (2.100), we choose a $T > t_0$ sufficiently large so that

$$
\frac{1}{\Gamma(\alpha+1)}\int_{T}^{\infty}s^{\alpha}\Big(\sum_{i=1}^{m}|P_i(s)|M_3\Big)ds \le 1 - c,
$$

where

$$M_3 = \max_{2(1-c)\le x\le 4}\{F_i(x) : 1 \le i \le m\}.$$

Let $C([t_0,\infty),\mathbb{R})$ be the set defined in the proof of Theorem 2.20. We define a closed, bounded and convex subset Ω of $C([t_0,\infty),\mathbb{R})$ as follows:

$$\Omega = \{x = x(t) \in C([t_0,\infty),\mathbb{R}) : 2(1-c) \le x(t) \le 4, t \ge t_0\},$$

and consider two maps $\mathcal{A}_1$ and $\mathcal{A}_2 : \Omega \to C([t_0,\infty),\mathbb{R})$ defined by

$$(\mathcal{A}_1 x)(t) = \begin{cases} 3 + c - cx(t-\tau), \ t \ge T, \\ (\mathcal{A}_1 x)(T), \qquad t_0 \le t \le T, \end{cases}$$

and

$$(\mathcal{A}_2 x)(t) = \begin{cases} \dfrac{1}{\Gamma(\alpha+1)} \displaystyle\int_t^\infty (s-t)^\alpha \Big(\sum_{i=1}^m P_i(s) F_i(x(s-\sigma_i))\Big) ds, \ t \ge T, \\ (\mathcal{A}_2 x)(T), \qquad t_0 \le t \le T. \end{cases}$$

As before, for any $x, y \in \Omega$ and $t \ge T$, we have

$$\begin{aligned} &(\mathcal{A}_1 x)(t) + (\mathcal{A}_2 y)(t) \\ &\le 3 + c - cx(t-\tau) + \frac{1}{\Gamma(\alpha+1)} \int_t^\infty (s-t)^\alpha \Big(\sum_{i=1}^m |P_i(s)||F_i(y(s-\sigma_i))|\Big) ds \\ &\le 3 + c + \frac{1}{\Gamma(\alpha+1)} \int_T^\infty s^\alpha \Big(\sum_{i=1}^m |P_i(s)| M_3\Big) ds \\ &\le 3 + c + 1 - c = 4, \end{aligned}$$

and

$$\begin{aligned} &(\mathcal{A}_1 x)(t) + (\mathcal{A}_2 y)(t) \\ &\ge 3 + c - cx(t-\tau) - \frac{1}{\Gamma(\alpha+1)} \int_t^\infty (s-t)^\alpha \Big(\sum_{i=1}^m |P_i(s)||F_i(y(s-\sigma_i))|\Big) ds \\ &\ge 3 + c - 4c - \frac{1}{\Gamma(\alpha+1)} \int_T^\infty s^\alpha \Big(\sum_{i=1}^m |P_i(s)| M_3\Big) ds \\ &\ge 3 + c - 4c - (1-c) = 2(1-c). \end{aligned}$$

In consequence, we get

$$2(1-c) \le (\mathcal{A}_1 x)(t) + (\mathcal{A}_2 y)(t) \le 4, \quad \text{for} \quad t \ge t_0.$$

This shows that $\mathcal{A}_1 x + \mathcal{A}_2 y \in \Omega$ for any $x, y \in \Omega$.

Proceeding as in the proof of Case I, we can establish that the mapping $\mathcal{A}_1$ is a contraction mapping on Ω and the mapping $\mathcal{A}_2$ is completely continuous. By Theorem 1.8, there is $x_0 \in \Omega$ such that $\mathcal{A}_1 x_0 + \mathcal{A}_2 x_0 = x_0$. Clearly, $x_0 = x_0(t)$ is a bounded positive solution of (2.99).

Case IV. $1 < c < \infty$. Again, by (2.100), we can choose a $T > t_0$ sufficiently large so that

$$\frac{1}{c\Gamma(\alpha+1)} \int_{T+\tau}^{\infty} s^{\alpha}\Big(\sum_{i=1}^{m} |P_i(s)| M_4\Big) ds \le c - 1,$$

where

$$M_4 = \max_{2(c-1) \le x \le 4c} \{F_i(x) : i = 1, 2, ..., m\}.$$

Let $C([t_0, \infty), \mathbb{R})$ be the set as considered in the proof of Theorem 2.20. Let Ω be a closed, bounded and convex subset of $C([t_0, \infty), \mathbb{R})$ defined by

$$\Omega = \{x = x(t) \in C([t_0, \infty), \mathbb{R}) : 2(c-1) \le x(t) \le 4c, t \ge t_0\}.$$

Define two maps $\mathcal{A}_1$ and $\mathcal{A}_2 : \Omega \to C([t_0, \infty), \mathbb{R})$ as follows:

$$(\mathcal{A}_1 x)(t) = \begin{cases} 3c + 1 - \dfrac{1}{c} x(t+\tau), \; t \ge T, \\ (\mathcal{A}_1 x)(T), \qquad\qquad t_0 \le t \le T, \end{cases}$$

and

$$(\mathcal{A}_2 x)(t) = \begin{cases} \dfrac{1}{c\Gamma(\alpha+1)} \displaystyle\int_{t+\tau}^{\infty} (s-t-\tau)^{\alpha}\Big(\sum_{i=1}^{m} P_i(s) F_i(x(s-\sigma_i))\Big) ds, \; t \ge T, \\ (\mathcal{A}_2 x)(T), \qquad\qquad t_0 \le t \le T. \end{cases}$$

In order to show that $\mathcal{A}_1 x + \mathcal{A}_2 y \in \Omega$ for any $x, y \in \Omega$ and $t \ge T$, we consider

$$\begin{aligned}
&(\mathcal{A}_1 x)(t) + (\mathcal{A}_2 y)(t) \\
&\le 3c + 1 - \frac{1}{c} x(t+\tau) \\
&\quad + \frac{1}{c} \frac{1}{\Gamma(\alpha+1)} \int_{t+\tau}^{\infty} (s-t-\tau)^{\alpha}\Big(\sum_{i=1}^{m} |P_i(s)||F_i(y(s-\sigma_i))|\Big) ds \\
&\le 3c + 1 + \frac{1}{c} \frac{1}{\Gamma(\alpha+1)} \int_{T+\tau}^{\infty} s^{\alpha}\Big(\sum_{i=1}^{m} (|P_i(s)| M_4\Big) ds \\
&\le 3c + 1 + (c-1) = 4c,
\end{aligned}$$

and

$$\begin{aligned}
&(\mathcal{A}_1 x)(t) + (\mathcal{A}_2 y)(t) \\
&\ge 3c + 1 - \frac{1}{c} x(t+\tau) \\
&\quad - \frac{1}{c} \frac{1}{\Gamma(\alpha+1)} \int_{t+\tau}^{\infty} (s-t)^{\alpha}\Big(\sum_{i=1}^{m} |P_i(s)||F_i(y(s-\sigma_i))|\Big) ds \\
&\ge 3c + 1 - 4 - \frac{1}{c} \frac{1}{\Gamma(\alpha+1)} \int_{T}^{\infty} s^{\alpha}\Big(\sum_{i=1}^{m} |P_i(s)| M_4\Big) ds \\
&\ge 3c - 3 - (c-1) = 2(c-1).
\end{aligned}$$

Hence we obtain

$$2(c-1) \leq \mathcal{A}_1 x(t) + \mathcal{A}_2 y(t) \leq 4c, \quad \text{for} \quad t \geq t_0,$$

which implies that $\mathcal{A}_1 x + \mathcal{A}_2 y \in \Omega$ for any $x, y \in \Omega$.

As in the proof of Case I, it can be shown that the mapping $\mathcal{A}_1$ is a contraction mapping on Ω and the mapping $\mathcal{A}_2$ is completely continuous. In consequence, the conclusion of Theorem 1.8 applies and there exists $x_0 \in \Omega$ such that $\mathcal{A}_1 x_0 + \mathcal{A}_2 x_0 = x_0$. It is easy to see that $x_0 = x_0(t)$ is a bounded positive solution of equation (2.99). The proof is completed. □

Remark 2.5. Minor adjustments are only necessary to discuss the neutral functional differential equation of the form

$${}_tD^{\alpha}_{+\infty}[x(t) + C(t)x(t-\tau)]' + F(t, x(\sigma_1(t)), ..., x(\sigma_m(t))) = f(t), \quad t \geq t_0,$$

where $\tau \in \mathbb{R}^+ = [0, \infty)$, $\sigma_i(t) \to \infty$ $(i = 1, 2, ..., m)$ as $t \to \infty$, $m \geq 1$ is an integer, and $F : [t_0, \infty) \times \mathbb{R} \times \cdots \times \mathbb{R} \to \mathbb{R}$ is continuous and bounded, $C, f \in C([t_0, \infty), \mathbb{R})$. So we omit the details.

Theorem 2.21. *Assume that* $c = -1$ *and that*

$$\sum_{j=0}^{\infty} \int_{t_0+j\tau}^{\infty} t^{\alpha} |P_i(t)| dt < \infty, \; i = 1, 2, ..., m. \tag{2.101}$$

Then equation (2.99) *has a bounded positive solution.*

Proof. By the condition (2.101), we can choose a sufficiently large $T > t_0$ so that

$$\frac{1}{\Gamma(\alpha+1)} \sum_{j=1}^{\infty} \int_{T+j\tau}^{\infty} s^{\alpha} \Big(\sum_{i=1}^{m} |P_i(s)| M_5 \Big) ds \leq 1,$$

where $M_5 = \max_{0 \leq x \leq 1} \{F_i(x) : 1 \leq i \leq m\}$.

We consider a closed, bounded and convex subset Ω of $C([t_0, \infty), \mathbb{R})$ as follows:

$$\Omega = \{x = x(t) \in C([t_0, \infty), \mathbb{R}) : 2 \leq x(t) \leq 4, t \geq t_0\}$$

and define a mapping $\mathcal{A} : \Omega \to C([t_0, \infty), \mathbb{R})$ as follow:

$$(\mathcal{A}x)(t) = \begin{cases} 3 - \dfrac{1}{\Gamma(\alpha+1)} \displaystyle\sum_{j=1}^{\infty} \int_{t+j\tau}^{\infty} (s-t-j\tau)^{\alpha} \\ \times \Big(\displaystyle\sum_{i=1}^{m} P_i(s) F_i(x(s-\sigma_i)) \Big) ds, & t \geq T, \\ (\mathcal{A}x)(T), & t_0 \leq t \leq T. \end{cases}$$

We first show that $\mathcal{A}\Omega \subset \Omega$. Indeed, for every $x \in \Omega$ and $t \geq T$, we get

$$(\mathcal{A}x)(t) \leq 3 + \frac{1}{\Gamma(\alpha+1)} \sum_{j=1}^{\infty} \int_{t+j\tau}^{\infty} (s-t-j\tau)^{\alpha} \Big(\sum_{i=1}^{m} |P_i(s)||F_i(x(s-\sigma_i))| \Big) ds$$

$$\leq 3 + \frac{1}{\Gamma(\alpha+1)} \sum_{j=1}^{\infty} \int_{T+j\tau}^{\infty} s^{\alpha} \Big(\sum_{i=1}^{m} |P_i(s)| M_5 \Big) ds$$
$$\leq 4,$$

and

$$(\mathcal{A}x)(t) \geq 3 - \frac{1}{\Gamma(\alpha+1)} \sum_{j=1}^{\infty} \int_{t+j\tau}^{\infty} (s-t-j\tau)^{\alpha} \Big(\sum_{i=1}^{m} |P_i(s)||F_i(x(s-\sigma_i))| \Big) ds$$
$$\geq 3 - \frac{1}{\Gamma(\alpha+1)} \sum_{j=1}^{\infty} \int_{T+j\tau}^{\infty} s^{\alpha} \Big(\sum_{i=1}^{m} |P_i(s)| M_5 \Big) ds$$
$$\geq 2.$$

Hence $\mathcal{A}\Omega \subset \Omega$.

We now show that $\mathcal{A}$ is continuous. Let $x_k = x_k(t) \in \Omega$ be such that $x_k(t) \to x(t)$ as $k \to \infty$. Since Ω is closed, $x = x(t) \in \Omega$. For $t \geq T$, we have

$$|(\mathcal{A}x_k)(t) - (\mathcal{A}x)(t)|$$
$$\leq \frac{1}{\Gamma(\alpha+1)} \sum_{j=1}^{\infty} \int_{T+j\tau}^{\infty} s^{\alpha} \Big(\sum_{i=1}^{m} |P_i(s)||F_i(x_k(s-\sigma_i)) - F_i(x(s-\sigma_i))| \Big) ds.$$

Noting that $|F_i(x_k(t-\sigma_i)) - F_i(x(t-\sigma_i))| \to 0$ as $k \to \infty$ for $i = 1, 2, ..., m$, and applying the Lebesgue dominated convergence theorem, we conclude that $\lim_{k\to\infty} ||(\mathcal{A}x_k)(t) - (\mathcal{A}x)(t)|| = 0$. This means that $\mathcal{A}$ is continuous.

In what follows, we show that $\mathcal{A}\Omega$ is relatively compact. By (2.101), for any $\varepsilon > 0$, take $T^* \geq T$ large enough so that

$$\frac{1}{\Gamma(\alpha+1)} \sum_{j=1}^{\infty} \int_{T^*+j\tau}^{\infty} s^{\alpha} \Big(M_5 \sum_{i=1}^{m} |P_i(s)| \Big) ds < \frac{\varepsilon}{2}.$$

Then, for $x \in \Omega$, $t_2 > t_1 \geq T^*$, we get

$$|(\mathcal{A}x)(t_2) - (\mathcal{A}x)(t_1)|$$
$$\leq \frac{1}{\Gamma(\alpha+1)} \sum_{j=1}^{\infty} \int_{t_2+j\tau}^{\infty} s^{\alpha} \Big(\sum_{i=1}^{m} |P_i(s)||F_i(x(s-\sigma_i))| \Big) ds$$
$$+ \frac{1}{\Gamma(\alpha+1)} \sum_{j=1}^{\infty} \int_{t_1+j\tau}^{\infty} s^{\alpha} \Big(\sum_{i=1}^{m} |P_i(s)||F_i(x(s-\sigma_i))| \Big) ds$$
$$\leq \frac{1}{\Gamma(\alpha+1)} \sum_{j=1}^{\infty} \int_{t_2+j\tau}^{\infty} s^{\alpha} \Big(M_5 \sum_{i=1}^{m} |P_i(s)| \Big) ds$$
$$+ \frac{1}{\Gamma(\alpha+1)} \sum_{j=1}^{\infty} \int_{t_1+j\tau}^{\infty} s^{\alpha} \Big(M_5 \sum_{i=1}^{m} |P_i(s)| \Big) ds$$
$$< \frac{\varepsilon}{2} + \frac{\varepsilon}{2} = \varepsilon.$$

For $T \le t_1 < t_2 \le T^*$, we choose a sufficiently large $J \in \mathbf{N}^+$ such that $T + j\tau \ge T^*$ as $j \ge J$. For $x \in \Omega$, we obtain

$$
\begin{aligned}
&|(\mathcal{A}x)(t_2) - (\mathcal{A}x)(t_1)| \\
&\le \frac{1}{\Gamma(\alpha+1)} \sum_{j=1}^{\infty} \int_{t_1+j\tau}^{t_2+j\tau} s^\alpha \Big(\sum_{i=1}^{m} |P_i(s)||F_i(x(s-\sigma_i))| \Big) ds \\
&\le \frac{1}{\Gamma(\alpha+1)} \left[\sum_{j=1}^{J} \int_{t_1+j\tau}^{t_2+j\tau} s^\alpha \Big(M_5 \sum_{i=1}^{m} |P_i(s)| \Big) ds \right. \\
&\quad \left. + \sum_{j=J+1}^{\infty} \int_{t_1+j\tau}^{t_2+j\tau} s^\alpha \Big(M_5 \sum_{i=1}^{m} |P_i(s)| \Big) ds \right] \\
&\le \frac{1}{\Gamma(\alpha+1)} \left[\max_{T+\tau \le s \le T^*+(J-1)\tau} \Big\{ s^\alpha \Big(M_5 \sum_{i=1}^{m} |P_i(s)| \Big) \Big\} J(t_2 - t_1) \right. \\
&\quad \left. + \sum_{j=1}^{\infty} \int_{T^*+j\tau}^{\infty} s^\alpha \Big(M_5 \sum_{i=1}^{m} |P_i(s)| \Big) ds \right].
\end{aligned}
$$

Then there exists a $\delta > 0$ such that

$$|(\mathcal{A}x)(t_2) - (\mathcal{A}x)(t_1)| < \varepsilon, \quad \text{if} \quad 0 < t_2 - t_1 < \delta.$$

For any $x \in \Omega$, $t_0 \le t_1 < t_2 \le T$, it is easy to see that

$$|(\mathcal{A}x)(t_2) - (\mathcal{A}x)(t_1)| = 0 < \varepsilon.$$

Therefore $\{\mathcal{A}x : x \in \Omega\}$ is uniformly bounded and equicontinuous on $[t_0, \infty)$, and hence $\mathcal{A}\Omega$ is relatively compact. By Theorem 1.5 (Schauder fixed point theorem), there exists $x_0 \in \Omega$ such that $\mathcal{A}x_0 = x_0$, that is,

$$
x_0(t) = \begin{cases}
3 - \dfrac{1}{\Gamma(\alpha+1)} \displaystyle\sum_{j=1}^{\infty} \int_{t+j\tau}^{\infty} (s-t-j\tau)^\alpha \\
\times \Big(\displaystyle\sum_{i=1}^{m} P_i(s) F_i(x_0(s-\sigma_i)) \Big) ds, & t \ge T, \\
x_0(T), & t_0 \le t \le T.
\end{cases}
$$

Then we have

$$x_0(t) - x_0(t-\tau) = \frac{1}{\Gamma(\alpha+1)} \int_t^\infty (s-t)^\alpha \Big(\sum_{i=1}^{m} P_i(s) F_i(x_0(s-\sigma_i)) \Big) ds, \quad t \ge T,$$

which implies that

$$[x_0(t) - x_0(t-\tau)]' = \frac{1}{\Gamma(\alpha)} \int_t^\infty (s-t)^{\alpha-1} \Big(\sum_{i=1}^{m} P_i(s) F_i(x_0(s-\sigma_i)) \Big) ds, \quad t \ge T.$$

It follows that $x_0 = x_0(t)$ is a bounded positive solution of equation (2.99). This completes the proof. □

Remark 2.6. When $F_i(x) \equiv x, P_i(t) \equiv p_i \in \mathbb{R}, i = 1, 2, ..., m$, equation (2.99) reduces to

$$ {}_tD^{\alpha}_{+\infty}[x(t) + cx(t-\tau)]' + \sum_{i=1}^{m} p_i x(t - \sigma_i) = 0, \quad t \geq t_0. \tag{2.102} $$

In this case, (2.100) and (2.101) cannot be satisfied. So, we provide an alternative sufficient condition for existence of nonoscillatory solutions of equation (2.99).

Theorem 2.22. *Assume that* $\alpha > 0$, c, τ, p_i, $\sigma_i \in \mathbb{R}$, $i = 1, 2, ..., m$. *If the characteristic equation of equation* (2.102)*:*

$$ \lambda^{\alpha+1} + c\lambda^{\alpha+1}e^{-\lambda\tau} = \sum_{i=1}^{m} p_i e^{-\lambda\sigma_i} \tag{2.103} $$

has a positive real root, then equation (2.102) *has a bounded positive solution.*

Proof. Let $\lambda_0 > 0$ be a real root of (2.103). Set $y(t) = e^{-\lambda_0 t}$. By using (2.103) and ${}_tD^{\alpha}_{+\infty}e^{-\lambda t} = \lambda^{\alpha}e^{-\lambda t}$, we get

$$ \begin{aligned} {}_tD^{\alpha}_{+\infty}[y(t) + cy(t-\tau)]' &= -(\lambda_0^{\alpha+1} + c\lambda_0^{\alpha+1}e^{-\lambda_0\tau})e^{-\lambda_0 t} \\ &= -\left(\sum_{i=1}^{m} p_i e^{-\lambda_0\sigma_i}\right)e^{-\lambda_0 t} \\ &= -\sum_{i=1}^{m} p_i y(t - \sigma_i). \end{aligned} $$

Clearly, $y(t)$ is a bounded positive solution of equation (2.102). The proof is completed. □

2.6.5 *Fractional Partial Functional Differential Equations*

In this subsection, we study the following fractional functional partial differential equation involving Riemann-Liouville fractional derivative

$$ \frac{\partial^{\alpha} u(x,t)}{\partial t^{\alpha}} = C(t)\triangle u + \sum_{i=1}^{n} P_i(x)u(x, t - \sigma_i) + R(x,t), \tag{2.104} $$

supplemented with the initial condition

$$ {}_0D_t^{-(1-\alpha)}u(x,t)\big|_{t\in[-\sigma,0]} = \varphi(x,t) \text{ for } x \in \Omega, \text{ where } \sigma = \max\{\sigma_i, i = 1, 2, \ldots, n\}, \tag{2.105} $$

and boundary conditions:

$$ \frac{\partial u(x,t)}{\partial N} = 0 \text{ on } (x,t) \in \partial\Omega \times [0, \infty), \tag{B1} $$

$$ u(x,t) = 0 \text{ on } (x,t) \in \partial\Omega \times [0, \infty), \tag{B2} $$

$$ \frac{\partial u(x,t)}{\partial N} + \nu u = 0 \text{ on } (x,t) \in \partial\Omega \times [0, \infty), \tag{B3} $$

where $0 < \alpha < 1$, $(x,t) \in \Omega \times (0,\infty) \equiv G$, Ω is a bounded domain in Euclidean $n-$dimensional space $\mathbb{R}^n$ with a piecewise smooth boundary $\partial\Omega$, $\triangle$ is the Laplacian in $\mathbb{R}^n$, $C \in C((0,\infty),(-\infty,0])$, $P_i \in C(\Omega,[0,\infty))$, $R(x,t) \in C(G,(-\infty,\infty))$, $\sigma_i \in [0,\infty), i = 1,2,\ldots,n$, N is the unit exterior normal vector to $\partial\Omega$ and $\nu(x,t)$ is a nonnegative continuous function on $\partial\Omega \times [0,\infty)$.

For $f : \mathbb{R}^n \times [0,\infty) \to \mathbb{R}^n$, Riemann-Liouville derivative with respect to time of the function f can be written as

$$\frac{\partial^\alpha f(x,t)}{\partial t^\alpha} = \frac{1}{\Gamma(1-\alpha)}\frac{\partial}{\partial t}\int_0^t \frac{f(x,t)}{(t-s)^\alpha}ds, \quad 0<\alpha<1.$$

A function $u(x,t)$ is said to be a solution of the problem (2.104) and (Bi) $(i = 1,2,3)$ if it satisfies (2.104) in the domain G and the boundary condition (Bi) $(i = 1,2,3)$.

The solution $u(x,t)$ of problem (2.104) and (Bi) $(i = 1,2,3)$ is said to be oscillatory in the domain G if for any positive number μ there exists a point $(x_1,t_1) \in \Omega \times [\mu,\infty)$ such that $u(x_1,t_1) = 0$ holds.

In the following, we will give the sufficient criteria for the oscillation of all solutions the equation (2.104) equipped with initial and Neumann, Dirichlet and Robin boundary conditions.

2.6.5.1 *Oscillation of Fractional ODEs*

To investigate oscillation of (2.104), let us study oscillation of the following fractional delay differential equation with Riemann-Liouville fractional derivative

$$_0D_t^\alpha x(t) + \sum_{i=1}^n p_i x(t-\tau_i) = f(t), \quad t > 0. \tag{2.106}$$

Without loss of generality, we assume the coefficients p_i of (2.106) are all nonzero and that $\tau_1 = \max\{\tau_1, ..., \tau_n\}$.

Definition 2.16. By a solution of (2.106) in $(0,\infty)$ with initial function $\varphi \in \mathrm{AC}[-\tau_1,0]$, we mean a function $x : [-\tau_1,\infty) \to R$ and $x \in \mathrm{AC}[0,b]$, for any $b > 0$, such that $x(t) = \varphi(t)$, $t \in [-\tau_1,0]$, $(_0D_t^\alpha x)(t)$ exists and $x(t)$ satisfies (2.106) in $(0,\infty)$. A solution $x(t)$ of equation (2.106) is called oscillatory if it has arbitrarily large number of zeros.

Lemma 2.15. *The solution of equation* (2.111) *has an exponent estimate*

$$x(t) = o(e^{q_0 t}) \quad (t > b > 0)$$

for a constant $q_0 > 0$.

Proof. Taking Riemann-Liouville integral of equation (2.111), we have

$$\begin{aligned} x(t) &= \frac{x_0}{\Gamma(\alpha)}t^{\alpha-1} - \sum_{i=1}^n \frac{p_i}{\Gamma(\alpha)}\int_0^t (t-s)^{\alpha-1}x(s-\tau_i)ds \\ &= \frac{x_0}{\Gamma(\alpha)}t^{\alpha-1} - \sum_{i=1}^n \frac{p_i}{\Gamma(\alpha)}F_i(t), \end{aligned} \tag{2.107}$$

where

$$F_i(t) = \int_0^t (t-s)^{\alpha-1} x(s-\tau_i)ds, \quad x_0 = {}_0D_t^{\alpha-1}x(0).$$

As $x(t) \in \mathrm{AC}[0,b]$, there exists a constant $M > 0$ such that $|x(t)| \leq M$ and $AC[-\tau_1, 0]$ is the Banach space with the norm $\|\cdot\|$. Let $\tau = \min\{\tau_i\}$ for $i \in \{1,2,...,n\}$. Then, for $t > b$, we have

$$\begin{aligned}
\|F_i(t)\| &\leq \int_0^t (t-s)^{\alpha-1}\|x(s-\tau_i)\|ds \\
&\leq \int_0^b (t-s)^{\alpha-1} \max_{s-\tau_i\leq\eta\leq s}\|x(\eta)\|ds + \int_b^t (t-s)^{\alpha-1} \max_{s-\tau_i\leq\eta\leq s}\|x(\eta)\|ds \\
&\leq \int_0^b (t-s)^{\alpha-1}(M+\|\varphi\|)ds + \int_b^t (t-s)^{\alpha-1}(\max_{b\leq\eta\leq s}\|x(\eta)\| + M + \|\varphi\|)ds \\
&\leq \frac{t^\alpha}{\alpha}(M+\|\varphi\|) + \int_b^t (t-s)^{\alpha-1}\max_{b\leq\eta\leq s}\|x(\eta)\|ds,
\end{aligned} \tag{2.108}$$

which, together with (2.107), yields

$$\begin{aligned}
\|x(t)\| &\leq \frac{b^{\alpha-1}}{\Gamma(\alpha)}|x_0| + \sum_{i=1}^n \frac{|p_i|}{\Gamma(\alpha)}\|F_i(t)\| \\
&\leq \frac{b^{\alpha-1}}{\Gamma(\alpha)}|x_0| + \frac{np}{\Gamma(\alpha)}\frac{t^\alpha}{\alpha}(M+\|\varphi\|) + \frac{np}{\Gamma(\alpha)}\int_b^t (t-s)^{\alpha-1}\max_{b\leq\eta\leq s}\|x(\eta)\|ds,
\end{aligned}$$

where $p = \max\{|p_i|\}$, for $i = 1,2,...,n$.

One can introduce nondecreasing function $m(t)$ as

$$m(t) = \frac{b^{\alpha-1}}{\Gamma(\alpha)}|x_0| + \frac{np}{\Gamma(\alpha)}\frac{t^\alpha}{\alpha}(M+\|\varphi\|).$$

By Lemma 2.9, there exists $k = k(\alpha)$ such that

$$\begin{aligned}
\|x(t)\| \leq \max_{b\leq s\leq t}\|x(s)\| &\leq m(t) + \frac{knp}{\Gamma(\alpha_0)}\int_b^t (t-s)^{\alpha_0-1}m(s)ds \\
&\leq m(t)\left(1 + \frac{knp}{\alpha_0\Gamma(\alpha_0)}t^{\alpha_0}\right).
\end{aligned} \tag{2.109}$$

Obviously, from (2.109) we infer that $x(t)$ has an exponent estimate. The proof is completed. □

Theorem 2.23. *If the equation*

$$P(\lambda) = \lambda^\alpha + \sum_{i=1}^n p_i e^{-\lambda\tau_i} = 0 \tag{2.110}$$

has no real roots, then every solution of

$${}_0D_t^\alpha x(t) + \sum_{i=1}^n p_i x(t-\tau_i) = 0, \quad t > 0 \tag{2.111}$$

is oscillatory.

Proof. On the contrary, let us assume that (2.111) has a positive solution $x(t)$ with Laplace transform $X(s)$. Taking the Laplace transform of both sides of (2.111), we have

$$\left(s^\alpha X(s) - {}_0D_t^{\alpha-1}x(0)\right) + \sum_{i=1}^{n} p_i \int_0^\infty e^{-st}x(t-\tau_i)dt = 0,$$

that is,

$$\left(s^\alpha X(s) - {}_0D_t^{\alpha-1}x(0)\right) + \sum_{i=1}^{n} p_i e^{-s\tau_i} X(s) + \sum_{i=1}^{n} p_i e^{-s\tau_i} \int_{-\tau_i}^{0} e^{-st}x(t)dt = 0.$$

Hence

$$(s^\alpha + \sum_{i=1}^{n} p_i e^{-s\tau_i})X(s) = {}_0D_t^{\alpha-1}x(0) - \sum_{i=1}^{n} p_i e^{-s\tau_i} \int_{-\tau_i}^{0} e^{-st}x(t)dt. \tag{2.112}$$

Let

$$P(s) = s^\alpha + \sum_{i=1}^{n} p_i e^{-s\tau_i}, \ x_0 = {}_0D_t^{\alpha-1}x(0),$$

$$\phi(s) = \sum_{i=1}^{n} \phi_i(s), \text{ where } \phi_i(s) = p_i e^{-s\tau_i} \int_{-\tau_i}^{0} e^{-st}x(t)dt.$$

Then, from (2.112), we get

$$P(s)X(s) = x_0 - \phi(s). \tag{2.113}$$

Since $P(s) = 0$ has no real roots, therefore $P(s) > 0$. Hence we obtain that p_1, the coefficient corresponding to the largest delay τ_1, is positive. Take $\varepsilon > 0$ be small enough so that $\tau_1 - \varepsilon > \tau_i$ for $i = 2, ..., n$. By positivity of $x(t)$ in AC$[-\tau_1, 0]$, there exists a constant $m > 0$ such that $x(t) \geq m$ and that

$$\begin{aligned}
\frac{e^{-s(\tau_1-\varepsilon)}}{\phi_1(s)} &= \frac{e^{-s(\tau_1-\varepsilon)}}{p_1 e^{-s\tau_1} \int_{-\tau_1}^{0} e^{-st}x(t)dt} \\
&\leq \frac{e^{s\varepsilon}}{p_1 m \int_{-\tau_1}^{0} e^{-st}dt} \\
&= \frac{s}{p_1 m \frac{e^{s\tau_1}-1}{e^{s\varepsilon}}} \\
&= \frac{1}{p_1 m \frac{\tau_1 e^{s\tau_1} e^{s\varepsilon} - \varepsilon e^{s\varepsilon}(e^{s\tau_1}-1)}{e^{2s\varepsilon}}} \\
&= \frac{e^{s\varepsilon}}{p_1 m(\tau_1 e^{s\tau_1} - \varepsilon(e^{s\tau_1}-1))} \to 0, \quad \text{as } s \to -\infty.
\end{aligned}$$

Thus we conclude that $\phi_1(s) > e^{-s(\tau_1-\varepsilon)} \to \infty$ as $s \to -\infty$ eventually. On the other hand, as $s \to -\infty$,

$$|\phi_i(s)| = o(e^{-s\tau_i}) = o(\phi_1(s))$$

for $i = 2, ..., n$. But both $P(s)$ and $X(s)$ are positive while $\phi(s) \to \infty$ as $s \to -\infty$. Hence (2.113) leads to a contradiction. The proof is completed. □

In Theorem 2.23, the characteristic equation (2.110) plays an important role in investigating the oscillation of equation (2.111). However, it remains an issue to determine whether (2.110) has a real root. In the following we derive some sufficient conditions for the oscillation of equation (2.111) which can easily be applied.

Corollary 2.8. *Assume that* $\alpha = p/q$, *where* p, q *are co-prime (that is,* $(p,q) = 1$*), and* $p_i\tau_i \geq 0$ *for* $i = 1, 2, ..., n$. *Then each of the following two conditions is sufficient for the oscillation of all solutions to equation* (2.111).

(i) $\left(\sum_{i=1}^{n} p_i\tau_i\right)\left(\sum_{i=1}^{n} p_i\right)^{\frac{1-\alpha}{\alpha}} > \frac{1}{e}$;

(ii) $\left(\prod_{i=1}^{n} p_i\right)^{\frac{1}{n}}\left(\sum_{i=1}^{n} \tau_i\right)\left(\sum_{i=1}^{n} p_i\right)^{\frac{1-\alpha}{\alpha}} > \frac{1}{e}$.

Proof. (i) As equation (2.110) has no positive real root. Assume, for the sake of contradiction, that (2.110) has a negative real root λ. Let $\bar{\lambda} = -\lambda > 0$. Then, from (2.110) we have

$$\bar{\lambda}^{\alpha} = \sum_{i=1}^{n} p_i e^{\bar{\lambda}\tau_i} \geq \sum_{i=1}^{n} p_i,$$

i.e.

$$\bar{\lambda} \geq \left(\sum_{i=1}^{n} p_i\right)^{\frac{1}{\alpha}} .$$

By making use of the inequality $e^x \geq xe$ for $x \geq 0$, we get

$$\begin{aligned}\bar{\lambda}^{\alpha} &= \sum_{i=1}^{n} p_i e^{\bar{\lambda}\tau_i} \geq \sum_{i=1}^{n} p_i\bar{\lambda}\tau_i e \\ &= \left(\sum_{i=1}^{n} p_i\tau_i\right)e\bar{\lambda}^{\alpha}\bar{\lambda}^{1-\alpha} \\ &\geq e\left(\sum_{i=1}^{n} p_i\tau_i\right)\left(\sum_{i=1}^{n} p_i\right)^{\frac{1-\alpha}{\alpha}}\bar{\lambda}^{\alpha},\end{aligned}$$

which implies that

$$\left(\sum_{i=1}^{n} p_i\tau_i\right)\left(\sum_{i=1}^{n} p_i\right)^{\frac{1-\alpha}{\alpha}} \leq \frac{1}{e}.$$

This is a contradiction. Hence, (2.110) has no negative real root and consequently the result follows as a consequence of Theorem 2.23.

(ii) In view of the arithmetic mean-geometric mean inequality:

$$\left(\prod_{i=1}^{n} p_i\right)^{\frac{1}{n}} \leq \frac{1}{n}\sum_{i=1}^{n} p_i ,$$

we have

$$\begin{aligned}\bar{\lambda}^{\alpha} &= \sum_{i=1}^{n} p_i e^{\bar{\lambda}\tau_i} \geq n\Big(\prod_{i=1}^{n} p_i e^{\bar{\lambda}\tau_i}\Big)^{\frac{1}{n}} \\ &= n\Big(\prod_{i=1}^{n} p_i\Big)^{\frac{1}{n}} \exp\Big(\frac{\bar{\lambda}}{n}\sum_{i=1}^{n}\tau_i\Big) \geq n\Big(\prod_{i=1}^{n} p_i\Big)^{\frac{1}{n}}\Big(\frac{\bar{\lambda}}{n}\sum_{i=1}^{n}\tau_i\Big)e \\ &= \Big(\prod_{i=1}^{n} p_i\Big)^{\frac{1}{n}}\Big(\sum_{i=1}^{n}\tau_i\Big)e\bar{\lambda}^{\alpha}\bar{\lambda}^{1-\alpha} \geq e\Big(\prod_{i=1}^{n} p_i\Big)^{\frac{1}{n}}\Big(\sum_{i=1}^{n}\tau_i\Big)\Big(\sum_{i=1}^{n} p_i\Big)^{\frac{1-\alpha}{\alpha}}\bar{\lambda}^{\alpha},\end{aligned}$$

which leads to a contradiction

$$\Big(\prod_{i=1}^{n} p_i\Big)^{\frac{1}{n}}\Big(\sum_{i=1}^{n}\tau_i\Big)\Big(\sum_{i=1}^{n} p_i\Big)^{\frac{1-\alpha}{\alpha}} \leq \frac{1}{e}.$$

Hence (2.110) has no negative real root and the result follows as an immediate application of Theorem 2.23. The proof is completed. □

We assume that $f(t) = o(e^{a_0 t})$ for some $a_0 > 0$. Then the Laplace transform $F(s)$ of $f(t)$ exists.

Theorem 2.24. *Let $a \in \mathbb{R}$ and assume that*

(H1) *Equation* (2.110) *has no roots in* $(0, \infty)$*;*
(H2) *a is the abscissa of convergence of $F(s)$, $F(s)$ has a singularity on* $\mathrm{Re}(s) = a$*, but $F(s)$ is analytic at $s = a$.*

Then every solution of (2.106) *is oscillatory.*

Proof. Suppose that (2.106) possesses an eventually positive solution $x(t)$ with Laplace transform $X(s)$ having abscissa of convergence b_0. Then $X(s)$ is analytic in the half-plane $\mathrm{Re}(s) > b_0$ and by Lemma 2.8, it cannot be analytically continued at $s = b_0$. That is, there is no complex neighborhood of b_0 on which we can find an analytic function which agrees with $X(s)$ for $\mathrm{Re}(s) > b_0$. Taking the Laplace transform of both sides of (2.106), we obtain

$$P(s)X(s) = x_0 - \phi(s) + F(s). \tag{2.114}$$

By analyticity, (2.114) holds for $\mathrm{Re}(s) > \max\{a, b_0\}$, and ϕ is an entire function. Now we can say $a > b_0$ is impossible because (2.114) and (H1), (H2) would imply a singularity of $X(s)$ in $\mathrm{Re}(s) > b_0$. On the other hand, $a \leq b_0$ is impossible because we could then use (H1), (H2), and (2.114) to analytically continue $X(s)$ at $s = b_0$. Thus (2.106) cannot have an eventually positive solution. This contradiction completes the proof. □

Lemma 2.16. *For any $c \in \mathbb{R}$, the Laplace transform $X_c(s)$ of $x_c(t)$ exists and has the same abscissa of convergence as $X(s)$.*

Proof. Obviously

$$\begin{aligned}X_c(s) &= \int_0^{\infty} e^{-st} x_c(t)dt \\ &= \int_0^{\infty} e^{-st} x(t+c)dt \\ &= e^{sc} \int_c^{\infty} e^{-st} x(t)dt \\ &= e^{sc}[X(s) - \int_0^c e^{-st} x(t)dt].\end{aligned} \tag{2.115}$$

Since the last integral defines an entire function of the complex variable s, we can see that both $X(s)$ and $X_c(s)$ converge or diverge for the same values of s, and have singularities at the same points. □

Theorem 2.25. *Suppose that*

(H3) *Equation* (2.110) *has no real roots;*
(H4) *the abscissa of convergence of* $F(s)$ *is* $-\infty$ *and, for some* $\varepsilon > 0$, $|F(s)| = o(e^{-s(\tau_i - \varepsilon)})$ *as* $s \to -\infty$. *Then every solution of* (2.106) *is oscillatory.*

Proof. Let us claim that (2.106) has a solution $x(t)$ such that for some $c \geq 0$, $x_c(t) > 0$ for $t \geq 0$. Let (2.106)$'$ denote equation (2.106) with f replaced by f_c. Then $x_c(t)$ is a positive solution of (2.106)$'$. It is easily checked by means of (2.115) that $F_c(s)$ also satisfies (H4). Since we are seeking a contradiction, we may as well assume that $x(t) > 0$ for $t \geq -\tau_1$. Then, in view of (2.114), and by Lemma 2.8, it follows that the abscissa of convergence of $X(s)$ is $-\infty$. As in the proof of Theorem 2.23, by (H3), (H4) and (2.114), we get $\lim\limits_{s \to -\infty} X(s) = -\infty$, which leads to the contradiction and completes the proof. □

2.6.5.2 *Boundary Value Problem* (2.104) *and* ($B1$)

Lemma 2.17. *Assume that*

(H5) $\int_{\Omega} R(x,t)dx \geq l_1 t^{-\alpha}$ *for* $t > 0$, *where* $l_1 > 0$ *is a constant.*

If problem (2.104) *and* ($B1$) *has an eventually positive solution, then the following fractional delay differential equation involving Riemann-Liouville fractional derivative*

$${}_0D_t^{\alpha} x(t) - \sum_{i=1}^{n} p_i x(t - \sigma_i) = 0, \quad t > 0 \tag{2.116}$$

has an eventually positive solution, where $p_i = \min\{P_i(x) : x \in \Omega\}$.

Proof. Without loss of generality, we assume that problem (2.104) and ($B1$) has $u(x,t) > 0$, and $u(x, t - \sigma_i) > 0$ in $\Omega \times (0, \infty)$ for some $t_1 > 0$.

Set

$$U(t) = \int_{\Omega} u(x,t)dx, \quad t \ge t_1,$$

and note that $U(t) > 0$ for $t > t_1$.

Integrating (2.104) with respect to x over the domain Ω, we get

$$\frac{\partial^{\alpha}}{\partial t^{\alpha}}\left[\int_{\Omega} u(x,t)dx\right] = C(t)\int_{\Omega} \triangle u dx + \sum_{i=1}^{n}\int_{\Omega} P_i(x)u(x,t-\sigma_i)dx + \int_{\Omega} R(x,t)dx. \tag{2.117}$$

Using the Green's formula and boundary condition $(B1)$, we obtain

$$\int_{\Omega} \triangle u dx = \int_{\partial\Omega} \frac{\partial u(x,t)}{\partial N} dS = 0, \quad t \ge t_1.$$

Then (2.117) becomes

$$\frac{\partial^{\alpha}}{\partial t^{\alpha}}\left[\int_{\Omega} u(x,t)dx\right] = \sum_{i=1}^{n}\int_{\Omega} P_i(x)u(x,t-\sigma_i)dx + \int_{\Omega} R(x,t)dx. \tag{2.118}$$

Let $r(t) = \int_{\Omega} R(x,t)dx$ and from the definitions of $U(t), p_i$ as above, (2.118) reduces to

$${}_0D_t^{\alpha}U(t) \ge \sum_{i=1}^{n} p_i U(t-\sigma_i) + r(t). \tag{2.119}$$

Integrating the inequality (2.119) from 0 to t and applying condition (H5), we have

$$\begin{aligned} U(t) &\ge \frac{U(0)}{\Gamma(\alpha)}t^{\alpha-1} + \sum_{i=1}^{n}\frac{p_i}{\Gamma(\alpha)}\int_0^t (t-s)^{\alpha-1}U(s-\sigma_i)ds + \frac{1}{\Gamma(\alpha)}\int_0^t (t-s)^{\alpha-1}r(s)ds \\ &\ge \frac{U(0)}{\Gamma(\alpha)}t^{\alpha-1} + \sum_{i=1}^{n}\frac{p_i}{\Gamma(\alpha)}\int_0^t (t-s)^{\alpha-1}U(s-\sigma_i)ds + \frac{l_1}{\Gamma(1-\alpha)}, \end{aligned} \tag{2.120}$$

where $U(0) = \int_{\Omega} \varphi(x,0)dx$. On the other hand, there exists a constant $t_1' > 0$ such that

$$\frac{|U(0)|}{\Gamma(\alpha)}t^{\alpha-1} \le \frac{l_1}{\Gamma(1-\alpha)}$$

for $t \ge t_1'$.

Let $\bar{t}_1 = \max\{t_1, t_1'\}$. Then, for $t > \bar{t}_1$, we get

$$U(t) \ge \sum_{i=1}^{n}\frac{p_i}{\Gamma(\alpha)}\int_0^t (t-s)^{\alpha-1}U(s-\sigma_i)ds. \tag{2.121}$$

Define a set $W = \{w(t) : w(t) \in C([\bar{t}_1 - \sigma, \infty), [0,1])\}$ and a mapping S on W by

$$(Sw)(t) = \begin{cases} \dfrac{1}{U(t)}\left[\displaystyle\sum_{i=1}^{n}\frac{p_i}{\Gamma(\alpha)}\int_0^t (t-s)^{\alpha-1}U(s-\sigma_i)w(s-\sigma_i)ds\right], & t \ge \bar{t}_1, \\ \dfrac{t-\bar{t}_1+\theta}{\theta}(Sw)(\bar{t}_1) + \left(1 - \dfrac{t-\bar{t}_1+\theta}{\theta}\right), & \bar{t}_1 - \sigma \le t < \bar{t}_1. \end{cases}$$

It easily follows from (2.121) that $SW \subset W$, and for any $w \in W$, we have $(Sw)(t) > 0$ for $\bar{t}_1 - \sigma \leq t < \bar{t}_1$.

Define a sequence of functions $\{w_n(t)\}$ in W by $w_0(t) = 1$, and $w_{n+1}(t) = Sw_n(t)$, $t \geq \bar{t}_1 - \sigma$. From (2.121), by induction, we have

$$0 \leq w_{n+1}(t) \leq w_n(t) \leq 1, \quad t \geq \bar{t}_1 - \sigma.$$

Then $\lim_{n\to\infty} w_n(t) = w(t)$, $t \geq \bar{t}_1 - \sigma$, exists, and $0 \leq w(t) \leq 1$. Further, in view of the Lebesgue dominated convergence theorem, we obtain

$$w(t) = \frac{1}{U(t)}\left[\sum_{i=1}^{n} \frac{p_i}{\Gamma(\alpha)} \int_0^t (t-s)^{\alpha-1} U(s-\sigma_i) w(s-\sigma_i) ds\right], \quad t \geq \bar{t}_1,$$

and

$$w(t) = \frac{t - \bar{t}_1 + \sigma}{\sigma} w(\bar{t}_1) + \left(1 - \frac{t - \bar{t}_1 + \sigma}{\sigma}\right) > 0, \ \bar{t}_1 - \sigma \leq t < \bar{t}_1.$$

Set $x(t) = U(t)w(t)$. Then $x(t)$ is a nonnegative solution of the following equation

$$x(t) = \sum_{i=1}^{n} \frac{p_i}{\Gamma(\alpha)} \int_0^t (t-s)^{\alpha-1} x(s-\sigma_i) ds, \ t \geq \bar{t}_1, \tag{2.122}$$

and $x(t) > 0$ for $\bar{t}_1 - \sigma \leq t < \bar{t}_1$. It is easy to infer that $x(t) > 0$ for $t \geq \bar{t}_1$ since $p_i > 0$ for $i = 1, 2, \ldots, n$. Therefore, $x(t) > 0$ for $t \geq \bar{t}_1 - \sigma$.

Note the integral equation (2.122) is equivalent to equation (2.116) with an initial condition $[{}_0D_t^{-(1-\alpha)} x(t)]_{t\in[-\sigma,0]} = 0$ for $t \in [-\sigma, 0]$. The proof is completed. □

Applying Theorem 2.23 and Lemma 2.17, we directly obtain the following results of oscillation.

Theorem 2.26. *Assume that condition (H5) holds, and if the equation*

$$P(\lambda) = \lambda^\alpha - \sum_{i=1}^{n} p_i e^{-\lambda\sigma_i} = 0 \tag{2.123}$$

has no real roots, then every solution of problem (2.104) *and (B1) is oscillatory.*

2.6.5.3 *Boundary Value Problem* (2.104) *and (B2)*

We firstly consider the following Dirichlet problem in the domain Ω:

$$\begin{cases} \Delta u + \eta u = 0 & \text{in } (x,t) \in \Omega \times [0,\infty), \\ u = 0 & \text{on } (x,t) \in \partial\Omega \times [0,\infty), \end{cases} \tag{2.124}$$

where η is a constant. It is well known (see Vladimirov, 1981) that the smallest eigenvalue η_1 of problem (2.124) is positive and corresponding eigenfunction $\Phi(x)$ is also positive on Ω.

Lemma 2.18. *Assume that*

(H6) $\int_\Omega \Phi(x) R(x,t) dx \geq l_2 t^{-\alpha}$ *for* $t > 0$, *where* $l_2 > 0$ *is a constant.*

If problem (2.104) *and* (B2) *has an eventually positive solution, then* (2.116) *has an eventually positive solution.*

Proof. Without loss of generality, we assume that problem (2.104) and ($B2$) has $u(x,t) > 0$ and $u(x,t-\sigma_i) > 0$ in $\Omega \times (0,\infty)$ for some $t_2 > 0$.

Define

$$V(t) = \int_\Omega u(x,t)\Phi(x)dx, \ t \geq t_2,$$

and notice that $V(t) > 0$ for $t \geq t_2$.

Integrating (2.104) with respect to t, we have

$$\begin{aligned} u(x,t) = &\frac{\varphi(x,0)}{\Gamma(\alpha)}t^{\alpha-1} + \frac{1}{\Gamma(\alpha)}\int_0^t (t-s)^{\alpha-1}C(s)\Delta u ds \\ &+\sum_{i=1}^n \frac{1}{\Gamma(\alpha)}\int_0^t (t-s)^{\alpha-1}P_i(x)u(x,s-\sigma_i)ds \\ &+\frac{1}{\Gamma(\alpha)}\int_0^t (t-s)^{\alpha-1}R(x,s)ds. \end{aligned} \tag{2.125}$$

Multiplying both sides of (2.125) by $\Phi(x)$, and integrating both sides with respect to $x \in \Omega$, we obtain

$$\begin{aligned} &\int_\Omega \Phi(x)u(x,t)dx \\ &= \frac{t^{\alpha-1}}{\Gamma(\alpha)}\int_\Omega \Phi(x)\varphi(x,0)dx + \frac{1}{\Gamma(\alpha)}\int_\Omega \Phi(x)\int_0^t (t-s)^{\alpha-1}C(s)\Delta u ds dx \\ &\quad+\sum_{i=1}^n \frac{1}{\Gamma(\alpha)}\int_\Omega \Phi(x)\int_0^t (t-s)^{\alpha-1}P_i(x)u(x,s-\sigma_i)dsdx \\ &\quad+\frac{1}{\Gamma(\alpha)}\int_\Omega \Phi(x)\int_0^t (t-s)^{\alpha-1}R(x,s)dsdx \\ &= \frac{t^{\alpha-1}}{\Gamma(\alpha)}\int_\Omega \Phi(x)\varphi(x,0)dx + \frac{1}{\Gamma(\alpha)}\int_0^t (t-s)^{\alpha-1}C(s)\int_\Omega \Phi(x)\Delta u dx ds \\ &\quad+\sum_{i=1}^n \frac{1}{\Gamma(\alpha)}\int_0^t (t-s)^{\alpha-1}\int_\Omega \Phi(x)P_i(x)u(x,s-\sigma_i)dxds \\ &\quad+\frac{1}{\Gamma(\alpha)}\int_0^t (t-s)^{\alpha-1}\int_\Omega \Phi(x)R(x,s)dxds \\ &\geq \frac{t^{\alpha-1}}{\Gamma(\alpha)}\int_\Omega \Phi(x)\varphi(x,0)dx + \frac{1}{\Gamma(\alpha)}\int_0^t (t-s)^{\alpha-1}C(s)\int_\Omega \Phi(x)\Delta u dx ds \\ &\quad+\sum_{i=1}^n \frac{p_i}{\Gamma(\alpha)}\int_0^t (t-s)^{\alpha-1}\int_\Omega \Phi(x)u(x,s-\sigma_i)dxds \\ &\quad+\frac{1}{\Gamma(\alpha)}\int_0^t (t-s)^{\alpha-1}\int_\Omega \Phi(x)R(x,s)dxds. \end{aligned} \tag{2.126}$$

Using Green's formula and boundary condition ($B2$), we have

$$\begin{aligned}\int_\Omega \Phi(x)\Delta u dx &= \int_{\partial\Omega}\left(\Phi(x)\frac{\partial u}{\partial N} - u\frac{\partial \Phi(x)}{\partial N}\right)dS + \int_\Omega u(x,t)\Delta\Phi(x)dx \\ &= -\eta_1\int_\Omega \Phi(x)u(x,t)dx, \quad t \geq t_2.\end{aligned} \tag{2.127}$$

From (2.126), (2.127) and condition (H6), we obtain

$$\begin{aligned}V(t) \geq\ & \frac{V(0)}{\Gamma(\alpha)}t^{\alpha-1} - \frac{\eta_1}{\Gamma(\alpha)}\int_0^t (t-s)^{\alpha-1}C(s)V(s)ds \\ &+ \sum_{i=1}^n \frac{p_i}{\Gamma(\alpha)}\int_0^t (t-s)^{\alpha-1}V(s-\sigma_i)dxds \\ &+ \frac{1}{\Gamma(\alpha)}\int_0^t (t-s)^{\alpha-1}\int_\Omega \Phi(x)R(x,s)dxds \\ \geq\ & \frac{V(0)}{\Gamma(\alpha)}t^{\alpha-1} + \sum_{i=1}^n \frac{p_i}{\Gamma(\alpha)}\int_0^t (t-s)^{\alpha-1}V(s-\sigma_i)dxds + \frac{l_2}{\Gamma(1-\alpha)},\end{aligned} \tag{2.128}$$

where $V(0) = \int_\Omega \Phi(x)\varphi(x,0)dx$. Then (2.128) is similar to (2.120). The rest of the proof is similar to that of Lemma 2.17, so we omit it. This completes the proof. □

Theorem 2.27. *Assume that condition (H6) holds, and that equation* (2.123) *has no real roots, then every solution of problem* (2.104) *and (B2) is oscillatory.*

2.6.5.4 *Boundary Value Problem* (2.104) *and (B3)*

Lemma 2.19. *Assume that condition (H5) holds. If problem* (2.104) *and (B3) has an eventually positive solution, then* (2.116) *has an eventually positive solution.*

Proof. Without loss of generality, we assume that problem (2.104) and ($B3$) has $u(x,t) > 0$ and $u(x, t-\sigma_i) > 0$ in $\Omega \times (0,\infty)$ for some $t_3 > 0$.

From the Green's formula and boundary condition ($B3$), it follows that

$$\int_\Omega \Delta u dx = -\int_{\partial\Omega} \nu u dS \leq 0, \tag{2.129}$$

where dS is the surface element on $\partial\Omega$. Combining (2.117) and (2.129), we get

$$\frac{\partial^\alpha}{\partial t^\alpha}\left[\int_\Omega u(x,t)dx\right] \geq \sum_{i=1}^n \int_\Omega P_i(x)u(x,t-\sigma_i)dx + \int_\Omega R(x,t)dx. \tag{2.130}$$

Define

$$U(t) = \int_\Omega u(x,t)dx,\ t \geq t_3,$$

and observe that $U(t) > 0$ for $t \geq t_3$. From (2.130) we have

$${}_0D_t^\alpha U(t) \geq \sum_{i=1}^n p_i U(t-\sigma_i) + r(t). \tag{2.131}$$

Integrating inequality (2.131) from 0 to t, we obtain (2.120). We omit the remaining part of the proof as it is similar to that of Lemma 2.17. This completes the proof. □

Theorem 2.28. *Assume that condition (H5) holds, and that equation* (2.123) *has no real roots, then every solution of problem* (2.104) *and (B3) is oscillatory.*

2.6.5.5 *Example*

In this subsection, we discuss an illustrative example.

Example 2.5. Consider the following fractional partial differential equation

$$\frac{\partial^\alpha u(x,t)}{\partial t^\alpha} = -\triangle u + (x^2+1)u(x,t-1) + t^{-\alpha}\sin x, \quad (x,t) \in (0,\pi)\times(0,\infty), \tag{2.132}$$

with the boundary conditions

$$-\frac{\partial u(0,t)}{\partial x} = \frac{\partial u(\pi,t)}{\partial x} = 0, \quad t \geq 0. \tag{2.133}$$

Here $0 < \alpha < 1$, $\Omega = (0,\pi)$, $R(x,t) = t^{-\alpha}\sin x$, $\sigma = 1$.

Obviously

$$\int_\Omega R(x,t)dx = \int_0^\pi t^{-\alpha}\sin x dx = 2t^{-\alpha},$$

and

$$P(x) > 1 = p, \quad (p\sigma)p^{\frac{1-\alpha}{\alpha}} = p^{\frac{1}{\alpha}} = 1 > \frac{1}{e}.$$

Thus all the conditions of Theorem 2.26 are satisfied. Therefore every solution of problem (2.132) and (2.133) is oscillatory in $(0,\pi)\times(0,\infty)$.

2.7 Notes and Remarks

The results in Section 2.2 are taken from Agarwal, Zhou and He, 2010. The main results in Section 2.3 are adopted from Zhou, Jiao and Li, 2009a. The material in Section 2.4 are due to Zhou, Jiao and Li, 2009b. The results in Section 2.5 are taken from Wang, Fečkan and Zhou, 2013c. The material in Section 2.6 are taken from Zhou, Alsaedi and Ahmad, 2018; Zhou, Ahmad and Alsaedi, 2019a; Zhou, Ahmad, Chen and Alsaedi, 2019.

Chapter 3

Fractional Ordinary Differential Equations in Banach Spaces

3.1 Introduction

In this chapter, we discuss the Cauchy problem of fractional ordinary differential equations in Banach spaces under hypotheses based on Carathéodory condition. The tools used include some classical and modern nonlinear analysis methods such as fixed point theory, measure of noncompactness method, topological degree method and Picard operators technique, etc.

Firstly, we give an example which show that the criteria on existence of solutions for the initial value problem of fractional differential equations in finite-dimensional spaces may not be true in infinite-dimensional cases. It is well known that Peano theorem of integer order ordinary differential equations is not true in infinite-dimensional Banach spaces. The first result in this direction was obtained by Dieudonne, 1950. Dieudonne, 1950, produced an example which showed that Peano theorem is not true in the space c_0 of sequences which converge to zero. In fact, Peano theorem of fractional differential equations is also not true in infinite-dimensional Banach spaces. In the following, we shall show that the existence result of nonlocal Cauchy problem for fractional abstract differential equations which has been obtained by N'Guerekata in 2009 is not true in the space c_0.

Example 3.1. Let $E = c_0 = \{z = (z_1, z_2, z_3, ...) : z_n \to 0 \text{ as } n \to \infty\}$ with the norm $\|z\| = \sup_{n\geq 1} |z_n|$ and $f(z) = 2(\sqrt{|z_1|}, \sqrt{|z_2|}, \sqrt{|z_3|}, ...)$ with $z = (z_1, z_2, z_3, ...) \in c_0$. Consider the nonlocal Cauchy problem for fractional differential equations given by

$$ {}_0^C D_t^q x(t) = f(x(t)), \quad x(0) = \xi, \ t \in (0, t_0] \tag{3.1}$$

where ${}_0^C D_t^q$ is Caputo fractional derivative of order $0 < q < 1$, $\xi = (1, 1/2^2, 1/3^2, ...) \in c_0$, $t_0 < (\frac{\Gamma(1+q)}{2})^{\frac{1}{q}}$.

It is obvious that $f : c_0 \to c_0$ is continuous. According to N'Guerekata, 2009, there exists a constant $k^* = \frac{\Gamma(1+q)}{\Gamma(1+q) - 2t_0^q}$, such that IVP (3.1) possesses at least one continuous solution $x \in C([0, t_0], c_0)$ and $x(t) = (x_1(t), x_2(t), x_3(t), ...) \in c_0$ on $[0, t_0]$ with $\sup_{t\in[0,t_0]} \|x(t)\| \leq k^*$. According to the definition of the norm of c_0, we can conclude that

$$ {}_0^C D_t^q x_n(t) = 2\sqrt{|x_n(t)|}, \quad x_n(0) = \frac{1}{n^2}, \quad t \in (0, t_0], \quad n = 1, 2, 3, ..., \tag{3.2} $$

where x_n satisfies that $x_n \in C([0, t_0], \mathbb{R})$ with $\sup_{t \in [0,t_0]} |x_n(t)| \leq k^*$.

Let us consider equation (3.2) which can be written as the following equivalent form

$$ x_n(t) = \frac{1}{n^2} + 2{}_0D_t^{-q}\sqrt{|x_n(t)|} = \frac{1}{n^2} + \frac{2}{\Gamma(q)} \int_0^t (t-s)^{q-1}\sqrt{|x_n(s)|}ds, \quad t \in [0, t_0]. \tag{3.3} $$

Since $(t-s)^{q-1} > 1$ with $s \in [0, t)$ for $t \in (0, t_0]$, we have by (3.3)

$$ x_n(t) \geq \frac{1}{n^2} + \frac{2}{\Gamma(q)} \int_0^t \sqrt{|x_n(s)|}ds, \quad t \in [0, t_0], \quad n = 1, 2, 3, ... \tag{3.4} $$

Assume that $y_n \in C([0, t_0], \mathbb{R})$ is a solution of the following integral equation

$$ y_n(t) = \frac{1}{4n^2} + \frac{2}{\Gamma(q)} \int_0^t \sqrt{|y_n(s)|}ds, \quad t \in [0, t_0], \quad n = 1, 2, 3, ... \tag{3.5} $$

Then, we get

$$ x_n(t) \geq y_n(t), \quad t \in [0, t_0], \quad n = 1, 2, 3, ... \tag{3.6} $$

In fact, suppose (for contraction) that the conclusion (3.6) is not true. Then, because of the continuity of x and y, and that $x_n(0) > y_n(0)$, it follows that there exists a $t_1 \in (0, t_0]$ such that

$$ x_n(t_1) = y_n(t_1), \quad x_n(t) > y_n(t) \quad t \in [0, t_1), \quad n = 1, 2, 3, ... \tag{3.7} $$

Then using (3.4) and (3.7), we get

$$ \begin{aligned} y_n(t_1) &= \frac{1}{4n^2} + \frac{2}{\Gamma(q)} \int_0^{t_1} \sqrt{|y_n(s)|}ds \\ &< \frac{1}{n^2} + \frac{2}{\Gamma(q)} \int_0^{t_1} \sqrt{|x_n(s)|}ds \\ &\leq x_n(t_1), \quad n = 1, 2, 3, ..., \end{aligned} $$

which is a contraction in view of (3.7). Hence the conclusion (3.6) is valid.

Since the integral (3.5) is equivalent to the following fractional IVP

$$ y_n'(t) = \frac{2}{\Gamma(q)}\sqrt{|y_n(t)|}, \quad y_n(0) = \frac{1}{4n^2}, \quad t \in [0, t_0], \quad n = 1, 2, 3, ..., \tag{3.8} $$

and noting $y_n(t) > 0$, $t \in [0, t_0]$, we can conclude that fractional IVP (3.8) has a continuous solution

$$ y_n(t) = \left(\frac{t}{\Gamma(q)} + \frac{1}{2n} \right)^2, \quad t \in [0, t_0], \quad n = 1, 2, 3, ..., $$

which means that

$$ x_n(t) \geq y_n(t) = \left(\frac{t}{\Gamma(q)} + \frac{1}{2n} \right)^2, \quad t \in [0, t_0], \quad n = 1, 2, 3, ... \tag{3.9} $$

Therefore, for $t \in (0, t_0]$, $\lim_{n\to\infty} x_n(t) \neq 0$ by (3.9), contracting $x(t) \in c_0$. Hence fractional IVP (3.1) has no nonlocal solution in c_0.

3.2 Cauchy Problems via Measure of Noncompactness Method

3.2.1 *Introduction*

In Section 3.2, we assume that X is a Banach space with the norm $|\cdot|$. Let $J \subset \mathbb{R}$. Denote $C(J, X)$ be the Banach space of continuous functions from J into X.

Let $r > 0$ and $\mathcal{C} = C([-r, 0], X)$ be the space of continuous functions from $[-r, 0]$ into X. For any element $z \in \mathcal{C}$, define the norm $\|z\|_* = \sup_{\theta \in [-r,0]} |z(\theta)|$.

Consider the initial value problem (fractional IVP) for fractional functional differential equation given by

$$\begin{cases} {}_0^C D_t^q x(t) = f(t, x_t), & t \in (0, a), \\ x_0 = \varphi \in \mathcal{C}, \end{cases} \tag{3.10}$$

where ${}_0^C D_t^q$ is Caputo fractional derivative of order $0 < q < 1$, f: $[0, a) \times \mathcal{C} \to X$ is a given function satisfying some assumptions and define x_t by $x_t(\theta) = x(t + \theta)$, for $\theta \in [-r, 0]$.

In this section, we shall discuss the existence of the solutions for fractional IVP (3.10) under assumptions that f satisfies Carathéodory condition and the condition on measure of noncompactness. Then, we give an example to illustrate the application of our abstract results.

Definition 3.1. A function $x \in C([-r, T], X)$ is a solution for fractional IVP (3.10) on $[-r, T]$ for $T \in (0, a)$ if

(i) the function $x(t)$ is absolutely continuous on $[0, T]$;
(ii) $x_0 = \varphi$;
(iii) x satisfies the equation in (3.10).

3.2.2 *Existence*

We are now ready to prove the existence of the solutions for fractional IVP (3.10) under the following hypotheses:

(H1) for almost all $t \in [0, a)$, the function $f(t, \cdot) : \mathcal{C} \to X$ is continuous and for each $z \in \mathcal{C}$, the function $f(\cdot, z) : [0, a) \to X$ is strongly measurable;
(H2) for each $\tau > 0$, there exist a constant $q_1 \in [0, q)$ and $m_1 \in L^{\frac{1}{q_1}}([0, a), \mathbb{R}^+)$ such that $|f(t, z)| \leq m_1(t)$ for all $z \in \mathcal{C}$ with $\|z\|_* \leq \tau$ and almost all $t \in [0, a)$;
(H3) there exist a constant $q_2 \in (0, q)$ and $m_2 \in L^{\frac{1}{q_2}}([0, a), \mathbb{R}^+)$ such that $\alpha(f(t, B)) \leq m_2(t)\alpha(B)$ for almost all $t \in [0, a)$ and B a bounded set in $\mathcal{C}$.

In order to prove the existence theorem, we need the following lemma.

Lemma 3.1. *Assume that the hypotheses (H1) and (H2) hold.* $x \in C([-r, T], X)$ *is a solution for fractional IVP* (3.10) *on* $[-r, T]$ *for* $T \in (0, a)$ *if and only if* x

satisfies the following relation

$$\begin{cases} x(\theta) = \varphi(\theta), & \text{for } \theta \in [-r, 0], \\ x(t) = \varphi(0) + \dfrac{1}{\Gamma(q)} \displaystyle\int_0^t (t-s)^{q-1} f(s, x_s) ds, & \text{for } t \in [0, T]. \end{cases} \tag{3.11}$$

Proof. Since x_t is continuous in $t \in [0, a)$, according to (H1), $f(t, x_t)$ is a measurable function in $[0, a)$. Direct calculation gives that $(t-s)^{q-1} \in L^{\frac{1}{1-q_1}}[0, t]$ for $t \in (0, a)$ and $q_1 \in [0, q)$. Let

$$b_1 = \frac{q-1}{1-q_1} \in (-1, 0), \qquad M = \|m_1\|_{L^{\frac{1}{q_1}}[0,a)}.$$

By using Hölder inequality and (H2), for $t \in (0, a)$, we obtain that

$$\begin{aligned} \int_0^t |(t-s)^{q-1} f(s, x_s)| ds &\leq \left(\int_0^t (t-s)^{\frac{q-1}{1-q_1}} ds \right)^{1-q_1} \|m_1\|_{L^{\frac{1}{q_1}}[0,t]} \\ &\leq \frac{M}{(1+b_1)^{1-q_1}} a^{(1+b_1)(1-q_1)}. \end{aligned}$$

Thus, $|(t-s)^{q-1} f(s, x_s)|$ is Lebesgue integrable with respect to $s \in [0, t)$ for all $t \in (0, a)$. From Lemma 1.6 (Bochner theorem), it follows that $(t-s)^{q-1} f(s, x_s)$ is Bochner integrable with respect to $s \in [0, t)$ for all $t \in (0, a)$.

Let $L(\tau, s) = (t-\tau)^{-q} |\tau - s|^{q-1} m_1(s)$. Since $L(\tau, s)$ is a nonnegative, measurable function on $D = [0, t] \times [0, t]$, then we have

$$\int_0^t \left(\int_0^t L(\tau, s) ds \right) d\tau = \int_D L(\tau, s) ds d\tau = \int_0^t \left(\int_0^t L(\tau, s) d\tau \right) ds$$

and

$$\begin{aligned} \int_D L(\tau, s) ds d\tau &= \int_0^t \left(\int_0^t L(\tau, s) ds \right) d\tau \\ &= \int_0^t (t-\tau)^{-q} \left(\int_0^t |\tau - s|^{q-1} m_1(s) ds \right) d\tau \\ &= \int_0^t (t-\tau)^{-q} \left(\int_0^\tau (\tau - s)^{q-1} m_1(s) ds \right) d\tau \\ &\quad + \int_0^t (t-\tau)^{-q} \left(\int_\tau^t (s-\tau)^{q-1} m_1(s) ds \right) d\tau \\ &\leq \frac{2M}{(1+b_1)^{1-q_1}} a^{(1+b_1)(1-q_1)} \int_0^t (t-\tau)^{-q} d\tau \\ &\leq \frac{2M}{(1-q)(1+b_1)^{1-q_1}} a^{(1+b_1)(1-q_1)+1-q}. \end{aligned}$$

Therefore, $L_1(\tau, s) = (t-\tau)^{-q} (\tau - s)^{q-1} f(s, x_s)$ is a Bochner integrable function on $D = [0, t] \times [0, t]$, then we have

$$\int_0^t d\tau \int_0^\tau L_1(\tau, s) ds = \int_0^t ds \int_s^t L_1(\tau, s) d\tau.$$

We now prove that

$$ {}_0D_t^q\Big({}_0D_t^{-q}f(t,x_t)\Big) = f(t,x_t), \quad \text{for } t\in(0,T], $$

where ${}_0D_t^q$ is Riemann-Liouville fractional derivative.

Indeed, we have

$$ \begin{aligned} {}_0D_t^q\Big({}_0D_t^{-q}f(t,x_t)\Big) &= \frac{1}{\Gamma(1-q)\Gamma(q)}\frac{d}{dt}\int_0^t (t-\tau)^{-q}\Big(\int_0^\tau (\tau-s)^{q-1}f(s,x_s)ds\Big)d\tau \\ &= \frac{1}{\Gamma(1-q)\Gamma(q)}\frac{d}{dt}\int_0^t d\tau\int_0^\tau L_1(\tau,s)ds \\ &= \frac{1}{\Gamma(1-q)\Gamma(q)}\frac{d}{dt}\int_0^t ds\int_s^t L_1(\tau,s)d\tau \\ &= \frac{1}{\Gamma(1-q)\Gamma(q)}\frac{d}{dt}\int_0^t f(s,x_s)ds\int_s^t (t-\tau)^{-q}(\tau-s)^{q-1}d\tau \\ &= \frac{d}{dt}\int_0^t f(s,x_s)ds \\ &= f(t,x_t), \quad \text{for } \ t\in(0,T]. \end{aligned} $$

If x satisfies the relation (3.11), then we can get that $x(t)$ is absolutely continuous on $[0,T]$. In fact, for any disjoint family of open intervals $\{(c_i,d_i)\}_{1\le i\le n}$ in $[0,T]$ with $\sum_{i=1}^n(d_i-c_i)\to 0$, we have

$$ \begin{aligned} &\sum_{i=1}^n |x(d_i)-x(c_i)| \\ &= \sum_{i=1}^n \frac{1}{\Gamma(q)}\left|\int_0^{d_i}(d_i-s)^{q-1}f(s,x_s)ds - \int_0^{c_i}(c_i-s)^{q-1}f(s,x_s)ds\right| \\ &\le \sum_{i=1}^n \frac{1}{\Gamma(q)}\left|\int_{c_i}^{d_i}(d_i-s)^{q-1}f(s,x_s)ds\right| \\ &\quad + \sum_{i=1}^n \frac{1}{\Gamma(q)}\left|\int_0^{c_i}(d_i-s)^{q-1}f(s,x_s)ds - \int_0^{c_i}(c_i-s)^{q-1}f(s,x_s)ds\right| \\ &\le \sum_{i=1}^n \frac{1}{\Gamma(q)}\int_{c_i}^{d_i}(d_i-s)^{q-1}m_1(s)ds \\ &\quad + \sum_{i=1}^n \frac{1}{\Gamma(q)}\int_0^{c_i}\left((c_i-s)^{q-1}-(d_i-s)^{q-1}\right)m_1(s)ds \\ &\le \sum_{i=1}^n \frac{1}{\Gamma(q)}\Big(\int_{c_i}^{d_i}(d_i-s)^{\frac{q-1}{1-q_1}}ds\Big)^{1-q_1}\|m_1\|_{L^{\frac{1}{q_1}}[0,T]} \\ &\quad + \sum_{i=1}^n \frac{1}{\Gamma(q)}\Big(\int_0^{c_i}(c_i-s)^{\frac{q-1}{1-q_1}}-(d_i-s)^{\frac{q-1}{1-q_1}}ds\Big)^{1-q_1}\|m_1\|_{L^{\frac{1}{q_1}}[0,T]} \\ &= \sum_{i=1}^n \frac{(d_i-c_i)^{(1+b_1)(1-q_1)}}{\Gamma(q)(1+b_1)^{1-q_1}}\|m_1\|_{L^{\frac{1}{q_1}}[0,T]} \end{aligned} $$

$$+\sum_{i=1}^{n}\frac{(c_i^{1+b_1}-d_i^{1+b_1}+(d_i-c_i)^{1+b_1})^{1-q_1}}{\Gamma(q)(1+b_1)^{1-q_1}}\|m_1\|_{L^{\frac{1}{q_1}}[0,T]}$$
$$\leq 2\sum_{i=1}^{n}\frac{(d_i-c_i)^{(1+b_1)(1-q_1)}}{\Gamma(q)(1+b_1)^{1-q_1}}\|m_1\|_{L^{\frac{1}{q_1}}[0,T]}\to 0.$$

Therefore, $x(t)$ is absolutely continuous on $[0,T]$, which implies that $x(t)$ is differentiable a.e. on $[0,T]$. According to the argument above and Definition 1.3, for $t\in(0,T]$, we have

$$\begin{aligned}
{}_0^CD_t^q x(t) &= {}_0^CD_t^q\left(\varphi(0)+\frac{1}{\Gamma(q)}\int_0^t (t-s)^{q-1}f(s,x_s)ds\right)\\
&= {}_0^CD_t^q\left(\frac{1}{\Gamma(q)}\int_0^t (t-s)^{q-1}f(s,x_s)ds\right)\\
&= {}_0^CD_t^q\left({}_0D_t^{-q}f(t,x_t)\right)\\
&= {}_0D_t^q\left({}_0D_t^{-q}f(t,x_t)\right)-[{}_0D_t^{-q}f(t,x_t)]_{t=0}\frac{t^{-q}}{\Gamma(1-q)}\\
&= f(t,x_t)-[{}_0D_t^{-q}f(t,x_t)]_{t=0}\frac{t^{-q}}{\Gamma(1-q)}\,.
\end{aligned}$$

Since $(t-s)^{q-1}f(s,x_s)$ is Lebesgue integrable with respect to $s\in[0,t)$ for all $t\in(0,T]$, we know that $[{}_0D_t^{-q}f(t,x_t)]_{t=0}=0$, which means that ${}_0^CD_t^q x(t)=f(t,x_t)$, for $t\in(0,T]$. Hence, $x\in C([-r,T],X)$ is a solution of fractional IVP (3.10). On the other hand, it is obvious that if $x\in C([-r,T],X)$ is a solution of fractional IVP (3.10), then x satisfies the relation (3.11), and this completes the proof. □

Theorem 3.1. *Assume that hypotheses (H1)-(H3) hold. Then, for every $\varphi\in\mathcal{C}$, there exists a solution $x\in C([-r,T],X)$ for fractional IVP* (3.10) *with some $T\in(0,a)$.*

Proof. Let $k>0$ be any number and we can choose $T\in(0,a)$ such that

$$\frac{T^{(1+b_1)(1-q_1)}}{\Gamma(q)(1+b_1)^{1-q_1}}\|m_1\|_{L^{\frac{1}{q_1}}[0,T]}\leq k \tag{3.12}$$

and

$$\frac{T^{(1+b_2)(1-q_2)}}{\Gamma(q)(1+b_2)^{1-q_2}}\|m_2\|_{L^{\frac{1}{q_2}}[0,T]}<1, \tag{3.13}$$

where $b_i=\frac{q-1}{1-q_i}\in(-1,0)$, $i=1,2$.

Consider the set B_k defined as follows

$$B_k=\left\{x\in C([-r,T],X)\,\middle|\, x_0=\varphi,\ \sup_{s\in[0,T]}|x(s)-\varphi(0)|\leq k\right\}.$$

Define the operator F on B_k as follows

$$\begin{cases} Fx(\theta) = \varphi(\theta), & \text{for } \theta \in [-r, 0], \\ Fx(t) = \varphi(0) + \dfrac{1}{\Gamma(q)} \displaystyle\int_0^t (t-s)^{q-1} f(t, x_s) ds, & \text{for } t \in [0, T], \end{cases}$$

where $x \in B_k$. We prove that the operator equation $x = Fx$ has a solution $x \in B_k$, which means that x is a solution of fractional IVP (3.10).

Firstly, we observe that for every $y \in B_k$, $(Fy)(t)$ is continuous on $t \in [-r, T]$ and for $t \in [0, T]$, by (3.12) and Hölder inequality, we have

$$\begin{aligned} |(Fy)(t) - \varphi(0)| &\le \frac{1}{\Gamma(q)} \int_0^t |(t-s)^{q-1} f(s, y_s)| ds \\ &\le \frac{1}{\Gamma(q)} \left(\int_0^t (t-s)^{\frac{q-1}{1-q_1}} ds \right)^{1-q_1} \|m_1\|_{L^{\frac{1}{q_1}}[0,T]} \\ &\le \frac{T^{(1+b_1)(1-q_1)}}{\Gamma(q)(1+b_1)^{1-q_1}} \|m_1\|_{L^{\frac{1}{q_1}}[0,T]} \\ &\le k, \end{aligned} \tag{3.14}$$

where $b_1 = \frac{q-1}{1-q_1} \in (-1, 0)$. Thus, $\sup_{t\in[0,T]} |(Fy)(t) - \varphi(0)| \le k$, which implies that $F : B_k \to B_k$.

Further, we prove that F is a continuous operator on B_k. Let $\{y^n\} \subseteq B_k$ with $y^n \to y$ on B_k. Then by (H1) and the fact that $y_t^n \to y_t$, as $n \to \infty$, $t \in [0, T]$, we have

$$f(s, y_s^n) \to f(s, y_s), \quad \text{a.e. } s \in [0, T], \quad \text{as } n \to \infty.$$

Noting that $(t-s)^{q-1}|f(s, y_s^n) - f(s, y_s)| \le (t-s)^{q-1} 2m_1(s)$, by Lebesgue dominated convergence theorem, as $n \to \infty$, we have

$$|(Fy^n)(t) - (Fy)(t)| \le \frac{1}{\Gamma(q)} \int_0^t (t-s)^{q-1} |f(s, y_s^n) - f(s, y_s)| ds \to 0.$$

Therefore $Fy^n \to Fy$ as $n \to \infty$ which implies that F is continuous.

For each $n \ge 1$, we define a sequence $\{x^n : n \ge 1\}$ in the following way

$$x^n(t) = \begin{cases} \varphi^0(t), & \text{for } t \in [-r, \frac{T}{n}], \\ \varphi(0) + \dfrac{1}{\Gamma(q)} \displaystyle\int_0^{t-\frac{T}{n}} (t-s)^{q-1} f(t, x_s^n) ds, & \text{for } t \in [\frac{T}{n}, T], \end{cases}$$

where $\varphi^0 \in C([-r, a), X)$ denotes the function defined by

$$\varphi^0(t) = \begin{cases} \varphi(t), & \text{for } t \in [-r, 0], \\ \varphi(0), & \text{for } t \in [0, a). \end{cases}$$

Using the similar method as we did in (3.14), we get that $x^n \in B_k$ for all $n \ge 1$.

Let $A = \{x^n : n \ge 1\}$. It follows that the set A is uniformly bounded. Further, we show that the set A is equicontinuous on $[-r, T]$.

If $-r \leq t_1 < t_2 \leq \frac{T}{n}$, then for each $x^n \in A$, we have $\lim_{t_1 \to t_2} |x^n(t_2) - x^n(t_1)| = \lim_{t_1 \to t_2} |\varphi^0(t_2) - \varphi^0(t_1)| = 0$ independently of $x^n \in A$. Next, if $-r \leq t_1 \leq \frac{T}{n} < t_2 \leq T$, then for each $x^n \in A$, by using Hölder inequality, we have

$$\begin{aligned}
&|x^n(t_2) - x^n(t_1)| \\
&\leq |\varphi(0) - \varphi^0(t_1)| + \left| \frac{1}{\Gamma(q)} \int_0^{t_2 - \frac{T}{n}} (t_2 - s)^{q-1} f(s, x_s^n) ds \right| \\
&\leq |\varphi(0) - \varphi^0(t_1)| + \frac{1}{\Gamma(q)} \int_0^{t_2 - \frac{T}{n}} (t_2 - s)^{q-1} m_1(s) ds \\
&\leq |\varphi(0) - \varphi^0(t_1)| + \frac{1}{\Gamma(q)} \left(\int_0^{t_2 - \frac{T}{n}} (t_2 - s)^{\frac{q-1}{1-q_1}} ds \right)^{1-q_1} \|m_1\|_{L^{\frac{1}{q_1}}[0,T]} \\
&= |\varphi(0) - \varphi^0(t_1)| + \frac{\left(t_2^{1+b_1} - (\frac{T}{n})^{1+b_1} \right)^{1-q_1}}{\Gamma(q)(1+b_1)^{1-q_1}} \|m_1\|_{L^{\frac{1}{q_1}}[0,T]}.
\end{aligned}$$

According to the definition of φ^0, and using the last inequality, we obtain that $|x^n(t_2) - x^n(t_1)| \to 0$ independently of $x^n \in A$, as $t_1 \to t_2$.

Finally, if $\frac{T}{n} \leq t_1 < t_2 \leq T$, then for each $x^n \in A$, by using Hölder inequality, we have

$$\begin{aligned}
&|x^n(t_2) - x^n(t_1)| \\
&= \left| \frac{1}{\Gamma(q)} \int_0^{t_2 - \frac{T}{n}} (t_2 - s)^{q-1} f(s, x_s^n) ds - \frac{1}{\Gamma(q)} \int_0^{t_1 - \frac{T}{n}} (t_1 - s)^{q-1} f(s, x_s^n) ds \right| \\
&\leq \left| \frac{1}{\Gamma(q)} \int_{t_1 - \frac{T}{n}}^{t_2 - \frac{T}{n}} (t_2 - s)^{q-1} f(s, x_s^n) ds \right| + \left| \frac{1}{\Gamma(q)} \int_0^{t_1 - \frac{T}{n}} (t_2 - s)^{q-1} f(s, x_s^n) ds \right. \\
&\quad \left. - \frac{1}{\Gamma(q)} \int_0^{t_1 - \frac{T}{n}} (t_1 - s)^{q-1} f(s, x_s^n) ds \right| \\
&\leq \frac{1}{\Gamma(q)} \int_{t_1 - \frac{T}{n}}^{t_2 - \frac{T}{n}} (t_2 - s)^{q-1} m_1(s) ds \\
&\quad + \frac{1}{\Gamma(q)} \int_0^{t_1 - \frac{T}{n}} \left((t_1 - s)^{q-1} - (t_2 - s)^{q-1} \right) m_1(s) ds \\
&\leq \frac{1}{\Gamma(q)} \left(\int_{t_1 - \frac{T}{n}}^{t_2 - \frac{T}{n}} (t_2 - s)^{\frac{q-1}{1-q_1}} ds \right)^{1-q_1} \|m_1\|_{L^{\frac{1}{q_1}}[0,T]} \\
&\quad + \frac{1}{\Gamma(q)} \left(\int_0^{t_1 - \frac{T}{n}} (t_1 - s)^{\frac{q-1}{1-q_1}} - (t_2 - s)^{\frac{q-1}{1-q_1}} ds \right)^{1-q_1} \|m_1\|_{L^{\frac{1}{q_1}}[0,T]} \\
&= \frac{\left((t_2 - t_1 + \frac{T}{n})^{1+b_1} - (\frac{T}{n})^{1+b_1} \right)^{1-q_1}}{\Gamma(q)(1+b_1)^{1-q_1}} \|m_1\|_{L^{\frac{1}{q_1}}[0,T]} \\
&\quad + \frac{\left(t_1^{1+b_1} - (\frac{T}{n})^{1+b_1} - t_2^{1+b_1} + (t_2 - t_1 + \frac{T}{n})^{1+b_1} \right)^{1-q_1}}{\Gamma(q)(1+b_1)^{1-q_1}} \|m_1\|_{L^{\frac{1}{q_1}}[0,T]}
\end{aligned}$$

$$\leq 2\frac{\left((t_2 - t_1 + \frac{T}{n})^{1+b_1} - (\frac{T}{n})^{1+b_1}\right)^{1-q_1}}{\Gamma(q)(1+b_1)^{1-q_1}} \|m_1\|_{L^{\frac{1}{q_1}}[0,T]}.$$

It is easy to see that the last inequality tends to zero independently of $x^n \in A$, as $t_1 \to t_2$, which means that the set A is equicontinuous.

Set $A(t) = \{x^n(t) : n \geq 1\}$ and $A_t = \{x_t^n : n \geq 1\}$ for any $t \in [0, T]$. By the properties (iv) and (vi) of the measure of noncompactness, for any fixed $t \in (0, T]$, we have

$$\alpha(A(t)) \leq \alpha\left(\left\{\frac{1}{\Gamma(q)}\int_0^t (t-s)^{q-1} f(s, x_s^n)ds : n \geq 1\right\}\right) + \alpha\left(\left\{\frac{1}{\Gamma(q)}\int_{t-\frac{T}{n}}^t (t-s)^{q-1} f(s, x_s^n)ds : n \geq 1\right\}\right).$$

For $\forall\ \epsilon > 0$, we can find δ sufficiently small such that

$$\frac{\delta^{(1+b_1)(1-q_1)}}{\Gamma(q)(1+b_1)^{1-q_1}}\|m_1\|_{L^{\frac{1}{q_1}}[0,T]} < \frac{\epsilon}{2}.$$

Therefore, for each $t \in (0, T]$, we can choose $N_\delta \geq 1$ such that $\frac{T}{n} \leq \delta$ for $n \geq N_\delta$. Then we obtain that

$$\begin{aligned} &\alpha\left(\left\{\frac{1}{\Gamma(q)}\int_{t-\frac{T}{n}}^t (t-s)^{q-1} f(s, x_s^n)ds : n \geq N_\delta\right\}\right) \\ &\leq \frac{2}{\Gamma(q)} \sup_{n \geq N_\delta} \int_{t-\frac{T}{n}}^t (t-s)^{q-1} m_1(s)ds \\ &< \epsilon, \end{aligned}$$

for each $t \in (0, T]$. Hence, by the properties (iii) and (v) of the measure of noncompactness, it follows that

$$\alpha\left(\left\{\frac{1}{\Gamma(q)}\int_{t-\frac{T}{n}}^t (t-s)^{q-1} f(s, x_s^n)ds : n \geq 1\right\}\right) < \epsilon.$$

Then, we obtain that

$$\alpha(A(t)) \leq \alpha\left(\left\{\frac{1}{\Gamma(q)}\int_0^t (t-s)^{q-1} f(s, x_s^n)ds : n \geq 1\right\}\right) + \epsilon,$$

for $t \in (0, T]$. By Proposition 1.18 and (H3), we have that

$$\begin{aligned} \alpha(A(t)) &\leq \frac{2}{\Gamma(q)}\int_0^t (t-s)^{q-1}\alpha(f(s, A_s))ds + \epsilon \\ &\leq \frac{2}{\Gamma(q)}\int_0^t (t-s)^{q-1} m_2(s)\alpha(A_s)ds + \epsilon, \end{aligned}$$

where $t \in (0, T]$. Since $x^n(\theta) = \varphi(\theta)$, $\theta \in [-r, 0]$, we have $\alpha(\{x^n(\theta) : n \geq 1\}) = 0$ for $\theta \in [-r, 0]$. Moreover, by Proposition 1.17, for $s \in [0, t]$ with $t \in (0, T]$, we deduce that

$$\alpha(A_s) = \max_{\theta \in [-r,0]} \alpha(\{x_s^n(\theta) : n \geq 1\}) \leq \sup_{s \in [0,t]} \alpha(\{x^n(s) : n \geq 1\}) = \sup_{s \in [0,t]} \alpha(A(s)).$$

Since ϵ is arbitrary, we have that

$$\alpha(A(t)) \leq \frac{2T^{(1+b_2)(1-q_2)}}{\Gamma(q)(1+b_2)^{1-q_2}}\|m_2\|_{L^{\frac{1}{q_2}}[0,T]} \sup_{s\in[0,t]} \alpha(A(s)),$$

where $t \in (0,T]$ and $b_2 = \frac{q-1}{1-q_2} \in (-1,0)$.

Since (3.13) and $x_0^n = \varphi$, we must have that $\alpha(A(t)) = 0$ for every $t \in [-r,T]$. Then, by Proposition 1.17, we have that $\alpha(A) = \sup_{t\in[-r,T]} \alpha(A(t)) = 0$. Therefore, A is a relatively compact subset of B_k. Then, there exists a subsequence if necessary, we may assume that the sequence $\{x^n\}_{n\geq 1}$ converges uniformly on $[-r,T]$ to a continuous function $x \in B_k$ with $x(\theta) = \varphi(\theta)$, $\theta \in [-r,0]$.

Moreover, for $t \in [0, \frac{T}{n}]$, we have

$$|(Fx^n)(t) - x^n(t)| \leq \frac{1}{\Gamma(q)}\int_0^{\frac{T}{n}} (t-s)^{q-1}|f(t,x_s^n)|ds \leq \frac{1}{\Gamma(q)}\int_0^{\frac{T}{n}} (t-s)^{q-1}m_1(s)ds$$

and for $t \in [\frac{T}{n}, T]$, we have

$$\begin{aligned}|(Fx^n)(t) - x^n(t)| &= \frac{1}{\Gamma(q)}\left|\int_0^t (t-s)^{q-1}f(t,x_s^n)ds - \int_0^{t-\frac{T}{n}} (t-s)^{q-1}f(t,x_s^n)ds\right| \\ &= \frac{1}{\Gamma(q)}\left|\int_{t-\frac{T}{n}}^t (t-s)^{q-1}f(t,x_s^n)ds\right| \\ &\leq \frac{1}{\Gamma(q)}\int_{t-\frac{T}{n}}^t (t-s)^{q-1}m_1(s)ds.\end{aligned}$$

Therefore, it follows that

$$\sup_{t\in[0,T]} |(Fx^n)(t) - x^n(t)| \to 0, \quad \text{as } n \to \infty. \tag{3.15}$$

Since

$$\begin{aligned}\sup_{t\in[0,T]} |(Fx)(t) - x(t)| \leq{}& \sup_{t\in[0,T]} |(Fx)(t) - (Fx^n)(t)| \\ &+ \sup_{t\in[0,T]} |(Fx^n)(t) - x^n(t)| + \sup_{t\in[0,T]} |x^n(t) - x(t)|,\end{aligned}$$

then, by (3.15) and the fact that F is a continuous operator, we obtain that $\sup_{t\in[0,T]} |(Fx)(t) - x(t)| = 0$. It follows that $x(t) = (Fx)(t)$ for every $t \in [0,T]$. Hence

$$x(t) = \begin{cases} \varphi(t), & \text{for } t \in [-r,0], \\ \varphi(0) + \dfrac{1}{\Gamma(q)}\displaystyle\int_0^t (t-s)^{q-1}f(t,x_s)ds, & \text{for } t \in [0,T] \end{cases}$$

solve fractional IVP (3.10), and this completes the proof. □

Corollary 3.1. *Assume that hypotheses (H1)-(H3) hold. Then, for every $\varphi \in \mathcal{C}$, there exist $T \in (0,a)$ and a sequence of continuous function $x^n : [-r,T] \to X$, such that*

(i) *$x^n(t)$ are absolutely continuous on $[0, T]$;*
(ii) *$x_0^n = \varphi$, for every $n \geq 1$, and*
(iii) *extracting a subsequence which is labeled in the same way such that $x^n(t) \to x(t)$ uniformly on $[-r, T]$ and $x : [-r, T] \to X$ is a solution for fractional IVP* (3.10).

We now give an example to illustrate the application of our abstract results.

Example 3.2. Consider the infinite system of fractional functional differential equations

$$\begin{cases} {}_0^C D_t^{\frac{1}{2}} x_n(t) = \dfrac{1}{nt^{1/3}} x_n^2(t-r), & \text{for } t \in (0, a), \\ x_n(\theta) = \varphi(\theta) = \dfrac{\theta}{n}, & \text{for } \theta \in [-r, 0],\ n = 1, 2, 3, \ldots . \end{cases} \tag{3.16}$$

Let $E = c_0 = \{x = (x_1, x_2, x_3, \ldots) : x_n \to 0\}$ with norm $|x| = \sup_{n \geq 1} |x_n|$. Then the infinite system (3.16) can be regarded as a fractional IVP of form (3.10) in E. In this situation, $q = \frac{1}{2}$, $x = (x_1, \ldots, x_n, \ldots)$, $x_t = x(t-r) = (x_1(t-r), \ldots, x_n(t-r), \ldots)$, $\varphi(\theta) = (\theta, \frac{\theta}{2}, \ldots, \frac{\theta}{n}, \ldots)$ for $\theta \in [-r, 0]$ and $f = (f_1, \ldots, f_n, \ldots)$, in which

$$f_n(t, x_t) = \frac{1}{nt^{1/3}} x_n^2(t-r). \tag{3.17}$$

It is obvious that conditions (H1) and (H2) are satisfied. Now, we check the condition (H3) and the argument is similar to Section 2.4. Let $t \in (0, a)$, $R > 0$ be given and $\{w^{(m)}\}$ be any sequence in $f(t, B)$, where $w^{(m)} = (w_1^{(m)}, \ldots, w_n^{(m)}, \ldots)$ and $B = \{z \in \mathcal{C} : \|z\|_* \leq R\}$ is a bounded set in $\mathcal{C}$. By (3.17), we have

$$0 \leq w_n^{(m)} \leq \frac{R^2}{nt^{1/3}}, \quad n, m = 1, 2, 3, \ldots . \tag{3.18}$$

So, $\{w_n^{(m)}\}$ is bounded and, by the diagonal method, we can choose a subsequence $\{m_i\} \subset \{m\}$ such that

$$w_n^{(m_i)} \to w_n, \quad \text{as } i \to \infty, \quad n = 1, 2, 3, \ldots, \tag{3.19}$$

which implies by virtue of (3.18) that

$$0 \leq w_n \leq \frac{R^2}{nt^{1/3}}, \quad n = 1, 2, 3, \ldots . \tag{3.20}$$

Hence $w = (w_1, \ldots, w_n, \ldots) \in c_0$. It is easy to see from (3.18)-(3.20) that

$$|w^{(m_i)} - w| = \sup_n |w_n^{(m_i)} - w_n| \to 0, \quad \text{as } i \to \infty.$$

Thus, we have proved that $f(t, B)$ is relatively compact in c_0 for $t \in (0, a)$, which means that $f(t, B) = 0$ for almost all $t \in [0, a)$ and B a bounded set in $\mathcal{C}$. Hence, the condition (H3) is satisfied. Finally, from Theorem 3.1, we can conclude that the infinite system (3.16) has a continuous solution.

3.3 Cauchy Problems via Topological Degree Method

3.3.1 *Introduction*

It is well known that the topological method is proved to be a powerful tool in the study of various problems which appear in nonlinear analysis. Particularly, a priori estimate method has been often used together with the Brouwer degree, the Leray-Schauder degree or the coincidence degree in order to prove the existence of solutions for some boundary value problems and bifurcation problems for nonlinear differential equations or nonlinear partial differential equations. See, for example, Fečkan, 2008; Mawhin, 1979.

In Section 3.3, we consider the following nonlocal problem via a coincidence degree for condensing mapping in a Banach space X

$$\begin{cases} {}_0^C D_t^q u(t) = f(t, u(t)), \quad t \in J := [0, T], \\ u(0) + g(u) = u_0, \end{cases} \tag{3.21}$$

where ${}_0^C D_t^q$ is Caputo fractional derivative of order $q \in (0, 1)$, u_0 is an element of X, $f : J \times X \to X$ is continuous. The nonlocal term $g : C(J, X) \to X$ is a given function, here $C(J, X)$ is the Banach space of all continuous functions from J into X with the norm $\|u\| := \sup_{t \in J} |u(t)|$ for $u \in C(J, X)$.

3.3.2 *Qualitative Analysis*

This subsection deals with existence of solutions for the nonlocal problem (3.21).

Definition 3.2. A function $u \in C^1(J, X)$ is said to be a solution of the nonlocal problem (3.21) if u satisfies the equation ${}_0^C D_t^q u(t) = f(t, u(t))$ a.e. on J, and the condition $u(0) + g(u) = u_0$.

Lemma 3.2. *A function $u \in C(J, X)$ is a solution of the fractional integral equation*

$$u(t) = u_0 - g(u) + \frac{1}{\Gamma(q)} \int_0^t (t - s)^{q-1} f(s, u(s)) ds, \tag{3.22}$$

if and only if u is a solution of the nonlocal problem (3.21).

We make some following assumptions:

(H1) for arbitrary $u, v \in C(J, X)$, there exists a constant $K_g \in [0, 1)$ such that

$$|g(u) - g(v)| \leq K_g \|u - v\|;$$

(H2) for arbitrary $u \in C(J, X)$, there exist $C_g, M_g > 0$, $q_1 \in [0, 1)$ such that

$$|g(u)| \leq C_g \|u\|^{q_1} + M_g;$$

(H3) for arbitrary $(t, u) \in J \times X$, there exist $C_f, M_f > 0$, $q_2 \in [0, 1)$ such that

$$|f(t, u)| \leq C_f |u|^{q_2} + M_f;$$

(H4) for any $r > 0$, there exists a constant $\beta_r > 0$ such that

$$\alpha(f(s, \mathcal{M})) \le \beta_r \alpha(\mathcal{M}),$$

for all $t \in J$, $\mathcal{M} \subset \mathfrak{B}_r := \{\|u\| \le r : u \in C(J, X)\}$ and

$$\frac{2T^q \beta_r}{\Gamma(q+1)} < 1.$$

Under the assumptions (H1)-(H4), we show that fractional integral equation (3.22) has at least one solution $u \in C(J, X)$.

Define operators

$$F : C(J,X) \to C(J,X), \quad (Fu)(t) = u_0 - g(u), \qquad t \in J,$$

$$G : C(J,X) \to C(J,X), \quad (Gu)(t) = \frac{1}{\Gamma(q)} \int_0^t (t-s)^{q-1} f(s, u(s)) ds, \quad t \in J,$$

$$\mathbb{T} : C(J,X) \to C(J,X), \quad \mathbb{T}u = Fu + Gu.$$

It is obvious that $\mathbb{T}$ is well defined. Then, fractional integral equation (3.22) can be written as the following operator equation

$$u = \mathbb{T}u = Fu + Gu.$$

Thus, the existence of a solution for the nonlocal problem (3.21) is equivalent to the existence of a fixed point for operator $\mathbb{T}$.

Lemma 3.3. *The operator $F : C(J, X) \to C(J, X)$ is Lipschitz with constant K_g. Consequently F is α-Lipschitz with the same constant K_g. Moreover, F satisfies the following growth condition:*

$$\|Fu\| \le |u_0| + C_g \|u\|^{q_1} + M_g, \tag{3.23}$$

for every $u \in C(J, X)$.

Proof. Using (H1), we have $\|Fu - Fv\| \le |g(u) - g(v)| \le K_g \|u - v\|$, for every $u, v \in C(J, X)$. By Proposition 1.23, F is α-Lipschitz with constant K_g. Relation (3.23) is a simple consequence of (H2). □

Lemma 3.4. *The operator $G : C(J, X) \to C(J, X)$ is continuous. Moreover, G satisfies the following growth condition:*

$$\|Gu\| \le \frac{T^q (C_f \|u\|^{q_2} + M_f)}{\Gamma(q+1)}, \tag{3.24}$$

for every $u \in C(J, X)$.

Proof. For that, let $\{u_n\}$ be a sequence of a bounded set $\mathfrak{B}_K \subseteq C(J, X)$ such that $u_n \to u$ in $\mathfrak{B}_K (K > 0)$. We have to show that $\|Gu_n - Gu\| \to 0$.

It is easy to see that $f(s, u_n(s)) \to f(s, u(s))$ as $n \to \infty$ due to the continuity of f. On the one hand, using (H3), we get for each $t \in J$, $(t-s)^{q-1} |f(s, u_n(s)) - f(s, u(s))| \le (t-s)^{q-1} 2(C_f K^{q_2} + M_f)$. On the other hand, using the fact that the function $s \to (t-s)^{q-1}2(C_f K^{q_2} + M_f)$ is integrable for $s \in [0,t]$, $t \in J$, Lebesgue dominated convergence theorem yields $\int_0^t (t-s)^{q-1}|f(s,u_n(s)) - f(s,u(s))|ds \to 0$ as $n \to \infty$. Then, for all $t \in J$,

$$|(Gu_n)(t) - (Gu)(t)| \le \frac{1}{\Gamma(q)} \int_0^t (t-s)^{q-1}|f(s,u_n(s)) - f(s,u(s))|ds \to 0, \quad n \to \infty.$$

Therefore, $Gu_n \to Gu$ as $n \to \infty$ which implies that G is continuous. Relation (3.24) is a simple consequence of (H3). □

Lemma 3.5. *The operator $G : C(J,X) \to C(J,X)$ is compact. Consequently G is α-Lipschitz with zero constant.*

Proof. In order to prove the compactness of G, we consider a bounded set $\mathcal{M} \subseteq C(J,X)$ and the key step is to show that $G(\mathcal{M})$ is relatively compact in $C(J,X)$.

Let $\{u_n\}$ be a sequence on $\mathcal{M} \subset \mathfrak{B}_K$, for every $u_n \in \mathcal{M}$. By relation (3.24), we have

$$\|Gu_n\| \le \frac{T^q(C_f K^{q_2} + M_f)}{\Gamma(q+1)} =: r,$$

for every $u_n \in \mathcal{M}$, so $G(\mathcal{M})$ is bounded in $\mathfrak{B}_r$.

Now we prove that $\{Gu_n\}$ is equicontinuous. For $0 \le t_1 < t_2 \le T$, we get

$$\begin{aligned}
&|(Gu_n)(t_1) - (Gu_n)(t_2)| \\
&\le \frac{1}{\Gamma(q)} \int_0^{t_1} \left((t_1-s)^{q-1} - (t_2-s)^{q-1}\right) |f(s,u_n(s))|ds \\
&\quad + \frac{1}{\Gamma(q)} \int_{t_1}^{t_2} (t_2-s)^{q-1}|f(s,u_n(s))|ds \\
&\le \frac{1}{\Gamma(q)} \int_0^{t_1} \left((t_1-s)^{q-1} - (t_2-s)^{q-1}\right) (C_f|u_n(s)|^{q_2} + M_f)ds \\
&\quad + \frac{1}{\Gamma(q)} \int_{t_1}^{t_2} (t_2-s)^{q-1} \left(C_f|u_n(s)|^{q_2} + M_f\right) ds \\
&\le \frac{(C_f K^{q_2} + M_f)}{\Gamma(q)} \left(\frac{t_1^q}{q} - \frac{t_2^q}{q} + \frac{(t_2-t_1)^q}{q} + \frac{(t_2-t_1)^q}{q} \right) \\
&\le \frac{2(C_f K^{q_2} + M_f)(t_2-t_1)^q}{\Gamma(q+1)}.
\end{aligned}$$

As $t_2 \to t_1$, the right-hand side of the above inequality tends to zero. Therefore $\{Gu_n\}$ is equicontinuous.

Consider a bounded set

$$\mathcal{M}(t) := \left\{ v_n(t) : v_n(t) = \frac{1}{\Gamma(q)} \int_0^t (t-s)^{q-1} f(s,v_n(s))ds \right\} \subset \mathfrak{B}_r.$$

Applying Proposition 1.17, we know that the function $t \to \alpha(\mathcal{M}(t))$ is continuous on J. Moreover,

$$(t-s)^{q-1}|f(s,v_n(s))| \le (t-s)^{q-1}(C_f r^{q_2} + M_f) \in L^1(J,\mathbb{R}_+), \ \textit{for} \ s \in [0,t], \ t \in J.$$

Using (H4) and Proposition 1.18, we have

$$\begin{aligned}\alpha(\mathcal{M}(t)) &\le \alpha\left(\left\{\frac{1}{\Gamma(q)}\int_0^t (t-s)^{q-1} f(s,\mathcal{M}(s))ds\right\}\right)\\ &\le \frac{2}{\Gamma(q)}\int_0^t (t-s)^{q-1}\alpha\left(f(s,\mathcal{M}(s))\right)ds\\ &\le \frac{2\beta_r}{\Gamma(q)}\int_0^t (t-s)^{q-1}\alpha(\mathcal{M}(s))ds,\end{aligned}$$

which implies that

$$\begin{aligned}\alpha(\mathcal{M}) &\le \left(\frac{2\beta_r}{\Gamma(q)}\int_0^t (t-s)^{q-1}ds\right)\alpha(\mathcal{M})\\ &\le \frac{2T^q\beta_r}{\Gamma(q+1)}\alpha(\mathcal{M})\\ &< \alpha(\mathcal{M}),\end{aligned}$$

due to the condition

$$\frac{2T^q\beta_r}{\Gamma(q+1)} < 1.$$

Then we can deduce that $\alpha(\mathcal{M}) = 0$. Therefore, $G(\mathcal{M})$ is a relatively compact subset of $C(J,X)$, and so, there exists a subsequence v_n which converge uniformly on J to some $v_* \in C(J,X)$ together with Arzela-Ascoli theorem. By Proposition 1.22, G is α-Lipschitz with zero constant. □

Theorem 3.2. *Assume that (H1)-(H4) hold, then the nonlocal problem* (3.21) *has at least one solution* $u \in C(J,X)$ *and the set of the solutions of system* (3.21) *is bounded in* $C(J,X)$.

Proof. Let $F, G, \mathbb{T} : C(J,X) \to C(J,X)$ be the operators defined in the beginning of this subsection. They are continuous and bounded. Moreover, F is α-Lipschitz with constant $K_g \in [0,1)$ and G is α-Lipschitz with zero constant (see Lemmas 3.3-3.5). Proposition 1.21 shows that $\mathbb{T}$ is a strict α-contraction with constant K_g.

Set

$$S_0 = \{u \in C(J,X) : \exists\, \lambda \in [0,1] \text{ such that } u = \lambda\mathbb{T}u\}.$$

Next, we prove that S_0 is bounded in $C(J,X)$. Consider $u \in S_0$ and $\lambda \in [0,1]$ such that $u = \lambda\mathbb{T}u$. It follows from (3.23) and (3.24) that

$$\begin{aligned}\|u\| &= \lambda\|\mathbb{T}u\| \le \lambda(\|Fu\| + \|Gu\|)\\ &\le |u_0| + C_g\|u\|^{q_1} + M_g + \frac{T^q(C_f\|u\|^{q_2} + M_f)}{\Gamma(q+1)}.\end{aligned} \tag{3.25}$$

This inequality (3.25), together with $q_1 < 1$ and $q_2 < 1$, shows that S_0 is bounded in $C(J,X)$. If not, we suppose by contradiction, $\rho := \|u\| \to \infty$. Dividing both sides of (3.25) by ρ, and taking $\rho \to \infty$, we have

$$1 \le \lim_{\rho\to\infty} \rho^{-1}\left(|u_0| + C_g\rho^{q_1} + M_g + \frac{T^q(C_f\rho^{q_2} + M_f)}{\Gamma(q+1)}\right) = 0.$$

This is a contradiction. Consequently, by Theorem 1.2 we deduce that $\mathbb{T}$ has at least one fixed point and the set of the fixed points of $\mathbb{T}$ is bounded in $C(J,X)$. □

Remark 3.1.

(i) If the growth condition (H2) is formulated for $q_1 = 1$, then the conclusions of Theorem 3.2 remain valid provided that $C_g < 1$;
(ii) if the growth condition (H3) is formulated for $q_2 = 1$, then the conclusions of Theorem 3.2 remain valid provided that $\frac{T^q C_f}{\Gamma(q+1)} < 1$;
(iii) if the growth conditions (H2) and (H3) are formulated for $q_1 = 1$ and $q_2 = 1$, then the conclusions of Theorem 3.2 remain valid provided that $C_g + \frac{T^q C_f}{\Gamma(q+1)} < 1$.

3.4 Cauchy Problems via Picard Operators Technique

3.4.1 *Introduction*

Assume that $(X, |\cdot|)$ is a Banach space, and $J := [0,T]$, $T > 0$. Let $C(J, X)$ be the Banach space of all continuous functions from J into X with the norm $\|x\| := \sup\{|x(t)| : t \in J\}$ for $x \in C(J,X)$.

Consider the following Cauchy problem of fractional differential equation

$$\begin{cases} {}_0^C D_t^q x(t) = f(t, x(t)), & \text{a.e. } t \in J, \\ x(0) = x_0 \in X, \end{cases} \tag{3.26}$$

where ${}_0^C D_t^q$ is Caputo fractional derivative of order $q \in (0,1)$, the function $f : J \times X \to X$ satisfies some assumptions that will be specified later.

To our knowledge, Picard operators and weak Picard operators technique due to Rus 1979, 1987, 1993, 2003; Rus and Muresan, 2000 have been used to study the existence for the solutions of some integer differential equations (see, Mureşan, 2004; Şerban, Rus and Petruşel, 2010). In the present section we consider suitable Bielecki norms in some convenient spaces and obtain existence, uniqueness and data dependence results for the solutions of the fractional Cauchy problem (3.26) via Picard operators and weak Picard operators technique.

Definition 3.3. A function $x \in C^1(J, X)$ is said to be a solution of the fractional Cauchy problem (3.26) if x satisfies the equation ${}_0^C D_t^q x(t) = f(t, x(t))$ a.e. on J, and the condition $x(0) = x_0$.

In Subsection 3.4.2, we give the existence, uniqueness and data dependence results for the solutions of (3.26) via Picard operator by the successive approximation method. In Subsection 3.4.3, we obtain the existence results for the solutions of (3.26) via weak Picard operator.

3.4.2 *Results via Picard Operators*

Consider a Banach space $(X, |\cdot|)$, let $\|\cdot\|_B$ and $\|\cdot\|_C$ be the Bielecki and Chebyshev norms on $C(J, X)$ defined by

$$\|x\|_B = \max_{t\in J} |x(t)|e^{-\tau t}(\tau > 0) \quad \text{and} \quad \|x\|_C = \max_{t\in J} |x(t)|$$

and denote by d_B and d_C their corresponding metrics. We consider the set

$$C_L^{q-q^*}(J,X)=\left\{x\in C(J,X):|x(t_1)-x(t_2|\le L|t_1-t_2|^{q-q^*}\text{ for all }t_1,t_2\in J\right\}$$

where $L>0$, $q^*\in(0,q)$, and

$$C_{\bar{L}}^{q}(J,X)=\{x\in C(J,X):|x(t_1)-x(t_2)|\le \bar{L}|t_1-t_2|^{q}\text{ for all }t_1,t_2\in J\}$$

where $\bar{L}>0$, and

$$C_{\bar{L}}^{q}(J,B_R)=\{x\in C(J,B_R):|x(t_1)-x(t_2)|\le \bar{L}|t_1-t_2|^{q}\text{ for all }t_1,t_2\in J\}$$

where $B_R=\{x\in X:|x|\le R\}$ with $R>0$.

If $d\in\{d_C,d_B\}$, then $(C(J,X),d)$, $(C_L^{q-q^*}(J,X),d)$, $(C_{\bar{L}}^{q}(J,X),d)$ and $(C_{\bar{L}}^{q}(J,B_R),d)$ are complete metric spaces.

Let $q_i\in(0,q)$, $i=1,2,3$ and the functions $m(t)\in L^{\frac{1}{q_1}}(J,\mathbb{R}_+)$, $\eta(t)\in L^{\frac{1}{q_2}}(J,\mathbb{R}_+)$, $\mu(t)\in L^{\frac{1}{q_3}}(J,\mathbb{R}_+)$ and $l(t)\in C(J,\mathbb{R}_+)$.

For brevity, let

$$M=\|m\|_{L^{\frac{1}{q_1}}J},\quad N=\|\eta\|_{L^{\frac{1}{q_2}}J},\quad V=\|\mu\|_{L^{\frac{1}{q_2}}J},\quad L_0=\max_{t\in J}\{l(t)\},$$
$$\beta=\frac{q-1}{1-q_1}\in(-1,0),\quad \gamma=\frac{q-1}{1-q_2}\in(-1,0),\quad \nu=\frac{q-1}{1-q_3}\in(-1,0).$$

Lemma 3.6. *A function $x\in C(J,X)$ is a solution of the fractional integral equation*

$$x(t)=x_0+\frac{1}{\Gamma(q)}\int_0^t(t-s)^{q-1}f(s,x(s))ds,\tag{3.27}$$

if and only if x is a solution of the fractional Cauchy problem (3.26).

Theorem 3.3. *Suppose the following conditions hold:*

(C1) *$f\in C(J\times X,X)$;*

(C2) *there exist a constant $q_1\in(0,q)$ and function $m(\cdot)\in L^{\frac{1}{q_1}}(J,\mathbb{R}_+)$ such that*

$$|f(t,x)|\le m(t)$$

for all $x\in X$ and all $t\in J$;

(C3) *there exists a constant $L>0$ such that*

$$L\ge\frac{2M}{\Gamma(q)(1+\beta)^{1-q_1}};$$

(C4) *there exists a function $l(\cdot)\in C(J,\mathbb{R}_+)$ such that*

$$|f(t,u_1)-f(t,u_2)|\le l(t)|u_1-u_2|$$

for all $u_i\in X$ $(i=1,2)$ and all $t\in J$;

(C5) *there exist constants q_1 and τ such that*

$$\frac{L_0}{\Gamma(q)}\frac{T^{(1+\beta)(1-q_1)}}{(1+\beta)^{1-q_1}}\left(\frac{q_1}{\tau}\right)^{q_1}<1.$$

Then the fractional Cauchy problem (3.26) *has a unique solution* x^* *in* $C_L^{q-q_1}(J,X)$, *and this solution can be obtained by the successive approximation method, starting from any element of* $C_L^{q-q_1}(J,X)$.

Proof. Consider the operator

$$A:(C_L^{q-q_1}(J,X),\|\cdot\|_B)\to(C_L^{q-q_1}(J,X),\|\cdot\|_B)$$

defined by

$$Ax(t)=x_0+\frac{1}{\Gamma(q)}\int_0^t(t-s)^{q-1}f(s,x(s))\,ds.$$

It is easy to see the operator A is well defined due to (C1).

Firstly, we check that $Ax\in C(J,X)$ for every $x\in C_L^{q-q_1}(J,X)$.

For any $\delta>0$, every $x\in C_L^{q-q_1}(J,X)$, by (C2) and Hölder inequality,

$$\begin{aligned}
&|(Ax)(t+\delta)-(Ax)(t)|\\
&\le\frac{1}{\Gamma(q)}\int_0^t\left((t-s)^{q-1}-(t+\delta-s)^{q-1}\right)|f(s,x(s))|ds\\
&\quad+\frac{1}{\Gamma(q)}\int_t^{t+\delta}(t+\delta-s)^{q-1}|f(s,x(s))|ds\\
&\le\frac{1}{\Gamma(q)}\int_0^t\left((t-s)^{q-1}-(t+\delta-s)^{q-1}\right)m(s)ds+\frac{1}{\Gamma(q)}\int_t^{t+\delta}(t+\delta-s)^{q-1}m(s)ds\\
&\le\frac{1}{\Gamma(q)}\left(\int_0^t\left((t-s)^{q-1}-(t+\delta-s)^{q-1}\right)^{\frac{1}{1-q_1}}ds\right)^{1-q_1}\left(\int_0^t(m(s))^{\frac{1}{q_1}}ds\right)^{q_1}\\
&\quad+\frac{1}{\Gamma(q)}\left(\int_t^{t+\delta}\left((t+\delta-s)^{q-1}\right)^{\frac{1}{1-q_1}}ds\right)^{1-q_1}\left(\int_t^{t+\delta}(m(s))^{\frac{1}{q_1}}ds\right)^{q_1}\\
&\le\frac{M}{\Gamma(q)}\left(\int_0^t\left((t-s)^{\beta}-(t+\delta-s)^{\beta}\right)ds\right)^{1-q_1}+\frac{M}{\Gamma(q)}\left(\int_t^{t+\delta}(t+\delta-s)^{\beta}ds\right)^{1-q_1}\\
&\le\frac{M}{\Gamma(q)(1+\beta)^{1-q_1}}\left(|t^{1+\beta}-(t+\delta)^{1+\beta}|+\delta^{1+\beta}\right)^{1-q_1}+\frac{M}{\Gamma(q)(1+\beta)^{1-q_1}}\delta^{(1+\beta)(1-q_1)}\\
&\le\frac{2M}{\Gamma(q)(1+\beta)^{1-q_1}}\delta^{(1+\beta)(1-q_1)}+\frac{M}{\Gamma(q)(1+\beta)^{1-q_1}}\delta^{(1+\beta)(1-q_1)}\\
&\le\frac{3M}{\Gamma(q)(1+\beta)^{1-q_1}}\delta^{(1+\beta)(1-q_1)}.
\end{aligned}$$

It is easy to see that the right-hand side of the above inequality tends to zero as $\delta\to0$. Therefore $Ax\in C(J,X)$.

Secondly, we show that $Ax\in C_L^{q-q_1}(J,X)$.

Without loss of generality, for any $t_1<t_2$, $t_1,t_2\in J$, applying (C2) and Hölder inequality, we have

$$|(Ax)(t_2)-(Ax)(t_1)|$$

$$
\begin{aligned}
&\leq \frac{1}{\Gamma(q)}\left|\int_0^{t_1}[(t_2-s)^{q-1}-(t_1-s)^{q-1}]f(s,x(s))ds+\int_{t_1}^{t_2}(t_2-s)^{q-1}f(s,x(s))ds\right| \\
&\leq \frac{1}{\Gamma(q)}\int_0^{t_1}\left((t_1-s)^{q-1}-(t_2-s)^{q-1}\right)|f(s,x(s))|ds \\
&\quad+\frac{1}{\Gamma(q)}\int_{t_1}^{t_2}(t_2-s)^{q-1}|f(s,x(s))|ds \\
&\leq \frac{1}{\Gamma(q)}\int_0^{t_1}\left((t_1-s)^{q-1}-(t_2-s)^{q-1}\right)m(s)ds+\frac{1}{\Gamma(q)}\int_{t_1}^{t_2}(t_2-s)^{q-1}m(s)ds \\
&\leq \frac{1}{\Gamma(q)}\left(\int_0^{t_1}\left((t_1-s)^{q-1}-(t_2-s)^{q-1}\right)^{\frac{1}{1-q_1}}ds\right)^{1-q_1}\left(\int_0^{t_1}(m(s))^{\frac{1}{q_1}}ds\right)^{q_1} \\
&\quad+\frac{1}{\Gamma(q)}\left(\int_{t_1}^{t_2}\left((t_2-s)^{q-1}\right)^{\frac{1}{1-q_1}}ds\right)^{1-q_1}\left(\int_{t_1}^{t_2}(m(s))^{\frac{1}{q_1}}ds\right)^{q_1} \\
&\leq \frac{M}{\Gamma(q)}\left(\int_0^{t_1}(t_1-s)^{\beta}-(t_2-s)^{\beta}ds\right)^{1-q_1}+\frac{M}{\Gamma(q)}\left(\int_{t_1}^{t_2}(t_2-s)^{\beta}ds\right)^{1-q_1} \\
&\leq \frac{M}{\Gamma(q)(1+\beta)^{1-q_1}}\left(t_1^{1+\beta}-t_2^{1+\beta}+(t_2-t_1)^{1+\beta}\right)^{1-q_1} \\
&\quad+\frac{M}{\Gamma(q)(1+\beta)^{1-q_1}}(t_2-t_1)^{(1+\beta)(1-q_1)} \\
&\leq \frac{2M}{\Gamma(q)(1+\beta)^{1-q_1}}|t_1-t_2|^{(1+\beta)(1-q_1)} \\
&\leq \frac{2M}{\Gamma(q)(1+\beta)^{1-q_1}}|t_1-t_2|^{q-q_1}.
\end{aligned}
$$

Similarly, for any $t_1 > t_2$, $t_1, t_2 \in J$, we also have the above inequality. This implies that Ax is belong to $C_L^{q-q_1}(J,X)$ due to (C3).

Thirdly, A is continuous.

For that, let $\{x_n\}$ be a sequence of B_R such that $x_n \to x$ as $n \to \infty$ in B_R. Then, $f(s,x_n(s)) \to f(s,x(s))$ as $n \to \infty$ due to (C1). On the one hand, by using (C2), we get for each $s \in [0,t]$, $|f(s,x_n(s)) - f(s,x(s))| \leq 2m(s)$. On the other hand, using the fact that the function $s \to 2(t-s)^{q-1}m(s)$ is integrable on $[0,t]$, Lebesgue dominated convergence theorem yields

$$\int_0^t (t-s)^{q-1}|f(s,x_n(s)) - f(s,x(s))|ds \to 0, \quad n \to \infty.$$

For all $t \in J$, we have

$$|(Ax_n)(t) - (Ax)(t)| \leq \frac{1}{\Gamma(q)}\int_0^t (t-s)^{q-1}|f(s,x_n(s)) - f(s,x(s))|ds.$$

Thus, $Ax_n \to Ax$ as $n \to \infty$ which implies that A is continuous.

Moreover, for all $x, z \in C_L^{q-q_1}(J,X)$, using (C4) and Hölder inequality we have

$$|(Ax)(t) - (Az)(t)| \leq \frac{1}{\Gamma(q)}\int_0^t (t-s)^{q-1}|f(s,x(s)) - f(s,z(s))|ds$$

$$\begin{aligned}
&\le \frac{1}{\Gamma(q)}\int_0^t (t-s)^{q-1} l(s)|x(s)-z(s)|ds \\
&\le \frac{1}{\Gamma(q)}\int_0^t (t-s)^{q-1} \max_{s\in[0,t]}\{l(s)\}\left(|x(s)-z(s)|e^{-\tau s}\right)e^{\tau s}ds \\
&\le \frac{L_0}{\Gamma(q)}\|x-z\|_B \int_0^t (t-s)^{q-1}e^{\tau s}ds \\
&\le \frac{L_0}{\Gamma(q)}\|x-z\|_B \left(\int_0^t (t-s)^{\beta}ds\right)^{1-q_1}\left(\int_0^t e^{\frac{\tau s}{q_1}}ds\right)^{q_1} \\
&\le \frac{L_0}{\Gamma(q)}\frac{T^{(1+\beta)(1-q_1)}}{(1+\beta)^{1-q_1}}\left(\frac{q_1}{\tau}\right)^{q_1} e^{\tau t}\|x-z\|_B.
\end{aligned}$$

It follows that

$$|(Ax)(t)-(Az)(t)|e^{-\tau t} \le \frac{L_0}{\Gamma(q)}\frac{T^{(1+\beta)(1-q_1)}}{(1+\beta)^{1-q_1}}\left(\frac{q_1}{\tau}\right)^{q_1}\|x-z\|_B$$

for all $t\in J$. So we have

$$\|Ax-Az\|_B \le \frac{L_0}{\Gamma(q)}\frac{T^{(1+\beta)(1-q_1)}}{(1+\beta)^{1-q_1}}\left(\frac{q_1}{\tau}\right)^{q_1}\|x-z\|_B$$

for all $x,\ z\in C_L^{q-q_1}(J,X)$. The operator A is of Lipschitz type with constant

$$L_A = \frac{L_0}{\Gamma(q)}\frac{T^{(1+\beta)(1-q_1)}}{(1+\beta)^{1-q_1}}\left(\frac{q_1}{\tau}\right)^{q_1} \tag{3.28}$$

and $0 < L_A < 1$ due to (C5). By applying Banach contraction mapping principle to this operator we obtain that A is a Picard operator. This completes the proof. □

Example 3.3. Consider the fractional Cauchy problem

$$\begin{cases} {}_0^CD_t^q x(t) = x(t), \quad q = \dfrac{1}{2}, \\ x(0) = 0 \in X \end{cases}$$

on $[0,1]$. Set $L_0 = 1$, $T = 1$, $q_1 = \frac{1}{3}$, then $\beta = -\frac{3}{4}$. Indeed

$$\frac{L_0}{\Gamma(q)}\frac{T^{(1+\beta)(1-q_1)}}{(1+\beta)^{1-q_1}}\left(\frac{q_1}{\tau}\right)^{q_1} < 1 \iff \frac{qL_0}{\Gamma(q+1)}\frac{T^{(1+\beta)(1-q_1)}}{(1+\beta)^{1-q_1}}\left(\frac{q_1}{\tau}\right)^{q_1} < 1,$$

which implies that we must choose a suitable $\tau_0 > 0$ such that $\frac{\frac{1}{2}}{\Gamma(\frac{3}{2})}\frac{1}{(\frac{1}{4})^{\frac{2}{3}}}\left(\frac{\frac{1}{3}}{\tau_0}\right)^{\frac{1}{3}} < 1$. Noting that $\Gamma(\frac{3}{2}) = \frac{\sqrt{\pi}}{2}$, for $\tau_0 = \frac{16}{9} > \frac{16}{3\sqrt{\pi^3}}$ we have the condition (C5) in Theorem 3.3.

Theorem 3.4. *Suppose the following conditions hold:*

(C1) $f \in C(J\times X, X)$*;*
(C2)′ *there exists a constant* $\bar{M} > 0$ *such that* $|f(t,x)| \le \bar{M}$ *for all* $x\in X$ *and all* $t \in J$*;*

(C3)′ *there exists a constant* $\bar{L} > 0$ *such that* $\bar{L} \geq \frac{2\bar{M}}{\Gamma(q+1)}$;
(C4)′ *there exists a constant* $\bar{L}_0 > 0$ *such that* $|f(t, u_1) - f(t, u_2)| \leq \bar{L}_0|u_1 - u_2|$ *for all* $u_i \in X$ $(i = 1, 2)$ *and all* $t \in J$;
(C5)′ *there exist constants* q_1 *and* τ *such that* $\bar{L}_{\bar{A}} = \frac{\bar{L}_0}{\Gamma(q)} \frac{T^{(1+\beta)(1-q_1)}}{(1+\beta)^{1-q_1}} \left(\frac{q_1}{\tau}\right)^{q_1} < 1$.

Then the fractional Cauchy problem (3.26) *has a unique solution* x^* *in* $C^q_{\bar{L}}(J, X)$, *and this solution can be obtained by the successive approximation method, starting from any element of* $C^q_{\bar{L}}(J, X)$.

Proof. Consider the following continuous operator

$$\bar{A} : (C^q_{\bar{L}}(J, X), \|\cdot\|_B) \to (C^q_{\bar{L}}(J, X), \|\cdot\|_B)$$

defined by

$$\bar{A}x(t) = x_0 + \frac{1}{\Gamma(q)} \int_0^t (t-s)^{q-1} f(s, x(s))\, ds.$$

As the proof in Theorem 3.3, applying the given conditions one can verify that

$$\|\bar{A}(x) - \bar{A}(z)\|_B \leq \frac{\bar{L}_0}{\Gamma(q)} \frac{T^{(1+\beta)(1-q_1)}}{(1+\beta)^{1-q_1}} \left(\frac{q_1}{\tau}\right)^{q_1} \|x - z\|_B$$

for all x, $z \in C^q_{\bar{L}}(J, X)$. So, the operator $\bar{A}$ is a Picard operator. □

Similarly, we can prove the following theorem.

Theorem 3.5. *Suppose the following conditions hold:*

(C1)′ $f \in C(J \times B_R, X)$;
(C2)″ *there exists a constant* $\bar{M}(R) > 0$ *such that* $|f(t, x)| \leq \bar{M}(R)$ *for all* $x \in B_R$ *and all* $t \in J$ *with* $R \geq |x_0| + \frac{\bar{M}(R)T^q}{\Gamma(q+1)}$;
(C3)″ *there exists a constant* $\bar{L} > 0$ *such that* $\bar{L} \geq \frac{2\bar{M}(R)}{\Gamma(q+1)}$;
(C4)″ *there exists a constant* $\bar{L}_0 > 0$ *such that* $|f(t, u_1) - f(t, u_2)| \leq \bar{L}_0|u_1 - u_2|$ *for all* $u_i \in B_R$ $(i = 1, 2)$ *and all* $t \in J$;
(C5)′ *there exist constants* q_1 *and* τ *such that* $\bar{L}_{\bar{A}} = \frac{\bar{L}_0}{\Gamma(q)} \frac{T^{(1+\beta)(1-q_1)}}{(1+\beta)^{1-q_1}} \left(\frac{q_1}{\tau}\right)^{q_1} < 1$.

Then the fractional Cauchy problem (3.26) *has a unique solution* x^* *in* $C^q_{\bar{L}}(J, B_R)$, *and this solution can be obtained by the successive approximation method, starting from any element of* $C^q_{\bar{L}}(J, B_R)$.

Consider the following fractional Cauchy problem

$$\begin{cases} {}_0^C D_t^q x(t) = g(t, x(t)), & t \in J, \\ x(0) = y_0 \in X, \end{cases} \tag{3.29}$$

where $g \in C(J \times X, X)$. By Lemma 3.6, a function $x \in C(J, X)$ is a solution of the fractional integral equation

$$x(t) = y_0 + \frac{1}{\Gamma(q)} \int_0^t (t-s)^{q-1} g(s, x(s))\, ds, \tag{3.30}$$

if and only if x is a solution of the fractional Cauchy problem (3.29).

Now, we consider both fractional integral equations (3.27) and (3.30).

Theorem 3.6. *Suppose the following:*

(D1) *all conditions in Theorem 3.3 are satisfied and* $x^* \in C_L^{q-q_1}(J,X)$ *is the unique solution of the fractional integral equation* (3.27)*;*
(D2) *with the same function* $m(\cdot)$ *as in Theorem 3.3,* $|g(t,x)| \leq m(t)$ *for all* $x \in X$ *and all* $t \in J$*;*
(D3) *with the same function* $l(\cdot)$ *as in Theorem 3.3,* $|g(t,u_1) - g(t,u_2)| \leq l(t)|u_1 - u_2|$ *for all* $u_i \in X$ $(i = 1,2)$ *and all* $t \in J$*;*
(D4) $L \geq \frac{2M}{\Gamma(q)(1+\beta)^{1-q_1}}$*;*
(D5) *there exists a constant* $q_2 \in (0,q)$ *and function* $\eta(\cdot) \in L^{\frac{1}{q_2}}(J,\mathbb{R}_+)$ *such that* $|f(t,u) - g(t,u)| \leq \eta(t)$ *for all* $u \in X$ *and all* $t \in J$.

If y^* *is the solution of the fractional integral equation* (3.30)*, then*

$$\|x^* - y^*\|_B \leq \frac{|x_0 - y_0| + \frac{NT^{(1+\gamma)(1-q_2)}}{\Gamma(q)(1+\gamma)^{1-q_2}}}{1 - L_A}, \tag{3.31}$$

where L_A *is given by* (3.28) *with* $\tau = \tau_0 > 0$ *such that* $0 < L_A < 1$.

Proof. Consider the following two operators

$$A,\ B : (C_L^{q-q_1}(J,X), \|\cdot\|_B) \to (C_L^{q-q_1}(J,X), \|\cdot\|_B)$$

defined by

$$Ax(t) = x_0 + \frac{1}{\Gamma(q)} \int_0^t (t-s)^{q-1} f(s, x(s))\, ds,$$
$$Bx(t) = y_0 + \frac{1}{\Gamma(q)} \int_0^t (t-s)^{q-1} g(s, x(s))\, ds$$

on J. We have

$$\begin{aligned} |Ax(t) - Bx(t)| &\leq |x_0 - y_0| + \frac{1}{\Gamma(q)} \int_0^t (t-s)^{q-1} |f(s,x(s)) - g(s,x(s))|ds \\ &\leq |x_0 - y_0| + \frac{1}{\Gamma(q)} \int_0^t (t-s)^{q-1} \eta(s) ds \\ &\leq |x_0 - y_0| + \frac{NT^{(1+\gamma)(1-q_2)}}{\Gamma(q)(1+\gamma)^{1-q_2}}, \end{aligned}$$

for $t \in J$. It follows that

$$\|Ax - Bx\|_B \leq |x_0 - y_0| + \frac{NT^{(1+\gamma)(1-q_2)}}{\Gamma(q)(1+\gamma)^{1-q_2}}.$$

So we can apply Theorem 1.15 to obtain (3.31) which completes the proof. □

Remark 3.2. All the results obtained in Theorem 3.3 hold even if the condition (C2) is replaced by the following:

(C2-E) there exist a constant $q_1 \in [0, q)$ and function $m(\cdot) \in L^{\frac{1}{q_1}}(J, \mathbb{R}_+)$ such that $|f(t,x)| \leq m(t)$ for all $x \in X$ and all $t \in J$.

In fact, we only need extend the space $L^p(J, \mathbb{R}_+)$ $(1 < p < \infty)$ to $L^p(J, \mathbb{R}_+)$ $(1 \leq p \leq \infty)$ where $L^p(J, \mathbb{R}_+)$ $(1 \leq p \leq \infty)$ be the Banach space of all Lebesgue measurable functions $\phi : J \to \mathbb{R}_+$ with $\|\phi\|_{L^p J} < \infty$.

3.4.3 *Results via Weakly Picard Operators*

Now, we consider another fractional integral equation

$$x(t) = x(0) + \frac{1}{\Gamma(q)} \int_0^t (t-s)^{q-1} f(s, x(s))\, ds \tag{3.32}$$

on J, where $f \in C(J \times X, X)$ is as in the fractional Cauchy problem (3.26).

Theorem 3.7. *Suppose that for the fractional integral equation* (3.32) *the same conditions as in Theorem 3.3 are satisfied. Then this equation has solutions in* $C_L^{q-q_1}(J, X)$. *If* $\mathcal{S} \subset C_L^{q-q_1}(J, X)$ *is its solutions set, then card* $\mathcal{S} =$ *card* X.

Proof. Consider the operator

$$A_* : (C_L^{q-q_1}(J, X), \|\cdot\|_B) \to (C_L^{q-q_1}(J, X), \|\cdot\|_B)$$

defined by

$$A_* x(t) = x(0) + \frac{1}{\Gamma(q)} \int_0^t (t-s)^{q-1} f(s, x(s))\, ds.$$

This is a continuous operator, but not a Lipschitz one. We can write

$$C_L^{q-q_1}(J, X) = \bigcup_{\alpha \in X} X_\alpha, \qquad X_\alpha = \left\{ x \in C_L^{q-q_1}(J, X) : x(0) = \alpha \right\}.$$

We have that X_α is an invariant set of A_* and we apply Theorem 3.3 to $A_*|_{X_\alpha}$. By using Theorem 1.3 we obtain that A_* is a weak Picard operator.

Consider the operator

$$A_*^\infty : C_L^{q-q_1}(J, X) \to C_L^{q-q_1}(J, X), \quad A_*^\infty x = \lim_{n \to \infty} A_*^n x.$$

From $A_*^{n+1}(x) = A_*(A_*^n(x))$ and the continuity of A_*, $A_*^\infty(x) \in F_{A_*}$. Then

$$A_*^\infty(C_L^{q-q_1}(J, X)) = F_{A_*} = \mathcal{S} \text{ and } \mathcal{S} \neq \varnothing.$$

So, *card* $\mathcal{S} =$ *card* X. □

Theorem 3.8. *Suppose that for the fractional integral equation* (3.32) *the same conditions as in Theorem 3.4 are satisfied. Then this equation has solutions in* $C_L^q(J, X)$. *If* $\mathcal{S} \subset C_L^q(J, X)$ *is its solutions set, then card* $\mathcal{S} =$ *card* X.

Proof. As the proof in Theorem 3.7, we need to consider the continuous operator (but not a Lipschitz one)

$$\bar{A}_* : (C^q_{\bar{L}}(J, X), \|\cdot\|_B) \to (C^q_{\bar{L}}(J, X), \|\cdot\|_B)$$

defined by

$$\bar{A}_* x(t) = x(0) + \frac{1}{\Gamma(q)} \int_0^t (t-s)^{q-1} f(s, x(s))\, ds.$$

We can write $C^q_{\bar{L}}(J, X) = \bigcup_{\alpha \in X} \bar{X}_\alpha$, $\bar{X}_\alpha = \left\{x \in C^q_{\bar{L}}(J, X) : x(0) = \alpha\right\}$. We have that $\bar{X}_\alpha$ is an invariant set of $\bar{A}_*$ and we apply Theorem 3.4 to $\bar{A}_*|_{X_\alpha}$. By using Theorem 1.3 we obtain that $\bar{A}_*$ is a weak Picard operator. Consider the operator $\bar{A}^\infty_* : C^q_{\bar{L}}(J, X) \to C^q_{\bar{L}}(J, X)$, $\bar{A}^\infty_*(x) = \lim_{n\to\infty} \bar{A}^n_*(x)$. From $\bar{A}^{n+1}_*(x) = \bar{A}_*(\bar{A}^n_*(x))$ and the continuity of $\bar{A}_*$, $\bar{A}^\infty_*(x) \in F_{\bar{A}_*}$. Then $\bar{A}^\infty_*(C^q_{\bar{L}}(J, X)) = F_{\bar{A}_*} = \mathcal{S}$ and $\mathcal{S} \neq \varnothing$. So, *card* $\mathcal{S}$ = *card* X. □

Similarly as above, we can prove the following result.

Theorem 3.9. *Suppose that for the fractional integral equation (3.32) the same conditions as in Theorem 3.5 are satisfied. Then this equation has solutions in* $C^q_{\bar{L}}(J, B_R)$. *If* $\mathcal{S} \subset C^q_{\bar{L}}(J, B_R)$ *is its solutions set, then card* $\mathcal{S}$ = *card* B_R.

In order to study data dependence for the solutions set of the fractional integral equation (3.32), we consider both (3.32) and the following fractional integral equation

$$x(t) = x(0) + \frac{1}{\Gamma(q)} \int_0^t (t-s)^{q-1} g(s, x(s))\, ds$$

on J where $g \in C(J \times X, X)$. Let $\mathcal{S}_1$ be the solutions set of this equation.

Theorem 3.10. *Suppose the following conditions:*

(E1) *there exists a function* $l(t) \in C(J, \mathbb{R}_+)$ *such that*

$$|f(t,u_1) - f(t,u_2)| \le l(t)|u_1 - u_2| \quad \text{and} \quad |g(t,u_1) - g(t,u_2)| \le l(t)|u_1 - u_2|$$

for all $u_i \in X$ $(i = 1, 2)$ *and all* $t \in J$;

(E2) *there exist* $q_1, q_3 \in (0, q)$ *and functions* $m(t) \in L^{\frac{1}{q_1}}(J, \mathbb{R}_+)$, $\mu(t) \in L^{\frac{1}{q_3}}(J, \mathbb{R}_+)$ *such that*

$$|f(t,x)| \le m(t) \quad \text{and} \quad |g(t,x)| \le \mu(t)$$

for all $x \in X$ *and all* $t \in J$;

(E3) *there exists a constant* $L > 0$ *such that*

$$L \ge \frac{2\max\{M, V\}}{\Gamma(q)\min\left\{(1+\beta)^{1-q_1}, (1+\nu)^{1-q_3}\right\}};$$

(E4) *there exist a constant* $q_2 \in (0, q)$ *and function* $\eta \in L^{\frac{1}{q_2}}(J, \mathbb{R}_+)$

$$|f(t,u) - g(t,u)| \le \eta(t)$$

for all $u \in X$ *and all* $t \in J$;

(E5) $\frac{L_0 T^q}{\Gamma(q+1)} < 1$.

Then

$$H_{\|\cdot\|_C}(\mathcal{S}, \mathcal{S}_1) \leq \frac{qNT^{(1+\gamma)(1-q_2)}}{(\Gamma(q+1) - L_0 T^q)(1+\gamma)^{1-q_2}}$$

where by $H_{\|\cdot\|_C}$ *we denote the Pompeiu-Hausdorff functional with respect to* $\|\cdot\|_C$ *on* $C_L^{q-q_1}(J,X)$.

Proof. Consider the operator

$$B_* : (C_L^{q-q_1}(J,X), \|\cdot\|_B) \to (C_L^{q-q_1}(J,X), \|\cdot\|_B)$$

defined by

$$B_* x(t) = x(0) + \frac{1}{\Gamma(q)} \int_0^t (t-s)^{q-1} g(s, x(s))\, ds, \text{ for } t \in J.$$

Because of (E1)-(E3), $A_*, B_* : (C_L^{q-q_1}(J,X), \|\cdot\|_B) \to (C_L^{q-q_1}(J,X), \|\cdot\|_B)$ are two orbitally continuous operators. Moreover, we have

$$\begin{aligned} |A_*^2 x(t) - A_* x(t)| &\leq \frac{L_0}{\Gamma(q)} \int_0^t (t-s)^{q-1} |A_* x(s) - x(s)| ds \\ &\leq \frac{L_0 T^q}{\Gamma(q+1)} \|A_* x - x\|_C, \end{aligned}$$

for all $x \in C_L^{q-q_1}(J,X)$. Similarly,

$$|B_*^2 x(t) - B_* x(t)| \leq \frac{L_0 T^q}{\Gamma(q+1)} \|B_* x - x\|_C$$

for all $x \in C_L^{q-q_1}(J,X)$. It follows that

$$\begin{aligned} \|A_*^2 x - A_* x\|_C &\leq \frac{L_0 T^q}{\Gamma(q+1)} \|A_* x - x\|_C, \\ \|B_*^2 x - B_* x\|_C &\leq \frac{L_0 T^q}{\Gamma(q+1)} \|B_* x - x\|_C. \end{aligned}$$

Because of (E4),

$$\begin{aligned} \|A_* x - B_* x\|_C &\leq \frac{1}{\Gamma(q)} \int_0^t (t-s)^{q-1} \eta(s) ds \\ &\leq \frac{NT^{(1+\gamma)(1-q_2)}}{\Gamma(q)(1+\gamma)^{1-q_2}}, \end{aligned}$$

for all $x \in C_L^{q-q_1}(J,X)$.

By (E5) and applying Theorem 1.16, we obtain

$$H_{\|\cdot\|_C}(F_{A_*}, F_{B_*}) \leq \frac{qNT^{(1+\gamma)(1-q_2)}}{(\Gamma(q+1) - L_0 T^q)(1+\gamma)^{1-q_2}}$$

and the theorem is proved. □

Theorem 3.11. *Suppose the following conditions:*

(E1)′ *there exists a constant $L_* > 0$ such that*

$$|f(t,u_1) - f(t,u_2)| \le L_*|u_1 - u_2| \ \text{ and } |g(t,u_1) - g(t,u_2)| \le L_*|u_1 - u_2|$$

for all $u_i \in X$ $(i = 1,2)$ and all $t \in J$;

(E2)′ *there exists a constant $M_* > 0$ such that*

$$|f(t,x)| \le M_* \ \text{ and } |g(t,x)| \le M_*$$

for all $x \in X$ and all $t \in J$;

(E3)′ *there exists a constant $\bar{L} > 0$ such that*

$$\bar{L} \ge \frac{2M_*}{\Gamma(q+1)};$$

(E4)′ *there exists a constant $\eta_* > 0$ such that*

$$|f(t,u) - g(t,u)| \le \eta_*$$

for all $u \in X$ and all $t \in J$;

(E5)′ $\frac{L_* T^q}{\Gamma(q+1)} < 1$.

Then we have

$$\bar{H}_{\|\cdot\|_C}(\mathcal{S}, \mathcal{S}_1) \le \frac{\eta_* T^q}{\Gamma(q+1) - L_* T^q}$$

where by $\bar{H}_{\|\cdot\|_C}$ we denote the Pompeiu-Hausdorff functional with respect to $\|\cdot\|_C$ on $C^q_{\bar{L}}(J,X)$.

Proof. Consider the operator

$$\bar{B}_* : (C^q_{\bar{L}}(J,X), \|\cdot\|_B) \to (C^q_{\bar{L}}(J,X), \|\cdot\|_B)$$

defined by

$$\bar{B}_* x(t) = x(0) + \frac{1}{\Gamma(q)} \int_0^t (t-s)^{q-1} g\,(s, x(s))\, ds, \ \text{ for } t \in J.$$

Applying (E1)′-(E3)′, $\bar{A}_*, \bar{B}_* : (C^{q-q_1}_{\bar{L}}(J,X), \|\cdot\|_B) \to (C^{q-q_1}_{\bar{L}}(J,X), \|\cdot\|_B)$ are two orbitally continuous operators. Moreover, we have

$$|\bar{A}^2_* x(t) - \bar{A}_* x(t)| \le \frac{L_* T^q}{\Gamma(q+1)} \|\bar{A}_*(x) - x\|_C,$$
$$|\bar{B}^2_* x(t) - \bar{B}_* x(t)| \le \frac{L_* T^q}{\Gamma(q+1)} \|\bar{B}_*(x) - x\|_C,$$

for all $x \in C^q_{\bar{L}}(J,X)$. It follows that

$$\|\bar{A}^2_*(x) - \bar{A}_*(x)\|_C \le \frac{L_* T^q}{\Gamma(q+1)} \|\bar{A}_*(x) - x\|_C,$$
$$\|\bar{B}^2_*(x) - \bar{B}_*(x)\|_C \le \frac{L_* T^q}{\Gamma(q+1)} \|\bar{B}_*(x) - x\|_C.$$

Because of (E4)$'$, we obtain

$$\|\bar{A}_*(x) - \bar{B}_*(x)\|_C \leq \frac{1}{\Gamma(q)} \int_0^t (t-s)^{q-1} \eta_* ds \leq \frac{\eta_* T^q}{\Gamma(q+1)},$$

for all $x \in C^q_{\bar{L}}(J, X)$.

By (E5)$'$ and applying Theorem 1.16, we obtain the result and the theorem is proved. □

Similarly, we can prove the following theorem.

Theorem 3.12. *Suppose the following:*

(E1)$''$ *there exists a constant $L_* > 0$ such that*

$$|f(t,u_1) - f(t,u_2)| \leq L_*|u_1 - u_2| \quad \text{and} \ |g(t,u_1) - g(t,u_2)| \leq L_*|u_1 - u_2|$$

for all $u_i \in B_R$ $(i = 1, 2)$ and all $t \in J$;

(E2)$''$ *there exists a constant $M_*(R) > 0$ such that*

$$|f(t,x)| \leq M_*(R) \quad \text{and} \ |g(t,x)| \leq M_*(R)$$

for all $x \in B_R$ and all $t \in J$ with

$$R \geq |x(0)| + \frac{M_*(R)T^q}{\Gamma(q+1)};$$

(E3)$''$ *there exists a constant $\bar{L} > 0$ such that*

$$\bar{L} \geq \frac{2M_*(R)}{\Gamma(q+1)};$$

(E4)$''$ *there exists a constant $\eta_* > 0$ such that*

$$|f(t,u) - g(t,u)| \leq \eta_*$$

for all $u \in B_R$ and all $t \in J$;

(E5)$''$ $\frac{L_* T^q}{\Gamma(q+1)} < 1$.

Then

$$\bar{H}_{\|\cdot\|_C}(\mathcal{S}, \mathcal{S}_1) \leq \frac{\eta_* T^q}{\Gamma(q+1) - L_* T^q}$$

where by $\bar{H}_{\|\cdot\|_C}$ we denote the Pompeiu-Hausdorff functional with respect to $\|\cdot\|_C$ on $C^q_{\bar{L}}(J, B_R)$.

3.5 Notes and Remarks

The results in Sections 3.1 and 3.2 are taken from Zhou, Jiao and Pečarić, 2013. The material in Section 3.3 is due to Wang, Zhou and Medved, 2012. The main results in Section 3.4 are adopted from Wang, Zhou and Wei, 2013.

Chapter 4

Fractional Abstract Evolution Equations

4.1 Introduction

The existence of mild solutions for the Cauchy problem of fractional evolution equations has been considered in several recent papers (see, e.g., Agarwal and Shmad, 2011; Belmekki and Benchohra, 2010; Chang, Kavitha and Mallika, 2009; Darwish, Henderson and Ntouyas, 2009; Hernandez, O'Regan and Balachandran, 2010; Hu, Ren and Sakthivel, 2009; Kumar and Sukavanam, 2012; Li, Peng and Jia, 2012; Shu, Lai and Chen, 2011; Wang, Chen and Xiao, 2012; Wang and Zhou, 2011; Wang, Fečkan and Zhou, 2011; Zhou and Jiao, 2010), much less is known about the fractional evolution equations with Riemann-Liouville fractional derivative.

In most of the existing articles, Schauder fixed point theorem, Krasnoselskii fixed point theorem or Darbo fixed point theorem, Kuratowski measure of noncompactness are employed to obtain the fixed points of the solution operator of the Cauchy problems under some restrictive conditions. In order to show that the solution operator is compact, a very common approach is to use Arzela-Ascoli theorem. However, it is difficult to check the relative compactness of the solution operator and the equicontinuity of certain family of functions which is given by the solution operator.

In this chapter, we discuss the existence of mild solutions of fractional abstract evolution equations. The suitable mild solutions of fractional evolution equations with Riemann-Liouville fractional derivative and Caputo fractional derivative are introduced respectively.

In Sections 4.2 and 4.3, by using the theory of Hausdorff measure of noncompactness, we investigate the existence of mild solutions for the Cauchy problems in the cases C_0-semigroup is compact and noncompact, respectively. In Section 4.4, the existence results of mild solutions of nonlocal problem of fractional evolution equations are presented. Section 4.5 concerns the existence of mild solutions for semilinear fractional evolution equations and optimal controls in the α-norm. Section 4.6 is devoted to study of the evolution equations with almost sectorial operator. In Sections 4.7 and 4.8, we study fractional evolution equations with Hilfer fractional derivative on a finite interval and an infinite interval respectively.

4.2 Evolution Equations with Riemann-Liouville Derivative

4.2.1 *Introduction*

Assume that X is a Banach space with the norm $|\cdot|$. Let $a \in \mathbb{R}^+$, $J = [0, a]$ and $J' = (0, a]$. Denote $C(J, X)$ as the Banach space of continuous functions from J into X with the norm

$$\|x\| = \sup_{t\in[0,a]} |x(t)|,$$

where $x \in C(J, X)$, and $B(X)$ be the space of all bounded linear operators from X to X with the norm $\|Q\|_{B(X)} = \sup\{|Q(x)| \mid |x| = 1\}$, where $Q \in B(X)$ and $x \in X$.

Consider the following nonlocal Cauchy problem of fractional evolution equation with Riemann-Liouville fractional derivative

$$\begin{cases} {}_0D_t^q x(t) = Ax(t) + f(t, x(t)), & \text{a.e. } t \in [0, a], \\ {}_0D_t^{q-1} x(0) + g(x) = x_0, \end{cases} \tag{4.1}$$

where ${}_0D_t^q$ is Riemann-Liouville fractional derivative of order q, ${}_0D_t^{q-1}$ is Riemann-Liouville fractional integral of order $1-q$, $0 < q < 1$, A is the infinitesimal generator of a C_0-semigroup of bounded linear operators $\{Q(t)\}_{t\geq 0}$ in Banach space X, $f : J\times X \to X$ is a given function, $g : C(J, X) \to L(J, X)$ is a given operator satisfying some assumptions and x_0 is an element of the Banach space X.

A strong motivation for investigating the nonlocal Cauchy problem (4.1) comes from physics. For example, fractional diffusion equations are abstract partial differential equations that involve fractional derivatives in space and time. They are useful to model anomalous diffusion, where a plume of particles spreads in a different manner than the classical diffusion equation predicts. The time fractional diffusion equation is obtained from the standard diffusion equation by replacing the first-order time derivative with a fractional derivative of order $q \in (0, 1)$, namely

$$\partial_t^q u(z, t) = Au(z, t), \quad t \geq 0, \quad z \in \mathbb{R}.$$

We can take $A = \partial_z^{\beta_1}$, for $\beta_1 \in (0, 1]$, or $A = \partial_z + \partial_z^{\beta_2}$ for $\beta_2 \in (1, 2]$, where $\partial_t^q, \partial_z^{\beta_1}, \partial_z^{\beta_2}$ are the fractional derivatives of order q, β_1, β_2 respectively. We refer the interested reader to Eidelman and Kochubei, 2004; Hanyga, 2002; Hayashi, Kaikina and Naumkin, 2005; Meerschaert, Benson, Scheffler *et al.*, 2002; Schneider and Wayes, 1989; Zaslavsky, 1994 and the references therein for more details.

The nonlocal conditions ${}_0D_t^{q-1}x(0) + g(x) = x_0$ and $x(0) + g(x) = x_0$ can be applied in physics with better effect than the classical initial conditions ${}_0D_t^{q-1}x(0) = x_0$ and $x(0) = x_0$ respectively. For example, $g(x)$ may be given by

$$g(x) = \sum_{i=1}^{m} c_i x(t_i),$$

where c_i $(i = 1, 2, \ldots, n)$ are given constants and $0 < t_1 < t_2 < \cdots < t_n \leq a$. Nonlocal conditions were initiated by Byszewski, 1991, which he proved the existence and uniqueness of mild solutions and classical solutions for nonlocal Cauchy problems. As remarked by Byszewski and Lakshmikantham, 1991, the nonlocal condition can be more useful than the standard initial condition to describe some physical phenomena.

In this section, we study the nonlocal Cauchy problems of fractional evolution equations with Riemann-Liouville fractional derivative by considering an integral equation which is given in terms of probability density. By using the theory of Hausdorff measure of noncompactness, we establish various existence theorems of mild solutions for the Cauchy problem (4.1) in the cases C_0-semigroup is compact and noncompact, respectively. Subsection 4.2.2 is devoted to obtain the appropriate definition on the mild solutions of the problem (4.1) by considering an integral equation which is given in terms of probability density. In Subsection 4.2.3, we give some preliminary lemmas. Subsection 4.2.4 provides various existence theorems of mild solutions for the Cauchy problem (4.1) in the case C_0-semigroup is compact. In Subsection 4.2.5, we establish various existence theorems of mild solutions for the Cauchy problem (4.1) in the case C_0-semigroup is noncompact.

4.2.2 *Definition of Mild Solutions*

The following lemma is the special case of Proposition 1.3.

Lemma 4.1. *(Zain and Tazali, 1982)*

(i) *Let $\xi, \eta \in \mathbb{R}$ such that $\eta > -1$. If $t > 0$, then*

$$ {}_0D_t^{-\xi} \frac{t^\eta}{\Gamma(\eta+1)} = \begin{cases} \dfrac{t^{\xi+\eta}}{\Gamma(\xi+\eta+1)}, & \text{if } \xi+\eta \neq -n \\ 0, & \text{if } \xi+\eta = -n \end{cases} \quad (n \in \mathbb{N}). $$

(ii) *Let $\xi > 0$ and $\varphi \in L((0,a), X)$. Define*

$$ G_\xi(t) = {}_0D_t^{-\xi}\varphi, \quad \text{for } t \in (0,a), $$

then

$$ {}_0D_t^{-\eta} G_\xi(t) = {}_0D_t^{-(\xi+\eta)}\varphi(t), \quad \eta > 0, \quad \text{almost all } t \in [0,a]. $$

Lemma 4.2. *The nonlocal Cauchy problem* (4.1) *is equivalent to the integral equation*

$$ x(t) = \frac{t^{q-1}}{\Gamma(q)}(x_0 - g(x)) + \frac{1}{\Gamma(q)}\int_0^t (t-s)^{q-1}[Ax(s) + f(s,x(s))]ds, \quad \text{for } t \in (0,a], \tag{4.2} $$

provided that the integral in (4.2) *exists.*

Proof. Suppose (4.2) is true, then

$$ {}_0D_t^{q-1}x(t) = {}_0D_t^{q-1}\left(\frac{t^{q-1}}{\Gamma(q)}(x_0 - g(x)) + \frac{1}{\Gamma(q)}\int_0^t (t-\tau)^{q-1}(Ax(\tau) + f(\tau, x(\tau)))d\tau\right), $$

applying Lemma 4.1 we obtain that

$$ {}_0D_t^{q-1}x(t) = x_0 - g(x) + \int_0^t (Ax(s) + f(s, x(s)))]ds, \text{ almost all } t \in [0, a], $$

and this proves that ${}_0D_t^{q-1}x(t)$ is absolutely continuous on $[0, a]$. Then we have

$$ {}_0D_t^q x(t) = \frac{d}{dt}\, {}_0D_t^{q-1}x(t) = Ax(t) + f(t, x(t)), \text{ almost all } t \in [0, a] $$

and

$$ {}_0D_t^{q-1}x(0) + g(x) = x_0. $$

The proof of the converse is given as follows.
Suppose (4.1) is true, then

$$ {}_0D_t^{-q}({}_0D_t^q x(t)) = {}_0D_t^{-q}(Ax(t) + f(t, x(t))). $$

Since

$$ \begin{aligned} {}_0D_t^{-q}({}_0D_t^q x(t)) &= x(t) - \frac{t^{q-1}}{\Gamma(q)}\, {}_0D_t^{q-1}x(0) \\ &= x(t) - \frac{t^{q-1}}{\Gamma(q)}(x_0 - g(x)), \quad \text{for } t \in (0, a], \end{aligned} $$

then we have

$$ \begin{aligned} x(t) &= \frac{t^{q-1}}{\Gamma(q)}(x_0 - g(x)) + {}_0D_t^{-q}(Ax(t) + f(t, x(t))) \\ &= \frac{t^{q-1}}{\Gamma(q)}(x_0 - g(x)) + \frac{1}{\Gamma(q)} \int_0^t (t-s)^{q-1}(Ax(s) + f(s, x(s)))ds, \quad \text{for } t \in (0, a]. \end{aligned} $$

The proof is completed. □

Before giving the definition of mild solution of (4.1), we firstly prove the following lemma.

Lemma 4.3. *If*

$$ x(t) = \frac{t^{q-1}}{\Gamma(q)}(x_0 - g(x)) + \frac{1}{\Gamma(q)} \int_0^t (t-s)^{q-1}[Ax(s) + f(s, x(s))]ds, \quad \textit{for } t > 0 \quad (4.3) $$

holds, then we have

$$ x(t) = t^{q-1}P_q(t)(x_0 - g(x)) + \int_0^t (t-s)^{q-1}P_q(t-s)f(s, x(s))ds, \quad \textit{for } t > 0, $$

where

$$ P_q(t) = \int_0^\infty q\theta M_q(\theta)Q(t^q\theta)d\theta. $$

Proof. Let $\lambda > 0$. Applying the Laplace transform

$$\nu(\lambda) = \int_0^\infty e^{-\lambda s} x(s) ds \quad \text{and} \quad \omega(\lambda) = \int_0^\infty e^{-\lambda s} f(s, x(s)) ds, \quad \text{for } \lambda > 0$$

to (4.3), we have

$$\begin{aligned}\nu(\lambda) &= \frac{1}{\lambda^q}(x_0 - g(x)) + \frac{1}{\lambda^q} A\nu(\lambda) + \frac{1}{\lambda^q}\,\omega(\lambda)\\ &= (\lambda^q I - A)^{-1}(x_0 - g(x)) + (\lambda^q I - A)^{-1}\omega(\lambda)\\ &= \int_0^\infty e^{-\lambda^q s} Q(s)(x_0 - g(x)) ds + \int_0^\infty e^{-\lambda^q s} Q(s)\omega(\lambda) ds,\end{aligned} \tag{4.4}$$

provided that the integrals in (4.4) exist, where I is the identity operator defined on X.

Set

$$\psi_q(\theta) = \frac{q}{\theta^{q+1}} M_q(\theta^{-q}),$$

whose Laplace transform is given by

$$\int_0^\infty e^{-\lambda\theta} \psi_q(\theta) d\theta = e^{-\lambda^q}, \quad \text{where } q \in (0, 1). \tag{4.5}$$

Using (4.5), we get

$$\begin{aligned}&\int_0^\infty e^{-\lambda^q s} Q(s)(x_0 - g(x)) ds = \int_0^\infty q t^{q-1} e^{-(\lambda t)^q} Q(t^q)(x_0 - g(x)) dt\\ &= \int_0^\infty \int_0^\infty q\psi_q(\theta) e^{-(\lambda t\theta)} Q(t^q) t^{q-1}(x_0 - g(x)) d\theta dt\\ &= \int_0^\infty \int_0^\infty q\psi_q(\theta) e^{-\lambda t} Q\left(\frac{t^q}{\theta^q}\right) \frac{t^{q-1}}{\theta^q}(x_0 - g(x)) d\theta dt\\ &= \int_0^\infty e^{-\lambda t}\left[q\int_0^\infty \psi_q(\theta) Q\left(\frac{t^q}{\theta^q}\right)\frac{t^{q-1}}{\theta^q}(x_0 - g(x)) d\theta\right] dt,\end{aligned} \tag{4.6}$$

$$\begin{aligned}&\int_0^\infty e^{-\lambda^q s} Q(s)\omega(\lambda) ds\\ &= \int_0^\infty \int_0^\infty q t^{q-1} e^{-(\lambda t)^q} Q(t^q) e^{-\lambda s} f(s, x(s)) ds dt\\ &= \int_0^\infty \int_0^\infty \int_0^\infty q\psi_q(\theta) e^{-(\lambda t\theta)} Q(t^q) e^{-\lambda s} t^{q-1} f(s, x(s)) d\theta ds dt\\ &= \int_0^\infty \int_0^\infty \int_0^\infty q\psi_q(\theta) e^{-\lambda(t+s)} Q\left(\frac{t^q}{\theta^q}\right)\frac{t^{q-1}}{\theta^q} f(s, x(s)) d\theta ds dt\\ &= \int_0^\infty e^{-\lambda t}\left[q\int_0^t \int_0^\infty \psi_q(\theta) Q\left(\frac{(t-s)^q}{\theta^q}\right)\frac{(t-s)^{q-1}}{\theta^q} f(s, x(s)) d\theta ds\right] dt.\end{aligned} \tag{4.7}$$

According to (4.6) and (4.7), we have

$$\begin{aligned}\nu(\lambda) = \int_0^\infty e^{-\lambda t}\Bigg[&q\int_0^\infty \psi_q(\theta) Q\left(\frac{t^q}{\theta^q}\right)\frac{t^{q-1}}{\theta^q}(x_0 - g(x)) d\theta\\ &+ q\int_0^t \int_0^\infty \psi_q(\theta) Q\left(\frac{(t-s)^q}{\theta^q}\right)\frac{(t-s)^{q-1}}{\theta^q} f(s, x(s)) d\theta ds\Bigg] dt.\end{aligned}$$

Now we can invert the last Laplace transform to get

$$\begin{aligned}x(t) &= q\int_0^\infty \theta t^{q-1}M_q(\theta)Q(t^q\theta)(x_0-g(x))d\theta\\ &\quad + q\int_0^t\int_0^\infty \theta(t-s)^{q-1}M_q(\theta)Q((t-s)^q\theta)f(s,x(s))d\theta ds\\ &= t^{q-1}P_q(t)(x_0-g(x)) + \int_0^t (t-s)^{q-1}P_q(t-s)f(s,x(s))ds.\end{aligned}$$

The proof is completed. □

Due to Lemma 4.3, we give the following definition of the mild solution of (4.1).

Definition 4.1. By the mild solution of the nonlocal Cauchy problem (4.1), we mean that the function $x \in C(J', X)$ which satisfies

$$x(t) = t^{q-1}P_q(t)(x_0-g(x)) + \int_0^t (t-s)^{q-1}P_q(t-s)f(s,x(s))ds, \quad \text{for } t \in (0,a].$$

Suppose that A is the infinitesimal generator of a C_0-semigroup $\{Q(t)\}_{t\geq 0}$ of uniformly bounded linear operators on Banach space X. This means that there exists $M > 1$ such that $M = \sup_{t\in[0,\infty)} \|Q(t)\|_{B(X)} < \infty$.

Proposition 4.1. *(Zhou and Jiao, 2010a) For any fixed $t > 0$, $P_q(t)$ is linear and bounded operator, i.e., for any $x \in X$*

$$|P_q(t)x| \leq \frac{M}{\Gamma(q)}|x|.$$

Proposition 4.2. *(Zhou and Jiao, 2010a) Operator $\{P_q(t)\}_{t>0}$ is strongly continuous, which means that, for $\forall\ x \in X$ and $0 < t' < t'' \leq a$, we have*

$$|P_q(t'')x - P_q(t')x| \to 0, \quad as\ t'' \to t'.$$

Proposition 4.3. *(Zhou and Jiao, 2010a) Assume that $\{Q(t)\}_{t>0}$ is compact operator. Then $\{P_q(t)\}_{t>0}$ is also compact operator.*

Proposition 4.4. *(Pazy, 1983) Assume that $\{Q(t)\}_{t>0}$ is compact operator. Then $\{Q(t)\}_{t>0}$ is equicontinuous.*

4.2.3 *Preliminary Lemmas*

Define

$$X^{(q)}(J') = \left\{x \in C(J', X) : \lim_{t\to 0+} t^{1-q}x(t) \text{ exists and is finite}\right\}.$$

For any $x \in X^{(q)}(J')$, let the norm $\|\cdot\|_q$ defined by

$$\|x\|_q = \sup_{t\in(0,a]}\{t^{1-q}|x(t)|\}.$$

Then $(X^{(q)}(J'), \|\cdot\|_q)$ is a Banach space.

For $r > 0$, define a closed subset $B_r^{(q)}(J') \subset X^{(q)}(J')$ as follows

$$B_r^{(q)}(J') = \{x \in X^{(q)}(J') : \|x\|_q \leq r\}.$$

Thus, $B_r^{(q)}(J')$ is a bounded closed and convex subset of $X^{(q)}(J')$.

Let $B(J)$ be the closed ball of the space $C(J, X)$ with radius r and center at 0, that is

$$B(J) = \{y \in C(J, X) : \|y\| \leq r\}.$$

Thus $B(J)$ is a bounded closed and convex subset of $C(J, X)$.

We introduce the following hypotheses:

(H0) $Q(t)(t > 0)$ is equicontinuous, i.e., $Q(t)$ is continuous in the uniform operator topology for $t > 0$;

(H1) for each $t \in J'$, the function $f(t, \cdot) : X \to X$ is continuous and for each $x \in X$, the function $f(\cdot, x) : J' \to X$ is strongly measurable;

(H2) there exists a function $m \in L(J', \mathbb{R}^+)$ such that

$${}_0D_t^{-q} m \in C(J', \mathbb{R}^+), \quad \lim_{t\to 0+} t^{1-q}{}_0D_t^{-q} m(t) = 0,$$

and

$$|f(t, x)| \leq m(t) \text{ for all } x \in B_r^{(q)}(J') \text{ and almost all } t \in [0, a];$$

(H3) there exists a constant $L \in (0, \frac{\Gamma(q)}{M})$ such that the operator $g : C(J', X) \to L(J', X)$ satisfies

$$|g(x_1) - g(x_2)| \leq L\|x_1 - x_2\|_q, \quad \text{for } x_1, x_2 \in B_r^{(q)}(J');$$

(H4) there exists a constant $r > 0$ such that

$$\frac{M}{\Gamma(q) - ML}\left(|x_0| + |g(0)| + \sup_{t\in(0,a]}\left\{t^{1-q}\int_0^t (t-s)^{q-1} m(s)ds\right\}\right) \leq r;$$

(H3)′ the operator $g : C(J', X) \to L(J', X)$ is a continuous and compact map, and there exist positive constants L_1, L_2 such that $L_1 \in (0, \frac{\Gamma(q)}{M})$ and $|g(x)| \leq L_1\|x\|_q + L_2$ for all $x \in B_r^{(q)}(J')$;

(H4)′ there exists a constant $r > 0$ such that

$$\frac{M}{\Gamma(q) - ML_1}\left(|x_0| + L_2 + \sup_{t\in(0,a]}\left\{t^{1-q}\int_0^t (t-s)^{q-1} m(s)ds\right\}\right) \leq r.$$

For any $x \in B_r^{(q)}(J')$, define an operator T as follows

$$(Tx)(t) = (T_1x)(t) + (T_2x)(t),$$

where

$$(T_1x)(t) = t^{q-1}P_q(t)(x_0 - g(x)), \qquad \text{for } t \in (0, a],$$

$$(T_2x)(t) = \int_0^t (t-s)^{q-1}P_q(t-s)f(s, x(s))ds, \quad \text{for } t \in (0, a].$$

It is easy to see that $\lim_{t\to 0^+} t^{1-q}(Tx)(t) = \frac{x_0 - g(x)}{\Gamma(q)}$. For any $y \in B(J)$, set

$$x(t) = t^{q-1}y(t), \quad \text{for } t \in (0,a].$$

Then, $x \in B_r^{(q)}(J')$. Define $\mathscr{T}$ as follows

$$(\mathscr{T}y)(t) = (\mathscr{T}_1 y)(t) + (\mathscr{T}_2 y)(t),$$

where

$$(\mathscr{T}_1 y)(t) = \begin{cases} t^{1-q}(T_1 x)(t), & \text{for } t \in (0,a], \\ \dfrac{x_0 - g(x)}{\Gamma(q)}, & \text{for } t = 0, \end{cases}$$

$$(\mathscr{T}_2 y)(t) = \begin{cases} t^{1-q}(T_2 x)(t), & \text{for } t \in (0,a], \\ 0, & \text{for } t = 0. \end{cases}$$

Obviously, x is a mild solution of (4.1) in $B_r^{(q)}(J')$ if and only if the operator equation $x = Tx$ has a solution $x \in B_r^{(q)}(J')$. Before giving the main results, we firstly prove the following lemmas.

Lemma 4.4. *Assume that (H0)-(H4) hold, then $\{\mathscr{T}y : \ y \in B(J)\}$ is equicontinuous.*

Proof. Claim I. $\{\mathscr{T}_1 y : \ y \in B(J)\}$ is equicontinuous.

For any $y \in B(J)$, let $x(t) = t^{q-1}y(t),\ t \in (0,a]$. Then $x \in B_r^{(q)}(J')$. For $t_1 = 0$, $0 < t_2 \le a$, we get

$$\begin{aligned} |(\mathscr{T}_1 y)(t_2) - (\mathscr{T}_1 y)(0)| &\le \left| P_q(t_2)(x_0 - g(x)) - \frac{x_0 - g(x)}{\Gamma(q)} \right| \\ &\le \left| \left(P_q(t_2) - \frac{1}{\Gamma(q)} \right)(x_0 - g(x)) \right| \\ &\le \left| \left(P_q(t_2) - \frac{1}{\Gamma(q)} \right) \right| (|x_0| + L\|x\|_q + |g(0)|) \\ &\le \left| \left(P_q(t_2) - \frac{1}{\Gamma(q)} \right) \right| (|x_0| + Lr + |g(0)|) \\ &\to 0, \quad \text{as } t_2 \to 0. \end{aligned}$$

For $0 < t_1 < t_2 \le a$, we get

$$\begin{aligned} |(\mathscr{T}_1 y)(t_2) - (\mathscr{T}_1 y)(t_1)| &\le |P_q(t_2)(x_0 - g(x)) - P_q(t_1)(x_0 - g(x))| \\ &\le |(P_q(t_2) - P_q(t_1))(x_0 - g(x))| \\ &\le |(P_q(t_2) - P_q(t_1))|(|x_0| + L\|x\|_q + |g(0)|) \\ &\le |(P_q(t_2) - P_q(t_1))|(|x_0| + Lr + |g(0)|) \\ &\to 0, \quad \text{as } t_2 \to t_1. \end{aligned}$$

Hence, $\{\mathscr{T}_1 y : \ y \in B(J)\}$ is equicontinuous.

Claim II. $\{\mathscr{T}_2 y : \ y \in B(J)\}$ is equicontinuous.

For any $y \in B(J)$, let $x(t) = t^{q-1}y(t),\ t \in (0,a]$. Then $x \in B_r^{(q)}(J')$. For $t_1 = 0$, $0 < t_2 \le a$, we get

$$\begin{aligned}|(\mathscr{T}_2y)(t_2) - (\mathscr{T}_2y)(0)| &= \left| t_2^{1-q} \int_0^{t_2} (t_2 - s)^{q-1} P_q(t_2 - s) f(s, x(s)) ds \right| \\ &\le \frac{M}{\Gamma(q)} t_2^{1-q} \int_0^{t_2} (t_2 - s)^{q-1} m(s) ds \\ &\to 0, \quad \text{as } t_2 \to 0.\end{aligned}$$

For $0 < t_1 < t_2 \le a$, we have

$$\begin{aligned}&|(\mathscr{T}_2y)(t_2) - (\mathscr{T}_2y)(t_1)| \\ &\le \left| \int_{t_1}^{t_2} t_2^{1-q}(t_2 - s)^{q-1} P_q(t_2 - s) f(s, x(s)) ds \right| \\ &\quad + \left| \int_0^{t_1} t_2^{1-q}(t_2 - s)^{q-1} P_q(t_2 - s) f(s, x(s)) ds \right. \\ &\quad \left. - \int_0^{t_1} t_1^{1-q}(t_1 - s)^{q-1} P_q(t_2 - s) f(s, x(s)) ds \right| \\ &\quad + \left| \int_0^{t_1} t_1^{1-q}(t_1 - s)^{q-1} P_q(t_2 - s) f(s, x(s)) ds \right. \\ &\quad \left. - \int_0^{t_1} t_1^{1-q}(t_1 - s)^{q-1} P_q(t_1 - s) f(s, x(s)) ds \right| \\ &\le \frac{M}{\Gamma(q)} \left| \int_{t_1}^{t_2} t_2^{1-q}(t_2 - s)^{q-1} m(s) ds \right| \\ &\quad + \frac{M}{\Gamma(q)} \int_0^{t_1} \left(t_1^{1-q}(t_1 - s)^{q-1} - t_2^{1-q}(t_2 - s)^{q-1} \right) m(s) ds \\ &\quad + \left| \int_0^{t_1} t_1^{1-q}(t_1 - s)^{q-1} \left(P_q(t_2 - s) f(s, x(s)) - P_q(t_1 - s) f(s, x(s)) \right) ds \right| \\ &\le I_1 + I_2 + I_3,\end{aligned}$$

where

$$\begin{aligned}I_1 &= \frac{M}{\Gamma(q)} \left| \int_0^{t_2} t_2^{1-q}(t_2 - s)^{q-1} m(s) ds - \int_0^{t_1} t_1^{1-q}(t_1 - s)^{q-1} m(s) ds \right|, \\ I_2 &= \frac{2M}{\Gamma(q)} \int_0^{t_1} \left(t_1^{1-q}(t_1 - s)^{q-1} - t_2^{1-q}(t_2 - s)^{q-1} \right) m(s) ds, \\ I_3 &= \left| \int_0^{t_1} t_1^{1-q}(t_1 - s)^{q-1} \left(P_q(t_2 - s) - P_q(t_1 - s) \right) f(s, x(s)) ds \right|.\end{aligned}$$

One can reduce that $\lim_{t_2 \to t_1} I_1 = 0$, since ${}_0D_t^{-q} m \in C(J', \mathbb{R}^+)$. Noting that

$$\left(t_1^{1-q}(t_1 - s)^{q-1} - t_2^{1-q}(t_2 - s)^{q-1} \right) m(s) \le t_1^{1-q}(t_1 - s)^{q-1} m(s),$$

and $\int_0^{t_1} t_1^{1-q}(t_1-s)^{q-1}m(s)ds$ exists ($s \in [0,t_1]$), then by Lebesgue dominated convergence theorem, we have

$$\int_0^{t_1} \left(t_1^{1-q}(t_1-s)^{q-1} - t_2^{1-q}(t_2-s)^{q-1}\right) m(s)ds \to 0, \quad \text{as } t_2 \to t_1,$$

then one can deduce that $\lim_{t_2\to t_1} I_2 = 0$.

For $\varepsilon > 0$ be enough small, we have

$$\begin{aligned} I_3 &\le \int_0^{t_1-\varepsilon} t_1^{1-q}(t_1-s)^{q-1}\|P_q(t_2-s)-P_q(t_1-s)\|_{B(X)}\ |f(s,x(s))|ds \\ &\quad + \int_{t_1-\varepsilon}^{t_1} t_1^{1-q}(t_1-s)^{q-1}\|P_q(t_2-s)-P_q(t_1-s)\|_{B(X)}\ |f(s,x(s))|ds \\ &\le t_1^{1-q}\int_0^{t_1}(t_1-s)^{q-1}m(s)ds \sup_{s\in[0,t_1-\varepsilon]} \|P_q(t_2-s)-P_q(t_1-s)\|_{B(X)} \\ &\quad + \frac{2M}{\Gamma(q)}\int_{t_1-\varepsilon}^{t_1} t_1^{1-q}(t_1-s)^{q-1}m(s)ds \\ &\le I_{31}+I_{32}+I_{33}, \end{aligned}$$

where

$$\begin{aligned} I_{31} &= \frac{r\Gamma(q)}{M}\sup_{s\in[0,t_1-\varepsilon]}\|P_q(t_2-s)-P_q(t_1-s)\|_{B(X)}, \\ I_{32} &= \frac{2M}{\Gamma(q)}\left|\int_0^{t_1} t_1^{1-q}(t_1-s)^{q-1}m(s)ds - \int_0^{t_1-\varepsilon}(t_1-\varepsilon)^{1-q}(t_1-\varepsilon-s)^{q-1}m(s)ds\right|, \\ I_{33} &= \frac{2M}{\Gamma(q)}\int_0^{t_1-\varepsilon}\left((t_1-\varepsilon)^{1-q}(t_1-\varepsilon-s)^{q-1} - t_1^{1-q}(t_1-s)^{q-1}\right)m(s)ds. \end{aligned}$$

By (H0), it is easy to see that $I_{31} \to 0$ as $t_2 \to t_1$. Similar to the proof that I_1, I_2 tend to zero, we get $I_{32} \to 0$ and $I_{33} \to 0$ as $\varepsilon \to 0$. Thus, I_3 tends to zero independently of $y \in B(J)$ as $t_2 \to t_1$, $\varepsilon \to 0$. Therefore, $|(\mathscr{T}_2 y)(t_2) - (\mathscr{T}_2 y)(t_1)|$ tends to zero independently of $y \in B(J)$ as $t_2 \to t_1$, which means that $\{\mathscr{T}_2 y : y \in B(J)\}$ is equicontinuous.

Therefore, $\{\mathscr{T} y : y \in B(J)\}$ is equicontinuous. □

Lemma 4.5. *Assume that (H1)-(H4) hold. Then $\mathscr{T}$ maps $B(J)$ into $B(J)$, and $\mathscr{T}$ is continuous in $B(J)$.*

Proof. Claim I. $\mathscr{T}$ maps $B(J)$ into $B(J)$. For any $y \in B(J)$, let $x(t) = t^{q-1}y(t)$. Then $x \in B_r^{(q)}(J')$.

For $t \in [0,a]$, by (H1)-(H4), we have

$$\begin{aligned} |(\mathscr{T}y)(t)| &\le |P_q(t)(x_0 - g(x))| + t^{1-q}\left|\int_0^t (t-s)^{q-1}P_q(t-s)f(s,x(s))ds\right| \\ &\le \frac{M}{\Gamma(q)}(|x_0| + L\|x\|_q + |g(0)|) + \frac{Mt^{1-q}}{\Gamma(q)}\int_0^t (t-s)^{q-1}|f(s,x(s))|ds \end{aligned}$$

$$\leq \frac{M}{\Gamma(q)}\left(|x_0| + Lr + |g(0)| + \sup_{t\in[0,a]}\left\{t^{1-q}\int_0^t (t-s)^{q-1}m(s)ds\right\}\right)$$

$$\leq r.$$

Hence, $\|\mathscr{T}y\| \leq r$, for any $y \in B(J)$.

Claim II. $\mathscr{T}$ is continuous in $B(J)$. For any $y_m, y \in B(J)$, $m = 1, 2, ...$, with $\lim_{m\to\infty} y_m = y$, we have

$$\lim_{m\to\infty} y_m(t) = y(t) \quad \text{and} \quad \lim_{m\to\infty} t^{q-1}y_m(t) = t^{q-1}y(t), \quad \text{for } t \in (0, a].$$

Then by (H1), we have

$$f(t, x_m(t)) = f(t, t^{q-1}y_m(t)) \to f(t, t^{q-1}y(t)) = f(t, x(t)), \quad \text{as } m \to \infty,$$

where $x_m(t) = t^{q-1}y_m(t)$ and $x(t) = t^{q-1}y(t)$.

On the one hand, using (H2), we get for each $t \in J'$,

$$(t-s)^{q-1}|f(s, x_m(s)) - f(s, x(s))| \leq (t-s)^{q-1}2m(s), \quad \text{a.e. in } [0, t].$$

On the other hand, the function $s \to (t-s)^{q-1}2m(s)$ is integrable for $s \in [0, t]$ and $t \in J$. By Lebesgue dominated convergence theorem, we get

$$\int_0^t (t-s)^{q-1}|f(s, x_m(s)) - f(s, x(s))|ds \to 0, \quad \text{as } m \to \infty.$$

For $t \in [0, a]$

$$\begin{aligned}
&|(\mathscr{T}y_m)(t) - (\mathscr{T}y)(t)| = |t^{1-q}(Tx_m(t) - Tx(t))| \\
&\leq |P_q(t)(g(x_m) - g(x))| + t^{1-q}\left|\int_0^t (t-s)^{q-1}P_q(t-s)(f(s, x_m(s)) - f(s, x(s)))ds\right| \\
&\leq \frac{ML}{\Gamma(q)}\|x_m - x\|_q + \frac{Mt^{1-q}}{\Gamma(q)}\int_0^t (t-s)^{q-1}|f(s, x_m(s)) - f(s, x(s))|ds \\
&\leq \frac{ML}{\Gamma(q)}\|y_m - y\| + \frac{Mt^{1-q}}{\Gamma(q)}\int_0^t (t-s)^{q-1}|f(s, x_m(s)) - f(s, x(s))|ds.
\end{aligned}$$

Therefore, $\mathscr{T}y_m \to \mathscr{T}y$ pointwise on J as $m \to \infty$, by which Lemma 4.4 implies that $\mathscr{T}y_m \to \mathscr{T}y$ uniformly on J as $m \to \infty$ and so $\mathscr{T}$ is continuous. □

Lemma 4.6. *Assume that (H0)-(H2), (H3)′ and (H4)′ hold. Then* $\{\mathscr{T}y : y \in B(J)\}$ *is equicontinuous.*

Proof. For any $y \in B(J)$, for $t_1 = 0$, $0 < t_2 \leq a$, then, we get

$$\begin{aligned}
&|(\mathscr{T}y)(t_2) - (\mathscr{T}y)(0)| \\
&\leq \left|P_q(t_2)(x_0 - g(x)) - \frac{x_0 - g(x)}{\Gamma(q)}\right| + \left|t_2^{1-q}\int_0^{t_2} (t_2-s)^{q-1}P_q(t_2-s)f(s, x(s))ds\right| \\
&\leq \left|P_q(t_2)(x_0 - g(x)) - \frac{x_0 - g(x)}{\Gamma(q)}\right| + \frac{M}{\Gamma(q)}t_2^{1-q}\int_0^{t_2} (t_2-s)^{q-1}m(s)ds \\
&\to 0, \quad \text{as } t_2 \to 0.
\end{aligned}$$

For any $y \in B(J)$ and $0 < t_1 < t_2 \le a$, we get

$$\begin{aligned}|(\mathscr{T}y)(t_2) - (\mathscr{T}y)(t_1)| &\le |(\mathscr{T}_1 y)(t_2) - (\mathscr{T}_1 y)(t_1)| + |(\mathscr{T}_2 y)(t_2) - (\mathscr{T}_2 y)(t_1)| \\ &\le |(P_q(t_2) - P_q(t_1))(x_0 - g(x))| + I_1 + I_2 + I_3,\end{aligned}$$

where I_1, I_2 and I_3 are defined as in the proof of Lemma 4.4. According to Proposition 4.2, we know that $|(\mathscr{T}y)(t_2) - (\mathscr{T}y)(t_1)|$ tends to zero independently of $y \in B(J)$ as $t_2 \to t_1$, which means that $\{\mathscr{T}y : y \in B(J)\}$ is equicontinuous. □

Lemma 4.7. *Assume that (H1), (H2), (H3)′ and (H4)′ hold. Then $\mathscr{T}$ maps $B(J)$ into $B(J)$, and $\mathscr{T}$ is continuous in $B(J)$.*

Proof. For any $y \in B(J)$, we have

$$|(\mathscr{T}y)(t)| \le \frac{|x_0| + L_1 r + L_2}{\Gamma(q)} \le r, \quad \text{for } t = 0$$

and

$$|(\mathscr{T}y)(t)| = t^{1-q}|(Tx)(t)| \le r, \quad \text{for } t \in (0, a].$$

Hence, $\|\mathscr{T}y\|_B \le r$, for any $y \in B(J)$. Using the similar argument as that we did in the proof of Lemma 4.5, we know that $\mathscr{T}$ is continuous in $B(J)$. □

4.2.4 *Compact Semigroup Case*

In the following, we suppose that the operator A generates a compact C_0-semigroup $\{Q(t)\}_{t\ge 0}$ on X, that is, for any $t > 0$, the operator $Q(t)$ is compact.

Theorem 4.1. *Assume that $Q(t)(t > 0)$ is compact. Furthermore assume that (H1)-(H4) hold. Then the nonlocal Cauchy problem* (4.1) *has at least one mild solution in $B_r^{(q)}(J')$.*

Proof. Obviously, x is a mild solution of (4.1) in $B_r^{(q)}(J')$ if and only if y is a fixed point of $y = \mathscr{T}y$ in $B(J)$, where $x(t) = t^{q-1}y(t)$. So, it is enough to prove that $y = \mathscr{T}y$ has a fixed point in $B(J)$.

For any $y_1, y_2 \in B(J)$, according to (H3), we have

$$\begin{aligned}|\mathscr{T}_1 y_1(t) - \mathscr{T}_1 y_2(t)| &= t^{1-q}|(T_1 x_1)(t) - (T_1 x_2)(t)| \\ &\le \frac{M}{\Gamma(q)}|g(x_1) - g(x_2)| \\ &\le \frac{ML}{\Gamma(q)}\|x_1 - x_2\|_q \\ &= \frac{ML}{\Gamma(q)}\|y_1 - y_2\|,\end{aligned}$$

which implies that $\|\mathscr{T}_1 y_1 - \mathscr{T}_1 y_2\| \le \frac{ML}{\Gamma(q)}\|y_1 - y_2\|$. Thus, we obtain that

$$\alpha(\mathscr{T}_1(B(J))) \le \frac{ML}{\Gamma(q)}\alpha(B(J)). \tag{4.8}$$

Next, we show that for any $t \in [0,a]$, $V(t) = \{(\mathscr{T}_2 y)(t), y \in B(J)\}$ is relatively compact in X. Obviously, $V(0)$ is relatively compact in X. Let $t \in (0,a]$ be fixed. For $\forall\ \varepsilon \in (0,t)$ and $\forall\ \delta > 0$, define an operator $\mathscr{T}_{\varepsilon,\delta}$ on $B(J)$ by the formula

$$\begin{aligned}(\mathscr{T}_{\varepsilon,\delta} y)(t) &= qt^{1-q}\int_0^{t-\varepsilon}\int_\delta^\infty \theta(t-s)^{q-1}M_q(\theta)Q((t-s)^q\theta)f(s,x(s))d\theta ds\\ &= qt^{1-q}Q(\varepsilon^q\delta)\int_0^{t-\varepsilon}\int_\delta^\infty \theta(t-s)^{q-1}M_q(\theta)Q((t-s)^q\theta-\varepsilon^q\delta)f(s,x(s))d\theta ds,\end{aligned}$$

where $x \in B_r^{(q)}(J')$. Then from the compactness of $Q(\varepsilon^q\delta)(\varepsilon^q\delta > 0)$, we obtain that the set $V_{\varepsilon,\delta}(t) = \{(\mathscr{T}_{\varepsilon,\delta} y)(t),\ y \in B(J)\}$ is relatively compact in X for $\forall\ \varepsilon \in (0,t)$ and $\forall\ \delta > 0$. Moreover, for every $y \in B(J)$, we have

$$\begin{aligned}&|(\mathscr{T}_2 y)(t) - (\mathscr{T}_{\varepsilon,\delta} y)(t)|\\ &\le \left|qt^{1-q}\int_0^t\int_0^\delta \theta(t-s)^{q-1}M_q(\theta)Q((t-s)^q\theta)f(s,x(s))d\theta ds\right|\\ &\quad + \left|qt^{1-q}\int_{t-\varepsilon}^t\int_\delta^\infty \theta(t-s)^{q-1}M_q(\theta)Q((t-s)^q\theta)f(s,x(s))d\theta ds\right|\\ &\le qMt^{1-q}\int_0^t (t-s)^{q-1}m(s)ds\int_0^\delta \theta M_q(\theta)d\theta\\ &\quad + qMt^{1-q}\int_{t-\varepsilon}^t (t-s)^{q-1}m(s)ds\int_0^\infty \theta M_q(\theta)d\theta\\ &\le qMt^{1-q}\int_0^t (t-s)^{q-1}m(s)ds\int_0^\delta \theta M_q(\theta)d\theta + \frac{M}{\Gamma(q)}t^{1-q}\int_{t-\varepsilon}^t (t-s)^{q-1}m(s)ds\\ &\to 0, \quad \text{as } \varepsilon \to 0,\ \delta \to 0.\end{aligned}$$

Therefore, there are relatively compact sets arbitrarily close to the set $V(t)$, $t > 0$. Hence the set $V(t)$, $t > 0$ is also relatively compact in X. Therefore, $\{(\mathscr{T}_2 y)(t),\ y \in B(J)\}$ is relatively compact by Arzela-Ascoli theorem. Thus, we have $\alpha(\mathscr{T}_2(B_r^{(q)}(J'))) = 0$. By (4.8), we have

$$\begin{aligned}\alpha(\mathscr{T}(B(J))) &\le \alpha(\mathscr{T}_1(B(J))) + \alpha(\mathscr{T}_2(B(J)))\\ &\le \frac{ML}{\Gamma(q)}\alpha(B(J)).\end{aligned}$$

Thus, the operator $\mathscr{T}$ is an α-contraction in $B(J)$. By Lemma 4.5, we know that $\mathscr{T}$ is continuous. Hence, Theorem 1.10 shows that $\mathscr{T}$ has a fixed point $y^* \in B(J)$. Let $x^*(t) = t^{q-1}y^*(t)$. Then x^* is a mild solution of (4.1). □

Theorem 4.2. *Assume that $Q(t)(t > 0)$ is compact. Furthermore assume that (H1), (H2), (H3)′ and (H4)′ hold. Then the nonlocal Cauchy problem* (4.1) *has at least one mild solution in $B_r^{(q)}(J')$.*

Proof. Since Proposition 4.4, $Q(t)(t > 0)$ is equicontinuous, which implies (H0) is satisfied. Then, by Lemmas 4.4-4.5, we know that $\mathscr{T} : B(J) \to B(J)$ is bounded, continuous and $\{\mathscr{T}y :\ y \in B(J)\}$ is equicontinuous.

According to the argument of Theorem 4.1, we only need prove that for any $t \in J$, the set $V_1(t) = \{(\mathscr{T}_1 y)(t) : y \in B(J)\}$ is relatively compact in X. Obviously, $V_1(0)$ is relatively compact in X. Let $0 < t \le a$ be fixed. For $\forall\ \delta > 0$, define an operator $\mathscr{T}_1^\delta$ on $B(J)$ by the formula

$$\begin{aligned}(\mathscr{T}_1^\delta y)(t) &= q\int_\delta^\infty \theta M_q(\theta)Q(t^q\theta)(x_0 - g(x))d\theta \\ &= qQ(t^q\delta)\int_\delta^\infty \theta M_q(\theta)Q(t^q\theta - t^q\delta)(x_0 - g(x))d\theta,\end{aligned}$$

where $x(t) = t^{q-1}y(t),\ t \in (0,a]$. From the compactness of $Q(t^q\delta)$ $(t^q\delta > 0)$, we obtain that the set $V_1^\delta(t) = \{(\mathscr{T}_1^\delta y)(t),\ y \in B(J)\}$ is relatively compact in X for $\forall\ \delta > 0$. Moreover, for any $y \in B(J)$, we have

$$\begin{aligned}|(\mathscr{T}_1 y)(t) - (\mathscr{T}_1^\delta y)(t)| &= \left|q\int_0^\delta \theta M_q(\theta)Q(t^q\theta)(x_0 - g(x))d\theta\right| \\ &\le qM(|x_0| + L_1 r + L_2)\int_0^\delta \theta M_q(\theta)d\theta.\end{aligned}$$

Therefore, there are relatively compact sets arbitrarily close to the set $V_1(t)$, $t > 0$. Hence the set $V_1(t)$, $t > 0$ is also relatively compact in X. Moreover, $\{\mathscr{T}y : y \in B(J)\}$ is uniformly bounded by Lemma 4.7. Therefore, $\{(\mathscr{T}y)(t),\ y \in B(J)\}$ is relatively compact by Arzela-Ascoli theorem. Hence, Theorem 1.10 shows that $\mathscr{T}$ has a fixed point $y^* \in B(J)$. Let $x^*(t) = t^{q-1}y^*(t)$. Then x^* is a mild solution of (4.1). □

Remark 4.1. If g is not a compact map, we use another method given in Zhu and Li, 2008 to consider the following integral equations

$$x(t) = t^{q-1}P_q\left(t + \frac{1}{n}\right)(x_0 - g(x)) + \int_0^t (t-s)^{q-1}P_q(t-s)f(s,x(s))ds, \quad t \in (0,a]. \tag{4.9}$$

For any $n \in \mathbb{N}$, noticing that the operator $Q(\frac{1}{n})$ is compact, one can easily derive the relative compactness of $V(0)$ and $V(t)(t > 0)$. Then, (4.9) has a solution in $B_r^{(q)}(J')$. By passing the limit, as $n \to \infty$, one obtains a mild solution of the nonlocal Cauchy problem (4.1). However, because $Q(t)$ is replaced by $Q(\frac{1}{n})$, one needs a more restrictive condition than (H4)$'$, such as

(H4)$''$ there exists a constant $r > 0$ such that

$$\frac{M_\varepsilon}{\Gamma(q)}\left(|x_0| + L_1 r + L_2 + \sup_{t\in(0,a]}\left\{t^{1-q}\int_0^t (t-s)^{q-1}m(s)ds\right\}\right) \le r,$$

where $M_\varepsilon = \sup_{t\in[0,a+\varepsilon]}\|Q(t)\|_{B(X)}$, ε is a small constant.

Remark 4.2. The condition (H2) of Theorems 4.1-4.2 can be replaced by the following condition.

(H2)′ There exist a constant $q_1 \in (0, q)$ and $m \in L^{\frac{1}{q_1}}(J, \mathbb{R}^+)$ such that

$$|f(t,x)| \le m(t) \ \text{ for all } x \in B_r^{(q)}(J') \ \text{ and almost all } t \in [0,a].$$

In fact, if (H2)′ holds, by using the Hölder inequality, for any $t_1, t_2 \in J'$ and $t_1 < t_2$, we obtain

$$\begin{aligned}
&|{}_0D_t^{-q}m(t_2) - {}_0D_t^{-q}m(t_1)| \\
&= \frac{1}{\Gamma(q)}\left|\int_0^{t_1}((t_2-s)^{q-1}-(t_1-s)^{q-1})m(s)ds + \int_{t_1}^{t_2}(t_2-s)^{q-1}m(s)ds\right| \\
&\le \frac{1}{\Gamma(q)}\left(\int_0^{t_1}((t_1-s)^{q-1}-(t_2-s)^{q-1})^{\frac{1}{1-q_1}}ds\right)^{1-q_1}\left(\int_0^{t_1}(m(s))^{\frac{1}{q_1}}ds\right)^{q_1} \\
&\quad + \frac{1}{\Gamma(q)}\left(\int_{t_1}^{t_2}((t_2-s)^{q-1})^{\frac{1}{1-q_1}}ds\right)^{1-q_1}\left(\int_{t_1}^{t_2}(m(s))^{\frac{1}{q_1}}ds\right)^{q_1} \\
&\le \frac{1}{\Gamma(q)}\left(\int_0^{t_1}((t_1-s)^{\frac{q-1}{1-q_1}}-(t_2-s)^{\frac{q-1}{1-q_1}})ds\right)^{1-q_1}\|m\|_{L^{\frac{1}{q_1}}} \\
&\quad + \frac{1}{\Gamma(q)}\left(\int_{t_1}^{t_2}(t_2-s)^{\frac{q-1}{1-q_1}}ds\right)^{1-q_1}\|m\|_{L^{\frac{1}{q_1}}} \\
&\le \frac{\|m\|_{L^{\frac{1}{q_1}}}}{\Gamma(q)}\left(\frac{1-q_1}{q-q_1}\right)^{1-q_1}\left(t_1^{\frac{q-q_1}{1-q_1}}+(t_2-t_1)^{\frac{q-q_1}{1-q_1}}-t_2^{\frac{q-q_1}{1-q_1}}\right)^{1-q_1} \\
&\quad + \frac{\|m\|_{L^{\frac{1}{q_1}}}}{\Gamma(q)}\left(\frac{1-q_1}{q-q_1}\right)^{1-q_1}\left((t_2-t_1)^{\frac{q-q_1}{1-q_1}}\right)^{1-q_1} \\
&\le \frac{2\|m\|_{L^{\frac{1}{q_1}}}}{\Gamma(q)}\left(\frac{1-q_1}{q-q_1}\right)^{1-q_1}(t_2-t_1)^{q-q_1} \to 0, \quad \text{as } t_2 \to t_1,
\end{aligned} \tag{4.10}$$

where

$$\|m\|_{L^{\frac{1}{q_1}}} = \left(\int_0^a (m(t))^{\frac{1}{q_1}}dt\right)^{q_1}.$$

Furthermore,

$$\begin{aligned}
&t^{1-q}\int_0^t (t-s)^{q-1}m(s)ds \\
&\le t^{1-q}\left(\int_0^t (t-s)^{\frac{q-1}{1-q_1}}ds\right)^{1-q_1}\left(\int_0^t (m(s))^{\frac{1}{q_1}}ds\right)^{q_1} \\
&\le \left(\frac{1-q_1}{q-q_1}\right)^{1-q_1} t^{1-q_1}\|m\|_{L^{\frac{1}{q_1}}} \\
&\to 0, \quad \text{as } t \to 0.
\end{aligned} \tag{4.11}$$

Thus, (4.10) and (4.11) mean that ${}_0D_t^{-q}m \in C(J', \mathbb{R}^+)$, and $\lim_{t\to 0^+} t^{1-q}{}_0D_t^{-q}m(t) = 0$. Hence, (H2) holds.

Example 4.1. Let $X = L^2([0,\pi],\mathbb{R})$. Consider the following fractional partial differential equations.

$$\begin{cases} \partial_t^q u(t,z) = \partial_z^2 u(t,z) + \partial_z G(t,u(t,z)), & z \in [0,\pi],\ t \in (0,a], \\ u(t,0) = u(t,\pi) = 0, & t \in (0,a], \\ u(0,z) + \sum_{i=0}^{n} \int_0^{\pi} k(z,y)u(t_i,y)dy = u_0(z), & z \in [0,\pi], \end{cases} \tag{4.12}$$

where ∂_t^q is Riemann-Liouville fractional partial derivative of order $0 < q < 1$, $a > 0$, G is a given function, n is a positive integer, $0 < t_0 < t_1 < \cdots < t_n \le a$, $u_0(z) \in X = L^2([0,\pi],\mathbb{R})$, $k(z,y) \in L^2([0,\pi] \times [0,\pi],\mathbb{R}^+)$.

We define an operator A by $Av = v''$ with the domain

$$D(A) = \{v(\cdot) \in X :\ v, v' \text{ absolutly continuous},\ v'' \in X,\ v(0) = v(\pi) = 0\}.$$

Then A generates a C_0-semigroup $\{Q(t)\}_{t\ge 0}$ which is compact, analytic and self-adjoint. Clearly the nonlocal Cauchy problem (4.2) and (H1) are satisfied.

The system (4.12) can be reformulated as the following nonlocal Cauchy problem in X

$$\begin{cases} {}_0D_t^q x(t) = Ax(t) + f(t,x(t)), & \text{almost all } t \in [0,a], \\ {}_0D_t^{q-1} x(0) + g(x) = x_0, \end{cases}$$

where $x(t) = u(t,\cdot)$, that is $x(t)(z) = u(t,z)$, $t \in (0,a]$, $z \in [0,\pi]$. The function $f : J' \times X \to X$ is given by

$$f(t,x(t))(z) = \partial_z G(t,u(t,z)),$$

and the operator $g : C(J',X) \to L(J',X)$ is given by

$$g(x)(z) = \sum_{i=0}^{n} K_g x(t_i)(z),$$

where $K_g v(z) = \int_0^{\pi} k(z,y)v(y)dy$, for $v \in X = L^2([0,\pi],\mathbb{R})$, $z \in [0,\pi]$.

We can take $q = 1/3$ and $f(t,x(t)) = t^{-1/4}\sin x(t)$, and choose

$$m(t) = t^{-1/4},\quad L = (n+1)\left(\int_0^{\pi}\int_0^{\pi} k^2(z,y)dydz\right)^{\frac{1}{2}}$$

and

$$r = \frac{M}{\Gamma(\frac{1}{3}) - ML}\left(|x_0| + g(0) + \frac{\Gamma(\frac{1}{3})\Gamma(\frac{3}{4})}{\Gamma(\frac{12}{13})} a^{\frac{3}{4}}\right).$$

Then, (H1)-(H4) are satisfied (noting that $K_g : X \to X$ is completely continuous). According to Theorem 4.1, system (4.12) has a mild solution in $B_r^{(1/3)}((0,a])$ provided that $\frac{ML}{\Gamma(1/3)} < 1$.

4.2.5 *Noncompact Semigroup Case*

If $Q(t)$ is noncompact, we give an assumption as follows.
(H5) There exists a constant $\ell > 0$ such that for any bounded $D \subset X$,

$$\alpha(f(t,D)) \le \ell\alpha(D).$$

Theorem 4.3. *Assume that (H0)-(H5) hold. Then the nonlocal Cauchy problem* (4.1) *has at least one mild solution in* $B_r^{(q)}(J')$.

Proof. By Lemmas 4.5-4.6, we know that $\mathscr{T}_2 : B(J) \to B(J)$ is bounded, continuous and $\{\mathscr{T}_2 y : y \in B(J)\}$ is equicontinuous. Next, we show that $\mathscr{T}_2$ is compact in a subset of $B(J)$.

For each bounded subset $B_0 \subset B(J)$, set

$$\mathscr{T}^1(B_0) = \mathscr{T}_2(B_0),\ \ \mathscr{T}^n(B_0) = \mathscr{T}_2\left(\overline{co}(\mathscr{T}^{n-1}(B_0))\right),\ n = 2, 3, \dots .$$

Then, from Propositions 1.18-1.20, for any $\varepsilon > 0$, there is a sequence $\{y_n^{(1)}\}_{n=1}^{\infty} \subset B_0$ such that

$$\begin{aligned}
\alpha(\mathscr{T}^1(B_0(t))) &= \alpha(\mathscr{T}_2(B_0(t)))\\
&\le 2\alpha\left(t^{1-q}\int_0^t (t-s)^{q-1}P_q(t-s)f(s,\{s^{q-1}y_n^{(1)}(s)\}_{n=1}^{\infty})ds\right) + \varepsilon\\
&\le \frac{4M}{\Gamma(q)}t^{1-q}\int_0^t (t-s)^{q-1}\alpha\left(f(s,\{s^{q-1}y_n^{(1)}(s)\}_{n=1}^{\infty})\right)ds + \varepsilon\\
&\le \frac{4M\ell}{\Gamma(q)}t^{1-q}\int_0^t (t-s)^{q-1}\alpha(\{s^{q-1}y_n^{(1)}(s)\}_{n=1}^{\infty})ds + \varepsilon\\
&\le \frac{4M\ell\alpha(B_0)}{\Gamma(q)}t^{1-q}\int_0^t (t-s)^{q-1}s^{q-1}ds + \varepsilon\\
&\le \frac{4M\ell\Gamma(q)t^q\alpha(B_0)}{\Gamma(2q)} + \varepsilon.
\end{aligned}$$

Since $\varepsilon > 0$ is arbitrary, we have

$$\alpha(\mathscr{T}^1(B_0(t))) \le \frac{4M\ell\Gamma(q)t^q}{\Gamma(2q)}\alpha(B_0).$$

From Propositions 1.18-1.20, for any $\varepsilon > 0$, there is a sequence $\{y_n^{(2)}\}_{n=1}^{\infty} \subset \overline{co}(\mathscr{T}^1(B_0))$ such that

$$\begin{aligned}
\alpha(\mathscr{T}^2(B_0(t))) &= \alpha(\mathscr{T}_2(\overline{co}(\mathscr{T}^1(B_0(t)))))\\
&\le 2\alpha\left(t^{1-q}\int_0^t (t-s)^{q-1}P_q(t-s)f(s,\{s^{q-1}y_n^{(2)}(s)\}_{n=1}^{\infty})ds\right) + \varepsilon\\
&\le \frac{4Mt^{1-q}}{\Gamma(q)}\int_0^t (t-s)^{q-1}\alpha\left(f(s,\{s^{q-1}y_n^{(2)}(s)\}_{n=1}^{\infty})\right)ds + \varepsilon\\
&\le \frac{4M\ell t^{1-q}}{\Gamma(q)}\int_0^t (t-s)^{q-1}\alpha(\{s^{q-1}y_n^{(2)}(s)\}_{n=1}^{\infty})ds + \varepsilon
\end{aligned}$$

$$\leq \frac{4M\ell t^{1-q}}{\Gamma(q)} \int_0^t (t-s)^{q-1} s^{q-1} \alpha(\{y_n^{(2)}(s)\}_{n=1}^{\infty}) ds + \varepsilon$$
$$\leq \alpha(B_0) \frac{(4M\ell)^2 t^{1-q}}{\Gamma(2q)} \int_0^t (t-s)^{q-1} s^{2q-1} ds + \varepsilon$$
$$= \frac{(4M\ell)^2 \Gamma(q)}{\Gamma(3q)} t^{2q} \alpha(B_0) + \varepsilon.$$

It can be shown, by mathematical induction, that for every $\bar{n} \in \mathbb{N}$,

$$\alpha(\mathscr{T}^{\bar{n}}(B_0(t))) \leq \frac{(4M\ell)^{\bar{n}} \Gamma(q)}{\Gamma((\bar{n}+1)q)} t^{\bar{n}q} \alpha(B_0).$$

Since

$$\lim_{\bar{n}\to\infty} \frac{(4M\ell a^q)^{\bar{n}} \Gamma(q)}{\Gamma((\bar{n}+1)q)} = 0,$$

there exists a positive integer $\hat{n}$ such that

$$\frac{(4M\ell)^{\hat{n}} \Gamma(q)}{\Gamma((\hat{n}+1)q)} t^{\hat{n}q} \leq \frac{(4M\ell a^q)^{\hat{n}} \Gamma(q)}{\Gamma((\hat{n}+1)q)} = k < 1.$$

Then

$$\alpha(\mathscr{T}^{\hat{n}}(B_0(t))) \leq k\alpha(B_0).$$

We know from Proposition 1.16, $\mathscr{T}^{\hat{n}}(B_0(t))$ is bounded and equicontinuous. Then, from Proposition 1.17, we have

$$\alpha(\mathscr{T}^{\hat{n}}(B_0)) = \max_{t\in[0,a]} \alpha(\mathscr{T}^{\hat{n}}(B_0(t))).$$

Hence

$$\alpha(\mathscr{T}^{\hat{n}}(B_0)) \leq k\alpha(B_0).$$

Let

$$D_0 = B(J),\ D_1 = \overline{co}(\mathscr{T}^{\hat{n}}(D)), \ldots,\ D_n = \overline{co}(\mathscr{T}^{\hat{n}}(D_{n-1})),\ n = 2, 3, \ldots .$$

Then, we can get

(i) $D_0 \supset D_1 \supset D_2 \supset \cdots \supset D_{n-1} \supset D_n \supset \cdots$;
(ii) $\lim_{n\to\infty} \alpha(D_n) = 0$.

Then $\hat{D} = \bigcap_{n=0}^{\infty} D_n$ is a nonempty, compact and convex subset in $B(J)$.

We prove $\mathscr{T}_2(\hat{D}) \subset \hat{D}$. Firstly, we show

$$\mathscr{T}_2(D_n) \subset D_n,\quad n = 0, 1, 2, \ldots . \tag{4.13}$$

From $\mathscr{T}^1(D_0) = \mathscr{T}_2(D_0) \subset D_0$, we know $\overline{co}(\mathscr{T}^1(D_0)) \subset D_0$. Therefore

$$\mathscr{T}^2(D_0) = \mathscr{T}_2(\overline{co}(\mathscr{T}^1(D_0))) \subset \mathscr{T}_2(D_0) = \mathscr{T}^1(D_0),$$

$$\mathscr{T}^3(D_0) = \mathscr{T}_2(\overline{co}(\mathscr{T}^2(D_0))) \subset \mathscr{T}_2(\overline{co}(\mathscr{T}^1(D_0))) = \mathscr{T}^2(D_0),$$

$$\vdots$$

$$\mathscr{T}^{\hat{n}}(D_0) = \mathscr{T}_2(\overline{co}(\mathscr{T}^{\hat{n}-1}(D_0))) \subset \mathscr{T}_2(\overline{co}(\mathscr{T}^{\hat{n}-2}(D_0))) = \mathscr{T}^{\hat{n}-1}(D_0).$$

Hence, $D_1 = \overline{co}(\mathscr{T}^{\hat{n}}(D_0)) \subset \overline{co}(\mathscr{T}^{\hat{n}-1}(D_0))$, so $\mathscr{T}(D_1) \subset \mathscr{T}(\overline{co}(\mathscr{T}^{\hat{n}-1}(D_0))) = \mathscr{T}^{\hat{n}}(D_0) \subset \overline{co}(\mathscr{T}^{\hat{n}}(D_0)) = D_1$. Employing the same method, we can prove $\mathscr{T}_2(D_n) \subset D_n (n = 0, 1, 2, ...)$. By (4.13), we get $\mathscr{T}_2(\hat{D}) \subset \bigcap_{n=0}^{\infty} \mathscr{T}_2(D_n) \subset \bigcap_{n=0}^{\infty} D_n = \hat{D}$. Then $\mathscr{T}_2(\hat{D})$ is compact. Hence, $\alpha(\mathscr{T}_2(\hat{D})) = 0$.

On the other hand, for any $y_1, y_2 \in \hat{D}$ and $t \in (0, a]$, according to (H3), we have

$$\begin{aligned}|\mathscr{T}_1 y_1(t) - \mathscr{T}_1 y_2(t)| &= t^{1-q}|(T_1x_1)(t) - (T_1x_2)(t)| \\ &\le \frac{M}{\Gamma(q)}|g(x_1) - g(x_2)| \\ &\le \frac{ML}{\Gamma(q)}\|x_1 - x_2\|_q \\ &= \frac{ML}{\Gamma(q)}\|y_1 - y_2\|,\end{aligned}$$

which implies that $\|\mathscr{T}_1 y_1 - \mathscr{T}_1 y_2\| \le \frac{ML}{\Gamma(q)}\|y_1 - y_2\|$. Thus, we obtain that

$$\alpha(\mathscr{T}_1(\hat{D})) \le \frac{ML}{\Gamma(q)}\alpha(\hat{D}). \tag{4.14}$$

By (4.14), we have

$$\begin{aligned}\alpha(\mathscr{T}(\hat{D})) &\le \alpha(\mathscr{T}_1(\hat{D})) + \alpha(\mathscr{T}_2(\hat{D})) \\ &\le \frac{ML}{\Gamma(q)}\alpha(\hat{D}).\end{aligned}$$

Thus, the operator $\mathscr{T}$ is an α-contraction in $\hat{D}$. By Lemma 4.5, we know that $\mathscr{T}$ is continuous. Hence, Theorem 1.10 shows that $\mathscr{T}$ has a fixed point $y^* \in B(J)$. Let $x^*(t) = t^{q-1}y^*(t)$. Then x^* is a mild solution of (4.1). □

Theorem 4.4. *Assume that (H0)-(H2), (H3)′, (H4)′ and (H5) hold, then the nonlocal Cauchy problem* (4.1) *has at least one mild solution in* $B_r^{(q)}(J')$.

Proof. Since $g(x)$ is compact and $P_q(t)$ is bounded, for every $t > 0$, $\{(\mathscr{T}_1 y)(t),\ y \in B(J)\}$ is relatively compact. Thus, we have $\alpha(\mathscr{T}_1(B(J))) = 0$.

By the proof of Theorem 4.3, we know that there exists a $\hat{D} \subset B(J)$ such that $\mathscr{T}_2(\hat{D})$ is relatively compact, i.e., $\alpha(\mathscr{T}_2(\hat{D})) = 0$. Hence, we have

$$\alpha(\mathscr{T}(\hat{D})) \le \alpha(\mathscr{T}_1(\hat{D})) + \alpha(\mathscr{T}_2(\hat{D})) = 0.$$

Hence, Theorem 1.10 shows that $\mathscr{T}$ has a fixed point $y^* \in B(J)$. Let $x^*(t) = t^{q-1}y^*(t)$. Then x^* is a mild solution of (4.1). □

4.3 Evolution Equations with Caputo Derivative

4.3.1 *Introduction*

Consider the following nonlocal Cauchy problems of fractional evolution equation with Caputo fractional derivative

$$\begin{cases} {}_0^C D_t^q x(t) = Ax(t) + f(t, x(t)), & \text{a.e. } t \in J := [0, a], \\ x(0) + g(x) = x_0, \end{cases} \tag{4.15}$$

where ${}_0^C D_t^q$ is Caputo fractional derivative of order q, $0 < q < 1$, A is the infinitesimal generator of a C_0-semigroup of bounded linear operators $\{Q(t)\}_{t \geq 0}$ in Banach space X, $f : J \times X \to X$, $g : C(J, X) \to L(J, X)$ are given operators satisfying some assumptions and x_0 is an element of the Banach space X.

In this section, by using the theory of Hausdorff measure of noncompactness and fixed point theorems, we study the nonlocal Cauchy problem (4.15) in the cases $Q(t)$ is compact and noncompact, respectively. Subsection 4.3.2 is devoted to obtaining the appropriate definition on the mild solutions of the problem (4.15) by considering a integral equation which is given in terms of probability density. In Subsection 4.3.3, we give some preliminary lemmas. Subsection 4.3.4 provides various existence theorems of mild solutions for the Cauchy problem (4.15) in the case $Q(t)$ is compact. In Subsection 4.3.5, we establish various existence theorems of mild solutions for the Cauchy problem (4.15) in the case $Q(t)$ is noncompact.

4.3.2 *Definition of Mild Solutions*

Lemma 4.8. *If*

$$x(t) = x_0 - g(x) + \frac{1}{\Gamma(q)} \int_0^t (t-s)^{q-1}[Ax(s) + f(s, x(s))]ds, \quad \textit{for } t \geq 0 \tag{4.16}$$

holds, then we have

$$x(t) = S_q(t)(x_0 - g(x)) + \int_0^t (t-s)^{q-1} P_q(t-s) f(s, x(s))ds, \quad \textit{for } t \geq 0,$$

where

$$S_q(t) = \int_0^\infty M_q(\theta) Q(t^q \theta) d\theta, \quad P_q(t) = \int_0^\infty q\theta M_q(\theta) Q(t^q \theta) d\theta.$$

Proof. Let $\lambda > 0$. Applying the Laplace transform

$$\nu(\lambda) = \int_0^\infty e^{-\lambda s} x(s) ds \text{ and } \omega(\lambda) = \int_0^\infty e^{-\lambda s} f(s, x(s)) ds, \quad \lambda > 0$$

to (4.16), we have

$$\begin{aligned} \nu(\lambda) &= \frac{1}{\lambda}(x_0 - g(x)) + \frac{1}{\lambda^q} A\nu(\lambda) + \frac{1}{\lambda^q}\,\omega(\lambda) \\ &= \lambda^{q-1}(\lambda^q I - A)^{-1}(x_0 - g(x)) + (\lambda^q I - A)^{-1}\omega(\lambda) \\ &= \lambda^{q-1} \int_0^\infty e^{-\lambda^q s} Q(s)(x_0 - g(x)) ds + \int_0^\infty e^{-\lambda^q s} Q(s)\omega(\lambda) ds, \end{aligned} \tag{4.17}$$

provided that the integrals in (4.17) exist, where I is the identity operator defined on X.

Using (4.5) and (4.17), we get

$$\begin{aligned}
&\lambda^{q-1}\int_0^\infty e^{-\lambda^q s}Q(s)(x_0-g(x))ds\\
&=\int_0^\infty q(\lambda t)^{q-1}e^{-(\lambda t)^q}Q(t^q)(x_0-g(x))dt\\
&=\int_0^\infty -\frac{1}{\lambda}\frac{d}{dt}\left(e^{-(\lambda t)^q}\right)Q(t^q)(x_0-g(x))dt \qquad (4.18)\\
&=\int_0^\infty\int_0^\infty \theta\psi_q(\theta)e^{-\lambda t\theta}Q(t^q)(x_0-g(x))d\theta dt\\
&=\int_0^\infty e^{-\lambda t}\left(\int_0^\infty \psi_q(\theta)Q\left(\frac{t^q}{\theta^q}\right)(x_0-g(x))d\theta\right)dt.
\end{aligned}$$

According to (4.7), (4.17) and (4.18), we have

$$\begin{aligned}
\nu(\lambda)=\int_0^\infty e^{-\lambda t}\bigg(&\int_0^\infty \psi_q(\theta)Q\left(\frac{t^q}{\theta^q}\right)(x_0-g(x))d\theta\\
&+q\int_0^t\int_0^\infty \psi_q(\theta)Q\left(\frac{(t-s)^q}{\theta^q}\right)f(s,x(s))\frac{(t-s)^{q-1}}{\theta^q}d\theta ds\bigg)dt.
\end{aligned}$$

Now we can invert the last Laplace transform to get

$$\begin{aligned}
x(t)&=\int_0^\infty \psi_q(\theta)Q\left(\frac{t^q}{\theta^q}\right)(x_0-g(x))d\theta\\
&\quad+q\int_0^t\int_0^\infty \psi_q(\theta)Q\left(\frac{(t-s)^q}{\theta^q}\right)f(s,x(s))\frac{(t-s)^{q-1}}{\theta^q}d\theta ds\\
&=\int_0^\infty M_q(\theta)Q(t^q\theta)(x_0-g(x))d\theta\\
&\quad+q\int_0^t\int_0^\infty \theta(t-s)^{q-1}M_q(\theta)Q((t-s)^q\theta)f(s,x(s))d\theta ds\\
&=S_q(t)(x_0-g(x))+\int_0^t(t-s)^{q-1}P_q(t-s)f(s,x(s))ds.
\end{aligned}$$

The proof is completed. □

Due to Lemma 4.8, we give the following definition of the mild solution of (4.15).

Definition 4.2. By the mild solution of the nonlocal Cauchy problem (4.15), we mean that the function $x\in C(J,X)$ which satisfies

$$x(t)=S_q(t)(x_0-g(x))+\int_0^t(t-s)^{q-1}P_q(t-s)f(s,x(s))ds,\quad \text{for } t\in[0,a].$$

Suppose that A is the infinitesimal generator of a C_0-semigroup $\{Q(t)\}_{t\geq 0}$ of uniformly bounded linear operators on Banach space X. This means that there exists $M>1$ such that $M=\sup_{t\in[0,\infty)}\|Q(t)\|_{B(X)}<\infty$.

Proposition 4.5. *(Zhou and Jiao, 2010a) For any fixed $t > 0$, $\{S_q(t)\}_{t>0}$ and $\{P_q(t)\}_{t>0}$ are linear and bounded operators, i.e., for any $x \in X$*

$$|S_q(t)x| \le M|x|, \quad |P_q(t)x| \le \frac{M}{\Gamma(q)}|x|.$$

Proposition 4.6. *(Zhou and Jiao, 2010a) Operators $\{S_q(t)\}_{t>0}$ and $\{P_q(t)\}_{t>0}$ are strongly continuous, which means that, for $\forall\ x \in X$ and $0 < t' < t'' \le a$, we have*

$$|S_q(t'')x - S_q(t')x| \to 0, \quad |P_q(t'')x - P_q(t')x| \to 0, \quad \textit{as } t'' \to t'.$$

Proposition 4.7. *(Zhou and Jiao, 2010a) Assume that $\{Q(t)\}_{t>0}$ is compact operator. Then $\{S_q(t)\}_{t>0}$ and $\{P_q(t)\}_{t>0}$ are also compact operators.*

Remark 4.3. Since $S_q(\cdot)$ and $P_q(\cdot)$ are associated with the q, there are no analogue of the semigroup property, i.e., $S_q(t+s) \ne S_q(t)S_q(s)$, $P_q(t+s) \ne P_q(t)P_q(s)$ for $t, s > 0$.

4.3.3 *Preliminary Lemmas*

For $r > 0$, let $B_r(J)$ be the closed ball of the space $C(J,X)$ with radius r and center at 0, that is,

$$B_r(J) = \{x \in C(J,X) : \ \|x\| \le r\},$$

where $\|x\| = \sup_{t\in[0,a]} |x(t)|$.

We introduce the following hypotheses:

(H0) $Q(t)(t > 0)$ is equicontinuous, i.e., $Q(t)$ is continuous in the uniform operator topology for $t > 0$;

(H1) for each $t \in J$, the function $f(t,\cdot) : X \to X$ is continuous and for each $x \in X$, the function $f(\cdot,x) : J \to X$ is strongly measurable;

(H2) there exists a function m such that

$${}_0D_t^{-q}m \in C(J,\mathbb{R}^+), \quad \lim_{t\to 0+} {}_0D_t^{-q}m(t) = 0$$

and

$$|f(t,x)| \le m(t) \ \text{ for all } x \in B_r(J) \text{ and almost all } t \in [0,a];$$

(H3) there exists a constant $L \in (0, \frac{1}{M})$, the operator $g : C(J,X) \to L(J,X)$ satisfies

$$|g(x_1) - g(x_2)| \le L\|x_1 - x_2\|, \ \text{ for } x_1, x_2 \in B_r(J);$$

(H4) there exists a constant $r > 0$ such that

$$\frac{M}{1-ML}\left(|x_0| + |g(0)| + \sup_{t\in[0,a]}\left\{\frac{1}{\Gamma(q)}\int_0^t (t-s)^{q-1}m(s)ds\right\}\right) \le r;$$

(H3)′ the operator $g : C(J, X) \to L(J, X)$ is a continuous and compact map, and there exist positive constants L_1, L_2 such that $|g(x)| \leq L_1\|x\| + L_2$ for all $x \in B_r(J)$;

(H4)′ there exists a constant $r > 0$ such that

$$\frac{M}{1 - ML_1}\left(|x_0| + L_2 + \sup_{t\in[0,a]}\left\{\frac{1}{\Gamma(q)}\int_0^t (t-s)^{q-1}m(s)ds\right\}\right) \leq r.$$

For any $x \in B_r(J)$, we define an operator T as follows

$$(Tx)(t) = (T_1x)(t) + (T_2x)(t),$$

where

$$(T_1x)(t) = S_q(t)(x_0 - g(x)), \qquad \text{for } t \in [0, a],$$

$$(T_2x)(t) = \int_0^t (t-s)^{q-1}P_q(t-s)f(s, x(s))ds, \quad \text{for } t \in [0, a].$$

Obviously, x is a mild solution of (4.15) in $B_r(J)$ if and only if the operator equation $x = Tx$ has a solution $x \in B_r(J)$.

Lemma 4.9. *Assume that (H0)-(H3) hold. Then* $\{T_2x : x \in B_r(J)\}$ *is equicontinuous.*

Proof. For any $x \in B_r(J)$, for $t_1 = 0$, $0 < t_2 \leq a$, by (H2), we get

$$\begin{aligned}|(T_2x)(t_2) - (T_2x)(0)| &= \left|\int_0^{t_2}(t_2-s)^{q-1}P_q(t_2-s)f(s,x(s))ds\right| \\ &\leq \frac{M}{\Gamma(q)}\int_0^{t_2}(t_2-s)^{q-1}m(s)ds \to 0, \quad \text{as } t_2 \to 0.\end{aligned}$$

For $0 < t_1 < t_2 \leq a$, we have

$$\begin{aligned}&|(T_2x)(t_2) - (T_2x)(t_1)| \\ &= \left|\int_0^{t_2}(t_2-s)^{q-1}P_q(t_2-s)f(s,x(s))ds - \int_0^{t_1}(t_1-s)^{q-1}P_q(t_1-s)f(s,x(s))ds\right| \\ &\leq \left|\int_{t_1}^{t_2}(t_2-s)^{q-1}P_q(t_2-s)f(s,x(s))ds\right| \\ &\quad + \left|\int_0^{t_1}(t_2-s)^{q-1}P_q(t_2-s)f(s,x(s))ds - \int_0^{t_1}(t_1-s)^{q-1}P_q(t_2-s)f(s,x(s))ds\right| \\ &\quad + \left|\int_0^{t_1}(t_1-s)^{q-1}P_q(t_2-s)f(s,x(s))ds - \int_0^{t_1}(t_1-s)^{q-1}P_q(t_1-s)f(s,x(s))ds\right| \\ &\leq \frac{M}{\Gamma(q)}\int_{t_1}^{t_2}(t_2-s)^{q-1}m(s)ds + \frac{M}{\Gamma(q)}\int_0^{t_1}\left((t_1-s)^{q-1} - (t_2-s)^{q-1}\right)m(s)ds \\ &\quad + \int_0^{t_1}(t_1-s)^{q-1}|P_q(t_2-s)f(s,x(s)) - P_q(t_1-s)f(s,x(s))|ds \\ &\leq \frac{M}{\Gamma(q)}\left|\int_0^{t_2}(t_2-s)^{q-1}m(s)ds - \int_0^{t_1}(t_1-s)^{q-1}m(s)ds\right|\end{aligned}$$

$$
\begin{aligned}
&+\frac{2M}{\Gamma(q)}\int_0^{t_1}\left((t_1-s)^{q-1}-(t_2-s)^{q-1}\right)m(s)ds\\
&+\int_0^{t_1}(t_1-s)^{q-1}\|P_q(t_2-s)-P_q(t_1-s)\|_{B(X)}m(s)ds\\
=:&\, I_1+I_2+I_3,
\end{aligned}
$$

Since ${}_0D_t^{-q}m \in C(J,\mathbb{R}^+)$, thus $I_1 \to 0$ as $t_2 \to t_1$.

For $t_1 < t_2$,

$$I_2 \le \frac{2M}{\Gamma(q)}\int_0^{t_1}(t_1-s)^{q-1}m(s)ds,$$

then by Lebesgue dominated convergence theorem, we have that $I_2 \to 0$ as $t_2 \to t_1$.

For $\varepsilon > 0$ be small enough, we have

$$
\begin{aligned}
I_3 \le& \int_0^{t_1-\varepsilon}(t_1-s)^{q-1}\|P_q(t_2-s)-P_q(t_1-s)\|_{B(X)}\,m(s)ds\\
&+\int_{t_1-\varepsilon}^{t_1}(t_1-s)^{q-1}\|P_q(t_2-s)-P_q(t_1-s)\|_{B(X)}\,m(s)ds\\
\le& \int_0^{t_1}(t_1-s)^{q-1}m(s)ds \sup_{s\in[0,t_1-\varepsilon]}\|P_q(t_2-s)-P_q(t_1-s)\|_{B(X)}\\
&+\frac{2M}{\Gamma(q)}\int_{t_1-\varepsilon}^{t_1}(t_1-s)^{q-1}m(s)ds\\
\le& \int_0^{t_1}(t_1-s)^{q-1}m(s)ds \sup_{s\in[0,t_1-\varepsilon]}\|P_q(t_2-s)-P_q(t_1-s)\|_{B(X)}\\
&+\frac{2M}{\Gamma(q)}\left|\int_0^{t_1}(t_1-s)^{q-1}m(s)ds-\int_0^{t_1-\varepsilon}(t_1-\varepsilon-s)^{q-1}m(s)ds\right|\\
&+\frac{2M}{\Gamma(q)}\int_0^{t_1-\varepsilon}[(t_1-\varepsilon-s)^{q-1}-(t_1-s)^{q-1}]m(s)ds.\\
=:&\, I_{31}+I_{32}+I_{33}.
\end{aligned}
$$

By (H0), it is easy to see that $I_{31} \to 0$ as $t_2 \to t_1$. Similar to the proof that I_1, I_2 tend to zero, we get $I_{32} \to 0$ and $I_{33} \to 0$ as $\varepsilon \to 0$. Thus, I_3 tends to zero independently of $x \in B_r(J)$ as $t_2 \to t_1$, $\varepsilon \to 0$. Therefore, $|(T_2x)(t_1) - (T_2x)(t_2)|$ tends to zero independently of $x \in B_r(J)$ as $t_2 \to t_1$, which means that $\{T_2x : x \in B_r(J)\}$ is equicontinuous. □

Lemma 4.10. *Assume that (H1)-(H4) hold. Then T maps $B_r(J)$ into $B_r(J)$, and T is continuous in $B_r(J)$.*

Proof. **Claim I.** T maps $B_r(J)$ into $B_r(J)$.

For any $x \in B_r(J)$ and $t \in J$, by using (H1)-(H4), we have

$$|(Tx)(t)| = |(T_1x)(t) + (T_2x)(t)|$$

$$
\begin{aligned}
&\le |S_q(t)(x_0 - g(x))| + \left| \int_0^t (t-s)^{q-1} P_q(t-s) f(s, x(s)) ds \right| \\
&\le M(|x_0| + L\|x - 0\| + |g(0)|) + \frac{M}{\Gamma(q)} \int_0^t (t-s)^{q-1} |f(s, x(s))| ds \\
&\le M\Bigg(|x_0| + Lr + |g(0)| + \sup_{t\in[0,a]} \left\{ \frac{1}{\Gamma(q)} \int_0^t (t-s)^{q-1} m(s) ds \right\} \Bigg) \\
&\le r.
\end{aligned}
$$

Hence, $\|Tx\| \le r$ for any $x \in B_r(J)$.

Claim II. T is continuous in $B_r(J)$.

For any $\{x_m\}_{m=1}^{\infty} \subseteq B_r(J), x \in B_r(J)$ with $\lim_{n\to\infty} \|x_m - x\| = 0$, by the condition (H1), we have

$$
\lim_{m\to\infty} f(s, x_m(s)) = f(s, x(s)).
$$

On the one hand, using (H2), we get for each $t \in J$, $(t-s)^{q-1}|f(s, x_m(s)) - f(s, x(s))| \le (t-s)^{q-1} 2m(s)$. On the other hand, the function $s \to (t-s)^{q-1} 2m(s)$ is integrable for $s \in [0, t]$ and $t \in J$. By Lebesgue dominated convergence theorem, we get

$$
\int_0^t (t-s)^{q-1} |f(s, x_m(s)) - f(s, x(s))| ds \to 0, \quad \text{as } m \to \infty.
$$

Then, for $t \in J$,

$$
\begin{aligned}
&|(Tx_m)(t) - (Tx)(t)| \\
&\le |S_q(t)(g(x_m) - g(x))| + \left| \int_0^t (t-s)^{q-1} P_q(t-s) [f(s, x_m(s)) - f(s, x(s))] ds \right| \\
&\le ML\|x_m - x\| + \frac{M}{\Gamma(q)} \int_0^t (t-s)^{q-1} |f(s, x_m(s)) - f(s, x(s))| ds.
\end{aligned}
$$

Therefore, $Tx_m \to Tx$ pointwise on J as $m \to \infty$, by which Lemma 4.9 implies that $Tx_m \to Tx$ uniformly on J as $m \to \infty$ and so T is continuous. □

Lemma 4.11. *Assume that there exists a constant $r > 0$ such that the conditions (H0)-(H2) and (H3)′ are satisfied. Then $\{Tx : x \in B_r(J)\}$ is equicontinuous.*

Proof. For any $x \in B_r(J)$ and $0 \le t_1 < t_2 \le a$, we get

$$
|(Tx)(t_2) - (Tx)(t_1)| \le |(S_q(t_2) - S_q(t_1))(x_0 - g(x))| + I_1 + I_2 + I_3,
$$

where I_1, I_2 and I_3 are defined as in the proof of Lemma 4.9. According to Proposition 4.6, we know that $|(Tx)(t_2) - (Tx)(t_1)|$ tends to zero independently of $x \in B_r(J)$ as $t_2 \to t_1$, which means that $\{Tx, x \in B_r(J)\}$ is equicontinuous. □

Lemma 4.12. *Assume that there exists a constant $r > 0$ such that the conditions (H1), (H2), (H3)′ and (H4)′ are satisfied. Then T maps $B_r(J)$ into $B_r(J)$, and T is continuous in $B_r(J)$.*

Proof. For any $x \in B_r(J)$ and $t \in J$, by using (H1), (H2), (H3)$'$ and (H4)$'$, we have

$$\begin{aligned}
|(Tx)(t)| &= |(T_1x)(t) + (T_2x)(t)| \\
&\leq |S_q(t)(x_0 - g(x))| + \left| \int_0^t (t-s)^{q-1} P_q(t-s) f(s, x(s)) ds \right| \\
&\leq M(|x_0| + L_1 r + L_2) + \frac{M}{\Gamma(q)} \int_0^t (t-s)^{q-1} |f(s, x(s))| ds \\
&\leq M\left(|x_0| + L_1 r + L_2 + \sup_{t\in[0,a]} \left\{ \frac{1}{\Gamma(q)} \int_0^t (t-s)^{q-1} m(s) ds \right\} \right) \\
&\leq r.
\end{aligned}$$

Hence, $\|Tx\| \leq r$ for any $x \in B_r(J)$. Using the similar argument as that we did in the proof of Lemma 4.10, we know that T is continuous in $B_r(J)$. □

4.3.4 *Compact Semigroup Case*

In the following, we suppose that the operator A generates a compact C_0-semigroup $\{Q(t)\}_{t\geq 0}$ on X, that is, for any $t > 0$, the operator $Q(t)$ is compact.

Theorem 4.5. *Assume that $Q(t)(t > 0)$ is compact. Furthermore assume that there exists a constant $r > 0$ such that the conditions (H1)-(H4) are satisfied. Then the nonlocal Cauchy problem* (4.15) *has at least one mild solution in $B_r(J)$.*

Proof. Since Proposition 4.4, $Q(t)(t > 0)$ is equicontinuous, which implies (H0) is satisfied.

For any $x_1, x_2 \in B_r(J)$ and $t \in J$, according to (H3), we have

$$\begin{aligned}
|(T_1x_1)(t) - (T_1x_2)(t)| &\leq M|g(x_1) - g(x_2)| \\
&\leq ML\|x_1 - x_2\|,
\end{aligned}$$

which implies that $\|T_1x_1 - T_1x_2\| \leq ML\|x_1 - x_2\|$. Thus, we obtain that

$$\alpha(T_1B_r(J)) \leq ML\alpha(B_r(J)). \tag{4.19}$$

Next, we show that $\{T_2x,\ x \in B_r(J)\}$ is relatively compact, i.e., $\alpha(T_2(B_r(J))) = 0$. It suffices to show that the family of functions $\{T_2x : x \in B_r(J)\}$ is uniformly bounded and equicontinuous, and for any $t \in J$, $\{(T_2x)(t) : x \in B_r(J)\}$ is relatively compact in X.

By Lemma 4.10, we have $\|T_2x\| \leq r$, for any $x \in B_r(J)$, which means that $\{T_2x : x \in B_r(J)\}$ is uniformly bounded. By Lemma 4.9, $\{T_2x : x \in B_r(J)\}$ is equicontinuous.

It remains to prove that for any $t \in J$, $V(t) = \{(T_2x)(t) : x \in B_r(J)\}$ is relatively compact in X.

Obviously, $V(0)$ is relatively compact in X. Let $0 < t \leq a$ be fixed. For $\forall\, \varepsilon \in (0, t)$ and $\forall\, \delta > 0$, define an operator T_ε^δ on $B_r(J)$ by the formula

$$(T_\varepsilon^\delta x)(t) = q \int_0^{t-\varepsilon} \int_\delta^\infty \theta(t-s)^{q-1} M_q(\theta) Q((t-s)^q\theta) f(s, x(s)) d\theta ds$$

$$= q\, Q(\varepsilon^q\delta) \int_0^{t-\varepsilon} \int_\delta^\infty \theta(t-s)^{q-1} M_q(\theta) Q((t-s)^q\theta - \varepsilon^q\delta) f(s, x(s)) d\theta ds,$$

where $x \in B_r(J)$. Then from the compactness of $Q(\varepsilon^q\delta)(\varepsilon^q\delta > 0)$, we obtain that the set $V_\varepsilon^\delta(t) = \{(T_\varepsilon^\delta x)(t) : \ x \in B_r(J)\}$ is relatively compact in X for $\forall\ \varepsilon \in (0, t)$ and $\forall\ \delta > 0$. Moreover, for any $x \in B_r(J)$, we have

$$\begin{aligned} |(T_2x)(t) - (T_\varepsilon^\delta x)(t)| &\le q\left| \int_0^t \int_0^\delta \theta(t-s)^{q-1} M_q(\theta) Q((t-s)^q\theta) f(s, x(s)) d\theta ds \right| \\ &\quad + q\left| \int_{t-\varepsilon}^t \int_\delta^\infty \theta(t-s)^{q-1} M_q(\theta) Q((t-s)^q\theta) f(s, x(s)) d\theta ds \right| \\ &\le qM \int_0^t (t-s)^{q-1} m(s) ds \int_0^\delta \theta M_q(\theta) d\theta \\ &\quad + \frac{M}{\Gamma(q)} \int_{t-\varepsilon}^t (t-s)^{q-1} m(s) ds \to 0, \quad \text{as } \varepsilon \to 0, \quad \delta \to 0. \end{aligned}$$

Therefore, there are relatively compact sets arbitrarily close to the set $V(t)$, $t > 0$. Hence the set $V(t),\ t > 0$ is also relatively compact in X.

Therefore, $\{(T_2x)(t) : \ x \in B_r(J)\}$ is relatively compact by Arzela-Ascoli theorem. Thus, we have $\alpha(T_2(B_r(J))) = 0$. By (4.19), we have

$$\alpha(T(B_r(J))) \le \alpha(T_1(B_r(J))) + \alpha(T_2(B_r(J))) \le ML\alpha(B_r(J)).$$

Thus, the operator T is an α-contraction in $B_r(J)$. By Lemma 4.10, we know that T is continuous. Hence, Theorem 1.10 shows that T has a fixed point in $B_r(J)$. Therefore, the nonlocal Cauchy problem (4.15) has a mild solution in $B_r(J)$. □

Theorem 4.6. *Assume that $Q(t)(t > 0)$ is compact. Furthermore assume that (H1), (H2), (H3)′ and (H4)′ hold. Then the nonlocal Cauchy problem* (4.15) *has at least a mild solution in $B_r(J)$.*

Proof. Since Proposition 4.4, $Q(t)(t > 0)$ is equicontinuous, which implies (H0) is satisfied. By Lemma 4.12, we have $\|Tx\| \le r$, for any $x \in B_r(J)$, which means that $\{Tx : \ x \in B_r(J)\}$ is uniformly bounded. By Lemmas 4.11-4.12, we know that T is continuous, $\{Tx : \ x \in B_r(J)\}$ is equicontinuous. It remains to prove that for $t \in J$, the set $\{(Tx)(t) : \ x \in B_r(J)\}$ is relatively compact in X.

According to the argument of Theorem 4.5, we only need to prove that for any $t \in J$, the set $V_1(t) = \{(T_1x)(t) : \ x \in B_r(J)\}$ is relatively compact in X.

Obviously, $V_1(0)$ is relatively compact in X. Let $0 < t \le a$ be fixed. For $\forall\ \delta > 0$, define an operator T_1^δ on $B_r(J)$ by the formula

$$\begin{aligned} (T_1^\delta x)(t) &= \int_\delta^\infty M_q(\theta) Q(t^q\theta)(x_0 - g(x)) d\theta \\ &= Q(t^q\delta) \int_\delta^\infty M_q(\theta) Q(t^q\theta - t^q\delta)(x_0 - g(x)) d\theta \end{aligned}$$

where $x \in B_r(J)$. From the compactness of $Q(t^q\delta)(t^q\delta > 0)$, we obtain that the set $V_1^\delta(t) = \{(T_1^\delta x)(t) : x \in B_r(J)\}$ is relatively compact in X for $\forall\ \delta > 0$. Moreover, for every $x \in B_r(J)$, we have

$$\begin{aligned}
&|(T_1x)(t) - (T_1^\delta x)(t)| \\
&= \left| \int_0^\infty M_q(\theta)Q(t^q\theta)(x_0 - g(x))d\theta - \int_\delta^\infty M_q(\theta)Q(t^q\theta)(x_0 - g(x))d\theta \right| \\
&= \left| \int_0^\delta M_q(\theta)Q(t^q\theta)(x_0 - g(x))d\theta \right| \\
&\le M(|x_0| + L_1 r + L_2) \int_0^\delta M_q(\theta)d\theta.
\end{aligned}$$

Therefore, there are relatively compact sets arbitrarily close to the set $V_1(t)$, $t > 0$. Hence the set $V_1(t)$, $t > 0$ is also relatively compact in X. Moreover, $\{Tx : x \in B_r(J)\}$ is uniformly bounded by Lemma 4.10. Therefore, $\{Tx : x \in B_r(J)\}$ is relatively compact by Arzela-Ascoli theorem. Therefore, $\alpha(T(B_r(J))) = 0$. Hence, Theorem 1.10 shows that T has a fixed point in $B_r(J)$, which means that the nonlocal Cauchy problem (4.15) has a mild solution. □

Remark 4.4. If g is not a compact mapping, we consider the following integral equations

$$x(t) = t^{q-1}P_q\left(t + \frac{1}{n}\right)(x_0 - g(x)) + \int_0^t (t-s)^{q-1}P_q(t-s)f(s, x(s))ds, \quad t \in (0, a]. \tag{4.20}$$

For any $n \in \mathbb{N}$, noticing that the operator $Q(\frac{1}{n})$ is compact, one can easily derive the relative compactness of $V(0)$ and $V(t)(t > 0)$. Then, (4.20) has a solution in $B_r^{(q)}(J')$. By passing the limit, as $n \to \infty$, one obtains a mild solution of the nonlocal Cauchy problem (4.15). However, because $Q(t)$ is replaced by $Q(\frac{1}{n})$, one needs a more restrictive condition than (H4)$'$, such as

(H4)$''$ there exists a constant $r > 0$ such that

$$M_\varepsilon\left(|x_0| + L_1 r + L_2 + \sup_{t\in[0,a]}\left\{\frac{1}{\Gamma(q)}\int_0^t (t-s)^{q-1}m(s)ds\right\}\right) \le r,$$

where $M_\varepsilon = \sup_{t\in[0,a+\varepsilon]} \|Q(t)\|_{B(X)}$, ε is a small constant.

Remark 4.5. The condition (H2) of Theorems 4.5-4.6 can be replaced by the following condition.

(H2)$'$ There exist a constant $q_1 \in (0, q)$ and $m \in L^{\frac{1}{q_1}}(J, \mathbb{R}^+)$ such that

$$|f(t,x)| \le m(t) \ \text{ for all } x \in B_r(J) \ \text{ and almost all } t \in [0, a].$$

We emphasize that (H2) is weaker than the condition (H2)$'$.

4.3.5 *Noncompact Semigroup Case*

If $Q(t)$ is noncompact, we give an assumption as follows.

(H5) There exists $\ell > 0$ such that for any bounded $D \subset X$,

$$\alpha(f(t, D)) \leq \ell\alpha(D).$$

Theorem 4.7. *Assume that (H0)-(H5) hold. Then the nonlocal Cauchy problem* (4.15) *has at least one mild solution in* $B_r(J)$.

Proof. By Lemmas 4.9-4.10, we know that $T : B_r(J) \to B_r(J)$ is bounded, continuous and $\{T_2x : x \in B_r(J)\}$ is equicontinuous. For each bounded subset $B_0 \subset B_r(J)$, set

$$T^1(B_0) = T_2(B_0),\ T^n(B_0) = T_2\left(\overline{co}(T^{n-1}(B_0))\right),\ n = 2, 3, \ldots .$$

Then for any $\varepsilon > 0$, there is a sequence $\{x_n^{(1)}\}_{n=1}^{\infty}$ such that

$$\begin{aligned}
\alpha(T^1(B_0(t))) &= \alpha(T_2(B_0(t))) \\
&\leq 2\alpha\left(\int_0^t (t-s)^{q-1}P_q(t-s)f(s, \{x_n^{(1)}(s)\}_{n=1}^{\infty})ds\right) + \varepsilon \\
&\leq \frac{4M}{\Gamma(q)}\int_0^t (t-s)^{q-1}\alpha\left(f(s, \{x_n^{(1)}(s)\}_{n=1}^{\infty})\right) ds + \varepsilon \\
&\leq \frac{4M\ell\alpha(B_0)}{\Gamma(q)}\int_0^t (t-s)^{q-1}ds + \varepsilon \\
&= \frac{4M\ell t^q\alpha(B_0)}{\Gamma(q+1)} + \varepsilon.
\end{aligned}$$

Since $\varepsilon > 0$ is arbitrary, we have

$$\alpha(T^1(B_0(t))) \leq \frac{4M\ell t^q}{\Gamma(q+1)}\alpha(B_0).$$

From Propositions 1.18-1.20, for any $\varepsilon > 0$, there is a sequence $\{x_n^{(2)}\}_{n=1}^{\infty} \subset \overline{co}(T^1(B_0))$ such that

$$\begin{aligned}
\alpha(T^2(B_0(t))) &= \alpha(T_2(\overline{co}(T^1(B_0(t))))) \\
&\leq 2\alpha\left(\int_0^t (t-s)^{q-1}P_q(t-s)f(s, \{x_n^{(2)}(s)\}_{n=1}^{\infty})ds\right) + \varepsilon \\
&\leq \frac{4M}{\Gamma(q)}\int_0^t (t-s)^{q-1}\alpha\left(f(s, \{x_n^{(2)}(s)\}_{n=1}^{\infty})\right) ds + \varepsilon \\
&\leq \frac{4M\ell}{\Gamma(q)}\int_0^t (t-s)^{q-1}\alpha(\{x_n^{(2)}(s)\}_{n=1}^{\infty})ds + \varepsilon \\
&\leq \frac{(4M\ell)^2\alpha(B_0)}{\Gamma(q)\Gamma(q+1)}\int_0^t (t-s)^{q-1}s^q ds + \varepsilon \\
&= \frac{(4M\ell)^2t^{2q}\alpha(B_0)}{\Gamma(2q+1)} + \varepsilon.
\end{aligned}$$

It can be shown, by mathematical induction, that for every $\bar{n} \in \mathbb{N}$,

$$\alpha(T^{\bar{n}}(B_0(t))) \le \frac{(4M\ell)^{\bar{n}} t^{\bar{n}q} \alpha(B_0)}{\Gamma(\bar{n}q+1)}.$$

Since

$$\lim_{\bar{n}\to\infty} \frac{(4M\ell a^q)^{\bar{n}}}{\Gamma(\bar{n}q+1)} = 0,$$

there exists a positive integer $\hat{n}$ such that

$$\frac{(4M\ell)^{\hat{n}} t^{\hat{n}q}}{\Gamma(\hat{n}q+1)} \le \frac{(4M\ell a^q)^{\hat{n}}}{\Gamma(\hat{n}q+1)} = k < 1.$$

Then

$$\alpha(T^{\hat{n}}(B_0(t))) \le k\alpha(B_0).$$

We know from Proposition 1.16, $T^{\hat{n}}(B_0(t))$ is bounded and equicontinuous, Then, from Proposition 1.17, we have

$$\alpha(T^{\hat{n}}(B_0)) = \max_{t\in[0,a]} \alpha(T^{\hat{n}}(B_0(t))).$$

Hence

$$\alpha(T^{\hat{n}}(B_0)) \le k\alpha(B_0).$$

By using the similar method as in the proof of Theorem 4.3, we can prove that there exists a $D \subset B_r(J)$ such that

$$\alpha(T_2(D)) = 0. \tag{4.21}$$

On the other hand, for any $x_1, x_2 \in D$ and $t \in J$, according to (H3), we have

$$\begin{aligned}|(T_1x_1)(t) - (T_1x_2)(t)| &\le M|g(x_1) - g(x_2)| \\ &\le ML\|x_1 - x_2\|,\end{aligned}$$

which implies that $\|T_1x_1 - T_1x_2\| \le ML\|x_1 - x_2\|$. Thus, we obtain that

$$\alpha(T_1D) \le ML\alpha(D). \tag{4.22}$$

By (4.21) and (4.22), we have

$$\alpha(T(D)) \le \alpha(T_1(D)) + \alpha(T_2(D)) \le ML\alpha(T(D)).$$

Thus, the operator T is an α-contraction in D. By Lemma 4.10, we know that T is continuous. Hence, Theorem 1.10 shows that T has a fixed point in $D \subset B_r(J)$. Therefore, the nonlocal Cauchy problem (4.15) has a mild solution in $B_r(J)$. □

Theorem 4.8. *Assume that (H0)-(H2), (H3)′, (H4)′ and (H5) hold, then the nonlocal Cauchy problem* (4.15) *has at least a mild solution in* $B_r(J)$.

Proof. By the proof of Theorem 4.7, we know that there exists a $D \subset B_r(J)$ such that $T_2(D)$ is relatively compact, i.e., $\alpha(T_2(D)) = 0$. Clearly, $\alpha(T_1(D)) = 0$, since $g(x)$ is compact and $S_q(t)$ is bounded. Hence, we have

$$\alpha(T(D)) \le \alpha(T_1(D)) + \alpha(T_2(D)) = 0.$$

Therefore, Theorem 1.10 shows that T has a fixed point in $D \subset B_r(J)$. Therefore, the nonlocal Cauchy problem (4.15) has a mild solution in $B_r(J)$. □

4.4 Nonlocal Problems for Evolution Equations

4.4.1 *Introduction*

The nonlocal condition has a better effect on the solution and is more precise for physical measurements than the classical condition alone. In this section, we discuss the existence of mild solutions of Cauchy problem for fractional evolution equations with nonlocal conditions

$$\begin{cases} {}_0^C D_t^\alpha x(t) = Ax(t) + f\left(t, x(t)\right), & \alpha \in (0,1),\ t \in J = [0,1], \\ x(0) = \sum_{k=1}^{m} a_k x(t_k), & k = 1, 2, ..., m, \end{cases} \tag{4.23}$$

where $A: \ D(A) \subset X \to X$ is the generator of a C_0-semigroup $\{Q(t)\}_{t\geq 0}$ on a Banach space X, $f: \ J \times X \to X$ is a given function and a_k $(k = 1, 2, ..., m)$ are real numbers with $\Sigma_{k=1}^m a_k \neq 1$ and t_k, $k = 1, 2, ..., m$ are given points satisfying $0 \leq t_1 \leq t_2 \leq \cdots \leq t_m < 1$.

In Subsection 4.4.2, a suitable definition of mild solution of the equation (4.23) is introduced by defining a bounded operator B. Meanwhile, two sufficient conditions are given to guarantee such B exists. In Subsection 4.4.3, we state two existence result, the first one relies on a growth condition on J and the other one relies on a growth condition involving two parts, one for $[0, t_m]$, and the other for $[t_m, 1]$.

4.4.2 *Definition of Mild Solutions*

Assume that P_α, S_α are defined as in Subsection 4.3.2. Suppose that there exists a bounded operator $B: X \to X$ given by

$$B = \left(I - \sum_{k=1}^{m} a_k S_\alpha(t_k)\right)^{-1}. \tag{4.24}$$

We can give two sufficient conditions to guarantee such B exists and is bounded.

Lemma 4.13. *The operator B defined in (4.24) exists and is bounded, if one of the following two conditions holds:*

(C1) *there exist some real numbers a_k such that*

$$M \sum_{k=1}^{m} |a_k| < 1, \tag{4.25}$$

where $M = \sup_{t\in(0,\infty)} \|Q(t)\|_{B(X)} < \infty$;

(C2) *$Q(t)$ is compact for each $t > 0$ and homogeneous linear nonlocal problems*

$$\begin{cases} {}_0^C D_t^\alpha x(t) = Ax(t), & \alpha \in (0,1),\ t \in J, \\ x(0) = \sum_{k=1}^{m} a_k x(t_k), \end{cases} \tag{4.26}$$

has no non-trivial mild solutions.

Proof. For (C1), it is easy to see

$$\left\|\sum_{k=1}^{m} a_k S_\alpha(t_k)\right\|_{B(X)} \leq M \sum_{k=1}^{m} |a_k| < 1,$$

where Proposition 4.5 and (4.25) are used. Thus by Neumann theorem, B defined by (4.24) exists and it is bounded.

For (C2), it is obvious that the mild solutions of the problem (4.26) is given by

$$x(t) = S_\alpha(t)x(0),$$

which implies that

$$x(0) = \sum_{k=1}^{m} a_k x(t_k) = \sum_{k=1}^{m} a_k S_\alpha(t_k)x(0).$$

By Proposition 4.7, $S_\alpha(t_k)$ is compact for each $t_k > 0$, $k = 1, 2, ..., m$. Then $\sum_{k=1}^{m} a_k S_\alpha(t_k)$ is also compact. Since the problem (4.26) has no non-trivial mild solutions, one can obtain the desired result via Fredholm alternative theorem. □

Now we introduce the following definition of mild solutions of the equation (4.23).

Definition 4.3. By a mild solution of the equation (4.23), we mean a function $x \in C(J, X)$ satisfying

$$x(t) = S_\alpha(t) \sum_{k=1}^{m} a_k B(g(t_k)) + g(t), \ t \in J, \tag{4.27}$$

where

$$g(t_k) = \int_0^{t_k} (t_k - s)^{\alpha-1} P_\alpha(t_k - s) f(s, x(s)) ds, \tag{4.28}$$

and

$$g(t) = \int_0^{t} (t - s)^{\alpha-1} P_\alpha(t - s) f(s, x(s)) ds, \ t \in J. \tag{4.29}$$

Remark 4.6. To explain the formula (4.27), we note that a mild solution of the fractional evolution equation (4.23) with the initial condition is just $x(t) = S_\alpha(t)x(0) + g(t)$, so taking into account also the nonlocal condition, we get

$$x(0) = \sum_{k=1}^{m} a_k S_\alpha(t_k)x(0) + \sum_{k=1}^{m} a_k g(t_k).$$

So $x(0) = \sum_{k=1}^{m} a_k B(g(t_k))$ and hence $x(t) = S_\alpha(t) \sum_{k=1}^{m} a_k B(g(t_k)) + g(t)$ which is just (4.27).

4.4.3 *Existence*

Our first existence result is based on the well-known Schaefer fixed point theorem.

In this subsection, we study our problem under the following assumptions:

(H1) $f : J \times X \to X$ satisfies the Carathéodory conditions;

(H2) there exists a function h such that ${}_0D_t^{-\alpha}h(t)$ exists for all $t \in J$ and ${}_0D_\cdot^{-\alpha}h(\cdot) \in C((0,1], \mathbb{R}^+)$ with $\lim_{t\to 0^+} {}_0D_t^{-\alpha}h(t) = 0$ and a nondecreasing continuous function $\Omega : \mathbb{R}^+ \to \mathbb{R}^+$ such that

$$|f(t,x)| \leq h(t)\Omega(|x|)$$

for all $x \in X$ and all $t \in J$;

(H3) the inequality

$$\limsup_{\rho\to\infty} \rho \left(M^2 B\Omega(\rho) \sum_{k=1}^{m} |a_k| {}_0D_t^{-\alpha}h(t_k) + M\Omega(\rho) \sup_{t\in J} {}_0D_t^{-\alpha}h(t) \right)^{-1} > 1$$

hold;

(H4) $Q(t)$ is compact for each $t > 0$.

We begin to consider the following problem

$$\begin{cases} {}_0^CD_t^\alpha x(t) = Ax(t) + \lambda f(t, x(t)), \quad \alpha \in (0,1], \ \lambda, t \in J, \\ x(0) = \sum_{k=1}^{m} a_k x(t_k). \end{cases} \tag{4.30}$$

Define an operator $F : C(J,X) \to C(J,X)$ as follows

$$(Fx)(t) = (F_1x)(t) + (F_2x)(t), \ t \in J,$$

where $F_i : C(J,X) \to C(J,X)$, $i = 1,2$ are given by the formulas

$$(F_1x)(t) = S_\alpha(t) \sum_{k=1}^{m} a_k B(g(t_k)), \ (F_2x)(t) = g(t),$$

where B is the operator defined in (4.24), $g(t_k)$ is defined in (4.28) and $g(t)$ is defined in (4.29).

Obviously, a mild solution of the equation (4.30) is a solution of the operator equation

$$x = \lambda Fx \tag{4.31}$$

and conversely. Thus, we can apply Schaefer fixed point theorem to derive the existence results of solutions of the equation (4.23).

Lemma 4.14. *Let x be any solution of the equation (4.31). Then, there exists $R^* > 0$ such that $\|x\|_C \leq R^*$ which is independent of the parameter $\lambda \in J$.*

Proof. Denote $R_0 := \|x\|$. Taking into accounts our conditions and Proposition 4.5 and Proposition 4.6, it follows from (4.27) that

$$\begin{aligned}|x(t)| &\le |(F_1x)(t)| + |(F_2x)(t)| \\ &\le M\sum_{k=1}^{m}|a_k|\|B\|_{B(X)}|g(t_k)| + |g(t)|,\ t\in J.\end{aligned} \tag{4.32}$$

Note that

$$\begin{aligned}|g(t_k)| &\le \int_0^{t_k}(t_k-s)^{\alpha-1}\|P_\alpha(t_k-s)\|_{B(X)}|f(s,x(s))|ds \\ &\le \frac{M}{\Gamma(\alpha)}\int_0^{t_k}(t_k-s)^{\alpha-1}h(s)\Omega(\|x\|)ds \\ &\le \frac{M\Omega(R_0)}{\Gamma(\alpha)}\int_0^{t_k}(t_k-s)^{\alpha-1}h(s)ds \\ &= M\Omega(R_0)\,{}_0D_{t_k}^{-\alpha}h(t_k),\ k=1,2,...,m,\end{aligned} \tag{4.33}$$

and

$$\begin{aligned}|g(t)| &\le \frac{M\Omega(R_0)}{\Gamma(\alpha)}\int_0^{t}(t-s)^{\alpha-1}h(s)ds \\ &= M\Omega(R_0)\sup_{t\in J}{}_0D_t^{-\alpha}h(t),\ t\in J.\end{aligned} \tag{4.34}$$

In view of (4.32)-(4.34), one can obtain

$$R_0 := \|x\| \le M^2\|B\|_{B(X)}\Omega(R_0)\sum_{k=1}^{m}|a_k|{}_0D_{t_k}^{-\alpha}h(t_k) + M\Omega(R_0)\sup_{t\in J}{}_0D_t^{-\alpha}h(t),\ t\in J,$$

which implies that

$$R_0\left(M^2\|B\|_{B(X)}\Omega(R_0)\sum_{k=1}^{m}|a_k|{}_0D_{t_k}^{-\alpha}h(t_k) + M\Omega(R_0)\sup_{t\in J}{}_0D_t^{-\alpha}h(t)\right)^{-1} \le 1. \tag{4.35}$$

However, it follows (H3) that there exists a $R^* > 0$ such that for all $R > R^*$ we can derive that

$$R\left(M^2\|B\|_{B(X)}\Omega(R)\sum_{k=1}^{m}|a_k|{}_0D_{t_k}^{-\alpha}h(t_k) + M\Omega(R)\sup_{t\in J}{}_0D_t^{-\alpha}h(t)\right)^{-1} > 1. \tag{4.36}$$

Now, comparing the equalities (4.35) and (4.36), we claim that $R_0 \le R^*$. As a result, we find that $\|x\| \le R^*$ which independents the parameter λ. This completes the proof. □

Let

$$\mathfrak{B}_{R^*} = \{x\in C(J,X) : \|x\| \le R^*\}.$$

Then $\mathfrak{B}_{R^*}$ is a bounded closed and convex subset in $C(J,X)$.

By Lemma 4.14, we can derive the following result.

Lemma 4.15. *The operator F maps $\mathfrak{B}_{R^*}$ into itself.*

Lemma 4.16. *The operator $F : \mathfrak{B}_{R^*} \to \mathfrak{B}_{R^*}$ is completely continuous.*

Proof. For our purpose, we only need to check that $F_i : \mathfrak{B}_{R^*} \to \mathfrak{B}_{R^*}$, $i = 1, 2$ is completely continuous. Firstly, by repeating the same producers of Lemma 4.10 and Theorem 4.5, one can obtain $F_2 : \mathfrak{B}_{R^*} \to \mathfrak{B}_{R^*}$ is completely continuous.

Secondly, one can check that $F_1 : \mathfrak{B}_{R^*} \to \mathfrak{B}_{R^*}$ is continuous (by (H1), (H2) and Proposition 4.5) and $F_1 : \mathfrak{B}_{R^*} \to \mathfrak{B}_{R^*}$ is compact in viewing of $S_\alpha(t)$ is compact for each $t > 0$ (by (H4) and Proposition 4.7). The proof is completed. □

Now, we can state the main result of this section.

Theorem 4.9. *Assume that (H1)-(H4) hold and the condition (C1) (or (C2)) is satisfied. Then the equation* (4.23) *has at least one solution $u \in C(J, X)$ and the set of the solutions of the equation* (4.23) *is bounded in $C(J, X)$.*

Proof. Obviously, the set $\{x \in C(J, X) : x = \lambda F x, 0 < \lambda < 1\}$ is bounded due to Lemma 4.15. Now we can apply Theorem 1.6 to derive that F has a fixed point in $\mathfrak{B}_{R^*}$ which is just the mild solution of the equation (4.23). This completes the proof. □

Our second existence result is based on O'Regan fixed point theorem.

In addition to (H1), (H4) and (C1) (or (C2)), motivated by Boucherif and Precup, 2003, 2007, we introduce the following two assumptions:

(H5) there exists a function h such that ${}_0D_t^{-\alpha}h(t)$ exists for all $t \in [0, t_m]$ and ${}_0D_\cdot^{-\alpha}h(\cdot) \in C((0, t_m], \mathbb{R}^+)$ with $\lim_{t\to 0^+} {}_0D_t^{-\alpha}h(t) = 0$ and a nondecreasing continuous function $\Omega : \mathbb{R}^+ \to \mathbb{R}^+$ such that

$$|f(t, x)| \leq h(t)\Omega(|x|)$$

for all $x \in X$, and for all $t \in [t_m, 1]$ there exists a function l such that ${}_{t_m}D_t^{-\alpha}l(t)$ exists and ${}_{t_m}D_\cdot^{-\alpha}l(\cdot) \in C([t_m, 1], \mathbb{R}^+)$ such that

$$|f(t, x)| \leq l(t), \tag{4.37}$$

for all $x \in X$. Moreover, Ω has the property

$$r > M\Omega(r)\left(\sum_{k=1}^{m} |a_k| \|B\|_{B(X)} + 1\right) \sup_{t\in[0,t_m]} {}_0D_t^{-\alpha}h(t) \tag{4.38}$$

for all $r > R_1^* > 0$;

(H6) there exists a function q such that ${}_{t_m}D_t^{-\alpha}q(t)$ exists for all $t \in [t_m, 1]$ and ${}_{t_m}D_\cdot^{-\alpha}q(\cdot) \in C([t_m, 1], \mathbb{R}^+)$ with $M \sup_{t\in[t_m,1]} {}_0D_t^{-\alpha}q(t) \leq 1$ and a nondecreasing continuous function $\Psi : \mathbb{R}^+ \to \mathbb{R}^+$ with $\Psi(r) < r$ for $r > 0$ such that

$$|f(t, x) - f(t, y)| \leq q(t)\Psi(|x - y|)$$

for all $t \in [t_m, 1]$ and $x, y \in X$.

Consider the equation (4.30) again and the equivalent equation

$$x = \lambda T x \tag{4.39}$$

where $T : C(J,X) \to C(J,X)$ is defined by

$$(Tx)(t) = (T_1 x)(t) + (T_2 x)(t),\ t \in J,$$

where $T_i : C(J,X) \to C(J,X)$, $i = 1,2$ given by

$$(T_1x)(t) = \begin{cases} S_\alpha(t)\displaystyle\sum_{k=1}^{m} a_k B(g(t_k)) + g(t), & t \in [0,t_m), \\ S_\alpha(t)\displaystyle\sum_{k=1}^{m} a_k B(g(t_k)) & \\ \quad + \displaystyle\int_0^{t_m} (t-s)^{\alpha-1} P_\alpha(t-s) f(s,x(s))ds, & t \in [t_m,1], \end{cases}$$

and

$$(T_2x)(t) = \begin{cases} 0, & t \in [0,t_m), \\ \displaystyle\int_{t_m}^{t} (t-s)^{\alpha-1} P_\alpha(t-s) f(s,x(s))ds, & t \in [t_m,1]. \end{cases}$$

We first prove that any solutions of the equation (4.39) have *a priori* bound.

Lemma 4.17. *Let x be any solution of the equation* (4.39). *Then, there exist $R_i^* > 0$ $i = 1,2$ such that $\|x\|_{C([0,t_m],X)} \le R_1^*$ and $\|x\|_{C([t_m,1],X)} \le R_2^*$. In other words, $\|x\| \le R^* = \max\{R_1^*, R_2^*\}$ which is independent of the parameter λ.*

Proof. Case I. We prove that there exists $R_1^* > 0$ such that $\|x\|_{C([0,t_m],X)} \le R_1^*$.

For $t \in [0,t_m]$ and $\lambda \in J$, denote $R_{[0,t_m]} := \|x\|_{C([0,t_m],X)}$, we have

$$\begin{aligned} |x(t)| &\le \lambda |(T_1x)(t)| + |(T_2x)(t)| \\ &\le M \sum_{k=1}^{m} |a_k| \|B\|_{B(X)} |g(t_k)| + |g(t)| \\ &\le M \sum_{k=1}^{m} |a_k| \|B\|_{B(X)} \frac{M}{\Gamma(\alpha)} \int_0^{t_k} (t_k - s)^{\alpha-1} h(s) \Omega(R_{[0,t_m]}) ds \\ &\quad + \frac{M}{\Gamma(\alpha)} \int_0^t (t-s)^{\alpha-1} h(s) \Omega(R_{[0,t_m]}) ds \\ &\le M\Omega(R_{[0,t_m]}) \left(\sum_{k=1}^{m} |a_k| \|B\|_{B(X)} + 1 \right) \sup_{t\in[0,t_m]} {}_0D_t^{-\alpha} h(t), \end{aligned}$$

which implies that

$$R_{[0,t_m]} \le M\Omega(R_{[0,t_m]}) \left(\sum_{k=1}^{m} |a_k| \|B\|_{B(X)} + 1 \right) \sup_{t\in[0,t_m]} {}_0D_t^{-\alpha} h(t).$$

From (4.38) we find that there exists $R_1^* \geq R_{[0,t_m]} > 0$ such that

$$\|x\|_{C([0,t_m],X)} \leq R_1^*.$$

Case II. We prove that there exists $R_2^* > 0$ such that $\|x\|_{C([t_m,1],X)} \leq R_2^*$. For $t \in [t_m, 1]$ and $\lambda \in J$, keeping in mind our assumptions, we find that

$$\begin{aligned}|x(t)| &\leq M \sum_{k=1}^{m} |a_k| \|B\|_{B(X)} \frac{M}{\Gamma(\alpha)} \int_0^{t_k} (t_k - s)^{\alpha-1} h(s) \Omega(R_1^*) ds \\ &\quad + \frac{M}{\Gamma(\alpha)} \int_0^{t_m} (t-s)^{\alpha-1} h(s) \Omega(R_1^*) ds + \frac{M}{\Gamma(\alpha)} \int_{t_m}^{t} (t-s)^{\alpha-1} h(s) ds \\ &\leq M\Omega(R_1^*) \left(\sum_{k=1}^{m} |a_k| \|B\|_{B(X)} + 1 \right) \sup_{t \in [0,t_m]} {}_0D_t^{-\alpha} h(t) \\ &\quad + M \sup_{t \in [t_m,1]} {}_{t_m}D_t^{-\alpha} l(t),\end{aligned}$$

which implies that

$$\|x\|_{C([t_m,1],X)} \leq R_2^*,$$

where

$$R_2^* = M \left[\Omega(R_1^*) \left(\sum_{k=1}^{m} |a_k| \|B\|_{B(X)} + 1 \right) \sup_{t \in [0,t_m]} {}_0D_t^{-\alpha} h(t) + \sup_{t \in [t_m,1]} {}_{t_m}D_t^{-\alpha} l(t) \right].$$

Let $R^* = \max\{R_1^*, R_2^*\}$. Then we find that any possible solutions of the equation (4.39) satisfy $\|x\| \leq R^*$ which are independent of the parameter λ. This completes the proof. □

Denote

$$\mathcal{D} = \{x \in C(J, X) : \ \|x\| < R^* + 1\}.$$

We can proceed as in the proof of Lemma 4.17 to derive the following result.

Lemma 4.18. *$T(\overline{\mathcal{D}})$ is bounded.*

One can proceed as in the proof of Lemma 4.16 to obtain the following result.

Lemma 4.19. *The operator $T_1 : \overline{\mathcal{D}} \to C(J, X)$ is completely continuous.*

Next, we show the following result.

Lemma 4.20. *The operator $T_2 : \overline{\mathcal{D}} \to C(J, X)$ is nonlinear contraction.*

Proof. It follows from the definition of T_2 we only need to show $T_2 : \overline{\mathcal{D}} \to C([t_m, 1], X)$ is a nonlinear contraction.

In fact, for any $x, y \in \overline{\mathcal{D}}$ and $t \in [t_m, 1]$, we have

$$|(T_2x)(t) - (T_2y)(t)| \leq \int_{t_m}^{t} (t-s)^{\alpha-1} \|P_\alpha(t-s)[f(s, x(s)) - f(s, y(s))]\| ds$$

$$\leq \frac{M}{\Gamma(\alpha)} \int_{t_m}^{t} (t-s)^{\alpha-1} q(s) \Psi(|x(s)-y(s)|) ds$$
$$\leq \frac{M\Psi(\|x-y\|)}{\Gamma(\alpha)} \int_{t_m}^{t} (t-s)^{\alpha-1} q(s) ds$$
$$\leq \left(M \sup_{t\in[t_m,1]} {}_{t_m}D_t^{-\alpha} q(t) \right) \Psi(\|x-y\|),$$

which implies that

$$\|T_2 x - T_2 y\| \leq \Psi(\|x-y\|).$$

This completes the proof. □

Now, we are ready to present the main result of this section.

Theorem 4.10. *Assume that (H1), (H4), (H5) and (H6) hold and the condition (C1) (or (C2)) is satisfied. Then the equation* (4.23) *has at least one solution* $u \in C(J, X)$.

Proof. By Lemma 4.17 we see that (ii) of Theorem 1.8 does not hold. Thus, there is no solution of the equation (4.39) with $x \in \partial\mathcal{D}$. Therefore, one can apply Theorem 1.8 to derive that T has a fixed point in $\mathcal{D}$ which is just the mild solution of the equation (4.23). This completes the proof. □

Remark 4.7. Theorem 4.10 also holds even if (H5) and (H6) are replaced by the following conditions:

(H5)′ condition (H5) is assumed without (4.37);
(H6)′ denoting $\delta := \lim_{r\to+\infty} \inf \frac{\Psi(r)}{r} \leq 1$, condition (H6) is assumed in addition with

$$M\delta \sup_{t\in[t_m,1]} {}_{t_m}D_t^{-\alpha} q(t) < 1.$$

Indeed, we can modify Case II in the proof of Lemma 4.17 as follows

$$|x(t)| \leq M \sum_{k=1}^{m} |a_k| \|B\|_{B(X)} \frac{M}{\Gamma(\alpha)} \int_0^{t_k} (t_k - s)^{\alpha-1} h(s) \Omega(R_1^*) ds$$
$$+ \frac{M}{\Gamma(\alpha)} \int_0^{t_m} (t-s)^{\alpha-1} h(s) \Omega(R_1^*) ds$$
$$+ \frac{M}{\Gamma(\alpha)} \int_{t_m}^{t} (t-s)^{\alpha-1} |f(s,0)| ds + \frac{M}{\Gamma(\alpha)} \int_{t_m}^{t} (t-s)^{\alpha-1} q(s) \Psi(|x(s)|) ds$$
$$\leq M\Omega(R_1^*) \left(\sum_{k=1}^{m} |a_k| \|B\|_{B(X)} + 1 \right) \sup_{t\in[0,t_m]} {}_{t_m}D_t^{-\alpha} h(t)$$
$$+ \frac{M \sup_{t\in[t_m,t]} |f(t,0)| (1-t_m)^{\alpha}}{\Gamma(\alpha+1)}$$

$$+\frac{M}{\Gamma(\alpha)}\int_{t_m}^{t}(t-s)^{\alpha-1}q(s)(\delta|x(s)|+\delta_1)ds,$$

for some $\delta_1 \geq 0$.

Then we have

$$R_2^* = \frac{1}{1-M\delta\sup_{t\in[t_m,1]}{}_{t_m}D_t^{-\alpha}q(t)}$$
$$\times\left[M\Omega(R_1^*)\left(\sum_{k=1}^{m}|a_k|\|B\|_{B(X)}+1\right)\sup_{t\in[0,t_m]}{}_{t_m}D_t^{-\alpha}h(t)\right.$$
$$\left.+\frac{M\sup_{t\in[t_m,t]}|f(t,0)|(1-t_m)^{\alpha}}{\Gamma(\alpha+1)}+M\delta_1\sup_{t\in[t_m,1]}{}_{t_m}D_t^{-\alpha}q(t)\right].$$

4.5 Optimal Controls of Fractional Evolution Equations

4.5.1 *Introduction*

In this section, we consider the following fractional evolution equations

$$\begin{cases} {}_0^CD_t^q x(t) = -Ax(t) + f(t, x(t)), & t \in J = [0,T],\ q \in (0,1), \\ x(0) = x_0, \end{cases} \tag{4.40}$$

where ${}_0^CD_t^q$ is the Caputo fractional derivative of order $0 < q < 1$, $-A : D(A) \to X$ is the infinitesimal generator of a compact analytic semigroup of uniformly bounded linear operators $\{T(t), t \geq 0\}$, $f : J \times X_\alpha \to X$ is specified later, where $X_\alpha = D(A^\alpha)(0 < \alpha < 1)$ is a Banach space with the norm $\|x\|_\alpha = \|A^\alpha x\|$ for $x \in X_\alpha$.

In the present section, we introduce a suitable α-mild solution for system (4.40). The introduced α-mild solution is associated with a probability density function and semigroup operator. Then we give some properties of two new linear operators associated with the probability density function and semigroup operator which are used throughout this section. We prove the existence of α-mild solutions for system (4.40) under some easy checked conditions. The main technique used here are fractional calculus, singular version Gronwall inequality visa Leray-Schauder fixed point theorem for compact maps. Further, we consider the Lagrange problem of systems governed by (4.40) and an existence result of optimal controls is presented.

The rest of this section is organized as follows. In Subsection 4.5.2, we give some preliminary results on the fractional powers of the generator of an analytic compact semigroup and introduce the α-mild solution of system (4.40). In Subsection 4.5.3, we study the existence and uniqueness of α-mild solutions for system (4.40). In Subsection 4.5.4, we introduce a class of admissible controls and an existence result of optimal controls for a Lagrange problem is proved. At last, an example is given to demonstrate the applicability of the result.

4.5.2 *Preliminaries*

We denote by X a Banach space with the norm $\|\cdot\|$ and $-A : D(A) \to X$ is the infinitesimal generator of a compact analytic semigroup of uniformly bounded linear operators $\{T(t), t \geq 0\}$. This means that there exists $M > 1$ such that $\|T(t)\| \leq M$. We assume without loss of generality that $0 \in \rho(A)$. This allows us to define the fractional power A^{α} for $0 < \alpha < 1$, as a closed linear operator on its domain $D(A^{\alpha})$ with inverse $A^{-\alpha}$ (see Pazy, 1983).

We have the following basic properties on A^{α}.

Theorem 4.11. *(Pazy, 1983)*

(i) $X_{\alpha} = D(A^{\alpha})$ *is a Banach space with the norm* $\|x\|_{\alpha} = \|A^{\alpha}x\|$ *for* $x \in X_{\alpha}$.
(ii) $T(t) : X \to X_{\alpha}$ *for each* $t > 0$.
(iii) $A^{\alpha}T(t)x = T(t)A^{\alpha}x$ *for each* $x \in X_{\alpha}$ *and* $t \geq 0$.
(iv) *For every* $t > 0$, $A^{\alpha}T(t)$ *is bounded on* X *and there exists* $N_{\alpha} > 0$ *and* $\nu > 0$ *such that*

$$\|A^{\alpha}T(t)\| \leq \frac{N_{\alpha}}{t^{\alpha}} e^{-\nu t} \leq \frac{N_{\alpha}}{t^{\alpha}}.$$

(v) $A^{-\alpha}$ *is a bounded linear operator in* X *with* $X_{\alpha} = Im\ (A^{-\alpha})$.
(vi) *If* $0 < \alpha \leq \beta < 1$, *then* $D(A^{\beta}) \hookrightarrow D(A^{\alpha})$.

Remark 4.8. Observe as in Liu and Chang, 2009 that by Theorem 4.11 (ii) and (iii), the restriction $T_{\alpha}(t)$ of $T(t)$ to X_{α} is exactly the part of $T(t)$ in X_{α}. Let $x \in X_{\alpha}$. Since $\|T(t)x\|_{\alpha} \leq \|A^{\alpha}T(t)x\| = \|T(t)A^{\alpha}x\| \leq \|T(t)\|\|A^{\alpha}x\| = \|T(t)\|\|x\|_{\alpha}$, and as t decreases to 0, $\|T(t)x - x\|_{\alpha} = \|A^{\alpha}T(t)x - A^{\alpha}x\| = \|T(t)A^{\alpha}x - A^{\alpha}x\| \to 0$, for all $x \in X_{\alpha}$, it follows that $\{T(t), t \geq 0\}$ is a family of strongly continuous semigroup on X_{α} and $\|T_{\alpha}(t)\| \leq \|T(t)\| \leq M$ for all $t \leq 0$.

In the sequel, we will also use $\|f\|_{L^p(J,R^+)}$ to denote the $L^p(J, R^+)$ norm of f whenever $f \in L^p(J, R^+)$ for some p with $1 < p < \infty$. We will set $\alpha \in (0,1)$ and denote by $\mathcal{C}_{\alpha}$, the Banach space $C(J, X_{\alpha})$ endowed with supnorm given by $\|x\|_{\infty} \equiv \sup_{t\in J} \|x\|_{\alpha}$, for $x \in \mathcal{C}_{\alpha}$.

Based on Subsection 4.3.2, we use the following definition of α-mild solution for the problem below.

Definition 4.4. By the α-mild solution of system (4.40), we mean that the function $x \in \mathcal{C}_{\alpha}$ which satisfies

$$x(t) = S(t)x_0 + \int_0^t (t-s)^{q-1} P(t-s) f(s, x(s))\, ds, \quad t \in J,$$

where

$$S(t) = \int_0^{\infty} M_q(\theta) T(t^q \theta) d\theta, \quad P(t) = \int_0^{\infty} q\theta M_q(\theta) T(t^q \theta) d\theta.$$

The following results will be used throughout this section.

Lemma 4.21. *The operators* S *and* P *have the following properties:*

(i) *For any fixed* $t \geq 0$, $S(t)$ *and* $P(t)$ *are linear and bounded operators, i.e., for any* $x \in X$,

$$\|S(t)x\| \leq M\|x\|, \quad \|P(t)x\| \leq \frac{M}{\Gamma(q)}\|x\|.$$

(ii) $\{S(t)\}_{t\geq 0}$ *and* $\{P(t)\}_{t\geq 0}$ *are strongly continuous.*

(iii) *Assume that* $\{T(t)\}_{t>0}$ *is compact operator. Then* $\{S(t)\}_{t>0}$ *and* $\{P(t)\}_{t>0}$ *are also compact operators.*

(iv) *For any* $x \in X$, $\beta \in (0,1)$ *and* $\alpha \in (0,1)$, *we have*

$$AP(t)x = A^{1-\beta}P(t)A^{\beta}x, \quad t \in J,$$

$$\|A^{\alpha}P(t)\| \leq \frac{N_{\alpha}q\Gamma(2-\alpha)}{\Gamma(1+q(1-\alpha))}t^{-\alpha q}, \quad 0 < t \leq T.$$

(v) *For fixed* $t \geq 0$ *and any* $x \in X_{\alpha}$, *we have*

$$\|S(t)x\|_{\alpha} \leq M\|x\|_{\alpha}, \quad \|P(t)x\|_{\alpha} \leq \frac{M}{\Gamma(q)}\|x\|_{\alpha}.$$

(vi) $S_{\alpha}(t)$ *and* $P_{\alpha}(t)$, $t > 0$ *are uniformly continuous, that is for each fixed* $t > 0$, *and* $\epsilon > 0$, *there exists* $h > 0$ *such that*

$$\|S_{\alpha}(t+\epsilon) - S_{\alpha}(t)\|_{\alpha} < \varepsilon, \quad \textit{for} \quad t+\epsilon \geq 0 \quad \textit{and} \quad |\epsilon| < h,$$

$$\|P_{\alpha}(t+\epsilon) - P_{\alpha}(t)\|_{\alpha} < \varepsilon, \quad \textit{for} \quad t+\epsilon \geq 0 \quad \textit{and} \quad |\epsilon| < h,$$

where

$$S_{\alpha}(t) = \int_0^{\infty} M_q(\theta)T_{\alpha}(t^q\theta)d\theta, \quad P_{\alpha}(t) = \int_0^{\infty} q\theta M_q(\theta)T_{\alpha}(t^q\theta)d\theta.$$

Proof. We only check (v) and (vi) as follows.

(v) For fixed $t \geq 0$ and any $x \in X_{\alpha}$, using (iii) of Theorem 4.11,

$$\begin{aligned}
\|S(t)x\|_{\alpha} &\leq \int_0^{\infty} M_q(\theta)\|A^{\alpha}T(t^q\theta)x\|d\theta \\
&\leq \int_0^{\infty} M_q(\theta)\|T(t^q\theta)\|\|A^{\alpha}x\|d\theta \\
&\leq M\int_0^{\infty} M_q(\theta)\|A^{\alpha}x\|d\theta \\
&= M\|x\|_{\alpha}
\end{aligned}$$

and

$$\begin{aligned}
\|P(t)x\|_{\alpha} &\leq \int_0^{\infty} q\theta M_q(\theta)\|A^{\alpha}T(t^q\theta)x\|d\theta \\
&\leq \int_0^{\infty} q\theta M_q(\theta)\|T(t^q\theta)\|\|A^{\alpha}x\|d\theta \\
&\leq M\int_0^{\infty} q\theta M_q(\theta)\|A^{\alpha}x\|d\theta
\end{aligned}$$

$$= \frac{M}{\Gamma(q)} \|x\|_\alpha.$$

(vi) For each fixed $t > 0$, and $h > \epsilon > 0$, one can obtain

$$\begin{aligned}
\|S_\alpha(t+\epsilon) - S_\alpha(t)\|_\alpha &\le \int_0^\infty M_q(\theta) \|T_\alpha((t+\epsilon)^q\theta) - T_\alpha(t^q\theta)\|_\alpha d\theta \\
&\le M \int_0^\infty M_q(\theta) \|T_\alpha((t+\epsilon)^q\theta - t^q\theta) - I\|_\alpha d\theta, \\
\|P_\alpha(t+\epsilon) - P_\alpha(t)\|_\alpha &\le \int_0^\infty q\theta M_q(\theta) \|T_\alpha((t+\epsilon)^q\theta) - T_\alpha(t^q\theta)\|_\alpha d\theta \\
&\le M \int_0^\infty q\theta M_q(\theta) \|T_\alpha((t+\epsilon)^q\theta - t^q\theta) - I\|_\alpha d\theta.
\end{aligned}$$

By Lemma 3.3 of Liu and Chang, 2009, we know that the uniformly continuous of $T_\alpha(t)$, $t > 0$. As a result,

$$\|T_\alpha((t+\epsilon)^q\theta - t^q\theta) - I\|_\alpha \to 0, \quad as \quad \epsilon \to 0.$$

It comes from

$$\int_0^\infty M_q(\theta) d\theta = 1 \quad and \quad \int_0^\infty \theta M_q(\theta) d\theta = \frac{1}{\Gamma(1+q)},$$

that $S_\alpha(t)$ and $P_\alpha(t)$, $t > 0$ is uniformly continuous. □

Lemma 4.22. *(Zeidler, 1990) For each $\psi \in L^p(J, X)$ with $1 \le p < +\infty$,*

$$\lim_{h\to 0} \int_0^T \|\psi(t+h) - \psi(t)\|^p dt = 0$$

where $\psi(s) = 0$ for s does not belong to J.

4.5.3 *Existence of α-Mild Solutions*

In this subsection, we give the existence of the α-mild solutions for system (4.40).

We make the following assumptions.

[**Hf**]: $f : J \times X_\alpha \to X$ satisfies:

(1) For each $x \in X_\alpha$, $t \to f(t, x)$ is measurable.

(2) For arbitrary x_1, $x_2 \in X_\alpha$ satisfying $\|x_1\|_\alpha$, $\|x_2\|_\alpha \le \rho$, there exists a constant $L_f(\rho) > 0$ such that

$$\|f(t, x_1) - f(t, x_2)\| \le L_f(\rho)\|x_1 - x_2\|_\alpha, \text{ for all } t \in J.$$

(3) There exists a constant $a_f > 0$ such that

$$\|f(t, x)\| \le a_f(1 + \|x\|_\alpha), \text{ for all } x \in X_\alpha \text{ and } t \in J.$$

Now we are ready to state and prove the main result in this subsection.

Theorem 4.12. *Assume that the condition [Hf] is satisfied. If $x_0 \in X_\alpha$ and $\alpha q < \frac{1}{2}$ for some $\frac{1}{2} < q < 1$, then system (4.40) has an unique α-mild solution on J.*

Proof. Define the function $F : \mathcal{C}_\alpha \to \mathcal{C}_\alpha$ by

$$(Fx)(t) = S(t)x_0 + \int_0^t (t-s)^{q-1}P(t-s)f(s, x(s))\, ds.$$

It is not difficult to verify that $Fx \in \mathcal{C}_\alpha$. In fact, for $0 \le t_1 < t_2 \le T$, by Lemma 4.21, Hölder inequality and $\alpha q < \frac{1}{2}$, one can deduce the following inequality

$$\begin{aligned}
&\|(Fx)(t_1) - (Fx)(t_2)\|_\alpha \\
&\le \|S(t_1)x_0 - S(t_2)x_0\|_\alpha \\
&\quad + \int_0^{t_1} (t_1-s)^{q-1}\|P(t_1-s)f(s,x(s)) - P(t_2-s)f(s,x(s))\|_\alpha ds \\
&\quad + \int_0^{t_1} |(t_1-s)^{q-1} - (t_2-s)^{q-1}|\|P(t_2-s)f(s,x(s))\|_\alpha ds \\
&\quad + \int_{t_1}^{t_2} (t_2-s)^{q-1}\|P(t_2-s)f(s,x(s))\|_\alpha ds \\
&\le \|S_\alpha(t_1) - S_\alpha(t_2)\|_\alpha \|x_0\|_\alpha \\
&\quad + \int_0^{t_1} (t_1-s)^{q-1}\|A^\alpha[P(t_2-s) - P(t_1-s)]\|\|f(s,x(s))\|ds \\
&\quad + \int_0^{t_1} |(t_1-s)^{q-1} - (t_2-s)^{q-1}|\|A^\alpha P(t_2-s)\|\|f(s,x(s))\|ds \\
&\quad + \int_{t_1}^{t_2} (t_2-s)^{q-1}\|A^\alpha P(t_2-s)\|\|f(s,x(s))\|ds \\
&\le \|S_\alpha(t_1) - S_\alpha(t_2)\|_\alpha \|x_0\|_\alpha \\
&\quad + \frac{qN_{\alpha+1}\Gamma(2-\alpha)}{\alpha\Gamma(1+q(1-\alpha))}\|f\|_{C(J,X)} \int_0^{t_1} (t_1-s)^{q-1}|(t_2-s)^{-q\alpha} - (t_1-s)^{-q\alpha}|ds \\
&\quad + \frac{N_\alpha q\Gamma(2-\alpha)}{\Gamma(1+q(1-\alpha))} \int_0^{t_1} |(t_1-s)^{q-1} - (t_2-s)^{q-1}|(t_2-s)^{-\alpha q}\|f(s,x(s))\|ds \\
&\quad + \frac{N_\alpha q\Gamma(2-\alpha)}{\Gamma(1+q(1-\alpha))} \int_{t_1}^{t_2} (t_2-s)^{q-\alpha q-1}\|f(s,x(s))\|ds \\
&\le \|S_\alpha(t_1) - S_\alpha(t_2)\|_\alpha \|x_0\|_\alpha \\
&\quad + \frac{qN_{\alpha+1}\Gamma(2-\alpha)}{\alpha\Gamma(1+q(1-\alpha))}\|f\|_{C(J,X)} \left(\int_0^{t_1} |(t_2-s)^{-q\alpha} - (t_1-s)^{-q\alpha}|^2 ds\right)^{\frac{1}{2}} \\
&\quad \times \left(\int_0^{t_1} (t_1-s)^{2(q-1)} ds\right)^{\frac{1}{2}} \\
&\quad + \frac{N_\alpha q\Gamma(2-\alpha)}{\Gamma(1+q(1-\alpha))} \left(\int_0^{t_1} |(t_1-s)^{q-1} - (t_2-s)^{q-1}|^2 ds\right)^{\frac{1}{2}} \\
&\quad \times \left(\int_0^{t_1} (t_2-s)^{-2\alpha q} ds\right)^{\frac{1}{2}} \|f\|_{C(J,X)}
\end{aligned}$$

$$
\begin{aligned}
&+\frac{N_\alpha q\Gamma(2-\alpha)}{\Gamma(1+q(1-\alpha))}\frac{1}{q(1-\alpha)}(t_2-t_1)^{q(1-\alpha)}\|f\|_{C(J,X)}\\
&\le \|S_\alpha(t_1)-S_\alpha(t_2)\|_\alpha\|x_0\|_\alpha\\
&+\sqrt{\frac{1}{2q-1}}t_1^{q-\frac{1}{2}}\frac{qN_{\alpha+1}\Gamma(2-\alpha)}{\alpha\Gamma(1+q(1-\alpha))}\|f\|_{C(J,X)}\\
&\times\left(\int_0^T|(t_2-s)^{-q\alpha}-(t_1-s)^{-q\alpha}|^2ds\right)^{\frac{1}{2}}\\
&+\frac{N_\alpha q\Gamma(2-\alpha)}{\Gamma(1+q(1-\alpha))}\left(\int_0^T|(t_1-s)^{q-1}-(t_2-s)^{q-1}|^2ds\right)^{\frac{1}{2}}\\
&\times\sqrt{\frac{1}{1-2\alpha q}}\left(t_2^{1-2\alpha q}-(t_2-t_1)^{1-2\alpha q}\right)^{\frac{1}{2}}\|f\|_{C(J,X)}\\
&+\frac{N_\alpha q\Gamma(2-\alpha)}{\Gamma(1+q(1-\alpha))}\frac{1}{q(1-\alpha)}(t_2-t_1)^{q(1-\alpha)}\|f\|_{C(J,X)},
\end{aligned}
$$

which implies that $Fx\in\mathcal{C}_\alpha$.

Then we proceed in four steps.

Step I. F is a continuous operator on $\mathcal{C}_\alpha$.

In fact, let x_1, $x_2\in\mathcal{C}_\alpha$ and $\|x_1-x_2\|_\infty\le 1$, then $\|x_1\|_\infty\le 1+\|x_2\|_\infty=\rho$, and

$$
\begin{aligned}
&\|(Fx_1)(t)-(Fx_2)(t)\|_\alpha\\
&\le\int_0^t(t-s)^{q-1}\|P(t-s)[f(s,x_1(s))-f(s,x_2(s))]\|_\alpha ds\\
&\le\int_0^t(t-s)^{q-1}\|A^\alpha P(t-s)\|\|f(s,x_1(s))-f(s,x_2(s))\|ds\\
&\le L_f(\rho)\frac{N_\alpha q\Gamma(2-\alpha)}{\Gamma(1+q(1-\alpha))}\int_0^t(t-s)^{q-1-\alpha q}\|x_1(s)-x_2(s)\|_\alpha ds\\
&\le L_f(\rho)\frac{N_\alpha q\Gamma(2-\alpha)}{\Gamma(1+q(1-\alpha))}\int_0^t(t-s)^{q-1-\alpha q}\|x_1(s)-x_2(s)\|_\alpha ds\\
&\le L_f(\rho)\frac{N_\alpha q\Gamma(2-\alpha)}{\Gamma(1+q(1-\alpha))}\left(\int_0^t(t-s)^{q-1-\alpha q}ds\right)\|x_1-x_2\|_\infty\\
&\le L_f(\rho)\frac{N_\alpha q\Gamma(2-\alpha)}{\Gamma(1+q(1-\alpha))}\frac{1}{q(1-\alpha)}t^{q(1-\alpha)}\|x_1-x_2\|_\infty.
\end{aligned}
$$

Therefore, it can easily been shown that

$$\|Fx_1-Fx_2\|_\infty\le L_f(\rho)\frac{N_\alpha q\Gamma(2-\alpha)}{\Gamma(1+q(1-\alpha))}\frac{1}{q(1-\alpha)}T^{q(1-\alpha)}\|x_1-x_2\|_\infty,$$

that is, F is continuous.

Step II. F is compact.

Let B is bounded subset of $\mathcal{C}_\alpha$, there exists a constant μ such that $\|x\|_\infty\le\mu$ for all $x\in B$. By [Hf](3), there exists a constant N such that $\|f(t,x(t))\|\le a_f(1+\mu)=$

N. Then FB is a bounded subset of $\mathcal{C}_\alpha$. In fact, let $x \in B$, using Lemma 4.21 (i) and (iv), one can obtain

$$\begin{aligned}\|(Fx)(t)\|_\alpha &\le \|S(t)x_0\|_\alpha + \int_0^t (t-s)^{q-1}\big\|P(t-s)f(s,x(s))\big\|_\alpha ds \\ &\le M\|x_0\|_\alpha + \int_0^t (t-s)^{q-1}\big\|A^\alpha P(t-s)\big\|\|f(s,x(s))\|ds \\ &\le M\|x_0\|_\alpha + \frac{N_\alpha q\Gamma(2-\alpha)}{\Gamma(1+q(1-\alpha))}N\int_0^t (t-s)^{q-\alpha q-1}ds \\ &\le M\|x_0\|_\alpha + \frac{N_\alpha q\Gamma(2-\alpha)}{\Gamma(1+q(1-\alpha))}N\frac{1}{q(1-\alpha)}t^{q(1-\alpha)}.\end{aligned}$$

Thus,

$$\|Fx\|_\infty \le M\mu + \frac{N_\alpha q\Gamma(2-\alpha)}{\Gamma(1+q(1-\alpha))}\frac{NT^{q(1-\alpha)}}{q(1-\alpha)},$$

which implies that FB is bounded.

Define $\Pi = FB$ and $\Pi(t) = \{(Fx)(t) \mid x \in B\}$ for $t \in J$. Clearly, $\Pi(0) = \{(Fx)(0) \mid x \in B\} = \{x_0\}$ is compact, and hence, it is only necessary to consider $t > 0$. For each $h \in (0,t), t \in (0,T]$, arbitrary $\delta > 0$, define

$$\Pi_{h,\delta}(t) = (F_{h,\delta}B)(t) = \Big\{(F_{h,\delta}x)(t) \mid x \in B\Big\}$$

where

$$\begin{aligned}&(F_{h,\delta}x)(t) \\ &= T(h^q\delta)\int_\delta^\infty M_q(\theta)T(t^q\theta - h^q\delta)x_0 d\theta \\ &\quad + T(h^q\delta)\int_0^{t-h}(t-s)^{q-1}\left(q\int_\delta^\infty \theta M_q(\theta)T((t-s)^q\theta - h^q\delta)d\theta\right)f(s,x(s))\,ds \\ &= \int_\delta^\infty M_q(\theta)T(t^q\theta)x_0 d\theta + q\int_0^{t-h}\int_\delta^\infty \theta(t-s)^{q-1}M_q(\theta)T((t-s)^q\theta)f(s,x(s))\,d\theta ds.\end{aligned}$$

Then the sets $\{(F_{h,\delta}x)(t) \mid x \in B\}$ are relatively compact in X_α since the operator $T(h^q\delta)$, $h^q\delta > 0$ is compact in X_α. It comes from the following inequalities

$$\begin{aligned}&\|(Fx)(t) - (F_{h,\delta}x)(t)\|_\alpha \\ &\le \left\|\int_0^\delta M_q(\theta)T(t^q\theta)x_0 d\theta\right\|_\alpha \\ &\quad + q\left\|\int_0^t\int_0^\delta \theta(t-s)^{q-1}M_q(\theta)T((t-s)^q\theta)f(s,x(s))\,d\theta ds\right\|_\alpha \\ &\quad + q\left\|\int_0^t\int_\delta^\infty \theta(t-s)^{q-1}M_q(\theta)T((t-s)^q\theta)f(s,x(s))\,d\theta ds\right. \\ &\quad \left.- \int_0^{t-h}\int_\delta^\infty \theta(t-s)^{q-1}M_q(\theta)T((t-s)^q\theta)f(s,x(s))\,d\theta ds\right\|_\alpha\end{aligned}$$

$$
\begin{aligned}
&\leq \int_0^\delta M_q(\theta)\|T(t^q\theta)x_0\|_\alpha d\theta \\
&\quad + q\int_0^t\int_0^\delta \theta(t-s)^{q-1}M_q(\theta)\|A^\alpha T((t-s)^q\theta)\|\|f(s,x(s))\|d\theta ds \\
&\quad + q\int_{t-h}^t\int_\delta^\infty \theta(t-s)^{q-1}M_q(\theta)\|A^\alpha T((t-s)^q\theta)\|\|f(s,x(s))\|d\theta ds \\
&\leq M\|x_0\|_\alpha\int_0^\delta M_q(\theta)d\theta + N_\alpha Nq\int_0^t\int_0^\delta \theta(t-s)^{q-1}M_q(\theta)((t-s)^q\theta)^{-\alpha}d\theta ds \\
&\quad + N_\alpha Nq\int_{t-h}^t\int_\delta^\infty \theta(t-s)^{q-1}M_q(\theta)((t-s)^q\theta)^{-\alpha}d\theta ds \\
&\leq M\|x_0\|_\alpha\int_0^\delta M_q(\theta)d\theta + N_\alpha Nq\int_0^t\int_0^\delta \theta^{1-\alpha}(t-s)^{q-q\alpha-1}M_q(\theta)d\theta ds \\
&\quad + N_\alpha Nq\int_{t-h}^t\int_\delta^\infty \theta^{1-\alpha}(t-s)^{q-q\alpha-1}M_q(\theta)d\theta ds \\
&\leq M\|x_0\|_\alpha\int_0^\delta M_q(\theta)d\theta + N_\alpha Nq\left(\int_0^t (t-s)^{q-q\alpha-1}ds\right)\int_0^\delta \theta^{1-\alpha}M_q(\theta)d\theta \\
&\quad + N_\alpha Nq\left(\int_{t-h}^t (t-s)^{q-q\alpha-1}ds\right)\int_0^\infty \theta^{1-\alpha}M_q(\theta)d\theta \\
&\leq M\|x_0\|_\alpha\int_0^\delta M_q(\theta)d\theta + N_\alpha Nq\left(\int_0^t (t-s)^{q-q\alpha-1}ds\right)\int_0^\delta \theta^{1-\alpha}M_q(\theta)d\theta \\
&\quad + N_\alpha Nq\frac{\Gamma(2-\alpha)}{\Gamma(1+q(1-\alpha))}\left(\int_{t-h}^t (t-s)^{q-q\alpha-1}ds\right)
\end{aligned}
$$

and

$$
\int_0^t (t-s)^{q-q\alpha-1}ds \leq \frac{1}{q(1-\alpha)}t^{q(1-\alpha)}, \quad \int_{t-h}^t (t-s)^{q-q\alpha-1}ds \leq \frac{1}{q(1-\alpha)}h^{q(1-\alpha)}
$$

that

$$
\begin{aligned}
&\|(Fx)(t)-(F_{h,\delta}x)(t)\|_\alpha \\
&\leq M\|x_0\|_\alpha\int_0^\delta M_q(\theta)d\theta + \frac{N_\alpha Nq}{q(1-\alpha)}T^{q(1-\alpha)}\int_0^\delta \theta^{1-\alpha}M_q(\theta)d\theta \\
&\quad + \frac{N_\alpha Nq\Gamma(2-\alpha)}{\Gamma(1+q(1-\alpha))}\frac{1}{q(1-\alpha)}h^{q(1-\alpha)}.
\end{aligned}
$$

Therefore, $\Pi(t) = \{(Fx)(t) \mid x \in B\}$ is relatively compact in X_α for all $t \in (0,T]$ and since it is compact at $t=0$ we have the relatively compactness in X_α for all $t \in J$.

Next, let us prove that $\Pi = FB$ is equicontinuous. For $T > h \geq 0$,

$$
\begin{aligned}
\|(Fx)(h)-(Fx)(0)\|_\alpha &\leq \|S_\alpha(h)-I\|_\alpha\|x_0\|_\alpha \\
&\quad + \frac{N_\alpha q\Gamma(2-\alpha)}{\Gamma(1+q(1-\alpha))}N\frac{1}{q(1-\alpha)}h^{q(1-\alpha)},
\end{aligned}
$$

and for $0 < s < t_1 < t_2 \leq T$,

$$\|(Fx)(t_1) - (Fx)(t_2)\|_\alpha \leq I_1 + I_2 + I_3 + I_4,$$

where

$$I_1 = \|S_\alpha(t_1) - S_\alpha(t_2)\|_\alpha \|x_0\|_\alpha,$$

$$I_2 = N\sqrt{\frac{1}{2q-1}} t_1^{q-\frac{1}{2}} \frac{qN_{\alpha+1}\Gamma(2-\alpha)}{\alpha\Gamma(1+q(1-\alpha))} \left(\int_0^T |(t_2-s)^{-q\alpha} - (t_1-s)^{-q\alpha}|^2 ds\right)^{\frac{1}{2}},$$

$$I_3 = N\frac{N_\alpha q\Gamma(2-\alpha)}{\Gamma(1+q(1-\alpha))} \left(\int_0^T |(t_1-s)^{q-1} - (t_2-s)^{q-1}|^2 ds\right)^{\frac{1}{2}}$$

$$\times \sqrt{\frac{1}{1-2\alpha q}} \left(t_2^{1-2\alpha q} - (t_2-t_1)^{1-2\alpha q}\right)^{\frac{1}{2}},$$

$$I_4 = N\frac{N_\alpha q\Gamma(2-\alpha)}{\Gamma(1+q(1-\alpha))} \frac{1}{q(1-\alpha)} (t_2-t_1)^{q(1-\alpha)}.$$

Now, we need to check that I_1, I_2, I_3, I_4 tend to 0 independently of $x \in B$ when $t_2 \to t_1$. In fact, let $x \in B$, one can deduce that $\lim_{t_2\to t_1} I_1 = 0$ and $\lim_{t_2\to t_1} I_2 = 0$ since by (iii) and (vi) of Lemma 4.21. Moreover, using the fact $|(t_1-s)^{q-1} - (t_2-s)^{q-1}| \to 0$ as $t_2 \to t_1$, and Lemma 4.22, we can deduce

$$\int_0^T |(t_1-s)^{q-1} - (t_2-s)^{q-1}|^2 ds \to 0, \text{ as } t_2 \to t_1.$$

Thus, $\lim_{t_2\to t_1} I_3 = 0$ since $\alpha q < \frac{1}{2}$. It is also easy to see $\lim_{t_2\to t_1} I_4 = 0$.

In summary, we have proven that FB is relatively compact, for $t \in J$, $\Pi = \{Fx \mid x \in B\}$ is a family of equicontinuous functions. Hence by the Arzela-Ascoli theorem, F is compact.

Step III. F has a fixed point in $\mathcal{C}_\alpha$.

According to Schauder fixed point theorem, it is sufficient to show that the set $\Sigma = \{x \in \mathcal{C}_\alpha \mid x = \sigma Fx, \sigma \in [0,1]\}$ is a bounded subset of $\mathcal{C}_\alpha$.

In fact, let $x \in \Sigma$,

$$\begin{aligned}
\|x(t)\|_\alpha &= \|\sigma(Fx)(t)\|_\alpha \\
&\leq \|S(t)x_0\|_\alpha + \int_0^t (t-s)^{q-1}\big\|P(t-s)f(s,x(s))\big\|_\alpha ds \\
&\leq M\|x_0\|_\alpha + \int_0^t (t-s)^{q-1}\big\|A^\alpha P(t-s)\big\|\|f(s,x(s))\|ds \\
&\leq M\|x_0\|_\alpha + \frac{a_f N_\alpha q\Gamma(2-\alpha)}{\Gamma(1+q(1-\alpha))}\int_0^t (t-s)^{q-\alpha q-1}(1+\|x(s)\|_\alpha)ds \\
&\leq M\|x_0\|_\alpha + \frac{a_f N_\alpha q\Gamma(2-\alpha)}{\Gamma(1+q(1-\alpha))}\frac{T^{q(1-\alpha)}}{q(1-\alpha)} \\
&\quad + \frac{a_f N_\alpha q\Gamma(2-\alpha)}{\Gamma(1+q(1-\alpha))}\int_0^t (t-s)^{q-\alpha q-1}\|x(s)\|_\alpha ds,
\end{aligned}$$

using the well known singular version Gronwall inequality, we can deduce that there exists a constant $M^* > 0$ such that $\|x\|_\infty \leq M^*$. Thus, Σ is a bounded subset of $\mathcal{C}_\alpha$. By Schauder fixed point theorem F has a fixed point $\mathcal{C}_\alpha$. Consequently, system (4.40) has at least one α-mild solution x on J.

Step IV. $x(\cdot)$ is unique.

Let $y(\cdot)$ be another α-mild solution of system (4.40) with the initial value y_0. It is not difficult to verify that there exists a constant $\rho > 0$ such that $\|x\|_\alpha, \|y\|_\alpha \leq \rho$. From

$$\begin{aligned}&\|x(t) - y(t)\|_\alpha\\ &\leq \|S(t)(x_0 - y_0)\|_\alpha + \int_0^t (t-s)^{q-1}\|P(t-s)(f(s,x(s)) - f(s,y(s)))\|_\alpha ds,\end{aligned}$$

we can deduce that

$$\begin{aligned}\|x(t) - y(t)\|_\alpha &\leq M\|(x_0 - y_0)\|_\alpha\\ &\quad + \frac{L_f(\rho)N_\alpha q\Gamma(2-\alpha)}{\Gamma(1+q(1-\alpha))}\int_0^t (t-s)^{q-\alpha q-1}\|x(s) - y(s)\|_\alpha ds.\end{aligned}$$

By singular version Gronwall inequality again, there exists a constant $\widetilde{M} > 0$ such that

$$\|x(t) - y(t)\|_\alpha \leq \widetilde{M}M\|x_0 - y_0\|_\alpha,$$

which yields the uniqueness of $x(\cdot)$. □

4.5.4 *Existence of Fractional Optimal Controls*

We suppose that Y is another separable reflexive Banach space from which the controls u take the value. We denote a class of nonempty closed and convex subsets of Y by $W_f(Y)$. The multifunction $\omega : J \longrightarrow W_f(Y)$ is measurable and $\omega(\cdot) \subset E$ where E is a bounded set of Y, the admissible control set $U_{ad} = S^p_\omega = \{u \in L^p(E) \mid u(t) \in \omega(t)\ a.e.\}$, $1 < p < \infty$. Then $U_{ad} \neq \emptyset$ (see Proposition 1.7 and Lemma 3.2 of Hu and Papageorgiou, 1997).

Consider the following controlled system

$$\begin{cases} D^q x(t) = -Ax(t) + f(t,x(t)) + C(t)u(t), t \in J, u \in U_{ad}, q \in (0,1),\\ x(0) = x_0.\end{cases} \tag{4.41}$$

Assumption **[HC]**: $C \in L_\infty(J, L(Y, X_\alpha))$.

It is easy to see that $Cu \in L^p(J, X_\alpha)$ for all $u \in U_{ad}$.

By Theorem 4.12, we have the following result.

Theorem 4.13. *In addition to assumptions of Theorem 4.12, suppose assumption [HC] holds. For every $u \in U_{ad}$ and $pq(1-\alpha) > 1$, system (4.41) has a α-mild solution corresponding to u given by*

$$\begin{aligned}x^u(t) = S(t)x_0 &+ \int_0^t (t-s)^{q-1}P(t-s)f(s,x(s))ds\\ &+ \int_0^t (t-s)^{q-1}P(t-s)C(s)u(s)ds.\end{aligned}$$

Proof. Compared with Theorem 4.12, the key step is to check the term containing control policy. Consider

$$\Psi(t) = \int_0^t (t-s)^{q-1} P(t-s) C(s) u(s) ds,$$

using Lemma 4.21(iv) and Hölder inequality again, we have

$$\begin{aligned}
\|\Psi(t)\|_\alpha &\le \left\| \int_0^t (t-s)^{q-1} P(t-s) C(s) u(s) ds \right\|_\alpha \\
&\le \int_0^t (t-s)^{q-1} \|A^\alpha P(t-s)\| \|C(s)u(s)\| ds \\
&\le \frac{\|C\|_\infty N_\alpha q \Gamma(2-\alpha)}{\Gamma(1+q(1-\alpha))} \int_0^t (t-s)^{q-q\alpha-1} \|u(s)\|_Y ds \\
&\le \frac{\|C\|_\infty N_\alpha q \Gamma(2-\alpha)}{\Gamma(1+q(1-\alpha))} \left(\int_0^t (t-s)^{\frac{p}{p-1}(q-q\alpha-1)} ds \right)^{\frac{p-1}{p}} \left(\int_0^t \|u(s)\|_Y^p ds \right)^{\frac{1}{p}} \\
&\le \frac{\|C\|_\infty N_\alpha q \Gamma(2-\alpha)}{\Gamma(1+q(1-\alpha))} \left(\frac{p-1}{pq(1-\alpha)-1} \right)^{\frac{p-1}{p}} T^{\frac{pq(1-\alpha)-1}{p-1}} \|u\|_{L^p(J,Y)},
\end{aligned}$$

where $\|C\|_\infty$ is the norm of operator C in Banach space $L_\infty(J, L(Y, X_\alpha))$. Thus, $\|(t-s)^{q-1}P(t-s)C(s)u(s)\|_\alpha$ is Lebesgue integrable with respect to $s \in [0,t]$ for all $t \in J$. From Lemma 1.6, it follows that $(t-s)^{q-1}P(t-s)C(s)u(s)$ is Bochner integral with respect to $s \in [0,t]$ for all $t \in J$. Hence $\Psi(\cdot) \in \mathcal{C}_\alpha$. Using Theorem 4.12, one can verify it immediately. □

Assumption [**HL**]:

(1) The functional $\mathcal{L} : J \times X_\alpha \times Y \longrightarrow \mathbb{R} \cup \{\infty\}$ is Borel measurable.
(2) $\mathcal{L}(t, \cdot, \cdot)$ is sequentially lower semicontinuous on $X_\alpha \times Y$ for almost all $t \in J$.
(3) $\mathcal{L}(t, x, \cdot)$ is convex on Y for each $x \in X_\alpha$ and almost all $t \in J$.
(4) There exist constants $d \ge 0$, $e > 0$, φ is nonnegative and $\varphi \in L^1(J, R)$ such that

$$\mathcal{L}(t, x, u) \ge \varphi(t) + d\|x\|_\alpha + e\|u\|_Y^p.$$

We consider the Lagrange problem (P):
Find $(x^0, u^0) \in C(J, X_\alpha) \times U_{ad}$ such that

$$J(x^0, u^0) \le J(x^u, u), \text{ for all } u \in U_{ad}$$

where

$$J(x^u, u) = \int_0^T \mathcal{L}(t, x^u(t), u(t)) dt,$$

x^u denotes the α-mild solution of system (4.41) corresponding to the control $u \in U_{ad}$.

In order to obtain the existence of fractional optimal controls we need the following important lemma.

Lemma 4.23. *Operator* $\mathscr{Q} : L^p(J,Y) \longrightarrow \mathcal{C}_\alpha$ *for some* $pq(1-\alpha) > 1$, *given by*

$$(\mathscr{Q}u)(\cdot) = \int_0^{\cdot} P(\cdot - s)C(s)u(s)ds$$

is strongly continuous.

Proof. Suppose that $\{u^n\} \subseteq L^p(J,Y)$ is bounded, we define $\Theta_n(t) = (\mathscr{Q}u^n)(t)$, $t \in J$. Similar to the proof of Theorem 4.13, one can know that for any fixed $t \in J$ and $pq(1-\alpha) > 1$, $\|\Theta_n(t)\|_\alpha$ is bounded. By Lemma 4.21, it is not difficult to verify that $\Theta_n(t)$ is compact in X_α and is also equicontinuous. Due to Arzela-Ascoli theorem again, $\{\Theta_n(t)\}$ is relatively compact in $\mathcal{C}_\alpha$. Obviously, $\mathscr{Q}$ is linear and continuous. Hence, $\mathscr{Q}$ is a strongly continuous operator. □

Now we can give the following result on existence of optimal controls for problem (P).

Theorem 4.14. *If the assumption [HL] and the assumptions of Theorem 4.13 hold, then the problem (P) admits at least one optimal pair.*

Proof. If $\inf\{J(x^u,u) \mid u \in U_{ad}\} = +\infty$, there is nothing to prove.

Assume that $\inf\{J(x^u,u) \mid u \in U_{ad}\} = \epsilon < +\infty$. Using assumption [HL], we have $\epsilon > -\infty$. By definition of infimum there exists a minimizing sequence feasible pair $\{(x^m,u^m)\} \subset A_{ad} \equiv \{(x,u) \mid x$ is a α-mild solution of system (4.41) corresponding to $u \in U_{ad}\}$, such that $J(x^m,u^m) \to \epsilon$ as $m \to +\infty$. Since $\{u^m\} \subseteq U_{ad}$, $m = 1,2,\ldots$, $\{u^m\}$ is bounded in $L^p(J,Y)$, there exists a subsequence, relabeled as $\{u^m\}$, and $u^0 \in L^p(J,Y)$ such that

$$u^m \xrightarrow{w} u^0 \text{ in } L^p(J,Y), \quad \text{as } m \to +\infty.$$

Since U_{ad} is closed and convex, thanks to Marzur Lemma, $u^0 \in U_{ad}$.

Suppose x^m (x^0) is the mild solution of system (4.41) corresponding to u^m (u^0). x^m and x^0 satisfy the following integral equation respectively

$$\begin{aligned} x^m(t) = S(t)x_0 &+ \int_0^t (t-s)^{q-1}P(t-s)f(s,x^m(s))ds \\ &+ \int_0^t (t-s)^{q-1}P(t-s)C(s)u^m(s)ds, \\ x^0(t) = S(t)x_0 &+ \int_0^t (t-s)^{q-1}P(t-s)f(s,x^0(s))ds \\ &+ \int_0^t (t-s)^{q-1}P(t-s)C(s)u^0(s)ds. \end{aligned}$$

It follows from the boundedness of $\{u^m\}$ $(\{u^0\})$ and Theorem 4.12, one can verify that there exists a positive number ρ such that $\|x^m\|_\infty, \|x^0\|_\infty \le \rho$.

For $t \in J$, we have

$$\|x^m(t) - x^0(t)\|_\alpha \le \|\eta_m^{(1)}(t)\|_\alpha + \|\eta_m^{(2)}(t)\|_\alpha$$

where

$$\eta_m^{(1)}(t) = \int_0^t (t-s)^{q-1} P(t-s)[f(s, x^m(s)) - f(s, x^0(s))]ds,$$

$$\eta_m^{(2)}(t) = \int_0^t (t-s)^{q-1} P(t-s)C(s)\left(u^m(s) - u^0(s)\right)ds.$$

Using Lemma 4.21(iv) and [Hf](2), one has

$$\|\eta_m^{(1)}(t)\|_\alpha \le \frac{L_f(\rho)N_\alpha q\Gamma(2-\alpha)}{\Gamma(1+q(1-\alpha))}\int_0^t (t-s)^{q-1-\alpha q}\|x_1(s) - x_2(s)\|_\alpha ds.$$

Using Lemma 4.23, one has

$$\eta_m^{(2)}(t) \xrightarrow{s} 0 \text{ in } X_\alpha, \text{ as } m \to \infty.$$

Thus,

$$\begin{aligned}&\|x^m(t) - x^0(t)\|_\alpha\\ &\le \|\eta_m^{(2)}(t)\|_\alpha + \frac{L_f(\rho)N_\alpha q\Gamma(2-\alpha)}{\Gamma(1+q(1-\alpha))}\int_0^t (t-s)^{q-1-\alpha q}\|x_1(s) - x_2(s)\|_\alpha ds.\end{aligned}$$

By virtue of singular version Gronwall inequality again, there exists $\widehat{M}^* > 0$ such that

$$\|x^m(t) - x^0(t)\|_\alpha \le \widehat{M}^* \|\eta_m^{(2)}(t)\|_\alpha$$

which yields that

$$x^m \longrightarrow x^0 \text{ in } C(J, X_\alpha), \text{ as } m \to \infty.$$

Since $C(J, X_\alpha) \hookrightarrow L^1(J, X_\alpha)$, using the assumption [HL] and Balder's theorem, we can obtain

$$\epsilon = \lim_{m\to\infty}\int_0^T \mathcal{L}(t, x^m(t), u^m(t))\,dt \ge \int_0^T \mathcal{L}\left(t, x^0(t), u^0(t)\right)dt = J\left(x^0, u^0\right) \ge \epsilon.$$

This show that J attains its minimum at $u^0 \in U_{ad}$. □

At last, an example is given to illustrate our theory. Consider the following problem:

$$\begin{cases} {}_0^C D_t^q x(t,y) - \Delta x(t,y) = x(t,y) + 2u(t,y), \ y \in \Omega, \ t \in [0,1], \ q = \frac{25}{26}, \\ x(t,y) \mid_{y\in\partial\Omega} = 0, \quad t > 0, \\ x(0,y) = 0 \end{cases} \tag{4.42}$$

where $\Omega \subset \mathbb{R}^3$ is a bounded domain, $\partial\Omega \in C^3$.

Define $X = Y = L^2(\Omega)$, $D(A) = H^2(\Omega)\bigcap H_0^1(\Omega)$, and $Ax = -\left(\frac{\partial^2 x}{\partial y_1^2} + \frac{\partial^2 x}{\partial y_2^2} + \frac{\partial^2 x}{\partial y_3^2}\right)$ for $x \in D(A)$. The admissible control set $U_{ad} = \{u \in Y \mid \|u\|_{L^2([0,1],Y)} \le 1\}$. By Sobolev embedded theorem, we can choose $\alpha = \frac{11}{24}$ then $X_{\frac{11}{24}} \hookrightarrow C^1(\overline{\Omega})$. Find the control $u(t,y)$ that minimizes the performance index

$$J(x,u) = \int_0^1\int_\Omega |x(t,y)|^2 dydt + \int_0^1\int_\Omega |u(t,y)|^2 dydt$$

subject to the problem (4.42).

Define $x(t)(y) = x(t,y)$, $C(t)u(t)(y) = 2u(t,y)$, $f(t,x(t))(y) = x(t)(y)$. Thus problem (4.42) can be rewritten as

$$\begin{cases} {}_0^C D_t^q x(t) = -Ax(t) + f(t,x(t)) + C(t)u(t), \ t \in [0,1], \ q \in (0,1), \\ x(0) = x_0 \end{cases} \tag{4.43}$$

with the cost function

$$J(u) = \int_0^1 \left(\|x(t)\|^2 + \|u(t)\|_E^2\right) dt.$$

It is not difficult to verify that $q\alpha = \frac{25}{26} \times \frac{4}{5} = \frac{275}{624} < \frac{1}{2}$ and $pq(1-\alpha) = 2 \times \frac{25}{26} \times \frac{13}{24} = \frac{25}{24} > 1$. Then it satisfies all the assumptions given in Theorem 4.14. Therefore, the problem (4.42) has at least one optimal pair.

4.6 Abstract Cauchy Problems with Almost Sectorial Operators

4.6.1 *Introduction*

Let X be a complex Banach space with norm $|\cdot|$. As usual, for a linear operator A, we denote by $D(A)$ the domain of A, by $\sigma(A)$ its spectrum, while $\rho(A) := \mathbb{C} - \sigma(A)$ is the resolvent set of A, and denote by the family $R(z;A) = (zI - A)^{-1}$, $z \in \rho(A)$ of bounded linear operators the resolvent of A. Moreover, we denote by $B(X,Y)$ the space of all bounded linear operators from Banach space X to Banach space Y with the usual operator norm $\|\cdot\|_{B(X,Y)}$, and we abbreviate this notation to $B(X)$ when $Y = X$, and write $\|T\|_{B(X)}$ as $\|T\|$ for every $T \in B(X)$.

When dealing with parabolic evolution equations, it is usually assumed that the partial differential operator in the linear part is a sectorial operator, stimulated by the fact that this class of operators appears very often in the applications. For example, one can find from Henry, 1981; Lunardi, 1995 and Pazy, 1983 that many elliptic differential operators equipped with homogeneous boundary conditions are sectorial when they are considered in Lebesgue spaces (e.g., L^p-space) or in the space of continuous functions. We here mention that the operator A_ε in Example 4.2, which acts on a domain of "dumb-bell with a thin handle", is sectorial on V_ϵ^p. However, as presented in Example 4.2 and Example 4.3, though the resolvent set of some partial differential operators considered in some special domains such as the limit "domain" of dumb-bell with a thin handle or in some spaces of more regular functions such as the space of Hölder continuous functions, contains a sector, but for which the resolvent operators do not satisfy the required estimate to be a sectorial operator.

Example 4.2. In this notation the "dumb-bell with a thin handle" has the form

$$\Omega_\varepsilon = D_1 \cup Q_\varepsilon \cup D_2 \quad (\varepsilon \in (0,1]; \text{ small}),$$

where D_1 and D_2 are mutually disjoint bounded domains in $\mathbb{R}^N (N \geq 2)$ with smooth boundaries, joined by a thin channel, Q_ε (which is not required to be cylindrical), which degenerates to a 1-dim line segment Q_0 as ε approaches zero. This implies that passing to the limit as $\varepsilon \to 0$, the limit "domain" of Ω_ε consists of the fixed part D_1, D_2 and the line segment Q_0. Without loss of generality, we may assume that $Q_0 = \{(x, 0, \ldots, 0); 0 < x < 1\}$. Let $P_0 = (0, 0, \ldots, 0), P_1 = (1, 0, \ldots, 0)$ be the points where the line segment touches the boundary of D_1 and D_2. Put $\Omega = D_1 \cup D_2$.

Firstly, consider the evolution equation of parabolic type equipped with Neumann boundary condition in the form

$$\begin{cases} u_t - \Delta u + u = f(u), & x \in \Omega_\epsilon,\ t > 0, \\ \dfrac{\partial u}{\partial n} = 0, & x \in \partial\Omega_\epsilon, \end{cases} \tag{4.44}$$

where Δ stands for the Laplacian operator with respect to the spatial variable $x \in \Omega_\epsilon$, $\partial\Omega_\epsilon$ is the boundary of Ω_ϵ, $\frac{\partial}{\partial n}$ denotes the outward normal derivative on $\partial\Omega_\epsilon$ and $f : \mathbb{R} \to \mathbb{R}$ is a nonlinearity. Let V_ϵ^p $(1 \le p < \infty)$ denote the family of spaces based on $L^p(\Omega_\epsilon)$, equipped with the norm

$$\|u\|_{V_\epsilon^p} = \left(\int_\Omega |u|^p + \frac{1}{\varepsilon^{N-1}} \int_{Q_\epsilon} |u|^p \right)^{1/p}.$$

Define the linear operator $A_\varepsilon : D(A_\varepsilon) \subset V_\epsilon^p \mapsto V_\epsilon^p$ by

$$D(A_\varepsilon) = \left\{ u \in W^{2,p}(\Omega_\epsilon) : \ \Delta u \in V_\epsilon^p,\ \frac{\partial u}{\partial n}\bigg|_{\partial\Omega_\epsilon} = 0 \right\},$$

$$A_\varepsilon u = -\Delta u + u, \quad u \in D(A_\varepsilon).$$

It follows from a standard argument that the operator A_ε generates an analytic semigroup on V_ϵ^p. Moreover, the following estimate holds

$$\|R(\lambda; -A_\varepsilon)\|_{B(L^p(\Omega_\epsilon))} \le \frac{C}{|\lambda|}, \quad \text{for } \lambda \in \Sigma'_\theta,$$

where $\Sigma'_\theta = \{\lambda \in \mathbb{C} : \ |\arg(\lambda - 1)| \le \theta\}$ with $\theta > \frac{\pi}{2}$, and C is a constant that does not depend on ε (see, e.g., Henry, 1981 and Pazy, 1983).

The limit problem of (4.44) as $\varepsilon \to 0$ is the following problem studied in Carvalho, Dlotko and Neseimento, 2008

$$\begin{cases} w_t - \Delta w + w = f(w), & x \in \Omega,\ t > 0, \\ \dfrac{\partial w}{\partial n} = 0, & x \in \partial\Omega, \\ v_t - \dfrac{1}{g}(g v_x)_x + v = f(v), & x \in Q_0 = (0, 1), \\ v(0) = w(P_0),\ v(1) = w(P_1), \end{cases}$$

where w is a function that lives in Ω and v lives in the line segment Q_0, the function $g : [0, 1] \to (0, \infty)$ is a smooth function related to the geometry of the channel Q_ε, more exactly, on the way the channel Q_ε collapses to the segment line Q_0. Observe that the vector (w, v) is continuous in the junction between Ω and Q_0 and the variable w does not depend on the variable v, but v depends on w.

We identify V_0^p with $L^p(\Omega) \oplus L_g^p(0,1)$ $(1 \leq p < \infty)$ endowed with the norm

$$\|(w,v)\|_{V_0^p} = \left(\int_\Omega |w|^p + \int_0^1 g|v|^p \right)^{1/p}.$$

Consider the operator $A_0 : D(A_0) \subset V_0^p \mapsto V_0^p$ defined by

$$D(A_0) = \{(w,v) \in V_0^P; w \in D(\Delta_\Omega), v \in L_g^p(0,1), w(P_0) = v(0), w(P_1) = v(1)\},$$

$$A_0(w,v) = \left(-\Delta w + w, -\frac{1}{g}(gv')' + v \right), \quad (w,v) \in V_0^p, \tag{4.45}$$

where Δ_Ω is the Laplace operator with homogeneous Neumann boundary conditions in $L^p(\Omega)$ and

$$D(\Delta_\Omega) = \left\{ u \in W^{2,p}(\Omega) : \left. \frac{\partial u}{\partial n} \right|_{\partial\Omega} = 0 \right\}.$$

As pointed out by Arrieta, Carvalho and Lozada-Cruz, 2009a, the operator A_0 defined by (4.45) is not a sectorial operator. Its spectrum is all real and, therefore, it is contained in a sector but the resolvent estimate is different from the case of sectorial operator. More precisely, the operator A_0 has the following properties (see also Arrieta, Carvalho and Lozada-Cruz, 2006, 2009a):

(a) the domain $D(A_0)$ is dense in V_0^P;
(b) if $p > \frac{N}{2}$, then A_0 is a closed operator;
(c) A_0 has compact resolvent;
(d) for some $\mu \in (0, \frac{\pi}{2})$, $\Sigma_\mu := \{\lambda \in \mathbb{C} \setminus \{0\} : \ |\arg \lambda| \leq \pi - \mu\} \cup \{0\} \subset \rho(-A_0)$, and for $\frac{N}{2} < q \leq p$, the following estimate holds:

$$\|R(\lambda; -A_0)\|_{B(V_0^q, V_0^p)} \leq \frac{C}{1 + |\lambda|^{\gamma'}}, \quad \lambda \in \Sigma_\mu, \tag{4.46}$$

for each $0 < \gamma' < 1 - \frac{N}{2q} - \frac{1}{2}(\frac{1}{q} - \frac{1}{p}) < 1$, where C is a positive constant.

Remark 4.9. In fact, it is easy to prove that the estimate (4.46) with $p = q > \frac{N}{2}$ is equivalent to

$$\|R(\lambda; -A_0)\|_{B(V_0^p)} \leq \frac{\widetilde{C}}{|\lambda|^{\gamma'}}, \quad \lambda \in \Sigma_\mu \setminus \{0\},$$

for $0 < \gamma' < 1 - \frac{N}{2p}$, where $\widetilde{C}$ is a positive constant.

We refer to Section 2 of Arrieta, Carvalho and Lozada-Cruz, 2006 for a complete and rigorous definition of the dumb-bell domain, and to Arrieta, 1995; Arrieta, Carvalho and Lozada-Cruz, 2006, 2009a,b; Dancer and Daners, 1997; Gadyl'shin, 2005; Jimbo, 1989 for related studies of partial differential equations involving dumb-bell domain.

Example 4.3. Assume that Ω is a bounded domain in $\mathbb{R}^N$ $(N \geq 1)$ with boundary $\partial\Omega$ of class C^{4m} $(m \in \mathbb{N})$. Let $C^l(\overline{\Omega})$, $l \in (0,1)$, denote the usual Banach space with norm $|\cdot|_l$. Consider the elliptic differential operator $A' : D(A') \subset C^l(\Omega) \mapsto C^l(\Omega)$ in the form

$$D(A') = \left\{u \in C^{2m+l}(\overline{\Omega}) : \; D^\beta u|_{\partial\Omega} = 0, |\beta| \leq m-1\right\},$$

$$A'u = \sum_{|\beta| \leq 2m} a_\beta(x) D^\beta u(x), \quad u \in D(A'),$$

where β is a multiindex in $(\mathbb{N} \bigcup \{0\})^n$,

$$|\beta| = \sum_{j=1}^n \beta_j, \quad D^\beta = \prod_{j=1}^n \left(-i\frac{\partial}{\partial x_j}\right)^{\beta_j}.$$

The coefficients $a_\beta : \overline{\Omega} \mapsto \mathbb{C}$ of A' are assumed to satisfy

(i) $a_\beta \in C^l(\overline{\Omega})$ for all $|\beta| \leq 2m$;
(ii) $a_\beta(x) \in \mathbb{R}$ for all $x \in \overline{\Omega}$ and $|\beta| = 2m$;
(iii) there exists a constant $M > 0$ such that

$$M^{-1}|\xi|^2 \leq \sum_{|\beta|=2m} a_\beta(x)\xi^\beta \leq M|\beta|^2, \quad \text{for all } \xi \in \mathbb{R}^N \text{ and } x \in \overline{\Omega}.$$

Then, the following statements hold.

(a) A' is not densely defined in $C^l(\overline{\Omega})$;
(b) there exist ν, $\varepsilon > 0$ such that

$$\sigma(A' + \nu) \subset S_{\frac{\pi}{2}-\varepsilon} = \left\{\lambda \in \mathbb{C} \setminus \{0\} : \; |\arg \lambda| \leq \frac{\pi}{2} - \varepsilon\right\} \cup \{0\},$$

$$\|R(\lambda; A' + \nu)\|_{B(C^l(\overline{\Omega}))} \leq \frac{C}{|\lambda|^{1-\frac{l}{2m}}}, \quad \lambda \in \mathbb{C} \setminus S_{\frac{\pi}{2}-\varepsilon};$$

(c) the exponent $\frac{l}{2m} - 1 \in (-1, 0)$ is sharp. In particular, the operator $A' + \nu$ is not sectorial.

Notice in particular that the Laplace operator satisfies the conditions (a)-(c) in Example 4.3. For more details we refer to Wahl, 1972.

Observe that from Example 4.2 and Remark 4.9, if $p > \frac{N}{2}$, then $A_0 \in \Theta_\mu^{-\gamma'}(V_0^p)$ for some $\gamma' \in (0, 1 - \frac{N}{2p})$ and $\mu \in (0, \frac{\pi}{2})$, that is, A_0 is an almost sectorial operator on V_0^p. Also, from Example 4.3 one can find that $(A' + \nu) \in \Theta_{\frac{\pi}{2}-\varepsilon}^{\frac{l}{2m}-1}(C^l(\overline{\Omega}))$, which implies that $A' + \nu$ is an almost sectorial operator on $C^l(\overline{\Omega})$.

In this section, motivated by the above consideration, we are interested in studying the Cauchy problem for the linear evolution equation

$$\begin{cases} {}_0^C D_t^\alpha u(t) + Au(t) = f(t), & t > 0, \\ u(0) = u_0, \end{cases} \tag{4.47}$$

as well as the Cauchy problem for the corresponding semilinear fractional evolution equation

$$\begin{cases} {}_0^C D_t^\alpha u(t) + Au(t) = f(t, u(t)), & t > 0, \\ u(0) = u_0 \end{cases} \tag{4.48}$$

in X, where ${}_0^C D_t^\alpha$, $0 < \alpha < 1$, is Caputo fractional derivative of order α and A is an almost sectorial operator, that is, $A \in \Theta_\omega^\gamma(X)$ $(-1 < \gamma < 0,\ 0 < \omega < \pi/2)$. The main purpose is to study the existence and uniqueness of mild solutions and classical solutions of Cauchy problems (4.47) and (4.48). To do this, we construct two operator families based on the generalized Mittag-Leffler functions and the resolvent operators associated with A, present deep anatomy on basic properties for these families consisting on the study of the compactness, and prove that, under natural assumptions, reasonable concept of solutions can be given to problems (4.47) and (4.48), which in turn is used to find solutions to the Cauchy problems.

Remark 4.10. We make no assumption on the density of the domain of A.

Remark 4.11.

(i) M. M. Dzhrbashyan and A. B. Nersessyan in Dzhrbashyan and Nersessyan, 1968 (see also Miller and Ross, 1993) showed that the solution of the Cauchy problem

$$\begin{cases} {}_0^C D_t^\alpha u(t) + \lambda u(t) = 0, & t > 0, \\ u(0) = 1, & 0 < \alpha < 1 \end{cases}$$

has the form $u(t) = E_\alpha(-\lambda t^\alpha)$, where E_α is the known Mittag-Leffler function. This result issues a warning to us that no matter how smooth the data u_0 is, it is inappropriate to define the mild solution of problem (4.47) as follows

$$u(t) = T(t)u_0 + \frac{1}{\Gamma(\alpha)} \int_0^t (t-s)^{\alpha-1} T(t-s) f(s) ds,$$

where $T(t)$ is the semigroup generated by A (see Remark 1.7(i)), though this fashion was used in some situations of previous research (see, e.g., Jaradat, Ao-Omari and Momani, 2008).

(ii) Let us point out that in the treatment of problems (4.47) and (4.48), one of the difficult points is to give reasonable concept of solutions (see also the works of Zhou and Jiao, 2010a; Hernandez, O'Regan and Balachandran, 2010). Another is that even though the operator A generates a semigroup $T(t)$ in X, it is not continuous at $t = 0$ for nonsmooth initial data u_0.

(iii) It is worth mentioning that if it is the case when A is a matrix (or even bounded linear operators) then Kilbas, Srivastava and Trujill, 2006, obtained an explicit representation of mild solution to problem (4.47).

Let us now give a short summary of this section, which is organized in a way close to that given by Carvalho, Dlotko and Nescimento, 2008. In Subsection 4.6.2, we construct a pair of families of operators and present deep anatomy on the properties for these families. Based on the families of operators defined in Subsection 4.6.2, a reasonable concept of solution is given in Subsection 4.6.3 to problems (4.47), which in turn is used to analyze the existence of mild solutions and classical solutions to the Cauchy problem. The corresponding semilinear problem (4.48) is studied in Subsection 4.6.4. We investigate the existence of mild solutions, and then the existence of classical solutions. Finally, based mainly in Carvalho, Dlotko and Nescimento, 2008; Periago and Straub, 2002, we present three examples in Subsection 4.6.5 to illustrate our results.

Remark 4.12. Let us note that results in this section can be easily extended to the case of (general) sectorial operators.

4.6.2 *Properties of Operators*

Throughout this subsection we let A be an operator in the class $\Theta_\omega^\gamma(X)$ and $-1 < \gamma < 0$, $0 < \omega < \pi/2$. In the sequel, we succeed in defining two families of operators based on the generalized Mittag-Leffler functions and the resolvent operators associated with A. They are two families of linear and bounded operators. In order to check the properties of the families, we need a third object, namely the semigroup associated with A. We stress that these families are used very frequently throughout the rest of this section. Below the letter C denotes various positive constants.

Define operator families $\{\mathcal{S}_\alpha(t)\}|_{t\in S^0_{\frac{\pi}{2}-\omega}}$, $\{\mathcal{P}_\alpha(t)\}|_{t\in S^0_{\frac{\pi}{2}-\omega}}$ by

$$\mathcal{S}_\alpha(t) := E_\alpha(-zt^\alpha)(A) = \frac{1}{2\pi i}\int_{\Gamma_\theta} E_\alpha(-zt^\alpha)R(z;A)dz,$$

$$\mathcal{P}_\alpha(t) := e_\alpha(-zt^\alpha)(A) = \frac{1}{2\pi i}\int_{\Gamma_\theta} e_\alpha(-zt^\alpha)R(z;A)dz,$$

where the integral contour $\Gamma_\theta := \{\mathbb{R}_+e^{i\theta}\}\cup\{\mathbb{R}_+e^{-i\theta}\}$, is oriented counter-clockwise and $\omega < \theta < \mu < \frac{\pi}{2} - |\arg t|$.

We need some basic properties of these families which are used further in this section.

Theorem 4.15. *For each fixed $t \in S^0_{\frac{\pi}{2}-\omega}$, $\mathcal{S}_\alpha(t)$ and $\mathcal{P}_\alpha(t)$ are linear and bounded operators on X. Moreover, there exists constants $C_s = C(\alpha,\gamma) > 0$, $C_p = C(\alpha,\gamma) > 0$ such that for all $t > 0$,*

$$\|\mathcal{S}_\alpha(t)\|_{B(X)} \le C_s t^{-\alpha(1+\gamma)}, \quad \|\mathcal{P}_\alpha(t)\|_{B(X)} \le C_p t^{-\alpha(1+\gamma)}. \tag{4.49}$$

Proof. Note, from the asymptotic expansion of $E_{\alpha,\beta}$ that for each fixed $t \in S^0_{\frac{\pi}{2}-\omega}$,

$$E_\alpha(-zt^\alpha),\ e_\alpha(-zt^\alpha) \in \mathcal{F}_0^\gamma(S_\mu^0).$$

Therefore, by (1.24), the operators families $\{\mathcal{S}_\alpha(t)\}|_{t\in S^0_{\frac{\pi}{2}-\omega}}$, $\{\mathcal{P}_\alpha(t)\}|_{t\in S^0_{\frac{\pi}{2}-\omega}}$ are well-defined, and for each $t \in S^0_{\frac{\pi}{2}-\omega}$, $\mathcal{S}_\alpha(t)$ and $\mathcal{P}_\alpha(t)$ are linear bounded operators on X. So, to prove the theorem, it is sufficient to prove that the estimates in (4.49) hold.

Let $T(t)$, $t \in S^0_{\frac{\pi}{2}-\omega}$, be the semigroup defined by (1.25). Then by (W4) and the Fubini Theorem, we get

$$\begin{aligned}\mathcal{S}_\alpha(t)x &= \frac{1}{2\pi i}\int_{\Gamma_\theta} E_\alpha(-zt^\alpha)R(z;A)xdz \\ &= \frac{1}{2\pi i}\int_0^\infty M_\alpha(\lambda)\int_{\Gamma_\theta} e^{-\lambda z t^\alpha}R(z;A)xdzd\lambda \\ &= \int_0^\infty M_\alpha(s)T(st^\alpha)xds, \quad t \in S^0_{\frac{\pi}{2}-\omega},\ x \in X.\end{aligned} \tag{4.50}$$

A similar argument shows that

$$\mathcal{P}_\alpha(t)x = \int_0^\infty \alpha s M_\alpha(s)T(st^\alpha)xds, \quad t \in S^0_{\frac{\pi}{2}-\omega},\ x \in X. \tag{4.51}$$

Hence, by (4.50), (4.51), Proposition 4.56 (iii), (W1) and (W3), we have

$$\begin{aligned}|\mathcal{S}_\alpha(t)x| &\le C_0\int_0^\infty M_\alpha(s)s^{-(1+\gamma)}t^{-\alpha(1+\gamma)}|x|ds \\ &\le C_0\frac{\Gamma(-\gamma)}{\Gamma(1-\alpha(1+\gamma))}t^{-\alpha(1+\gamma)}|x|, \quad t>0,\ x\in X, \\ |\mathcal{P}_\alpha(t)x| &\le \alpha C_0\int_0^\infty M_\alpha(s)s^{-\gamma}t^{-\alpha(1+\gamma)}|x|ds \\ &\le \alpha C_0\frac{\Gamma(1-\gamma)}{\Gamma(1-\alpha\gamma)}t^{-\alpha(1+\gamma)}|x| \quad t>0,\ x\in X.\end{aligned}$$

Therefore, the estimates in (4.49) hold. This completes the proof. □

From now on, we frequently use the representations (4.50) and (4.51) for operators $\mathcal{S}_\alpha(t)$ and $\mathcal{P}_\alpha(t)$, respectively.

Theorem 4.16. *For $t > 0$, $\mathcal{S}_\alpha(t)$ and $\mathcal{P}_\alpha(t)$ are continuous in the uniform operator topology. Moreover, for every $r > 0$, the continuity is uniform on $[r,\infty)$.*

Proof. Let $\epsilon > 0$ be given. For every $r > 0$, it follows from (W3) that we may choose $\delta_1, \delta_2 > 0$ such that

$$\frac{2C_0}{r^{\alpha(1+\gamma)}}\int_0^{\delta_1}\Psi_\alpha(s)s^{-(1+\gamma)}ds \le \frac{\epsilon}{3}, \quad \frac{2C_0}{r^{\alpha(1+\gamma)}}\int_{\delta_2}^\infty \Psi_\alpha(s)s^{-(1+\gamma)}ds \le \frac{\epsilon}{3}. \tag{4.52}$$

Then we deduce, by Proposition 1.25(i), that there exists a positive constant δ such that

$$\int_{\delta_1}^{\delta_2} M_\alpha(s)\|T(t_1^\alpha s) - T(t_2^\alpha s)\|_{B(X)}ds \le \frac{\epsilon}{3}, \tag{4.53}$$

for $t_1, t_2 \ge r$ and $|t_1 - t_2| < \delta$.

On the other hand, using (4.52), (4.53) and Theorem 4.15, we get

$$\begin{aligned}|\mathcal{S}_\alpha(t_1)x-\mathcal{S}_\alpha(t_2)x| &\le \int_0^{\delta_1} M_\alpha(s)\left(\|T(t_1^\alpha s)\|_{B(X)}+\|T(t_2^\alpha s)\|_{B(X)}\right)|x|ds\\
&\quad+\int_{\delta_1}^{\delta_2} M_\alpha(s)\|T(t_1^\alpha s)-T(t_2^\alpha s)\|_{B(X)}|x|ds\\
&\quad+\int_{\delta_2}^{\infty} M_\alpha(s)\left(\|T(t_1^\alpha s)\|_{B(X)}+\|T(t_2^\alpha s)\|_{B(X)}\right)|x|ds\\
&\le \frac{2C_0}{r^{\alpha(1+\gamma)}}\int_0^{\delta_1}\Psi_\alpha(s)s^{-(1+\gamma)}|x|ds\\
&\quad+\int_{\delta_1}^{\delta_2} M_\alpha(s)\|T(t_1^\alpha s)-T(t_2^\alpha s)\|_{B(X)}|x|ds\\
&\quad+\frac{2C_0}{r^{\alpha(1+\gamma)}}\int_{\delta_2}^{\infty}\Psi_\alpha(s)s^{-(1+\gamma)}|x|ds\\
&\le \epsilon|x|,\quad \text{for any } x\in X,\end{aligned}$$

that is,

$$\|\mathcal{S}_\alpha(t_1)-\mathcal{S}_\alpha(t_2)\|_{B(X)}\le\epsilon,$$

which implies that $\mathcal{S}_\alpha(t)$ is uniformly continuous on $[r,\infty)$ in the uniform operator topology and hence, by the arbitrariness of $r>0$, $\mathcal{S}_\alpha(t)$ is continuous in the uniform operator topology for $t>0$. A similar argument enables us to give the characterization of continuity on $\mathcal{P}_\alpha(t)$. This completes the proof. □

Theorem 4.17. *Let $0<\beta<1-\gamma$. Then*

(i) *The range $R(\mathcal{P}_\alpha(t))$ of $\mathcal{P}_\alpha(t)$ for $t>0$, is contained in $D(A^\beta)$;*
(ii) *$\mathcal{S}'_\alpha(t)x=-t^{\alpha-1}A\mathcal{P}_\alpha(t)x$ ($x\in X$), and $\mathcal{S}'_\alpha(t)x$ for $x\in D(A)$ is locally integrable on $(0,\infty)$;*
(iii) *for all $x\in D(A)$ and $t>0$, $|A\mathcal{S}_\alpha(t)x|\le Ct^{-\alpha(1+\gamma)}|Ax|$, here C is a constant depending on γ,α.*

Proof. It follows from Proposition 1.25(iv) that for all $x\in X$, $t>0$, $T(t)x\in D(A^\beta)$ with $\beta>0$. Therefore, in view of (4.51), Proposition 1.25(iv) and (W3) we have

$$\begin{aligned}|A^\beta\mathcal{P}_\alpha(t)x| &\le \int_0^\infty \alpha sM_\alpha(s)\|A^\beta T(t^\alpha s)\|_{B(X)}|x|ds\\
&\le \alpha C't^{-\alpha(\gamma+\beta+1)}\int_0^\infty M_\alpha(s)s^{-(\beta+\gamma)}ds|x|\\
&\le \alpha C'\frac{\Gamma(1-\beta-\gamma)}{\Gamma(1-\alpha(\beta+\gamma+1))}t^{-\alpha(1+\beta+\gamma)}|x|,\end{aligned}$$

which implies that the assertion (i) holds.

From (i), it is easy to see that for all $x\in X$,

$$\mathcal{S}'_\alpha(t)x=-t^{\alpha-1}A\mathcal{P}_\alpha(t)x.$$

Moreover, for every $x \in D(A)$, one has by Proposition 4.56(iv),

$$|t^{\alpha-1}A\mathcal{P}_\alpha(t)x| \leq t^{\alpha-1}\int_0^\infty \alpha s M_\alpha(s)\|T(t^\alpha s)\|_{B(X)}|Ax|ds \\ \leq \alpha C_0 \frac{\Gamma(1-\gamma)}{\Gamma(1-\alpha\gamma)} t^{-\alpha\gamma-1}|Ax|.$$

Since $-\alpha\gamma-1 > -1$, this shows that $\mathcal{S}'_\alpha(t)x$ for each $x \in D(A)$, is locally integrable on $(0,\infty)$, that is, (ii) is true.

Moreover, Proposition 1.25(iv) and (4.50) imply that

$$|A\mathcal{S}_\alpha(t)x| \leq C_0 t^{-\alpha(1+\gamma)} \int_0^\infty M_\alpha(s)s^{-1-\gamma}ds|Ax| \\ \leq C_0 \frac{\Gamma(-\gamma)}{\Gamma(1-\alpha(1+\gamma))} t^{-\alpha(1+\gamma)}|Ax|, \quad x \in D(A).$$

This means that (iii) holds, and completes the proof. □

Remark 4.13. Particularly, from the proof of Theorem 4.17(i), we can conclude that

$$\|A\mathcal{P}_\alpha(t)\|_{B(X)} \leq Ct^{-\alpha(2+\gamma)},$$

where C is a constant depending on γ, α. Moreover, using a similar argument with that in Theorem 4.16, we have that $A\mathcal{P}_\alpha(t)$ for $t > 0$ is continuous in the uniform operator topology.

Theorem 4.18. *The following properties hold.*

(i) *Let $\beta > 1+\gamma$. For all $x \in D(A^\beta)$, $\lim_{t\to 0; t>0} \mathcal{S}_\alpha(t)x = x$;*
(ii) *for all $x \in D(A)$,$(\mathcal{S}_\alpha(t) - I)x = \int_0^t -s^{\alpha-1}A\mathcal{P}_\alpha(s)xds$;*
(iii) *for all $x \in D(A)$, $t > 0$, ${}_0D_t^\alpha \mathcal{S}_\alpha(t)x = -A\mathcal{S}_\alpha(t)x$;*
(iv) *for all $t > 0$, $\mathcal{S}_\alpha(t) = {}_0D_t^{\alpha-1}(t^{\alpha-1}\mathcal{P}_\alpha(t))$.*

Proof. For any $x \in X$, note by (4.50) and (W_3) that

$$\mathcal{S}_\alpha(t)x - x = \int_0^\infty \Psi_\alpha(s)(T(t^\alpha s)x - x)ds.$$

On the other hand, by Proposition 1.25(v) it follows that $D(A^\beta) \subset \Sigma_T$ in view of $\beta > 1+\gamma$. Therefore, we deduce, using Proposition 1.25(iii), that for any $x \in D(A^\beta)$, there exits a function $\eta(s) \in L^1(0,+\infty)$ depending on $\Psi_\alpha(s)$ such that

$$\|\Psi_\alpha(s)(T(t^\alpha s)x - x)\|_{B(X)} \leq \eta(s).$$

Hence, by means of Lebesgue dominated convergence theorem we obtain that

$$\mathcal{S}_\alpha(t)x - x \to 0, \quad \text{as } t \to 0,$$

that is, the assertion (i) remains true.

From (i) and Theorem 4.17(ii) we get for all $x \in D(A)$,

$$(\mathcal{S}_\alpha(t) - I)x = \lim_{s\to 0}(\mathcal{S}_\alpha(t)x - \mathcal{S}_\alpha(s)x) = \int_0^t -\lambda^{\alpha-1}A\mathcal{P}_\alpha(\lambda)xd\lambda,$$

which implies that the assertion (ii) holds.

To prove (iii), first it is easy to see that $\frac{1}{\varphi_0} \in \mathcal{F}(S_\mu^0)$ and the operator $\varphi_0(A)$ is injective. Taking $x \in D(A)$, by Proposition 4.55(iii) one has

$$\mathcal{S}_\alpha(t)x = E_\alpha(-zt^\alpha)(A)x = (E_\alpha(-zt^\alpha)\varphi_0)(A)\left(\frac{1}{\varphi_0}\right)(A)x.$$

Moreover, by (1.14) we have $\sup_{z\to\infty}|zt^\alpha E_\alpha(-zt^\alpha)| < \infty$, which implies that

$$|zE_\alpha(-zt^\alpha)(1+z)^{-1}| \le C|z|^{-1}t^{-\alpha}, \quad \text{as } z \to \infty,$$

where C is a constant which is independent of t. Consequently,

$$-zE_\alpha(-zt^\alpha)(1+z)^{-1} \in \mathcal{F}_0^\gamma(S_\mu^0). \tag{4.54}$$

Notice also that

$${}_0^CD_t^\alpha E_\alpha(-zt^\alpha)(1+z)^{-1}R(z;A) = (-z)E_\alpha(-zt^\alpha)(1+z)^{-1}R(z;A).$$

Combining Proposition 1.24(ii) and (4.54), we get

$$\begin{aligned}{}_0^CD_t^\alpha((E_\alpha(-zt^\alpha)(1+z^\beta)^{-1})(A)) &= \frac{1}{2\pi i}\int_{\Gamma_\theta}(-z)E_\alpha(-zt^\alpha)(1+z)^{-1}R(z;A)dz\\ &= (-z)(A)\big(E_\alpha(-zt^\alpha)(1+z)^{-1}\big)(A)\\ &= -A\big(E_\alpha(-zt^\alpha)(1+z)^{-1}\big)(A).\end{aligned}$$

Hence, we obtain

$$\begin{aligned}{}_0D_t^\alpha\mathcal{S}_\alpha(t)x &= -A\big(E_\alpha(-zt^\alpha)(1+z)^{-1}\big)(A)(1+z)(A)x\\ &= -A(E_\alpha(-zt^\alpha))(A)x\\ &= -A\mathcal{S}_\alpha(t)x.\end{aligned}$$

This proves (iii).

For (iv), by a similar argument with (iii), one can prove that $t^{\alpha-1}e_\alpha(-zt^\alpha)$ belongs to $\mathcal{F}_0^\gamma(S_\mu^0)$ for $t > 0$ and hence

$${}_0D_t^{\alpha-1}(t^{\alpha-1}\mathcal{P}_\alpha(t)) = {}_0D_t^{\alpha-1}((t^{\alpha-1}e_\alpha(-zt^\alpha)(A)) = (E_\alpha(-zt^\alpha))(A) = \mathcal{S}_\alpha(t),$$

in view of

$${}_0D_t^{\alpha-1}\big(t^{\alpha-1}e_\alpha(-zt^\alpha)\big) = E_\alpha(-zt^\alpha).$$

This completes the proof. □

Before proceeding with our theory further, we present the following result.

Lemma 4.24. *If $R(\lambda;-A)$ is compact for every $\lambda > 0$, then $T(t)$ is compact for every $t > 0$.*

Proof. Note first that as a consequence of Theorem 3.13 in Periago and Straub, 2002, for every $\lambda \in \mathbb{C}$ with $Re(\lambda) > 0$, $R(\lambda; -A) = \int_0^\infty e^{-\lambda s} T(s) ds$ defines a bounded linear operator on X. Therefore, we obtain

$$\lambda R(\lambda; -A)T(t) - T(t) = \lambda \int_0^\infty e^{-\lambda s}(T(t+s) - T(t))ds. \tag{4.55}$$

Let $\epsilon > 0$ be given. For every $\lambda > 0$ and $t > 0$, it follows from Theorem 4.16 that there exists a $\nu > 0$ such that

$$\sup_{s\in[0,\nu]} \|T(s+t) - T(t)\|_{B(X)} \le \frac{\epsilon}{2}.$$

So

$$\lambda \int_0^\nu e^{-s\lambda} \|T(t+s) - T(t)\|_{B(X)} ds \le \frac{\epsilon}{2}. \tag{4.56}$$

On the other hand, by Proposition 1.25(iii) we get

$$\begin{aligned} \lambda \left\| \int_\nu^\infty e^{-s\lambda}(T(s+t) - T(t))ds \right\|_{B(X)} &\le \lambda C \int_\nu^\infty e^{-s\lambda}((t+s)^{-1-\gamma} + t^{-\gamma-1})ds \\ &\le 2Ct^{-\gamma-1}e^{-\lambda\nu}, \end{aligned}$$

which implies that there exits a $\lambda_0 > 0$ such that

$$\lambda \left\| \int_\nu^\infty e^{-s\lambda}(T(s+t) - T(t))ds \right\|_{B(X)} \le \frac{\epsilon}{2}, \quad \lambda \ge \lambda_0. \tag{4.57}$$

Thus, for all $\lambda \ge \lambda_0$, using (4.55), (4.56) and (4.57) we deduce that

$$\begin{aligned} \|\lambda R(\lambda; -A)T(t) - T(t)\|_{B(X)} &\le \lambda \int_0^\nu e^{-s\lambda} \|T(t+s) - T(t)\|_{B(X)} ds \\ &\quad + \lambda \int_\nu^\infty e^{-s\lambda} \|T(s+t) - T(t)\|_{B(X)} ds \\ &\le \epsilon. \end{aligned}$$

It follows from the arbitrariness of $\nu > 0$ that

$$\lim_{\lambda\to\infty} \|\lambda R(\lambda; -A)T(t) - T(t)\|_{B(X)} = 0.$$

Since $\lambda R(\lambda; -A)T(t)$ is compact for every $\lambda > 0$ and $t > 0$, $T(t)$ is compact for every $t > 0$. □

With the help of this lemma we now show the following result.

Theorem 4.19. *If $R(\lambda; -A)$ is compact for every $\lambda > 0$, then $\mathcal{S}_\alpha(t)$, $\mathcal{P}_\alpha(t)$ are compact for every $t > 0$.*

Proof. Let $\epsilon > 0$ be arbitrary. Put

$$\varsigma_\epsilon(t) = \int_\epsilon^\infty \Psi_\alpha(s)T(st^\alpha - \epsilon t^\alpha)ds, \quad \zeta_\epsilon(t) = \int_\epsilon^\infty \Psi_\alpha(s)T(st^\alpha)ds.$$

Then, one has $\zeta_\epsilon(t) = T(\epsilon t^\alpha)\varsigma_\epsilon(t)$, and it is easy to prove that for every $t > 0$, $\varsigma_\epsilon(t)$ is bounded linear operators on X. Therefore, from the compactness of $T(t), t > 0$, we see that $\zeta_\epsilon(t)$ is compact for every $t > 0$.

On the other hand, note that

$$\|\zeta_\epsilon(t) - \mathcal{S}_\alpha(t)\|_{B(X)} \leq \left\| \int_0^\epsilon \Psi_\alpha(s)T(st^\alpha)ds \right\|_{B(X)} \leq C_0 t^{-\alpha(1+\gamma)} \int_0^\epsilon \Psi_\alpha(s)s^{-1-\gamma}ds.$$

Hence, it follows from the compactness of $\zeta_\epsilon(t), t > 0$ that $\mathcal{S}_\alpha(t)$ is compact for every $t > 0$. By a similar technique we can conclude that $\mathcal{P}_\alpha(t)$ is compact for every $t > 0$. The proof is completed. □

4.6.3 *Linear Problems*

Let $A \in \Theta_\omega^\gamma(X)$ with $-1 < \gamma < 0$ and $0 < \omega < \pi/2$. We discuss the existence and uniqueness of mild solutions and classical solutions for inhomogeneous linear abstract Cauchy problem (4.47). We assume the following condition.
($\mathbf{H}^*$) $u \in C([0,T],X)$, $g_{1-\alpha} * u \in C^1((0,T],X)$, $u(t) \in D(A)$ for $t \in (0,T]$, $Au \in L^1((0,T),X)$, and u satisfies (4.47).

Then, by Definitions 1.1 and 1.3, one can rewrite (4.47) as

$$u(t) = u_0 - \frac{1}{\Gamma(\alpha)} \int_0^t (t-s)^{\alpha-1}Au(s)ds + \frac{1}{\Gamma(\alpha)} \int_0^t (t-s)^{\alpha-1}f(s)ds, \qquad (4.58)$$

for $t \in [0,T]$.

Before presenting the definition of mild solution of problem (4.47), we first prove the following lemma.

Lemma 4.25. *If $u : [0,T] \to X$ is a function satisfying the assumption (H^*), then $u(t)$ satisfies the following integral equation*

$$u(t) = \mathcal{S}_\alpha(t)u_0 + \int_0^t (t-s)^{\alpha-1}\mathcal{P}_\alpha(t-s)f(s)ds, \quad t \in (0,T].$$

Proof. Note that the Laplace transform of a abstract function $f \in L^1(\mathbb{R}^+,X)$ is defined by $\bar{f}(\lambda) = \int_0^\infty e^{-\lambda t}f(t)dt$, $\lambda > 0$. Applying Laplace transform to (4.58) we get $\bar{u}(\lambda) = \frac{u_0}{\lambda} - \frac{1}{\lambda^\alpha}A\bar{u}(\lambda) + \frac{\bar{f}(\lambda)}{\lambda^\alpha}$, that is,

$$\bar{u}(\lambda) = \lambda^{\alpha-1}(\lambda^\alpha + A)^{-1}u_0 + (\lambda^\alpha + A)^{-1}\bar{f}(\lambda).$$

On the other hand, using Proposition 1.26 and (W2) we deduce that

$$\begin{aligned}
&\lambda^{\alpha-1}(\lambda^\alpha + A)^{-1}u_0 + (\lambda^\alpha + A)^{-1}\bar{f}(\lambda) \\
&= \lambda^{\alpha-1}\int_0^\infty e^{-\lambda^\alpha t}T(t)u_0 dt + \int_0^\infty e^{-\lambda^\alpha t}T(t)\bar{f}(\lambda)dt
\end{aligned}$$

$$
\begin{aligned}
&= \int_0^\infty \frac{d}{d\lambda} e^{-(\lambda t)^\alpha} T(t^\alpha) u_0 dt + \int_0^\infty \int_0^\infty \alpha t^{\alpha-1} e^{-(\lambda t)^\alpha t} T(t^\alpha) f(s) e^{-s\lambda} ds dt \\
&= \int_0^\infty \int_0^\infty \frac{\alpha t}{\tau^\alpha} \Psi_\alpha\left(\frac{1}{\tau^\alpha}\right) e^{-\lambda t\tau} T(t^\alpha) u_0 d\tau dt \\
&\quad + \int_0^\infty \int_0^\infty \int_0^\infty \frac{\alpha}{\tau^{2\alpha}} t^{\alpha-1} \Psi\left(\frac{1}{\tau^\alpha}\right) e^{-\lambda t} T(\frac{t^\alpha}{\tau^\alpha}) f(s) e^{-s\lambda} d\tau ds dt \\
&= \int_0^\infty \int_0^\infty \frac{\alpha}{\tau^{\alpha+1}} \Psi_\alpha\left(\frac{1}{\tau^\alpha}\right) e^{-\lambda t} T(\frac{t^\alpha}{\tau^\alpha}) u_0 d\tau dt \\
&\quad + \int_0^\infty \int_0^\infty \int_0^\infty \alpha\tau t^{\alpha-1} \Psi(\tau) T(t^\alpha \tau) f(s) e^{-(s+t)\lambda} d\tau ds dt \\
&= \int_0^\infty e^{-\lambda t} \int_0^\infty \Psi_\alpha(\tau) T(t^\alpha \tau) u_0 d\tau \\
&\quad + \int_0^\infty e^{-t\lambda} \int_0^t (t-s)^{\alpha-1} f(s) \left(\int_0^\infty \alpha\tau \Psi(\tau) T((t-s)^\alpha \tau) d\tau \right) ds dt \\
&= \int_0^\infty e^{-\lambda t} \mathcal{S}_\alpha(t) u_0 dt + \int_0^\infty e^{-\lambda t} \int_0^t (t-s)^\alpha \mathcal{P}_\alpha(t-s) f(s) ds dt \\
&= \int_0^\infty e^{-\lambda t} \left(\mathcal{S}_\alpha(t) u_0 + \int_0^t (t-s)^{\alpha-1} \mathcal{P}_\alpha(t-s) f(s) ds \right) dt.
\end{aligned}
$$

This implies that

$$
\bar{u}(\lambda) = \int_0^\infty e^{-\lambda t} \left(\mathcal{S}_\alpha(t) u_0 + \int_0^t (t-s)^{\alpha-1} \mathcal{P}_\alpha(t-s) f(s) ds \right) dt.
$$

Now using the uniqueness of the Laplace transform (cf. Theorem 1.1.6 of Xiao and Liang, 1998), we deduce that the assertion of lemma holds. This completes this proof. □

Motivated by Lemma 4.25, we adopt the following concept of mild solution to problem (4.47).

Definition 4.5. By a mild solution of problem (4.47), we mean a function $u \in C((0,T], X)$ satisfying

$$
u(t) = \mathcal{S}_\alpha(t) u_0 + \int_0^t (t-s)^{\alpha-1} \mathcal{P}_\alpha(t-s) f(s) ds, \quad t \in (0,T].
$$

Remark 4.14. It is to be noted that

(a) unlike the case of strongly continuous operator semigroups, we do not require the mild solution of problem (4.47) to be continuous at $t = 0$. Moreover, in general, since the operator $\mathcal{S}_\alpha(t)$ is singular at $t = 0$, solutions to problem (4.47) are assumed to have the same kind of singularity at $t = 0$ as the operator $\mathcal{S}_\alpha(t)$. This is the case, for instance, if f $\equiv 0$ so that we have that $u(t) = \mathcal{S}_\alpha(t) u_0$, which presents a discontinuity at the initial time;

(b) when $u_0 \in D(A^\beta)$, $\beta > 1+\gamma$, it follows from Theorem 4.18(i) that the mild solution is continuous at $t = 0$.

For $f \in L^1((0,T),X)$, the initial problem (4.47) has a unique mild solution for every $u_0 \in X$. We now are interested in imposing further condition on f and u_0 so that the mild solution becomes a classical solution. To this end we first introduce the following definition.

Definition 4.6. By a classical solution to problem (4.47), we mean a function $u(t) \in C([0,T],X)$ with ${}_0^C D_t^\alpha u(t) \in C((0,T],X)$, which, for all $t \in (0,T]$, takes values in $D(A)$ and satisfies (4.47).

We are now ready to state our main result in this subsection.

Theorem 4.20. *Let $A \in \Theta_\omega^\gamma(X)$ with $0 < \omega < \frac{\pi}{2}$. Suppose that $f(t) \in D(A)$ for all $0 < t \le T$, $Af(t) \in L^\infty((0,T),X)$, and $f(t)$ is Hölder continuous with an exponent $\theta' > \alpha(1+\gamma)$, that is,*

$$|f(t) - f(s)| \le K|t-s|^{\theta'}, \quad \textit{for all } 0 < t, s \le T.$$

Then, for every $u_0 \in D(A)$, there exists a classical solution to problem (4.47) *and this solution is unique.*

Proof. For $u_0 \in D(A)$, let $u(t) = \mathcal{S}_\alpha(t)u_0$ $(t > 0)$. Then it follows from Theorem 4.18(i, iii) that $u(t)$ is a classical solution of the following problem

$$\begin{cases} {}_0^C D_t^\alpha u(t) + Au(t) = 0, & 0 < t \le T, \\ u(0) = u_0. \end{cases} \tag{4.59}$$

Moreover, from Lemma 4.25, it is easy to see that $u(t)$ is the only solution to problem (4.59). Put

$$w(t) = \int_0^t (t-s)^{\alpha-1}\mathcal{P}_\alpha(t-s)f(s)ds, \quad 0 < t \le T.$$

Then from the assumptions on f and Theorem 4.15 we obtain

$$\begin{aligned} |Aw(t)| &\le \int_0^t (t-s)^{\alpha-1}\|\mathcal{P}_\alpha(t-s)\|_{B(X)}\|Af(t)\|_{L^\infty(0,T)}ds \\ &\le C_p\|Af(t)\|_{L^\infty(0,T)}\frac{1}{-\alpha\gamma}t^{-\gamma\alpha}, \end{aligned}$$

which implies that $w(t) \in D(A)$ for all $0 < t \le T$.

Next, we show ${}_0^C D_t^\alpha w(t) \in C((0,T],X)$. Since $w(0) = 0$ and hence

$${}_0^C D_t^\alpha w(t) = {}_0 D_t^1 {}_0 D_t^{\alpha-1} w(t) = \frac{d}{dt}\int_0^t \mathcal{S}_\alpha(t-s)f(s)ds, \tag{4.60}$$

in view of Propositions 1.4, 1.8, 1.9 and Theorem 4.18(iv). Let

$$v(t) = \int_0^t \mathcal{S}_\alpha(t-s)f(s)ds,$$

it remains to prove $v(t) \in C^1((0,T],X)$. Let $h > 0$ and $h \le T - t$. Then it is easy to verify the identity

$$\frac{v(t+h)-v(t)}{h} = \int_0^t \frac{\mathcal{S}_\alpha(t+h-s)-\mathcal{S}_\alpha(t-s)}{h} f(s)ds + \frac{1}{h}\int_t^{t+h} \mathcal{S}_\alpha(t+h-s)f(s)ds.$$

Again by the assumptions on f and Theorem 4.15, we have, for $t > 0$ fixed,

$$|(t-s)^{\alpha-1}A\mathcal{P}_\alpha(t-s)f(s)| \le C_p(t-s)^{-\alpha\gamma-1}|Af(s)| \in L^1((0,T),X),$$

for all $s \in [0,t)$. Therefore, using Theorem 4.17(ii) and the Lebesgue dominated convergence theorem we get

$$\begin{aligned}
&\lim_{h\to 0}\int_0^t \frac{\mathcal{S}_\alpha(t+h-s)-\mathcal{S}_\alpha(t-s)}{h} f(s)ds \\
&= \int_0^t (t-s)^{\alpha-1}(-A)\mathcal{P}_\alpha(t-s)f(s)ds \qquad (4.61)\\
&= -Aw(t).
\end{aligned}$$

Furthermore, note that

$$\begin{aligned}
&\frac{1}{h}\int_t^{t+h} \mathcal{S}_\alpha(t+h-s)f(s)ds \\
&= \frac{1}{h}\int_0^h \mathcal{S}_\alpha(s)f(t+h-s)ds \\
&= \frac{1}{h}\int_0^h \mathcal{S}_\alpha(s)(f(t+h-s)-f(t-s))ds \\
&\quad + \frac{1}{h}\int_0^h \mathcal{S}_\alpha(s)(f(t-s)-f(t))ds + \frac{1}{h}\int_0^h \mathcal{S}_\alpha(s)f(t)ds.
\end{aligned}$$

From Theorem 4.15 and the Hölder continuity on f we have

$$\frac{1}{h}\left|\int_0^h \mathcal{S}_\alpha(s)(f(t+h-s)-f(t-s))ds\right| \le \frac{C_sKh^{\theta'-\alpha(1+\gamma)}}{1-\alpha(1+\gamma)},$$

and

$$\frac{1}{h}\left|\int_0^h \mathcal{S}_\alpha(s)(f(t-s)-f(t))ds\right| \le \frac{C_sKh^{\theta'-\alpha(1+\gamma)}}{1+\theta-\alpha(1+\gamma)}.$$

And since $f(t) \in D(A)$ $(0 < t \le T)$, $\lim_{h\to 0}\frac{1}{h}\int_0^h \mathcal{S}_\alpha(s)f(t)ds = f(t)$ in view of Theorem 4.18(i). Hence,

$$\frac{1}{h}\int_t^{t+h} \mathcal{S}_\alpha(t+h-s)f(s)ds \to f(t), \quad \text{as } h \to 0^+. \qquad (4.62)$$

Combining (4.61) and (4.62) we deduce that v is differentiable from the right at t and $v'_+(t) = f(t) - Aw(t)$, $t \in (0,T]$. By a similar argument with the above one has that v is differentiable from the left at t and $v'_-(t) = f(t) - Aw(t),\ t \in (0,T]$. Next, we prove $Aw(t) \in C((0,T],X)$. To the end, let $Aw(t) = I_1(t) + I_2(t)$, where

$$I_1(t) = \int_0^t (t-s)^{\alpha-1} A\mathcal{P}_\alpha(t-s)\big(f(s) - f(t)\big)ds,$$
$$I_2(t) = \int_0^t A(t-s)^{\alpha-1}\mathcal{P}_\alpha(t-s)f(t)ds.$$

By Theorem 4.18(ii), we obtain

$$I_2(t) = -(\mathcal{S}_\alpha(t) - I)f(t).$$

So, by the assumption of f and Theorem 4.16 note that $I_2(t)$ is continuous for $0 < t \leq T$. To prove the same conclusion for $I_1(t)$, we let $0 < h \leq T - t$ and write

$$\begin{aligned}
&I_1(t+h) - I_1(t)\\
&= \int_0^t \left((t+h-s)^{\alpha-1} A\mathcal{P}_\alpha(t+h-s) - (t-s)^{\alpha-1}A\mathcal{P}_\alpha(t-s)\right)\big(f(s) - f(t)\big)ds\\
&+ \int_0^t (t+h-s)^{\alpha-1}A\mathcal{P}_\alpha(t+h-s)(f(t) - f(t+h))ds\\
&+ \int_t^{t+h} (t+h-s)^{\alpha-1}A\mathcal{P}_\alpha(t+h-s)\big(f(s) - f(t+h)\big)ds\\
&=: h_1(t) + h_2(t) + h_3(t).
\end{aligned}$$

For $h_1(t)$, on the one hand, it follows from Theorem 4.16 that

$$\begin{aligned}
&\lim_{h\to 0}(t+h-s)^{\alpha-1}A\mathcal{P}_\alpha(t+h-s)(f(s) - f(t))\\
&= (t-s)^{\alpha-1}A\mathcal{P}_\alpha(t-s)(f(s) - f(t)).
\end{aligned}$$

On the other hand, for $t \in (0,T]$ fixed, by Remark 4.13 and the assumption on f, we get

$$\begin{aligned}
&|(t+h-s)^{\alpha-1}A\mathcal{P}_\alpha(t+h-s)\big(f(s) - f(t)\big)|\\
&\leq C'_p K(t+h-s)^{-\alpha(1+\gamma)-1}(t-s)^{\theta'}\\
&\leq C'_p K(t-s)^{(\theta'-\alpha-\alpha\gamma)-1} \in L^1((0,t),X).
\end{aligned}$$

Thus, by mean of the Lebesgue dominated convergence theorem one has

$$\begin{aligned}
&\lim_{h\to 0}\int_0^t (t+h-s)^{\alpha-1}A\mathcal{P}_\alpha(t+h-s)\big(f(s) - f(t)\big)ds\\
&= \int_0^t (t-s)^{\alpha-1}A\mathcal{P}_\alpha(t-s)\big(f(s) - f(t)\big)ds,
\end{aligned}$$

which implies that $h_1(t) \to 0$ as $h \to 0^+$.

For $h_2(t)$, using Theorem 4.17(i), Remark 4.13,

$$\int_0^t (t+h-s)^{\alpha-1}\|A\mathcal{P}_\alpha(t+h-s)\|_{B(X)}|f(t) - f(t+h)|ds$$

$$\leq \int_0^t C_p' K(t+h-s)^{-\alpha(1+\gamma)-1} h^{\theta'} ds$$
$$= \frac{C_p' K h^{\theta'}}{\alpha(1+\gamma)} (h^{-\alpha(1+\gamma)} - (h+t)^{-\alpha(1+\gamma)}).$$

This yields $h_2(t) \to 0$ as $h \to 0^+$.

Moreover, $h_3(t) \to 0$ as $h \to 0^+$ by the following estimate

$$\left| \int_t^{t+h} (t+h-s)^{\alpha-1} \mathcal{P}_\alpha(t+h-s)\big(Af(s) - Af(t+h)\big) ds \right|$$
$$\leq \frac{2C_p}{-\alpha\gamma} \|Af(s)\|_{L^\infty(0,T)} h^{-\alpha\gamma}$$

in view of $Af(s) \in L^\infty((0,T), X)$ and Theorem 4.16.

The same reasoning establishes $I_1(t-h) - I_1(h) \to 0$ as $h \to 0^+$. Consequently, $Aw \in C((0,T], X)$, which implies that $v' \in C((0,T], X)$, provided that f is continuous on $(0,T]$. Then, by (4.60) we have ${}_0^C D_t^\alpha w \in C((0,T], X)$. Hence, we prove that $u + w$ is a classical solution to problem (4.47), and Lemma 4.25 implies that it is unique. This completes the proof. □

4.6.4 *Nonlinear Problems*

In this subsection we apply the theory developed in the previous sections to nonlinear abstract Cauchy problem (4.48).

Definition 4.7. By a mild solution to problem (4.48), we mean a function $u \in C((0,T], X)$ satisfying

$$u(t) = \mathcal{S}_\alpha(t) u_0 + \int_0^t (t-s)^{\alpha-1} \mathcal{P}_\alpha(t-s) f(s, u(s)) ds, \quad t \in (0,T].$$

Theorem 4.21. *Let $A \in \Theta_\omega^\gamma(X)$ with $-1 < \gamma < -\frac{1}{2}$ and $0 < \omega < \frac{\pi}{2}$. Suppose that the nonlinear mapping $f : (0,T] \times X \to X$ is continuous with respect to t and there exist constants $M, N > 0$ such that*

$$|f(t,x) - f(t,y)| \leq M\left(1 + |x|^{\nu-1} + |y|^{\nu-1}\right) |x-y|,$$

$$|f(t,x)| \leq N(1 + |x|^\nu),$$

for all $t \in (0,T]$ and for each $x, y \in X$, where ν is a constant in $[1, -\frac{\gamma}{1+\gamma})$. Then, for every $u_0 \in X$, there exists a $T_0 > 0$ such that the problem (4.48) has a unique mild solution defined on $(0, T_0]$.

Proof. For fixed $r > 0$, we introduce the metric space

$$F_r(T, u_0) = \{u \in C((0,T], X) : \ \rho_T(u, \mathcal{S}_\alpha(t) u_0) \leq r\},$$

$$\rho_T(u_1, u_2) = \sup_{t \in (0,T]} |u_1(t) - u_2(t)|.$$

It is not difficult to see that, with this metric, $F_r(T, u_0)$ is a complete metric space. Take $L = T^{\alpha(1+\gamma)}r + C_s|u_0|$, then for any $u \in F_r(T, u_0)$, we have

$$|s^{\alpha(1+\gamma)}u(s)| \leq s^{\alpha(1+\gamma)}|u - \mathcal{S}_\alpha(t)u_0| + s^{\alpha(1+\gamma)}|\mathcal{S}_\alpha(t)u_0| \leq L.$$

Choose $0 < T_0 \leq T$ such that

$$C_pN\frac{T_0^{-\alpha\gamma}}{-\alpha\gamma} + C_pNL^\nu T_0^{-\alpha(\nu(1+\gamma)+\gamma)}B(-\gamma\alpha, 1 - \nu\alpha(1+\gamma)) \leq r, \tag{4.63}$$

$$MC_p\frac{T_0^{-\alpha\gamma}}{-\alpha\gamma} + 2L^{\rho-1}T_0^{-\alpha(\gamma+(1+\gamma)(\nu-1))}B(-\alpha\gamma, 1 - \alpha(1+\gamma)(\nu-1)) \leq \frac{1}{2}, \tag{4.64}$$

where $B(\eta_1, \eta_2)$ with $\eta_i > 0$, $i = 1, 2$, denotes the Beta function. Assume that $u_0 \in X$. Consider the mapping Γ^α given by

$$(\Gamma^\alpha u)(t) = \mathcal{S}_\alpha(t)u_0 + \int_0^t (t-s)^{\alpha-1}\mathcal{P}_\alpha(t-s)f(s, u(s))ds, \quad u \in F_r(T_0, u_0).$$

By the assumptions on f, Theorem 4.15 and Theorem 4.16, we see that $(\Gamma^\alpha u)(t) \in C((0,T], X)$ and

$$\begin{aligned}
|(\Gamma^\alpha u)(t) - \mathcal{S}_\alpha(t)u_0| &\leq C_pN\int_0^t (t-s)^{-\alpha\gamma-1}(1 + |u(s)|^\nu)ds \\
&\leq C_pN\frac{T_0^{-\alpha\gamma}}{-\alpha\gamma} + \int_0^t C_pNL^\nu(t-s)^{-\alpha\gamma-1}s^{-\nu\alpha(1+\gamma)}ds \\
&\leq C_pN\frac{T_0^{-\alpha\gamma}}{-\alpha\gamma} + C_pNL^\nu T_0^{-\alpha(\nu(1+\gamma)+\gamma)}B(-\gamma\alpha, 1 - \nu\alpha(1+\gamma)) \\
&\leq r,
\end{aligned}$$

in view of (4.63). So, Γ^α maps $F_r(T_0, u_0)$ into itself. Next, for any $u, v \in F_r(T_0, u_0)$, by the assumptions on f and Theorem 4.15, we have

$$\begin{aligned}
&|(\Gamma^\alpha u)(t) - (\Gamma^\alpha v)(t)| \\
&= \left|\int_0^t (t-s)^{\alpha-1}\mathcal{P}_\alpha(t-s)\big(f(s,u(s)) - f(s,v(s))\big)ds\right| \\
&\leq C_pM\int_0^t (t-s)^{-\alpha\gamma-1}\left(1 + |u(s)|^{\rho-1} + |v(s)|^{\rho-1}\right)|u(s) - v(s)|ds \\
&\leq C_pM\rho_t(u,v)\int_0^t (t-s)^{-\alpha\gamma-1}\left(1 + 2L^{\nu-1}s^{-\alpha(\nu-1)(1+\gamma)}\right)ds \\
&\leq 2L^{\rho-1}T_0^{-\alpha(\gamma+(1+\gamma)(\nu-1))}B(-\alpha\gamma, 1-\alpha(1+\gamma)(\nu-1))\rho_{T_0}(u,v) \\
&\quad + MC_p\frac{T_0^{-\alpha\gamma}}{-\alpha\gamma}\rho_{T_0}(u,v).
\end{aligned}$$

This yields that Γ^α is a contraction on $F_r(T_0, u_0)$ due to (4.64). So, Γ_α has a unique fixed point $u \in F_r(T_0, u_0)$ in view of Banach contraction mapping principle, this means that u is a mild solution to problem (4.48) defined on $(0, T_0]$. The proof is completed. □

By a similar argument with the proof of Theorem 4.21 we have:

Corollary 4.1. *Assume that* $A \in \Theta^{\gamma}_{\omega}(X)$ *with* $-1 < \gamma < -\frac{2}{3}$ *and* $0 < \omega < \frac{\pi}{2}$. *Suppose in addition that the nonlinear mapping* $f : (0, T] \times X^{\beta} \to X$, $\beta \in (1+\gamma, -1-2\gamma)$, *is continuous with respect to* t *and there exist constants* $M, N > 0$ *such that*

$$|f(t,x) - f(t,y)| \le M\left(1 + |x|_{\beta}^{\nu-1} + |y|_{\beta}^{\nu-1}\right)|x-y|_{\beta},$$

$$|f(t,x)| \le N\left(1 + |x|_{\beta}^{\nu}\right),$$

for all $t \in (0,T]$ *and for each* $x, y \in X^{\beta}$, *where* ν *is a constant in* $[1, -\frac{\gamma+\beta}{1+\gamma})$. *Then, for every* $u_0 \in X^{\beta}$, *there exists a* $T_0 > 0$ *such that the problem* (4.48) *has a unique mild solution* $u \in C((0,T_0], X^{\beta})$.

Remark 4.15. If $A \in \Theta^{\gamma}_{\omega}(X)$ with $-1 < \gamma < 0$ and $0 < \omega < \frac{\pi}{2}$, then we can derive the local existence and uniqueness of mild solutions to problem (4.48), under the conditions:

(i) $u_0 \in X^{\beta}$ with $\beta > 1 + \gamma$;
(ii) the nonlinear mapping $f : [0,T] \times X \to X$ is continuous with respect to t and there exists a continuous function $L_f(\cdot) : \mathbb{R}^+ \to \mathbb{R}^+$ such that

$$|f(t,x) - f(t,y)| \le L_f(r)|x-y|,$$

for all $0 \le t \le T$ and for each $x, y \in X$ satisfying $|x|, |y| \le r$.

Indeed, for $r > \frac{C_p T_0^{-\alpha\gamma}}{-\alpha\gamma} \sup_{t\in[0,T]} |f(t,u_0)|$ fixed, we may choose $0 < T_0 \le T$ such that

$$\sup_{t\in[0,T_0]} |(\mathcal{S}_{\alpha}(t) - I)u_0| + \frac{C_p T_0^{-\alpha\gamma}}{-\alpha\gamma}\left(L_f(r)r + \sup_{t\in[0,T_0]} |f(t,u_0)|\right) < r \tag{4.65}$$

in view of Theorem 4.18(i). Assume that the map Γ^{α} is defined the same as in Theorem 4.21 and the space $F_r(T_0, u_0)$ is replaced by the following Banach space:

$$F_r'(T_0,u_0) = \left\{ u \in C([0,T_0], X) : \ u(0) = u_0 \ \text{ and } \sup_{t\in[0,T_0]} |u - u_0| \le r \right\}.$$

Then, it is easy to verify, thanks to the assumptions on f and (4.65), that Γ^{α} maps $F_r'(T_0,u_0)$ into itself and is a contraction on $F_r'(T_0,u_0)$, which implies that the problem (4.48) has a unique mild solution defined on $[0,T_0]$.

Since $1 > 1+\gamma$ $(-1 < \gamma < -\frac{1}{2})$, $X^1 = D(A)$ is a Banach space endowed with the graph norm $|x|_{X^1} = |Ax|$, for $x \in X^1$.

The following is the existence of X^1-smooth solutions.

Theorem 4.22. *Let* $A \in \Theta^{\gamma}_{\omega}(X)$ *with* $-1 < \gamma < -\frac{1}{2}$ *and* $0 < \omega < \frac{\pi}{2}$ *and* $u_0 \in X^1$. *Assume that there exists a continuous function* $M_f(\cdot) : \mathbb{R}^+ \to \mathbb{R}^+$ *and a constant* $N_f > 0$ *such that the nonlinear mapping* $f : (0,T] \times X^1 \to X^1$ *satisfies*

$$|f(t,x) - f(t,y)|_{X^1} \le M_f(r)|x-y|_{X^1},$$

$$|f(t, \mathcal{S}_\alpha(t)u_0)|_{X^1} \le N_f\big(1 + t^{-\alpha(1+\gamma)}|u_0|_{X^1}\big),$$

for all $0 < t \le T$ *and for each* $x, y \in X^1$ *satisfying* $\sup_{t\in(0,T]} |x(t) - \mathcal{S}_\alpha(t)u_0|_{X^1} \le r$, $\sup_{t\in(0,T]} |y(t) - \mathcal{S}_\alpha(t)u_0|_{X^1} \le r$. *Then there exists a* $T_0 > 0$ *such that the problem* (4.48) *has a unique mild solution defined on* $(0, T_0]$.

Proof. For $u_0 \in X^1$ and $r > 0$, set

$$F_r''(T, u_0) = \Big\{u \in C((0,T], X^1) : \sup_{t\in(0,T]} |u - \mathcal{S}_\alpha(t)u_0|_{X^1} \le r\Big\}.$$

For any $u \in F_r''(T, u_0)$, by the assumptions on f and Theorem 4.15 we have

$$\begin{aligned}
&|(\Gamma^\alpha u)(t) - \mathcal{S}_\alpha(t)u_0|_{X^1} \\
&\le \int_0^t (t-s)^{\alpha-1}\|\mathcal{P}_\alpha(t-s)\|_{B(X)}|f(s,u(s)) - f(s,\mathcal{S}_\alpha(t)u_0)|_{X^1}ds \\
&\quad + \int_0^t (t-s)^{\alpha-1}\|\mathcal{P}_\alpha(t-s)\|_{B(X)}|f(s,\mathcal{S}_\alpha(t)u_0))|_{X^1}ds \\
&\le C_p\int_0^t (t-s)^{-\alpha\gamma-1}\big(M_f(r)r + N_f + N_f s^{-\alpha(1+\gamma)}|u_0|\big)ds \\
&\le C_p(M_f(r)r + N_f)\frac{T^{-\alpha\gamma}}{-\alpha\gamma} + C_pN_fT^{-\alpha(1+2\gamma)}\beta(-\gamma\alpha, 1-\alpha(1+\gamma))|u_0|.
\end{aligned}$$

Using this result, it follows from an analogous idea with Theorem 4.21 that the claim of theorem follows. Here we omit the details. □

Next, we derive mild solutions under the condition of compactness on the resolvent of A.

Theorem 4.23. *Let* $A \in \Theta_\omega^\gamma(X)$ *with* $-1 < \gamma < 0$ *and* $0 < \omega < \frac{\pi}{2}$. *Let*

(H1) $R(\lambda; -A)$ *is compact for every* $\lambda > 0$;

(H2) $f : [0,T] \times X \to X$ *is a Carathéodory function and for any* $r > 0$, *there exists a function* $m_r(t) \in L^p((0,T), \mathbb{R}^+)$ *with* $p > -\frac{1}{\alpha\gamma}$ *such that*

$$|f(t,x)| \le m_r(t), \quad \text{and } \liminf_{r\to+\infty} \frac{|m_r(t)|_{L^p(0,T)}}{r} = \sigma < \infty$$

for a.e. $t \in [0,T]$ *and all* $x \in X$ *satisfying* $|x| \le r$.

Then for every $u_0 \in D(A^\beta)$ *with* $\beta > 1 + \gamma$, *the problem* (4.48) *has at least one mild solution, provided that*

$$C_p\sigma\left(\frac{T^{1-(1+\alpha\gamma)q}}{1-(1+\alpha\gamma)q}\right)^{1/q} < 1, \tag{4.66}$$

where $q = p/(p-1)$.

Proof. Assume that $u_0 \in D(A^\beta)$. On $C([0,T],X)$ define the map

$$(\Gamma^\alpha u)(t) = \mathcal{S}_\alpha(t)u_0 + \int_0^t (t-s)^{\alpha-1}\mathcal{P}_\alpha(t-s)f(s,u(s))ds.$$

From our assumptions it is easy to see that Γ_μ is well defined and maps $C([0,T],X)$ into itself. Put

$$\Omega_r = \{u \in C([0,T],X): \ \|u\| \le r, \quad \text{for all } 0 \le t \le T\},$$

for $r > 0$ as selected below. We seek for solutions in Ω_r. We claim that there exists an integer $r > 0$ such that Γ^α maps Ω_r into Ω_r. In fact, if this is not the case, then for each $r > 0$, there would exist $u^r \in \Omega_r$ and $t^r \in [0,T]$ such that $\|(\Gamma^\alpha u^r)(t^r)\| > r$. On the other hand, by (H2) and Theorem 4.15 we get

$$\begin{aligned}
r &< |(\Gamma^\alpha u^r)(t^r)| \\
&\le |\mathcal{S}_\alpha(t^r)u_0| + \int_0^{t^r} |(t^r-s)^{\alpha-1}\mathcal{P}_\alpha(t^r-s)f(s,u(s))|ds \\
&\le \sup_{t\in[0,T]} |\mathcal{S}_\alpha(t)u_0| + \int_0^{t^r} C_p(t^r-s)^{-1-\alpha\gamma}m_r(s)ds \\
&\le \sup_{t\in[0,T]} |\mathcal{S}_\alpha(t)u_0| + C_p\left(\int_0^{t^r} s^{-(1+\alpha\gamma)q}ds\right)^{\frac{1}{q}}\left(\int_0^{t^r} m_r^p(s)ds\right)^{\frac{1}{p}} \\
&\le \sup_{t\in[0,T]} |\mathcal{S}_\alpha(t)u_0| + C_p\|m_r\|_{L^p(0,T)}\left(\frac{T^{1-(1+\alpha\gamma)\,q}}{1-(1+\alpha\gamma)q}\right)^{\frac{1}{q}},
\end{aligned}$$

where $q = p/(p-1)$. Dividing on both sides by r and taking the lower limit as $r \to \infty$, one has

$$1 \le C_p\sigma\left(\frac{T^{1-(1+\alpha\gamma)\,q}}{1-(1+\alpha\gamma)q}\right)^{1/q},$$

which contradicts (4.66). Hence for some positive integer r, $\Gamma^\alpha(\Omega_r) \subset \Omega_r$.

The rest of the proof is divided into three steps.

Claim I. Γ^α is continuous on Ω_r.

Take $\{u_n\}_{n=1}^\infty \subset \Omega_r$ with $u_n \to u$ as $n \to \infty$ in $C([0,T],X)$. Then by the continuity of f with respect to the second argument we deduce that

$$f(s,u_n(s)) \to f(s,u(s)), \quad \text{as } n \to \infty, \quad \text{a.e. } s \in [0,T].$$

Moreover, observe from (H2) and Theorem 4.15, that for fixed $0 < t \le T$,

$$(t-s)^{\alpha-1}|\mathcal{P}_\alpha(t-s)f(s,u_n(s))| \le C_p(t-s)^{-1-\alpha\gamma}m_r(s).$$

Thus, by means of Lebesgue dominated convergence theorem we obtain that

$$\int_0^t (t-s)^{\alpha-1}\|\mathcal{P}_\alpha(t-s)\|_{B(X)}|f(s,u_n(s)) - f(s,u(s))|ds \to 0, \quad \text{as } n \to \infty,$$

which means that $\lim_{n\to\infty} \|\Gamma^\alpha u_n - \Gamma^\alpha u\|_\infty = 0$, that is, Γ^α is continuous on Ω_r.

Claim II. $P = \{(\Gamma^\alpha u): \ u \in \Omega_r\}$ is equicontinuous.

For $0 < t_1 < t_2 \le T$ and $\delta > 0$ small enough, we have

$$|(\Gamma^\alpha u)(t_1) - (\Gamma^\alpha u)(t_2)| \le I_1 + I_2 + I_3 + I_4 + I_5,$$

where

$$\begin{aligned}
I_1 &= |\mathcal{S}_\alpha(t_1)u_0 - \mathcal{S}_\alpha(t_2)u_0|,\\
I_2 &= \int_{t_1}^{t_2} (t_2-s)^{\alpha-1}|\mathcal{P}_\alpha(t_2-s)f(s,u(s))|ds,\\
I_3 &= \int_0^{t_1-\delta} (t_1-s)^{\alpha-1}\|\mathcal{P}_\alpha(t_2-s)-\mathcal{P}_\alpha(t_1-s)\|_{B(X)}|f(s,u(s))|ds,\\
I_4 &= \int_{t_1-\delta}^{t_1} (t_1-s)^{\alpha-1}\|\mathcal{P}_\alpha(t_2-s)-\mathcal{P}_\alpha(t_1-s)\|_{B(X)}|f(s,u(s))|ds,\\
I_5 &= \int_0^{t_1} |(t_2-s)^{\alpha-1}-(t_1-s)^{\alpha-1}|\|\mathcal{P}_\alpha(t_2-s)\|_{B(X)}|f(s,u(s))|ds.
\end{aligned}$$

From Theorem 4.16 and Theorem 4.18(i) it is easy to see that $I_1 \to 0$ when $t_1 \to t_2$. Moreover, using (H2) and Theorem 4.15 we get

$$\begin{aligned}
I_2 &\le C_p\Big(\frac{(t_2-t_1)^{1-(1+\alpha\gamma)q}}{1-(1+\alpha\gamma)q}\Big)^{1/q}\|m_r\|_{L^p(0,T)},\\
I_3 &\le \sup_{s\in[0,t_1-\delta]}\|\mathcal{P}_\alpha(t_2-s)-\mathcal{P}_\alpha(t_1-s)\|_{B(X)}\Big(\int_0^{t_1-\delta}(t_1-s)^{q\alpha-q}qds\Big)^{1/q}\|m_r\|_{L^p(0,T)}\\
&\le \sup_{s\in[0,t_1-\delta]}\|\mathcal{P}_\alpha(t_2-s)-\mathcal{P}_\alpha(t_1-s)\|_{B(X)}\Big(\frac{t_1^{1+q(\alpha-1)}-\delta^{1+q(\alpha-1)}}{1+q(\alpha-1)}\Big)^{1/q}\|m_r\|_{L^p(0,T)},\\
I_4 &\le C_p\int_{t_1-\delta}^{t_1}(t_1-s)^{\alpha-1}\cdot 2(t_1-s)^{-\alpha(\gamma+1)}m_r(s)ds\\
&\le 2C_p\frac{\delta^{1-(1+\alpha\gamma)q}}{1-(1+\alpha\gamma)q}\|m_r\|_{L^p(0,T)},\\
I_5 &\le \int_0^{t_1} C_p\big((t_1-s)^{\alpha-1}-(t_2-s)^{\alpha-1}\big)(t_2-s)^{-\alpha(1+\gamma)}m_r(s)ds\\
&\le \int_0^{t_1} C_p\big((t_1-s)^{-\gamma\alpha-1}-(t_2-s)^{-\alpha\gamma-1}\big)m_r(s)ds\\
&\le C_p\Big(\int_0^{t_1}(t_1-s)^{-q(\gamma\alpha+1)}-(t_2-s)^{-q(\alpha\gamma+1)}ds\Big)^{1/q}\|m_r\|_{L^p(0,T)}\\
&= C_p\Big(\frac{(t_2-t_1)^{1-(1+\alpha\gamma)q}}{1-(1+\alpha\gamma)q}+\frac{t_1^{1-(1+\alpha\gamma)q}-t_2^{1-(1+\alpha\gamma)q}}{1-(1+\alpha\gamma)q}\Big)^{1/q}\|m_r\|_{L^p(0,T)}.
\end{aligned}$$

It follows from Theorem 4.16 that I_i $(i = 2,3,4,5)$ tends to zero independent of $u \in \Omega_r$ as $t_2 - t_1 \to 0$, $\delta \to 0$. Hence, we can conclude that

$$|(\Gamma^\alpha u)(t_1) - (\Gamma^\alpha u)(t_2)| \to 0, \quad \text{as } t_2 - t_1 \to 0,$$

and the limit is independent of $u \in \Omega_r$. For the case when $0 = t_1 < t_2 \leq T$, since

$$\int_0^{t_2} (t_2 - s)^{\alpha-1} |P(t_2 - s) f(s, u(s))| ds \leq C_p \Big(\frac{t_2^{1-q(\alpha\gamma+1)}}{1 - q(\alpha\gamma + 1)} \Big)^{1/q} \|m_r\|_{L^p(0,T)},$$

in view of (H2) and Theorem 4.15, $|(\Gamma^\alpha u)(t_2)|$ can be made small when t_2 is small independently of $u \in \Omega_r$. Thus, we prove that the assertion in Claim II holds.

Claim III. For each $t \in [0, T]$, $\{(\Gamma^\alpha u)(t); u \in \Omega_r\}$ is precompact in X.

For the case when $t = 0$, it is not difficult to see that $\{(\Gamma^\alpha u)(0) : u \in \Omega_r\} = \{u_0 : u \in \Omega_r\}$ is compact. Let $t \in (0, T]$ be fixed and $\epsilon, \delta > 0$. For $u \in \Omega_r$, define the map $\Gamma^\alpha_{\epsilon,\delta}$ by

$$(\Gamma^\alpha_{\epsilon,\delta} u)(t) = \mathcal{S}_\alpha(t) u_0 + \int_0^{t-\epsilon} \int_\delta^\infty \alpha\tau (t - s)^{\alpha-1} M_\alpha(\tau) T((t - s)^\alpha \tau) f(s, u(s)) d\tau ds.$$

Since A has compact resolvent, $\{T(t)\}_{t>0}$ is compact in view of Theorem 4.19. Thus, for each $t \in (0, T]$, $\{(\mathcal{F}_{\epsilon,\delta} u)(t) : u \in \Omega_r, \ \delta > 0, \ 0 < \epsilon < t\}$ is precompact in X. On the other hand, using (H2) and Theorem 4.15, a direct calculation yields

$$\begin{aligned}
&|(\Gamma^\alpha u)(t) - (\Gamma^\alpha_{\epsilon,\delta} u)(t)| \\
&\leq \left| \int_0^t \int_0^\delta \alpha\tau (t - s)^{\alpha-1} M_\alpha(\tau) T((t - s)^\alpha \tau) f(s, u(s)) d\tau ds \right| \\
&\quad + \left| \int_{t-\epsilon}^t \int_\delta^\infty \alpha\tau (t - s)^{\alpha-1} M_\alpha(\tau) T((t - s)^\alpha \tau) f(s, u(s)) d\tau ds \right| \\
&\leq \int_0^t C_p (t - s)^{-1-\alpha\gamma} m_r(s) ds \int_0^\delta \tau^{-\gamma} M_\alpha(\tau) d\tau \\
&\quad + \int_{t-\epsilon}^t C_p (t - s)^{-1-\alpha\gamma} m_r(s) ds \int_\delta^\infty \tau^{-\gamma} M_\alpha(\tau) d\tau \\
&\leq C_p \Big(\frac{T^{1-(1+\alpha\gamma)q}}{1 - (1 + \alpha\gamma) q} \Big)^{1/q} \|m_r\|_{L^p(0,T)} \int_0^\delta \tau^{-\gamma} M_\alpha(\tau) d\tau \\
&\quad + C_p \Big(\frac{\epsilon^{1-(1+\alpha\gamma)q}}{1 - (1 + \alpha\gamma) q} \Big)^{1/q} \|m_r\|_{L^p(0,T)} \frac{\Gamma(1 - \gamma)}{\Gamma(1 - \gamma\alpha)}.
\end{aligned}$$

Using the total boundedness we have that for each $t \in (0, T]$ $\{(\Gamma^\alpha u)(t) : u \in \Omega_r\}$ is precompact in X. Therefore, for each $t \in [0, T]$, $\{(\Gamma^\alpha u)(t) : u \in \Omega_r\}$ is precompact in X.

Finally, by Claims I-III and Arzela-Ascoli theorem, we conclude that Γ^α is a compact operator. Hence, Γ^α has a fixed point, which gives rise to a mild solution. This completes the proof. □

Theorem 4.24. *Let $A \in \Theta^\gamma_\omega(X)$ with $0 < \omega < \frac{\pi}{2}$ and $-1 < \gamma < -\frac{1}{2}$. Suppose that there exists a continuous function $M'_f(\cdot) : \mathbb{R}^+ \to \mathbb{R}^+$ and a constant $\kappa > \alpha(1 + \gamma)$ such that the nonlinear mapping $f : [0, T] \times X \to X$ satisfies*

$$|f(t, x) - f(s, y)| \leq M'_f(r)(|t - s|^\kappa + |x - y|),$$

for all $0 \leq t \leq T$ *and* $x, y \in X$ *satisfying* $|x|, |y| \leq r$. *In addition, let the assumptions of Theorem 4.22 be satisfies and* u *be a mild solution corresponding to* u_0, *defined on* $[0, T_0]$. *Then* u *is in fact the unique classical solution to problem* (4.48), *existing on* $[0, T_0]$, *provided that* $u_0 \in D(A)$ *with* $Au_0 \in D(A^\beta)$, $\beta > (1+\gamma)$.

Proof. In order to prove that u is a classical solution, by Theorem 4.20 and the condition on f, we only have to verify that u is Hölder continuous with an exponent $\varsigma > \alpha(1+\gamma)$ on $(0, T_0]$. For fixed $t \in (0, T_0]$, take $0 < h < 1$ such that $h + t \leq T_0$. We estimate the difference

$$\begin{aligned}|u(t+h) - u(t)| &\leq |\mathcal{S}_\alpha(t+h)u_0 - \mathcal{S}_\alpha(t)u_0| \\ &\quad + \left| \int_0^h (t+h-s)^{\alpha-1}\mathcal{P}(t+h-s)f(s,u(s))ds \right| \\ &\quad + \left| \int_0^t (t-s)^{\alpha-1}\mathcal{P}(t-s)\big(f(s+h,u(s+h)) - f(s,u(s))\big)ds \right| \\ &= I_1 + I_2 + I_3.\end{aligned}$$

According to Theorem 4.15, Theorem 4.17(ii) and the assumptions on f we obtain

$$\begin{aligned}I_1 &= \left| \int_0^t -s^{\alpha-1}A\mathcal{P}_\alpha(s)u_0 ds \right| \leq \frac{C_p}{-\alpha\gamma}\big((t+h)^{-\alpha\gamma} - t^{-\alpha\gamma}\big), \\ I_3 &\leq M'C_p \int_0^t (t-s)^{-\alpha\gamma-1}(|h|^\kappa + |u(s+h) - u(s)|)ds \\ &\leq \frac{M'C_p}{-\alpha\gamma}T_0^{-\alpha\gamma}h^\kappa + M'C_p \int_0^t (t-s)^{-\alpha\gamma-1}|u(s+h) - u(s)|ds.\end{aligned}$$

Put $N_2 = \sup_{t\in(0,T_0)} |f(t, u(t))|$. Then, it follows from Theorem 4.15 that

$$\begin{aligned}I_2 &\leq C_p \int_0^h (t+h-s)^{-\alpha\gamma-1}|f(s,u(s))|ds \\ &\leq \frac{C_pN_2}{-\alpha\gamma}\big((t+h)^{-\alpha\gamma} - t^{-\alpha\gamma}\big).\end{aligned}$$

Collecting these estimates and using the inequality $(t+h)^{-\alpha\gamma} - t^{-\alpha\gamma} \leq h^{-\alpha\gamma}$ $(0 < -\alpha\gamma < 1)$ we have

$$\begin{aligned}&|u(t+h) - u(t)| \\ &\leq \frac{C_pN_2 + C_p}{-\alpha\gamma}\big((t+h)^{-\alpha\gamma} - t^{-\alpha\gamma}\big) + \frac{M_p'}{-\alpha\gamma}T_0^{-\alpha\gamma}h^\kappa \\ &\quad + M'C_p \int_0^t (t-s)^{-\alpha\gamma-1}|u(s+h) - u(s)|ds \\ &\leq \frac{C_pN_2 + C_p + M'C_p}{-\alpha\gamma}h^\varsigma + M'C_p \int_0^t (t-s)^{-\alpha\gamma-1}|u(s+h) - u(s)|ds,\end{aligned}$$

where $\varsigma = \min\{\kappa, -\alpha\gamma\} > \alpha(\gamma+1)$. Now, it follows from the usual Gronwall inequality that u has Hölder continuity on $(0, T_0]$. This completes the proof of theorem. □

4.6.5 *Applications*

In this subsection, we present three examples (Examples 4.4-4.6) motivated from physics, which do not aim at generality but indicate how our theorems can be applied to concrete problems. Examples 4.4 and 4.5 are inspired directly from the work of Carvalho, Dlotko and Nescimento, 2008, and they describe anomalous diffusion on fractals (physical objects of fractional dimension, like some amorphous semiconductors or strongly porous materials; see Anh and Leonenko, 2001; Metzler and Klafter, 2000 and references therein). Example 4.4 is the limit problem of certain fractional diffusion equations in complex systems on domains of "dumb-bell with a thin handle"(see, e.g., Anh and Leonenko, 2001; Metzler and Klafter, 2000). Example 4.5 displays anomalous dynamical behavior of anomalous transport processes (see, e.g., Anh and Leonenko, 2001; Metzler and Klafter, 2000). Example 4.6 is a modified fractional Schrödinger equation with fractional Laplacians whose physical background is statistical physics and fractional quantum mechanics (see, e.g., Hu and Kallianpur, 2000; Podlubny, 1999). We refer the reader to Kirane, Laskri and Tatar, 2005 and references therein for more research results related to fractional Laplacians.

Example 4.4. Consider the system of fractional partial differential equations in the form

$$\begin{cases} {}_0^C D_t^\alpha w - \Delta w + w = f(w), & x \in \Omega,\ t > 0, \\ \dfrac{\partial w}{\partial n} = 0, & x \in \partial\Omega, \\ {}_0^C D_t^\alpha v - \dfrac{1}{g}(g v_x)_x + v = f(v), & x \in (0,1), \\ v(0) = w(P_0)\,, v(1) = w(P_1), & \\ w(x,0) = w_0(x) \quad x \in \Omega, & v(x,0) = v_0(x), \quad x \in (0,1), \end{cases} \tag{4.67}$$

where $\Omega = D_1 \cup D_2$ and D_1 and D_2 are mutually disjoint bounded domains in $\mathbb{R}^N (N \geq 2)$ with smooth boundaries, joined by the line segment Q_0, and ${}_0^C D_t^\alpha$, $0 < \alpha < 1$, is the regularized Caputo fractional derivative of order α, that is,

$$({}_0^C D_t^\alpha u)(t,x) = \frac{1}{\Gamma(1-\alpha)} \left(\frac{\partial}{\partial t} \int_0^t (t-s)^{-\alpha} u(s,x) ds - t^{-\alpha} u(0,x) \right). \tag{4.68}$$

When $\alpha = 1$, we regard (4.67) as the limit problem of (4.44) as $\varepsilon \to 0$, which is described in more detail in Example 4.2. Here, our objective is to show that system (4.67) is well posed in $V_0^p = L^p(\Omega) \oplus L_g^p(0,1)$ $(1 \leq p < \infty)$.

Let the operators $A_0 : D(A_0) \subset V_0^p \mapsto V_0^p$ be defined by

$$D(A_0) = \{(w,v) \in V_0^P :\ w \in D(\Delta_\Omega), v \in L_g^p(0,1), w(P_0) = v(0), w(P_1) = v(1)\},$$

$$A_0(w,v) = \left(-\Delta w + w, -\frac{1}{g}(g v')' + v \right), \quad (w,v) \in V_0^p,$$

where Δ_Ω is the Laplace operator with homogeneous Neumann boundary conditions in $L^p(\Omega)$ and

$$D(\Delta_\Omega) = \left\{ u \in W^{2,p}(\Omega) : \left. \frac{\partial u}{\partial n} \right|_{\partial\Omega} = 0 \right\}.$$

From Example 4.2, if $p > \frac{N}{2}$, then $A_0 \in \Theta_\mu^{-\gamma'}(V_0^p)$ for some $\gamma' \in (0, 1 - \frac{N}{2p})$ and $\mu \in (0, \frac{\pi}{2})$. Therefore, system (4.67) can be seen as an abstract evolution equation in the form

$$\begin{cases} {}_0^C D_t^\alpha u + A_0 u = f(u), & t > 0, \\ u(0) = u_0 = (w_0, v_0) \in V_0^p. \end{cases} \tag{4.69}$$

We assume that the nonlinearity $f : \mathbb{R} \to \mathbb{R}$ is globally Lipschitz continuous. It can define a Nemitskiĭ operator from V_0^p into itself by $f(w, v) = (f_\Omega(w), f_I(v))$ with $f_\Omega(w)(x) = f(w(x))$, $x \in \Omega$ and $f_I(v)(x) = f(v(x))$, $x \in (0, 1)$ such that

$$|f(u) - f(u')|_{V_0^p} \leq L''(r)|u - u'|_{V_0^p},$$

for all $u, u' \in V_0^p$ satisfying $|u|_{V_0^p}, |u'|_{V_0^p} \leq r$. Hence, from Remark 4.15, (4.69) (that is, (4.67)) has a unique mild solution provided that $u_0 \in D(A_0^\beta)$ with $\beta > 1 - \gamma'$ (in particular, $u_0 \in D(A_0)$).

Example 4.5. Let Ω be a bounded domain in $\mathbb{R}^N$ $(N \geq 1)$ with boundary $\partial\Omega$ of class C^4. Consider the fractional initial-boundary value problem of form

$$\begin{cases} ({}_0^C D_t^\alpha u)(t, x) - \Delta u(t, x) = f(u(t, x)), & t > 0,\ x \in \Omega, \\ u|_{\partial\Omega} = 0, \\ u(0, x) = u_0(x), & x \in \Omega, \end{cases} \tag{4.70}$$

in the space $C^l(\overline{\Omega})$ $(0 < l < 1)$, where Δ stands for the Laplacian operator with respect to the spatial variable and ${}_0^C D_t^\alpha$, representing the regularized Caputo fractional derivative of order α $(0 < \alpha < 1)$, is given by (4.68). Set

$$\widetilde{A} = -\Delta, \quad D(\widetilde{A}) = \{u \in C^{2+l}(\overline{\Omega}) : \ u = 0 \ \text{ on } \partial\Omega\}.$$

It follows from Example 4.3 that there exist ν, $\varepsilon > 0$ such that $\widetilde{A} + \nu \in \Theta_{\frac{\pi}{2}-\varepsilon}^{\frac{l}{2}-1}(C^l(\overline{\Omega}))$. Then, problem (4.70) can be written abstractly as

$${}_0^C D_t^\alpha u(t) + \widetilde{A} u(t) = f(u), \quad t > 0.$$

With respect to the nonlinearity f, we assume that $f : \mathbb{R} \to \mathbb{R}$ is continuously differentiable and satisfies the condition

$$|f(x) - f(y)| \leq \frac{k(r)}{r}|x - y|, \quad |x|, |y| \leq r, \tag{4.71}$$

for any $r > 0$. It defines a Nemitskiĭ operator from $C^l(\overline{\Omega})$ into itself by $f(u)(x) = f(u(x))$ with

$$|f(u) - f(v)|_{C^l(\overline{\Omega})} \leq k(r)|u - v|_{C^l(\overline{\Omega})}, \quad |v|_{C^l(\overline{\Omega})}, |u|_{C^l(\overline{\Omega})} \leq r.$$

Noting $\frac{l}{2}-1\in(-1,-\frac{1}{2})$, we then obtain the following conclusion: (i) according to Remark 4.15, (4.70) has a unique mild solution for each $u_0\in D(\widetilde{A}^{\beta})$ with $\beta>\frac{l}{2}$. Moreover, (ii) if f',f'' are continuously differentiable functions satisfying the condition (4.71), then one finds that the Nemitskiĭ operator satisfies the assumptions of Theorem 4.22 and Theorem 4.24, which implies that for each $u_0\in D(\widetilde{A})$ with $\widetilde{A}u_0\in D(\widetilde{A}^{\beta})$ $(\beta>\frac{l}{2})$, the corresponding mild solution to (4.70) is also a unique classical solution.

Example 4.6. Consider the following fractional Cauchy problem

$$\begin{cases}({}_0^CD_t^{\alpha}y)(t,x)+(-i\Delta+\sigma)^{1/2}u(t,x)=f(u(t,x)), & t>0,\ x\in\mathbb{R}^2,\\ u(0,x)=u_0(x), & x\in\mathbb{R}^2,\end{cases}\tag{4.72}$$

in $L^3(\mathbb{R}^2)$, where $\sigma>0$ is a suitable constant, $i\Delta$ is the Schrödinger operator and ${}_0^CD_t^{\alpha}(0<\alpha<1)$ is given by (4.68). Let

$$\widehat{A}=(-i\Delta+\sigma)^{1/2},\quad D(\widehat{A})=W^{1,3}(\mathbb{R}^2)\quad \text{(a Sobolev space)}.$$

Then $i\Delta$ generates a β-times integrated semigroup $S^{\beta}(t)$ with $\beta=\frac{5}{12}$ on $L^3(\mathbb{R}^2)$ such that $\|S^{\beta}(t)\|_{B(L^3(\mathbb{R}^2))}\le\widehat{M}t^{\beta}$ for all $t\ge0$ and some constants $\widehat{M}>0$ (see Neerven and Straub, 1998). Therefore, by virtue of Theorem 1.3.5 and Definition 1.3.1 for $C=I$ of Xiao and Liang, 1998, we deduce that the operator $-i\Delta+\sigma$ belongs to $\Theta_{\frac{\pi}{2}}^{\beta-1}(L^3(\mathbb{R}^2))$, which denotes the family of all linear closed operators $A:D(A)\subset L^3(\mathbb{R}^2)\to L^3(\mathbb{R}^2)$ satisfying

$$\sigma(A)\subset S_{\frac{\pi}{2}}=\{z\in\mathbb{C}\setminus\{0\}:\ |\arg z|\le\frac{\pi}{2}\}\cup\{0\},$$

and for every $\frac{\pi}{2}<\mu<\tau$ there exists a constant C_μ such that $\|R(z;A)\|\le C_\mu|z|^{\beta-1}$. Thus, it follows from Proposition 3.6 of Periago and Straub, 2002 that $\widehat{A}\in\Theta_{\omega}^{-1+2\beta}(L^3(\mathbb{R}^2))$ for some $0<\omega<\frac{2}{\pi}$. Moreover, the system (4.72) can be rewritten as follows:

$$\begin{cases}({}_0^CD_t^{\alpha}y)(t,x)+\widehat{A}u=f(u), & t>0,\\ u(0,x)=u_0\in L^3(\mathbb{R}^2).\end{cases}$$

Assume that $f:\mathbb{C}\to\mathbb{C}$ is globally Lipschitz continuous. Then we have a Nemitskiĭ operator from $L^3(\mathbb{R}^2)$ to itself given by $f(u)(x)=f(u(x))$, and for a constant $\widehat{L}(r)$ and all $u,v\in L^3(\mathbb{R}^2)$ such that $|u|_{L^3(\mathbb{R}^2)}\le r$ and $|v|_{L^3(\mathbb{R}^2)}\le r$. Consequently, it follows from Remark 4.15 that (4.72) has a unique mild solution provided $u_0\in D(\widehat{A})^{\tau}$ with $\tau>\frac{5}{6}$.

4.7 Evolution Equations with Hilfer Derivative

4.7.1 *Introduction*

In this section, we consider the Cauchy problem of fractional evolution equations with an almost sectorial operator

$$\begin{cases}{}^HD_{0+}^{\mu,\nu}x(t)=Ax(t)+f(t,x(t)), & t\in(0,T],\\ {}_0D_t^{-(1-\mu)(1-\nu)}x(0)=x_0,\end{cases}\tag{4.73}$$

where ${}^H D_{0+}^{\mu,\nu}$ is the Hilfer fractional derivative of order $0 < \mu < 1$ and type $0 \le \nu \le 1$, ${}_0D_t^{-(1-\mu)(1-\nu)}$ is Riemann-Liouville fractional integral of order $(1-\mu)(1-\nu)$, A is an almost sectorial operator in Banach space X, that is, $A \in \Theta_\omega^{-k}(X)$ ($0 < k < 1$ and $0 < \omega < \frac{\pi}{2}$). We denote the semigroup associated with A by $\{Q(t)\}_{t\ge 0}$. Let $g : [0,T] \times X \to X$ be a function to be defined later, $x_0 \in X, T \in (0,\infty)$.

The Hilfer fractional derivative is a natural generalization of Caputo derivative and Riemann-Liouville derivative. It is obvious that fractional differential equations with Hilfer derivatives include fractional differential equations with a Riemann-Liouville derivative or Caputo derivative as special cases. In this section, we will prove two existence theorems of mild solutions for (4.73) in the cases that the semigroup associated with the almost sectorial operator is compact as well as noncompact. In Subsection 4.7.2, we will construct three families of operators and present some properties for these families. In Subsection 4.7.3, we will give some useful lemmas before proving the main results. In Subsection 4.7.4, we will show some new existence results of mild solutions for Cauchy problem (4.73).

4.7.2 *Preliminaries*

Assume that X is a Banach space with the norm $|\cdot|$. Let J be a finite interval of $\mathbb{R}$. By $C(J,X)$ we denote the Banach space of all continuous functions from J to X with the norm $\|u\| = \sup_{t\in J}|u(t)| < \infty$. We denote by $\mathcal{L}(X)$ the space of all bounded linear operators from X to X with the usual operator norm $\|\cdot\|_{\mathcal{L}(X)}$.

Let A be a linear operator from X to itself. Denote by $D(A)$ the domain of A, by $\sigma(A)$ its spectrum, while $\rho(A) := \mathbb{C} - \sigma(A)$ is the resolvent set of A. Let $S_\lambda^0 = \{z \in \mathbb{C}\backslash\{0\} : |\arg z| < \lambda\}$ be the open sector for $0 < \lambda < \pi$, and S_λ be its closure, i.e., $S_\lambda = \{z \in \mathbb{C}\backslash\{0\} : |\arg z| \le \lambda\} \cup \{0\}$.

Definition 4.8. Let $0 < k < 1$ and $0 < \omega < \frac{\pi}{2}$. We denote $\Theta_\omega^{-k}(X)$ as a family of all closed linear operators $A : D(A) \subset X \to X$ such that
(i) $\sigma(A) \subset S_\omega = \{z \in \mathbb{C} \setminus \{0\} : |\arg z| \le \omega\} \cup \{0\}$ and
(ii) for any $\lambda \in (\omega, \pi)$, there exists C_λ such that

$$\|R(z;A)\|_{\mathcal{L}(X)} \le C_\lambda |z|^{-k}, \text{ for all } z \in \mathbb{C} \setminus S_\lambda,$$

where $R(z;A) = (zI - A)^{-1}, z \in \rho(A)$ is the resolvent operator of A. The linear operator A will be called an almost sectorial operator on X if $A \in \Theta_\omega^{-k}(X)$.

Define the power of A as

$$A^\beta = \frac{1}{2\pi i}\int_{\Gamma_\rho} z^\beta R(z;A)dz, \quad \beta > 1-k,$$

where $\Gamma_\rho = \{\mathbb{R}^+ e^{i\rho}\}\bigcup\{\mathbb{R}^+ e^{-i\rho}\}$ is an appropriate path oriented counterclockwise and $\omega < \rho < \lambda$. Then, the linear power space $X_\beta := D(A^\beta)$ can be defined and X_β is a Banach space with the graph norm $\|x\|_\beta = |A^\beta x|$, $x \in D(A^\beta)$.

Next, let us introduce the semigroup associated with A. We denote the semigroup associated with A by $\{Q(t)\}_{t\geq 0}$. For $t \in S^0_{\frac{\pi}{2}-\omega}$

$$Q(t) = e^{-tz}(A) = \frac{1}{2\pi i}\int_{\Gamma_\rho} e^{-tz}R(z;A)dz,$$

where the integral contour $\Gamma_\rho = \{\mathbb{R}^+e^{i\rho}\}\bigcup\{\mathbb{R}^+e^{-i\rho}\}$ is oriented counter-clockwise and $\omega < \rho < \lambda < \frac{\pi}{2} - |\arg t|$, forms an analytic semigroup of growth order $1-k$.

Define the Wright function $M_\mu(\theta)$ by (see Definition 1.9)

$$M_\mu(\theta) = \sum_{n=1}^{\infty} \frac{(-\theta)^{n-1}}{(n-1)!\Gamma(1-\mu n)}, \quad 0<\mu<1,\ \theta\in\mathbb{C},$$

with the following property

$$\int_0^\infty \theta^\delta M_\mu(\theta)d\theta = \frac{\Gamma(1+\delta)}{\Gamma(1+\mu\delta)}, \quad \text{for } \delta\geq 0.$$

Lemma 4.26. *(Gu, 2015) The problem* (4.73) *is equivalent to the integral equation*

$$\begin{aligned} x(t) = {} & \frac{x_0}{\Gamma(\nu(1-\mu)+\mu)}t^{-(1-\mu)(1-\nu)} \\ & + \frac{1}{\Gamma(\mu)}\int_0^t (t-s)^{\mu-1}[Ax(s)+f(s,x(s))]ds, \quad t\in(0,T]. \end{aligned} \tag{4.74}$$

Lemma 4.27. *Assume that $x(t)$ satisfies integral equation* (4.74). *Then*

$$x(t) = S_{\mu,\nu}(t)x_0 + \int_0^t K_\mu(t-s)f(s,x(s))ds, \quad t\in(0,T],$$

where

$$S_{\mu,\nu}(t) = {}_0D_t^{-\nu(1-\mu)}K_\mu(t),\ K_\mu(t) = t^{\mu-1}P_\mu(t),\ \text{and } P_\mu(t) = \int_0^\infty \mu\theta M_\mu(\theta)Q(t^\mu\theta)d\theta.$$

Proof. This proof is similar to Gu, 2015, so we omit it. □

In view of Lemma 4.27, we have the following definition.

Definition 4.9. If $x\in C((0,T],X)$ satisfies

$$x(t) = S_{\mu,\nu}(t)x_0 + \int_0^t K_\mu(t-s)f(s,x(s))ds, \quad t\in(0,T],$$

then $x(t)$ is called a mild solution of the Cauchy problem (4.73).

Lemma 4.28. *(Jaiwal, 2022) If $\{Q(t)\}_{t>0}$ is a compact operator, then $\{S_{\mu,\nu}(t)\}_{t>0}$ and $\{P_\mu(t)\}_{t>0}$ are also compact operators.*

Lemma 4.29. *(Zhou, 2016) Let $\beta > 1-k$. For all $x\in D(A^\beta)$, we have* $\lim_{t\to 0+} P_\mu(t)x = \frac{x}{\Gamma(\mu)}$.

Lemma 4.30. *Assume that $\{Q(t)\}_{t>0}$ is a compact operator. Then $\{Q(t)\}_{t>0}$ is equicontinuous.*

Lemma 4.31. *For any fixed $t > 0$, $P_\mu(t)$, $K_\mu(t)$ and $S_{\mu,\nu}(t)$ are linear operators, and for any $x \in X$,*

$$|P_\mu(t)x| \leq L_1 t^{\mu(k-1)}|x|, \ |K_\mu(t)x| \leq L_1 t^{\mu k-1}|x|, \text{ and } |S_{\mu,\nu}(t)x| \leq L_2 t^{-1+\nu-\mu\nu+\mu k}|x|,$$

where

$$L_1 = \frac{C_0\Gamma(k)}{\Gamma(\mu k)}, \quad L_2 = \frac{C_0\Gamma(k)}{\Gamma(\nu(1-\mu)+\mu k)}.$$

Proof. By

$$\int_0^\infty \theta^\delta M_\mu(\theta)d\theta = \frac{\Gamma(1+\delta)}{\Gamma(1+\mu\delta)}, \text{ for } \delta \geq 0,$$

we have

$$\begin{aligned} |P_\mu(t)x| &= \left| \int_0^\infty \mu\theta M_\mu(\theta)Q(t^\mu\theta)xd\theta \right| \\ &\leq \mu C_0 \int_0^\infty M_\mu(\theta)\theta^k t^{\mu(k-1)}|x|d\theta \\ &\leq L_1 t^{\mu(k-1)}|x|, \text{ for } t \in (0,T] \text{ and } x \in X. \end{aligned}$$

Moreover, for $t \in (0,T]$ and $x \in X$,

$$|K_\mu(t)x| = |t^{\mu-1}P_\mu(t)x| \leq L_1 t^{\mu k-1}|x|,$$

and

$$\begin{aligned} |S_{\mu,\nu}(t)x| = |{}_0D_t^{-\nu(1-\mu)}K_\mu(t)x| &= \left| \frac{1}{\Gamma(\nu(1-\mu))} \int_0^t (t-s)^{\nu(1-\mu)-1}K_\mu(s)xds \right| \\ &\leq \frac{C_0\Gamma(k)}{\Gamma(\mu k)\Gamma(\nu(1-\mu))} \int_0^t (t-s)^{\nu(1-\mu)-1}s^{\mu k-1}|x|ds \\ &\leq L_2 t^{-1+\nu-\mu\nu+\mu k}|x|. \end{aligned}$$

This completes the proof. □

Lemma 4.32. *(Jaiwal, 2022) Assume that $\{Q(t)\}_{t>0}$ is equicontinuous. Then $\{P_\mu(t)\}_{t>0}$, $\{K_\mu(t)\}_{t>0}$ and $\{S_{\mu,\nu}(t)\}_{t>0}$ are strongly continuous, that is, for any $x \in X$ and $t'' > t' > 0$,*

$$\begin{aligned} &|P_\mu(t')x - P_\mu(t'')x| \to 0, \ |K_\mu(t')x - K_\mu(t'')x| \to 0, \\ &|S_{\mu,\nu}(t')x - S_{\mu,\nu}(t'')x| \to 0, \quad \text{as } t'' \to t'. \end{aligned}$$

4.7.3 *Some Lemmas*

Throughout this section, we assume that $A \in \Theta_\omega^{-k}(X)$, $0 < k < 1$ and $0 < \omega < \frac{\pi}{2}$. Furthermore, we suppose that $x_0 \in D(A^\beta)$ with $\beta > 1-k$.

We introduce the following hypotheses:

(H1) $Q(t)$ is continuous in the uniform operator topology for $t > 0$, i.e., $\{Q(t)\}_{t>0}$ is equicontinuous;

(H2) the map $t \to f(t,x)$ is measurable for all $x \in X$ and the map $x \to f(t,x)$ is continuous for a.e. $t \in [0,T]$;

(H3) there exists a function $m \in L((0,T],\mathbb{R}^+)$ satisfying

$$ {}_0D_t^{-\mu k} m \in C((0,T],\mathbb{R}^+), \quad \lim_{t\to 0+} t^{1-\nu+\mu\nu-\mu k}\, {}_0D_t^{-\mu k} m(t) = 0 $$

and $|f(t,x)| \le m(t)$, for a.e. $t \in [0,T]$ and any $x \in X$;

(H4) there exists a constant $r > 0$ such that

$$ L_2|x_0| + L_1 \sup_{t\in[0,T]} \left\{ t^{1-\nu+\mu\nu-\mu k} \int_0^t (t-s)^{\mu k-1} m(s)ds \right\} \le r, $$

where

$$ L_1 = \frac{C_0\Gamma(k)}{\Gamma(\mu k)}, \quad L_2 = \frac{C_0\Gamma(k)}{\Gamma(\nu(1-\mu)+\mu k)}. $$

Let

$$ C_\mu((0,T],X) = \{x \in C((0,T],X) : \lim_{t\to 0+} t^{1-\nu+\mu\nu-\mu k}|x(t)| \text{ exists and is finite}\}, $$

with the norm

$$ \|x\|_\mu = \sup_{t\in(0,T]} \{t^{1-\nu+\mu\nu-\mu k}|x(t)|\}. $$

Then $(C_\mu((0,T],X), \|\cdot\|_\mu)$ is a Banach space.

For any $x \in C_\mu((0,T],X)$, define an operator $\mathcal{T}$ as follows

$$ (\mathcal{T}x)(t) = (\mathcal{T}_1 x)(t) + (\mathcal{T}_2 x)(t), $$

where

$$ (\mathcal{T}_1 x)(t) = S_{\mu,\nu}(t)x_0, \quad (\mathcal{T}_2 x)(t) = \int_0^t K_\mu(t-s)f(s,x(s))ds, \quad \text{for } t \in (0,T]. $$

Clearly, the problem (4.73) has a mild solution $x^* \in C_\mu((0,T],X)$ if and only if $\mathcal{T}$ has a fixed point $x^* \in C_\mu((0,T],X)$.

It is easy to show that

$$ \lim_{t\to 0+} t^{1-\nu+\mu\nu-\mu k} S_{\mu,\nu}(t)x_0 = 0. \tag{4.75} $$

In fact,

$$ \begin{aligned} t^{1-\nu+\mu\nu-\mu k} S_{\mu,\nu}(t)x_0 &= \frac{t^{1-\nu+\mu\nu-\mu k}}{\Gamma(\nu(1-\mu))} \int_0^t (t-s)^{\nu(1-\mu)-1} s^{\mu-1} P_\mu(s)x_0 ds \\ &= \frac{1}{\Gamma(\nu(1-\mu))} \int_0^1 (1-z)^{\nu(1-\mu)-1} z^{\mu-1} t^{\mu(1-k)} P_\mu(tz)x_0 dz. \end{aligned} $$

By Lemma 4.29, $\lim_{t\to 0+} t^{\mu(1-k)} P_\mu(tz)x_0 = 0$ and $\int_0^1 (1-z)^{\nu(1-\mu)-1} z^{\mu-1} dz$ exists, so (4.75) holds.

In addition, from Lemma 4.31 and (H3), we have

$$\left|t^{1-\nu+\mu\nu-\mu k}\int_0^t K_\mu(t-s)f(s,x(s))ds\right| \le L_1 t^{1-\nu+\mu\nu-\mu k}\int_0^t (t-s)^{\mu k-1}m(s)ds \to 0, \quad \text{as } t\to 0. \tag{4.76}$$

For any $u \in C([0,T],X)$, set

$$x(t) = t^{-1+\nu-\mu\nu+\mu k}u(t), \quad t\in (0,T].$$

Clearly, $x \in C_\mu((0,T],X)$. Define an operator $\mathcal{F}$ as follows

$$(\mathcal{F}u)(t) = (\mathcal{F}_1 u)(t) + (\mathcal{F}_2 u)(t),$$

where

$$(\mathcal{F}_1 u)(t) = \begin{cases} t^{1-\nu+\mu\nu-\mu k}(\mathcal{T}_1 x)(t), & \text{for } t\in(0,T],\\ 0, & \text{for } t=0,\end{cases}$$

$$(\mathcal{F}_2 u)(t) = \begin{cases} t^{1-\nu+\mu\nu-\mu k}(\mathcal{T}_2 x)(t), & \text{for } t\in(0,T],\\ 0, & \text{for } t=0.\end{cases}$$

Let

$$\Omega_r = \{u \in C([0,T],X): \ \|u\|\le r\},$$

and

$$\widetilde{\Omega}_r = \{x \in C_\mu((0,T],X): \ \|x\|_\mu \le r\}.$$

Clearly, Ω_r and $\widetilde{\Omega}_r$ are nonempty, convex and closed subsets of $C([0,T],X)$ and $C_\mu((0,T],X)$, respectively.

Before giving the main results, we first prove the following lemmas.

Lemma 4.33. *Assume that (H1)-(H4) hold. Then, the set* $\{\mathcal{F}u: \ u\in\Omega_r\}$ *is equicontinuous.*

Proof. Step I. We first prove that $\{\mathcal{F}_1 u: \ u\in\Omega_r\}$ is equicontinuous.

For $t_1=0$, $t_2\in(0,T]$, by (4.75), we obtain

$$\left|(\mathcal{F}_1u)(t_2)-(\mathcal{F}_1u)(0)\right| \le \left|{t_2}^{1-\nu+\mu\nu-\mu k}S_{\mu,\nu}(t_2)x_0 - 0\right| \to 0, \quad \text{as } \ t_2\to 0.$$

For any $t_1,t_2\in(0,T]$ and $t_1<t_2$, we have

$$\begin{aligned}\left|(\mathcal{F}_1u)(t_2)-(\mathcal{F}_1u)(t_1)\right| &\le \left|{t_2}^{1-\nu+\mu\nu-\mu k}S_{\mu,\nu}(t_2)x_0 - {t_1}^{1-\nu+\mu\nu-\mu k}S_{\mu,\nu}(t_1)x_0\right|\\ &\le \left|{t_2}^{1-\nu+\mu\nu-\mu k}\right|\left|S_{\mu,\nu}(t_2)x_0 - S_{\mu,\nu}(t_1)x_0\right|\\ &\quad + \left|{t_2}^{1-\nu+\mu\nu-\mu k} - {t_1}^{1-\nu+\mu\nu-\mu k}\right|\left|S_{\mu,\nu}(t_1)x_0\right|\\ &\to 0, \quad \text{as } \ t_2\to t_1.\end{aligned}$$

Hence, $\{\mathcal{F}_1 u: \ u\in\Omega_r\}$ is equicontinuous.

Step II. We prove that $\{\mathcal{F}_2 u: \ u\in\Omega_r\}$ is equicontinuous.

Let $x(t) = t^{-1+\nu-\mu\nu+\mu k}u(t)$, for any $u \in \Omega_r$, $t \in (0, T]$. Then $x \in \widetilde{\Omega}_r$.
For $t_1 = 0$, $0 < t_2 < T$, by (4.76), we have

$$\big|(\mathcal{F}_2 u)(t_2) - (\mathcal{F}_2 u)(0)\big| = \left|{t_2}^{1-\nu+\mu\nu-\mu k}\int_0^{t_2} K_\mu(t_2 - s)f(s, x(s))ds\right|$$
$$\to 0, \quad \text{as} \quad t_2 \to 0.$$

For $0 < t_1 < t_2 \le T$, we get

$$\begin{aligned}
&|(\mathcal{F}_2 u)(t_2) - (\mathcal{F}_2 u)(t_1)| \\
&\le \left|{t_1}^{1-\nu+\mu\nu-\mu k}\int_{t_1}^{t_2}(t_2 - s)^{\mu-1}P_\mu(t_2 - s)f(s, x(s))ds\right| \\
&\quad + \left|{t_1}^{1-\nu+\mu\nu-\mu k}\int_0^{t_1}\big((t_2 - s)^{\mu-1} - (t_1 - s)^{\mu-1}\big)P_\mu(t_2 - s)f(s, x(s))ds\right| \\
&\quad + \left|{t_1}^{1-\nu+\mu\nu-\mu k}\int_0^{t_1}(t_1 - s)^{\mu-1}\big(P_\mu(t_2 - s) - P_\mu(t_1 - s)\big)f(s, x(s))ds\right| \\
&\quad + |{t_2}^{1-\nu+\mu\nu-\mu k} - {t_1}^{1-\nu+\mu\nu-\mu k}|\left|\int_0^{t_2}(t_2 - s)^{\mu-1}P_\mu(t_2 - s)f(s, x(s))ds\right| \\
&\le I_1 + I_2 + I_3 + I_4,
\end{aligned}$$

where

$$I_1 = L_1{t_1}^{1-\nu+\mu\nu-\mu k}\left|\int_0^{t_2}(t_2 - s)^{\mu k-1}m(s)ds - \int_0^{t_1}(t_1 - s)^{\mu k-1}m(s)ds\right|,$$

$$I_2 = 2L_1{t_1}^{1-\nu+\mu\nu-\mu k}\int_0^{t_1}\big((t_1 - s)^{\mu-1} - (t_2 - s)^{\mu-1}\big)(t_2 - s)^{\mu(k-1)}m(s)ds,$$

$$I_3 = {t_1}^{1-\nu+\mu\nu-\mu k}\left|\int_0^{t_1}(t_1 - s)^{\mu-1}\big(P_\mu(t_2 - s) - P_\mu(t_1 - s)\big)f(s, x(s))ds\right|,$$

$$I_4 = |{t_2}^{1-\nu+\mu\nu-\mu k} - {t_1}^{1-\nu+\mu\nu-\mu k}|\left|L_1\int_0^{t_2}(t_2 - s)^{\mu k-1}m(s)ds\right|.$$

One can deduce that $\lim_{t_2 \to t_1} I_1 = 0$, since ${}_0D_t^{-\mu k}m \in C((0, T], \mathbb{R}^+)$. Noting that

$$((t_1 - s)^{\mu-1} - (t_2 - s)^{\mu-1})(t_2 - s)^{\mu(k-1)}m(s) \le (t_1 - s)^{\mu k-1}m(s), \quad \text{for } s \in [0, t_1),$$

then by Lebesgue dominated convergence theorem, we have

$$\int_0^{t_1}\big((t_1 - s)^{\mu-1} - (t_2 - s)^{\mu-1}\big)(t_2 - s)^{\mu(k-1)}m(s)ds \to 0, \quad \text{as } t_2 \to t_1,$$

which implies $I_2 \to 0$ as $t_2 \to t_1$.

By (H3), for $\varepsilon > 0$, we have

$$\begin{aligned}
I_3 \le {t_1}^{1-\nu+\mu\nu-\mu k}&\int_0^{t_1-\varepsilon}(t_1 - s)^{\mu-1}\|P_\mu(t_2 - s) - P_\mu(t_1 - s)\|_{\mathcal{L}(X)}|f(s, x(s))|ds \\
&+ {t_1}^{1-\nu+\mu\nu-\mu k}\left|\int_{t_1-\varepsilon}^{t_1}(t_1 - s)^{\mu-1}\big(P_\mu(t_2 - s) - P_\mu(t_1 - s)\big)f(s, x(s))ds\right|
\end{aligned}$$

$$\leq {t_1}^{1-\nu+\mu\nu-\mu k}\int_0^{t_1}(t_1-s)^{\mu-1}m(s)ds\sup_{s\in[0,t_1-\varepsilon]}\|P_\mu(t_2-s)-P_\mu(t_1-s)\|_{\mathcal{L}(X)}$$
$$+2L_1{t_1}^{1-\nu+\mu\nu-\mu k}\int_{t_1-\varepsilon}^{t_1}(t_1-s)^{\mu k-1}m(s)ds$$
$$\leq I_{31}+I_{32}+I_{33},$$

where

$$I_{31}={t_1}^{1-\nu+\mu\nu-\mu k}\int_0^{t_1}(t_1-s)^{\mu-1}m(s)ds\sup_{s\in[0,t_1-\varepsilon]}\|P_\mu(t_2-s)-P_\mu(t_1-s)\|_{\mathcal{L}(X)},$$
$$I_{32}=2L_1{t_1}^{1-\nu+\mu\nu-\mu k}\left|\int_0^{t_1}(t_1-s)^{\mu k-1}m(s)ds-\int_0^{t_1-\varepsilon}(t_1-\varepsilon-s)^{\mu k-1}m(s)ds\right|,$$
$$I_{33}=2L_1{t_1}^{1-\nu+\mu\nu-\mu k}\int_0^{t_1-\varepsilon}((t_1-\varepsilon-s)^{\mu k-1}-(t_1-s)^{\mu k-1})m(s)ds.$$

By (H1) and Lemma 4.32, it is easy to see that $I_{31}\to 0$ as $t_2\to t_1$. Similar to the proof that I_1, I_2 tend to zero, we get $I_{32}\to 0$ and $I_{33}\to 0$ as $\varepsilon\to 0$. Thus, I_3 tends to zero as $t_2\to t_1$. Clearly, $I_4\to 0$ as $t_2\to t_1$.

Therefore, $\{\mathcal{F}_2u:\ u\in\Omega_r\}$ is equicontinuous. Furthermore, $\{\mathcal{F}u:\ u\in\Omega_r\}$ is equicontinuous. □

Lemma 4.34. *Assume that (H2)-(H4) hold. Then* $\mathcal{F}\Omega_r\subset\Omega_r$.

Proof. Let $x(t)=t^{-1+\nu-\mu\nu+\mu k}u(t)$, for $u\in\Omega_r$, $t\in(0,T]$. Then $x\in\widetilde{\Omega}_r$.

From Lemma 4.33, we know that $\mathcal{F}\Omega_r\subset C([0,T],X)$. For $t>0$ and any $u\in\Omega_r$, by (H4), we have

$$|(\mathcal{F}u)(t)|\leq\left|t^{1-\nu+\mu\nu-\mu k}S_{\mu,\nu}(t)x_0\right|+\left|t^{1-\nu+\mu\nu-\mu k}\int_0^t K_\mu(t-s)f(s,x(s))ds\right|$$
$$\leq L_2|x_0|+L_1t^{1-\nu+\mu\nu-\mu k}\int_0^t(t-s)^{\mu k-1}m(s)ds\leq r.$$

For $t=0$, we have $|(\mathcal{F}u)(0)|=0<r$. Therefore, $\mathcal{F}\Omega_r\subset\Omega_r$. □

Lemma 4.35. *Assume that (H2)-(H4) hold. Then* $\mathcal{F}$ *is continuous.*

Proof. Let $\{u_n\}_{n=1}^\infty$ be a sequence in Ω_r which is convergent to $u\in\Omega_r$. Consequently,

$$\lim_{n\to\infty}u_n(t)=u(t),\ \text{and}\ \lim_{n\to\infty}t^{-1+\nu-\mu\nu+\mu k}u_n(t)=t^{-1+\nu-\mu\nu+\mu k}u(t),\ \text{for}\ \ t\in(0,T].$$

Let $x(t)=t^{-1+\nu-\mu\nu+\mu k}u(t)$, $x_n(t)=t^{-1+\nu-\mu\nu+\mu k}u_n(t)$, $t\in(0,T]$. Then $x,x_n\in\widetilde{\Omega}_r$. In view of (H2), we have

$$\lim_{n\to\infty}f(t,x_n(t))=\lim_{n\to\infty}f(t,t^{-1+\nu-\mu\nu+\mu k}u_n(t))=f(t,t^{-1+\nu-\mu\nu+\mu k}u(t))=f(t,x(t)).$$

For each $t \in (0,T]$, $(t-s)^{\mu k-1}|f(s,x_n(s)) - f(s,x(s))| \le 2(t-s)^{\mu k-1}m(s)$. By Lebesgue dominated convergence theorem, we obtain

$$\int_0^t (t-s)^{\mu k-1}|f(s,x_n(s)) - f(s,x(s))|ds \to 0, \quad \text{as} \quad n \to \infty.$$

Thus, for $t \in [0,T]$,

$$\begin{aligned}
&\big|(\mathcal{F}u_n)(t) - (\mathcal{F}u)(t)\big| \\
&\le t^{1-\nu+\mu\nu-\mu k} \int_0^t |K_\mu(t-s)(f(s,x_n(s)) - f(s,x(s)))|ds \\
&\le L_1 t^{1-\nu+\mu\nu-\mu k} \int_0^t (t-s)^{\mu k-1}|f(s,x_n(s)) - f(s,x(s))|ds \to 0, \quad \text{as} \quad n \to \infty.
\end{aligned}$$

Therefore, $\|\mathcal{F}u_n - \mathcal{F}u\| \to 0$ as $n \to \infty$. Hence, $\mathcal{F}$ is continuous. The proof is completed. □

4.7.4 *Existence Results*

Theorem 4.25. *Assume that $Q(t)(t > 0)$ is compact. Furthermore suppose that (H2)-(H4) hold. Then the Cauchy problem (4.73) has at least one mild solution in $\widetilde{\Omega}_r$.*

Proof. Clearly, the problem (4.73) exists a mild solution $x \in \widetilde{\Omega}_r$ if and only if the operator $\mathcal{F}$ has a fixed point $u \in \Omega_r$, where $u(t) = t^{1-\nu+\mu\nu-\mu k}x(t)$. Hence, we only need to prove that the operator $\mathcal{F}$ has a fixed point in Ω_r. From Lemmas 4.34 and 4.35, we know that $\mathcal{F}\Omega_r \subset \Omega_r$ and $\mathcal{F}$ is continuous. In view of Lemma 4.33, the set $\{\mathcal{F}u : u \in \Omega_r\}$ is equicontinuous. It remains to prove that for $t \in [0,T]$, $\{(\mathcal{F}u)(t) : u \in \Omega_r\}$ is relatively compact in X. Clearly, $\{(\mathcal{F}u)(0) : u \in \Omega_r\}$ is relatively compact in X. We only consider the case $t > 0$. For any $\varepsilon \in (0,t)$ and $\delta > 0$, define $\mathcal{F}_{\varepsilon,\delta}$ on Ω_r as follows

$$\begin{aligned}
(\mathcal{F}_{\varepsilon,\delta}u)(t) &:= t^{1-\nu+\mu\nu-\mu k}(\mathcal{T}_{\varepsilon,\delta}x)(t) \\
&:= t^{1-\nu+\mu\nu-\mu k}\bigg(S_{\mu,\nu}(t)x_0 + \int_0^{t-\varepsilon}\int_\delta^\infty \mu\theta(t-s)^{\mu-1}M_\mu(\theta) \\
&\quad \times Q((t-s)^\mu\theta)f(s,x(s))d\theta ds\bigg).
\end{aligned}$$

Thus,

$$\begin{aligned}
(\mathcal{F}_{\varepsilon,\delta}u)(t) = t^{1-\nu+\mu\nu-\mu k}\bigg(&S_{\mu,\nu}(t)x_0 + Q(\varepsilon^\mu\delta)\int_0^{t-\varepsilon}\int_\delta^\infty \mu\theta(t-s)^{\mu-1}M_\mu(\theta) \\
&\times Q((t-s)^\mu\theta - \varepsilon^\mu\delta)f(s,x(s))d\theta ds\bigg).
\end{aligned}$$

By Lemma 4.28, we know that $S_{\mu,\nu}(t)$ is compact because $Q(t)$ is compact for $t > 0$. Furthermore, $Q(\varepsilon^\mu\delta)$ is compact, then the set $\{(\mathcal{F}_{\varepsilon,\delta}u)(t),\ u \in \Omega_r\}$ is relatively compact in X for any $\varepsilon \in (0,t)$ and for any $\delta > 0$. Moreover, for every $u \in \Omega_r$, we find

$$\begin{aligned}
&\Big|(\mathcal{F}u)(t) - (\mathcal{F}_{\varepsilon,\delta}u)(t)\Big| \\
&\le t^{1-\nu+\mu\nu-\mu k}\Big|\int_0^t\int_0^\delta \mu\theta(t-s)^{\mu-1}M_\mu(\theta)Q((t-s)^\mu\theta)f(s,x(s))d\theta ds\Big| \\
&\quad + t^{1-\nu+\mu\nu-\mu k}\Big|\int_{t-\varepsilon}^t\int_\delta^\infty \mu\theta(t-s)^{\mu-1}M_\mu(\theta)Q((t-s)^\mu\theta)f(s,x(s))d\theta ds\Big| \\
&\le \mu C_0 t^{1-\nu+\mu\nu-\mu k}\int_0^t (t-s)^{\mu k-1}|f(s,x(s))|ds\int_0^\delta \theta^k M_\mu(\theta)d\theta \\
&\quad + \mu C_0 t^{1-\nu+\mu\nu-\mu k}\int_{t-\varepsilon}^t (t-s)^{\mu k-1}|f(s,x(s))|ds\int_0^\infty \theta^k M_\mu(\theta)d\theta \\
&\le \mu C_0 t^{1-\nu+\mu\nu-\mu k}\int_0^t (t-s)^{\mu k-1}m(s)ds\int_0^\delta \theta^k M_\mu(\theta)d\theta \\
&\quad + \mu C_0 t^{1-\nu+\mu\nu-\mu k}\int_{t-\varepsilon}^t (t-s)^{\mu k-1}m(s)ds\int_0^\infty \theta^k M_\mu(\theta)d\theta \\
&\to 0, \quad \text{as } \varepsilon \to 0,\ \delta \to 0.
\end{aligned}$$

Therefore, $\{(\mathcal{F}u)(t):\ u \in \Omega_r\}$ is also a relatively compact set in X for $t \in [0,T]$. Thus, $\{\mathcal{F}u:\ u \in \Omega_r\}$ is relatively compact by Arzela-Ascoli theorem. Hence, $\mathcal{F}$ is a completely continuous operator. Schauder fixed point theorem shows that $\mathcal{F}$ has at least a fixed point $u^* \in \Omega_r$. Let $x^*(t) = t^{-1+\nu-\mu\nu+\mu k}u^*(t)$. Thus,

$$x^*(t) = S_{\mu,\nu}(t)x_0 + \int_0^t K_\mu(t-s)f(s,x^*(s))ds, \quad t \in (0,T],$$

which implies that x^* is a mild solution of (4.73) in $\widetilde{\Omega}_r$. The proof is completed. □

In the case that $Q(t)$ is noncompact for $t > 0$, we give an assumption as follows:

(H5) there exists a constant $K > 0$ such that for any bounded $D \subseteq X$,

$$\alpha_1(f(t,D)) \le Kt^{1-\nu+\mu\nu-\mu k}\alpha_1(D), \quad \text{for a.e. } t \in [0,T],$$

where α_1 is the Kuratowski measure of noncompactness.

Theorem 4.26. *Assume that (H1)-(H5) hold. Then the Cauchy problem* (4.73) *has at least one mild solution in* $\widetilde{\Omega}_r$.

Proof. Let $u_0(t) = t^{1-\nu+\mu\nu-\mu k}S_{\mu,\nu}(t)x_0$ for all $t \in [0,T]$ and $u_{n+1} = \mathcal{F}u_n$, $n = 0,1,2,\ldots$. By Lemma 4.34, $\mathcal{F}u_n \in \Omega_r$, for $u_n \in \Omega_r$. Consider set $\mathcal{V} = \{\mathcal{F}u_n) :\ u_n \in \Omega_r\}_{n=0}^\infty$, and we will prove set $\mathcal{V}$ is relatively compact. In view of Lemma 4.33, the set $\mathcal{V}$ is equicontinuous. We only need to prove $\mathcal{V}(t) = \{(\mathcal{F}u_n)(t),\ u_n \in \Omega_r\}_{n=0}^\infty$ is relatively compact in X for $t \in [0,T]$.

By the properties of measure of noncompactness, for any $t \in [0,T]$ we have

$$\begin{aligned}\alpha_1(\{u_n(t)\}_{n=0}^{\infty}) &= \alpha_1(\{u_0(t)\} \cup \{u_n(t)\}_{n=1}^{\infty}) \\ &= \alpha_1(\{u_n(t)\}_{n=1}^{\infty}) = \alpha_1(\mathcal{V}(t)).\end{aligned} \tag{4.77}$$

Let $x_n(t) = t^{-1+\nu-\mu\nu+\mu k} u_n(t)$, $t \in (0,T]$, $n = 0,1,2,\ldots$. By the condition (H5) and Proposition 1.18, we have

$$\begin{aligned}&\alpha_1(\mathcal{V}(t)) \\ &= \alpha_1(\{(\mathcal{F}u_n)(t)\}_{n=0}^{\infty}) \\ &= \alpha_1\left(\left\{t^{1-\nu+\mu\nu-\mu k} S_{\mu,\nu}(t)x_0 + t^{1-\nu+\mu\nu-\mu k} \int_0^t K_\mu(t-s) f(s, x_n(s))ds\right\}_{n=0}^{\infty}\right) \\ &= \alpha_1\left(\left\{t^{1-\nu+\mu\nu-\mu k} \int_0^t K_\mu(t-s) f(s, x_n(s))ds\right\}_{n=0}^{\infty}\right) \\ &\le 2L_1 t^{1-\nu+\mu\nu-\mu k} \int_0^t (t-s)^{\mu k-1} \alpha_1\left(f(s, \{s^{-1+\nu-\mu\nu+\mu k} u_n(s)\}_{n=0}^{\infty})\right) ds \\ &\le 2L_1 K T^{1-\nu+\mu\nu-\mu k} \int_0^t (t-s)^{\mu k-1} s^{1-\nu+\mu\nu-\mu k} \alpha_1\left(\{s^{-1+\nu-\mu\nu+\mu k} u_n(s)\}_{n=0}^{\infty}\right) ds \\ &\le 2L_1 K T^{1-\nu+\mu\nu-\mu k} \int_0^t (t-s)^{\mu k-1} \alpha_1(\{u_n(s)\}_{n=0}^{\infty}) ds.\end{aligned}$$

In view of (4.77), we obtain

$$\alpha_1(\mathcal{V}(t)) \le 2L_1 K T^{1-\nu+\mu\nu-\mu k} \int_0^t (t-s)^{\mu k-1} \alpha_1(\mathcal{V}(s)) ds.$$

Therefore, by the inequality in Henry, 1981 (p. 188), we obtain that $\alpha_1(\mathcal{V}(t)) = 0$, then $\mathcal{V}(t)$ is relatively compact. Consequently, it follows from Arzela-Ascoli theorem that set $\mathcal{V}$ is relatively compact, i.e., there exists a convergent subsequence of $\{u_n\}_{n=0}^{\infty}$. With no confusion, let $\lim_{n\to\infty} u_n = u^* \in \Omega_r$.

Thus, by continuity of the operator $\mathcal{F}$, we have

$$u^* = \lim_{n\to\infty} u_n = \lim_{n\to\infty} \mathcal{F}u_{n-1} = \mathcal{F}\left(\lim_{n\to\infty} u_{n-1}\right) = \mathcal{F}u^*.$$

Let $x^*(t) = t^{-1+\nu-\mu\nu+\mu k} u^*(t)$. Thus, x^* is a mild solution of (4.73) in $\widetilde{\Omega}_r$. The proof is completed. □

In the following, we prove the existence and uniqueness of a mild solution of the Cauchy problem (4.73).

(H6) There exists a function $L \in C([0,T], \mathbb{R}^+)$ such that $I_{0+}^{\mu k} L \in C([0,T], \mathbb{R}^+)$,

$$|f(t, x_1(t)) - f(t, x_2(t))| \le L(t)\|x_1 - x_2\|_\mu, \text{ for any } x_1, x_2 \in \widetilde{\Omega}_r,$$

and

$$\sup_{t\in[0,T]} \left\{L_1 T^{1-\nu+\mu\nu-\mu k} \int_0^t (t-s)^{\mu k-1} L(s) ds\right\} \le l_0 < 1.$$

Theorem 4.27. *Assume that the conditions (H2)-(H4) and (H6) hold. Then the Cauchy problem* (4.73) *has a unique mild solution in* $\widetilde{\Omega}_r$.

Proof. From Lemma 4.34, we know that $\mathcal{F}\Omega_r \subset \Omega_r$. For any $u_1, u_2 \in \Omega_r$, $t \in [0, T]$, we have

$$\begin{aligned}
&\big|(\mathcal{F}u_1)(t) - (\mathcal{F}u_2)(t)\big| \\
&\leq T^{1-\nu+\mu\nu-\mu k} \int_0^t |K_\mu(t-s)(f(s, x_1(s)) - f(s, x_2(s)))|ds \\
&\leq L_1 T^{1-\nu+\mu\nu-\mu k} \int_0^t (t-s)^{\mu k-1} |f(s, x_1(s)) - f(s, x_2(s))|ds \\
&\leq L_1 T^{1-\nu+\mu\nu-\mu k} \int_0^t (t-s)^{\mu k-1} L(s)\|x_1 - x_2\|_\mu ds \\
&\leq l_0 \|u_1 - u_2\|.
\end{aligned}$$

Thus

$$\|(\mathcal{F}u_1) - (\mathcal{F}u_2)\| \leq l_0 \|u_1 - u_2\|,$$

which implies that $\mathcal{F}$ is a contraction mapping. In view of the contraction mapping principle, $\mathcal{F}$ has the unique fixed point $u^* \in \Omega_r$. Let $x^*(t) = t^{-1+\nu-\mu\nu+\mu k} u^*(t)$. Thus, x^* is a unique mild solution of (4.73) in $\widetilde{\Omega}_r$. The proof is completed. □

4.8 Infinite Interval Problems with Hilfer Derivative

4.8.1 *Introduction*

Consider the Cauchy problem of fractional evolution equations on an infinite interval

$$\begin{cases} {}_0^H D_t^{\mu,\nu} x(t) = Ax(t) + f(t, x(t)), & t \in (0, \infty), \\ {}_0 D_t^{-(1-\mu)(1-\nu)} x(0) = x_0, \end{cases} \tag{4.78}$$

where ${}_0^H D_t^{\mu,\nu}$ is the Hilfer fractional derivative of order $0 < \mu < 1$ and type $0 \leq \nu \leq 1$, ${}_0 D_t^{-(1-\mu)(1-\nu)}$ is Riemann-Liouville integral of order $(1-\mu)(1-\nu)$, A is the infinitesimal generator of a strongly continuous semigroup of bounded linear operators (i.e., C_0-semigroup) $\{Q(t)\}_{t\geq 0}$ in Banach space X, $f : [0, \infty) \times X \to X$ is a function to be defined later.

Most of the results involve the existence of solutions for fractional evolution equations on a finite interval $[0, T]$, where $T \in (0, \infty)$. It seems that there are few works concerned with fractional evolution equations on an infinite interval. The Arzela-Ascoli theorem and various fixed point theorems are widely used to study the existence of solutions. It is well known that the classical Arzela-Ascoli theorem is powerful technique to give a necessary and sufficient condition for judging the relative compactness of a family of abstract continuous functions, while it is limited to finite closed interval.

In this section, by using the generalized Arzela-Ascoli theorem and some new techniques, we give sufficient conditions of the existence for global mild solutions when the semigroup is compact as well as noncompact. In particular, we do not need to assume that the $f(t,\cdot)$ satisfies the Lipschitz condition.

4.8.2 *Preliminaries*

Assume that X is a Banach space with the norm $|\cdot|$. Let J be an infinite interval of $\mathbb{R}$. By $C(J,X)$ we denote the space of all continuous functions from J to X with the norm $\|u\|_0 = \sup_{t\in J}|u(t)| < \infty$. We denote by $\mathcal{L}(X)$ the space of all bounded linear operators from X to X with the usual operator norm $\|\cdot\|_{\mathcal{L}(X)}$.

Lemma 4.36. *(K. M. Furati, 2012) The Cauchy problem* (4.78) *is equivalent to the integral equation*

$$\begin{aligned} x(t) = &\frac{x_0}{\Gamma(\nu(1-\mu)+\mu)} t^{(\nu-1)(1-\mu)} \\ &+ \frac{1}{\Gamma(\mu)}\int_0^t (t-s)^{\mu-1}[Ax + f(s,x(s))]ds, \ t\in(0,\infty). \end{aligned} \tag{4.79}$$

The Wright function $M_\mu(\theta)$ is defined by (see Definition 1.9)

$$M_\mu(\theta) = \sum_{n=1}^{\infty} \frac{(-\theta)^{n-1}}{(n-1)!\Gamma(1-\mu n)}, \quad 0<\mu<1, \ \theta\in\mathbb{C},$$

which satisfies the following equality.

$$\int_0^\infty \theta^\delta M_\mu(\theta)d\theta = \frac{\Gamma(1+\delta)}{\Gamma(1+\mu\delta)}, \quad \text{for } \theta \geq 0.$$

Lemma 4.37. *If integral equation* (4.79) *holds, then we have*

$$x(t) = S_{\nu,\mu}(t)x_0 + \int_0^t K_\mu(t-s)f(s,x(s))ds, \ t\in(0,\infty), \tag{4.80}$$

where

$$K_\mu(t) = t^{\mu-1}P_\mu(t), \ P_\mu(t) = \int_0^\infty \mu\theta M_\mu(\theta)Q(t^\mu\theta)d\theta, \ S_{\nu,\mu}(t) = {}_0D_t^{-\nu(1-\mu)}K_\mu(t).$$

Proof. Let $\lambda > 0$. Applying the Laplace transform

$$\chi(\lambda) = \int_0^\infty e^{-\lambda s}x(s)ds \ \text{ and } \ \omega(\lambda) = \int_0^\infty e^{-\lambda s}f(s,x(s))ds$$

to (4.79), we have

$$\begin{aligned} \chi(\lambda) &= \lambda^{(1-\nu)(1-\mu)-1}x_0 + \frac{1}{\lambda^\mu}A\chi(\lambda) + \frac{1}{\lambda^\mu}\,\omega(\lambda) \\ &= \lambda^{\nu(\mu-1)}(\lambda^\mu I - A)^{-1}x_0 + (\lambda^\mu I - A)^{-1}\omega(\lambda) \\ &= \lambda^{\nu(\mu-1)}\int_0^\infty e^{-\lambda^\mu s}Q(s)x_0 ds + \int_0^\infty e^{-\lambda^\mu s}Q(s)\omega(\lambda)ds, \end{aligned} \tag{4.81}$$

provided that the integrals in (4.81) exist, where I is the identity operator defined on X.

Let

$$\psi_\mu(\theta) = \frac{\mu}{\theta^{\mu+1}} M_\mu(\theta^{-\mu}),$$

whose Laplace transform is given by

$$\int_0^\infty e^{-\lambda\theta}\psi_\mu(\theta)d\theta = e^{-\lambda^\mu}, \quad \text{where } \mu \in (0,1). \tag{4.82}$$

Using (2.4), we have

$$\begin{aligned}\int_0^\infty e^{-\lambda^\mu s}Q(s)x_0 ds &= \int_0^\infty \mu t^{\mu-1}e^{-(\lambda t)^\mu}Q(t^\mu)x_0 dt \\ &= \int_0^\infty\int_0^\infty \mu\psi_\mu(\theta)e^{-(\lambda t\theta)}Q(t^\mu)t^{\mu-1}x_0 d\theta dt \\ &= \int_0^\infty\int_0^\infty \mu\psi_\mu(\theta)e^{-\lambda t}Q\left(\frac{t^\mu}{\theta^\mu}\right)\frac{t^{\mu-1}}{\theta^\mu}x_0 d\theta dt \\ &= \int_0^\infty e^{-\lambda t}\left[\mu\int_0^\infty \psi_\mu(\theta)Q\left(\frac{t^\mu}{\theta^\mu}\right)\frac{t^{\mu-1}}{\theta^\mu}x_0 d\theta\right]dt \\ &= \int_0^\infty e^{-\lambda t}t^{\mu-1}P_\mu(t)x_0 dt,\end{aligned} \tag{4.83}$$

and

$$\begin{aligned}&\int_0^\infty e^{-\lambda^\mu s}Q(s)\omega(\lambda)ds \\ &= \int_0^\infty\int_0^\infty \mu t^{\mu-1}e^{-(\lambda t)^\mu}Q(t^\mu)e^{-\lambda s}f(s,x(s))dsdt \\ &= \int_0^\infty\int_0^\infty\int_0^\infty \mu\psi_\mu(\theta)e^{-(\lambda t\theta)}Q(t^\mu)e^{-\lambda s}t^{\mu-1}f(s,x(s))d\theta dsdt \\ &= \int_0^\infty\int_0^\infty\int_0^\infty \mu\psi_\mu(\theta)e^{-\lambda(t+s)}Q\left(\frac{t^\mu}{\theta^\mu}\right)\frac{t^{\mu-1}}{\theta^\mu}f(s,x(s))d\theta dsdt \\ &= \int_0^\infty e^{-\lambda t}\left[\mu\int_0^t\int_0^\infty \psi_\mu(\theta)Q\left(\frac{(t-s)^\mu}{\theta^\mu}\right)\frac{(t-s)^{\mu-1}}{\theta^\mu}f(s,x(s))d\theta ds\right]dt \\ &= \int_0^\infty e^{-\lambda t}\left[\int_0^t (t-s)^{\mu-1}P_\mu(t-s)f(s,x(s))ds\right]dt.\end{aligned} \tag{4.84}$$

Since the Laplace inverse transform of $\lambda^{\nu(\mu-1)}$ is

$$\mathcal{L}^{-1}(\lambda^{\nu(\mu-1)}) = \begin{cases}\dfrac{t^{\nu(1-\mu)-1}}{\Gamma(\nu(1-\mu))}, & 0 < \nu \leq 1, \\ \delta(t), & \nu = 0,\end{cases}$$

where $\delta(t)$ is the Delta function, by (4.81), (4.83) and (4.84), for $t \in (0, \infty)$ we obtain

$$\begin{aligned} x(t) &= \big(\mathcal{L}^{-1}(\lambda^{\nu(\mu-1)}) * K_\mu(t)\big)x_0 + \int_0^t K_\mu(t-s)f(s,x(s))ds \\ &= \big({}_0D_t^{-\nu(1-\mu)}K_\mu(t)\big)x_0 + \int_0^t K_\mu(t-s)f(s,x(s))ds \\ &= S_{\nu,\mu}(t)x_0 + \int_0^t K_\mu(t-s)f(s,x(s))ds. \end{aligned} \tag{4.85}$$

This completes the proof. □

Due to Lemma 4.37, we give the following definition of the mild solution of (4.78).

Definition 4.10. By the mild solution of the Cauchy problem (4.78), we mean that the function $x \in C((0,\infty), X)$ which satisfies

$$x(t) = S_{\nu,\mu}(t)x_0 + \int_0^t K_\mu(t-s)f(s,x(s))ds, \ t \in (0,\infty).$$

Throughout this section, we introduce the following hypotheses:

(H0) $Q(t)$ is continuous in the uniform operator topology for $t > 0$, and $\{Q(t)\}_{t\geq 0}$ is uniformly bounded, i.e., there exists $M > 1$ such that $\sup_{t\in[0,+\infty)} |Q(t)| < M$.

Lemma 4.38. *(Jaiwal, 2022) If $\{Q(t)\}_{t>0}$ is a compact operator, then $\{S_{\nu,\mu}(t)\}_{t>0}$ and $\{P_\mu(t)\}_{t>0}$ are also compact operators.*

Lemma 4.39. *(Zhou and Jiao, 2010a) Assume that $\{Q(t)\}_{t>0}$ is a compact operator. Then $\{Q(t)\}_{t>0}$ is equicontinuous.*

Lemma 4.40. *Under assumption (H0), $P_\mu(t)$ is continuous in the uniform operator topology for $t > 0$.*

Proof. For any $t > 0$, $h > 0$ and $x \in X$, we have

$$|P_\mu(t+h)x - P_\mu(t)x| = \left| \int_0^\infty \mu\theta M_\mu(\theta)[Q((t+h)^\mu\theta) - Q(t^\mu\theta)]xd\theta \right|.$$

Since

$$\left| \int_0^\infty \mu\theta M_\mu(\theta)[Q((t+h)^\mu\theta) - Q(t^\mu\theta)]xd\theta \right| \leq 2M \int_0^\infty \mu\theta M_\mu(\theta)d\theta|x| = \frac{2M}{\Gamma(\mu)}|x|,$$

then by Lebesgue dominated convergence theorem, we have

$$|P_\mu(t+h)x - P_\mu(t)x| \to 0 \quad \text{indepently of } t \text{ and } x, \text{ as } h \to 0.$$

Therefore, $P_\mu(t)$ is continuous in the uniform operator topology for $t > 0$. This completes the proof. □

Lemma 4.41. *Under assumption (H0), for any fixed $t > 0$, $\{K_\mu(t)\}_{t>0}$ and $\{S_{\nu,\mu}(t)\}_{t>0}$ are linear operators, and for any $x \in X$*

$$|K_\mu(t)x| \leq \frac{Mt^{\mu-1}}{\Gamma(\mu)}|x| \quad \text{and} \quad |S_{\nu,\mu}(t)x| \leq \frac{Mt^{(\nu-1)(\mu-1)}}{\Gamma(\nu(1-\mu)+\mu)}|x|.$$

Proof. From the equality

$$\int_0^\infty \theta^\delta M_\mu(\theta)d\theta = \frac{\Gamma(1+\delta)}{\Gamma(1+\mu\delta)},$$

we know that

$$|P_\mu(t)x| = \left|\int_0^\infty \mu\theta M_\mu(\theta)Q(t^\mu\theta)xd\theta\right| \leq \frac{M}{\Gamma(\mu)}|x|, \quad \text{for } t \in [0,\infty) \text{ and } x \in X,$$

then we have

$$|K_\mu(t)x| \leq \frac{Mt^{\mu-1}}{\Gamma(\mu)}|x|, \quad \text{for } t \in (0,\infty) \text{ and } x \in X.$$

For $t \in (0,\infty)$ and $x \in X$,

$$\begin{aligned}
|S_{\nu,\mu}(t)x| &= \left|{}_0D_t^{-\nu(1-\mu)}K_\mu(t)x\right| \\
&= \left|\frac{1}{\Gamma(\nu(1-\mu))}\int_0^t (t-s)^{\nu(1-\mu)-1}K_\mu(s)xds\right| \\
&= \left|\frac{1}{\Gamma(\nu(1-\mu))}\int_0^t (t-s)^{\nu(1-\mu)-1}s^{\mu-1}P_\mu(s)xds\right| \\
&= \left|\frac{t^{(\nu-1)(1-\mu)}}{\Gamma(\nu(1-\mu))}\int_0^1 (1-s)^{\nu(1-\mu)-1}s^{\mu-1}P_\mu(ts)xds\right| \\
&\leq \frac{t^{(\nu-1)(1-\mu)}M}{\Gamma(\nu(1-\mu))\Gamma(\mu)}\int_0^1 (1-s)^{\nu(1-\mu)-1}s^{\mu-1}ds|x| \\
&= \frac{t^{(\nu-1)(1-\mu)}M}{\Gamma(\nu(1-\mu)+\mu)}|x|.
\end{aligned} \tag{4.86}$$

This completes the proof. □

Lemma 4.42. *Under assumption (H0), $\{K_\mu(t)\}_{t>0}$ and $\{S_{\nu,\mu}(t)\}_{t>0}$ are strongly continuous, which means that, for any $x \in X$ and $0 < t' < t'' \leq b$, we have*

$$|K_\mu(t')x - K_\mu(t'')x| \to 0 \quad \text{and} \quad |S_{\nu,\mu}(t')x - S_{\nu,\mu}(t'')x| \to 0, \quad \text{as } t'' \to t'.$$

Proof. By Lemma 4.40, we know that$\{P_\mu(t)\}_{t>0}$ is strongly continuous, then we easily obtain $\{K_\mu(t)\}_{t>0}$ is also strongly continuous.

For any $x \in X$ and $0 < t_1 < t_2 \leq b$, we have

$$\begin{aligned}
&|S_{\nu,\mu}(t_2)x - S_{\nu,\mu}(t_1)x| \\
&= \frac{1}{\Gamma(\nu(1-\mu))}\left|\int_0^{t_2}(t_2-s)^{\nu(1-\mu)-1}K_\mu(s)xds - \int_0^{t_1}(t_1-s)^{\nu(1-\mu)-1}K_\mu(s)xds\right| \\
&= \frac{1}{\Gamma(\nu(1-\mu))} \\
&\quad \times \left|\int_0^{t_2}(t_2-s)^{\nu(1-\mu)-1}s^{\mu-1}P_\mu(s)xds - \int_0^{t_1}(t_1-s)^{\nu(1-\mu)-1}s^{\mu-1}P_\mu(s)xds\right| \\
&\leq \frac{1}{\Gamma(\nu(1-\mu))}\left|\int_{t_1}^{t_2}(t_2-s)^{\nu(1-\mu)-1}s^{\mu-1}P_\mu(s)xds\right| \\
&\quad + \frac{1}{\Gamma(\nu(1-\mu))}\left|\int_0^{t_1}\left((t_2-s)^{\nu(1-\mu)-1}-(t_1-s)^{\nu(1-\mu)-1}\right)s^{\mu-1}P_\mu(s)xds\right| \\
&\leq \frac{Mt_1^{\mu-1}}{\Gamma(\nu(1-\mu))\Gamma(\mu)}\frac{1}{\nu(1-\mu)}(t_2-t_1)^{\nu(1-\mu)}|x| \\
&\quad + \frac{M}{\Gamma(\nu(1-\mu))\Gamma(\mu)}\left|\int_0^{t_1}\left((t_2-s)^{\nu(1-\mu)-1}-(t_1-s)^{\nu(1-\mu)-1}\right)s^{\mu-1}ds\right||x|.
\end{aligned} \tag{4.87}$$

Since

$$\left|\int_0^{t_1}\left((t_2-s)^{\nu(1-\mu)-1}-(t_1-s)^{\nu(1-\mu)-1}\right)s^{\mu-1}ds\right| \leq 2\int_0^{t_1}(t_1-s)^{\nu(1-\mu)-1}s^{\mu-1}ds$$

exists, then by Lebesgue dominated convergence theorem, we have

$$\left|\int_0^{t_1}\left((t_2-s)^{\nu(1-\mu)-1}-(t_1-s)^{\nu(1-\mu)-1}\right)s^{\mu-1}ds\right| \to 0, \quad \text{as } t_2 \to t_1.$$

Consequently, we have

$$|S_{\nu,\mu}(t_2)x - S_{\nu,\mu}(t_1)x| \to 0, \quad \text{as } t_2 \to t_1,$$

i.e., $\{S_{\nu,\mu}(t)\}_{t>0}$ is strongly continuous. This completes the proof. □

4.8.3 *Lemmas*

Let

$$C_1([0,\infty),X) = \left\{u \in C([0,\infty),X) : \lim_{t\to\infty}\frac{|u(t)|}{1+t} = 0\right\}.$$

Then, $C_1([0,\infty),X)$ is a Banach space with the norm $\|u\| = \sup_{t\in[0,\infty)}|u(t)|/(1+t) < \infty$.

We introduce the following hypotheses:

(H1) $f(t,\cdot)$ is Lebesgue measurable with respect to t on $[0,\infty)$. $f(\cdot,x)$ is continuous with respect to x on X.

(H2) There exists a function $m : (0, \infty) \to (0, \infty)$ such that
$${}_0D_t^{-\mu} m(t) \in C((0,\infty),(0,\infty)), \quad |f(t,x)| \le m(t), \quad \text{for all } x \in X, \ t \in (0,\infty),$$
and
$$\lim_{t\to 0+} t^{(1-\mu)(1-\nu)}{}_0D_t^{-\mu} m(t) = 0, \quad \lim_{t\to\infty} \frac{t^{(1-\mu)(1-\nu)}}{1+t}{}_0D_t^{-\mu} m(t) = 0.$$

Let
$$C_\mu((0,\infty),X) = \Big\{x \in C((0,\infty),X) : \lim_{t\to 0+} t^{(1-\mu)(1-\nu)}|x(t)| \text{ exists and is finite,}$$
$$\lim_{t\to\infty} \frac{t^{(1-\mu)(1-\nu)}|x(t)|}{1+t} = 0\Big\}.$$

Then $(C_\mu((0,\infty),X), \|\cdot\|_\mu)$ is a Banach space with the norm
$$\|x\|_\mu = \sup_{t\in[0,\infty)} \frac{t^{(1-\mu)(1-\nu)}|x(t)|}{1+t}.$$

For any $x \in C_\mu((0,\infty),X)$, define an operator Ψ as follows
$$(\Psi x)(t) = (\Psi_1 x)(t) + (\Psi_2 x)(t),$$
where
$$(\Psi_1 x)(t) = S_{\nu,\mu}(t)x_0, \quad (\Psi_2 x)(t) = \int_0^t K_\mu(t-s)f(s,x(s))ds, \quad \text{for } t \in (0,\infty).$$

For any $u \in C_1([0,\infty),X)$, set
$$x(t) = t^{-(1-\mu)(1-\nu)}u(t), \quad \text{for } t \in (0,\infty).$$
Then, $x \in C_\mu((0,\infty),X)$. Define an operator Φ as follows
$$(\Phi u)(t) = (\Phi_1 u)(t) + (\Phi_2 u)(t),$$
where
$$(\Phi_1 u)(t) = \begin{cases} t^{(1-\mu)(1-\nu)}(\Psi_1 x)(t), & \text{for } t \in (0,\infty), \\ \dfrac{x_0}{\Gamma(\nu(1-\mu)+\mu)}, & \text{for } t = 0, \end{cases}$$
$$(\Phi_2 u)(t) = \begin{cases} t^{(1-\mu)(1-\nu)}(\Psi_2 x)(t), & \text{for } t \in (0,\infty), \\ 0, & \text{for } t = 0. \end{cases}$$

Obviously, $x \in C_\mu((0,\infty),X)$ is a mild solution of (4.78) if and only if the operator equation $x = \Psi x$ has a solution $x \in C_\mu((0,\infty),X)$.

In view of (H2), we have
$$\lim_{t\to 0+} \frac{t^{(1-\mu)(1-\nu)}}{1+t}{}_0D_t^{-\mu} m(t) = 0, \quad \lim_{t\to\infty} \frac{t^{(1-\mu)(1-\nu)}}{1+t}{}_0D_t^{-\mu} m(t) = 0.$$
Thus, there exists a constant $r > 0$ such that
$$\sup_{t\in[0,\infty)} \left\{ \frac{L|x_0|}{\Gamma(\nu(1-\mu)+\mu)} + \frac{Lt^{(1-\mu)(1-\nu)}}{1+t}{}_0D_t^{-\mu} m(t) \right\} \le r,$$

i.e.,

$$\sup_{t\in[0,\infty)}\left\{\frac{L|x_0|}{\Gamma(\nu(1-\mu)+\mu)}+\frac{L}{\Gamma(\mu)}\frac{t^{(1-\mu)(1-\nu)}}{1+t}\int_0^t(t-s)^{\mu-1}m(s)ds\right\}\le r. \quad (4.88)$$

Let

$$\Omega_r=\{u\in C_1([0,\infty),X):\ \|u\|\le r\},\quad \widetilde{\Omega}_r=\{x\in C_\mu((0,\infty),X):\ \|x\|_\mu\le r\}.$$

Clearly, Ω_r is a nonempty, convex, and closed subset of $C_1([0,\infty),X)$, and $\widetilde{\Omega}_r$ is a nonempty, convex, and closed subset of $C_\mu((0,\infty),X)$.

Let

$$V:=\big\{v: v(t)=(\Phi u)(t)/(1+t),\ u\in\Omega_r\big\}.$$

Lemma 4.43. *Assume that (H0), (H1) and (H2) hold. Then the set V is equicontinuous.*

Proof. Step I. We first prove that $\big\{v : v(t) = (\Phi_1 u)(t)/(1+t),\ u\in\Omega_r\big\}$ is equicontinuous.

As $\lim_{t\to 0+}P_\mu(t)x_0 = x_0/\Gamma(\mu)$, we find

$$\begin{aligned}\lim_{t\to 0+}t^{(1-\mu)(1-\nu)}S_{\nu,\mu}(t)x_0 &= \lim_{t\to 0+}\frac{t^{(1-\mu)(1-\nu)}}{\Gamma(\nu(1-\mu))}\int_0^t(t-s)^{\nu(1-\mu)-1}s^{\mu-1}P_\mu(s)x_0ds\\ &= \lim_{t\to 0+}\frac{1}{\Gamma(\nu(1-\mu))}\int_0^1(1-z)^{\nu(1-\mu)-1}z^{\mu-1}P_\mu(tz)x_0dz\\ &= \frac{1}{\Gamma(\nu(1-\mu))\Gamma(\mu)}\int_0^1(1-z)^{\nu(1-\mu)-1}z^{\mu-1}x_0dz\\ &= \frac{x_0}{\Gamma(\nu(1-\mu)+\mu)}.\end{aligned}$$

Hence, for $t_1=0$, $t_2\in(0,\infty)$, we obtain

$$\begin{aligned}&\left|\frac{(\Phi_1u)(t_2)}{1+t_2}-(\Phi_1u)(0)\right|\\ &\le\left|\frac{1}{1+t_2}{t_2}^{(1-\mu)(1-\nu)}S_{\nu,\mu}(t_2)x_0-\frac{x_0}{\Gamma(\nu(1-\mu)+\mu)}\right|\\ &\to 0,\quad \text{as } t_2\to 0.\end{aligned}$$

For any $t_1,t_2\in(0,\infty)$ and $t_1<t_2$, we have

$$\begin{aligned}&\left|\frac{(\Phi_1u)(t_2)}{1+t_2}-\frac{(\Phi_1u)(t_1)}{1+t_1}\right|\\ &=\left|\frac{{t_2}^{(1-\mu)(1-\nu)}S_{\nu,\mu}(t_2)x_0}{1+t_2}-\frac{{t_1}^{(1-\mu)(1-\nu)}S_{\nu,\mu}(t_1)x_0}{1+t_1}\right|\\ &\le\left|\frac{{t_2}^{(1-\mu)(1-\nu)}S_{\nu,\mu}(t_2)x_0}{1+t_2}-\frac{{t_2}^{(1-\mu)(1-\nu)}S_{\nu,\mu}(t_2)x_0}{1+t_1}\right|\\ &\quad+\left|\frac{{t_2}^{(1-\mu)(1-\nu)}S_{\nu,\mu}(t_2)x_0}{1+t_1}-\frac{{t_1}^{(1-\mu)(1-\nu)}S_{\nu,\mu}(t_1)x_0}{1+t_1}\right|\end{aligned}$$

$$\begin{aligned}
&\leq \left|t_2{}^{(1-\mu)(1-\nu)}S_{\nu,\mu}(t_2)x_0\right|\frac{|t_2-t_1|}{(1+t_2)(1+t_1)} \\
&\quad + \left|t_2{}^{(1-\mu)(1-\nu)}S_{\nu,\mu}(t_2)x_0 - t_1{}^{(1-\mu)(1-\nu)}S_{\nu,\mu}(t_1)x_0\right|\frac{1}{1+t_1} \\
&\leq \left|t_2{}^{(1-\mu)(1-\nu)}S_{\nu,\mu}(t_2)x_0\right|\frac{|t_2-t_1|}{(1+t_2)(1+t_1)} \\
&\quad + \left|t_2{}^{(1-\mu)(1-\nu)}\right|\left|S_{\nu,\mu}(t_2)x_0 - S_{\nu,\mu}(t_1)x_0\right|\frac{1}{1+t_1} \\
&\quad + \left|t_2{}^{(1-\mu)(1-\nu)} - t_1{}^{(1-\mu)(1-\nu)}\right|\left|S_{\nu,\mu}(t_1)x_0\right|\frac{1}{1+t_1} \\
&\to 0, \quad \text{as } t_2 \to t_1.
\end{aligned}$$

Hence, $\{v : v(t) = (\Phi_1 u)(t)/(1+t),\ u \in \Omega_r\}$ is equicontinuous.

Step II. We prove that $\{v : v(t) = (\Phi_2 u)(t)/(1+t),\ u \in \Omega_r\}$ is equicontinuous.

Let $x(t) = t^{-(1-\mu)(1-\nu)}u(t)$, for any $u \in \Omega_r$, $t \in (0,\infty)$. Then $x \in \widetilde{\Omega}_r$. For $\varepsilon > 0$, in view of (H2), there exists $T > 0$ such that

$$\frac{L}{\Gamma(\mu)}\frac{t^{(1-\mu)(1-\nu)}}{1+t}\int_0^t (t-s)^{\mu-1}m(s)ds < \frac{\varepsilon}{2}, \quad \text{for } t > T. \tag{4.89}$$

For $t_1, t_2 > T$, in virtue of (H2) and (4.89), we find

$$\begin{aligned}
\left|\frac{(\Phi_2 u)(t_2)}{1+t_2} - \frac{(\Phi_2 u)(t_1)}{1+t_1}\right| &\leq \left|\frac{t_2{}^{(1-\mu)(1-\nu)}}{1+t_2}\int_0^{t_2} K_\mu(t_2-s)f(s,x(s))ds\right| \\
&\quad + \left|\frac{t_1{}^{(1-\mu)(1-\nu)}}{1+t_1}\int_0^{t_1} K_\mu(t_1-s)f(s,x(s))ds\right| \\
&\leq \frac{L}{\Gamma(\mu)}\frac{t_2{}^{(1-\mu)(1-\nu)}}{1+t_2}\int_0^{t_2}(t_2-s)^{\mu-1}m(s)ds \\
&\quad + \frac{L}{\Gamma(\mu)}\frac{t_1{}^{(1-\mu)(1-\nu)}}{1+t_1}\int_0^{t_1}(t_1-s)^{\mu-1}m(s)ds \\
&< \varepsilon.
\end{aligned}$$

When $t_1 = 0$, $0 < t_2 < T$, we have

$$\begin{aligned}
\left|\frac{(\Phi_2 u)(t_2)}{1+t_2} - (\Phi_2 u)(0)\right| &= \left|\frac{t_2{}^{(1-\mu)(1-\nu)}}{1+t_2}\int_0^{t_2} K_\mu(t_2-s)f(s,x(s))ds\right| \\
&\leq \frac{L}{\Gamma(\mu)}\frac{t_2{}^{(1-\mu)(1-\nu)}}{1+t_2}\int_0^{t_2}(t_2-s)^{\mu-1}m(s)ds \\
&\to 0, \quad \text{as } t_2 \to 0.
\end{aligned}$$

For $0 < t_1 < t_2 \leq T$, we find

$$\begin{aligned}
&\left|\frac{(\Phi_2 u)(t_2)}{1+t_2} - \frac{(\Phi_2 u)(t_1)}{1+t_1}\right| \\
&\leq \left|\frac{t_1{}^{(1-\mu)(1-\nu)}}{1+t_1}\int_{t_1}^{t_2}(t_2-s)^{\mu-1}P_\mu(t_2-s)f(s,x(s))ds\right|
\end{aligned}$$

$$
\begin{aligned}
&+\left|\frac{t_1{}^{(1-\mu)(1-\nu)}}{1+t_1}\int_0^{t_1}\Big((t_2-s)^{\mu-1}-(t_1-s)^{\mu-1}\Big)P_\mu(t_2-s)f(s,x(s))ds\right|\\
&+\left|\frac{t_1{}^{(1-\mu)(1-\nu)}}{1+t_1}\int_0^{t_1}(t_1-s)^{\mu-1}\Big(P_\mu(t_2-s)-P_\mu(t_1-s)\Big)f(s,x(s))ds\right|\\
&+\left|\frac{t_2{}^{(1-\mu)(1-\nu)}}{1+t_2}-\frac{t_1{}^{(1-\mu)(1-\nu)}}{1+t_1}\right|\left|\int_0^{t_2}(t_2-s)^{\mu-1}P_\mu(t_2-s)f(s,x(s))ds\right|\\
&\le I_1+I_2+I_3+I_4,
\end{aligned}
$$

where

$$
\begin{aligned}
I_1&=\frac{L}{\Gamma(\mu)}\frac{t_1{}^{(1-\mu)(1-\nu)}}{1+t_1}\left|\int_0^{t_2}(t_2-s)^{\mu-1}m(s)ds-\int_0^{t_1}(t_1-s)^{\mu-1}m(s)ds\right|,\\
I_2&=\frac{2L}{\Gamma(\mu)}\frac{t_1{}^{(1-\mu)(1-\nu)}}{1+t_1}\int_0^{t_1}\Big((t_1-s)^{\mu-1}-(t_2-s)^{\mu-1}\Big)m(s)ds,\\
I_3&=\frac{t_1{}^{(1-\mu)(1-\nu)}}{1+t_1}\left|\int_0^{t_1}(t_1-s)^{\mu-1}\Big(P_\mu(t_2-s)-P_\mu(t_1-s)\Big)f(s,x(s))ds\right|,\\
I_4&=\left|\frac{t_2{}^{(1-\mu)(1-\nu)}}{1+t_2}-\frac{t_1{}^{(1-\mu)(1-\nu)}}{1+t_1}\right|\frac{L}{\Gamma(\mu)}\int_0^{t_2}(t_2-s)^{\mu-1}m(s)ds.
\end{aligned}
$$

One can deduce that $\lim_{t_2\to t_1} I_1=0$, as ${}_0D_t^{-\mu}m(t)\in C((0,\infty),(0,\infty))$. Noting that

$$((t_1-s)^{\mu-1}-(t_2-s)^{\mu-1})m(s)\le (t_1-s)^{\mu-1}m(s),\quad \text{for } s\in[0,t_1),$$

then by Lebesgue dominated convergence theorem, we find

$$\int_0^{t_1}\Big((t_1-s)^{\mu-1}-(t_2-s)^{\mu-1}\Big)m(s)ds\to 0,\quad \text{as } t_2\to t_1,$$

so, $I_2\to 0$ as $t_2\to t_1$.

For $\varepsilon>0$ be enough small, we have

$$
\begin{aligned}
I_3\le\;&\frac{t_1{}^{(1-\mu)(1-\nu)}}{1+t_1}\int_0^{t_1-\varepsilon}(t_1-s)^{\mu-1}\big\|P_\mu(t_2-s)-P_\mu(t_1-s)\big\|_{\mathcal{L}(X)}|f(s,x(s))|ds\\
&+\frac{t_1{}^{(1-\mu)(1-\nu)}}{1+t_1}\int_{t_1-\varepsilon}^{t_1}(t_1-s)^{\mu-1}\big\|P_\mu(t_2-s)-P_\mu(t_1-s)\big\|_{\mathcal{L}(X)}|f(s,x(s))|ds\\
\le\;&\frac{t_1{}^{(1-\mu)(1-\nu)}}{1+t_1}\int_0^{t_1-\varepsilon}(t_1-s)^{\mu-1}m(s)ds\sup_{s\in[0,t_1-\varepsilon]}\big\|P_\mu(t_2-s)-P_\mu(t_1-s)\big\|_{\mathcal{L}(X)}\\
&+\frac{2L}{\Gamma(\mu)}\frac{t_1{}^{(1-\mu)(1-\nu)}}{1+t_1}\int_{t_1-\varepsilon}^{t_1}(t_1-s)^{\mu-1}m(s)ds\\
\le\;&I_{31}+I_{32}+I_{33},
\end{aligned}
$$

where

$$I_{31}=\frac{t_1{}^{(1-\mu)(1-\nu)}}{1+t_1}\int_0^{t_1-\varepsilon}(t_1-s)^{\mu-1}m(s)ds\sup_{s\in[0,t_1-\varepsilon]}\big\|P_\mu(t_2-s)-P_\mu(t_1-s)\big\|_{\mathcal{L}(X)},$$

$$I_{32} = \frac{2L}{\Gamma(\mu)} \frac{t_1{}^{(1-\mu)(1-\nu)}}{1+t_1} \left| \int_0^{t_1} (t_1 - s)^{\mu-1} m(s)ds - \int_0^{t_1-\varepsilon} (t_1 - \varepsilon - s)^{\mu-1} m(s)ds \right|,$$

$$I_{33} = \frac{2L}{\Gamma(\mu)} \frac{t_1{}^{(1-\mu)(1-\nu)}}{1+t_1} \int_0^{t_1-\varepsilon} ((t_1 - \varepsilon - s)^{\mu-1} - (t_1 - s)^{\mu-1}) m(s)ds.$$

By (H0) and Lemma 4.42, it is easy to see that $I_{31} \to 0$ as $t_2 \to t_1$. Similar to the proof that I_1, I_2 tend to zero, we obtain $I_{32} \to 0$ and $I_{33} \to 0$ as $\varepsilon \to 0$. Thus, I_3 tends to zero as $t_2 \to t_1$. Clearly, $I_4 \to 0$ as $t_2 \to t_1$.

For $0 < t_1 < T < t_2$, if $t_2 \to t_1$, then $t_2 \to T$ and $t_1 \to T$. Thus, for $u \in \Omega_r$

$$\begin{aligned}
&\left| \frac{(\Phi_2 u)(t_2)}{1+t_2} - \frac{(\Phi_2 u)(t_1)}{1+t_1} \right| \\
&\leq \left| \frac{(\Phi_2 u)(t_2)}{1+t_2} - \frac{(\Phi_2 u)(T)}{1+T} \right| + \left| \frac{(\Phi_2 u)(T)}{1+T} - \frac{(\Phi_2 u)(t_1)}{1+t_1} \right| \to 0, \quad \text{as } t_2 \to t_1.
\end{aligned}$$

Consequently,

$$\left| \frac{(\Phi_2 u)(t_2)}{1+t_2} - \frac{(\Phi_2 u)(t_1)}{1+t_1} \right| \to 0, \quad \text{as } t_2 \to t_1.$$

Therefore, $\{v : v(t) = (\Phi_2 u)(t)/(1+t),\ u \in \Omega_r\}$ is equicontinuous. Furthermore, V is equicontinuous. □

Lemma 4.44. *Assume that (H1) and (H2) hold. Then,* $\lim_{t\to\infty} |(\Phi u)(t)|/(1+t) = 0$ *uniformly for* $u \in \Omega_r$.

Proof. In fact, for any $u \in \Omega_r$, by (H2) and Lemma 4.41, we find

$$\begin{aligned}
\frac{|(\Phi u)(t)|}{1+t} &\leq \left| \frac{t^{(1-\mu)(1-\nu)}}{1+t} S_{\nu,\mu}(t) x_0 \right| + \left| \frac{t^{(1-\mu)(1-\nu)}}{1+t} \int_0^t K_\mu(t-s) f(s, x(s)) ds \right| \\
&\leq \frac{L|x_0|}{\Gamma(\nu(1-\mu)+\mu)(1+t)} + \frac{L}{\Gamma(\mu)} \frac{t^{(1-\mu)(1-\nu)}}{1+t} \int_0^t (t-s)^{\mu-1} m(s)ds, \quad t > 0.
\end{aligned} \tag{4.90}$$

By (H2), we derive

$$\frac{|(\Phi u)(t)|}{1+t} \to 0, \quad \text{as } t \to \infty,$$

which implies that $\lim_{t\to\infty} |(\Phi u)(t)|/(1+t) = 0$ uniformly for $u \in \Omega_r$. This completes the proof. □

Lemma 4.45. *Assume that (H1) and (H2) hold. Then* $\Phi\Omega_r \subset \Omega_r$.

Proof. From Lemmas 4.43 and 4.44, we know that $\Phi\Omega_r \subset C_1([0,\infty), X)$. For $t > 0$ and any $u \in \Omega_r$, by (4.88) and (4.90), we have

$$\frac{|(\Phi u)(t)|}{1+t} \leq \frac{L|x_0|}{\Gamma(\nu(1-\mu)+\mu)} + \frac{L}{\Gamma(\mu)} \frac{t^{(1-\mu)(1-\nu)}}{1+t} \int_0^t (t-s)^{\mu-1} m(s)ds \leq r.$$

For $t = 0$, we find

$$|(\Phi u)(0)| = \frac{x_0}{\Gamma(\nu(1-\mu)+\mu)} \leq \frac{Lx_0}{\Gamma(\nu(1-\mu)+\mu)} \leq r.$$

Therefore, $\Phi\Omega_r \subset \Omega_r$. □

Lemma 4.46. *Suppose that (H1) and (H2) hold. Then Φ is continuous.*

Proof. Indeed, let $\{u_n\}_{n=1}^{\infty}$ be a sequence in Ω_r which is convergent to $u \in \Omega_r$. Consequently,

$$\lim_{n\to\infty} u_n(t) = u(t), \text{ and } \lim_{n\to\infty} t^{-(1-\mu)(1-\nu)}u_n(t) = t^{-(1-\mu)(1-\nu)}u(t), \text{ for } t \in (0,\infty).$$

Let $x(t) = t^{-(1-\mu)(1-\nu)}u(t)$, $x_n(t) = t^{-(1-\mu)(1-\nu)}u_n(t)$, $t \in (0,\infty)$. Then $x, x_n \in \widetilde{\Omega}_r$. In view of (H1), we have

$$\lim_{n\to\infty} f(t, x_n(t)) = \lim_{n\to\infty} f(t, t^{-(1-\mu)(1-\nu)}u_n(t)) = f(t, t^{-(1-\mu)(1-\nu)}u(t)) = f(t, x(t)).$$

On the one hand, using (H2), we get for each $t \in (0,\infty)$,

$$(t-s)^{\mu-1}|f(s, x_n(s)) - f(s, x(s))| \le 2(t-s)^{\mu-1}m(s), \quad \text{a.e. in } [0,t).$$

On the other hand, the function $s \to 2(t-s)^{\mu-1}m(s)$ is integrable for $s \in [0,t)$, $t \in [0,\infty)$. By Lebesgue dominated convergence theorem, we obtain

$$\int_0^t (t-s)^{\mu-1}|f(s, x_n(s)) - f(s, x(s))|ds \to 0, \quad \text{as } n \to \infty.$$

Thus, for $t \in [0,\infty)$,

$$\begin{aligned}
&\left|\frac{(\Phi u_n)(t)}{1+t} - \frac{(\Phi u)(t)}{1+t}\right| \\
&\le \frac{t^{(1-\mu)(1-\nu)}}{1+t}\int_0^t |K_\mu(t-s)(f(s, x_n(s)) - f(s, x(s)))|ds \\
&\le \frac{L}{\Gamma(\mu)}\frac{t^{(1-\mu)(1-\nu)}}{1+t}\int_0^t (t-s)^{\mu-1}|f(s, x_n(s)) - f(s, x(s))|ds \to 0, \quad \text{as } n \to \infty.
\end{aligned}$$

Therefore, $\|\Phi u_n - \Phi u\| \to 0$ as $n \to \infty$. Hence, Φ is continuous. The proof is completed. □

4.8.4 *Existence on Infinite Interval*

Theorem 4.28. *Assume that $Q(t)(t > 0)$ is compact. Furthermore suppose that (H1) and (H2) hold. Then the Cauchy problem* (4.78) *has at least one mild solution.*

Proof. Clearly, the problem (4.78) exists a mild solution $x \in \widetilde{\Omega}_r$ if and only if the operator Φ has a fixed point $u \in \Omega_r$, where $u(t) = t^{(1-\mu)(1-\nu)}x(t)$. Hence, we only need to prove that the operator Φ has a fixed point in Ω_r. From Lemmas 4.45 and 4.46, we know that $\Phi\Omega_r \subset \Omega_r$ and Φ is continuous. In order to prove that Φ is a completely continuous operator, we need to prove that $\Phi\Omega_r$ is a relatively compact set. In view of Lemmas 4.43 and 4.44, the set $V = \{v : v(t) = (\Phi u)(t)/(1+t),\ u \in \Omega_r\}$ is equicontinuous on $[0,h]$ for any $h > 0$, and $\lim_{t\to\infty} |(\Phi u)(t)|/(1+t) = 0$ uniformly for $u \in \Omega_r$. According to Lemma 1.4, we only need to prove $V(t) = \{v(t) : v(t) = (\Phi u)(t)/(1+t),\ u \in \Omega_r\}$ is relatively compact in X for $t \in [0,\infty)$.

Obviously, $V(0)$ is relatively compact in X. We only consider the case $t > 0$. For $\forall\ \varepsilon \in (0,t)$ and $\delta > 0$, define $\Phi_{\varepsilon,\delta}$ on Ω_r as follows:

$$(\Phi_{\varepsilon,\delta}u)(t) := t^{(1-\mu)(1-\nu)}(\Psi_{\varepsilon,\delta}x)(t)$$
$$= t^{(1-\mu)(1-\nu)}\left(S_{\nu,\mu}(t)x_0 + \int_0^{t-\varepsilon}\int_\delta^\infty \mu\theta(t-s)^{\mu-1}M_\mu(\theta)Q((t-s)^\mu\theta)f(s,x(s))d\theta ds\right).$$

Thus,

$$\frac{(\Phi_{\varepsilon,\delta}u)(t)}{1+t} = \frac{t^{(1-\mu)(1-\nu)}}{1+t}\left(S_{\nu,\mu}(t)x_0\right.$$
$$\left. + Q(\varepsilon^\mu\delta)\int_0^{t-\varepsilon}\int_\delta^\infty \mu\theta(t-s)^{\mu-1}M_\mu(\theta)Q((t-s)^\mu\theta - \varepsilon^\mu\delta)f(s,x(s))d\theta ds\right).$$

By Lemma 4.38, we know that $S_{\nu,\mu}(t)$ is compact because $Q(t)$ is compact for $t > 0$. Further, $Q(\varepsilon^\mu\delta)$ is compact, then the set $\{\frac{(\Phi_{\varepsilon,\delta}u)(t)}{1+t},\ u \in \Omega_r\}$ is relatively compact in X for any $\varepsilon \in (0,t)$ and for any $\delta > 0$. Moreover, for every $u \in \Omega_r$, we find

$$\left|\frac{(\Phi u)(t)}{1+t} - \frac{(\Phi_{\varepsilon,\delta}u)(t)}{1+t}\right|$$
$$\leq \frac{t^{(1-\mu)(1-\nu)}}{1+t}\left|\int_0^t\int_0^\delta \mu\theta(t-s)^{\mu-1}M_\mu(\theta)Q((t-s)^\mu\theta)f(s,x(s))d\theta ds\right|$$
$$+ \frac{t^{(1-\mu)(1-\nu)}}{1+t}\left|\int_{t-\varepsilon}^t\int_\delta^\infty \mu\theta(t-s)^{\mu-1}M_\mu(\theta)Q((t-s)^\mu\theta)f(s,x(s))d\theta ds\right|$$
$$\leq \frac{\mu L t^{(1-\mu)(1-\nu)}}{1+t}\int_0^t (t-s)^{\mu-1}m(s)ds\int_0^\delta \theta M_\mu(\theta)d\theta$$
$$+ \frac{\mu L t^{(1-\mu)(1-\nu)}}{1+t}\int_{t-\varepsilon}^t (t-s)^{\mu-1}m(s)ds\int_0^\infty \theta M_\mu(\theta)d\theta$$
$$\to 0, \quad \text{as } \varepsilon \to 0,\ \delta \to 0.$$

Thus, $V(t)$ is also a relatively compact set in X for $t \in [0,\infty)$. Therefore, Schauder fixed point theorem implies that Φ has at least a fixed point $u^* \in \Omega_r$. Let $x^*(t) = t^{-(1-\mu)(1-\nu)}u^*(t)$. Thus,

$$x^*(t) = S_{\nu,\mu}(t)x_0 + \int_0^t K_\mu(t-s)f(s,x^*(s))ds,\ t \in (0,\infty),$$

which implies that $x^* \in \widetilde{\Omega}_r$ is a mild solution of (4.78). The proof is completed. □

In the case that $Q(t)$ is noncompact for $t > 0$, we need the following hypothesis:

(H3) there exists a constant $K > 0$ such that for any bounded set $D \subseteq X$,

$$\alpha_1(f(t,D)) \leq Kt^{(1-\mu)(1-\nu)}\alpha_1(D), \quad \text{for a.e. } t \in [0,\infty),$$

where α_1 is the Kuratowski measure of noncompactness.

Theorem 4.29. *Assume that (H0), (H1), (H2) and (H3) hold. Then the Cauchy problem* (4.78) *has at least one mild solution.*

Proof. Let $u_0(t) = t^{(1-\mu)(1-\nu)}S_{\nu,\mu}(t)x_0$ for all $t \in [0,\infty)$ and $u_{n+1} = \Phi u_n$, $n = 0,1,2,\ldots$. By Lemma 4.45, $\Phi u_n \in \Omega_r$, for $u_n \in \Omega_r$, $n = 0,1,2,\ldots$. Consider set $\mathcal{V} = \{v_n : v_n(t) = (\Phi u_n)(t)/(1+t),\ u_n \in \Omega_r\}_{n=0}^{\infty}$, and we will prove set $\mathcal{V}$ is relatively compact.

In view of Lemmas 4.43 and 4.44, the set $\mathcal{V}$ is equicontinuous and $\lim_{t\to\infty} |(\Phi u_n)(t)|/(1+t) = 0$ uniformly for $u_n \in \Omega_r$. According to Lemma 1.4, we only need to prove $\mathcal{V}(t) = \{v_n(t) : v_n(t) = (\Phi u_n)(t)/(1+t),\ u_n \in \Omega_r\}_{n=0}^{\infty}$ is relatively compact in X for $t \in [0,\infty)$.

Let $x_n(t) = t^{-(1-\mu)(1-\nu)}u_n(t)$, $t \in (0,\infty)$, $n = 0,1,2,\ldots$. By the condition (H3) and Proposition 1.19, we have

$$\begin{aligned}
\alpha_1(\mathcal{V}(t)) &= \alpha_1\Big(\Big\{\frac{(\Phi u_n)(t)}{1+t}\Big\}_{n=0}^{\infty}\Big)\\
&= \alpha_1\Big(\Big\{\frac{t^{(1-\mu)(1-\nu)}}{1+t}S_{\nu,\mu}(t)x_0 + \frac{t^{(1-\mu)(1-\nu)}}{1+t}\int_0^t K_\mu(t-s)f(s,x_n(s))ds\Big\}_{n=0}^{\infty}\Big)\\
&= \alpha_1\Big(\Big\{\frac{t^{(1-\mu)(1-\nu)}}{1+t}\int_0^t K_\mu(t-s)f(s,x_n(s))ds\Big\}_{n=0}^{\infty}\Big)\\
&\le \frac{2L}{\Gamma(\mu)}\frac{t^{(1-\mu)(1-\nu)}}{1+t}\int_0^t (t-s)^{\mu-1}\alpha_1\left(f(s,\{s^{-(1-\mu)(1-\nu)}u_n(s)\}_{n=0}^{\infty})\right)ds\\
&\le \frac{2LK}{\Gamma(\mu)}\frac{t^{(1-\mu)(1-\nu)}}{1+t}\int_0^t (t-s)^{\mu-1}s^{(1-\mu)(1-\nu)}\alpha_1\left(\{s^{-(1-\mu)(1-\nu)}u_n(s)\}_{n=0}^{\infty}\right)ds\\
&\le \frac{2LK}{\Gamma(\mu)}\frac{t^{(1-\mu)(1-\nu)}}{1+t}\int_0^t (t-s)^{\mu-1}(1+s)\alpha_1\Big(\Big\{\frac{u_n(s)}{1+s}\Big\}_{n=0}^{\infty}\Big)ds.
\end{aligned}$$

On the other hand, by the properties of measure of noncompactness, for any $t \in [0,\infty)$ we have

$$\alpha_1\Big(\Big\{\frac{u_n(t)}{1+t}\Big\}_{n=0}^{\infty}\Big) = \alpha_1\Big(\Big\{\frac{u_0(t)}{1+t}\Big\} \cup \Big\{\frac{u_n(t)}{1+t}\Big\}_{n=1}^{\infty}\Big) = \alpha_1\Big(\Big\{\frac{u_n(t)}{1+t}\Big\}_{n=1}^{\infty}\Big) = \alpha_1(\mathcal{V}(t)).$$

Thus

$$\alpha_1(\mathcal{V}(t)) \le \frac{2LKM^*}{\Gamma(\mu)}\int_0^t (t-s)^{\mu-1}(1+s)\alpha_1(\mathcal{V}(s))ds, \tag{4.91}$$

where $M^* = \max_{t\in[0,\infty)}\{\frac{t^{(1-\mu)(1-\nu)}}{1+t}\}$. From (4.91), we know that

$$\alpha_1(\mathcal{V}(t)) \le \frac{4LKM^*}{\Gamma(\mu)}\int_0^t (t-s)^{\mu-1}\alpha_1(\mathcal{V}(s))ds,$$

or

$$\alpha_1(\mathcal{V}(t)) \le \frac{4LKM^*}{\Gamma(\mu)}\int_0^t (t-s)^{\mu-1}s\alpha_1(\mathcal{V}(s))ds$$

holds. Therefore, by the inequality in Henry, 1981, we obtain that $\alpha_1(\mathcal{V}(t)) = 0$, then $\mathcal{V}(t)$ is relatively compact. Consequently, it follows from Lemma 1.4 that set $\mathcal{V}$

is relatively compact, i.e., there exists a convergent subsequence of $\{u_n\}_{n=0}^{\infty}$. With no confusion, let $\lim_{n\to\infty} u_n = u^*$, $u^* \in \Omega_r$.

Thus, by continuity of the operator Φ, we have

$$u^* = \lim_{n\to\infty} u_n = \lim_{n\to\infty} \Phi u_{n-1} = \Phi\left(\lim_{n\to\infty} u_{n-1}\right) = \Phi u^*.$$

Let $x^*(t) = t^{-(1-\mu)(1-\nu)}u^*(t)$. Thus, $x^* \in \widetilde{\Omega}_r$ is a mild solution of (4.78). The proof is completed. □

By Theorems 4.28 and 4.29, we have the following corollaries.

Corollary 4.2. *Assume that $Q(t)$ is compact for $t > 0$ and (H1) holds. Furthermore suppose that*

(H2)′ *there exists a function $m : (0,\infty) \to (0,\infty)$ and $\alpha \in (0,1), M > 0$ such that*

$$ {}_0D_t^{-\mu}m(t) \in C((0,\infty),(0,\infty)), \quad t^{(1-\mu)(1-\nu)}{}_0D_t^{-\mu}m(t) \le Mt^{\alpha},$$

and

$$|f(t,x)| \le m(t), \quad \textit{for all} \;\; x \in X, \quad t \in (0,\infty).$$

Then the Cauchy problem (4.78) *has at least one mild solution.*

Corollary 4.3. *Assume that (H0), (H1), (H2)′ and (H3) hold. Then the Cauchy problem* (4.78) *has at least one mild solution.*

Example 4.7. Let $X = L^2([0,\pi],\mathbb{R})$. Consider the following fractional partial differential equations on infinite interval

$$\begin{cases} {}_0^HD_t^{\mu,\nu}x(t,z) = \partial_z^2 x(t,z) + t^{-\eta}, & z \in [0,\pi],\ t > 0, \\ x(t,0) = x(t,\pi) = 0, & t > 0 \\ {}_0D_t^{-(1-\nu)(1-\mu)}x(0,z) = x_0(z), & z \in [0,\pi]. \end{cases} \tag{4.92}$$

We define an operator A by $Av = v''$ with the domain

$$D(A) = \{v \in X :\ v, v'' \text{ are absolutely continuous},\ v'' \in X,\ v(0) = v(\pi) = 0\}.$$

Then A generates a compact, analytic, self-adjoint semigroup $\{T(t)\}_{t>0}$. Then problem (4.92) can be rewritten as follows

$$\begin{cases} {}_0^HD_t^{\mu,\nu}x(t) = Ax(t) + f_1(t,x(t)), & t > 0, \\ {}_0D_t^{-(1-\nu)(1-\mu)}x(0) = x_0, \end{cases} \tag{4.93}$$

where $f_1(t,x) := t^{-\eta}$ for $\eta \in (\mu, 1-\nu+\nu\mu)$ satisfies (H1), and $|f_1(t,x(t))| \le t^{-\eta}$, $t \in (0,\infty)$. Let $m(t) = t^{-\eta}$, for $t > 0$. Then

$$ {}_0D_t^{-\mu}m(t) = \frac{\Gamma(1-\eta)}{\Gamma(1+\mu-\eta)}t^{\mu-\eta}, \quad t^{(1-\mu)(1-\nu)}{}_0D_t^{-\mu}m(t) = \frac{\Gamma(1-\eta)}{\Gamma(1+\mu-\eta)}t^{\alpha},$$

where $\alpha = 1-\nu+\mu\nu-\eta \in (0,1)$. This means that the condition (H2)′ is satisfied. By Corollary 4.2, the problem (4.92) has at least a mild solution.

4.9 Notes and Remarks

The results in Section 4.2 are taken from Zhou, Zhang and Shen, 2013. The material in Section 4.3 due to Zhou, Shen and Zhang, 2013. The main results in Section 4.4 is taken from Wang, Zhou and Fečkan, 2014. The results in Section 4.5 are adopted from Wang and Zhou, 2011a. The contents of Section 4.6 are taken from Wang, Chen and Xiao, 2012. The results in Section 4.7 are taken from Zhou, Li and Zhou, 2022. The material in Section 4.8 due to Zhou, 2022.

Chapter 5

Fractional Impulsive Differential Equations

5.1 Introduction

The theory of impulsive differential equations has recently years been an object of increasing interest because of its wide applicability in biology, in medicine and in more and more fields. The reason for this applicability arises from the fact that impulsive differential problems is an appropriate model for describing process which at certain moments change their state rapidly and which cannot be described using the classical differential problems. For a wide bibliography and exposition on this object see for instance the monographs of Benchohra, Henderson and Ntouyas, 2006; Bainov and Simeonov, 1993; Lakshmikantham, Bainov and Simeonov, 1989; Yang, 2001 and the papers of Abada, Benchohra and Hammouche, 2009; Ahmed, 2003 and 2007; Akhmet, 2005; Fan and Li, 2010; Fan, 2010; Liang, Liu and Xiao, 2009; Liu, 1999; Battelli and Fečkan, 1997; Mophou, 2010; Nieto and O'Regan, 2009; Wang, Xiang and Peng, 2009; Wang and Wei, 2010; Wei, Xiang and Peng, 2006; Wei, Hou and Teo, 2006.

Recently, a number of papers have been written on Cauchy problems, boundary value problems and nonlocal problems for fractional impulsive differential equations, one can see Ahmad *et al.*, 2009 and 2010; Bench, 2009; Agarwal *et al.*, 2010; Ahmad and Wang, 2010; Balachandran, 2010; Tian and Bai, 2010; Cao and Chen, 2010; Wang *et al.*, 2010; Wang, Ahmad and Zhang, 2010; Wang, 2011; Wang, Zhang and Song, 2011; Yang and Chen, 2011; Cao and Chen, 2012 and the references therein.

However, Fečkan, Zhou and Wang, 2012; Kosmatov, 2012, point out on the error in former solutions for some fractional impulsive differential equations by constructing a counterexample and establish a general framework to seek a natural solution for fractional impulsive differential equations.

Section 5.2 is concerned with the existence and uniqueness of solutions for fractional impulsive initial value equations. In Section 5.3, we give some sufficient conditions for the existence of the solutions for fractional impulsive boundary value equations, and use a new generalized singular Gronwall inequality to obtain the data dependence. In Section 5.4, we establish the existence results of solutions for fractional impulsive Langevin equations by utilizing boundedness, continuity, monotonicity and nonnegative of Mittag-Leffler function and fixed point methods. Section 5.5 is devoted to the existence of PC-mild solutions for Cauchy problems and nonlocal problems for fractional impulsive evolution equations.

5.2 Impulsive Initial Value Problems

5.2.1 *Introduction*

Consider the Cauchy problems for the following fractional impulsive differential equations

$$\begin{cases} {}_0^CD_t^q u(t) = f(t, u(t)), & t \in J' := J\backslash\{t_1, ..., t_m\}, \quad J := [0, T], \\ u(t_k^+) = u(t_k^-) + y_k, & y_k \in \mathbb{R}, \ k = 1, 2, ..., m, \\ u(0) = u_0, \end{cases} \tag{5.1}$$

where ${}_0^CD_t^q$ is Caputo fractional derivative of order $q \in (0, 1)$ with the lower limit zero, $u_0 \in \mathbb{R}$, $f : J \times \mathbb{R} \to \mathbb{R}$ is jointly continuous, and t_k satisfy $0 = t_0 < t_1 < \cdots < t_m < t_{m+1} = T$, $u(t_k^+) = \lim_{\epsilon \to 0^+} u(t_k + \epsilon)$ and $u(t_k^-) = \lim_{\epsilon \to 0^-} u(t_k + \epsilon)$ represent the right and left limits of $u(t)$ at $t = t_k$.

In Subsection 5.2.2, we introduce the definition of a solution of the problem (5.1). Subsection 5.2.3 is concerned with the existence and uniqueness of solutions for (5.1).

5.2.2 *Formula of Solutions*

Note that

$$u(t) = u_0 - \frac{1}{\Gamma(q)} \int_0^a (a - s)^{q-1} h(s) ds + \frac{1}{\Gamma(q)} \int_0^t (t - s)^{q-1} h(s) ds,$$

solves the Cauchy problems

$$\begin{cases} {}_0^CD_t^q u(t) = h(t), & t \in J, \\ u(0) = u_0 - \dfrac{1}{\Gamma(q)} \displaystyle\int_0^a (a - s)^{q-1} h(s) ds. \end{cases}$$

One can obtain the following result immediately.

Lemma 5.1. *Let $q \in (0, 1)$ and $h : J \to \mathbb{R}$ be continuous. A function $u \in C(J, \mathbb{R})$ is a solution of the fractional integral equation*

$$u(t) = u_0 - \frac{1}{\Gamma(q)} \int_0^a (a - s)^{q-1} h(s) ds + \frac{1}{\Gamma(q)} \int_0^t (t - s)^{q-1} h(s) ds,$$

if and only if u is a solution of the following fractional Cauchy problems

$$\begin{cases} {}_0^CD_t^q u(t) = h(t), & t \in J, \\ u(a) = u_0, & a > 0. \end{cases}$$

As a consequence of Lemma 5.1, we have the following result which is useful in what follows.

Lemma 5.2. *Let $q \in (0,1)$ and $h : J \to \mathbb{R}$ be continuous. A function u is a solution of the fractional integral equation*

$$u(t) = \begin{cases} u_0 + \frac{1}{\Gamma(q)} \int_0^t (t-s)^{q-1} h(s)ds, & \textit{for } t \in [0, t_1), \\ u_0 + y_1 + \frac{1}{\Gamma(q)} \int_0^t (t-s)^{q-1} h(s)ds, & \textit{for } t \in (t_1, t_2), \\ u_0 + y_1 + y_2 + \frac{1}{\Gamma(q)} \int_0^t (t-s)^{q-1} h(s)ds, & \textit{for } t \in (t_2, t_3), \\ \vdots & \\ u_0 + \sum_{i=1}^{m} y_i + \frac{1}{\Gamma(q)} \int_0^t (t-s)^{q-1} h(s)ds, & \textit{for } t \in (t_m, T], \end{cases} \tag{5.2}$$

if and only if u is a solution of the following impulsive problem

$$\begin{cases} {}_0^C D_t^q u(t) = h(t), & t \in (0, T], \\ u(t_k^+) = u(t_k^-) + y_k, & k = 1, 2, ..., m, \\ u(0) = u_0. \end{cases} \tag{5.3}$$

Proof. Assume u satisfies (5.3). If $t \in [0, t_1]$, then

$${}_0^C D_t^q u(t) = h(t), \quad t \in (0, t_1] \text{ with } u(0) = u_0. \tag{5.4}$$

Integrating the expression (5.4) from 0 to t by virtue of Definition 1.1, one can obtain

$$u(t) = u_0 + \frac{1}{\Gamma(q)} \int_0^t (t-s)^{q-1} h(s)ds.$$

If $t \in (t_1, t_2]$, then

$${}_0^C D_t^q u(t) = h(t), \ t \in (t_1, t_2] \text{ with } u(t_1^+) = u(t_1^-) + y_1.$$

By Lemma 5.1, one obtain

$$\begin{aligned} u(t) &= u(t_1^+) - \frac{1}{\Gamma(q)} \int_0^{t_1} (t_1 - s)^{q-1} h(s)ds + \frac{1}{\Gamma(q)} \int_0^t (t-s)^{q-1} h(s)ds \\ &= u(t_1^-) + y_1 - \frac{1}{\Gamma(q)} \int_0^{t_1} (t_1 - s)^{q-1} h(s)ds + \frac{1}{\Gamma(q)} \int_0^t (t-s)^{q-1} h(s)ds \\ &= u_0 + y_1 + \frac{1}{\Gamma(q)} \int_0^t (t-s)^{q-1} h(s)ds. \end{aligned}$$

If $t \in (t_2, t_3]$, then using again Lemma 5.1, we get

$$u(t) = u(t_2^+) - \frac{1}{\Gamma(q)} \int_0^{t_2} (t_2 - s)^{q-1} h(s)ds + \frac{1}{\Gamma(q)} \int_0^t (t-s)^{q-1} h(s)ds$$

$$= u(t_2^-) + y_2 - \frac{1}{\Gamma(q)}\int_0^{t_2}(t_2 - s)^{q-1}h(s)ds + \frac{1}{\Gamma(q)}\int_0^t (t-s)^{q-1}h(s)ds$$
$$= u_0 + y_1 + y_2 + \frac{1}{\Gamma(q)}\int_0^t (t-s)^{q-1}h(s)ds.$$

If $t \in (t_k, t_{k+1}]$, then again from Lemma 5.1 we get

$$u(t) = u_0 + \sum_{i=1}^{k} y_i + \frac{1}{\Gamma(q)}\int_0^t (t-s)^{q-1}h(s)ds.$$

Conversely, assume that u satisfies (5.2). If $t \in (0, t_1]$ then $u(0) = u_0$ and using the fact that ${}_0^C D_t^q$ is the left inverse of ${}_0D_t^{-q}$ we get (5.4). If $t \in (t_k, t_{k+1}]$, $k = 1, 2, ..., m$ and using the fact of Caputo fractional derivative of a constant is equal to zero, we obtain ${}_0^C D_t^q u(t) = h(t)$, $t \in (t_k, t_{k+1}]$ and $u(t_k^+) = u(t_k^-) + y_k$. This completes the proof. □

Definition 5.1. A function $u \in PC^1(J, \mathbb{R})$ is said to be a solution of the problem (5.1) if u satisfies the integral equation

$$u(t) = \begin{cases} u_0 + \dfrac{1}{\Gamma(q)}\displaystyle\int_0^t (t-s)^{q-1} f(s, u(s))ds, & \text{for } t \in [0, t_1), \\ u_0 + y_1 + \dfrac{1}{\Gamma(q)}\displaystyle\int_0^t (t-s)^{q-1} f(s, u(s))ds, & \text{for } t \in (t_1, t_2), \\ u_0 + y_1 + y_2 + \dfrac{1}{\Gamma(q)}\displaystyle\int_0^t (t-s)^{q-1} f(s, u(s))ds, & \text{for } t \in (t_2, t_3), \\ \vdots & \\ u_0 + \displaystyle\sum_{i=1}^{m} y_i + \dfrac{1}{\Gamma(q)}\displaystyle\int_0^t (t-s)^{q-1} f(s, u(s))ds, & \text{for } t \in (t_m, T]. \end{cases}$$

Theorem 5.1. *(Ye, Gao and Ding, 2007) Suppose $\beta > 0$, $\widetilde{a}(t)$ is a nonnegative function locally integrable on $[0, T)$ and $\widetilde{g}(t)$ is a nonnegative, nondecreasing continuous function defined on $\widetilde{g}(t) \le M$, $t \in [0, T)$, and suppose $y(t)$ is nonnegative and locally integrable on $[0, T)$ with*

$$y(t) \le \widetilde{a}(t) + \widetilde{g}(t)\int_0^t (t-s)^{\beta-1}y(s)ds, \ t \in [0, T).$$

Then

$$y(t) \le \widetilde{a}(t) + \int_0^t \left(\sum_{n=1}^{\infty} \frac{(\widetilde{g}(t)\Gamma(\beta))^n}{\Gamma(n\beta)}(t-s)^{n\beta-1}\widetilde{a}(s)\right) ds, \ t \in [0, T).$$

Theorem 5.2. *Under the hypothesis of Theorem 5.1, let $\widetilde{a}(t)$ be a nondecreasing function on $[0, T)$. Then we have*

$$y(t) \le \widetilde{a}(t)E_\beta(\widetilde{g}(t)\Gamma(\beta)t^\beta),$$

where E_β is the Mittag-Leffler function.

Remark 5.1. There exists a constant $M_{\widetilde{g}}^* > 0$ independent of $\widetilde{a}$ such that

$$y(t) \le M_{\widetilde{g}}^* \widetilde{a}, \quad \text{for all } 0 \le t < T.$$

5.2.3 *Existence*

This subsection deals with the existence and uniqueness of solutions for the problem (5.1). Before stating and proving the main results, we introduce the following hypotheses.

(H1) $f : J \times \mathbb{R} \to \mathbb{R}$ is jointly continuous;
(H2) there exists $q_1 \in (0, q)$ and a real function $m(\cdot) \in L^{\frac{1}{q_1}}(J, \mathbb{R})$ such that $|f(t, u)| \leq m(t)$, for all $u \in \mathbb{R}$;
(H3) there exists $q_2 \in (0, q)$ and a real function $h(\cdot) \in L^{\frac{1}{q_2}}(J, \mathbb{R})$ such that $|f(t, u_1) - f(t, u_2)| \leq h(t)|u_1 - u_2|$, for all $u_1, u_2 \in \mathbb{R}$.

For brevity, let

$$\gamma = \frac{T^q}{\Gamma(q+1)}, \quad \beta = \frac{q-1}{1-q_1}, \quad \alpha = \frac{q-1}{1-q_2}.$$

Theorem 5.3. *Assume that (H1)-(H3) hold. If*

$$\frac{T^{(1+\alpha)(1-q_2)}\|h\|_{L^{\frac{1}{q_2}}J}}{\Gamma(q)(1+\alpha)^{1-q_2}} < 1, \tag{5.5}$$

then the problem (5.1) *has a unique solution on* J.

Proof. Transform the problem (5.1) into a fixed point problem. Consider the operator $F : PC(J, \mathbb{R}) \to PC(J, \mathbb{R})$ defined by

$$(Fu)(t) = u_0 + \sum_{i=1}^{k} y_i + \frac{1}{\Gamma(q)} \int_0^t (t-s)^{q-1} f(s, u(s))ds. \tag{5.6}$$

It is obvious that F is well defined due to (H1).

Claim I. $Fu \in PC(J, \mathbb{R})$ for every $u \in PC(J, \mathbb{R})$.

If $t \in [0, t_1]$, then for every $u \in C([0, t_1], \mathbb{R})$ and any $\delta > 0$, by using Hölder inequality, we get

$$\begin{aligned}
&|(Fu)(t+\delta) - (Fu)(t)| \\
&\leq \frac{1}{\Gamma(q)} \left| \int_0^t \left((t+\delta-s)^{q-1} - (t-s)^{q-1}\right) f(s, u(s))ds \right| \\
&\quad + \frac{1}{\Gamma(q)} \left| \int_t^{t+\delta} (t+\delta-s)^{q-1} f(s, u(s))ds \right| \\
&\leq \frac{1}{\Gamma(q)} \left(\int_0^t \left((t+\delta-s)^{q-1} - (t-s)^{q-1}\right)^{\frac{1}{1-q_1}} ds \right)^{1-q_1} \left(\int_0^t (m(s))^{\frac{1}{q_1}} ds \right)^{q_1} \\
&\quad + \frac{1}{\Gamma(q)} \left(\int_t^{t+\delta} \left((t+\delta-s)^{q-1}\right)^{\frac{1}{1-q_1}} ds \right)^{1-q_1} \left(\int_t^{t+\delta} (m(s))^{\frac{1}{q_1}} ds \right)^{q_1},
\end{aligned}$$

which implies that

$$|Fu(t+\delta) - Fu(t)| \le \frac{2\delta^{(1+\beta)(1-q_1)}\|m\|_{L^{\frac{1}{q_1}}[0,t_1]}}{\Gamma(q)(1+\beta)^{1-q_1}}.$$

It is easy to see that the right-hand side of the above inequality tends to zero as $\delta \to 0$. Thus, $Fu \in C([0,t_1],\mathbb{R})$.

If $t \in (t_1,t_2]$, then for every $u \in C((t_1,t_2],\mathbb{R})$ and any $\delta > 0$, repeating the same process, one can obtain

$$|Fu(t+\delta) - Fu(t)| \le \frac{2\delta^{(1+\beta)(1-q_1)}\|m\|_{L^{\frac{1}{q_1}}[t_1,t_2]}}{\Gamma(q)(1+\beta)^{1-q_1}},$$

which implies that $Fu \in C((t_1,t_2],\mathbb{R})$.

If $t \in (t_k,t_{k+1}]$, $k = 1,2,...,m$, then for every $u \in C((t_k,t_{k+1}],\mathbb{R})$ and any $\delta > 0$, repeating the same process again, one can obtain

$$|Fu(t+\delta) - Fu(t)| \le \frac{2\delta^{(1+\beta)(1-q_1)}\|m\|_{L^{\frac{1}{q_1}}[t_k,t_{k+1}]}}{\Gamma(q)(1+\beta)^{1-q_1}},$$

which implies that $Fu \in C((t_k,t_{k+1}],\mathbb{R})$.

From the above discussion, we must have $Fu \in PC(J,\mathbb{R})$ for every $u \in PC(J,\mathbb{R})$.

Claim II. F is a contraction operator on $PC(J,\mathbb{R})$.

In fact, for arbitrary $u, v \in PC(J,\mathbb{R})$, we get

$$\begin{aligned}
|(Fu)(t) - (Fv)(t)| &\le \frac{1}{\Gamma(q)}\int_0^t (t-s)^{q-1}|f(s,u_1(s)) - f(s,u_2(s))|ds \\
&\le \frac{1}{\Gamma(q)}\int_0^t (t-s)^{q-1}h(s)|u_1(s) - u_2(s)|ds \\
&\le \frac{\|h\|_{L^{\frac{1}{q_2}}J}}{\Gamma(q)}\left(\int_0^t \left((t-s)^{q-1}\right)^{\frac{1}{1-q_2}} ds\right)^{1-q_2} \|u_1 - u_2\|_{PC} \\
&= \frac{T^{(1+\alpha)(1-q_2)}\|h\|_{L^{\frac{1}{q_2}}J}}{\Gamma(q)(1+\alpha)^{1-q_2}}\|u_1 - u_2\|_{PC}.
\end{aligned}$$

Thus, F is a contraction mapping on $PC(J,\mathbb{R})$ due to the condition (5.5). By applying Banach contraction mapping principle we know that the operator F has a unique fixed point on $PC(J,\mathbb{R})$. Therefore, the problem (5.1) has a unique solution

$$u(t) = \begin{cases}
u_0 + \dfrac{1}{\Gamma(q)}\displaystyle\int_0^t (t-s)^{q-1}f(s,u(s))ds, & \text{for } t \in [0,t_1), \\
u_0 + y_1 + \dfrac{1}{\Gamma(q)}\displaystyle\int_0^t (t-s)^{q-1}f(s,u(s))ds, & \text{for } t \in (t_1,t_2), \\
u_0 + y_1 + y_2 + \dfrac{1}{\Gamma(q)}\displaystyle\int_0^t (t-s)^{q-1}f(s,u(s))ds, & \text{for } t \in (t_2,t_3), \\
\vdots & \\
u_0 + \displaystyle\sum_{i=1}^m y_i + \dfrac{1}{\Gamma(q)}\displaystyle\int_0^t (t-s)^{q-1}f(s,u(s))ds, & \text{for } t \in (t_m,T].
\end{cases}$$

This completes the proof. □

The second result is based on Schaefer fixed point theorem.

Now, we replace (H2) into the following linear growth condition:

(H2)$'$ there exists a constant $L > 0$ such that

$$|f(t,u)| \leq L(1+|u|), \quad \text{for each } t \in J \text{ and all } u \in \mathbb{R}.$$

Theorem 5.4. *Assume that (H1) and (H2)$'$ hold. Then the problem* (5.1) *has at least one solution.*

Proof. Transform the problem (5.1) into a fixed point problem. Consider the operator $F : PC(J,\mathbb{R}) \to PC(J,\mathbb{R})$ defined as (5.6). For the sake of convenience, we subdivide the proof into several steps.

Claim I. F is continuous.

Let $\{u_n\}$ be a sequence such that $u_n \to u$ in $PC(J,\mathbb{R})$. Then for each $t \in J$, we have

$$\begin{aligned} |(Fu_n)(t) - (Fu)(t)| &\leq \frac{1}{\Gamma(q)} \int_0^t (t-s)^{q-1} |f(s,u_n(s)) - f(s,u(s))| ds \\ &\leq \frac{T^q}{\Gamma(q+1)} \|f(\cdot,u_n(\cdot)) - f(\cdot,u(\cdot))\|_{PC}. \end{aligned}$$

Due to (H1), f is jointly continuous, then we have

$$\|Fu_n - Fu\|_{PC} \to 0, \quad \text{as } n \to \infty.$$

Claim II. F maps bounded sets into bounded sets in $PC(J,\mathbb{R})$.

Indeed, it is enough to show that for any $\eta^* > 0$, there exists a $\ell > 0$ such that for each $u \in B_{\eta^*} = \{y \in PC(J,\mathbb{R}) : \|u\|_{PC} \leq \eta^*\}$, we have $\|Fu\|_{PC} \leq \ell$.

For each $t \in J$, we get

$$\begin{aligned} |(Fu)(t)| &\leq |u_0| + \sum_{i=1}^{m} |y_i| + \frac{1}{\Gamma(\alpha)} \int_0^t (t-s)^{q-1} |f(s,u(s))| ds \\ &\leq |u_0| + \sum_{i=1}^{m} |y_i| + \frac{LT^q \eta^*}{\Gamma(q+1)}, \end{aligned}$$

which implies that

$$\|Fy\|_{PC} \leq |u_0| + \sum_{i=1}^{m} |y_i| + \frac{LT^q \eta^*}{\Gamma(q+1)} =: \ell.$$

Claim III. F maps bounded sets into equicontinuous sets of $PC(J,\mathbb{R})$.

For interval $[0,t_1]$, $0 \leq s_1 < s_2 \leq t_1$, $u \in B_{\eta^*}$. Using (H2)$'$, we have

$$\begin{aligned} |(Fu)(s_2) - (Fu)(s_1)| \leq{}& \frac{1}{\Gamma(q)} \int_0^{s_1} \left((s_1-s)^{q-1} - (s_2-s)^{q-1}\right) |f(s,u(s))| ds \\ &+ \frac{1}{\Gamma(q)} \int_{s_1}^{s_2} (s_2-s)^{q-1} |f(s,y(s))| ds \end{aligned}$$

$$
\begin{aligned}
&\leq \frac{L}{\Gamma(q)}\int_0^{s_1}\left((s_1-s)^{q-1}-(s_2-s)^{q-1}\right)(1+|u(s)|)ds\\
&\quad+\frac{L}{\Gamma(q)}\int_{s_1}^{s_2}(s_2-s)^{q-1}(1+|u(s)|)ds\\
&\leq \frac{L(1+\eta^*)}{\Gamma(q)}\int_0^{s_1}(s_1-s)^{q-1}-(s_2-s)^{q-1}ds\\
&\quad+\frac{L(1+\eta^*)}{\Gamma(q)}\int_{s_1}^{s_2}(s_2-s)^{q-1}ds\\
&\leq \frac{L(1+\eta^*)}{\Gamma(q+1)}\left(|s_1^q-s_2^q|+2(s_2-s_1)^q\right)\\
&\leq \frac{3L(1+\eta^*)(s_2-s_1)^q}{\Gamma(q+1)}.
\end{aligned}
$$

As $s_2 \to s_1$, the right-hand side of the above inequality tends to zero, therefore F is equicontinuous on interval $[0, t_1]$.

In general, for the time interval $(t_k, t_{k+1}]$, we similarly obtain the following inequality

$$
|(Fu)(s_2)-(Fu)(s_1)| \leq \frac{3L(1+\eta^*)(s_2-s_1)^q}{\Gamma(q+1)} \to 0, \text{ as } s_2 \to s_1.
$$

This yields that F is equicontinuous on interval $(t_k, t_{k+1}]$.

As a consequence of Claim I-III together with PC-type Arzela-Ascoli theorem (see Lemma 1.3 in the case of $X = \mathbb{R}$), we can conclude that $F : B_{\eta^*} \to B_{\eta^*}$ is continuous and completely continuous.

Claim IV. *A priori* bound.

Now it remains to show that the set

$$
E(F) = \{u \in PC(J,\mathbb{R}) : u = \lambda Fu, \text{ for some } \lambda \in (0,1)\}
$$

is bounded.

Let $u \in E(F)$, then $u = \lambda Fu$ for some $\lambda \in (0, 1)$.

Without loss of generality, for the time interval $t \in (t_k, t_{k+1}]$,

$$
\begin{aligned}
|u(t)| &\leq |u_0| + \sum_{i=1}^{k}|y_i| + \frac{1}{\Gamma(q)}\int_0^t (t-s)^{q-1}|f(s,u(s))|ds\\
&\leq |u_0| + \sum_{i=1}^{k}|y_i| + \frac{LT^q}{\Gamma(q+1)} + \frac{L}{\Gamma(q)}\int_0^t (t-s)^{q-1}|u(s)|ds.
\end{aligned}
$$

By Lemma 5.1, there exists a $M_k^* > 0$ such that

$$
|u(t)| \leq M_k^*, \ t \in (t_k, t_{k+1}].
$$

Set $M^* = \max_{1\leq k\leq m} M_k^*$. Thus for every $t \in J$, we have

$$
\|u\|_{PC} \leq M^*.
$$

This shows that the set $E(F)$ is bounded.

As a consequence of Schaefer fixed point theorem, we deduce that F has a fixed point which is a solution of the problem (5.1). The proof is completed. □

Remark 5.2. Let the assumptions of Theorem 5.4 hold. If f is uniformly Lipschtiz continuous with respect to the second variable, then one can obtain the uniqueness of solutions by virtue of Lemma 5.1 again.

In the following theorem we apply the nonlinear alternative of Leray-Schauder type in which the condition (H2)$'$ is weakened.

(H2)$''$ There exists a constant $q_3 \in (0,q)$ such that real valued function $\phi_f(t) \in L^{\frac{1}{q_3}}(J,\mathbb{R})$ and there exists a L^1-integrable and nondecreasing $\psi : [0,+\infty) \to (0,+\infty)$ such that
$$|f(t,u)| \leq \phi_f(t)\psi(|u|) \text{ for each } t \in J \text{ and all } u \in \mathbb{R};$$

(H4) the following inequality
$$r\left(\chi + \frac{\psi(r)T^{q-q_3}(1-q_3)^{1-q_3}\vartheta}{\Gamma(q)(q-q_3)^{1-q_3}}\right)^{-1} > 1$$
has at least one positive solution, where $\chi = |u_0| + \sum_{i=1}^{m}|y_i|$ and $\vartheta = \|\phi_f\|_{L^{\frac{1}{q_3}}J}$.

Theorem 5.5. *Assume that (H1), (H2)$''$ and (H4) hold. Then the problem* (5.1) *has at least one solution.*

Proof. Consider the operator F defined in Theorem 5.4. It can be easily shown that F is continuous and completely continuous. Repeating the same process in Claim IV in Theorem 5.4, using (H2)$''$ and Hölder inequality again, for each $t \in J$, we have
$$\begin{aligned}
|u(t)| &\leq |(Fu)(t)| \\
&\leq |u_0| + \sum_{i=1}^{m}|y_i| + \frac{1}{\Gamma(q)}\int_0^t (t-s)^{q-1}\phi_f(s)\psi(|u(s)|)ds \\
&\leq |u_0| + \sum_{i=1}^{m}|y_i| + \frac{\psi(\|u\|_{PC})}{\Gamma(q)}\int_0^t (t-s)^{q-1}\phi_f(s)ds \\
&\leq |u_0| + \sum_{i=1}^{m}|y_i| + \frac{\psi(\|u\|_{PC})}{\Gamma(q)}\left(\int_0^t (t-s)^{\frac{q-1}{1-q_3}}ds\right)^{1-q_3}\left(\int_0^t (\phi_f(s))^{\frac{1}{q_3}}ds\right)^{q_3} \\
&\leq \chi + \frac{\psi(\|u\|_{PC})T^{q-q_3}(1-q_3)^{1-q_3}\vartheta}{\Gamma(q)(q-q_3)^{1-q_3}}.
\end{aligned}$$
Thus
$$\frac{r}{|u_0| + \sum_{i=1}^{m}|y_i| + \frac{\psi(r)T^{q-q_3}(1-q_3)^{1-q_3}\vartheta}{\Gamma(q)(q-q_3)^{1-q_3}}} \leq 1.$$
By (H4), there exists a $N^* > 0$ such that $\|u\|_{PC} \neq N^*$.

Let $U = \{u \in PC(J,\mathbb{R}) : \|u\|_{PC} < N^*\}$. The operator $F : \overline{U} \to PC(J,\mathbb{R})$ is continuous and completely continuous. From the choice of U, there is no $u \in \partial U$ such that $u = \lambda^* F(u)$, $\lambda^* \in [0,1]$. As a consequence of the nonlinear alternative of Leray-Schauder type, we deduce that F has a fixed point $u \in \overline{U}$, which implies that the problem (5.1) has at least one solution $u \in PC(J,\mathbb{R})$. This completes the proof. □

5.3 Impulsive Boundary Value Problems

5.3.1 *Introduction*

In the present section, we consider the boundary value problems for the following fractional impulsive differential equations

$$\begin{cases} {}_0^C D_t^q u(t) = f(t, u(t)), & t \in J' := J \setminus \{t_1, ..., t_m\}, \ J := [0,1], \\ \Delta u(t_k) = y_k, \ \Delta u'(t_k) = \bar{y}_k, & k = 1, 2, ..., m, \\ u(0) = 0, \ u'(1) = 0, \end{cases} \tag{5.7}$$

where ${}_0^C D_t^q$ is Caputo fractional derivative of order $q \in (1,2)$ with the lower limit zero, t_k satisfy $0 = t_0 < t_1 < \cdots < t_m < t_{m+1} = 1$, and $y_k, \bar{y}_k \in \mathbb{R}$.

In Subsection 5.3.2, we give a formula of solutions to the problem (5.7). Subsection 5.3.3 is concerned with the existence and uniqueness of solutions for (5.7).

5.3.2 *Formula of Solutions*

In this subsection, we give a formula of solutions to boundary problem for impulsive fractional differential equations

$$\begin{cases} {}_0^C D_t^q u(t) = h(t), & t \in J', \ q \in (1,2), \\ \Delta u(t_k) = y_k, \ \Delta u'(t_k) = \bar{y}_k, & k = 1, 2, ..., m, \\ u(0) = 0, \ u'(1) = 0, \end{cases} \tag{5.8}$$

where $y_k, \bar{y}_k \in \mathbb{R}$.

Lemma 5.3. *Let $q \in (1,2)$ and $h : J \to \mathbb{R}$ be continuous. A function u given by*

$$u(t) = \begin{cases} \dfrac{1}{\Gamma(q)} \displaystyle\int_0^t (t-s)^{q-1} h(s) ds \\ \quad - \left(\dfrac{1}{\Gamma(q-1)} \displaystyle\int_0^1 (1-s)^{q-2} h(s) ds + \sum_{k=1}^m \bar{y}_k \right) t, & \text{for } t \in [0, t_1), \\ \dfrac{1}{\Gamma(q)} \displaystyle\int_0^t (t-s)^{q-1} h(s) ds + \bar{y}_1 (t - t_1) + y_1 \\ \quad - \left(\dfrac{1}{\Gamma(q-1)} \displaystyle\int_0^1 (1-s)^{q-2} h(s) ds + \sum_{k=1}^m \bar{y}_k \right) t, & \text{for } t \in (t_1, t_2), \\ \vdots \\ \dfrac{1}{\Gamma(q)} \displaystyle\int_0^t (t-s)^{q-1} h(s) ds + \sum_{i=1}^k \bar{y}_i (t - t_i) + \sum_{i=1}^k y_i \\ \quad - \left(\dfrac{1}{\Gamma(q-1)} \displaystyle\int_0^1 (1-s)^{q-2} h(s) ds + \sum_{k=1}^m \bar{y}_k \right) t, & \text{for } t \in (t_k, t_{k+1}], \\ & k = 1, 2, ..., m, \end{cases} \tag{5.9}$$

is a unique solution of the impulsive problem (5.8).

Proof. A general solution u of the first equation of (5.8) on each interval (t_k, t_{k+1}) $(k = 0, 1, ..., m)$ is given by

$$u(t) = \frac{1}{\Gamma(q)} \int_0^t (t-s)^{q-1} h(s) ds + a_k + b_k t, \quad \text{for } t \in (t_k, t_{k+1}), \tag{5.10}$$

where $t_0 = 0$ and $t_{m+1} = 1$.

Then, we have

$$u'(t) = \frac{1}{\Gamma(q-1)} \int_0^t (t-s)^{q-2} h(s) ds + b_k, \quad \text{for } t \in (t_k, t_{k+1}). \tag{5.11}$$

Applying the boundary conditions of (5.8), we find that

$$a_0 = 0, \quad b_m = -\frac{1}{\Gamma(q-1)} \int_0^1 (1-s)^{q-2} h(s) ds. \tag{5.12}$$

Next, using the right impulsive condition of (5.8), we derive

$$b_k = b_{k-1} + \bar{y}_k,$$

which by (5.12) implies

$$b_j = -\frac{1}{\Gamma(q-1)} \int_0^1 (1-s)^{q-2} h(s) ds - \sum_{k=j+1}^{m} \bar{y}_k, \quad j = 0, 1, 2, ..., m-1. \tag{5.13}$$

Furthermore, using the left impulsive condition of (5.8), we derive

$$a_k + b_k t_k = a_{k-1} + b_{k-1} t_k + y_k,$$

which is equivalent to

$$a_k = a_{k-1} + (b_{k-1} - b_k) t_k + y_k = a_{k-1} + y_k - \bar{y}_k t_k,$$

so by (5.12) we obtain

$$a_j = \sum_{k=1}^{j} (y_k - \bar{y}_k t_k), \quad j = 1, 2, ..., m. \tag{5.14}$$

Hence for $j = 1, 2, ..., m$, (5.13) and (5.14) imply

$$\begin{aligned} a_j + b_j t &= \sum_{k=1}^{j} (y_k - \bar{y}_k t_k) + \left(-\frac{1}{\Gamma(q-1)} \int_0^1 (1-s)^{q-2} h(s) ds - \sum_{k=j+1}^{m} \bar{y}_k \right) t \\ &= \sum_{k=1}^{j} \bar{y}_k (t - t_k) + \sum_{k=1}^{j} y_k - \left(\frac{1}{\Gamma(q-1)} \int_0^1 (1-s)^{q-2} h(s) ds + \sum_{k=1}^{m} \bar{y}_k \right) t. \end{aligned} \tag{5.15}$$

Now it is clear that (5.10), (5.12) and (5.15) imply (5.9).

Conversely, assume that u satisfies (5.9). By a direct computation, it follows that the solution given by (5.9) satisfies (5.8). This completes the proof. □

Motivated by the above results, we give the following concept of the solution for the problem (5.7).

Definition 5.2. A function $u \in PC^1(J, \mathbb{R})$ is said to be a solution of the problem (5.7) if u satisfies the integral equation

$$u(t) = \begin{cases} \frac{1}{\Gamma(q)} \int_0^t (t-s)^{q-1} h(s) ds \\ \quad - \left(\frac{1}{\Gamma(q-1)} \int_0^1 (1-s)^{q-2} f(s, u(s)) ds + \sum_{k=1}^{m} \bar{y}_k \right) t, & \text{for } t \in [0, t_1), \\ \frac{1}{\Gamma(q)} \int_0^t (t-s)^{q-1} h(s) ds + \bar{y}_1 (t - t_1) + y_1 \\ \quad - \left(\frac{1}{\Gamma(q-1)} \int_0^1 (1-s)^{q-2} h(s) ds + \sum_{k=1}^{m} \bar{y}_k \right) t, & \text{for } t \in (t_1, t_2), \\ \vdots \\ \frac{1}{\Gamma(q)} \int_0^t (t-s)^{q-1} f(s, u(s)) ds + \sum_{i=1}^{k} \bar{y}_i (t - t_i) + \sum_{i=1}^{k} y_i \\ \quad - \left(\frac{1}{\Gamma(q-1)} \int_0^1 (1-s)^{q-2} f(s, u(s)) ds + \sum_{k=1}^{m} \bar{y}_k \right) t, & \text{for } t \in (t_k, t_{k+1}], \\ & k = 1, 2, ..., m. \end{cases}$$

Moreover, we need the following known results.

Lemma 5.4. *(Wang, Xiang and Peng, 2009) Let $u \in C(J, \mathbb{R})$ satisfy the following inequality:*

$$|u(t)| \leq a + b \int_0^t |u(\theta)|^{\lambda_1} d\theta + c \int_0^1 |u(\theta)|^{\lambda_2} d\theta, \ t \in J,$$

where $\lambda_1 \in [0, 1], \lambda_2 \in [0, 1), a, b, c \geq 0$ are constants. Then there exists a constant $M^* > 0$ such that

$$|u(t)| \leq M^*.$$

Remark 5.3. For $\lambda_1 < 1$ we can take M^* to be the unique positive solution of $M^* = a + bM^{*\lambda_1} + cM^{*\lambda_2}$. Using the classical Gronwall inequality, for $\lambda_1 = 1$ we can take M^* to be the unique positive solution of $M^* = \left(a + cM^{*\lambda_2}\right) e^b$.

Using Lemma 5.4, we can obtain the following generalized Gronwall inequality with mixed integral term.

Lemma 5.5. *Let $u \in C(J, \mathbb{R})$ satisfy the following inequality:*

$$|u(t)| \leq a + b \int_0^t (t-s)^{q-1} |u(s)|^{\lambda_1} ds + c \int_0^1 (1-s)^{q-2} |u(s)|^{\lambda_2} ds, \tag{5.16}$$

where $q \in (1,2)$, $a, b, c \geq 0$ *are constants,* $\lambda_1 \in [0, 1 - \frac{1}{p}]$, $\lambda_2 \in [0, 1 - \frac{1}{p})$, *and for some* $p > 1$ *such that* $p(q-2) + 1 > 0$. *Then there exists a constant* $M_* > 0$ *such that*

$$|u(t)| \leq M_*.$$

Proof. It follows from (5.16) and Hölder inequality that

$$\begin{aligned}|u(t)| &\leq a + b\left(\int_0^t (t-s)^{p(q-1)}ds\right)^{\frac{1}{p}}\left(\int_0^t |u(s)|^{\frac{\lambda_1 p}{p-1}}ds\right)^{\frac{p-1}{p}} \\ &\quad + c\left(\int_0^1 (1-s)^{p(q-2)}ds\right)^{\frac{1}{p}}\left(\int_0^1 |u(s)|^{\frac{\lambda_2 p}{p-1}}ds\right)^{\frac{p-1}{p}} \\ &\leq a + b\left(\frac{1}{p(q-1)+1}\right)^{\frac{1}{p}}\int_0^t |u(s)|^{\frac{\lambda_1 p}{p-1}}ds + c\left(\frac{1}{p(q-2)+1}\right)^{\frac{1}{p}}\int_0^1 |u(s)|^{\frac{\lambda_2 p}{p-1}}ds \\ &\leq a + b\int_0^t |u(s)|^{\frac{\lambda_1 p}{p-1}}ds + c\left(\frac{1}{p(q-2)+1}\right)^{\frac{1}{p}}\int_0^1 |u(s)|^{\frac{\lambda_2 p}{p-1}}ds.\end{aligned}$$

Applying Lemma 5.4, there exists a constant $M_* > 0$ such that

$$|u(t)| \leq M_*.$$

The proof is completed. □

Remark 5.4. Constant M_* can be determined by using Remark 5.3.

5.3.3 *Existence*

This subsection deals with the existence and uniqueness of solutions for the problem (5.7).

Theorem 5.6. *Let* $f : J \times \mathbb{R} \to \mathbb{R}$ *be a continuous function mapping. Assume that there exists a positive constant* L *such that*

(A1) $|f(t,u) - f(t,v)| \leq L|u - v|$, *for all* $t \in J$, $u, v \in \mathbb{R}$,

with $L \leq \frac{\Gamma(1+q)}{2(1+q)}$. *Then the problem* (5.7) *has a unique solution on* J.

Proof. Setting $\sup_{t\in J} |f(t,0)| = M$ and

$$B_r = \left\{u \in PC^1(J, \mathbb{R}) : \|u\|_{PC^1} \leq r\right\},$$

where

$$r \geq 2\left(\frac{1+q}{\Gamma(1+q)}M + \sum_{i=1}^m |\bar{y}_i| + 2\sum_{i=1}^m |y_i|\right).$$

Define an operator $F : B_r \to PC^1(J, \mathbb{R})$ by

$$(Fu)(t) = \frac{1}{\Gamma(q)}\int_0^t (t-s)^{q-1} f(s, u(s))ds + \sum_{i=1}^k \bar{y}_i(t - t_i) + \sum_{i=1}^k y_i$$

$$-\left(\frac{1}{\Gamma(q-1)}\int_0^1(1-s)^{q-2}f(s,u(s))ds+\sum_{i=1}^{m}\bar{y}_i\right)t,$$
$$t\in(t_k,t_{k+1}],\ k=0,1,2,...,m.$$

It is obvious that F is well defined due to $f : J\times\mathbb{R}\to\mathbb{R}$ is jointly continuous and maps bounded subsets of $J\times\mathbb{R}$ to bounded subsets of $\mathbb{R}$.

Claim I. $FB_r\subset B_r$.

For $u\in B_r$, $t\in J'$, we have

$$\begin{aligned}
&|(Fu)(t)|\\
&=\left|\frac{1}{\Gamma(q)}\int_0^t(t-s)^{q-1}f(s,u(s))ds+\sum_{i=1}^{k}\bar{y}_i(t-t_i)+\sum_{i=1}^{k}y_i\right.\\
&\quad\left.-\left(\frac{1}{\Gamma(q-1)}\int_0^1(1-s)^{q-2}f(s,u(s))ds+\sum_{i=1}^{m}\bar{y}_i\right)t\right|\\
&\le\left|\frac{1}{\Gamma(q)}\int_0^t(t-s)^{q-1}f(s,u(s))ds\right|+\left|\frac{1}{\Gamma(q-1)}\int_0^1(1-s)^{q-2}f(s,u(s))ds\right|\\
&\quad+\sum_{i=1}^{m}|\bar{y}_i|+2\sum_{i=1}^{m}|y_i|\\
&\le\frac{1}{\Gamma(q)}\int_0^t(t-s)^{q-1}|f(s,u(s))-f(s,0)|ds+\frac{1}{\Gamma(q)}\int_0^t(t-s)^{q-1}|f(s,0)|ds\\
&\quad+\frac{1}{\Gamma(q-1)}\int_0^1(1-s)^{q-2}|f(s,u(s))-f(s,0)|ds\\
&\quad+\frac{1}{\Gamma(q-1)}\int_0^1(1-s)^{q-2}|f(s,0)|ds+\sum_{i=1}^{m}|\bar{y}_i|+2\sum_{i=1}^{m}|y_i|\\
&\le L\frac{1+q}{\Gamma(1+q)}r+M\frac{1+q}{\Gamma(1+q)}+\sum_{i=1}^{m}|\bar{y}_i|+2\sum_{i=1}^{m}|y_i|\\
&\le r.
\end{aligned}$$

Claim II. F is a contraction mapping.

For $u,v\in B_r$ and for each $t\in J'$, we obtain

$$\begin{aligned}
&|(Fu)(t)-(Fv)(t)|\\
&=\left|\frac{1}{\Gamma(q)}\int_0^t(t-s)^{q-1}f(s,u(s))ds-\frac{t}{\Gamma(q-1)}\int_0^1(1-s)^{q-2}f(s,u(s))ds\right.\\
&\quad\left.-\left(\frac{1}{\Gamma(q)}\int_0^t(t-s)^{q-1}f(s,v(s))ds-\frac{t}{\Gamma(q-1)}\int_0^1(1-s)^{q-2}f(s,v(s))ds\right)\right|\\
&\le\frac{1}{\Gamma(q)}\int_0^t(t-s)^{q-1}|f(s,u(s))-f(s,v(s))|ds
\end{aligned}$$

$$
\begin{aligned}
&+\frac{1}{\Gamma(q-1)}\int_0^1(1-s)^{q-2}|f(s,u(s))-f(s,v(s))|ds\\
&\le\left(\frac{L}{\Gamma(q)}\int_0^t(t-s)^{q-1}ds\right)\|u-v\|_{PC^1}+\left(\frac{L}{\Gamma(q-1)}\int_0^1(1-s)^{q-2}ds\right)\|u-v\|_{PC^1}\\
&\le L\frac{1+q}{\Gamma(1+q)}\|u-v\|_{PC^1}\\
&\le\frac{1}{2}\|u-v\|_{PC^1},
\end{aligned}
$$

which implies that

$$\|Fu-Fv\|_{PC^1}\le\frac{1}{2}\|u-v\|_{PC^1}.$$

Therefore F is a contraction.

Thus, the conclusion of theorem follows by Banach contraction mapping principle. The proof is completed. □

Now we are ready to state and prove the following existence result.

Theorem 5.7. *Let $f: J\times\mathbb{R}\to\mathbb{R}$ be a continuous function mapping with $|f(t,u)|\le\mu(t)$, for all $(t,u)\in J\times\mathbb{R}$ where $\mu\in L^{\frac{1}{\sigma}}(J,\mathbb{R})$ and $\sigma\in(0,q-1)$. Then the problem (5.7) has at least one solution on J.*

Proof. Let's choose

$$r\ge\frac{\|\mu\|_{L^{\frac{1}{\sigma}}J}}{\Gamma(q)(\frac{q-\sigma}{1-\sigma})^{1-\sigma}}+\frac{\|\mu\|_{L^{\frac{1}{\sigma}}J}}{\Gamma(q-1)(\frac{q-\sigma-1}{1-\sigma})^{1-\sigma}}+2\sum_{i=1}^m|\bar{y}_i|+\sum_{i=1}^m|y_i|,$$

and denote

$$B_r=\{u\in PC^1(J,\mathbb{R}):\|u\|_{PC^1}\le r\}.$$

We define the operators P and Q on B_r as

$$
\begin{aligned}
(Pu)(t)&=\frac{1}{\Gamma(q)}\int_0^t(t-s)^{q-1}f(s,u(s))ds-\left(\frac{1}{\Gamma(q-1)}\int_0^1(1-s)^{q-2}f(s,u(s))ds\right)t,\\
(Qu)(t)&=\sum_{i=1}^k\bar{y}_i(t-t_i)+\sum_{i=1}^k y_i-\sum_{i=1}^m\bar{y}_i t.
\end{aligned}
$$

For any $u,v\in B_r$ and $t\in J$, using the estimation condition on f and Hölder inequality,

$$
\begin{aligned}
\int_0^t\left|(t-s)^{q-1}f(s,u(s))\right|ds&\le\left(\int_0^t(t-s)^{\frac{q-1}{1-\sigma}}ds\right)^{1-\sigma}\left(\int_0^t(\mu(s))^{\frac{1}{\sigma}}ds\right)^{\sigma}\\
&\le\frac{\|\mu\|_{L^{\frac{1}{\sigma}}J}}{(\frac{q-\sigma}{1-\sigma})^{1-\sigma}},\\
\int_0^1\left|(1-s)^{q-2}f(s,u(s))\right|ds&\le\left(\int_0^1(1-s)^{\frac{q-2}{1-\sigma}}ds\right)^{1-\sigma}\left(\int_0^1(\mu(s))^{\frac{1}{\sigma}}ds\right)^{\sigma}
\end{aligned}
$$

$$\leq \frac{\|\mu\|_{L^{\frac{1}{\sigma}}J}}{(\frac{q-\sigma-1}{1-\sigma})^{1-\sigma}}.$$

Therefore,

$$\|Pu+Qv\|_{PC^1} \leq \frac{\|\mu\|_{L^{\frac{1}{\sigma}}J}}{\Gamma(q)(\frac{q-\sigma}{1-\sigma})^{1-\sigma}} + \frac{\|\mu\|_{L^{\frac{1}{\sigma}}J}}{\Gamma(q-1)(\frac{q-\sigma-1}{1-\sigma})^{1-\sigma}} + 2\sum_{i=1}^{m}|\bar{y}_i| + \sum_{i=1}^{m}|y_i| \leq r.$$

Thus $Pu+Qv \in B_r$. It is obvious that Q is a contraction with the constant zero. On the other hand, the continuity of f implies that the operator P is continuous. Also, P is uniformly bounded on B_r since

$$\|Pu\|_{PC^1} \leq \frac{\|\mu\|_{L^{\frac{1}{\sigma}}J}}{\Gamma(q)(\frac{q-\sigma}{1-\sigma})^{1-\sigma}} + \frac{\|\mu\|_{L^{\frac{1}{\sigma}}J}}{\Gamma(q-1)(\frac{q-\sigma-1}{1-\sigma})^{1-\sigma}} \leq r.$$

Now we need to prove the compactness of the operator P.

Letting $\Omega = J \times B_r$, we can define $\sup_{(t,x)\in\Omega}|f(t,u)| = f_{\max}$, and consequently for any $t_k < \tau_2 < \tau_1 \leq t_{k+1}$ we have

$$\begin{aligned}
&|(Pu)(\tau_2)-(Pu)(\tau_1)| \\
&= \left|\frac{1}{\Gamma(q)}\int_0^{\tau_2}(\tau_2-s)^{q-1}f(s,u(s))ds - \frac{\tau_2}{\Gamma(q-1)}\int_0^1(1-s)^{q-2}f(s,u(s))ds\right. \\
&\quad \left. -\left(\frac{1}{\Gamma(q)}\int_0^{\tau_1}(\tau_1-s)^{q-1}f(s,u(s))ds - \frac{\tau_1}{\Gamma(q-1)}\int_0^1(1-s)^{q-2}f(s,u(s))ds\right)\right| \\
&\leq \left|\frac{1}{\Gamma(q)}\int_0^{\tau_2}\left((\tau_2-s)^{q-1}-(\tau_1-s)^{q-1}\right)f(s,u(s))ds\right. \\
&\quad \left. +\frac{1}{\Gamma(q)}\int_{\tau_2}^{\tau_1}(\tau_1-s)^{q-1}f(s,u(s))ds\right| + \left|\frac{\tau_2-\tau_1}{\Gamma(q-1)}\int_0^1(1-s)^{q-2}f(s,u(s))ds\right| \\
&\leq f_{\max}\left(\frac{2(\tau_1-\tau_2)^q+\tau_1^q-\tau_2^q}{\Gamma(1+q)} + \frac{\tau_1-\tau_2}{\Gamma(q)}\right),
\end{aligned}$$

which tends to zero as $\tau_2 \to \tau_1$. This yields that P is equicontinuous on interval $(t_k, t_{k+1}]$. So P is relatively compact on B_r.

Hence, by PC-type Arzela-Ascoli theorem (see Lemma 1.3 in the case of $X=\mathbb{R}$), P is compact on B_r. Thus all the assumptions of Theorem 1.7 are satisfied and the conclusion of Theorem 1.7 implies that the problem (5.7) has at least one solution on J. The proof is completed. □

In order to obtain the data dependence of solutions, we revise (A1) to the assumption:

(A2) there exist $L>0$ and $\lambda \in (0, 1-\frac{1}{p})$ where $p(q-2)+1>0$ with $p>1$ such that

$$|f(t,u)-f(t,v)| \leq L|u-v|^{\lambda}, \text{ for each } t \in J, \text{ and all } u,v \in \mathbb{R}.$$

Further, we give the following data dependence result.

Theorem 5.8. *Assume that the conditions of Theorem 5.7 and the additional condition (A2) hold. Let $v(\cdot)$ be another solution of the problem (5.7) with impulsive conditions $\Delta v(t_k) = y_k$, $\Delta v'(t_k) = \bar{y}_k$, $k = 1, 2, ..., m$, and boundary value conditions $v(0) = 0$, $v'(1) = 0$. Then there exists a constant $M_* > 0$ such that $\|u - v\|_{PC^1} \leq M_*$.*

Proof. By Theorem 5.7, the problem (5.7) has a solution $u(\cdot)$ in $PC^1(J, X)$. Keeping in mind our conditions, $v(\cdot)$ be another solution of the problem (5.7) with impulsive conditions $\Delta v(t_k) = y_k$, $\Delta v'(t_k) = \bar{y}_k$, $k = 1, 2, ..., m$, and boundary value conditions $v(0) = 0$, $v'(1) = 0$. Note the condition (A2), we obtain

$$|u(t) - v(t)| \leq \frac{L}{\Gamma(q)} \int_0^t (t-s)^{q-1} |u(s) - v(s)|^\lambda ds + \frac{L}{\Gamma(q-1)} \int_0^1 (1-s)^{q-2} |u(s) - v(s)|^\lambda ds.$$

By Lemma 5.5, we obtain $\|u - v\|_{PC^1} \leq M_*$. This completes the proof. □

Remark 5.5. Under the assumptions of Theorem 5.8, we do not obtain the uniqueness of the solutions.

Remark 5.6. By Remark 5.4 we see that M_* is the unique positive solution of

$$M_* = \frac{L}{\Gamma(q)} M_*^{\frac{\lambda p}{p-1}} + \frac{L}{\Gamma(q-1)} \left(\frac{1}{p(q-2)+1} \right)^{\frac{1}{p}} M_*^{\frac{\lambda p}{p-1}},$$

so

$$M_* = \left(\frac{L}{\Gamma(q)} + \frac{L}{\Gamma(q-1)} \left(\frac{1}{p(q-2)+1} \right)^{\frac{1}{p}} \right)^{\frac{1}{1-\frac{\lambda p}{p-1}}}.$$

5.4 Impulsive Langevin Equations

5.4.1 *Introduction*

In 1908, Langevin introduced a concept of a equation of motion of a Brownian particle which is named after Langevin equation, thereafter, Langevin is regarded as one of the founder of the subject of stochastic differential equations. Langevin equations have been widely used to describe stochastic problems in physics, chemistry and electrical engineering. For example, Brownian motion is well described by the Langevin equation when the random fluctuation force is assumed to be white noise. For systems in complex media, standard Langevin equation does not provide the correct description of the dynamics. As a results, various generalizations of Langevin equations have been offered to describe dynamical processes in a fractal medium. One such generalization is the generalized Langevin equation which incorporates the fractal and memory properties with a dissipative memory kernel into the Langevin equation. This gives rise to study Langevin equation with fractional derivatives, see for instance Mainardi and Pironi, 1996; Lutz, 2001; Fa, 2006, 2007; Kobolev and Romanov, 2000; Picozzi and West, 2002; Bazzani, Bassi and Turchetti, 2003; Lim, Li and Teo, 2008; Ahmad and Eloe, 2010; Ahmad, Nieto, Alsaedi *et al.*, 2012; Sandev, Tomovski and Dubbeldam, 2011; Sandev, Metzler and Tomovski, 2012, and the references therein.

It is remarkable that many evolution processes are characterized by the fact that at certain moments of time they experience a change of state abruptly. These processes are subject to short term perturbations whose duration is negligible in comparison with the duration of the processes. Consequently, it is natural to assume that these perturbations act instantaneously, that is in the form of impulses. In particular, when we want to describe fractional Langevin equations subject to abrupt changes as well as other phenomena such as earthquake, it is nature to use fractional impulsive Langevin equations to describe such problems. Thus, we offer to study the following fractional impulsive Langevin equations

$$\begin{cases} {}_0^CD_t^\beta({}_0^CD_t^\alpha+\lambda)x(t)=f(t,x(t)), & t\in J':=J\setminus\{t_1,...,t_m\}, J:=[0,1], \\ \Delta x(t_k):=x(t_k^+)-x(t_k^-)=I_k, & I_k\in\mathbb{R}, \\ x(0)=0,\ x(\eta_k)=0,\ x(1)=0, & \eta_k\in(t_k,t_{k+1}),\ k=0,1,...,m-1, \end{cases} \tag{5.17}$$

where $f: J\times\mathbb{R}\to\mathbb{R}$ is a given function, $0<\alpha,\beta<1$ with $0<\alpha+\beta<1$, $\lambda>0$, $0=t_0<t_1<t_2<\cdots<t_m<t_{m+1}=1$, $x(t_k^+)=\lim_{\epsilon\to0^+}x(t_k+\epsilon)$ and $x(t_k^-)=\lim_{\epsilon\to0^-}x(t_k+\epsilon)$ represent the right and left limits of $x(t)$ at $t=t_k$, the constants I_k denotes the size of the jump.

Moreover, we also study the following nonlinear impulsive problems:

$$\begin{cases} {}_0^CD_t^\beta({}_0^CD_t^\alpha+\lambda)x(t)=f(t,x(t)), & t\in J', \\ \Delta x(t_k):=x(t_k^+)-x(t_k^-)=I_k(x(t_k^-)), & \\ x(0)=0,\ x(\eta_k)=0,\ x(1)=0, & \eta_k\in(t_k,t_{k+1}),\ k=0,1,2,...,m-1, \end{cases} \tag{5.18}$$

where the nonlinear impulsive terms $I_k:\mathbb{R}\to\mathbb{R}$ are specified latter.

Subsection 5.4.2 is devoted to giving the formula of solutions for the linear Langevin equations and some basic properties of classical and generalized Mittag-Leffler functions, then proceed to obtain the general solutions of the linear fractional impulsive Langevin equations. In Subsection 5.4.3, we deal with the existence and uniqueness of solution for the problem (5.17), and extend the existence results to problem (5.18).

5.4.2 *Formula of Solutions*

Firstly, we study the following linear Langevin equations with two different fractional derivatives

$${}_0^CD_t^\beta({}_0^CD_t^\alpha+\lambda)x(t)=f(t),\ t\in J_i,\ i=0,1,2,...,m, \tag{5.19}$$

where $J_0:=[0,t_1]$, $J_i:=(t_i,t_{i+1}]$, $i=1,2,...,m-1$, $J_m:=(t_m,1]$.

Lemma 5.6. *For $q > 0$, the general solution of fractional differential equation ${}_0^C D_t^q u(t) = 0$ is given by*

$$u(t) = c_0 + c_1 t + c_2 t^2 + \cdots + c_{n-1} t^{n-1}$$

where $c_i \in \mathbb{R}, i = 0, 1, 2, \ldots, n-1$ $(n = [q]+1)$ and $[q]$ denotes the integer part of the real number q.

Remark 5.7. In view of Lemma 5.6, it follows that

$${}_0 D_t^{-q} ({}_0^C D_t^q u)(t) = u(t) + c_0 + c_1 t + c_2 t^2 + \cdots + c_{n-1} t^{n-1}$$

where $c_i \in \mathbb{R}, i = 0, 1, 2, \ldots, n-1$, $n = [q]+1$.

It is remarkable that Ahmad, Nieto and Alsaedi *et al.*, 2012 studied existence of solutions of linear Langevin equations with two different fractional derivatives by finding a fixed point of a suitable fractional integral equation (see Lemma 2.1 in Ahmad, Nieto and Alsaedi *et al.*), however, the general presentation of solutions for such equations have not been deduced. Here, we try to find a general solution the equation (5.19).

Lemma 5.7. *A function $u \in C(J_i, \mathbb{R})$, $i = 0, 1, 2, ..., m$, is a solution of the equation (5.19) if and only if u is a solution of the integral equation*

$$\begin{aligned} x(t) &= E_\alpha(-t^\alpha \lambda) b_i - \frac{1}{\lambda} \left(1 - E_\alpha(-t^\alpha \lambda)\right) a_i \\ &\quad + \int_0^t (t-z)^{\alpha+\beta-1} E_{\alpha,\alpha+\beta}(-(t-z)^\alpha \lambda) f(z) dz. \end{aligned} \tag{5.20}$$

Proof. In view of Remark 5.7 and by integrating the equation (5.19) from zero to t we have

$$({}_0^C D_t^\alpha + \lambda) x(t) = \int_0^t \frac{(t-s)^{\beta-1}}{\Gamma(\beta)} f(s) ds - a_i, \ i = 0, 1, ..., m,$$

where a_i are constants.

By adopting the same idea and techniques in Zhou and Jiao, 2010, the general solution of

$$({}_0^C D_t^\alpha + \lambda) x(t) = h(t) \tag{5.21}$$

is

$$x(t) = \mathscr{T}(t) b_i + \int_0^t (t-s)^{\alpha-1} \mathscr{S}(t-s) h(s) ds, \tag{5.22}$$

where

$$\mathscr{T}(t) := \int_0^\infty M_\alpha(\theta) e^{-t^\alpha \theta \lambda} d\theta, \quad \mathscr{S}(t) := \alpha \int_0^\infty \theta M_\alpha(\theta) e^{-t^\alpha \theta \lambda} d\theta.$$

Here M_α is the Wright function (see Definition 1.9). Meanwhile, the solution of the equation (5.21) have been considered in the monograph Kilbas, Srivastava and Trujillo, 2006, (see pp. 140-141, (3.1.32)-(3.1.34)), and it is given by the following expression:

$$x(t) = E_\alpha(-t^\alpha \lambda) b_i + \int_0^t (t-s)^{\alpha-1} e_\alpha(-(t-s)^\alpha \lambda) h(s) ds. \tag{5.23}$$

Combined (5.22) and (5.23), we can rewrite $\mathscr{T}(t) = E_\alpha(-t^\alpha \lambda)$, $\mathscr{S}(t) = e_\alpha(-t^\alpha \lambda)$. Note $e_\alpha(z) = \alpha E'_\alpha(z)$ and so

$$(t-s)^{\alpha-1} e_\alpha(-(t-s)^\alpha \lambda) = \frac{d}{ds}\left(\frac{1}{\lambda} E_\alpha(-(t-s)^\alpha \lambda)\right).$$

This yields that

$$\int_0^t (t-s)^{\alpha-1} e_\alpha(-(t-s)^\alpha \lambda) ds = \frac{1}{\lambda}\left(1 - E_\alpha(-t^\alpha \lambda)\right).$$

So the final formula of solution of the equation (5.19) should be

$$\begin{aligned} x(t) &= E_\alpha(-t^\alpha \lambda) b_i + \int_0^t (t-s)^{\alpha-1} e_\alpha(-(t-s)^\alpha \lambda) \\ &\quad \times \left(\int_0^s \frac{(s-z)^{\beta-1}}{\Gamma(\beta)} f(z) dz - a_i\right) ds \\ &= E_\alpha(-t^\alpha \lambda) b_i - \left(\int_0^t (t-s)^{\alpha-1} e_\alpha(-(t-s)^\alpha \lambda) ds\right) a_i \\ &\quad + \frac{1}{\Gamma(\beta)} \int_0^t \int_0^s (s-z)^{\beta-1} (t-s)^{\alpha-1} e_\alpha(-(t-s)^\alpha \lambda) f(z) dz ds \\ &= E_\alpha(-t^\alpha \lambda) b_i - \frac{1}{\lambda}\left(1 - E_\alpha(-t^\alpha \lambda)\right) a_i \\ &\quad + \frac{1}{\Gamma(\beta)} \int_0^t f(z) \int_z^t (t-s)^{\alpha-1} (s-z)^{\beta-1} e_\alpha(-(t-s)^\alpha \lambda) ds dz \\ &= E_\alpha(-t^\alpha \lambda) b_i - \frac{1}{\lambda}\left(1 - E_\alpha(-t^\alpha \lambda)\right) a_i \\ &\quad + \int_0^t (t-z)^{\alpha+\beta-1} E_{\alpha,\alpha+\beta}(-(t-z)^\alpha \lambda) f(z) dz, \end{aligned}$$

where we use the fact in Theorem 4 of Prabhakar, 1971. Hence

$$\begin{aligned} x(t) = E_\alpha(-t^\alpha \lambda) b_i &- \frac{1}{\lambda}\left(1 - E_\alpha(-t^\alpha \lambda)\right) a_i \\ &+ \int_0^t (t-z)^{\alpha+\beta-1} E_{\alpha,\alpha+\beta}(-(t-z)^\alpha \lambda) f(z) dz, \end{aligned}$$

where $E_{\alpha,\alpha+\beta}$ is the generalized Mittag-Leffler function:

$$E_{\alpha,\alpha+\beta}(z) = \sum_{k=0}^{\infty} \frac{z^k}{\Gamma(k\alpha+\alpha+\beta)} = \frac{1}{\Gamma(\beta)} \sum_{k=0}^{\infty} \frac{z^k}{\Gamma(k\alpha+\alpha)} B(k\alpha+\alpha, \beta)$$

$$
\begin{aligned}
&= \frac{1}{\Gamma(\beta)} \sum_{k=0}^{\infty} \frac{z^k}{\Gamma(k\alpha+\alpha)} \int_0^1 u^{k\alpha+\alpha-1}(1-u)^{\beta-1}du \\
&= \frac{1}{\Gamma(\beta)} \int_0^1 \sum_{k=0}^{\infty} \frac{(uz)^k}{\Gamma(k\alpha+\alpha)} u^{\alpha-1}(1-u)^{\beta-1}du \\
&= \frac{1}{\Gamma(\beta)} \int_0^1 u^{\alpha-1}(1-u)^{\beta-1} e_\alpha(uz)du,
\end{aligned}
$$

with B denoting Beta function, which is coincided with

$$
\begin{aligned}
E_{\alpha,\alpha+\beta}(-z) &= \frac{1}{\Gamma(\beta)} \int_0^1 u^{\alpha-1}(1-u)^{\beta-1} e_\alpha(-uz)du \\
&= \frac{\alpha}{\Gamma(\beta)} \int_0^1 u^{\alpha-1}(1-u)^{\beta-1} \int_0^\infty \theta M_\alpha(\theta) e^{-uz\theta} d\theta du \\
&= \frac{\alpha}{\Gamma(\beta)} \int_0^\infty \theta M_\alpha(\theta) \left(\int_0^1 u^{\alpha-1}(1-u)^{\beta-1} e^{-uz\theta} du \right) d\theta \\
&= \frac{\Gamma(\alpha+1)}{\Gamma(\alpha+\beta)} \int_0^\infty \theta M_\alpha(\theta) {}_1F_1(\alpha, \alpha+\beta, -z\theta) d\theta,
\end{aligned}
\tag{5.24}
$$

where

$$
{}_1F_1(\alpha, \alpha+\beta, z) = \frac{1}{B(\alpha,\beta)} \int_0^1 u^{\alpha-1}(1-u)^{\beta-1} e^{uz} du
$$

is the hypergeometric function (see, Seaborn, 1991). □

It is well known that classical and generalized Mittag-Leffler functions have played important role in the study of fractional ordinary differential equation with constant coefficients Bonilla, 2007, and fractional diffusion equations Atkinson and Osseiran, 2011. However, it seems that the known properties of these special functions are not explicit and complete. For example, the literature usually address that the classical and generalized Mittag-Leffler functions are boundedness, but not give a explicit boundedness. Meanwhile, other important properties such as continuity, monotonicity, nonnegative and etc seems have not been systematically reported. Here, we try to revisit some basic properties of classical and generalized Mittag-Leffler functions by using one-side probability density function.

Lemma 5.8. *Let $0 < \alpha$, $\beta < 1$. The functions E_α, e_α and $E_{\alpha,\alpha+\beta}$ are nonnegative and have the following properties:*

(i) *For any $\lambda > 0$ and $t \in J$,*

$$
E_\alpha(-t^\alpha\lambda) \le 1, \; e_\alpha(-t^\alpha\lambda) \le \frac{1}{\Gamma(\alpha)}, \; E_{\alpha,\alpha+\beta}(-t^\alpha\lambda) \le \frac{1}{\Gamma(\alpha+\beta)}.
$$

Moreover, $E_\alpha(0) = 1$, $e_\alpha(0) = \frac{1}{\Gamma(\alpha)}$, $E_{\alpha,\alpha+\beta}(0) = \frac{1}{\Gamma(\alpha+\beta)}$.

(ii) *For any* $\lambda > 0$ *and* $t_1, t_2 \in J$,

$$\begin{aligned} E_\alpha(-t_2^\alpha\lambda) &\to E_\alpha(-t_1^\alpha\lambda) \ \textit{as}\ t_2 \to t_1, \\ e_\alpha(-t_2^\alpha\lambda) &\to e_\alpha(-t_1^\alpha\lambda) \ \textit{as}\ t_2 \to t_1, \\ E_{\alpha,\alpha+\beta}(-t_2^\alpha\lambda) &\to E_{\alpha,\alpha+\beta}(-t_1^\alpha\lambda) \ \textit{as}\ t_2 \to t_1. \end{aligned} \tag{5.25}$$

Or rather,

$$\begin{aligned} |E_\alpha(-t_2^\alpha\lambda) - E_\alpha(-t_1^\alpha\lambda)| &:= O(|t_2 - t_1|^\alpha) \ \textit{as}\ t_2 \to t_1, \\ |e_\alpha(-t_2^\alpha\lambda) - e_\alpha(-t_1^\alpha\lambda)| &:= O(|t_2 - t_1|^\alpha) \ \textit{as}\ t_2 \to t_1, \\ |E_{\alpha,\alpha+\beta}(-t_2^\alpha\lambda) - E_{\alpha,\alpha+\beta}(-t_1^\alpha\lambda)| &:= O(|t_2 - t_1|^\alpha) \ \textit{as}\ t_2 \to t_1. \end{aligned}$$

(iii) *For any* $\lambda > 0$, $t_1, t_2 \in J$ *and* $t_1 \le t_2$,

$$\begin{aligned} E_\alpha(-t_2^\alpha\lambda) &\le E_\alpha(-t_1^\alpha\lambda), \\ e_\alpha(-t_2^\alpha\lambda) &\le e_\alpha(-t_1^\alpha\lambda), \\ E_{\alpha,\alpha+\beta}(-t_2^\alpha\lambda) &\le E_{\alpha,\alpha+\beta}(-t_1^\alpha\lambda). \end{aligned} \tag{5.26}$$

(iv) *For any* $\lambda > 0$ *and* $t_* > 0$,

$$1 - E_\alpha(-t_*^\alpha\lambda) > 0.$$

Proof. (i) For any $\lambda > 0$ and $t \in J$,

$$\begin{aligned} E_\alpha(-t^\alpha\lambda) &\le \int_0^\infty M_\alpha(\theta)e^{-t^\alpha\theta\lambda}d\theta \le \int_0^\infty M_\alpha(\theta)d\theta = E_\alpha(0) = 1, \\ e_\alpha(-t^\alpha\lambda) &\le \alpha\int_0^\infty \theta M_\alpha(\theta)e^{-t^\alpha\theta\lambda}d\theta \le \alpha\int_0^\infty \theta M_\alpha(\theta)d\theta = e_\alpha(0) = \frac{1}{\Gamma(\alpha)}, \\ E_{\alpha,\alpha+\beta}(-t^\alpha\lambda) &\le \frac{1}{\Gamma(\alpha)\Gamma(\beta)}\int_0^1 u^{\alpha-1}(1-u)^{\beta-1}du = E_{\alpha,\alpha+\beta}(0) = \frac{1}{\Gamma(\alpha+\beta)}. \end{aligned}$$

(ii) We only check the result (5.25) for $0 \le t_1 < t_2 \le 1$. In fact, using the inequality $t_2^\alpha - t_1^\alpha \le (t_2 - t_1)^\alpha$ and Lagrange mean value theorem, we have

$$\begin{aligned} |e_\alpha(-t_2^\alpha\lambda) - e_\alpha(-t_1^\alpha\lambda)| &\le \alpha\int_0^\infty \theta M_\alpha(\theta)\left|e^{-t_2^\alpha\theta\lambda} - e^{-t_1^\alpha\theta\lambda}\right|d\theta \\ &= \alpha\int_0^\infty \theta M_\alpha(\theta)\left(\int_0^1 e^{-\eta t_2^\alpha\theta\lambda-(1-\eta)t_1^\alpha\theta\lambda}d\eta\right)|t_2^\alpha - t_1^\alpha|\theta\lambda d\theta \\ &\le |t_2 - t_1|^\alpha\alpha\lambda\int_0^\infty \theta^2 M_\alpha(\theta)d\theta \\ &\le |t_2 - t_1|^\alpha\frac{2\alpha\lambda}{\Gamma(1+2\alpha)} \\ &= O(|t_2 - t_1|^\alpha), \ \text{as}\ t_2 \to t_1. \end{aligned}$$

Next, one can use the formula (5.24) via the above facts and Beta function to derive

$$|E_{\alpha,\alpha+\beta}(-t_2^\alpha\lambda) - E_{\alpha,\alpha+\beta}(-t_1^\alpha\lambda)|$$

$$
\begin{aligned}
&\leq \frac{1}{\Gamma(\beta)} \int_0^1 u^{\alpha-1}(1-u)^{\beta-1}|e_\alpha(-t_2^\alpha \lambda u) - e_\alpha(-t_1^\alpha \lambda u)|du \\
&\leq |t_2 - t_1|^\alpha \frac{2\alpha\lambda B(\alpha,\beta)}{\Gamma(1+2\alpha)\Gamma(\beta)} \\
&= O(|t_2 - t_1|^\alpha), \quad \text{as } t_2 \to t_1.
\end{aligned}
$$

(iii) Similar to the proof in (ii), for any $t_1, t_2 \in J$ and $t_1 \leq t_2$, we obtain

$$
\begin{aligned}
e_\alpha(-t_1^\alpha \lambda) - e_\alpha(-t_2^\alpha \lambda) &= \alpha \int_0^\infty \theta M_\alpha(\theta) \left(e^{-t_1^\alpha \theta\lambda} - e^{-t_2^\alpha \theta\lambda}\right) d\theta \\
&= \alpha\lambda(t_2^\alpha - t_1^\alpha) \int_0^\infty \theta^2 M_\alpha(\theta) \left(\int_0^1 e^{-\eta t_2^\alpha \theta\lambda - (1-\eta) t_1^\alpha \theta\lambda} d\eta \right) d\theta \\
&\geq 0.
\end{aligned}
$$

Next, one can use the formula (5.24) via the above facts and Beta function to derive the result (5.26).

(iv) In fact,

$$
E_\alpha(-t_*^\alpha \lambda) = \int_0^\infty M_\alpha(\theta) e^{-t_*^\alpha \theta\lambda} d\theta < \int_0^\infty M_\alpha(\theta) d\theta = 1,
$$

where we use the fact $e^{-t_*^\alpha \theta\lambda} < 1$ for $\theta \in (0,\infty)$ and $\lambda > 0$. The proof is completed. □

Secondly, we deduce the general solutions of the following linear fractional impulsive Langevin equations

$$
\begin{cases}
{}_0^C D_t^\beta ({}_0^C D_t^\alpha + \lambda) x(t) = f(t), & t \in J', \\
\Delta x(t_k) := x(t_k^+) - x(t_k^-) = I_k, & I_k \in \mathbb{R}, \\
x(0) = 0, \ x(\eta_k) = 0, \ x(1) = 0, & \eta_k \in (t_k, t_{k+1}), \ k = 0, 1, 2, ..., m-1.
\end{cases} \tag{5.27}
$$

For brevity, we denote

$$
\begin{aligned}
(Ff)(t) &:= \int_0^t (t-z)^{\alpha+\beta-1} E_{\alpha,\alpha+\beta}(-(t-z)^\alpha \lambda) f(z) dz, \quad t \in J, \\
(T_0 f)(t) &:= -\frac{1 - E_\alpha(-t^\alpha \lambda)}{1 - E_\alpha(-\eta_0^\alpha \lambda)} (Ff)(\eta_0), \quad t \in J_0, \\
(T_i f)(t) &:= \frac{E_\alpha(-t^\alpha \lambda) - E_\alpha(-\eta_i^\alpha \lambda)}{E_\alpha(-t_i^\alpha \lambda) - E_\alpha(-\eta_i^\alpha \lambda)} [(T_{i-1} f)(t_i) \\
&\quad + (Ff)(\eta_i) + I_i] - (Ff)(\eta_i), \quad t \in J_i, \ i = 1, 2, ..., m-1, \\
(T_m f)(t) &:= \frac{E_\alpha(-t^\alpha \lambda) - E_\alpha(-\lambda)}{E_\alpha(-t_m^\alpha \lambda) - E_\alpha(-\lambda)} [(T_{m-1} f)(t_m) \\
&\quad + (Ff)(1) + I_m] - (Ff)(1), \quad t \in J_m.
\end{aligned}
$$

Clearly, by Lemma 5.8(iii), the above symbols are well defined.

Lemma 5.9. *A general solution x of the equation* (5.27) *on the interval J is given by*

$$x(t) = \begin{cases} (Ff)(t) + (T_0 f)(t), & \text{for } t \in J_0, \\ (Ff)(t) + (T_1 f)(t), & \text{for } t \in J_1, \\ \vdots \\ (Ff)(t) + (T_i f)(t), & \text{for } t \in J_i, \\ \vdots \\ (Ff)(t) + (T_m f)(t), & \text{for } t \in J_m. \end{cases} \tag{5.28}$$

Proof. For $t \in J_0$, integrating both sides of the first equation of (5.27), one can obtain that (see (5.20))

$$\begin{aligned} x(t) = {} & E_\alpha(-t^\alpha \lambda) b_0 - \frac{1}{\lambda}\left(1 - E_\alpha(-t^\alpha \lambda)\right) a_0 \\ & + \int_0^t (t-z)^{\alpha+\beta-1} E_{\alpha,\alpha+\beta}(-(t-z)^\alpha \lambda) f(z) dz. \end{aligned} \tag{5.29}$$

Using the conditions $x(0) = 0$ and $x(\eta_0) = 0$, we get

$$\begin{aligned} a_0 &= \frac{\lambda}{1 - E_\alpha(-\eta_0^\alpha \lambda)} \int_0^{\eta_0} (\eta_0 - z)^{\alpha+\beta-1} E_{\alpha,\alpha+\beta}(-(\eta_0 - z)^\alpha \lambda) f(z) dz, \\ b_0 &= 0. \end{aligned} \tag{5.30}$$

Submitting (5.30) to (5.29), we obtain

$$\begin{aligned} x(t) &= -\frac{1 - E_\alpha(-t^\alpha \lambda)}{1 - E_\alpha(-\eta_0^\alpha \lambda)} \int_0^{\eta_0} (\eta_0 - z)^{\alpha+\beta-1} E_{\alpha,\alpha+\beta}(-(\eta_0 - z)^\alpha \lambda) f(z) dz \\ &\quad + \int_0^t (t-z)^{\alpha+\beta-1} E_{\alpha,\alpha+\beta}(-(t-z)^\alpha \lambda) f(z) dz, \\ x(t_1) &= (Ff)(t_1) + (T_0 f)(t_1). \end{aligned}$$

For $t \in J_1$, integrating both sides of the first equation of (5.27), one can obtain that

$$\begin{aligned} x(t) = {} & E_\alpha(-t^\alpha \lambda) b_1 - \frac{1}{\lambda}\left(1 - E_\alpha(-t^\alpha \lambda)\right) a_1 \\ & + \int_0^t (t-z)^{\alpha+\beta-1} E_{\alpha,\alpha+\beta}(-(t-z)^\alpha \lambda) f(z) dz. \end{aligned}$$

Since

$$\begin{aligned} x(t_1^+) &= E_\alpha(-t_1^\alpha \lambda) b_1 - \frac{1}{\lambda}\left(1 - E_\alpha(-t_1^\alpha \lambda)\right) a_1 + (Ff)(t_1), \\ x(t_1^-) &= (Ff)(t_1) + (T_0 f)(t_1), \end{aligned}$$

from $x(t_1^+) = x(t_1^-) + I_1, x(\eta_1) = 0$, it follows

$$E_\alpha(-t_1^\alpha\lambda)b_1 - \frac{1}{\lambda}\left(1 - E_\alpha(-t_1^\alpha\lambda)\right)a_1 = (T_0f)(t_1) + I_1,$$
$$E_\alpha(-\eta_1^\alpha\lambda)b_1 - \frac{1}{\lambda}\left(1 - E_\alpha(-\eta_1^\alpha\lambda)\right)a_1 + (Ff)(\eta_1) = 0$$

and solving the above equations, we can get

$$a_1 = \frac{\lambda E_\alpha(-\eta_1^\alpha\lambda)}{E_\alpha(-t_1^\alpha\lambda) - E_\alpha(-\eta_1^\alpha\lambda)}[(T_0f)(t_1) + (Ff)(\eta_1) + I_1] + \lambda(Ff)(\eta_1),$$
$$b_1 = -(Ff)(\eta_1) + \frac{1 - E_\alpha(-\eta_1^\alpha\lambda)}{E_\alpha(-t_1^\alpha\lambda) - E_\alpha(-\eta_1^\alpha\lambda)}[(T_0f)(t_1) + (Ff)(\eta_1) + I_1].$$

Hence, we obtain

$$x(t) = (Ff)(t) + (T_1f)(t), \qquad \text{for } t \in J_2,$$
$$x(t_2) = (Ff)(t_2) + (T_1f)(t_2).$$

Repeating the above methods on the subinterval J_i, $i = 2, 3, ..., m-1$ respectively, one can obtain that $x(t) = (Ff)(t) + (T_if)(t)$ for $t \in J_i$.

Finally, for $t \in J_m$, integrating both side of the first equation of (5.27) again, one can obtain that

$$x(t) = E_\alpha(-t^\alpha\lambda)b_m - \frac{1}{\lambda}\left(1 - E_\alpha(-t^\alpha\lambda)\right)a_m + \int_0^t (t-z)^{\alpha+\beta-1}E_{\alpha,\alpha+\beta}(-(t-z)^\alpha\lambda)f(z)dz.$$

Note that $x(t_m^+) = x(t_m^-) + I_m = (Fx)(t_m) + (T_{m-1}x)(t_m) + I_m$, $x(1) = 0$, one can obtain

$$a_m = \frac{\lambda E_\alpha(-\lambda)}{E_\alpha(-t_m^\alpha\lambda) - E_\alpha(-\lambda)}[(T_{m-1}f)(t_m) + (Ff)(1) + I_m] + \lambda(Ff)(1),$$
$$b_m = \frac{1 - E_\alpha(-\lambda)}{E_\alpha(-t_m^\alpha\lambda) - E_\alpha(-\lambda)}[(T_{m-1}f)(t_m) + (Ff)(1) + I_m] - (Ff)(1).$$

Then, we get

$$x(t) = (Ff)(t) + (T_mf)(t), \quad \text{for } t \in J_m.$$

This completes the proof. □

Remark 5.8. If we denote

$$(T_0f)(t) := -\frac{1 - E_\alpha(-t^\alpha\lambda)}{1 - E_\alpha(-\eta_0^\alpha\lambda)}(Ff)(\eta_0), \ t \in J_0,$$
$$(T_if)(t) := [(T_{i-1}f)(t_i) + I_i + M_i(t_i)(Ff)(\eta_i)]\frac{E_\alpha(-t^\alpha\lambda) - E_\alpha(-\eta_i^\alpha\lambda)}{E_\alpha(-t_i^\alpha\lambda) - E_\alpha(-\eta_i^\alpha\lambda)} - M_i(t)(Ff)(\eta_i), \ t \in J_i, \ i = 1, 2, ..., m-1,$$
$$(T_mf)(t) := [(T_{m-1}f)(t_m) + I_m + M_m(t_m)(Ff)(1)]\frac{E_\alpha(-t^\alpha\lambda) - E_\alpha(-\lambda)}{E_\alpha(-t_m^\alpha\lambda) - E_\alpha(-\lambda)}$$

$$
\begin{aligned}
&- M_m(t)(Ff)(1),\ t \in J_m,\\
M_i(t) &:= \frac{1 - E_\alpha(-t^\alpha \lambda)}{1 - E_\alpha(-\eta_i^\alpha \lambda)},\ t \in J_i,\ i = 1, 2, ..., m-1,\\
M_m(t) &:= \frac{1 - E_\alpha(-t^\alpha \lambda)}{1 - E_\alpha(-\lambda)},
\end{aligned}
$$

then an alternative formula of general solutions of the equation (5.27) on J is given by

$$
x(t) = \begin{cases} (Ff)(t) + (T_0 f)(t), & \text{for } t \in J_0,\\ (Ff)(t) + (T_1 f)(t), & \text{for } t \in J_1,\\ \vdots & \\ (Ff)(t) + (T_i f)(t), & \text{for } t \in J_i,\\ \vdots & \\ (Ff)(t) + (T_m f)(t), & \text{for } t \in J_m. \end{cases} \tag{5.31}
$$

By directly computation, one can verify that (5.31) is coincided with (5.28). But, (5.28) seems more suitable than (5.31).

5.4.3 *Existence*

This subsection deals with the existence and uniqueness of solution for the problem (5.17). A number of papers have been recently written on fractional impulsive initial and boundary value problems. However, both Fečkan, Zhou and Wang, 2012 and Kosmatov, 2012, point out on the error in former solutions for some impulsive fractional differential equations by construct a counterexample and establish a general framework to seek a nature solution for such problems. Motivated by Fečkan, Zhou and Wang, 2012 and Kosmatov, 2012, we define what it means for the problem (5.17) to have a solution.

Definition 5.3. A function $x \in PC(J, \mathbb{R})$ whose Caputo fractional derivative existing on J is said to be a solution of the problem (5.17) if $x(t)$ satisfies

$$
x(t) = \begin{cases} (Fx)(t) + (T_0 x)(t), & \text{for } t \in J_0,\\ (Fx)(t) + (T_1 x)(t), & \text{for } t \in J_1,\\ \vdots & \\ (Fx)(t) + (T_i x)(t), & \text{for } t \in J_i,\\ \vdots & \\ (Fx)(t) + (T_m x)(t), & \text{for } t \in J_m, \end{cases}
$$

where

$$
(Fx)(t) = \int_0^t (t-z)^{\alpha+\beta-1} E_{\alpha,\alpha+\beta}(-(t-z)^\alpha \lambda) f(z, x(z)) dz,\ t \in J,
$$

$$(T_0x)(t) = -\frac{1-E_\alpha(-t^\alpha\lambda)}{1-E_\alpha(-\eta_0^\alpha\lambda)}(Fx)(\eta_0),\ t\in J_0,$$

$$(T_ix)(t) = \frac{E_\alpha(-t^\alpha\lambda)-E_\alpha(-\eta_i^\alpha\lambda)}{E_\alpha(-t_i^\alpha\lambda)-E_\alpha(-\eta_i^\alpha\lambda)}[(T_{i-1}x)(t_i)+(Fx)(\eta_i)+I_i]-(Fx)(\eta_i),$$
$$t\in J_i,\ i=1,2,...,m-1,$$

$$(T_mx)(t) = \frac{E_\alpha(-t^\alpha\lambda)-E_\alpha(-\lambda)}{E_\alpha(-t_m^\alpha\lambda)-E_\alpha(-\lambda)}[(T_{m-1}x)(t_m)+(Fx)(1)+I_m]-(Fx)(1),$$
$$t\in J_m.$$

Before stating the main results, we introduce the following hypotheses.

(H1) $f: J\times\mathbb{R}\to\mathbb{R}$ is jointly continuous;

(H2) there exists a function $n(\cdot)\in L^{\frac{1}{q_1}}(J,\mathbb{R}^+)$ such that $|f(t,x)|\le n(t)$ for all $t\in J$ and all $x\in\mathbb{R}$, where $q_1\in(0,\alpha+\beta)$;

(H3) there exists a function $h(\cdot)\in L^{\frac{1}{q_2}}(J,\mathbb{R}^+)$ such that $|f(t,x)-f(t,y)|\le h(t)|x-y|$ for all $t\in J$ and all $x,y\in\mathbb{R}$, where $q_2\in(0,\alpha+\beta)$.

Consider an operator $N: PC(J,\mathbb{R})\to PC(J,\mathbb{R})$ defined by

$$(Nx)(t) = \begin{cases} (Fx)(t)+(T_0x)(t), & \text{for } t\in J_0,\\ (Fx)(t)+(T_1x)(t), & \text{for } t\in J_1,\\ \vdots & \\ (Fx)(t)+(T_ix)(t), & \text{for } t\in J_i,\\ \vdots & \\ (Fx)(t)+(T_mx)(t), & \text{for } t\in J_m. \end{cases} \tag{5.32}$$

It is obvious that N is well defined due to (H1). Then, we can transform existence of solutions of the problem (5.17) into a fixed point problem of the operator N.

For brevity, denote $p_1 = \frac{\alpha+\beta-1}{1-q_1}$, $p_2 = \frac{\alpha+\beta-1}{1-q_2}$.

We are ready to state the first existence and uniqueness result in this subsection.

Theorem 5.9. *Assume that (H1)-(H3) hold. If*

$$M_F + M_m < 1, \tag{5.33}$$

then the problem (5.17) *has a unique solution, where*

$$M_F = \frac{\|h\|_{L^{\frac{1}{q_2}}J}}{\Gamma(\alpha+\beta)(1+p_2)^{1-q_2}}, \tag{5.34}$$

and

$$M_m = \frac{1-E_\alpha(-\lambda)}{E_\alpha(-t_m^\alpha\lambda)-E_\alpha(-\lambda)}(M_F+M_{m-1})+M_F, ...,\ M_0 = \frac{M_F}{1-E_\alpha(-\eta_0^\alpha\lambda)}.$$

Proof. We verify that N defined by (5.32) is a contraction mapping. We divide our proof into two steps.

Claim I. $Nx \in PC(J,\mathbb{R})$ for every $x \in PC(J,\mathbb{R})$.

If $t \in J_0$, then for every $x \in C(J_0,\mathbb{R})$ and any $\delta > 0$, $0 < t < t+\delta \leq t_1$, by (H2), Lemma 5.8 and Hölder inequality, we get

$$\begin{aligned}
&|(Nx)(t+\delta)-(Nx)(t)| \\
&\leq |(Fx)(t+\delta)-(Fx)(t)| + |(T_0x)(t+\delta)-(T_0x)(t)| \\
&\leq \left|\int_0^{t+\delta}(t+\delta-z)^{\alpha+\beta-1}E_{\alpha,\alpha+\beta}(-(t+\delta-z)^\alpha\lambda)f(z,x(z))dz\right. \\
&\quad \left.-\int_0^t (t-z)^{\alpha+\beta-1}E_{\alpha,\alpha+\beta}(-(t-z)^\alpha\lambda)f(z,x(z))dz\right| \\
&\quad +\left|\frac{E_\alpha(-(t+\delta)^\alpha\lambda)-E_\alpha(-t^\alpha\lambda)}{1-E_\alpha(-\eta_0^\alpha\lambda)}(Fx)(\eta_0)\right| \\
&\leq \int_0^t (t+\delta-z)^{\alpha+\beta-1}|E_{\alpha,\alpha+\beta}(-(t+\delta-z)^\alpha\lambda)-E_{\alpha,\alpha+\beta}(-(t-z)^\alpha\lambda)|n(z)dz \\
&\quad +\int_0^t \left|(t+\delta-z)^{\alpha+\beta-1}-(t-z)^{\alpha+\beta-1}\right| |E_{\alpha,\alpha+\beta}(-(t-z)^\alpha\lambda)|n(z)dz \\
&\quad +\int_t^{t+\delta}(t+\delta-z)^{\alpha+\beta-1}|E_{\alpha,\alpha+\beta}(-(t+\delta-z)^\alpha\lambda)|n(z)dz \\
&\quad +\frac{|E_\alpha(-(t+\delta)^\alpha\lambda)-E_\alpha(-t^\alpha\lambda)|}{1-E_\alpha(-\eta_0^\alpha\lambda)}|(Fx)(\eta_0)| \\
&\leq O(\delta^\alpha)\left(\int_0^t (t+\delta-z)^{\frac{\alpha+\beta-1}{1-q_1}}dz\right)^{1-q_1}\left(\int_0^t n(z)^{\frac{1}{q_1}}dz\right)^{q_1} \\
&\quad +\frac{1}{\Gamma(\alpha+\beta)}\left(\int_0^t \left((t-z)^{\alpha+\beta-1}-(t+\delta-z)^{\alpha+\beta-1}\right)^{\frac{1}{1-q_1}}dz\right)^{1-q_1} \\
&\quad \times\left(\int_0^t n(z)^{\frac{1}{q_1}}dz\right)^{q_1}+\frac{1}{\Gamma(\alpha+\beta)}\left(\int_t^{t+\delta}(t+\delta-z)^{\frac{\alpha+\beta-1}{1-q_1}}dz\right)^{1-q_1} \\
&\quad \times\left(\int_t^{t+\delta}n(z)^{\frac{1}{q_1}}dz\right)^{q_1}+O(\delta^\alpha)\frac{|(Fx)(\eta_0)|}{1-E_\alpha(-\eta_0^\alpha\lambda)} \\
&\leq O(\delta^\alpha)\left(\frac{|(Fx)(\eta_0)|}{1-E_\alpha(-\eta_0^\alpha\lambda)}+\frac{\|n\|_{L^{\frac{1}{q_1}}J_0}}{(1+p_1)^{1-q_1}}\right)+\frac{2\delta^{(1+p_1)(1-q_1)}\|n\|_{L^{\frac{1}{q_1}}J_0}}{\Gamma(\alpha+\beta)(1+p_1)^{1-q_1}} \\
&\to 0, \quad \text{as } \delta\to 0,
\end{aligned}$$

where we use the facts

$$\int_0^t (t+\delta-z)^{\frac{\alpha+\beta-1}{1-q_1}}dz = \frac{(t+\delta)^{1+p_1}-\delta^{1+p_1}}{1+p_1}\leq\frac{1}{1+p_1},$$

$$\int_0^t \left((t-z)^{\alpha+\beta-1}-(t+\delta-z)^{\alpha+\beta-1}\right)^{\frac{1}{1-q_1}}dz$$

$$\leq \int_0^t \left((t-z)^{\frac{\alpha+\beta-1}{1-q_1}} - (t+\delta-z)^{\frac{\alpha+\beta-1}{1-q_1}}\right) dz$$
$$\leq \frac{\delta^{1+p_1}}{1+p_1},$$
$$\int_t^{t+\delta} (t+\delta-z)^{\frac{\alpha+\beta-1}{1-q_1}} dz = \frac{\delta^{1+p_1}}{1+p_1},$$

and

$$|(Fx)(\eta_0)| \leq \frac{1}{\Gamma(\alpha+\beta)} \left(\int_0^{\eta_0} (\eta_0 - z)^{\frac{\alpha+\beta-1}{1-q_1}} dz\right)^{1-q_1} \left(\int_0^{\eta_0} n(z)^{\frac{1}{q_1}} dz\right)^{q_1}$$
$$\leq \frac{\|n\|_{L^{\frac{1}{q_1}} J_0}}{\Gamma(\alpha+\beta)(1+p_1)^{1-q_1}}.$$

Thus we obtain $Nx \in C(J_0, \mathbb{R})$.

If $t \in J_1$, then for every $x \in C(J_1, \mathbb{R})$ and any $\delta > 0$, $t_1 < t < t+\delta \leq t_2$, repeating the above process, one can obtain

$$|(Nx)(t+\delta) - (Nx)(t)|$$
$$\leq O(\delta^\alpha) \left(\frac{(1 - E_\alpha(-\eta_1^\alpha \lambda))(|(Fx)(\eta_1)| + |(T_0x)(t_1)| + |I_1|)}{E_\alpha(-t_1^\alpha \lambda) - E_\alpha(-\eta_1^\alpha \lambda)} + \frac{\|n\|_{L^{\frac{1}{q_1}} J_1}}{(1+p_1)^{1-q_1}}\right)$$
$$+ \frac{2\delta^{(1+p_1)(1-q_1)} \|n\|_{L^{\frac{1}{q_1}} J_1}}{\Gamma(\alpha+\beta)(1+p_1)^{1-q_1}} \to 0$$

as $\delta > 0$, where we use the fact

$$|(T_0x)(t_1)| \leq \frac{1}{1 - E_\alpha(-\eta_0^\alpha \lambda)} |(Fx)(\eta_0)|,$$

and

$$|(Fx)(\eta_1)| \leq \frac{\|n\|_{L^{\frac{1}{q_1}} J_1}}{\Gamma(\alpha+\beta)(1+p_1)^{1-q_1}}.$$

Thus $Nx \in C(J_1, \mathbb{R})$.

With the same argument, one can verify that $Nx \in C(J_i, \mathbb{R})$, for every $x \in C(J_i, \mathbb{R}), i = 2, ..., m$.

From the above fact, we can conclude that $Nx \in PC(J, \mathbb{R})$, for every $x \in PC(J, \mathbb{R})$.

Claim II. N is a contraction mapping on $PC(J, \mathbb{R})$.

For arbitrary $x, y \in PC(J,\mathbb{R})$ and $t \in J$, by (H3), Lemma 5.8 and Hölder inequality, we get

$$\begin{aligned}
&|(Fx)(t) - (Fy)(t)| \\
&\leq \frac{1}{\Gamma(\alpha+\beta)} \int_0^t (t-z)^{\alpha+\beta-1} |f(z,x(z)) - f(z,y(z))| dz \\
&\leq \frac{1}{\Gamma(\alpha+\beta)} \left(\int_0^t (t-z)^{\frac{\alpha+\beta-1}{1-q_2}} dz \right)^{1-q_2} \left(\int_0^t h(z)^{\frac{1}{q_2}} dz \right)^{q_2} \|x-y\|_{PC} \qquad (5.35) \\
&\leq \frac{\|h\|_{L^{\frac{1}{q_2}} J}}{\Gamma(\alpha+\beta)(1+p_2)^{1-q_2}} \|x-y\|_{PC} \\
&=: M_F \|x-y\|_{PC}.
\end{aligned}$$

If $t \in J_0$, for arbitrary $x, y \in C(J_0,\mathbb{R})$, we get

$$\begin{aligned}
|(T_0x)(t) - (T_0y)(t)| &\leq \frac{1}{1-E_\alpha(-\eta_0^\alpha \lambda)} |(Fx)(\eta_0) - (Fy)(\eta_0)| \\
&\leq \frac{1}{1-E_\alpha(-\eta_0^\alpha \lambda)} M_F \|x-y\|_{PC} \\
&=: M_0 \|x-y\|_{PC}.
\end{aligned}$$

Thus, $\|Nx - Ny\|_{PC} \leq (M_F + M_0) \|x-y\|_{PC}$.

If $t \in J_1$, for arbitrary $x, y \in C(J_1,\mathbb{R})$, we get

$$\begin{aligned}
&|(T_1x)(t) - (T_1y)(t)| \\
&\leq \frac{1-E_\alpha(-\eta_1^\alpha \lambda)}{E_\alpha(-t_1^\alpha \lambda) - E_\alpha(-\eta_1^\alpha \lambda)} |(Fx)(\eta_1) - (Fy)(\eta_1) + (T_0x)(t_1) - (T_0y)(t_1)| \\
&\quad + |(Fx)(\eta_1) - (Fy)(\eta_1)| \\
&\leq \left(\frac{1-E_\alpha(-\eta_1^\alpha \lambda)}{E_\alpha(-t_1^\alpha \lambda) - E_\alpha(-\eta_1^\alpha \lambda)} (M_F + M_0) + M_F \right) \|x-y\|_{PC} \\
&=: M_1 \|x-y\|_{PC}.
\end{aligned}$$

Thus, $\|Nx - Ny\|_{PC} \leq (M_F + M_1)\|x-y\|_{PC}$.

If $t \in J_i$, for arbitrary $x, y \in C(J_i,\mathbb{R})$, $i = 2, ..., m-1$, with the same argument, we get $\|Nx - Ny\|_{PC} \leq (M_F + M_i)\|x-y\|_{PC}$.

If $t \in J_m$, for arbitrary $x, y \in C(J_m,\mathbb{R})$, we get

$$\begin{aligned}
&|(T_mx)(t) - (T_my)(t)| \\
&\leq \frac{1-E_\alpha(-\lambda)}{E_\alpha(-t_m^\alpha \lambda) - E_\alpha(-\lambda)} |(Fx)(1) - (Fy)(1) + (T_{m-1}x)(t_m) - (T_{m-1}y)(t_m)| \\
&\quad + |(Fx)(1) - (Fy)(1)| \\
&\leq \left(\frac{1-E_\alpha(-\lambda)}{E_\alpha(-t_m^\alpha \lambda) - E_\alpha(-\lambda)} (M_F + M_{m-1}) + M_F \right) \|x-y\|_{PC} \\
&=: M_m \|x-y\|_{PC}.
\end{aligned}$$

Thus, $\|Nx - Ny\|_{PC} \leq (M_F + M_m)\|x-y\|_{PC}$.

Moreover, it is easy to see $M_0 < M_1 < M_2 < \cdots < M_m$. Due to the condition (5.33), N has a unique fixed point on $PC(J,\mathbb{R})$ by Banach contraction mapping principle. This completes the proof. □

Our second result is based on the well-known fixed point theorem due to Krasnoselskii (see Theorem 1.7).

Theorem 5.10. *Assume the conditions (H1)-(H3) hold. If* $M_F < 1$, *then the problem* (5.17) *has at least a solution on* $PC(J,\mathbb{R})$.

Proof. Setting $B_r = \{x \in PC(J,\mathbb{R}) : \|x\|_{PC} \le r\}$, where $r \ge \overline{M_F} + \overline{M_m}$, and $\overline{M_m}$, $\overline{M_F}$ are finite positive constants defined by

$$\overline{M_m} = \overline{M_F} + \frac{1 - E_\alpha(-\lambda)}{E_\alpha(-t_m^\alpha \lambda) - E_\alpha(-\lambda)}(|I_m| + \overline{M_{m-1}} + \overline{M_F}), ..., \overline{M_0} = \frac{\overline{M_F}}{1 - E_\alpha(-\eta_0^\alpha \lambda)},$$

and

$$\overline{M_F} := \frac{\|n\|_{L^{\frac{1}{q_1}} J}}{\Gamma(\alpha+\beta)(1+p_1)^{1-q_1}}.$$

Claim I. $(Fx)(t) + (T_i y)(t) \in B_r$ for any $t \in J_i$ and $x, y \in B_r$.

By the Claim I of Theorem 5.9, $(Fx)(t)$ and $(T_i x)(t)$ are obviously continuous in J_i for every $x \in B_r$.

For every $x, y \in B_r$ and $t \in J_0$, by (H2), Lemma 5.8 and Hölder inequality again, we get

$$\begin{aligned}
|(Fx)(t) + (T_0 y)(t)| &\le \left| \int_0^t (t-z)^{\alpha+\beta-1} E_{\alpha,\alpha+\beta}(-(t-z)^\alpha \lambda) f(z, x(z)) dz \right| \\
&\quad + \frac{1 - E_\alpha(-t^\alpha \lambda)}{1 - E_\alpha(-\eta_0^\alpha \lambda)} |(Fy)(\eta_0)| \\
&\le \frac{\|n\|_{L^{\frac{1}{q_1}} J_0}}{\Gamma(\alpha+\beta)(1+p_1)^{1-q_1}} + \frac{\|n\|_{L^{\frac{1}{q_1}} J_0}}{\Gamma(\alpha+\beta)(1+p_1)^{1-q_1}} \frac{1}{1 - E_\alpha(-\eta_0^\alpha \lambda)} \\
&\le \overline{M_F} + \overline{M_0} \le r.
\end{aligned}$$

For every $x, y \in B_r$ and $t \in J_i$, $i = 1, 2, ..., m-1$, by (H2), Lemma 5.8 and Hölder inequality, we have

$$\begin{aligned}
&|(Fx)(t) + (T_i y)(t)| \\
&\le \left| \int_0^t (t-z)^{\alpha+\beta-1} E_{\alpha,\alpha+\beta}(-(t-z)^\alpha \lambda) f(z, x(z)) dz \right| \\
&\quad + \left| \frac{E_\alpha(-t^\alpha \lambda) - E_\alpha(-\eta_i^\alpha \lambda)}{E_\alpha(-t_i^\alpha \lambda) - E_\alpha(-\eta_i^\alpha \lambda)} [(T_{i-1} y)(t_i) + (Fy)(\eta_i) + I_i] - (Fy)(\eta_i) \right| \\
&\le \frac{\|n\|_{L^{\frac{1}{q_1}} J_i}}{\Gamma(\alpha+\beta)(1+p_1)^{1-q_1}} \\
&\quad + \frac{1 - E_\alpha(-\eta_i^\alpha \lambda)}{E_\alpha(-t_i^\alpha \lambda) - E_\alpha(-\eta_i^\alpha \lambda)} \left(|I_i| + |(T_{i-1} x)(t_i)| + \frac{\|n\|_{L^{\frac{1}{q_1}} J_i}}{\Gamma(\alpha+\beta)(1+p_1)^{1-q_1}} \right)
\end{aligned}$$

$$
\begin{aligned}
&+\frac{\|n\|_{L^{\frac{1}{q_1}}J_i}}{\Gamma(\alpha+\beta)(1+p_1)^{1-q_1}}\\
&\le \overline{M_F}+\left(\overline{M_F}+\frac{1-E_\alpha(-\eta_i^\alpha\lambda)}{E_\alpha(-t_i^\alpha\lambda)-E_\alpha(-\eta_i^\alpha\lambda)}(|I_i|+\overline{M_{i-1}}+\overline{M_F})\right)\\
&=: \overline{M_F}+\overline{M_i}\le r.
\end{aligned}
$$

For every $x, y \in B_r$ and $t \in J_m$, after a similar computation we obtain

$$
\begin{aligned}
|(Fx)(t)+(T_my)(t)| &\le \overline{M_F}+\left(\overline{M_F}+\frac{1-E_\alpha(-\lambda)}{E_\alpha(-t_m^\alpha\lambda)-E_\alpha(-\lambda)}(|I_m|+\overline{M_{m-1}}+\overline{M_F})\right)\\
&:= \overline{M_F}+\overline{M_m}\le r.
\end{aligned}
$$

Clearly, $\overline{M_m} > \overline{M_{m-1}} > \cdots > \overline{M_0}$. Due to the definition of the ball B_r, we must have $Fx + T_iy \in B_r$ for any $t \in J_i$ and $x, y \in B_r$.

Claim II. F is a contraction mapping on B_r.

By (5.35) we have $\|Fx - Fy\|_{PC} \le M_F\|x - y\|_{PC}$. The assumption $M_F < 1$ implies that F is a contraction mapping.

Claim III. T_i is a completely continuous operator on $B_r|_{J_i}$, $i = 0, 1, 2, ..., m$.

Similar to the Claim I of Theorem 5.9, one can easy to verify that T_i is continuous and $\{T_ix : x \in B_r\}$ is an equicontinuous set. Moreover, $\{T_ix : x \in B_r\}$ is uniformly bounded. Thus, T_i is a completely continuous operator on $B_r|_{J_i}, i = 0, 1, 2, ..., m$ due to Arzela-Ascoli theorem.

Applying Theorem 1.7, the problem (5.17) has at least a solution on $PC(J, \mathbb{R})$. The proof is completed. □

To end this section, we extend the above existence results to the equation (5.18). Now, we denote

$$
\begin{aligned}
&(Ff)(t) := \int_0^t (t-z)^{\alpha+\beta-1}E_{\alpha,\alpha+\beta}(-(t-z)^\alpha\lambda)f(z,x(z))dz, && t\in J,\\
&(\overline{T_0f})(t) := -\frac{1-E_\alpha(-t^\alpha\lambda)}{1-E_\alpha(-\eta_0^\alpha\lambda)}(Ff)(\eta_0), && t\in J_0,\\
&(\overline{T_if})(t) := \frac{E_\alpha(-t^\alpha\lambda)-E_\alpha(-\eta_i^\alpha\lambda)}{E_\alpha(-t_i^\alpha\lambda)-E_\alpha(-\eta_i^\alpha\lambda)}\big[(\overline{T_{i-1}f})(t_i)\\
&\qquad +(Ff)(\eta_i)+I_i(x(t_k^-))\big]-(Ff)(\eta_i), && t\in J_i,\ i=1,2,...,m-1,\\
&(\overline{T_mf})(t) := \frac{E_\alpha(-t^\alpha\lambda)-E_\alpha(-\lambda)}{E_\alpha(-t_m^\alpha\lambda)-E_\alpha(-\lambda)}\big[(\overline{T_{m-1}f})(t_m)\\
&\qquad +(Ff)(1)+I_m(x(t_k^-))\big]-(Ff)(1), && t\in J_m.
\end{aligned}
$$

Using the same method as in Lemma 5.9, one can obtain the following result immediately.

Lemma 5.10. *A general solution x of the equation (5.18) on the interval J is given by*

$$x(t) = \begin{cases} (Ff)(t) + \overline{(T_0 f)}(t), & \text{for } t \in J_0, \\ (Ff)(t) + \overline{(T_1 f)}(t), & \text{for } t \in J_1, \\ \vdots \\ (Ff)(t) + \overline{(T_i f)}(t), & \text{for } t \in J_i, \\ \vdots \\ (Ff)(t) + \overline{(T_m f)}(t), & \text{for } t \in J_m. \end{cases}$$

We make a necessary assumption on the nonlinear impulsive terms.

(H4) There exist constants $L > 0$ and $M > 0$ such that $|I_k(x) - I_k(y)| \leq L|x - y|$, with $|I_k(x)| \leq M$, for all $x, y \in \mathbb{R}$, $k = 1, 2, \ldots, m$.

Now the reader can apply the same methods as in the above theorems to obtain the following existence results. So we omit details of the proof here.

Theorem 5.11. *Assume the assumptions (H1)-(H4) hold. If*

$$M_F + \widetilde{M_m} < 1,$$

then the problem (5.18) has a unique solution, where M_F is defined in (5.34) and

$$\widetilde{M_m} = \frac{1 - E_\alpha(-\lambda)}{E_\alpha(-t_m^\alpha \lambda) - E_\alpha(-\lambda)} (M_F + \widetilde{M_{m-1}} + L) + M_F, \ldots,$$
$$M_0 = \frac{1}{1 - E_\alpha(-\eta_0^\alpha \lambda)} M_F.$$

Theorem 5.12. *Assume the assumptions (H1)-(H4) hold. If $M_F < 1$, then the problem (5.18) has at least a solution on $PC(J, \mathbb{R})$.*

5.5 Impulsive Evolution Equations

5.5.1 *Introduction*

Consider the nonlocal Cauchy problems for fractional impulsive evolution equations:

$$\begin{cases} {}_0^C D_t^\alpha x(t) = Ax(t) + f(t, x(t)), & t \in J, \ t \neq t_k, \\ x(0) = x_0 + g(x), \\ x(t_k^+) = x(t_k^-) + y_k, & k = 1, 2, ..., \delta, \end{cases} \tag{5.36}$$

where ${}_0^C D_t^\alpha$ is Caputo fractional derivative of order α, A: $D(A) \subseteq X \to X$ is the generator of a C_0-semigroup $\{Q(t)\}_{t \geq 0}$ on a Banach space X, f: $J \times X \to X$ is continuous, x_0, y_k are the element of X, g is a given function, $0 = t_0 < t_1 < t_2 < \cdots < t_\delta < t_{\delta+1} = b$, $x(t_k^+) = \lim_{h \to 0^+} = x(t_k + h)$ and $x(t_k^-) = x(t_k)$ represent respectively the right and left limits of $x(t)$ at $t = t_k$.

In Subsection 5.5.2, we give the definition of mild solution of problem (5.36). Subsection 5.5.3 is devoted to the existence and uniqueness results under the different assumptions on nonlinear term.

5.5.2 *Cauchy Problems*

In this subsection, we introduce a concept of solutions for our problems. We first consider an nonhomogeneous impulsive linear fractional equation of the form

$$\begin{cases} {}_0^C D_t^\alpha x(t) = Ax(t) + h(t), & \alpha \in (0,1),\ t \in J = [0,b],\ t \neq t_k, \\ x(0) = x_0, \\ x(t_k^+) = x(t_k^-) + y_k, & k = 1,2,...,\delta, \end{cases} \tag{5.37}$$

where $h \in PC(J, X)$. We observe that $x(\cdot)$ can be decomposed to $v(\cdot) + w(\cdot)$ where v is the continuous mild solution for

$$\begin{cases} {}_0^C D_t^\alpha v(t) = Av(t) + h(t), & t \in J, \\ v(0) = x_0, \end{cases} \tag{5.38}$$

on J, and w is the PC-mild solution for

$$\begin{cases} {}_0^C D_t^\alpha w(t) = Aw(t), & t \in J,\ t \neq t_k, \\ w(0) = 0, \\ w(t_k^+) = w(t_k^-) + y_k, & k = 1,2,...,\delta. \end{cases} \tag{5.39}$$

Indeed, by adding together (5.38) with (5.39), it follows (5.37). Note v is continuous, so $v(t_k^+) = v(t_k^-)$, $k = 1,2,...,\delta$. On the other hand, any solution of (5.37) can be decomposed to (5.38) and (5.39).

A mild solution of (5.38) is given by

$$v(t) = S_\alpha(t)x_0 + \int_0^t (t-s)^{\alpha-1} P_\alpha(t-s)h(s)ds,\ t \in J,$$

where

$$S_\alpha(t) = \int_0^\infty M_\alpha(\theta)Q(t^\alpha\theta)d\theta, \quad P_\alpha(t) = \int_0^\infty \alpha\theta M_\alpha(\theta)Q(t^\alpha\theta)d\theta.$$

Now we rewrite system (5.39) in the equivalent integral equation

$$w(t) = \begin{cases} \dfrac{1}{\Gamma(\alpha)}\displaystyle\int_0^t (t-s)^{\alpha-1}Aw(s)ds, & \text{for } t \in [0,t_1], \\ y_1 + \dfrac{1}{\Gamma(\alpha)}\displaystyle\int_0^t (t-s)^{\alpha-1}Aw(s)ds, & \text{for } t \in (t_1,t_2], \\ y_1 + y_2 + \dfrac{1}{\Gamma(\alpha)}\displaystyle\int_0^t (t-s)^{\alpha-1}Aw(s)ds, & \text{for } t \in (t_2,t_3], \\ \vdots \\ \displaystyle\sum_{i=1}^{\delta} y_i + \dfrac{1}{\Gamma(\alpha)}\displaystyle\int_0^t (t-s)^{\alpha-1}Aw(s)ds, & \text{for } t \in (t_\delta,b]. \end{cases} \tag{5.40}$$

Then the above equation (5.40) can be expressed as

$$w(t) = \sum_{i=1}^{\delta} \chi_i(t) y_i + \frac{1}{\Gamma(\alpha)} \int_0^t (t-s)^{\alpha-1} A w(s) ds, \quad \text{for } t \in J, \tag{5.41}$$

where

$$\chi_i(t) = \begin{cases} 0, & \text{for } t \in [0, t_i), \\ 1, & \text{for } t \in [t_i, b] \cup (b, \infty). \end{cases}$$

We adopt the idea used in Section 4.3 and apply the Laplace transform for (5.41) to get

$$u(\lambda) = \sum_{i=1}^{\delta} \frac{e^{-t_i \lambda}}{\lambda} y_i + \frac{1}{\lambda^\alpha} A u(\lambda),$$

which implies

$$u(\lambda) = \sum_{i=1}^{\delta} e^{-t_i \lambda} \lambda^{\alpha-1} \left(\lambda^\alpha I - A\right)^{-1} y_i.$$

Note that the Laplace transform of $S_\alpha(t) y_i$ is $\lambda^{\alpha-1} \left(\lambda^\alpha I - A\right)^{-1} y_i$. Thus we can derive the mild solution of (5.39) as

$$w(t) = \sum_{i=1}^{\delta} \chi_i(t) S_\alpha(t - t_i) y_i.$$

Summarizing, the mild solution of (5.37) is given by

$$x(t) = S_\alpha(t) x_0 + \sum_{i=1}^{\delta} \chi_i(t) S_\alpha(t - t_i) y_i + \int_0^t (t-s)^{\alpha-1} P_\alpha(t-s) h(s) ds,$$

i.e.,

$$x(t) = \begin{cases} S_\alpha(t) x_0 + \int_0^t (t-s)^{\alpha-1} P_\alpha(t-s) h(s) ds, & \text{for } t \in [0, t_1], \\ S_\alpha(t) x_0 + S_\alpha(t - t_1) y_1 + \int_0^t (t-s)^{\alpha-1} P_\alpha(t-s) h(s) ds, & \text{for } t \in (t_1, t_2], \\ \vdots & \\ S_\alpha(t) x_0 + \sum_{i=1}^{\delta} S_\alpha(t - t_i) y_i + \int_0^t (t-s)^{\alpha-1} P_\alpha(t-s) h(s) ds, & \text{for } t \in (t_\delta, b]. \end{cases}$$

By using the above results, we can introduce the following definition of the mild solution for system (5.36).

Definition 5.4. By a PC-mild solution of the system (5.36) we mean that a function $x \in PC(J, X)$ which satisfies the following integral equation

$$x(t) = \begin{cases} S_\alpha(t)x_0 + \displaystyle\int_0^t (t-s)^{\alpha-1} P_\alpha(t-s) f(s, x(s))\, ds, & \text{for } t \in [0, t_1], \\ S_\alpha(t)x_0 + S_\alpha(t-t_1)y_1 & \\ \quad + \displaystyle\int_0^t (t-s)^{\alpha-1} P_\alpha(t-s) f(s, x(s))\, ds, & \text{for } t \in (t_1, t_2], \\ \vdots & \\ S_\alpha(t)x_0 + \displaystyle\sum_{i=1}^{\delta} S_\alpha(t-t_i)y_i & \\ \quad + \displaystyle\int_0^t (t-s)^{\alpha-1} P_\alpha(t-s) f(s, x(s))\, ds, & \text{for } t \in (t_\delta, b]. \end{cases}$$

5.5.3 *Nonlocal Problems*

In this subsection, we derive some existence and uniqueness results concerning the PC-mild solution for system (5.36) under the different assumptions on f.

Case I. f is Lipschitz.

Let us list the following hypotheses:

(HA) A is the infinitesimal generator of a compact semigroup $\{T(t)\}_{t\geq 0}$ in X;

(HF1) f: $J \times X \to X$ is continuous and there exists a constant $q_1 \in (0, \alpha)$ and a real-valued function $L_f(t) \in L^{\frac{1}{q_1}}(J, \mathbb{R}^+)$ such that

$$|f(t,x) - f(t,y)| \leq L_f(t)|x - y|, \ t \in J, \ x, y \in X.$$

For brevity, let us take

$$T^* = \left[\left(\frac{1-q_1}{\alpha - q_1} \right) b^{\frac{\alpha - q_1}{1-q_1}} \right]^{1-q_1} \|L_f\|_{L^{\frac{1}{q_1}} J}.$$

Theorem 5.13. *Let (HA) and (HF1) be satisfied. Then for every $x_0 \in X$, the system* (5.36) *has a unique PC-mild solution on J provided that*

$$0 < \frac{\alpha M T^*}{\Gamma(1+\alpha)} < 1. \tag{5.42}$$

Proof. Let $x_0 \in X$ be fixed. Define an operator T on $PC(J,X)$ by

$$(Tx)(t) = \begin{cases} S_\alpha(t)x_0 + \int_0^t (t-s)^{\alpha-1} P_\alpha(t-s) f(s, x(s))\, ds, & \text{for } t \in [0, t_1], \\ S_\alpha(t)x_0 + S_\alpha(t-t_1)y_1 & \\ \quad + \int_0^t (t-s)^{\alpha-1} P_\alpha(t-s) f(s, x(s))\, ds, & \text{for } t \in (t_1, t_2], \\ \vdots & \\ S_\alpha(t)x_0 + \sum_{i=1}^{\delta} S_\alpha(t-t_i)y_i & \\ \quad + \int_0^t (t-s)^{\alpha-1} P_\alpha(t-s) f(s, x(s))\, ds, & \text{for } t \in (t_\delta, b]. \end{cases}$$

By our assumptions and Lemma 1.6, T is well defined on $PC(J,X)$.

Claim I. $Tx \in PC(J,X)$ for $x \in PC(J,X)$.

For $0 \le \tau < t \le t_1$, taking into account the imposed assumptions and applying Proposition 4.5, we obtain

$$\begin{aligned} &|(Tx)(t) - (Tx)(\tau)| \\ &\le |S_\alpha(t)x_0 - S_\alpha(\tau)x_0| + \int_\tau^t (t-s)^{\alpha-1} |P_\alpha(t-s) f(s, x(s))|\, ds \\ &\quad + \int_0^\tau (t-s)^{\alpha-1} |P_\alpha(t-s) f(s, x(s)) - P_\alpha(\tau-s) f(s, x(s))|\, ds \\ &\quad + \int_0^\tau |(t-s)^{\alpha-1} - (\tau-s)^{\alpha-1}|\, |P_\alpha(\tau-s) f(s, x(s))|\, ds \\ &\le \|S_\alpha(t) - S_\alpha(\tau)\|_{B(X)} |x_0| + \frac{\alpha M}{\Gamma(1+\alpha)} \int_\tau^t (t-s)^{\alpha-1} |f(s, x(s))|\, ds \\ &\quad + \sup_{s \in [0,\tau]} \|P_\alpha(t-s) - P_\alpha(\tau-s)\|_{B(X)} \int_0^\tau (t-s)^{\alpha-1} |f(s, x(s))|\, ds \\ &\quad + \frac{\alpha M \|f\|_{C([0,t_1],X)}}{\Gamma(1+\alpha)} \left| \int_0^\tau (\tau-s)^{\alpha-1} ds - \int_0^\tau (t-s)^{\alpha-1} ds \right| \\ &\le \|S_\alpha(t) - S_\alpha(\tau)\|_{B(X)} |x_0| \\ &\quad + \frac{t_1^\alpha \|f\|_{PC}}{\alpha} \sup_{s \in [0,\tau]} \|P_\alpha(t-s) - P_\alpha(\tau-s)\|_{B(X)} \\ &\quad + \frac{3M \|f\|_{PC} (t-\tau)^\alpha}{\Gamma(1+\alpha)}, \end{aligned}$$

where we use the inequality $t^\alpha - \tau^\alpha \le (t-\tau)^\alpha$. Keeping in mind of Proposition 4.7, the first and second terms tend to zero as $t \to \tau$. Moreover, it is obvious that the last terms tends to zero too as $t \to \tau$. Thus, we can deduce that $Tx \in C([0,t_1],X)$.

For $t_1 \le \tau < t < t_2$, keeping in mind our assumptions and applying Proposition 4.5 again, we have

$$|(Tx)(t) - (Tx)(\tau)|$$

$$
\begin{aligned}
&\leq \|S_\alpha(t) - S_\alpha(\tau)\||x_0| + \|S_\alpha(t-t_1) - S_\alpha(\tau - t_1)\|_{B(X)}|y_1| \\
&\quad + \frac{t_2^\alpha \|f\|_{PC}}{\alpha} \sup_{s\in[0,\tau]} \|P_\alpha(t-s) - P_\alpha(\tau - s)\|_{B(X)} \\
&\quad + \frac{3M\|f\|_{PC}(t-\tau)^\alpha}{\Gamma(1+\alpha)}.
\end{aligned}
$$

As $t \to \tau$, the right hand side of the above inequality tends to zero. Thus, we can deduce that $Tx \in C((t_1, t_2], X)$.

Similarly, we can also obtain that $Tx \in C((t_2, t_3], X), ..., Tx \in C((t_\delta, b], X)$. That is, $Tx \in PC(J, X)$.

Claim II. T is contraction on $PC(J, X)$.

For each $t \in [0, t_1]$, it comes from our assumptions and Proposition 4.5 that

$$
\begin{aligned}
&|(Tx)(t) - (Ty)(t)| \\
&\leq \frac{\alpha M}{\Gamma(1+\alpha)} \int_0^t (t-s)^{\alpha-1} L_f(s)|x(s) - y(s)|ds \\
&\leq \frac{\alpha M \|x-y\|_{PC}}{\Gamma(1+\alpha)} \int_0^t (t-s)^{\alpha-1} L_f(s)ds \\
&\leq \frac{\alpha M \|x-y\|_{PC}}{\Gamma(1+\alpha)} \left(\int_0^t (t-s)^{\frac{\alpha-1}{1-q_1}} ds \right)^{1-q_1} \|L_f\|_{L^{\frac{1}{q_1}}[0,t_1]} \\
&\leq \frac{\alpha M \|x-y\|_{PC}}{\Gamma(1+\alpha)} \left[\left(\frac{1-q_1}{\alpha - q_1} \right) t_1^{\frac{\alpha - q_1}{1-q_1}} \right]^{1-q_1} \|L_f\|_{L^{\frac{1}{q_1}}[0,t_1]}.
\end{aligned}
$$

In general, for each $t \in (t_k, t_{k+1}]$, using our assumptions and Proposition 4.5 again,

$$
\begin{aligned}
&|(Tx)(t) - (Ty)(t)| \\
&\leq \frac{\alpha M \|x-y\|_{PC}}{\Gamma(1+\alpha)} \left[\left(\frac{1-q_1}{\alpha - q_1} \right) t_{k+1}^{\frac{\alpha - q_1}{1-q_1}} \right]^{1-q_1} \|L_f\|_{L^{\frac{1}{q_1}}[t_k,t_{k+1}]}.
\end{aligned}
$$

Thus,

$$
\|Tx - Ty\|_{PC} \leq \frac{\alpha M T^*}{\Gamma(1+\alpha)} \|x-y\|_{PC}.
$$

Hence, the condition (5.42) allows us to conclude in view of Banach contraction mapping principle, that T has a unique fixed point $x \in PC(J, X)$ which is just the unique PC-mild solution of system (5.36). □

Case II. *f* is not Lipschitz.

We make the following assumptions.

(C1) f: $J \times X \to X$ is continuous and maps a bounded set into a bounded set;

(C2) for each $x_0 \in X$, there exists a constant $r > 0$ such that

$$
M \left(|x_0| + \sum_{k=1}^{\delta} |y_k| + \frac{b^\alpha}{\Gamma(1+\alpha)} \sup_{s\in J, \phi \in Y_\Gamma} |f(s, \phi(s))| \right) \leq r,
$$

where

$$
Y_\Gamma = \{ \phi \in PC(J, X) \mid \|\phi\| \leq r \text{ for } t \in J \}.
$$

Theorem 5.14. *Suppose that (HA), (C1) and (C2) are satisfied. Then for every* $x_0 \in X$*, the system* (5.36) *has at least a PC-mild solution on* J.

Proof. Let $x_0 \in X$ be fixed. We introduce that map

$$T : PC(J, X) \to PC(J, X)$$

by

$$(Tv)(t) = (T_1 v)(t) + (T_2 v)(t),$$

where

$$(T_1 v)(t) = S_\alpha(t)x_0 + \int_0^t (t-s)^{\alpha-1} P_\alpha(t-s) f(s, v(s))\, ds, \quad t \in J \setminus \{t_1, t_2, ..., t_\delta\},$$

and

$$(T_2 v)(t) = \begin{cases} 0, & t \in [0, t_1], \\ \displaystyle\sum_{i=1}^{k} S_\alpha(t - t_i) y_i, & t \in (t_k, t_{k+1}],\ k = 1, ..., \delta. \end{cases} \tag{5.43}$$

For each $t \in [0, t_1]$, $v \in Y_\Gamma$,

$$\begin{aligned} |(Tv)(t)| &\le |(T_1 v)(t)| + |(T_2 v)(t)| \\ &\le M|x_0| + \frac{b^\alpha M}{\Gamma(1+\alpha)} \sup_{s \in J, \phi \in Y_\Gamma} |f(s, \phi(s))|. \end{aligned}$$

For each $t \in (t_k, t_{k+1}]$, $v \in Y_\Gamma$,

$$\begin{aligned} |(Tv)(t)| &\le |(T_1 v)(t)| + |(T_2 v)(t)| \\ &\le M|x_0| + M \sum_{k=1}^{\delta} |y_k| + \frac{b^\alpha M}{\Gamma(1+\alpha)} \sup_{s \in J, \phi \in Y_\Gamma} |f(s, \phi(s))|. \end{aligned}$$

Noting that the condition (C2), we see that $T : Y_\Gamma \to Y_\Gamma$.

Claim I. T is a continuous mapping from Y_Γ to Y_Γ.

In order to derive the continuity of T, we only check that T_1 and T_2 are all continuous.

For this purpose, we assume that $v_n \to v$ in Y_Γ. It comes from the continuity of f that $(\cdot - s)^{\alpha-1} f(s, v_n(s)) \to (\cdot - s)^{\alpha-1} f(s, v(s))$, as $n \to \infty$. Noting that

$$(t-s)^{\alpha-1} |f(s, v_n(s)) - f(s, v(s))| \le (t-s)^{\alpha-1} \sup_{s \in J, \phi \in Y_\Gamma} |f(s, \phi(s))|,$$

for $s \in [0, t]$, $t \in J$, by means of Lebesgue dominated convergence theorem, we obtain that

$$\int_0^t (t-s)^{\alpha-1} |f(s, v_n(s)) - f(s, v(s))| ds \to 0, \text{ as } n \to \infty.$$

It is easy to see that for each $t \in J$,

$$|(T_1 v_n)(t) - (T_1 v)(t)| \le \frac{\alpha M}{\Gamma(1+\alpha)} \int_0^t (t-s)^{\alpha-1} |f(s, v_n(s)) - f(s, v(s))| ds$$

$$\to 0, \text{ as } n \to \infty.$$

Thus, T_1 is continuous. On the other hand, it is obvious that T_2 is continuous.

Since T_1 and T_2 are continuous, T is continuous.

Claim II. T is a compact operator, or T_1 and T_2 are compact operators.

The compactness of T_2 is clear since it is a constant map (see (5.43)).

Now we prove the compactness of T_1. For each $t \in J$, the set $\{S_\alpha(t)x_0\}$ is precompact in X since $S_\alpha(t)$, $t > 0$ is compact.

Also, for each $t \in J$, arbitrary $b > h > 0$, $\varepsilon > 0$, the set

$$\left\{T(h^\alpha\varepsilon)\int_0^{t-h}(t-s)^{q-1}\left(\alpha\int_\varepsilon^\infty \theta M_\alpha(\theta)T((t-s)^\alpha\theta - h^\alpha\varepsilon)d\theta\right)\right.$$
$$\left.\times f(s, v(s))\,ds \mid v \in Y_\Gamma\right\}$$
$$= \left\{\alpha\int_0^{t-h}\int_\varepsilon^\infty \theta(t-s)^{\alpha-1}M_\alpha(\theta)T((t-s)^\alpha\theta)f(s, v(s))\,d\theta ds \mid v \in Y_\Gamma\right\}$$

is precompact in X since $T(h^\alpha\varepsilon)$ is compact.

Proceeding as in the proof of Theorem 3.1 in our previous work Zhou and Jiao, 2010b, one can obtain

$$\alpha\int_0^{t-h}\int_\varepsilon^\infty \theta(t-s)^{\alpha-1}M_\alpha(\theta)T((t-s)^\alpha\theta)f(s, v(s))\,d\theta ds$$
$$\to \alpha\int_0^{t}\int_0^\infty \theta(t-s)^{\alpha-1}M_\alpha(\theta)T((t-s)^\alpha\theta)f(s, v(s))\,d\theta ds,$$

as $h \to 0$, $\varepsilon \to 0$.

Thus, we can conclude that

$$\left\{\int_0^t (t-s)^{\alpha-1}P_\alpha(t-s)f(s, v(s))\,ds \mid v \in Y_\Gamma\right\}$$
$$= \left\{\alpha\int_0^{t}\int_0^\infty \theta(t-s)^{\alpha-1}M_\alpha(\theta)T((t-s)^\alpha\theta)f(s, v(s))\,d\theta ds \mid v \in Y_\Gamma\right\}$$

is precompact in X.

Therefore, the set

$$\left\{S_\alpha(t)x_0 + \sum_{i=1}^k S_\alpha(t-t_i)y_i + \int_0^t (t-s)^{\alpha-1}P_\alpha(t-s)f(s, v(s))\,ds \mid v \in Y_\Gamma\right\}$$

is precompact in X.

Thus, for each $t \in J$, $\{(T_1v)(t) \mid v \in Y_\Gamma\}$ is precompact in X.

Next, we show the equicontinuity of $\mathcal{M} = \{(T_1v)(\cdot) \mid v \in Y_\Gamma\}$.

The equicontinuity of $\{S_\alpha(t)x_0 \mid\ t \in J \setminus \{t_1, t_2, ..., t_\delta\}\}$, can be shown using the fact of $S_\alpha(\cdot)$ is continuous.

Now, we only need to check the equicontinuity of the second term in $\mathcal{M}$.

For $t \in J$, let $0 \leq t' < t'' \leq t_1$, set

$$I_1 = \left| \int_{t'}^{t''} (t'' - s)^{\alpha-1} P_\alpha(t'' - s) f(s, v(s)) ds \right|,$$

$$I_2 = \left| \int_0^{t'} \left((t'' - s)^{\alpha-1} - (t' - s)^{\alpha-1} \right) P_\alpha(t'' - s) f(s, v(s)) ds \right|,$$

$$I_3 = \left| \int_0^{t'} (t' - s)^{\alpha-1} \left(P_\alpha(t'' - s) - P_\alpha(t' - s) \right) f(s, v(s)) ds \right|.$$

After some computation, we have

$$\left| \int_0^{t''} (t'' - s)^{\alpha-1} P_\alpha(t'' - s) f(s, v(s)) ds - \int_0^{t'} (t' - s)^{\alpha-1} P_\alpha(t' - s) f(s, v(s)) ds \right|$$
$$\leq I_1 + I_2 + I_3.$$

Now repeating the previous discussion in Theorem 3.1 of Zhou and Jiao, 2010, we derive that I_1, I_2, I_2 tend to zero as $t'' \to t'$.

Accordingly, we see that the functions in $\mathcal{M}$ are equicontinuous. Therefore, T_1 is a compact operator by Arzela-Ascoli theorem, and hence T is also a compact operator. Now, Schauder fixed point theorem implies that T has a fixed point, which gives rise to a PC-mild solution. □

To end this section, we make the following assumptions.

(D1) f: $J \times X \to X$ is continuous and there exists a function $m(\cdot) \in L^\infty(J, \mathbb{R}^+)$ such that

$$|f(t, x)| \leq m(t), \text{ for all } x \in X \text{ and } t \in J.$$

Theorem 5.15. *Suppose that (HA) and (D1) are satisfied. Then system* (5.36) *has at least a PC-mild solution on J.*

Proof. We defined that $T : PC(J, X) \to PC(J, X)$ as in Theorem 5.14 by $(Tv)(t) = (T_1 v)(t) + (T_2 v)(t)$. Then we proceed in several steps.

Claim I. T is a continuous mapping from $PC(J, X)$ to $PC(J, X)$.

Let $\{v_n\}$ be a sequence in $PC(J, X)$ such that $v_n \to v$ in $PC(J, X)$. It comes from (D1) that $(\cdot - s)^{\alpha-1} f(s, v_n(s)) \to (\cdot - s)^{\alpha-1} f(s, v(s))$, as $n \to \infty$, and note that

$$(t - s)^{\alpha-1} |f(s, v_n(s)) - f(s, v(s))| \leq 2m(s)(t - s)^{\alpha-1} \in L^1(J, \mathbb{R}^+),$$

for $s \in [0, t]$, $t \in J$. Similar to the discussion in Theorem 5.14, one can prove that T is a continuous mapping from $PC(J, X)$ to $PC(J, X)$.

Claim II. T maps bounded sets into bounded sets in $PC(J, X)$.

So, let us prove that for any $r > 0$ there exits a $M^* > 0$ such that for each $v \in B_r = \{v \in PC(J, X) \mid \|v\|_{PC} \leq r\}$, we have $\|Tv\|_{PC} \leq M^*$.

Indeed, for any $v \in B_r$,

$$|(Tv)(t)| \leq |(T_1 v)(t)| + |(T_2 v)(t)|$$

$$\leq M|x_0| + M\sum_{i=1}^{\delta}|y_i| + \frac{b^\alpha M}{\Gamma(1+\alpha)}\|m\|_{L^\infty J},$$

which implies

$$\|Tv\|_{PC} \leq M|x_0| + M\sum_{i=1}^{\delta}|y_i| + \frac{b^\alpha M}{\Gamma(1+\alpha)}\|m\|_{L^\infty J} \equiv M^*.$$

Claim III. T is a compact operator.

In order to verify that T is a compact operator, one can repeat the same process in Claim II of Theorem 5.14 only need replace $\sup_{s\in J,\phi\in Y_\Gamma}\|f(s,\phi(s))\|$ by $\|m\|_{L^\infty J}$.

Claim IV. The set $\Theta = \{x \in PC(J,X) \mid x = \lambda Tx, \lambda \in [0,1]\}$ is bounded.

Let $v \in \Theta$. Then $v(t) = \lambda(Tv)(t)$ for some $\lambda \in [0,1]$. Thus, for $t \in J$, directly calculation implies that $\|Tv\|_{PC} \leq M^*$. Hence, we deduce that Θ is a bounded set.

Since we have already proven that T is continuous and compact, thanks to the Schaefer fixed point theorem, T has a fixed point which is a PC-mild solution of system (5.36) on J. □

Remark 5.9. In the assumption (D1), the condition $m(\cdot) \in L^\infty(J,\mathbb{R}^+)$ can be replaced by $m(\cdot) \in L^{\frac{1}{q_2}}(J,\mathbb{R}^+)$ where $\frac{1}{q_2} \in [0,\alpha)$.

5.6 Notes and Remarks

The material in Section 5.2 due to Fečkan, Zhou and Wang, 2012. The results in Section 5.3 are adopted from Wang, Zhou and Fečkan, 2012. The main results of Section 5.4 are from Wang, Fečkan and Zhou, 2013. The material in Section 5.5 due to Wang, Fečkan and Zhou, 2011.

Chapter 6

Fractional Boundary Value Problems

6.1 Introduction

Critical point theory and variational methods are crucial in the study of many mathematical models of real-world problems. We realized that critical point theory, which has been mostly developed by specialist in ordinary differential equations, partial differential equations, differential topology, optimization, should be made more popular among people working in fractional differential equations.

The main purpose of this chapter is to present a new approach via critical point theory to study the existence of solutions for the boundary value problem of fractional differential equations. In Section 6.2, we consider the existence of solutions for fractional boundary value problems by using the critical point theory. Section 6.3 is devoted to the existence of multiple solutions to the boundary value problem which arises from studying the steady fractional advection dispersion equation. In Section 6.4, according to variational methods, we investigate the multiplicity results for the solutions for boundary value problem.

6.2 Solutions for BVP with Left and Right Fractional Integrals

6.2.1 *Introduction*

In this section, we consider the fractional boundary value problem (BVP for short) of the following form

$$\begin{cases} \dfrac{d}{dt}\left(\dfrac{1}{2}\,{}_0D_t^{-\beta}(u'(t)) + \dfrac{1}{2}\,{}_tD_T^{-\beta}(u'(t))\right) + \nabla F(t, u(t)) = 0, \quad \text{a.e. } t \in [0, T], \\ u(0) = u(T) = 0, \end{cases} \tag{6.1}$$

where ${}_0D_t^{-\beta}$ and ${}_tD_T^{-\beta}$ are the left and right Riemann-Liouville fractional integrals of order $0 \le \beta < 1$, respectively, F: $[0, T] \times \mathbb{R}^N \to \mathbb{R}$ is a given function satisfying some assumptions and $\nabla F(t, x)$ is the gradient of F at x. In particular, if $\beta = 0$, BVP (6.1) reduces to the standard second order BVP.

Physical models containing fractional differential operators have recently renewed attention from scientists which is mainly due to applications as models for physical phenomena exhibiting anomalous diffusion. A strong motivation for investigating the BVP (6.1) comes from fractional advection dispersion equation (ADE for short). A fractional ADE is a generalization of the classical ADE in which the second-order derivative is replaced with a fractional-order derivative. In contrast to the classical ADE, the fractional ADE has solutions that resemble the highly skewed and heavy-tailed breakthrough curves observed in field and laboratory studies (see, Benson, Schumer, Meerschaert *et al.*, 2001; Benson, Wheatcraft and Meerschaert, 2000a), in particular in contaminant transport of ground-water flow (see, Benson, Wheatcraft and Meerschaert, 2000b). Benson *et al.* stated that solutes moving through a highly heterogeneous aquifer violations violates the basic assumptions of local second-order theories because of large deviations from the stochastic process of Brownian motion.

Let $\phi(t,x)$ represents the concentration of a solute at a point x at time t in an arbitrary bounded connected set $\Omega \subset \mathbb{R}^N$. According to Benson, Wheatcraft and Meerschaert, 2000a; Fix and Roop, 2004, the N-dimensional form of the fractional ADE can be written as

$$\frac{\partial \phi}{\partial t} = -\nabla(v\phi) - \nabla(\nabla^{-\beta}(-k\nabla\phi)) + f, \quad \text{in } \Omega, \tag{6.2}$$

where v is a constant mean velocity, k is a constant dispersion coefficient, $v\phi$ and $-k\nabla\phi$ denote the mass flux from advection and dispersion respectively. The components of $\nabla^{-\beta}$ in (6.2) are linear combination of the left and right Riemann-Liouville fractional integral operators

$$(\nabla^{-\beta}(-k\nabla\phi))_i = (q \ {}_{-\infty}D_{x_i}^{-\beta} + (1-q) \ {}_{x_i}D_{+\infty}^{-\beta})\Big(-k\frac{\partial \phi}{\partial x_i}\Big), \quad i = 1, ..., N, \tag{6.3}$$

where $q \in [0,1]$ describes the skewness of the transport process, and $\beta \in [0,1)$ is the order of the left and right Liouville-Weyl fractional integral operators on the real line (see Definition 1.5). This equation may be interpreted as stating that the mass flux of a particle is related to the negative gradient via a combination of the left and right fractional integrals. Equation (6.3) is physically interpreted as a Fick's law for concentrations of particles with a strong nonlocal interaction.

For discussions of equation (6.2), see Benson, Wheatcraft and Meerschaert, 2000b; Fix and Roop, 2004. When $\beta = 0$, the dispersion operators in (6.2) are identical and the classical ADE is recovered. In a more general version of (6.2), k is replaced by a symmetric positive definite matrix.

A special case of the fractional ADE (equation (6.2)) describes symmetric transitions. In this case, $\nabla^{-\beta}$ is equivalent to the symmetric operator

$$(\nabla^{-\beta})_i = \frac{1}{2} \ {}_{-\infty}D_{x_i}^{-\beta} + \frac{1}{2} \ {}_{x_i}D_{+\infty}^{-\beta}, \quad i = 1, ..., N. \tag{6.4}$$

Combining (6.2) and (6.4) gives the mass balance equation for advection and symmetric fractional dispersion.

The fractional ADE has been studied in one dimension (see, e.g., Benson, Wheatcraft and Meerschaert, 2000b), and in three dimension (see Lu, Molz and

Fix, 2002), over infinite domains by using the Fourier transform of fractional differential operators to determine a classical solution. Variational methods, especially the Galerkin approximation has been investigated to find the solutions of BVP (see, e.g., Fix and Roop, 2004) and fractional ADE (see, e.g., Ervin and Roop, 2006) on a finite domain by establishing some suitable fractional derivative spaces. A Lagrangian structure for some partial differential equations is obtained by using the fractional embedding theory of continuous Lagrangian systems (see, Cresson, 2010).

We note that for nonlinear BVP, some fixed point theorems were already applied successfully to investigate the existence of solutions (see, e.g., Agarwal, Benchohra and Hamani, 2010; Ahmad and Nieto, 2009; Benchohra, Hamani and Ntouyas, 2009; Zhang, 2010). However, it seems that fixed point theorem is not appropriate for discussing BVP (6.1) since the equivalent integral equation is not easy to be obtained. On the other hand, there is another effective approach, calculus of variation, which proved to be very useful in determining the existence of solutions for integer order differential equation provided that equation with certain boundary conditions possesses a variational structure on some suitable Sobolev spaces, for example, one can refer to Corvellec, Motreanu and Saccon, 2010; Li, Liang and Zhang, 2005; Mawhin and Willem, 1989; Rabinowitz, 1986; Tang and Wu, 2010 and the references therein for detailed discussions.

However, to the best of author's knowledge, there are few results on the solutions to BVP which were established by the critical point theory, since it is often very difficult to establish a suitable space and variational functional for fractional differential equations with some boundary conditions. These difficulties are mainly caused by the following properties of fractional integral and fractional derivative operators. These are:

(i) the composition rule in general fails to be satisfied by fractional integral and fractional derivative operators (e.g., Lemma 2.21 in Kilbas, Srivastava and Trujillo, 2006);

(ii) the fractional integral is a singular integral operator and fractional derivative operator is non-local (see Definitions 1.1, 1.2 and 1.3), and

(iii) the adjoint of a fractional differential operator is not the negative of itself (e.g., Lemma 2.7 in Kilbas, Srivastava and Trujillo, 2006).

It should be mentioned here that the fractional variational principles were started to be investigated deeply. The fractional calculus of variations was introduced by Riewe, 1996, where he presented a new approach to mechanics that allows one to obtain the equations for a nonconservative system using certain functionals. Klimek, 2002, gave another approach by considering fractional derivatives, and corresponding Euler-Lagrange equations were obtained, using both Lagrangian and Hamiltonian formalisms. Agrawal, 2002, presented Euler-Lagrange equations for unconstrained and constrained fractional variational problems, and as a continuation of Agrawal's work, the generalized mechanics are considered to obtain the

Hamiltonian formulation for the Lagrangian depending on fractional derivative of coordinates (see, Rabei, Nawafleh, Hijjawi *et al.*, 2007). The recent book by Malinowska and Torres, 2012, provides a broad introduction to the important subject of fractional calculus of variations.

In Section 6.2, we investigate the existence of solutions for BVP (6.1). The technical tool is the critical point theory. In Subsection 6.2.2, we develop a fractional derivative space and some propositions are proven which aid in our analysis, and in Subsection 6.2.3, we shall exhibit a variational structure for BVP (6.1). The results presented in Subsections 6.2.2 and 6.2.3 are basic, but crucial to limpidly reveal that under some suitable assumptions, the critical points of the variational functional defined on a suitable Hilbert space are the solutions of BVP (6.1). In Subsection 6.2.4, we introduce some critical point theorems. Also, various criteria on the existence of solutions for BVP (6.1) is established.

As it was already mentioned, if $\beta = 0$, then BVP (6.1) reduces to the standard second order BVP of the following form

$$\begin{cases} u''(t) + \nabla F(t, u(t)) = 0, & \text{a.e. } t \in [0, T], \\ u(0) = u(T) = 0, \end{cases}$$

where $F\colon\ [0,T] \times \mathbb{R}^N \to \mathbb{R}$ is a given function and $\nabla F(t, x)$ is the gradient of F at x. Although many excellent results have been worked out on the existence of solutions for second order BVP (e.g., Li, Liang and Zhang, 2005; Rabinowitz, 1986), it seems that no similar results were obtained in the literature for fractional BVP. The present results in Section 6.2 are to show that the critical point theory is an effective approach to tackle the existence of solutions for fractional BVP.

6.2.2 *Fractional Derivative Space*

Let us recall that for any fixed $t \in [0, T]$ and $1 \le p < \infty$,

$$\|u\|_{L^p[0,t]} = \left(\int_0^t |u(\xi)|^p d\xi \right)^{\frac{1}{p}}, \quad \|u\|_{L^p} = \left(\int_0^T |u(t)|^p dt \right)^{\frac{1}{p}} \quad \text{and } \|u\| = \max_{t \in [0,T]} |u(t)|.$$

The following result yields the boundedness of the Riemann-Liouville fractional integral operators from the space $L^p([0,T], \mathbb{R}^N)$ to the space $L^p([0,T], \mathbb{R}^N)$, where $1 \le p < \infty$. It should be mentioned here that the similar results have been presented in Fix and Roop, 2004; Kilbas, Srivastava and Trujillo, 2006; Samko, Kilbas and Marichev, 1993.

Lemma 6.1. *Let $0 < \alpha \le 1$ and $1 \le p < \infty$. For any $f \in L^p([0,T], \mathbb{R}^N)$, we have*

$$\|{}_0D_\xi^{-\alpha} f\|_{L^p[0,t]} \le \frac{t^\alpha}{\Gamma(\alpha+1)} \|f\|_{L^p[0,t]}, \quad \textit{for } \xi \in [0,t], \quad t \in [0,T]. \tag{6.5}$$

Proof. Inspired by the proof of Young theorem in Adams, 1975, we can prove (6.5).

In fact, if $p = 1$, we have

$$\begin{aligned}\|{}_0D_\xi^{-\alpha} f\|_{L^1[0,t]} &= \frac{1}{\Gamma(\alpha)}\left|\int_0^t \int_0^\xi (\xi-\tau)^{\alpha-1} f(\tau)d\tau d\xi\right| \\ &\le \frac{1}{\Gamma(\alpha)}\int_0^t \int_0^\xi (\xi-\tau)^{\alpha-1} |f(\tau)|d\tau d\xi \\ &= \frac{1}{\Gamma(\alpha)}\int_0^t |f(\tau)|d\tau \int_\tau^t (\xi-\tau)^{\alpha-1} d\xi \\ &= \frac{1}{\Gamma(\alpha+1)}\int_0^t |f(\tau)|(t-\tau)^{\alpha} d\tau \\ &\le \frac{t^\alpha}{\Gamma(\alpha+1)}\|f\|_{L^1[0,t]}, \quad \text{for } t\in[0,T].\end{aligned} \tag{6.6}$$

Now, suppose that $1 < p < \infty$ and $g \in L^q([0,T],\mathbb{R}^N)$, where $\frac{1}{p}+\frac{1}{q}=1$. We have

$$\begin{aligned}&\left|\int_0^t g(\xi)\int_0^\xi (\xi-\tau)^{\alpha-1} f(\tau)d\tau d\xi\right| \\ &= \left|\int_0^t g(\xi)\int_0^\xi \tau^{\alpha-1} f(\xi-\tau)d\tau d\xi\right| \\ &\le \int_0^t |g(\xi)|\int_0^\xi \tau^{\alpha-1} |f(\xi-\tau)|d\tau d\xi \\ &= \int_0^t \tau^{\alpha-1}d\tau \int_\tau^t |g(\xi)||f(\xi-\tau)|d\xi \\ &\le \int_0^t \tau^{\alpha-1}d\tau\left(\int_\tau^t |g(\xi)|^q d\xi\right)^{\frac{1}{q}}\left(\int_\tau^t |f(\xi-\tau)|^p d\xi\right)^{\frac{1}{p}} \\ &\le \frac{t^\alpha}{\alpha}\|f\|_{L^p[0,t]}\|g\|_{L^q[0,t]}, \quad \text{for } t\in[0,T].\end{aligned} \tag{6.7}$$

For any fixed $t \in [0,T]$, consider the functional $H_{\xi * f} : L^q([0,T],\mathbb{R}^N) \to \mathbb{R}$

$$H_{\xi*f}(g) = \int_0^t \left(\int_0^\xi (\xi-\tau)^{\alpha-1} f(\tau)d\tau\right)g(\xi)d\xi. \tag{6.8}$$

According to (6.7), it is obvious that $H_{\xi*f} \in (L^q([0,T],\mathbb{R}^N))^*$, where $(L^q([0,T],\mathbb{R}^N))^*$ denotes the dual space of $L^q([0,T],\mathbb{R}^N)$. Therefore, by (6.7), (6.8) and Riesz representation theorem, there exists $h \in L^p([0,T],\mathbb{R}^N)$ such that

$$\int_0^t h(\xi)g(\xi)d\xi = \int_0^t \left(\int_0^\xi (\xi-\tau)^{\alpha-1} f(\tau)d\tau\right)g(\xi)d\xi \tag{6.9}$$

and

$$\|h\|_{L^p[0,t]} \le \frac{t^\alpha}{\alpha}\|f\|_{L^p[0,t]} \tag{6.10}$$

for all $g \in L^q([0,T],\mathbb{R}^N)$. Hence, we have by (6.9)

$$\frac{1}{\Gamma(\alpha)}h(\xi) = \frac{1}{\Gamma(\alpha)}\int_0^\xi (\xi-\tau)^{\alpha-1} f(\tau)d\tau = {}_0D_\xi^{-\alpha} f(\xi), \quad \text{for } \xi \in [0,t],$$

which means that

$$\|{}_0D_\xi^{-\alpha}f\|_{L^p[0,t]} = \frac{1}{\Gamma(\alpha)}\|h\|_{L^p[0,t]} \leq \frac{t^\alpha}{\Gamma(\alpha+1)}\|f\|_{L^p[0,t]} \tag{6.11}$$

according to (6.10). Combining (6.6) and (6.11), we obtain the inequality (6.5). □

In order to establish a variational structure for BVP (6.1), it is necessary to construct appropriate function spaces. Denote by $C_0^\infty([0,T],\mathbb{R}^N)$ the set of all functions $h \in C^\infty([0,T],\mathbb{R}^N)$ with $h(0) = h(T) = 0$. According to Lemma 6.1, for any $h \in C_0^\infty([0,T],\mathbb{R}^N)$ and $1 < p < \infty$, we have $h \in L^p([0,T],\mathbb{R}^N)$ and ${}_0^CD_t^\alpha h \in L^p([0,T],\mathbb{R}^N)$. Therefore, one can construct a set of space $E_0^{\alpha,p}$, which depend on L^p-integrability of Caputo fractional derivative of a function.

Definition 6.1. Let $0 < \alpha \leq 1$ and $1 < p < \infty$. The fractional derivative space $E_0^{\alpha,p}$ is defined by the closure of $C_0^\infty([0,T],\mathbb{R}^N)$ with respect to the norm

$$\|u\|_{\alpha,p} = \left(\int_0^T |u(t)|^p dt + \int_0^T |{}_0^CD_t^\alpha u(t)|^p dt\right)^{\frac{1}{p}}, \quad \forall u \in E_0^{\alpha,p}.$$

Remark 6.1.

(i) It is obvious that the fractional derivative space $E_0^{\alpha,p}$ is the space of functions $u \in L^p([0,T],\mathbb{R}^N)$ having an α-order Caputo fractional derivative ${}_0^CD_t^\alpha u \in L^p([0,T],\mathbb{R}^N)$ and $u(0) = u(T) = 0$.

(ii) For any $u \in E_0^{\alpha,p}$, noting the fact that $u(0) = 0$, we have ${}_0^CD_t^\alpha u(t) = {}_0D_t^\alpha u(t)$, $t \in [0,T]$ according to Proposition 1.1.

(iii) It is easy to verify that $E_0^{\alpha,p}$ is a reflexive and separable Banach space.

Proposition 6.1. *Let $0 < \alpha \leq 1$ and $1 < p < \infty$. The fractional derivative space $E_0^{\alpha,p}$ is a reflexive and separable Banach space.*

Proof. In fact, owing to $L^p([0,T],\mathbb{R}^N)$ be reflexive and separable, the Cartesian product space

$$L_2^p([0,T],\mathbb{R}^N) = L^p([0,T],\mathbb{R}^N) \times L^p([0,T],\mathbb{R}^N)$$

is also a reflexive and separable Banach space with respect to the norm

$$\|v\|_{L_2^p} = \left(\sum_{i=1}^2 \|v_i\|_{L^p[0,T]}^p\right)^{\frac{1}{p}}, \tag{6.12}$$

where $v = (v_1, v_2) \in L_2^p([0,T],\mathbb{R}^N)$.

Consider the space $\Omega = \{(u, {}_0^CD_t^\alpha u) : u \in E_0^{\alpha,p}\}$, which is a closed subset of $L_2^p([0,T],\mathbb{R}^N)$ as $E_0^{\alpha,p}$ is closed. Therefore, Ω is also a reflexive and separable Banach space with respect to the norm (6.12) for $v = (v_1, v_2) \in \Omega$.

We form the operator $A : E_0^{\alpha,p} \to \Omega$ as follows

$$A : u \to (u,\ {}_0^CD_t^\alpha u), \quad \forall\, u \in E_0^{\alpha,p}.$$

It is obvious that

$$\|u\|_{\alpha,p} = \|Au\|_{L_2^p},$$

which means that the operator $A : u \to (u, {}_0^CD_t^\alpha u)$ is an isometric isomorphic mapping and the space $E_0^{\alpha,p}$ is isometric isomorphic to the space Ω. Thus $E_0^{\alpha,p}$ is a reflexive and separable Banach space, and this completes the proof. □

Applying Proposition 1.9 and Lemma 6.1, we now can give the following useful estimates.

Proposition 6.2. *Let $0 < \alpha \leq 1$ and $1 < p < \infty$. For all $u \in E_0^{\alpha,p}$, we have*

$$\|u\|_{L^p[0,T]} \leq \frac{T^\alpha}{\Gamma(\alpha+1)} \|{}_0^CD_t^\alpha u\|_{L^p[0,T]}. \tag{6.13}$$

Moreover, if $\alpha > \frac{1}{p}$ and $\frac{1}{p} + \frac{1}{q} = 1$, then

$$\|u\| \leq \frac{T^{\alpha-\frac{1}{p}}}{\Gamma(\alpha)((\alpha-1)q+1)^{\frac{1}{q}}} \|{}_0^CD_t^\alpha u\|_{L^p[0,T]}. \tag{6.14}$$

Proof. For any $u \in E_0^{\alpha,p}$, according to (1.11) and noting the fact that $u(0) = 0$, we have that

$${}_0D_t^{-\alpha}({}_0^CD_t^\alpha u(t)) = u(t), \quad t \in [0,T].$$

Therefore, in order to prove inequalities (6.13) and (6.14), we only need to prove that

$$\|{}_0D_t^{-\alpha}({}_0^CD_t^\alpha u)\|_{L^p[0,T]} \leq \frac{T^\alpha}{\Gamma(\alpha+1)} \|{}_0^CD_t^\alpha u\|_{L^p[0,T]}, \tag{6.15}$$

where $0 < \alpha \leq 1$ and $1 < p < \infty$, and

$$\|{}_0D_t^{-\alpha}({}_0^CD_t^\alpha u)\| \leq \frac{T^{\alpha-\frac{1}{p}}}{\Gamma(\alpha)((\alpha-1)q+1)^{\frac{1}{q}}} \|{}_0^CD_t^\alpha u\|_{L^p[0,T]}, \tag{6.16}$$

where $\alpha > \frac{1}{p}$ and $\frac{1}{p} + \frac{1}{q} = 1$.

Firstly, we note that ${}_0^CD_t^\alpha u \in L^p([0,T], \mathbb{R}^N)$, the inequality (6.15) follows from (6.5) directly.

We are now in a position to prove (6.16). For $\alpha > \frac{1}{p}$, choose q such that $\frac{1}{p} + \frac{1}{q} = 1$. $\forall\, u \in E_0^{\alpha,p}$, we have

$$\begin{aligned}
|{}_0D_t^{-\alpha}({}_0^CD_t^\alpha u(t))| &= \frac{1}{\Gamma(\alpha)} \left| \int_0^t (t-s)^{\alpha-1}\, {}_0^CD_s^\alpha u(s)ds \right| \\
&\leq \frac{1}{\Gamma(\alpha)} \left(\int_0^t (t-s)^{(\alpha-1)q} ds \right)^{\frac{1}{q}} \|{}_0^CD_t^\alpha u\|_{L^p[0,T]} \\
&\leq \frac{T^{\frac{1}{q}+\alpha-1}}{\Gamma(\alpha)((\alpha-1)q+1)^{\frac{1}{q}}} \|{}_0^CD_t^\alpha u\|_{L^p[0,T]} \\
&= \frac{T^{\alpha-\frac{1}{p}}}{\Gamma(\alpha)((\alpha-1)q+1)^{\frac{1}{q}}} \|{}_0^CD_t^\alpha u\|_{L^p[0,T]},
\end{aligned}$$

and this completes the proof. □

According to (6.13), we can consider $E_0^{\alpha,p}$ with respect to the norm

$$\|u\|_{\alpha,p} = \|{}_0^C D_t^\alpha u\|_{L^p[0,T]} = \left(\int_0^T |{}_0^C D_t^\alpha u(t)|^p dt\right)^{\frac{1}{p}} \tag{6.17}$$

in the following analysis.

Proposition 6.3. *Let $0 < \alpha \leq 1$ and $1 < p < \infty$. Assume that $\alpha > \frac{1}{p}$ and the sequence $\{u_k\}$ converges weakly to u in $E_0^{\alpha,p}$, i.e., $u_k \rightharpoonup u$. Then $u_k \to u$ in $C([0,T],\mathbb{R}^N)$, i.e., $\|u - u_k\| \to 0$, as $k \to \infty$.*

Proof. If $\alpha > \frac{1}{p}$, then by (6.14) and (6.17), the injection of $E_0^{\alpha,p}$ into $C([0,T],\mathbb{R}^N)$, with its natural norm $\|\cdot\|$, is continuous, i.e., if $u_k \to u$ in $E_0^{\alpha,p}$, then $u_k \to u$ in $C([0,T],\mathbb{R}^N)$.

Since $u_k \rightharpoonup u$ in $E_0^{\alpha,p}$, it follows that $u_k \rightharpoonup u$ in $C([0,T],\mathbb{R}^N)$. In fact, for any $h \in (C([0,T],\mathbb{R}^N))^*$, if $u_k \to u$ in $E_0^{\alpha,p}$, then $u_k \to u$ in $C([0,T],\mathbb{R}^N)$, and thus $h(u_k) \to h(u)$. Therefore, $h \in (E_0^{\alpha,p})^*$, which means that $(C([0,T],\mathbb{R}^N))^* \subseteq (E_0^{\alpha,p})^*$.

Hence, if $u_k \rightharpoonup u$ in $E_0^{\alpha,p}$, then for any $h \in (C([0,T],\mathbb{R}^N))^*$, we have $h \in (E_0^{\alpha,p})^*$, and thus $h(u_k) \to h(u)$, i.e., $u_k \rightharpoonup u$ in $C([0,T],\mathbb{R}^N)$.

By the Banach-Steinhaus theorem, $\{u_k\}$ is bounded in $E_0^{\alpha,p}$ and, hence, in $C([0,T],\mathbb{R}^N)$. We are now in a position to prove that the sequence $\{u_k\}$ is equi-uniformly continuous.

Let $\frac{1}{p} + \frac{1}{q} = 1$ and $0 \leq t_1 < t_2 \leq T$. $\forall\ f \in L^p([0,T],\mathbb{R}^N)$, by using Hölder inequality and noting that $\alpha > \frac{1}{p}$, we have

$$\begin{aligned}
&|{}_0D_{t_1}^{-\alpha} f(t_1) - {}_0D_{t_2}^{-\alpha} f(t_2)| \\
&= \frac{1}{\Gamma(\alpha)}\left|\int_0^{t_1} (t_1 - s)^{\alpha-1} f(s)ds - \int_0^{t_2} (t_2 - s)^{\alpha-1} f(s)ds\right| \\
&\leq \frac{1}{\Gamma(\alpha)}\left|\int_0^{t_1} (t_1 - s)^{\alpha-1} f(s)ds - \int_0^{t_1} (t_2 - s)^{\alpha-1} f(s)ds\right| \\
&\quad + \frac{1}{\Gamma(\alpha)}\left|\int_{t_1}^{t_2} (t_2 - s)^{\alpha-1} f(s)ds\right| \\
&\leq \frac{1}{\Gamma(\alpha)}\int_0^{t_1} \left((t_1 - s)^{\alpha-1} - (t_2 - s)^{\alpha-1}\right) |f(s)|ds \\
&\quad + \frac{1}{\Gamma(\alpha)}\int_{t_1}^{t_2} (t_2 - s)^{\alpha-1}|f(s)|ds
\end{aligned} \tag{6.18}$$

$$
\begin{aligned}
&\leq \frac{1}{\Gamma(\alpha)}\left(\int_0^{t_1}\left((t_1-s)^{\alpha-1}-(t_2-s)^{\alpha-1}\right)^q ds\right)^{\frac{1}{q}}\|f\|_{L^p[0,T]} \\
&\quad+\frac{1}{\Gamma(\alpha)}\left(\int_{t_1}^{t_2}(t_2-s)^{(\alpha-1)q}ds\right)^{\frac{1}{q}}\|f\|_{L^p[0,T]} \\
&\leq \frac{1}{\Gamma(\alpha)}\left(\int_0^{t_1}(t_1-s)^{(\alpha-1)q}-(t_2-s)^{(\alpha-1)q}ds\right)^{\frac{1}{q}}\|f\|_{L^p[0,T]} \\
&\quad+\frac{1}{\Gamma(\alpha)}\left(\int_{t_1}^{t_2}(t_2-s)^{(\alpha-1)q}ds\right)^{\frac{1}{q}}\|f\|_{L^p[0,T]} \\
&= \frac{\|f\|_{L^p[0,T]}}{\Gamma(\alpha)(1+(\alpha-1)q)^{\frac{1}{q}}}\left(t_1^{(\alpha-1)q+1}-t_2^{(\alpha-1)q+1}+(t_2-t_1)^{(\alpha-1)q+1}\right)^{\frac{1}{q}} \\
&\quad+\frac{\|f\|_{L^p[0,T]}}{\Gamma(\alpha)(1+(\alpha-1)q)^{\frac{1}{q}}}\left((t_2-t_1)^{(\alpha-1)q+1}\right)^{\frac{1}{q}} \\
&\leq \frac{2\|f\|_{L^p[0,T]}}{\Gamma(\alpha)(1+(\alpha-1)q)^{\frac{1}{q}}}(t_2-t_1)^{\alpha-1+\frac{1}{q}} \\
&= \frac{2\|f\|_{L^p[0,T]}}{\Gamma(\alpha)(1+(\alpha-1)q)^{\frac{1}{q}}}(t_2-t_1)^{\alpha-\frac{1}{p}}.
\end{aligned}
$$

Therefore, the sequence $\{u_k\}$ is equi-uniformly continuous since, for $0 \leq t_1 < t_2 \leq T$, by applying (6.18) and in view of (6.17), we have

$$
\begin{aligned}
|u_k(t_1)-u_k(t_2)| &= \left|{}_0D_{t_1}^{-\alpha}\left({}_0^CD_{t_1}^{\alpha}u_k(t_1)\right)-{}_0D_{t_2}^{-\alpha}\left({}_0^CD_{t_2}^{\alpha}u_k(t_2)\right)\right| \\
&\leq \frac{2(t_2-t_1)^{\alpha-\frac{1}{p}}}{\Gamma(\alpha)(1+(\alpha-1)q)^{\frac{1}{q}}}\left\|{}_0^CD_t^{\alpha}u_k\right\|_{L^p[0,T]} \\
&= \frac{2(t_2-t_1)^{\alpha-\frac{1}{p}}}{\Gamma(\alpha)(1+(\alpha-1)q)^{\frac{1}{q}}}\|u_k\|_{\alpha,p} \\
&\leq c(t_2-t_1)^{\alpha-\frac{1}{p}},
\end{aligned}
$$

where $\frac{1}{p}+\frac{1}{q}=1$ and $c \in \mathbb{R}^+$ is a constant. By Arzela-Ascoli theorem, $\{u_k\}$ is relatively compact in $C([0,T],\mathbb{R}^N)$. By the uniqueness of the weak limit in $C([0,T],\mathbb{R}^N)$, every uniformly convergent subsequence of $\{u_k\}$ converges uniformly on $[0,T]$ to u. The proof is completed. □

6.2.3 *Variational Structure*

In this subsection, we establish a variational structure which enables us to reduce the existence of solutions of BVP (6.1) to the one of critical points of corresponding functional defined on the space $E_0^{\alpha,p}$ with $p=2$ and $\frac{1}{2}<\alpha\leq 1$.

First of all, making use of Proposition 1.4, for any $u \in AC([0,T],\mathbb{R}^N)$, BVP

(6.1) transforms to

$$\begin{cases} \dfrac{d}{dt}\left(\dfrac{1}{2}\, {}_0D_t^{-\frac{\beta}{2}}({}_0D_t^{-\frac{\beta}{2}}u'(t)) + \dfrac{1}{2}\, {}_tD_T^{-\frac{\beta}{2}}({}_tD_T^{-\frac{\beta}{2}}u'(t))\right) + \nabla F(t,u(t)) = 0, \\ u(0) = u(T) = 0, \end{cases} \tag{6.19}$$

for almost every $t \in [0,T]$, where $\beta \in [0,1)$.

Furthermore, in view of Definition 1.3 and Proposition 1.2, it is obvious that $u \in AC([0,T],\mathbb{R}^N)$ is a solution of BVP (6.19) if and only if u is a solution of the following problem

$$\begin{cases} \dfrac{d}{dt}\left(\dfrac{1}{2}\, {}_0D_t^{\alpha-1}({}_0^CD_t^{\alpha}u(t)) - \dfrac{1}{2}\, {}_tD_T^{\alpha-1}({}_t^CD_T^{\alpha}u(t))\right) + \nabla F(t,u(t)) = 0, \\ u(0) = u(T) = 0, \end{cases} \tag{6.20}$$

for almost every $t \in [0,T]$, where $\alpha = 1 - \frac{\beta}{2} \in (\frac{1}{2}, 1]$. Therefore, we seek a solution u of BVP (6.20) which, of course, corresponds to the solution u of BVP (6.1) provided that $u \in AC([0,T],\mathbb{R}^N)$.

Let us denote by

$$D^{\alpha}(u(t)) = \frac{1}{2}\, {}_0D_t^{\alpha-1}({}_0^CD_t^{\alpha}u(t)) - \frac{1}{2}\, {}_tD_T^{\alpha-1}({}_t^CD_T^{\alpha}u(t)). \tag{6.21}$$

We are now in a position to give a definition of the solution of BVP (6.20).

Definition 6.2. A function $u \in AC([0,T],\mathbb{R}^N)$ is called a solution of BVP (6.20) if

(i) $D^{\alpha}(u(t))$ is derivable for almost every $t \in [0,T]$, and
(ii) u satisfies (6.20).

In the sequel, we treat BVP (6.20) in the Hilbert space $E^{\alpha} = E_0^{\alpha,2}$ with the corresponding norm $\|u\|_{\alpha} = \|u\|_{\alpha,2}$ which we defined in (6.17).

Consider the functional $u \to -\int_0^T ({}_0^CD_t^{\alpha}u(t), {}_t^CD_T^{\alpha}u(t))dt$ on E^{α}. The following estimate is useful for our further discussion.

Proposition 6.4. *If* $\frac{1}{2} < \alpha \leq 1$, *then for any* $u \in E^{\alpha}$, *we have*

$$|\cos(\pi\alpha)|\|u\|_{\alpha}^2 \leq -\int_0^T ({}_0^CD_t^{\alpha}u(t), {}_t^CD_T^{\alpha}u(t))dt \leq \frac{1}{|\cos(\pi\alpha)|}\|u\|_{\alpha}^2. \tag{6.22}$$

Proof. Let $u \in E^{\alpha}$ and $\tilde{u}$ be the extension of u by zero on $\mathbb{R} \setminus [0,T]$. Then $\mathrm{supp}(\tilde{u}) \subseteq [0,T]$. However, as the left and right fractional derivatives are nonlocal,

$$\mathrm{supp}({}_{-\infty}D_t^{\alpha}\tilde{u}) \subseteq [0,\infty) \quad \text{and} \quad \mathrm{supp}({}_tD_{+\infty}^{\alpha}\tilde{u}) \subseteq (-\infty,T].$$

Nonetheless, the product $({}_{-\infty}D_t^{\alpha}\tilde{u}, {}_tD_{+\infty}^{\alpha}\tilde{u})$ has support in $[0,T]$.

On the other hand, according to Theorem 2.3 and Lemma 2.4 in Ervin and Roop, 2006, we have

$$\begin{aligned} \int_{-\infty}^{\infty} ({}_{-\infty}D_t^{\alpha}\tilde{u}(t), {}_tD_{+\infty}^{\alpha}\tilde{u}(t))dt &= \cos(\pi\alpha)\int_{-\infty}^{\infty} |{}_{-\infty}D_t^{\alpha}\tilde{u}(t)|^2 dt \\ &= \cos(\pi\alpha)\int_{-\infty}^{\infty} |{}_tD_{+\infty}^{\alpha}\tilde{u}(t)|^2 dt, \end{aligned} \tag{6.23}$$

where $_{-\infty}D_t^\alpha$ and $_tD_{+\infty}^\alpha$ are Liouville-Weyl fractional derivatives on the real line (see Definition 1.5). Helpful in establishing (6.23) is the Fourier transform of Liouville-Weyl fractional derivative on the real line (see, Podlubny, 1999). Hence, according to Remark 6.1, (6.23) and noting that $\cos(\pi\alpha) \in [-1, 0)$ as $\alpha \in (\frac{1}{2}, 1]$, we have

$$\begin{aligned}
-\int_0^T ({}_0^CD_t^\alpha u(t), {}_t^CD_T^\alpha u(t))dt &= -\int_0^T ({}_0D_t^\alpha u(t), {}_tD_T^\alpha u(t))dt \\
&= -\int_0^T ({}_{-\infty}D_t^\alpha \tilde{u}(t), {}_tD_{+\infty}^\alpha \tilde{u}(t))dt \\
&= -\int_{-\infty}^\infty ({}_{-\infty}D_t^\alpha \tilde{u}(t), {}_tD_{+\infty}^\alpha \tilde{u}(t))dt \\
&= -\cos(\pi\alpha) \int_{-\infty}^\infty |{}_{-\infty}D_t^\alpha \tilde{u}(t)|^2 dt \\
&= -\cos(\pi\alpha) \int_0^\infty |{}_0D_t^\alpha \tilde{u}(t)|^2 dt \\
&\geq -\cos(\pi\alpha) \int_0^T |{}_0D_t^\alpha u(t)|^2 dt \\
&= |\cos(\pi\alpha)| \int_0^T |{}_0^CD_t^\alpha u(t)|^2 dt \\
&= |\cos(\pi\alpha)| \|u\|_\alpha^2.
\end{aligned} \tag{6.24}$$

On the other hand, by using Young inequality, we obtain

$$\begin{aligned}
\left|\int_0^T ({}_0^CD_t^\alpha u(t), {}_t^CD_T^\alpha u(t))dt\right| &= \left|\int_0^T ({}_0D_t^\alpha u(t), {}_tD_T^\alpha u(t))dt\right| \\
&\leq \int_0^T \frac{1}{\sqrt{2\varepsilon}} |{}_0D_t^\alpha u(t)| \sqrt{2\varepsilon}\, |{}_tD_T^\alpha u(t)| dt \\
&\leq \frac{1}{4\varepsilon} \int_0^T |{}_0D_t^\alpha u(t)|^2 dt + \varepsilon \int_0^T |{}_tD_T^\alpha u(t)|^2 dt \\
&= \frac{1}{4\varepsilon} \int_0^T |{}_0^CD_t^\alpha u(t)|^2 dt + \varepsilon \int_0^\infty |{}_tD_{+\infty}^\alpha \tilde{u}(t)|^2 dt \\
&\leq \frac{1}{4\varepsilon} \|u\|_\alpha^2 + \varepsilon \int_{-\infty}^\infty |{}_tD_{+\infty}^\alpha \tilde{u}(t)|^2 dt \\
&= \frac{1}{4\varepsilon} \|u\|_\alpha^2 + \frac{\varepsilon}{|\cos(\pi\alpha)|} \left|\int_{-\infty}^\infty ({}_{-\infty}D_t^\alpha \tilde{u}(t), {}_tD_{+\infty}^\alpha \tilde{u}(t))dt\right| \\
&= \frac{1}{4\varepsilon} \|u\|_\alpha^2 + \frac{\varepsilon}{|\cos(\pi\alpha)|} \left|\int_0^T ({}_0D_t^\alpha u(t), {}_tD_T^\alpha u(t))dt\right| \\
&= \frac{1}{4\varepsilon} \|u\|_\alpha^2 + \frac{\varepsilon}{|\cos(\pi\alpha)|} \left|\int_0^T ({}_0^CD_t^\alpha u(t), {}_t^CD_T^\alpha u(t))dt\right|.
\end{aligned}$$

Therefore, by taking $\varepsilon = |\cos(\pi\alpha)|/2$, we have

$$\left|\int_0^T ({}_0^CD_t^\alpha u(t), {}_t^CD_T^\alpha u(t))dt\right| \leq \frac{1}{|\cos(\pi\alpha)|} \|u\|_\alpha^2. \tag{6.25}$$

The inequality (6.22) follows then from (6.24) and (6.25), and the proof is completed. □

Remark 6.2. According to (6.22) and (6.23), for any $u \in E^\alpha$, it is obvious that

$$\begin{aligned}
\int_0^T |{}_t^C D_T^\alpha u(t)|^2 dt &\leq \int_{-\infty}^{\infty} |{}_t D_{+\infty}^\alpha \tilde{u}(t)|^2 dt \\
&= -\int_0^T \frac{({}_0^C D_t^\alpha u(t), {}_t^C D_T^\alpha u(t))}{|\cos(\pi\alpha)|} dt \\
&\leq \frac{1}{|\cos(\pi\alpha)|^2} \|u\|_\alpha^2,
\end{aligned}$$

which means that ${}_t^C D_T^\alpha u \in L^2([0,T], \mathbb{R}^N)$.

In the following, we establish a variational structure on E^α with $\alpha \in (\frac{1}{2}, 1]$. Also, we show that the critical points of that functional are indeed solutions of BVP (6.20), and therefore, are solutions of BVP (6.1).

Theorem 6.1. *Let $L : [0,T] \times \mathbb{R}^N \times \mathbb{R}^N \times \mathbb{R}^N \to \mathbb{R}$ be defined by*

$$L(t, x, y, z) = -\frac{1}{2}(y, z) - F(t, x),$$

where $F : [0,T] \times \mathbb{R}^N \to \mathbb{R}$ satisfies the following assumption:

(A) *$F(t,x)$ is measurable in t for each $x \in \mathbb{R}^N$, continuously differentiable in x for almost every $t \in [0,T]$ and there exist $m_1 \in C(\mathbb{R}^+, \mathbb{R}^+)$ and $m_2 \in L^1([0,T], \mathbb{R}^+)$ such that*

$$|F(t,x)| \leq m_1(|x|) m_2(t), \quad |\nabla F(t,x)| \leq m_1(|x|) m_2(t)$$

for all $x \in \mathbb{R}^N$ and a.e. in $t \in [0,T]$.

If $\frac{1}{2} < \alpha \leq 1$, then the functional defined by

$$\begin{aligned}
\varphi(u) &= \int_0^T L(t, u(t), {}_0^C D_t^\alpha u(t), {}_t^C D_T^\alpha u(t)) dt \\
&= \int_0^T \Big(-\frac{1}{2}({}_0^C D_t^\alpha u(t), {}_t^C D_T^\alpha u(t)) - F(t, u(t)) \Big) dt
\end{aligned} \tag{6.26}$$

is continuously differentiable on E^α, and $\forall u, v \in E^\alpha$, we have

$$\begin{aligned}
\langle \varphi'(u), v \rangle &= \int_0^T \big(D_x L(t, u(t), {}_0^C D_t^\alpha u(t), {}_t^C D_T^\alpha u(t)), v(t)\big)\, dt \\
&\quad + \int_0^T \big(D_y L(t, u(t), {}_0^C D_t^\alpha u(t), {}_t^C D_T^\alpha u(t)), {}_0^C D_t^\alpha v(t)\big)\, dt \\
&\quad + \int_0^T \big(D_z L(t, u(t), {}_0^C D_t^\alpha u(t), {}_t^C D_T^\alpha u(t)), {}_t^C D_T^\alpha v(t)\big)\, dt \qquad (6.27) \\
&= -\int_0^T \frac{1}{2} \big(({}_0^C D_t^\alpha u(t), {}_t^C D_T^\alpha v(t)) + ({}_t^C D_T^\alpha u(t), {}_0^C D_t^\alpha v(t))\big)\, dt \\
&\quad - \int_0^T (\nabla F(t, u(t)), v(t)) dt.
\end{aligned}$$

Proof. First, we note that for a.e. $t \in [0,T]$ and every $[x,y,z] \in \mathbb{R}^N \times \mathbb{R}^N \times \mathbb{R}^N$, one has

$$|L(t,x,y,z)| \leq m_1(|x|)m_2(t) + \frac{1}{4}(|y|^2 + |z|^2), \tag{6.28}$$

$$|D_x L(t,x,y,z)| \leq m_1(|x|)m_2(t), \tag{6.29}$$

$$|D_y L(t,x,y,z)| \leq \frac{1}{2}|z| \text{ and } |D_z L(t,x,y,z)| \leq \frac{1}{2}|y|. \tag{6.30}$$

Then, inspired by the proof of Theorem 1.4 in Mawhin and Willem, 1989, it suffices to prove that at every point u,φ has a directional derivative $\varphi'(u) \in (E^\alpha)^*$ given by (6.27) and that the mapping

$$\varphi' : E^\alpha \to (E^\alpha)^*, \quad u \to \varphi'(u)$$

is continuous.

1) It follows easily from Remark 6.2 and (6.28) that φ is everywhere finite on E^α. Let us define, for u and v fixed in E^α, $t \in [0,T]$, $\lambda \in [-1,1]$,

$$G(\lambda,t) = L(t, u(t) + \lambda v(t), {}_0^C D_t^\alpha u(t) + \lambda\, {}_0^C D_t^\alpha v(t), {}_t^C D_T^\alpha u(t) + \lambda\, {}_t^C D_T^\alpha v(t))$$

and

$$\psi(\lambda) = \int_0^T G(\lambda,t)dt = \varphi(u + \lambda v).$$

We shall apply Leibniz formula of differentiation under integral sign to ψ. By (6.29) and (6.30), we have

$$\begin{aligned}
&|D_\lambda G(\lambda,t)| \\
&= \left|(D_x L(t, u(t) + \lambda v(t), {}_0^C D_t^\alpha u(t) + \lambda\, {}_0^C D_t^\alpha v(t), {}_t^C D_T^\alpha u(t) + \lambda\, {}_t^C D_T^\alpha v(t)), v(t))\right| \\
&\quad + \left|(D_y L(t, u(t) + \lambda v(t), {}_0^C D_t^\alpha u(t) + \lambda\, {}_0^C D_t^\alpha v(t),\right. \\
&\quad \left. {}_t^C D_T^\alpha u(t) + \lambda\, {}_t^C D_T^\alpha v(t)), {}_0^C D_t^\alpha v(t))\right| \\
&\quad + \left|(D_z L(t, u(t) + \lambda v(t), {}_0^C D_t^\alpha u(t) + \lambda\, {}_0^C D_t^\alpha v(t),\right. \\
&\quad \left. {}_t^C D_T^\alpha u(t) + \lambda\, {}_t^C D_T^\alpha v(t)), {}_t^C D_T^\alpha v(t))\right| \\
&\leq m_1(|u(t) + \lambda v(t)|)m_2(t)|v(t)| + \frac{1}{2}|{}_t^C D_T^\alpha u(t) + \lambda\, {}_t^C D_T^\alpha v(t)||{}_0^C D_t^\alpha v(t)| \\
&\quad + \frac{1}{2}\left|{}_0^C D_t^\alpha u(t) + \lambda\, {}_0^C D_t^\alpha v(t)\right|\left|{}_t^C D_T^\alpha v(t)\right| \\
&\leq m_0 m_2(t)|v(t)| + \frac{1}{2}\left|{}_t^C D_T^\alpha u(t)\right|\left|{}_0^C D_t^\alpha v(t)\right| + \frac{1}{2}|{}_0^C D_t^\alpha u(t)||{}_t^C D_T^\alpha v(t)| \\
&\quad + \left|{}_0^C D_t^\alpha v(t)||{}_t^C D_T^\alpha v(t)\right|,
\end{aligned}$$

where

$$m_0 = \max_{(\lambda,t)\in[-1,1]\times[0,T]} m_1(|u(t) + \lambda v(t)|).$$

Since $m_2 \in L^1([0,T], \mathbb{R}^+)$, v is continuous on $[0,T]$, and in view of Remark 6.2, we have

$$|D_\lambda G(\lambda,t)| \leq d(t),$$

where $d \in L^1([0,T],\mathbb{R}^+)$. Thus Leibniz formula is applicable and

$$\begin{aligned}\frac{d}{d\lambda}\psi(0) &= \int_0^T D_\lambda G(0,t)dt \\ &= \int_0^T (D_xL(t,u(t),{}_0^CD_t^\alpha u(t),{}_t^CD_T^\alpha u(t)),v(t))dt \\ &\quad + \int_0^T (D_yL(t,u(t),{}_0^CD_t^\alpha u(t),{}_t^CD_T^\alpha u(t)),{}_0^CD_t^\alpha v(t))dt \\ &\quad + \int_0^T (D_zL(t,u(t),{}_0^CD_t^\alpha u(t),{}_t^CD_T^\alpha u(t)),{}_t^CD_T^\alpha v(t))dt.\end{aligned}$$

Moreover,

$$\begin{aligned}\left|D_xL(t,u(t),{}_0^CD_t^\alpha u(t),{}_t^CD_T^\alpha u(t))\right| &\le m_1(|u(t)|)m_2(t), \\ \left|D_yL(t,u(t),{}_0^CD_t^\alpha u(t),{}_t^CD_T^\alpha u(t))\right| &\le \frac{1}{2}v\left|{}_t^CD_T^\alpha u(t)\right|\end{aligned}$$

and

$$|D_zL(t,u(t),{}_0^CD_t^\alpha u(t),{}_t^CD_T^\alpha u(t))| \le \frac{1}{2}|{}_0^CD_t^\alpha u(t)|.$$

Thus, by Remark 6.2 and (6.14),

$$\begin{aligned}\langle\varphi'(u),v\rangle &= \int_0^T (D_xL(t,u(t),{}_0^CD_t^\alpha u(t),{}_t^CD_T^\alpha u(t)),v(t))dt \\ &\quad + \int_0^T (D_yL(t,u(t),{}_0^CD_t^\alpha u(t),{}_t^CD_T^\alpha u(t)),{}_0^CD_t^\alpha v(t))dt \\ &\quad + \int_0^T (D_zL(t,u(t),{}_0^CD_t^\alpha u(t),{}_t^CD_T^\alpha u(t)),{}_t^CD_T^\alpha v(t))dt \\ &\le c_1\|v\| + c_2\|{}_0^CD_t^\alpha v(t)\|_{L^2} + c_3\|{}_t^CD_T^\alpha v(t)\|_{L^2} \\ &\le c_1\|v\| + c_2\|v\|_\alpha + \frac{c_3}{|\cos(\pi\alpha)|}\|v\|_\alpha \\ &\le c_4\|v\|_\alpha,\end{aligned}$$

where c_1, c_2, c_3 and c_4 are some positive constants. Therefore, φ has, at u, a directional derivative $\varphi'(u) \in (E^\alpha)^*$ given by (6.27).

2) By a theorem of Krasnoselskii, (6.29) and (6.30) imply that the mapping from E^α into $L^1([0,T],\mathbb{R}^N) \times L^2([0,T],\mathbb{R}^N) \times L^2([0,T],\mathbb{R}^N)$ defined by

$$u \to (D_xL(\cdot,u,{}_0^CD_t^\alpha u,{}_t^CD_T^\alpha u), D_yL(\cdot,u,{}_0^CD_t^\alpha u,{}_t^CD_T^\alpha u), D_zL(\cdot,u,{}_0^CD_t^\alpha u,{}_t^CD_T^\alpha u))$$

is continuous, so that φ' is continuous from E^α into $(E^\alpha)^*$, and the proof is completed. □

Theorem 6.2. *Let $\frac{1}{2} < \alpha \le 1$ and φ be defined by (6.26). If condition (A) is satisfied and $u \in E^\alpha$ is a solution of corresponding Euler equation $\varphi'(u) = 0$, then u is a solution of BVP (6.20) which, of course, corresponding to the solution of BVP (6.1).*

Proof. By Theorem 6.1 and Proposition 1.10, we have

$$
\begin{aligned}
0 &= \langle \varphi'(u), v\rangle \\
&= -\int_0^T \frac{1}{2}[({}_0^C D_t^\alpha u(t), {}_t^C D_T^\alpha v(t)) + ({}_t^C D_T^\alpha u(t), {}_0^C D_t^\alpha v(t))]dt \\
&\quad - \int_0^T (\nabla F(t,u(t)), v(t))dt \\
&= \int_0^T \left(\frac{1}{2}({}_0 D_t^{\alpha-1}({}_0^C D_t^\alpha u(t)), v'(t)) - \frac{1}{2}({}_t D_T^{\alpha-1}({}_t^C D_T^\alpha u(t)), v'(t))\right) dt \\
&\quad - \int_0^T (\nabla F(t,u(t)), v(t))dt
\end{aligned} \tag{6.31}
$$

for all $v \in E^\alpha$.

Let us define $w \in C([0,T], \mathbb{R}^N)$ by

$$
w(t) = \int_0^t \nabla F(s, u(s))ds, \quad t \in [0,T],
$$

so that

$$
\int_0^T (w(t), v'(t))dt = \int_0^T \left(\int_0^t (\nabla F(s,u(s)), v'(t))ds\right)dt.
$$

By the Fubini theorem and noting that $v(T) = 0$, we obtain

$$
\begin{aligned}
\int_0^T (w(t), v'(t))dt &= \int_0^T \left(\int_s^T (\nabla F(s,u(s)), v'(t))dt\right)ds \\
&= \int_0^T (\nabla F(s,u(s)), v(T) - v(s))ds \\
&= -\int_0^T (\nabla F(s,u(s)), v(s))ds.
\end{aligned}
$$

Hence, by (6.31) we have, for every $v \in E^\alpha$,

$$
\int_0^T \left(\frac{1}{2}\, {}_0 D_t^{\alpha-1}({}_0^C D_t^\alpha u(t)) - \frac{1}{2}\, {}_t D_T^{\alpha-1}({}_t^C D_T^\alpha u(t)) + w(t), v'(t)\right)dt = 0. \tag{6.32}
$$

If (e_j) denotes the canonical basis of $\mathbb{R}^N$, we can choose $v \in E^\alpha$ such that

$$
v(t) = \sin\frac{2k\pi t}{T} e_j \quad \text{or} \quad v(t) = e_j - \cos\frac{2k\pi t}{T} e_j, \qquad k = 1, 2, \ldots \text{ and } j = 1, \ldots, N.
$$

The theory of Fourier series and (6.32) imply that

$$
\frac{1}{2}\, {}_0 D_t^{\alpha-1}({}_0^C D_t^\alpha u(t)) - \frac{1}{2}\, {}_t D_T^{\alpha-1}({}_t^C D_T^\alpha u(t)) + w(t) = C,
$$

a.e. $t \in [0,T]$, for some $C \in \mathbb{R}^N$. According to the definition of $w \in C([0,T], \mathbb{R}^N)$, we have

$$
\frac{1}{2}\, {}_0 D_t^{\alpha-1}({}_0^C D_t^\alpha u(t)) - \frac{1}{2}\, {}_t D_T^{\alpha-1}({}_t^C D_T^\alpha u(t)) = -\int_0^t \nabla F(s,u(s))ds + C,
$$

a.e. $t \in [0,T]$, for some $C \in \mathbb{R}^N$.

In view of $\nabla F(\cdot, u(\cdot)) \in L^1([0,T], \mathbb{R}^N)$, we shall identify the equivalence class $D^\alpha(u(t))$ given by (6.21) and its continuous representation

$$\begin{aligned} D^\alpha(u(t)) &= \frac{1}{2}\,{}_0D_t^{\alpha-1}({}_0^CD_t^\alpha u(t)) - \frac{1}{2}\,{}_tD_T^{\alpha-1}({}_t^CD_T^\alpha u(t)) \\ &= -\int_0^t \nabla F(s, u(s))ds + C \end{aligned} \tag{6.33}$$

for $t \in [0, T]$.

Therefore, it follows from (6.33) and a classical result of Lebesgue theory that $-\nabla F(\cdot, u(\cdot))$ is the classical derivative of $D^\alpha(u(t))$ a.e. on $[0, T]$ which means that (i) in Definition 6.2 is verified.

Since $u \in E^\alpha$ implies that $u \in AC([0,T], \mathbb{R}^N)$, it remains to show that u satisfies (6.20). In fact, according to (6.33), we can get that

$$\frac{d}{dt}D^\alpha(u(t)) = \frac{d}{dt}\Big(\frac{1}{2}\,{}_0D_t^{\alpha-1}({}_0^CD_t^\alpha u(t)) - \frac{1}{2}\,{}_tD_T^{\alpha-1}({}_t^CD_T^\alpha u(t))\Big) = -\nabla F(t, u(t)).$$

Moreover, $u \in E^\alpha$ implies that $u(0) = u(T) = 0$, and therefore (6.1) is verified. The proof is completed. □

From now on, φ given by (6.26) is considered as a functional on E^α with $\frac{1}{2} < \alpha \leq 1$.

6.2.4 *Existence under Ambrosetti-Rabinowitz Condition*

According to Theorem 6.2, we know that in order to find solutions of BVP (6.1), it suffices to obtain the critical points of functional φ given by (6.26). We need to use some critical point theorems.

First, we use Theorem 1.14 to consider the existence of solutions for BVP (6.1). Assume that condition (A) is satisfied. Recall that, in our setting in (6.26), the corresponding functional φ on E^α given by

$$\varphi(u) = \int_0^T \Big(-\frac{1}{2}({}_0^CD_t^\alpha u(t), {}_t^CD_T^\alpha u(t)) - F(t, u(t))\Big)dt$$

is continuously differentiable according to Theorem 6.1 and is also weakly lower semi-continuous functional on E^α as the sum of a convex continuous function (see Theorem 1.2 in Mawhin and Willem, 1989) and of a weakly continuous one (see Proposition 1.2 in Mawhin and Willem, 1989).

In fact, according to Proposition 6.3, if $u_k \rightharpoonup u$ in E^α, then $u_k \to u$ in $C([0,T], \mathbb{R}^N)$. Therefore, $F(t, u_k(t)) \to F(t, u(t))$ a.e. $t \in [0, T]$. By Lebesgue dominated convergence theorem, we have $\int_0^T F(t, u_k(t))dt \to \int_0^T F(t, u(t))dt$, which means that the functional $u \to \int_0^T F(t, u(t))dt$ is weakly continuous on E^α. Moreover, the following lemma implies that the functional $u \to -\int_0^T [({}_0^CD_t^\alpha u(t), {}_t^CD_T^\alpha u(t))/2]dt$ is convex and continuous on E^α.

Lemma 6.2. *Let $\frac{1}{2} < \alpha \leq 1$ and condition (A) be satisfied. If $u \in E^{\alpha}$, then the functional $H : E^{\alpha} \to \mathbb{R}^N$ denoted by*

$$H(u) = -\frac{1}{2} \int_0^T ({}_0^C D_t^{\alpha} u(t), {}_t^C D_T^{\alpha} u(t))dt$$

is convex and continuous on E^{α}.

Proof. The continuity follows from (6.22) and (6.17) directly. We are now in a position to prove the convexity of H.

Let $\lambda \in (0,1)$, $u, v \in E^{\alpha}$ and $\tilde{u}, \tilde{v}$ be the extension of u and v by zero on $\mathbb{R}/[0,T]$ respectively. Since Caputo fractional derivative operator is linear operator, we have by Remark 6.2 and (6.23) that

$$\begin{aligned}
&H((1-\lambda)u + \lambda v)\\
&= -\frac{1}{2} \int_0^T ({}_0^C D_t^{\alpha}((1-\lambda)u(t) + \lambda v(t)), {}_t^C D_T^{\alpha}((1-\lambda)u(t) + \lambda v(t)))dt\\
&= -\frac{1}{2} \int_0^T ({}_0 D_t^{\alpha}((1-\lambda)u(t) + \lambda v(t)), {}_t D_T^{\alpha}((1-\lambda)u(t) + \lambda v(t)))dt\\
&= -\frac{1}{2} \int_{-\infty}^{\infty} ({}_{-\infty} D_t^{\alpha}((1-\lambda)\tilde{u}(t) + \lambda \tilde{v}(t)), {}_t D_{+\infty}^{\alpha}((1-\lambda)\tilde{u}(t) + \lambda \tilde{v}(t)))dt\\
&= \frac{|\cos(\pi\alpha)|}{2} \int_{-\infty}^{\infty} |{}_{-\infty} D_t^{\alpha}((1-\lambda)\tilde{u}(t) + \lambda \tilde{v}(t))|^2 dt\\
&\leq \frac{|\cos(\pi\alpha)|}{2} \int_{-\infty}^{\infty} \left((1-\lambda)|{}_{-\infty} D_t^{\alpha}\tilde{u}(t)|^2 + \lambda|{}_{-\infty} D_t^{\alpha}\tilde{v}(t)|^2\right) dt\\
&= \int_{-\infty}^{\infty} \left(-\frac{1-\lambda}{2}({}_{-\infty} D_t^{\alpha}\tilde{u}(t), {}_t D_{+\infty}^{\alpha}\tilde{u}(t)) - \frac{\lambda}{2}({}_{-\infty} D_t^{\alpha}\tilde{v}(t), {}_t D_{+\infty}^{\alpha}\tilde{v}(t)) \right) dt\\
&= \int_0^T \left(-\frac{1-\lambda}{2}({}_0^C D_t^{\alpha} u(t), {}_t^C D_T^{\alpha} u(t)) - \frac{\lambda}{2}({}_0^C D_t^{\alpha} v(t), {}_t^C D_T^{\alpha} v(t)) \right) dt\\
&= (1-\lambda)H(u) + \lambda H(v),
\end{aligned}$$

which implies that H is a convex functional defined on E^{α}. This completes the proof. □

According to the arguments above, if φ is coercive, by Theorem 1.14, φ has a minimum so that BVP (6.1) is solvable. It remains to find conditions under which φ is coercive on E^{α}, i.e. $\lim_{\|u\|_{\alpha} \to \infty} \varphi(u) = +\infty$, for $u \in E^{\alpha}$. We shall see that it suffices to require that $F(t,x)$ is bounded by a function for a.e., $t \in [0,T]$ and all $x \in \mathbb{R}^N$.

Theorem 6.3. *Let $\alpha \in (\frac{1}{2}, 1]$ and assume that F satisfies condition (A). If*

$$|F(t,x)| \leq \bar{a}|x|^2 + \bar{b}(t)|x|^{2-\gamma} + \bar{c}(t), \quad t \in [0,T],\ x \in \mathbb{R}^N, \tag{6.34}$$

where $\bar{a} \in [0, |\cos(\pi\alpha)|\Gamma^2(\alpha+1)/2T^{2\alpha})$, $\gamma \in (0,2)$, $\bar{b} \in L^{2/\gamma}([0,T],\mathbb{R})$, and $\bar{c} \in L^1([0,T],\mathbb{R})$, then BVP (6.1) has at least one solution which minimizes φ on E^{α}.

Proof. According to arguments above, our problem reduces to prove that φ is coercive on E^α. For $u \in E^\alpha$, it follows from (6.22), (6.34) and (6.13) that

$$\begin{aligned}
\varphi(u) &= -\frac{1}{2}\int_0^T ({}_0^C D_t^\alpha u(t), {}_t^C D_T^\alpha u(t))dt - \int_0^T F(t,u(t))dt \\
&\geq \frac{|\cos(\pi\alpha)|}{2}\int_0^T |{}_0^C D_t^\alpha u(t)|^2 dt - \bar{a}\int_0^T |u(t)|^2 dt \\
&\quad - \int_0^T \bar{b}(t)|u(t)|^{2-\gamma}dt - \int_0^T \bar{c}(t)dt \\
&= \frac{|\cos(\pi\alpha)|}{2}\|u\|_\alpha^2 - \bar{a}\|u\|_{L^2}^2 - \int_0^T \bar{b}(t)|u(t)|^{2-\gamma}dt - \bar{c}_1 \\
&\geq \frac{|\cos(\pi\alpha)|}{2}\|u\|_\alpha^2 - \bar{a}\|u\|_{L^2}^2 - \left(\int_0^T |\bar{b}(t)|^{2/\gamma}dt\right)^{\gamma/2}\left(\int_0^T |u(t)|^2 dt\right)^{1-\gamma/2} - \bar{c}_1 \\
&= \frac{|\cos(\pi\alpha)|}{2}\|u\|_\alpha^2 - \bar{a}\|u\|_{L^2}^2 - \bar{b}_1\|u\|_{L^2}^{2-\gamma} - \bar{c}_1 \\
&\geq \frac{|\cos(\pi\alpha)|}{2}\|u\|_\alpha^2 - \frac{\bar{a}T^{2\alpha}}{\Gamma^2(\alpha+1)}\|u\|_\alpha^2 - \bar{b}_1\left(\frac{T^\alpha}{\Gamma(\alpha+1)}\right)^{2-\gamma}\|u\|_\alpha^{2-\gamma} - \bar{c}_1 \\
&= \left(\frac{|\cos(\pi\alpha)|}{2} - \frac{\bar{a}T^{2\alpha}}{\Gamma^2(\alpha+1)}\right)\|u\|_\alpha^2 - \bar{b}_1\left(\frac{T^\alpha}{\Gamma(\alpha+1)}\right)^{2-\gamma}\|u\|_\alpha^{2-\gamma} - \bar{c}_1,
\end{aligned}$$

where $\bar{b}_1 = \left(\int_0^T |\bar{b}(t)|^{2/\gamma}dt\right)^{\gamma/2}$ and $\bar{c}_1 = \int_0^T \bar{c}(t)dt$.

Noting that $\bar{a} \in [0, |\cos(\pi\alpha)|\Gamma^2(\alpha+1)/2T^{2\alpha})$ and $\gamma \in (0,2)$, we have

$$\varphi(u) = +\infty \quad \text{as } \|u\|_\alpha \to \infty,$$

and hence φ is coercive, which completes the proof. □

Our task is now to use Theorem 1.15 (Mountain pass theorem) to find a nonzero critical point of functional φ on E^α.

Theorem 6.4. *Let $\alpha \in (\frac{1}{2}, 1]$ and suppose that F satisfies condition (A). If*

- **(A1)** *$F \in C([0,T] \times \mathbb{R}^N, \mathbb{R})$ and there exists $\mu \in [0, \frac{1}{2})$ and $M > 0$ such that $0 < F(t,x) \leq \mu(\nabla F(t,x), x)$ for all $x \in \mathbb{R}^N$ with $|x| \geq M$ and $t \in [0,T]$;*
- **(A2)** *$\limsup_{|x|\to 0} F(t,x)/|x|^2 < |\cos(\pi\alpha)|\Gamma^2(\alpha+1)/2T^{2\alpha}$ uniformly for $t \in [0,T]$ and $x \in \mathbb{R}^N$;*

are satisfied, then BVP (6.1) has at least one nonzero solution on E^α.

Proof. We will verify that φ satisfies all conditions of Theorem 1.15.

First, we will prove that φ satisfies (PS) condition. Since $F(t,x)-\mu(\nabla F(t,x),x)$ is continuous for $t \in [0,T]$ and $|x| \leq M$, there exists $c \in \mathbb{R}^+$, such that

$$F(t,x) \leq \mu(\nabla F(t,x), x) + c, \quad t \in [0,T], \quad |x| \leq M.$$

By condition (A1), we obtain

$$F(t,x) \leq \mu(\nabla F(t,x), x) + c, \quad t \in [0,T], \quad x \in \mathbb{R}^N. \tag{6.35}$$

Let $\{u_k\} \subset E^\alpha$, $|\varphi(u_k)| \leq K$, $k = 1, 2, ...,$ $\varphi'(u_k) \to 0$. Notice that

$$\langle \varphi'(u_k), u_k \rangle = -\int_0^T [({}_0^C D_t^\alpha u_k(t), {}_t^C D_T^\alpha u_k(t)) + (\nabla F(t, u_k(t)), u_k(t))]dt. \quad (6.36)$$

It follows from (6.35), (6.36) and (6.22) that

$$\begin{aligned} K \geq \varphi(u_k) &= -\frac{1}{2}\int_0^T ({}_0^C D_t^\alpha u_k(t), {}_t^C D_T^\alpha u_k(t))dt - \int_0^T F(t, u_k(t))dt \\ &\geq -\frac{1}{2}\int_0^T ({}_0^C D_t^\alpha u_k(t), {}_t^C D_T^\alpha u_k(t))dt - \mu \int_0^T (\nabla F(t, u_k(t)), u_k(t))dt - cT \\ &= \left(\mu - \frac{1}{2}\right)\int_0^T ({}_0^C D_t^\alpha u_k(t), {}_t^C D_T^\alpha u_k(t))dt + \mu\langle \varphi'(u_k), u_k\rangle - cT \\ &\geq \left(\frac{1}{2} - \mu\right)|\cos(\pi\alpha)|\|u_k\|_\alpha^2 - \mu\|\varphi'(u_k)\|_\alpha\|u_k\|_\alpha - cT, \quad k = 1, 2, \dots . \end{aligned}$$

Since $\varphi'(u_k) \to 0$, there exists $N_0 \in \mathbb{N}$ such that

$$K \geq \left(\frac{1}{2} - \mu\right)|\cos(\pi\alpha)|\|u_k\|_\alpha^2 - \|u_k\|_\alpha - cT, \quad k > N_0,$$

and this implies that $\{u_k\} \subset E^\alpha$ is bounded. Since E^α is a reflexive space, going to a subsequence if necessary, we may assume that $u_k \rightharpoonup u$ weakly in E^α, thus we have

$$\begin{aligned} \langle \varphi'(u_k) - \varphi'(u), u_k - u\rangle &= \langle \varphi'(u_k), u_k - u\rangle - \langle \varphi'(u), u_k - u\rangle \\ &\leq \|\varphi'(u_k)\|_\alpha\|u_k - u\|_\alpha - \langle \varphi'(u), u_k - u\rangle \quad (6.37) \\ &\to 0, \quad \text{as } k \to \infty. \end{aligned}$$

Moreover, according to (6.14) and Proposition 6.3, we have u_k is bounded in $C([0,T], \mathbb{R}^N)$ and $\|u_k - u\| \to 0$ as $k \to \infty$. Hence, we have

$$\int_0^T \nabla F(t, u_k(t))dt \to \int_0^T \nabla F(t, u(t))dt, \quad \text{as } k \to \infty. \quad (6.38)$$

Noting that

$$\begin{aligned} &\langle \varphi'(u_k) - \varphi'(u), u_k - u\rangle \\ &= -\int_0^T ({}_0^C D_t^\alpha(u_k(t) - u(t)), {}_t^C D_T^\alpha(u_k(t) - u(t)))dt \\ &\quad - \int_0^T \Big((\nabla F(t, u_k(t)) - \nabla F(t, u(t))), (u_k(t) - u(t))\Big)dt \\ &\geq |\cos(\pi\alpha)|\|u_k - u\|_\alpha^2 - \left|\int_0^T (\nabla F(t, u_k(t)) - \nabla F(t, u(t)))dt\right|\|u_k - u\|. \end{aligned}$$

Combining (6.37) and (6.38), it is easy to verify that $\|u_k - u\|_\alpha^2 \to 0$ as $k \to \infty$, and hence that $u_k \to u$ in E^α. Thus, we obtain the desired convergence property.

From $\limsup_{|x|\to 0} F(t,x)/|x|^2 < |\cos(\pi\alpha)|\Gamma^2(\alpha+1)/2T^{2\alpha}$ uniformly for $t \in [0,T]$, there exists $\epsilon \in (0, |\cos(\pi\alpha)|)$ and $\delta > 0$ such that $F(t,x) \leq (|\cos(\pi\alpha)| - \epsilon)(\Gamma^2(\alpha+1)/2T^{2\alpha})|x|^2$ for all $t \in [0,T]$ and $x \in \mathbb{R}^N$ with $|x| \leq \delta$.

Let $\rho = \frac{\Gamma(\alpha)((\alpha-1)/2+1)^{\frac{1}{2}}}{T^{\alpha-\frac{1}{2}}}\delta$ and $\sigma = \epsilon\rho^2/2 > 0$. Then it follows from (6.14) that

$$\|u\| \leq \frac{T^{\alpha-\frac{1}{2}}}{\Gamma(\alpha)((\alpha-1)/2+1)^{\frac{1}{2}}}\|u\|_\alpha = \delta$$

for all $u \in E^\alpha$ with $\|u\|_\alpha = \rho$. Therefore, we have

$$\begin{aligned}
\varphi(u) &= -\frac{1}{2}\int_0^T ({}_0^C D_t^\alpha u(t), {}_t^C D_T^\alpha u(t))dt - \int_0^T F(t, u(t))dt \\
&\geq \frac{|\cos(\pi\alpha)|}{2}\|u\|_\alpha^2 - (|\cos(\pi\alpha)| - \epsilon)\frac{\Gamma^2(\alpha+1)}{2T^{2\alpha}}\int_0^T |u(t)|^2 dt \\
&\geq \frac{|\cos(\pi\alpha)|}{2}\|u\|_\alpha^2 - \frac{1}{2}(|\cos(\pi\alpha)| - \epsilon)\|u\|_\alpha^2 \\
&= \frac{1}{2}\epsilon\|u\|_\alpha^2 \\
&= \sigma
\end{aligned}$$

for all $u \in E^\alpha$ with $\|u\|_\alpha = \rho$. This implies (ii) in Theorem 1.15 is satisfied.

It is obvious from the definition of φ and (A2) that $\varphi(0) = 0$, and therefore, it suffices to show that φ satisfies (iii) in Theorem 1.15.

Since $0 < F(t, x) \leq \mu(\nabla F(t, x), x)$ for all $x \in \mathbb{R}^N$ and $|x| \geq M$, a simple regularity argument then shows that there exists $r_1, r_2 > 0$ such that

$$F(t, x) \geq r_1|x|^{1/\mu} - r_2, \quad x \in \mathbb{R}^N, \quad t \in [0, T].$$

For any $u \in E^\alpha$ with $u \neq 0$, $\kappa > 0$ and noting that $\mu \in [0, \frac{1}{2})$ and (6.22), we have

$$\begin{aligned}
\varphi(\kappa u) &= -\frac{1}{2}\int_0^T ({}_0^C D_t^\alpha \kappa u(t), {}_t^C D_T^\alpha \kappa u(t))dt - \int_0^T F(t, \kappa u(t))dt \\
&\leq \frac{\kappa^2}{2|\cos(\pi\alpha)|}\|u\|_\alpha^2 - r_1\int_0^T |\kappa u(t)|^{1/\mu}dt + r_2 T \\
&= \frac{\kappa^2}{2|\cos(\pi\alpha)|}\|u\|_\alpha^2 - r_1\kappa^{1/\mu}\|u\|_{L^{1/\mu}}^{1/\mu} + r_2 T \\
&\to -\infty
\end{aligned}$$

as $\kappa \to \infty$. Then there exists a sufficiently large κ_0 such that $\varphi(\kappa_0 u) \leq 0$. Hence (iii) in Theorem 1.15 holds.

Lastly noting that $\varphi(0) = 0$ while for our critical point u, $\varphi(u) \geq \sigma > 0$. Hence u is a nontrivial weak solution of BVP (6.1), and this completes the proof. □

Corollary 6.1. $\forall\ \alpha \in (\frac{1}{2}, 1]$, *suppose that F satisfies conditions (A) and (A1). If*

(A2)′ $F(t, x) = o(|x|^2)$, *as $|x| \to 0$ uniformly for $t \in [0, T]$ and $x \in \mathbb{R}^N$*

is satisfied, then BVP (6.1) has at least one nonzero solution on E^α.

6.2.5 *Superquadratic Case*

Under the usual Ambrosetti-Rabinowitz condition, it is easy to show that the energy functional associated with the system has the Mountain Pass geometry and satisfies the (PS) condition. However, the A.R. condition is so strong that many potential functions can not satisfy it, then the problem becomes more delicate and complicated.

Assume that $F : [0,T] \times \mathbb{R}^N \to \mathbb{R}$ satisfies the condition (A) which is assumed as in Subsection 6.2.3.

In the following, we introduce the function space E^α, where $\alpha \in (\frac{1}{2}, 1]$. For $u \in E^\alpha$, where

$$E^\alpha := \left\{u \in L^2(0,T;\mathbb{R}^N) : \, {}_0^C D_t^\alpha u \in L^2(0,T;\mathbb{R}^N)\right\}$$

is a reflexive Banach space with the norm defined by

$$\|u\|_\alpha = \|{}_0^C D_t^\alpha u\|_{L_2}$$

and

$$\|u\| = \max_{t\in[0,\,T]} |u(t)|.$$

It follows from Theorem 6.1 that the functional φ on E^α given by

$$\varphi(u) = \int_0^T \left(-\frac{1}{2}\left({}_0^C D_t^\alpha u(t), \, {}_t^C D_T^\alpha u(t)\right) - F(t, u(t))\right) dt$$

is continuously differentiable on E^α. Moreover, we have

$$\begin{aligned}\langle \varphi'(u), v\rangle = &-\int_0^T \frac{1}{2}\left(\left({}_0^C D_t^\alpha u(t), \, {}_t^C D_T^\alpha v(t)\right) + \left({}_t^C D_T^\alpha u(t), \, {}_0^C D_t^\alpha v(t)\right)\right) dt \\ &- \int_0^T \left(\nabla F(t, u(t)), v(t)\right) dt.\end{aligned}$$

Recall that a sequence $\{u_n\} \subset E^\alpha$ is said to be a (C) sequence of φ if $\varphi(u_n)$ is bounded and $(1 + \|u_n\|_\alpha)\|\varphi(u_n)\|_\alpha \to 0$ as $n \to \infty$. The functional φ satisfies condition (C) if every (C) sequence of φ has a convergent subsequence. This condition is due to Cerami, 1978.

For the superquadratic case, we make the following assumptions:

(A3) $\lim\limits_{|x|\to 0} \frac{F(t,x)}{|x|^2} = 0$, $\liminf\limits_{|x|\to\infty} \frac{F(t,x)}{|x|^2} \geq L > \frac{\pi^2}{|\cos(\pi\alpha)|\Gamma^2(2-\alpha)T^{2\alpha}(3-2\alpha)}$ uniformly for some $L > 0$ and a.e. $t \in [0,T]$;

(A4) $\limsup\limits_{|x|\to+\infty} \frac{F(t,x)}{|x|^r} \leq M < +\infty$ uniformly for some $M > 0$ and a.e. $t \in [0,T]$;

(A5) $\liminf\limits_{|x|\to+\infty} \frac{(\nabla F(t,x),x) - 2F(t,x)}{|x|^\mu} \geq Q > 0$ uniformly for some $Q > 0$ and a.e. $t \in [0,T]$, where $r > 2$ and $\mu > r - 2$.

We will first establish the following lemma.

Lemma 6.3. *Assume (A), (A4), (A5) hold, then the functional φ satisfies condition (C).*

Proof. Let $\{u_n\} \subset E^\alpha$ is a (C) sequence of φ, that is $\varphi(u_n)$ is bounded and $(1+\|u_n\|_\alpha)\|\varphi'(u_n)\|_\alpha \to 0$ as $n \to \infty$. Then there exists M_0 such that

$$|\varphi(u_n)| \le M_0 \quad \text{and} \quad (1+\|u_n\|_\alpha)\|\varphi'(u_n)\|_\alpha \le M_0, \tag{6.39}$$

for all $n \in \mathbb{N}$.

By (A4), there exist positive constants B_1 and M_1 such that

$$F(t,x) \le B_1|x|^r$$

for all $|x| \ge M_1$ and a.e. $t \in [0,T]$.

It follows from (A) that

$$|F(t,x)| \le \max_{s\in[0,\, M_1]} a(s)b(t)$$

for all $|x| \le M_1$ and a.e. $t \in [0,T]$. Therefore, we obtain

$$F(t,x) \le B_1|x|^r + \max_{s\in[0,\, M_1]} a(s)b(t), \tag{6.40}$$

for all $x \in \mathbb{R}^N$ and a.e. $t \in [0,T]$.

Combining (6.22) and (6.40), we get

$$\begin{aligned}\frac{|\cos(\pi\alpha)|}{2}\|u_n\|_\alpha^2 &\le \varphi(u_n) + \int_0^T F(t,u_n(t))dt \\ &\le M_0 + \max_{s\in[0,M_1]} a(s)\int_0^T b(t)dt + B_1\int_0^T |u_n(t)|^r dt.\end{aligned} \tag{6.41}$$

On the other hand, by (A5), there exist $\eta > 0$ and $M_2 > 0$ such that

$$\big(\nabla F(t,x),x\big) - 2F(t,x) \ge \eta|x|^\mu$$

for a.e. $t \in [0,T]$ and $|x| \ge M_2$.

By (A), we have

$$|\,(\nabla F(t,x),x) - 2F(t,x)| \le (2+M_2) \max_{s\in[0,\, M_2]} a(s)b(t)$$

for all $|x| \le M_2$ and a.e. $t \in [0,T]$.

Therefore, we obtain

$$(\nabla F(t,x),x) - 2F(t,x) \ge \eta|x|^\mu - (2+M_2) \max_{s\in[0,\, M_2]} a(s)b(t), \tag{6.42}$$

for all $x \in \mathbb{R}^N$ and a.e. $t \in [0,T]$.

It follows from (6.39) and (6.42) that

$$\begin{aligned}3M_0 &\ge 2\varphi(u_n) - \langle\varphi'(u_n),u_n\rangle \\ &= 2\int_0^T \Big[-\frac{1}{2}({}_0^CD_t^\alpha u_n(t),\, {}_t^CD_T^\alpha u_n(t)) - F(t,u_n(t))\Big]dt \\ &\quad - \int_0^T \Big[-({}_0^CD_t^\alpha u_n(t), {}_t^CD_T^\alpha u_n(t)) - (\nabla F(t,u_n(t)),u_n(t))\Big]dt\end{aligned}$$

$$= \int_0^T \left[(\nabla F(t, u_n(t)), u_n(t)) - 2F(t, u_n(t))\right] dt$$
$$\geq \eta \int_0^T |u_n(t)|^\mu dt - (2 + M_2) \max_{s \in [0,\, M_2]} a(s) \int_0^T b(t) dt,$$

thus, $\int_0^T |u_n(t)|^\mu dt$ is bounded.

If $\mu > r$, then

$$\int_0^T |u_n(t)|^r dt \leq T^{\frac{\mu - r}{\mu}} \left(\int_0^T |u_n(t)|^\mu dt \right)^{r/\mu},$$

which combining (6.41) implies that $\|u_n\|_\alpha$ is bounded.

If $\mu \leq r$, then

$$\int_0^T |u_n(t)|^r dt \leq \|u_n\|_\infty^{r-\mu} \int_0^T |u_n(t)|^\mu dt \leq C_1^{r-\mu} \|u_n\|_\alpha^{r-\mu} \int_0^T |u_n(t)|^\mu dt,$$

where

$$C_1 := \frac{T^{\alpha - \frac{1}{2}}}{\Gamma(\alpha)(2\alpha - 1)^{\frac{1}{2}}}$$

by (6.14).

Since $\mu > r - 2$, it follows from (6.41) that $\|u_n\|_\alpha$ is bounded too. Thus $\|u_n\|_\alpha$ is bounded in E^α.

By Proposition 6.3, the sequence $\{u_n\}$ has a subsequence, also denoted by $\{u_n\}$, such that

$$u_n \rightharpoonup u \ \text{ in } \ E^\alpha \quad \text{and} \quad u_n \to u \ \text{ in } \ C([0, T], \mathbb{R}^N).$$

Then we obtain $u_n \to u$ in E^α by use of the same argument of Theorem 6.4. The proof of Lemma 6.3 is completed. □

We state our first existence result as follows.

Theorem 6.5. *Assume that (A3)-(A5) hold and that $F(t, x)$ satisfies the condition (A). Then BVP (6.1) has at least one solution on E^α.*

Proof. By (A3), there exist $\epsilon_1 \in (0, |\cos(\pi\alpha)|)$ and $\delta > 0$ such that

$$F(t, x) \leq (|\cos(\pi\alpha)| - \epsilon_1) \frac{\Gamma^2(\alpha + 1)}{2T^{2\alpha}} |x|^2$$

for a.e. $t \in [0, T]$ and $x \in \mathbb{R}^N$ with $|x| \leq \delta$.

Let

$$\rho = \frac{\Gamma(\alpha)(2(\alpha - 1) + 1)^{\frac{1}{2}}}{T^{\alpha - \frac{1}{2}}} \delta \quad \text{and} \quad \sigma = \frac{\epsilon_1 \rho^2}{2} > 0.$$

Then it follows from (6.14) that

$$\|u\| \leq \frac{T^{\alpha - \frac{1}{2}}}{\Gamma(\alpha)(2(\alpha - 1) + 1)^{\frac{1}{2}}} \|u\|_\alpha = \delta$$

for all $u \in E^\alpha$ with $\|u\|_\alpha = \rho$.

Therefore, we have

$$\begin{aligned}\varphi(u) &= \int_0^T \Big[-\frac{1}{2}\left({}_0^CD_t^\alpha u(t),\, {}_t^CD_T^\alpha u(t)\right) - F(t,u(t))\Big] dt \\ &\geq \frac{|\cos(\pi\alpha)|}{2}\|u\|_\alpha^2 - (|\cos(\pi\alpha)| - \epsilon_1)\frac{\Gamma^2(\alpha+1)}{2T^{2\alpha}}\int_0^T |u(t)|^2 dt \\ &\geq \frac{|\cos(\pi\alpha)|}{2}\|u\|_\alpha^2 - \frac{|\cos(\pi\alpha)| - \epsilon_1}{2}\|u\|_\alpha^2 \\ &= \frac{\epsilon_1}{2}\|u\|_\alpha^2 \\ &= \sigma\end{aligned}$$

for all $u \in E^\alpha$ with $\|u\|_\alpha = \rho$. This implies that (ii) in Theorem 1.15 is satisfied.

It is obvious from the definition of φ and (A3) that $\varphi(0) = 0$, and therefore, it suffices to show that φ satisfies (iii) in Theorem 1.15.

By (A3), there exist $\epsilon_2 > 0$ and $M_3 > 0$ such that

$$F(t,x) > \left(\frac{\pi^2}{|\cos(\pi\alpha)|\Gamma^2(2-\alpha)T^{2\alpha}(3-2\alpha)} + \epsilon_2\right)|x|^2$$

for all $|x| \geq M_3$ and a.e. $t \in [0,T]$.

It follows from (A) that

$$|F(t,x)| \leq \max_{s\in[0,\, M_3]} a(s)b(t),$$

for all $|x| \leq M_3$ and a.e. $t \in [0,T]$.

Therefore, we obtain

$$\begin{aligned}F(t,x) \geq &\left(\frac{\pi^2}{|\cos(\pi\alpha)|\Gamma^2(2-\alpha)T^{2\alpha}(3-2\alpha)} + \epsilon_2\right)(|x|^2 - M_3^2) \\ &- \max_{s\in[0,\, M_3]} a(s)b(t),\end{aligned} \tag{6.43}$$

for all $x \in \mathbb{R}^N$ and a.e. $t \in [0,T]$.

Choosing $u_0 = \left(\frac{T}{\pi}\sin\frac{\pi t}{T}, 0, ..., 0\right) \in E^\alpha$, then

$$\|u_0\|_{L^2}^2 = \frac{T^3}{2\pi^2} \text{ and } \|u_0\|_\alpha^2 \leq \frac{T^{3-2\alpha}}{\Gamma^2(2-\alpha)(3-2\alpha)}. \tag{6.44}$$

For $\varsigma > 0$ and noting that (6.43) and (6.44), we have

$$\begin{aligned}\varphi(\varsigma u_0) &= \int_0^T \Big[-\frac{1}{2}\,({}_0^C D_t^\alpha \varsigma u_0(t), {}_t^C D_T^\alpha \varsigma u_0(t)) - F(t, \varsigma u_0(t))\Big] dt \\ &\leq \frac{\varsigma^2}{2|\cos(\pi\alpha)|}\|u_0\|_\alpha^2 \\ &\quad - \left(\frac{\varsigma^2\pi^2}{|\cos(\pi\alpha)|T^{2\alpha}\Gamma^2(2-\alpha)(3-2\alpha)} + \varsigma^2\epsilon_2\right)\int_0^T |u_0(t)|^2 dt + C_2 \\ &\leq \frac{\varsigma^2}{2|\cos(\pi\alpha)|}\cdot\frac{T^{3-2\alpha}}{\Gamma^2(2-\alpha)(3-2\alpha)} \\ &\quad - \frac{\varsigma^2\pi^2}{|\cos(\pi\alpha)|T^{2\alpha}\Gamma^2(2-\alpha)(3-2\alpha)}\cdot\frac{T^3}{2\pi^2} - \frac{\varsigma^2\epsilon_2 T^3}{2\pi^2} + C_2 \\ &\to -\infty\end{aligned}$$

as $\varsigma \to \infty$, where C_2 is a positive constant. Then there exists a sufficiently large ς_0 such that $\varphi(\varsigma_0 u_0) \leq 0$. Hence (iii) in Theorem 1.15 holds.

Finally, noting that $\varphi(0) = 0$ while for critical point u, $\varphi(u) \geq \sigma > 0$. Hence u is a nontrivial solution of BVP (6.1), and this completes the proof. □

We give an example to illustrate our results.

Example 6.1. In BVP (6.1), let

$$F(t, x) = \ln(1 + 2|x|^2)|x|^2.$$

These show that all conditions of Theorem 6.5 are satisfied, where

$$r = 2.5, \quad \mu = 2.$$

By Theorem 6.5, BVP (6.1) has at least one solution $u \in E^\alpha$.

6.2.6 *Asymptotically Quadratic Case*

For the asymptotically quadratic case, we assume:

(A4)′ $\limsup\limits_{|x|\to+\infty} \frac{F(t,x)}{|x|^2} \leq M < +\infty$ uniformly for some $M > 0$ and a.e. $t \in [0, T]$;

(A6) there exists $\tau(t) \in L^1([0, T], \mathbb{R}^+)$ such that $(\nabla F(t, x), x) - 2F(t, x) \geq \tau(t)$ for all $x \in \mathbb{R}^N$ and a.e. $t \in [0, T]$;

(A7) $\lim\limits_{|x|\to+\infty} [(\nabla F(t, x), x) - 2F(t, x)] = +\infty$ for a.e. $t \in [0, T]$.

Theorem 6.6. *Assume that $F(t, x)$ satisfies (A), (A3), (A4)′, (A6) and (A7). Then BVP (6.1) has at least one solution on E^α.*

The following lemmas are needed in the proof of Theorem 6.6.

Lemma 6.4. *Assume that (A7) holds. Then for any $\varepsilon > 0$, there exists a subset $E_\varepsilon \subset [0, T]$ with $\alpha([0, T]\backslash E_\varepsilon) < \varepsilon$ such that*

$$\lim_{|x|\to\infty} [(\nabla F(t, x), x) - 2F(t, x)] = +\infty$$

uniformly for $t \in E_\varepsilon$.

The proof is similar to that of Lemma 2 in Tang and Wu, 2001, and is omitted.

Lemma 6.5. *Assume that (A), (A4)′, (A6) and (A7) hold. Then the functional* φ *satisfies condition (C).*

Proof. Suppose that $\{u_n\} \subset E^\alpha$ is a (C) sequence of φ, that is $\varphi(u_n)$ is bounded and $(1 + \|u_n\|_\alpha)\|\varphi'(u_n)\|_\alpha \to 0$ as $n \to \infty$. Then we have

$$\liminf_{n\to\infty} [\langle \varphi'(u_n), u_n\rangle - 2\varphi(u_n)] > -\infty,$$

which implies that

$$\limsup_{n\to\infty} \int_0^T [(\nabla F(t, u_n),\ u_n) - 2F(t, u_n)]dt < +\infty. \tag{6.45}$$

We only need to show that $\{u_n\}$ is bounded in E^α. If $\{u_n\}$ is unbounded, we may assume, without loss of generality, that $\|u_n\|_\alpha \to \infty$ as $n \to \infty$. Put $z_n = \frac{u_n}{\|u_n\|_\alpha}$, we then have $\|z_n\|_\alpha = 1$. Going to a sequence if necessary, we assume that $z_n \rightharpoonup z$ in E^α, $z_n \to z$ in $C([0, T], \mathbb{R}^N)$ and $L^2([0, T], \mathbb{R}^N)$.

By (A2)′, it follows that there exist constants $B_2 > 0$ and $M_4 > 0$ such that

$$F(t, x) \le B_2|x|^2$$

for all $|x| \ge M_4$ and a.e. $t \in [0, T]$.

By condition (A), it follows that

$$|F(t, x)| \le \max_{s\in[0,\, M_4]} a(s)b(t)$$

for all $|x| \le M_4$ and a.e. $t \in [0, T]$. Therefore, we obtain

$$F(t, x) \le B_2|x|^2 + \max_{s\in[0,\, M_4]} a(s)b(t)$$

for all $x \in \mathbb{R}^N$ and a.e. $t \in [0, T]$. Therefore, we have

$$\begin{aligned}\varphi(u) &= \int_0^T \Big[-\frac{1}{2}\left({}_0^cD_t^\alpha u(t),\ {}_t^cD_T^\alpha u(t)\right) - F(t, u(t))\Big]dt \\ &\ge \frac{|\cos(\pi\alpha)|}{2}\|u\|_\alpha^2 - B_2\int_0^T |u|^2 dt - \max_{s\in[0,\, M_4]} a(s)\int_0^T b(t)dt,\end{aligned}$$

from which, it follows that

$$\frac{\varphi(u_n)}{\|u_n\|_\alpha^2} \ge \frac{|\cos(\pi\alpha)|}{2} - B_2\|z_n\|_{L_2}^2 - \frac{1}{\|u_n\|_\alpha^2}\max_{s\in[0,\, M_4]} a(s)\int_0^T b(t)dt.$$

Passing to the limit in the last inequality, we get

$$\frac{|\cos(\pi\alpha)|}{2} - B_2\|z\|_{L_2}^2 \le 0,$$

which yields $z \ne 0$. Therefore, there exists a subset $E \subset [0, T]$ with $\alpha(E) > 0$ such that $z(t) \ne 0$ on E.

By virtue of Lemma 6.4, for $\varepsilon = \frac{1}{2}\alpha(E) > 0$, we can choose a subset $E_\varepsilon \subset [0,T]$ with $\alpha([0,T]\backslash E_\varepsilon) < \varepsilon$ such that

$$\lim_{|x|\to\infty} [(\nabla F(t,x),x) - 2F(t,x)] = +\infty, \tag{6.46}$$

uniformly for $t \in E_\varepsilon$.

We assert that $\alpha(E\bigcap E_\varepsilon) > 0$. If not, $\alpha(E\bigcap E_\varepsilon) = 0$.

Since $E = (E\bigcap E_\varepsilon)\bigcup(E\backslash E_\varepsilon)$, it follows that

$$\begin{aligned}0 < \alpha(E) &= \alpha\Big(E\bigcap E_\varepsilon\Big) + \alpha(E\backslash E_\varepsilon)\\ &\le \alpha([0,T]\backslash E_\varepsilon)\\ &< \varepsilon = \frac{1}{2}\alpha(E),\end{aligned}$$

which leads to a contradiction and establishes the assertion.

By (A6), we obtain

$$\begin{aligned}&\int_0^T [(\nabla F(t,u_n),u_n) - 2F(t,u_n)]dt\\ &= \int_{E\bigcap E_\varepsilon} [(\nabla F(t,u_n),u_n) - 2F(t,u_n)]dt\\ &\quad + \int_{[0,T]\backslash(E\bigcap E_\varepsilon)} [(\nabla F(t,u_n),u_n) - 2F(t,u_n)]dt\\ &\ge \int_{E\bigcap E_\varepsilon} [(\nabla F(t,u_n),u_n) - 2F(t,u_n)]dt - \int_0^T |\tau(t)|dt.\end{aligned} \tag{6.47}$$

By (6.46), (6.47) and Fatou lemma, it follows that

$$\lim_{n\to\infty} \int_0^T [(\nabla F(t,u_n),u_n) - 2F(t,u_n)]dt = +\infty,$$

which contradicts (6.45). This contradiction shows that $\|u_n\|_\alpha$ is bounded in E^α and this completes the proof. □

Theorem 6.7. *Assume that $F(t,x)$ satisfies (A), (A3), (A4)$'$ and the following conditions:*

(A6)$'$ *there exists $\tau(t) \in L^1(0,T;\mathbb{R}^+)$ such that $(\nabla F(t,x),x) - 2F(t,x) \le \tau(t)$ for all $x \in \mathbb{R}^N$ and a.e. $t \in [0,T]$;*

(A7)$'$ $\lim_{|x|\to+\infty} [(\nabla F(t,x),x) - 2F(t,x)] = -\infty$ *for a.e. $t \in [0,T]$.*

Then BVP (6.1) has at least one solution on E^α.

By virtue of Lemma 6.4 and Lemma 6.5, similar to Theorem 6.5, we can complete the proof of Theorem 6.6 by using the similar proof of Theorem 6.5. Theorem 6.7 can be proved similarly.

We give an example to illustrate our results.

Example 6.2. In BVP (6.1), let $T = 2\pi$ and $F(t,x) = \kappa f(x)(2+\sin t)\arctan|x|^2$, where $\kappa > 0$ and $f(x)$ will be specified below.

Let $f(x) = |x|^2 + \ln(1+|x|^2)$. Noting that $0 \le \ln(1+|x|^2) \le |x|^2$, we see that (A) and (A4)′ hold. It is also easy to see that (A3) hold for

$$\kappa > \frac{(2\pi)^{1-2\alpha}}{|\cos(\pi\alpha)|\Gamma^2(2-\alpha)(3-2\alpha)}.$$

Furthermore, we have

$$(\nabla f(x), x) - 2f(x) = \frac{2|x|^2}{1+|x|^2} - 2\ln(1+|x|^2) \to -\infty$$

as $|x| \to +\infty$. Therefore, we have

$$\begin{aligned}&(\nabla F(t,x), x) - 2F(t,x)\\ &= \kappa\frac{2|x|^2}{1+|x|^4} f(x)(2+\sin t) + \kappa[(\nabla f(x), x) - 2f(x)](2+\sin t)\arctan|x|^2\\ &\to -\infty\end{aligned}$$

uniformly for all $t \in [0, 2\pi]$ as $|x| \to +\infty$. Thus (A6)′ and (A7)′ hold. By virtue of Theorem 6.7, we conclude that BVP (6.1) has at least one solution on E^α.

If $f(x) = |x|^2 - \ln(1+|x|^2)$, then exact the same conclusions as above hold true by Theorem 6.6.

6.3 Multiple Solutions for BVP with Parameters

6.3.1 *Introduction*

In this section, we study the existence of three solutions to BVP of the form

$$\begin{cases}\dfrac{d}{dt}\Big(\dfrac{1}{2}{}_0D_t^{-\beta}(u'(t)) + \dfrac{1}{2}{}_tD_T^{-\beta}(u'(t))\Big) + \lambda\nabla F(t,u(t)) = 0, & t\in[0,T],\\ u(0) = u(T) = 0,\end{cases} \tag{6.48}$$

where $T > 0$, $\lambda > 0$ is a parameter, $0 \le \beta < 1$, ${}_0D_t^{-\beta}$ and ${}_tD_T^{-\beta}$ are the left and right Riemann-Liouville fractional integrals of order β, respectively, $N \ge 1$ is an integer, $F : [0,T]\times\mathbb{R}^N \to \mathbb{R}$ is a given function such that $F(t,\mathbf{x})$ is measurable in t for each $\mathbf{x} = (x_1,\ldots,x_N) \in \mathbb{R}^N$ and continuously differentiable in $\mathbf{x}$ for a.e. $t \in [0,T]$, $F(t,0,\ldots,0) \equiv 0$ on $[0,T]$, and $\nabla F(t,\mathbf{x}) = (\partial F/\partial x_1,\ldots,\partial F/\partial x_N)$ is the gradient of F at $\mathbf{x}$. By a solution of (6.48), we mean an absolutely continuous function $u : [0,T] \to \mathbb{R}^N$ such that $u(t)$ satisfies both equation for a.e. $t \in [0,T]$ and the boundary conditions in (6.48). We notice that when $\beta = 0$, problem (6.48) has the form

$$\begin{cases}u''(t) + \lambda\nabla F(t,u(t)) = 0, & t\in[0,T],\\ u(0) = u(T) = 0,\end{cases} \tag{6.49}$$

which has been extensively studied.

The equation in (6.48) is motivated by the steady fractional advection dispersion equation studied in Ervin and Roop, 2006,

$$-D\,a\,(p\,{}_0D_t^{-\beta} + q\,{}_tD_T^{-\beta})Du + b(t)Du + c(t)u = f, \tag{6.50}$$

where D represents a single spatial derivative, $0 \le p, q \le 1$ satisfying $p + q = 1$, $a > 0$ is a constant, and b, c, f are functions satisfying some suitable conditions. The interest in (6.50) arises from its application as a model for physical phenomena exhibiting anomalous diffusion; i.e., diffusion not accurately modeled by the usual advection dispersion equation. Anomalous diffusion has been used in modeling turbulent flow (see, Carreras, Lynch and Zaslavsky, 2001; Shlesinger, West and Klafter, 1987), and chaotic dynamics of classical conservative systems (see, Zaslavsky, Stevens and Weitzner, 1993). The reader may find more background information and applications on (6.50) in Benson, Wheatcraft and Meerschaert, 2000a; Ervin and Roop, 2006.

Example 6.3. When $N = 1$, problem (6.48) reduces to the scalar BVP

$$\begin{cases} \dfrac{d}{dt}\Big(\dfrac{1}{2}{}_0D_t^{-\beta}(u'(t)) + \dfrac{1}{2}{}_tD_T^{-\beta}(u'(t))\Big) + \lambda f(t, u(t)) = 0, \quad t \in [0, T], \\ u(0) = u(T) = 0, \end{cases} \tag{6.51}$$

where $f : [0, T] \times \mathbb{R} \to \mathbb{R}$ is such that $f(t, x)$ is measurable in t for each $x \in \mathbb{R}$ and continuous in x for a.e. $t \in [0, T]$.

It is clear that the equation in (6.51) is of the special form of (6.50) with $D = d/dt$, $a = 1$, $p = q = \frac{1}{2}$, $b(t) = c(t) = 0$, and $f = \lambda f(t, u)$.

We also notice that since (6.50) is the steady fractional advection dispersion equation, it has no dependence on the time variable and it just depends on the space variable t (here, the notation t stands for the space variable in (6.50)). Since the space we studied is one dimensional and has the form of an interval, say $[0, T]$, the boundary conditions in the space reduce to the conditions at the two endpoints $t = 0$ and $t = T$ of the interval. In Subsection 6.3.2, we discuss the existence of Dirichlet type boundary conditions.

6.3.2 *Existence*

For $0 \le \beta < 1$ given in (6.48), let $\alpha = 1 - \frac{\beta}{2} \in (\frac{1}{2}, 1]$ and define

$$\rho_\alpha = \frac{16N}{T^2\Gamma^2(2-\alpha)}\left(\frac{1}{3-2\alpha}\left(\frac{T}{4}\right)^{3-2\alpha} + \int_{T/4}^{3T/4} g^2(t)dt + \int_{3T/4}^{T} h^2(t)dt\right), \tag{6.52}$$

where

$$g(t) = t^{1-\alpha} - (t - T/4)^{1-\alpha}, \tag{6.53}$$

$$h(t) = t^{1-\alpha} - (t - T/4)^{1-\alpha} - (t - 3T/4)^{1-\alpha}. \tag{6.54}$$

In the remainder of this section, for some $c, d, l, m, p \in \mathbb{R}$, let the bold letters $\mathbf{c}$, $\mathbf{d}$, $\mathbf{l}$, $\mathbf{m}$, and $\mathbf{p}$ be the constant vectors in $\mathbb{R}^N$ defined by

$$\mathbf{c} = (c, ..., c), \quad \mathbf{d} = (d, ..., d), \quad \mathbf{l} = (l, ..., l), \quad \mathbf{m} = (m, ..., m), \quad \mathbf{p} = (p, ..., p),$$

and any other bold letter, such as $\mathbf{x}$, is used to denote an arbitrary vector in $\mathbb{R}^N$.

Let E^α be the space of functions $u \in L^2([0,T], \mathbb{R}^N)$ having an α-order Caputo fractional derivatives ${}_0^C D_t^\alpha u \in L^2([0,T], \mathbb{R}^N)$ and $u(0) = u(T) = 0$. Then, by Remark 6.1(i) and Proposition 6.1, E^α is a reflexive and separable Banach space with the norm

$$\|u\|_\alpha = \Big(\int_0^T |u(t)|^2 dt + \int_0^T |{}_0^C D_t^\alpha u(t)|^2 dt \Big)^{\frac{1}{2}}, \quad \text{for any } u \in E^\alpha.$$

We see that the norm $\|u\|_\alpha$ is equivalent to the norm defined as the follow

$$\|u\|_\alpha = \Big(\int_0^T |{}_0^C D_t^\alpha u(t)|^2 dt \Big)^{\frac{1}{2}}, \quad \text{for any } u \in E^\alpha.$$

We recall the norms

$$\|u\|_{L^2} = \Big(\int_0^T |u(t)|^2 dt \Big)^{\frac{1}{2}} \quad \text{and} \quad \|u\| = \max_{t\in[0,T]} |u(t)|.$$

For $u \in E^\alpha$, let the functionals Φ and Ψ be defined as follows

$$\Phi(u) = -\frac{1}{2} \int_0^T \big({}_0^C D_t^\alpha u(t), {}_t^C D_T^\alpha u(t)\big) dt, \tag{6.55}$$

$$\Psi(u) = \int_0^T F(t, u(t)) dt. \tag{6.56}$$

Then, by Theorem 6.1, we see that Φ and Ψ are continuously differentiable, and for any $u, v \in E^\alpha$, we have

$$\langle \Phi'(u), v \rangle = -\frac{1}{2} \int_0^T \big[\big({}_0^C D_t^\alpha u(t), {}_t^C D_T^\alpha v(t)\big) + \big({}_t^C D_T^\alpha u(t), {}_0^C D_t^\alpha v(t)\big) \big] dt, \tag{6.57}$$

$$\langle \Psi'(u), v \rangle = \int_0^T \big(\nabla F(t, u(t)), v(t)\big) dt.$$

Parts (i) and (ii) of Lemma 6.6 below are taken from Lemma 6.2 and Theorem 6.2, respectively.

Lemma 6.6. *We have that*

- **(i)** *The functional Φ is convex and continuous on E^α.*
- **(ii)** *If $u \in E^\alpha$ is a critical point of the functional $\Phi - \lambda\Psi$, then u is a solution of BVP* (6.48).

We now state the results of this subsection.

Theorem 6.8. *Assume that there exist four positive constants c, d, l and m, with*

$$d < m \quad \text{and} \quad c < \frac{T^{\alpha-\frac{1}{2}} \rho_\alpha^{\frac{1}{2}} d}{\Gamma(\alpha)(2\alpha - 1)^{\frac{1}{2}}} < |\cos(\pi\alpha)| l < |\cos(\pi\alpha)| m, \tag{6.58}$$

such that

$$F(t,\mathbf{x}) \geq 0, \quad \textit{for } (t,\mathbf{x}) \in [0,T] \times [-m,m]^N, \tag{6.59}$$

$$\max_{|\mathbf{x}|\leq c} F(t,\mathbf{x}) \leq F(t,\mathbf{c}), \quad \max_{|\mathbf{x}|\leq l} F(t,\mathbf{x}) \leq F(t,\mathbf{l}), \quad \max_{|\mathbf{x}|\leq m} F(t,\mathbf{x}) \leq F(t,\mathbf{m}), \tag{6.60}$$

$$\frac{\int_0^T F(t,\mathbf{c})dt}{c^2} < \frac{\Gamma^2(\alpha)\cos^2(\pi\alpha)(2\alpha-1)}{T^{2\alpha-1}\rho_\alpha d^2}\left(\int_{T/4}^{3T/4} F(t,\mathbf{d})dt - \int_0^T F(t,\mathbf{c})dt\right), \tag{6.61}$$

$$\frac{\int_0^T F(t,\mathbf{l})dt}{l^2} < \frac{\Gamma^2(\alpha)\cos^2(\pi\alpha)(2\alpha-1)}{T^{2\alpha-1}\rho_\alpha d^2}\left(\int_{T/4}^{3T/4} F(t,\mathbf{d})dt - \int_0^T F(t,\mathbf{c})dt\right), \tag{6.62}$$

$$\frac{\int_0^T F(t,\mathbf{m})dt}{m^2-l^2} < \frac{\Gamma^2(\alpha)\cos^2(\pi\alpha)(2\alpha-1)}{T^{2\alpha-1}\rho_\alpha d^2}\left(\int_{T/4}^{3T/4} F(t,\mathbf{d})dt - \int_0^T F(t,\mathbf{c})dt\right). \tag{6.63}$$

Then, for each $\lambda \in (\underline{\lambda}, \bar{\lambda})$, *the system* (6.48) *has at least three solutions* u_1, u_2 *and* u_3 *such that* $\max_{t\in[0,T]} |u_1(t)| < c$, $\max_{t\in[0,T]} |u_2(t)| < l$, *and* $\max_{t\in[0,T]} |u_3(t)| < m$, *where*

$$\underline{\lambda} = \frac{\rho_\alpha d^2}{2|\cos(\pi\alpha)|\left(\int_{T/4}^{3T/4} F(t,\mathbf{d})dt - \int_0^T F(t,\mathbf{c})dt\right)} \tag{6.64}$$

and

$$\begin{aligned}\bar{\lambda} = \min\Bigg\{&\frac{\Gamma^2(\alpha)(2\alpha-1)|\cos(\pi\alpha)|c^2}{2T^{2\alpha-1}\int_0^T F(t,\mathbf{c})dt}, \frac{\Gamma^2(\alpha)(2\alpha-1)|\cos(\pi\alpha)|l^2}{2T^{2\alpha-1}\int_0^T F(t,\mathbf{l})dt},\\ &\frac{\Gamma^2(\alpha)(2\alpha-1)|\cos(\pi\alpha)|(m^2-l^2)}{2T^{2\alpha-1}\int_0^T F(t,\mathbf{m})dt}\Bigg\}.\end{aligned} \tag{6.65}$$

Proof. For any $x \in \mathbb{R}$, let $p(x) = \max\{-m, \min\{x,m\}\}$. For any $\mathbf{x} = (x_1,\ldots,x_N) \in E^\alpha$, let $\tilde{F}(t,\mathbf{x}) = F(t,\tilde{\mathbf{x}})$, where $\tilde{\mathbf{x}} = (p(x_1),\ldots,p(x_N))$. Then, $\tilde{F}(t,\mathbf{x})$ is measurable in t for each $\mathbf{x} \in \mathbb{R}^N$ and continuously differentiable in $\mathbf{x}$ for a.e. $t \in [0,T]$, and $\tilde{F}(t,0,\ldots,0) = 0$ on $[0,T]$. Note that $-m \leq p(u_i) \leq m$ for any $u = (u_1,\ldots,u_N) \in E^\alpha$ and $i = 1,\ldots,N$. Then, (6.59) implies that

$$\tilde{F}(t,u) \geq 0, \quad \text{for } (t,u) \in [0,T] \times E^\alpha. \tag{6.66}$$

Note that $d < m$ and $c < l < m$ by (6.58). Then, we have

$$\begin{aligned}&\tilde{F}(t,\mathbf{x}) = F(t,\mathbf{x}), \quad \text{for } (t,\mathbf{x}) \in [0,T] \times \mathbb{R}^N \text{ with } |\mathbf{x}| < m,\\ &\tilde{F}(t,\mathbf{c}) = F(t,\mathbf{c}), \quad \tilde{F}(t,\mathbf{d}) = F(t,\mathbf{d}),\\ &\tilde{F}(t,\mathbf{l}) = F(t,\mathbf{l}), \quad \tilde{F}(t,\mathbf{m}) = F(t,\mathbf{m}).\end{aligned} \tag{6.67}$$

Let the continuously differentiable functional Φ be given by (6.55) and the functional $\tilde{\Psi}$ be defined by

$$\tilde{\Psi}(u) = \int_0^T \tilde{F}(t,u(t))dt, \quad \text{for } u \in E^\alpha. \tag{6.68}$$

Then, by Proposition 6.4 and (6.55), we have

$$\frac{1}{2}|\cos(\pi\alpha)|\,\|u\|_\alpha^2 \le \Phi(u) \le \frac{1}{2|\cos(\pi\alpha)|}\|u\|_\alpha^2, \quad \text{for } u \in E^\alpha. \tag{6.69}$$

Moreover, $\tilde{\Psi}$ is continuously differentiable, and for any $u, v \in E^\alpha$, in view of (6.66), we have

$$\tilde{\Psi}(u) \ge 0 \quad \text{and} \quad \langle \tilde{\Psi}'(u), v\rangle = \int_0^T (\nabla \tilde{F}(t, u(t)), v(t))dt. \tag{6.70}$$

In the following, we will apply Theorem 1.19 with $X = E^\alpha$ to the functionals Φ and $\tilde{\Psi}$.

We first show that some basic assumptions of Theorem 1.19 are satisfied. The convexity and coercivity of Φ follow from Lemma 6.6(i) and (6.69), respectively. For any $u, v \in E^\alpha$, from Proposition 6.4 and (6.57),

$$\begin{aligned}
&\langle \Phi'(u) - \Phi'(v), u - v\rangle \\
&= -\frac{1}{2}\int_0^T \Big[\Big({}_0^CD_t^\alpha u(t), {}_t^CD_T^\alpha(u(t) - v(t))\Big) + \Big({}_t^CD_T^\alpha u(t), {}_0^CD_t^\alpha(u(t) - v(t))\Big)\Big]dt \\
&\quad + \frac{1}{2}\int_0^T \Big[\Big({}_0^CD_t^\alpha v(t), {}_t^CD_T^\alpha(u(t) - v(t))\Big) + \Big({}_t^CD_T^\alpha v(t), {}_0^CD_t^\alpha(u(t) - v(t))\Big)\Big]dt \\
&= -\int_0^T \Big({}_0^CD_t^\alpha(u(t) - v(t)), {}_t^CD_T^\alpha(u(t) - v(t))\Big)dt \\
&\ge |\cos(\pi\alpha)|\,\|u - v\|_\alpha^2.
\end{aligned}$$

Thus, Φ' is uniformly monotone. Hence, by Theorem 26.A(d) in Zeidler, 1990, $(\Phi')^{-1} : (E^\alpha)^* \to E^\alpha$ exists and is continuous. Suppose that $u_n \rightharpoonup u \in E^\alpha$. Then, by Proposition 6.3 $u_n \to u$ in $C([0,T], \mathbb{R}^N)$. Since $\tilde{F}(t, \mathbf{x})$ is continuously differentiable in $\mathbf{x}$ for a.e. $t \in [0, 1]$, from the derivative formula in (6.70), we have $\tilde{\Psi}'(u_n) \to \tilde{\Psi}'(u)$, i.e., $\tilde{\Psi}'$ is strongly continuous. Therefore, $\tilde{\Psi}'$ is a compact operator by Proposition 26.2 in Zeidler, 1990.

Next, note that the facts that $\tilde{F}(t, 0, ..., 0) = 0$ on $[0, T]$ and the inequality in (6.70), from Proposition 6.4, (6.55) and (6.68), we see that conditions (i) and (ii) of Theorem 1.19 are satisfied.

Now, we show that condition (iii) of Theorem 1.19 holds. For $i = 1, ..., N$, let

$$w_i(t) = \begin{cases} \dfrac{4d}{T}t, & t \in [0, T/4), \\ d, & t \in [T/4, 3T/4], \\ \dfrac{4d}{T}(T - t), & t \in (3T/4, T], \end{cases}$$

and $w(t) = (w_1(t), ..., w_N(t))$. Then, $w \in E^\alpha$ and

$${}_0^CD_t^\alpha w_i(t) = \frac{4d}{T\Gamma(2-\alpha)}\begin{cases} t^{1-\alpha}, & t \in [0, T/4), \\ g(t), & t \in [T/4, 3T/4], \\ h(t), & t \in (3T/4, T], \end{cases} \tag{6.71}$$

where $g(t)$ and $h(t)$ are defined by (6.53) and (6.54). From (6.52) and (6.71),

$$\begin{aligned}
&\int_0^T |{}_0^C D_t^\alpha w(t)|^2 dt \\
&= N\bigg(\int_0^T |{}_0^C D_t^\alpha w_1(t)|^2 dt + \int_{T/4}^{3T/4} |{}_0^C D_t^\alpha w_1(t)|^2 dt + \int_{3T/4}^{T} |{}_0^C D_t^\alpha w_1(t)|^2 dt\bigg) \\
&= \frac{16Nd^2}{T^2\Gamma^2(2-\alpha)}\bigg(\int_0^{T/4} t^{2-2\alpha} dt + \int_{T/4}^{3T/4} g^2(t)dt + \int_{3T/4}^{T} |h(t)|^2 dt\bigg) \\
&= \frac{16Nd^2}{T^2\Gamma^2(2-\alpha)}\bigg(\frac{1}{3-2\alpha}\left(\frac{T}{4}\right)^{3-2\alpha} + \int_{T/4}^{3T/4} g^2(t)dt + \int_{3T/4}^{T} h^2(t)dt\bigg) \\
&= \rho_\alpha d^2.
\end{aligned}$$

Then, $\|w\|_\alpha^2 = \rho_\alpha d^2$. Thus, from (6.69) with $u = w$,

$$\frac{1}{2}|\cos(\pi\alpha)|\rho_\alpha d^2 \le \Phi(w) \le \frac{1}{2|\cos(\pi\alpha)|}\rho_\alpha d^2. \tag{6.72}$$

Let

$$\begin{aligned}
r_1 &= \frac{\Gamma^2(\alpha)(2\alpha-1)|\cos(\pi\alpha)|}{2T^{2\alpha-1}}c^2, \quad r_2 = \frac{\Gamma^2(\alpha)(2\alpha-1)|\cos(\pi\alpha)|}{2T^{2\alpha-1}}l^2, \\
r_3 &= \frac{\Gamma^2(\alpha)(2\alpha-1)|\cos(\pi\alpha)|}{2T^{2\alpha-1}}(m^2 - l^2).
\end{aligned} \tag{6.73}$$

Then, from (6.58) and (6.72), we have $r_1 < \Phi(w) < r_2$ and $r_3 > 0$. For any $u \in E^\alpha$, from the first inequality in (6.69), we see that $\|u\|_\alpha^2 \le 2\Phi(u)/|\cos(\pi\alpha)|$. Then, by (6.14) and (6.17), we have

$$\|u\|^2 \le \frac{T^{2\alpha-1}}{\Gamma^2(\alpha)(2\alpha-1)}\|u\|_\alpha^2 \le \frac{2T^{2\alpha-1}\Phi(u)}{\Gamma^2(\alpha)(2\alpha-1)|\cos(\pi\alpha)|}.$$

Thus, by (6.73), we have the following implications

$$\begin{aligned}
&\Phi(u) < r_1 \Rightarrow \|u\| < c, \\
&\Phi(u) < r_2 \Rightarrow \|u\| < l, \\
&\Phi(u) < r_2 + r_3 \Rightarrow \|u\| < m.
\end{aligned} \tag{6.74}$$

This, together with (6.60) and (6.67), implies

$$\sup_{u\in\Phi^{-1}(-\infty,r_1)} \int_0^T \tilde{F}(t,u(t))dt \le \int_0^T \max_{|\mathbf{x}|\le c} F(t,\mathbf{x})dt \le \int_0^T F(t,\mathbf{c})dt, \tag{6.75}$$

$$\sup_{u\in\Phi^{-1}(-\infty,r_2)} \int_0^T \tilde{F}(t,u(t))dt \le \int_0^T \max_{|\mathbf{x}|\le l} F(t,\mathbf{x})dt \le \int_0^T F(t,\mathbf{l})dt,$$

$$\sup_{u\in\Phi^{-1}(-\infty,r_2+r_3)} \int_0^T \tilde{F}(t,u(t))dt \le \int_0^T \max_{|\mathbf{x}|\le m} F(t,\mathbf{x})dt \le \int_0^T F(t,\mathbf{m})dt.$$

Let φ, β, γ and α be defined by (1.20)-(1.23). Then, taking into account the fact that $0 \in \Phi^{-1}(-\infty, r_i)$, $i = 1, 2$, from (6.68) and (6.73), it follows that

$$\varphi(r_1) \le \frac{\sup_{u\in\Phi^{-1}(-\infty,r_1)} \tilde{\Psi}(u)}{r_1} \le \frac{2T^{2\alpha-1}\int_0^T F(t,\mathbf{c})dt}{\Gamma^2(\alpha)(2\alpha-1)|\cos(\pi\alpha)|c^2}, \tag{6.76}$$

$$\varphi(r_2) \le \frac{\sup_{u\in\Phi^{-1}(-\infty,r_2)} \tilde{\Psi}(u)}{r_2} \le \frac{2T^{2\alpha-1}\int_0^T F(t,\mathbf{l})dt}{\Gamma^2(\alpha)(2\alpha-1)|\cos(\pi\alpha)|l^2}, \tag{6.77}$$

$$\gamma(r_2,r_3) = \frac{\sup_{u\in\Phi^{-1}(-\infty,r_2+r_3)} \tilde{\Psi}(u)}{r_3} \le \frac{2T^{2\alpha-1}\int_0^T F(t,\mathbf{m})dt}{\Gamma^2(\alpha)(2\alpha-1)|\cos(\pi\alpha)|(m^2-l^2)}. \tag{6.78}$$

On the other hand, in view of the fact that $w(t) = \mathbf{d} < \mathbf{m}$ on $[T/4, 3T/4]$ and from (6.66) and (6.67),

$$\int_0^T \tilde{F}(t,w(t))dt \ge \int_{T/4}^{3T/4} \tilde{F}(t,w(t))dt = \int_{T/4}^{3T/4} \tilde{F}(t,\mathbf{d})dt.$$

Note that $w \in \Phi^{-1}[r_1, r_2)$, from (1.21) and (6.75), we obtain

$$\begin{aligned}
\beta(r_1,r_2) &\ge \inf_{u\in\Phi^{-1}(-\infty,r_1)} \frac{\tilde{\Psi}(w)-\tilde{\Psi}(u)}{\Phi(w)-\Phi(u)} \\
&\ge \inf_{u\in\Phi^{-1}(-\infty,r_1)} \frac{\tilde{\Psi}(w)-\tilde{\Psi}(u)}{\Phi(w)} \\
&\ge \frac{\int_{T/4}^{3T/4} \tilde{F}(t,\mathbf{d})dt - \int_0^T \tilde{F}(t,\mathbf{c})dt}{\Phi(w)}.
\end{aligned}$$

By (6.72), $1/\Phi(w) \ge 2|\cos(\pi\alpha)|/(\rho_\alpha d^2)$. Then

$$\beta(r_1,r_2) \ge \frac{2|\cos(\pi\alpha)|}{\rho_\alpha d^2}\left(\int_{T/4}^{3T/4} \tilde{F}(t,\mathbf{d})dt - \int_0^T \tilde{F}(t,\mathbf{c})dt\right). \tag{6.79}$$

For $\underline{\lambda}$ and $\bar{\lambda}$ defined by (6.64) and (6.65), from (6.61)-(6.63) and (6.76)-(6.79), we have

$$\begin{aligned}
\varphi(r_1) &< \frac{1}{\bar{\lambda}} < \frac{1}{\underline{\lambda}} < \beta(r_1,r_2), \\
\varphi(r_2) &< \frac{1}{\bar{\lambda}} < \frac{1}{\underline{\lambda}} < \beta(r_1,r_2), \\
\gamma(r_2,r_3) &< \frac{1}{\bar{\lambda}} < \frac{1}{\underline{\lambda}} < \beta(r_1,r_2).
\end{aligned}$$

In view of (1.23), $\alpha(r_1,r_2,r_3) < 1/\bar{\lambda} < 1/\underline{\lambda} < \beta(r_1,r_2)$; i.e., condition (iii) of Theorem 1.19 holds. Hence, all the assumptions of Theorem 1.19 are satisfied. Then, by Theorem 1.19, for each $\lambda \in (\underline{\lambda}, \bar{\lambda})$, the functional $\Phi - \lambda\tilde{\Psi}$ has three distinct critical points u_1, u_2 and u_3 such that $u_1 \in \Phi^{-1}(-\infty, r_1)$, $u_2 \in \Phi^{-1}[r_1, r_2)$, and $u_3 \in \Phi^{-1}(-\infty, r_2 + r_3)$. From (6.74), we have

$$\|u_1\| < c, \quad \|u_2\| < l, \quad \|u_3\| < m.$$

Then, in view of (6.65), (6.67) and (6.68), we have $\tilde{\Psi}(u) = \Psi(u)$. Therefore, u_1, u_2 and u_3 are three distinct critical points of the functional $\Phi - \lambda\Psi$. Thus, by Proposition 6.6(ii), u_1, u_2 and u_3 are three distinct solutions of (6.48). This completes the proof of the theorem. □

The following results are consequences of Theorem 6.8. In particular, Corollaries 6.2 and 6.4 give some conditions for the system (6.49) to have at least three solutions, and Corollary 6.3 provide some relatively simpler existence criteria for the system (6.48).

Corollary 6.2. *Assume that there exist four positive constants c, d, l and m, with*

$$c < (8N)^{\frac{1}{2}} d < l < m,$$

such that (6.59) *and* (6.60) *hold, and*

$$\frac{\int_0^T F(t,\mathbf{c})dt}{c^2} < \frac{1}{8Nd^2}\Big(\int_{T/4}^{3T/4} F(t,\mathbf{d})dt - \int_0^T F(t,\mathbf{c})dt\Big),$$
$$\frac{\int_0^T F(t,\mathbf{l})dt}{l^2} < \frac{1}{8Nd^2}\Big(\int_{T/4}^{3T/4} F(t,\mathbf{d})dt - \int_0^T F(t,\mathbf{c})dt\Big),$$
$$\frac{\int_0^T F(t,\mathbf{m})dt}{m^2 - l^2} < \frac{1}{8Nd^2}\Big(\int_{T/4}^{3T/4} F(t,\mathbf{d})dt - \int_0^T F(t,\mathbf{c})dt\Big).$$

Then, for each $\lambda \in (\underline{\lambda}_1, \bar{\lambda}_1)$, system (6.49) *has at least three solutions u_1, u_2, and u_3 such that $\max_{t\in[0,T]} |u_1(t)| < c$, $\max_{t\in[0,T]} |u_2(t)| < l$, and $\max_{t\in[0,T]} |u_3(t)| < m$, where*

$$\underline{\lambda}_1 = \frac{4Nd^2}{T\Big(\int_{T/4}^{3T/4} F(t,\mathbf{d})dt - \int_0^T F(t,\mathbf{c})dt\Big)},$$
$$\bar{\lambda}_1 = \min\left\{\frac{c^2}{2T\int_0^T F(t,\mathbf{c})dt}, \frac{l^2}{2T\int_0^T F(t,\mathbf{l})dt}, \frac{m^2 - l^2}{2T\int_0^T F(t,\mathbf{m})dt}\right\}.$$

Proof. When $\alpha = 1$, from (6.52), we have $\rho_\alpha = 8N/T$. Then, under the assumptions of Corollary 6.2, it is easy to see that all the conditions of Theorem 6.8 hold for $\alpha = 1$. Note that the system (6.49) is a special case of the system (6.48) with $\alpha = 1$. The conclusion then follows directly from Theorem 6.8. The proof is completed. □

Corollary 6.3. *Assume that there exist three positive constants c, d and p, with*

$$d < p \text{ and } c < \frac{T^{\alpha-\frac{1}{2}}\rho_\alpha^{\frac{1}{2}} d}{\Gamma(\alpha)(2\alpha-1)^{\frac{1}{2}}} < \frac{|\cos(\pi\alpha)|p}{\sqrt{2}}, \tag{6.80}$$

such that

$$F(t,\mathbf{x}) \geq 0 \quad \text{for } (t,\mathbf{x}) \in [0,T] \times [-p,p]^N, \tag{6.81}$$

$$\max_{|\mathbf{x}|\leq c} F(t,\mathbf{x}) \leq F(t,\mathbf{c}), \quad \max_{|\mathbf{x}|\leq p/\sqrt{2}} F(t,\mathbf{x}) \leq F\left(t, \frac{\mathbf{p}}{\sqrt{2}}\right), \quad \max_{|\mathbf{x}|\leq p} F(t,\mathbf{x}) \leq F(t,\mathbf{p}), \tag{6.82}$$

$$\frac{\int_0^T F(t,\mathbf{c})dt}{c^2} < \frac{\Gamma^2(\alpha)\cos^2(\pi\alpha)(2\alpha-1)}{T^{2\alpha-1}\rho_\alpha d^2(1+\cos^2(\pi\alpha))} \int_{T/4}^{3T/4} F(t,\mathbf{d})dt, \tag{6.83}$$

$$\frac{\int_0^T F(t,\mathbf{p})dt}{p^2} < \frac{\Gamma^2(\alpha)\cos^2(\pi\alpha)(2\alpha-1)}{2T^{2\alpha-1}\rho_\alpha d^2(1+\cos^2(\pi\alpha))}\int_{T/4}^{3T/4} F(t,\mathbf{d})dt. \tag{6.84}$$

Then, for each $\lambda \in (\underline{\lambda}_2, \bar{\lambda}_2)$, *system* (6.48) *has at least three solutions* u_1, u_2, *and* u_3 *such that* $\max_{t\in[0,T]}|u_1(t)| < c$, $\max_{t\in[0,T]}|u_2(t)| < p/\sqrt{2}$, *and* $\max_{t\in[0,T]}|u_3(t)| < p$, *where*

$$\underline{\lambda}_2 = \frac{\rho_\alpha d^2(1+\cos^2(\pi\alpha))}{2|\cos(\pi\alpha)|\int_{T/4}^{3T/4} F(t,\mathbf{d})dt}, \tag{6.85}$$

$$\bar{\lambda}_2 = \min\left\{\frac{\Gamma^2(\alpha)(2\alpha-1)|\cos(\pi\alpha)|c^2}{2T^{2\alpha-1}\int_0^T F(t,\mathbf{c})dt}, \frac{\Gamma^2(\alpha)(2\alpha-1)|\cos(\pi\alpha)|p^2}{4T^{2\alpha-1}\int_0^T F(t,\mathbf{p})dt}\right\}. \tag{6.86}$$

Proof. Let $l = p/\sqrt{2}$ and $m = p$. Then, from (6.80)-(6.82), we see that (6.58)-(6.60) hold. By (6.82) and (6.84), we have

$$\begin{aligned}\frac{\int_0^T F(t,\mathbf{l})dt}{l^2} &= \frac{2\int_0^T F(t,\mathbf{p}/\sqrt{2})dt}{p^2} \le \frac{2\int_0^T F(t,\mathbf{p})dt}{p^2}\\ &< \frac{\Gamma^2(\alpha)\cos^2(\pi\alpha)(2\alpha-1)}{T^{2\alpha-1}\rho_\alpha d^2(1+\cos^2(\pi\alpha))}\int_{T/4}^{3T/4} F(t,\mathbf{d})dt,\end{aligned} \tag{6.87}$$

and

$$\begin{aligned}\frac{\int_0^T F(t,\mathbf{m})dt}{m^2-l^2} &= \frac{2\int_0^T F(t,\mathbf{p})dt}{p^2}\\ &< \frac{\Gamma^2(\alpha)\cos^2(\pi\alpha)(2\alpha-1)}{T^{2\alpha-1}\rho_\alpha d^2(1+\cos^2(\pi\alpha))}\int_{T/4}^{3T/4} F(t,\mathbf{d})dt.\end{aligned} \tag{6.88}$$

Note from (6.80) it follows that

$$\frac{\Gamma^2(\alpha)(2\alpha-1)}{T^{2\alpha-1}\rho_\alpha d^2} < \frac{1}{c^2}.$$

Combining this inequality with (6.83), we obtain

$$\begin{aligned}&\frac{\Gamma^2(\alpha)\cos^2(\pi\alpha)(2\alpha-1)}{T^{2\alpha-1}\rho_\alpha d^2}\left(\int_{T/4}^{3T/4} F(t,\mathbf{d})dt - \int_0^T F(t,\mathbf{c})dt\right)\\ &> \frac{\Gamma^2(\alpha)\cos^2(\pi\alpha)(2\alpha-1)}{T^{2\alpha-1}\rho_\alpha d^2}\int_{T/4}^{3T/4} F(t,\mathbf{d})dt - \frac{\cos^2(\pi\alpha)}{c^2}\int_0^T F(t,\mathbf{c})dt\\ &> \frac{\Gamma^2(\alpha)\cos^2(\pi\alpha)(2\alpha-1)}{T^{2\alpha-1}\rho_\alpha d^2}\int_{T/4}^{3T/4} F(t,\mathbf{d})dt\\ &\quad - \frac{\Gamma^2(\alpha)\cos^4(\pi\alpha)(2\alpha-1)}{T^{2\alpha-1}\rho_\alpha d^2(1+\cos^2(\pi\alpha))}\int_{T/4}^{3T/4} F(t,\mathbf{c})dt\\ &= \frac{\Gamma^2(\alpha)\cos^2(\pi\alpha)(2\alpha-1)}{T^{2\alpha-1}\rho_\alpha d^2(1+\cos^2(\pi\alpha))}\int_{T/4}^{3T/4} F(t,\mathbf{d})dt.\end{aligned} \tag{6.89}$$

By (6.83) and (6.87)-(6.89), we see that (6.61)-(6.62) hold. From (6.64), (6.65), (6.85), (6.86) and (6.89), we have $\underline{\lambda} < \underline{\lambda}_2$ and $\bar{\lambda} = \bar{\lambda}_2$. Therefore, the conclusion now follows from Theorem 6.8. The proof is completed. □

Corollary 6.4. *Assume that there exist three positive constants c, d and p, with*

$$c < (8N)^{\frac{1}{2}} d < \frac{p}{\sqrt{2}}, \tag{6.90}$$

such that (6.81) *and* (6.82) *hold, and*

$$\frac{\int_0^T F(t, \mathbf{c})dt}{c^2} < \frac{1}{16Nd^2} \int_{T/4}^{3T/4} F(t, \mathbf{d})dt, \tag{6.91}$$

and

$$\frac{\int_0^T F(t, \mathbf{p})dt}{p^2} < \frac{1}{32Nd^2} \int_{T/4}^{3T/4} F(t, \mathbf{d})dt. \tag{6.92}$$

Then, for each $\lambda \in (\underline{\lambda}_3, \bar{\lambda}_3)$, system (6.49) *has at least three solutions u_1, u_2, and u_3 such that $\max_{t\in[0,T]} |u_1(t)| < c$, $\max_{t\in[0,T]} |u_2(t)| < p/\sqrt{2}$, and $\max_{t\in[0,T]} |u_3(t)| < p$, where*

$$\underline{\lambda}_3 = \frac{8Nd^2}{T\int_{T/4}^{3T/4} F(t, \mathbf{d})dt},$$

$$\bar{\lambda}_3 = \min\left\{\frac{c^2}{2T\int_0^T F(t, \mathbf{c})dt}, \frac{p^2}{4T\int_0^T F(t, \mathbf{p})dt}\right\}.$$

Proof. When $\alpha = 1$, from (6.52), we have $\rho_\alpha = 8N/T$. Under the assumptions of Corollary 6.4, it is easy to see that all the conditions of Corollary 6.3 hold for $\alpha = 1$. Note that system (6.49) is a special case of system (6.48) with $\alpha = 1$. The conclusion then follows directly from Corollary 6.3. The proof is completed. □

Remark 6.3. We want to point out that when F does not depend on t, (6.91) and (6.92) reduce to

$$\frac{F(\mathbf{c})}{c^2} < \frac{F(\mathbf{d})}{32Nd^2} \quad \text{and} \quad \frac{F(\mathbf{p})}{p^2} < \frac{F(\mathbf{d})}{64Nd^2}, \tag{6.93}$$

and $\underline{\lambda}_3$ and $\bar{\lambda}_3$ become

$$\underline{\lambda}_3 = \frac{16Nd^2}{T^2F(\mathbf{d})} \quad \text{and} \quad \bar{\lambda}_3 = \min\left\{\frac{c^2}{2T^2F(\mathbf{c})}, \frac{p^2}{4T^2F(\mathbf{p})}\right\}. \tag{6.94}$$

Remark 6.4. We observe that, in our results, no asymptotic condition on F is needed and only local conditions on F are imposed to guarantee the existence of solutions. Moreover, in the conclusions of the above results, one of the three solutions may be trivial since $\nabla F(t, 0, ..., 0)$ may be zero.

In the remainder of this subsection, we give two examples to illustrate the applicability of our results.

Example 6.4. Let $T > 0$. For $(t, x, y) \in [0, T] \times \mathbb{R}^2$, let $F(t, x, y) = tG(x, y)$, where $G : \mathbb{R}^2 \to \mathbb{R}$ satisfies that $G(-x, -y) = G(x, y)$, and that for $x \in [0, \infty)$ and $y \in \mathbb{R}$,

$$G(x,y) = \begin{cases} x^3 + |y|^3, & 0 \le x \le 1,\ 0 \le |y| \le 1, \\ x^3 + 2|y|^{3/2} - 1, & 0 \le x \le 1,\ |y| > 1, \\ 2x^{3/2} + |y|^3 - 1, & x > 1,\ 0 \le |y| \le 1, \\ 2x^{3/2} + 2|y|^{3/2} - 2, & x > 1,\ |y| > 1. \end{cases} \tag{6.95}$$

It is easy to verify that $F : [0, T] \times \mathbb{R}^2 \to \mathbb{R}$ is measurable in t for $(x, y) \in \mathbb{R}^2$ and continuously differentiable in x and y for $t \in [0, T]$, and $F(t, 0, 0) \equiv 0$ on $[0, T]$.

Let $0 \le \beta < 1$, $\alpha = 1 - \frac{\beta}{2} \in (\frac{1}{2}, 1]$, ρ_α be defined by (6.52), and $u(t) = (u_1(t), u_2(t))$. We claim that for each

$$\lambda \in \left(\frac{\rho_\alpha (1 + \cos^2(\pi\alpha))}{T^2 |\cos(\pi\alpha)|}, \infty \right),$$

the system

$$\begin{cases} \dfrac{d}{dt}\Big(\dfrac{1}{2}\, {}_0D_t^{-\beta}(u'(t)) + \dfrac{1}{2}\, {}_tD_T^{-\beta}(u'(t)) \Big) + \lambda \nabla F(t, u(t)) = 0, \quad t \in [0, T], \\ u(0) = u(T) = 0 \end{cases} \tag{6.96}$$

has at least three solutions.

In fact, system (6.96) is a special case of system (6.48) with $N = 2$. For $0 < c < 1$ and $p > 1$, in view of (6.95), we have

$$\frac{\int_0^T F(t, c, c)dt}{c^2} = \frac{2c^3 \int_0^T t dt}{c^2} = T^2 c, \tag{6.97}$$

$$\frac{\int_0^T F(t, p, p)dt}{p^2} = \frac{(4p^{3/2} - 2) \int_0^T t dt}{p^2} = \frac{T^2 (2p^{3/2} - 1)}{p^2}. \tag{6.98}$$

Choose $d = 1$. Then,

$$\int_{T/4}^{3T/4} F(t, d, d)dt = 2 \int_{T/4}^{3T/4} t dt = \frac{1}{2} T^2. \tag{6.99}$$

By (6.97)-(6.99), we see that there exist $0 < c^* < 1$ and $p^* > 1$ such that (6.80), (6.83) and (6.84) hold for any $0 < c < c^*$ and $p > p^*$. Moreover, (6.81) and (6.82) hold for any $c, p > 0$. Finally, note from (6.85) and (6.86) that

$$\underline{\lambda}_2 = \frac{\rho_\alpha (1 + \cos^2(\pi\alpha))}{T^2 |\cos(\pi\alpha)|},$$

$$\bar{\lambda}_2 \to \infty \quad \text{as} \quad c \to 0^+ \quad \text{and} \quad p \to \infty.$$

Then, the claim follows from Corollary 6.3.

Example 6.5. Let $F:\mathbb{R}^2\to\mathbb{R}$ satisfy that $F(-x,-y)=F(x,y)$, and that for $x\in[0,\infty)$ and $y\in\mathbb{R}$,

$$F(x,y)=\begin{cases} x^3, & 0\le x\le 1,\ 0\le |y|\le 1,\\ x^3+2|y|^{3/2}-3|y|+1, & 0\le x\le 1,\ |y|>1,\\ 2x^{3/2}-1, & x>1,\ 0\le |y|\le 1,\\ 2x^{3/2}+2|y|^{3/2}-3|y|, & x>1,\ |y|>1. \end{cases} \tag{6.100}$$

It is easy to verify that $F:\mathbb{R}^2\to\mathbb{R}$ is continuously differentiable in x and y and $F(0,0)=0$.

Let $T>0$ and $u(t)=(u_1(t),u_2(t))$. We claim that for each $\lambda\in(32/T^2,\infty)$, the system

$$\begin{cases} u''(t)+\lambda\nabla F(u(t))=0, & t\in[0,T],\\ u(0)=u(T)=0 \end{cases} \tag{6.101}$$

has at least three solutions. In fact, the system (6.101) is a special case of the system (6.49) with $N=2$. For $0<c<1$ and $p>1$, from (6.100), we have

$$\frac{F(c,c)}{c^2}=\frac{c^3}{c^2}=c, \tag{6.102}$$

$$\frac{F(p,p)}{p^2}=\frac{4p^{3/2}-3p}{p^2}=\frac{4p^{\frac{1}{2}}-3}{p}. \tag{6.103}$$

Choose $d=1$. Then

$$\frac{F(d,d)}{32Nd^2}=\frac{1}{64}\quad\text{and}\quad\frac{F(d,d)}{64Nd^2}=\frac{1}{128}. \tag{6.104}$$

By (6.102)-(6.104), we see that there exist $0<c^*<1$ and $p^*>1$ such that (6.90) and (6.93) hold for any $0<c<c^*$ and $p>p^*$. Moreover, (6.81) and (6.82) hold for any $c,p>0$. Finally, note from (6.94) that

$$\underline{\lambda}_3=\frac{32}{T^2}\quad\text{and}\quad\bar{\lambda}_3\to\infty,\quad\text{as}\ \ c\to 0^+\ \text{and}\ \ p\to\infty.$$

Then, the claim follows from Corollary 6.4 and Remark 6.3.

Remark 6.5. As noted in Remark 6.4, one of the three solutions in the conclusions of the above examples may be trivial.

6.4 Infinite Solutions for BVP with Left and Right Fractional Integrals

6.4.1 *Introduction*

In this section, we consider BVP (6.1), i.e.,

$$\begin{cases} \dfrac{d}{dt}\left(\dfrac{1}{2}\,{}_0D_t^{-\beta}(u'(t))+\dfrac{1}{2}\,{}_tD_T^{-\beta}(u'(t))\right)+\nabla F(t,u(t))=0, & \text{a.e. } t\in[0,T],\\ u(0)=u(T)=0, \end{cases}$$

where ${}_0D_t^{-\beta}$ and ${}_tD_T^{-\beta}$ are the left and right Riemann-Liouville fractional integrals of order $0 \leq \beta < 1$ respectively. Assume that $F : [0,T] \times \mathbb{R}^N \to \mathbb{R}$ satisfies the condition (A) which is assumed as in Subsection 6.2.3.

In particular, if $\beta = 0$, BVP (6.1) reduces to the standard second-order BVP.

In the Subsection 6.4.2, using variational methods we prove the multiplicity results for the solutions of problem (6.1).

6.4.2 *Existence*

Making use of the Proposition 1.4 and Definition 1.3, for any $u \in AC([0,T], \mathbb{R}^N)$, BVP (6.1) is equivalent to (6.20).

In the following, we will treat BVP (6.20) in the Hilbert space $E^\alpha = E_0^{\alpha,2}$ with the corresponding norm $\|u\|_\alpha = \|u\|_{\alpha,2}$.

As E^α is a reflexive and separable Banach space, then there are $e_j \in E^\alpha$ and $e_j^* \in (E^\alpha)^*$ such that

$$E^\alpha = \overline{\text{span}\{e_j : j = 1, 2, ...\}} \quad \text{and} \quad (E^\alpha)^* = \overline{\text{span}\{e_j^* : j = 1, 2, ...\}}.$$

For $k = 1, 2, ...,$ denote

$$X_j := \text{span}\{e_j\}, \quad Y_k := \bigoplus_{j=1}^{k} X_j, \quad Z_k := \overline{\bigoplus_{j=k}^{\infty} X_j}.$$

Theorem 6.9. *Assume that $F(t,x)$ satisfies the condition (A), and suppose the following conditions hold:*

(A1) *there exist $\kappa > 2$ and $r > 0$ such that*

$$\kappa F(t,x) \leq (\nabla F(t,x), x)$$

for a.e. $t \in [0,T]$ and all $|x| \geq r$ in $\mathbb{R}^N$;

(A2) *there exist positive constants $\mu > 2$ and $Q > 0$ such that*

$$\limsup_{|x| \to +\infty} \frac{F(t,x)}{|x|^\mu} \leq Q$$

uniformly for a.e. $t \in [0,T]$;

(A3) *there exist $\mu' > 2$ and $Q' > 0$ such that*

$$\limsup_{|x| \to +\infty} \frac{F(t,x)}{|x|^{\mu'}} \geq Q'$$

uniformly for a.e. $t \in [0,T]$;

(A4) *$F(t,x) = F(t,-x)$ for $t \in [0,T]$ and all x in $\mathbb{R}^N$.*

Then BVP (6.1) has infinite solutions $\{u_n\}$ on E^α for every positive integer n such that $\|u_n\|_\infty \to \infty$, as $n \to \infty$.

Proof. Let $\{u_n\} \subset E^\alpha$ such that $\varphi(u_n)$ is bounded and $\varphi'(u_n) \to 0$ as $n \to \infty$, where $\varphi(u)$ and $\varphi'(u)$ are defined by (6.26) and (6.27) respectively. First we prove $\{u_n\}$ is a bounded sequence, otherwise, $\{u_n\}$ would be unbounded sequence, passing to a subsequence, still denoted by $\{u_n\}$, such that $\|u_n\|_\alpha \geq 1$ and $\|u_n\|_\alpha \to \infty$, as $n \to \infty$.

Noting that

$$\langle \varphi'(u_n), u_n \rangle = -\int_0^T \left(({}_0^C D_t^\alpha u_n(t), {}_t^C D_T^\alpha u_n(t)) + (\nabla F(t, u_n(t)), u_n(t)) \right) dt.$$

In view of the condition (A1) and (6.22) that

$$\begin{aligned}
&\varphi(u_n) - \frac{1}{\kappa}\langle \varphi'(u_n), u_n \rangle \\
&= \left(\frac{1}{\kappa} - \frac{1}{2}\right) \int_0^T ({}_0^C D_t^\alpha u_n(t), {}_t^C D_T^\alpha u_n(t)) dt \\
&\quad + \int_0^T \left(\frac{1}{\kappa}(\nabla F(t, u_n(t)), u_n(t)) - F(t, u_n(t)) \right) dt \\
&\geq \left(\frac{1}{2} - \frac{1}{\kappa}\right) |\cos(\pi\alpha)| \|u_n\|_\alpha^2 \\
&\quad + \left(\int_{\Omega_1} + \int_{\Omega_2} \right) \left(\frac{1}{\kappa}(\nabla F(t, u_n(t)), u_n(t)) - F(t, u_n(t)) \right) dt \\
&\geq \left(\frac{1}{2} - \frac{1}{\kappa}\right) |\cos(\pi\alpha)| \|u_n\|_\alpha^2 - C_1,
\end{aligned}$$

where $\Omega_1 := \{t \in [0, T] : |u_n(t)| \leq r\}$, $\Omega_2 := [0, T] \backslash \Omega_1$ and C_1 is a positive constant.

Since $\varphi(u_n)$ is bounded, there exists a positive constant C_2, such that $|\varphi(u_n)| \leq C_2$. Hence, we have

$$\begin{aligned}
C_2 \geq \varphi(u_n) &\geq \left(\frac{1}{2} - \frac{1}{\kappa}\right) |\cos(\pi\alpha)| \|u_n\|_\alpha^2 + \frac{1}{\kappa}\langle \varphi'(u_n), u_n \rangle - C_1 \\
&\geq \left(\frac{1}{2} - \frac{1}{\kappa}\right) |\cos(\pi\alpha)| \|u_n\|_\alpha^2 - \frac{1}{\kappa}\|\varphi'(u_n)\|_\alpha \|u_n\|_\alpha - C_1,
\end{aligned} \tag{6.105}$$

so $\{u_n\}$ is a bounded sequence in E^α by (6.105).

Since E^α is a reflexive space, going to a subsequence if necessary, we may assume that $u_n \rightharpoonup u$ weakly in E^α, thus we have

$$\begin{aligned}
\langle \varphi'(u_n) - \varphi'(u), u_n - u \rangle &= \langle \varphi'(u_n), u_n - u \rangle - \langle \varphi'(u), u_n - u \rangle \\
&\leq \|\varphi'(u_n)\|_\alpha \|u_n - u\|_\alpha - \langle \varphi'(u), u_n - u \rangle \\
&\to 0, \quad \text{as } n \to \infty.
\end{aligned} \tag{6.106}$$

Moreover, according to (6.14) and Proposition 6.3, we have u_n is bounded in $C([0, T], \mathbb{R}^N)$ and $\|u_n - u\| \to 0$ as $n \to \infty$.

Observing that

$$
\begin{aligned}
&\langle \varphi'(u_n) - \varphi'(u), u_n - u\rangle \\
&= -\int_0^T ({}_0^C D_t^\alpha (u_n(t) - u(t)), {}_t^C D_T^\alpha (u_n(t) - u(t)))dt \\
&\quad - \int_0^T (\nabla F(t, u_n(t)) - \nabla F(t, u(t)), u_n(t) - u(t))dt \\
&\geq |\cos(\pi\alpha)| \|u_n(t) - u(t)\|_\alpha^2 \\
&\quad - \left|\int_0^T (\nabla F(t, u_n(t)) - \nabla F(t, u(t)))dt\right| \|u_n(t) - u(t)\|.
\end{aligned}
$$

Combining this with (6.106), it is easy to verify that $\|u_n(t)-u(t)\|_\alpha \to 0$ as $n \to \infty$, and hence that $u_n \to u$ in E^α. Thus, $\{u_n\}$ admits a convergent subsequence.

For any $u \in Y_k$, let

$$
\|u\|_* := \left(\int_0^T |u(t)|^{\mu'} dt\right)^{1/\mu'}, \tag{6.107}
$$

and it is easy to verify that $\|\cdot\|_*$ define by (6.107) is a norm of Y_k. Since all the norms of a finite dimensional normed space are equivalent, so there exists positive constant C_3 such that

$$
C_3\|u\|_\alpha \leq \|u\|_* \quad \text{for } u \in Y_k. \tag{6.108}
$$

In view of (A3), there exist two positive constants M_1 and C_4 such that

$$
F(t, x) \geq M_1|x|^{\mu'}, \tag{6.109}
$$

for a.e. $t \in [0, T]$ and $|x| \geq C_4$.

It follows from (6.22), (6.108) and (6.109) that

$$
\begin{aligned}
\varphi(u) &= -\int_0^T \frac{1}{2}({}_0^C D_t^\alpha u(t), {}_t^C D_T^\alpha u(t))dt - \int_0^T F(t, u(t))dt \\
&\leq \frac{1}{|2\cos(\pi\alpha)|}\|u\|_\alpha^2 - \int_{\Omega_3} F(t, u(t))dt - \int_{\Omega_4} F(t, u(t))dt \\
&\leq \frac{1}{|2\cos(\pi\alpha)|}\|u\|_\alpha^2 - M_1\int_{\Omega_3} |u(t)|^{\mu'} dt - \int_{\Omega_4} F(t, u(t))dt \\
&= \frac{1}{|2\cos(\pi\alpha)|}\|u\|_\alpha^2 - M_1\int_0^T |u(t)|^{\mu'} dt + M_1\int_{\Omega_4} |u(t)|^{\mu'} dt - \int_{\Omega_4} F(t, u(t))dt \\
&\leq \frac{1}{|2\cos(\pi\alpha)|}\|u\|_\alpha^2 - C_3^{\mu'} M_1\|u\|_\alpha^{\mu'} + C_5,
\end{aligned}
$$

where $\Omega_3 := \{t \in [0, T] : \ |u(t)| \geq C_4\}$, $\Omega_4 := [0, T]\backslash\Omega_3$ and C_5 is a positive constant.

Since $\mu' > 2$, then there exist positive constants d_k such that

$$
\varphi(u) \leq 0, \quad \text{for } u \in Y_k, \quad \text{and } \|u\|_\alpha \geq d_k. \tag{6.110}
$$

For any $u \in Z_k$, let

$$\|u\|_\mu := \left(\int_0^T |u(t)|^\mu dt\right)^{1/\mu} \quad \text{and} \quad \beta_k := \sup_{\substack{u\in Z_k\\ \|u\|_\alpha=1}} \|u\|_\mu,$$

then we conclude $\beta_k \to 0$ as $k \to \infty$.

In fact, it is obvious that $\beta_k \geq \beta_{k+1} > 0$, so $\beta_k \to \beta$ as $k \to \infty$. For every $k \in \mathbb{N}$, there exists $u_k \in Z_k$ such that

$$\|u_k\|_\alpha = 1 \;\text{ and } \|u_k\|_\mu > \beta_k/2. \tag{6.111}$$

As E^α is reflexive, $\{u_k\}$ has a weakly convergent subsequence, still denoted by $\{u_k\}$, such that $u_k \rightharpoonup u$. We claim $u = 0$.

In fact, for any $f_m \in \{f_n : n = 1, 2, ...\}$, we have $f_m(u_k) = 0$, when $k > m$, so

$$f_m(u_k) \to 0, \quad \text{as } k \to \infty$$

for any $f_m \in \{f_n : n = 1, 2, ...\}$, therefore $u = 0$.

By Proposition 6.3, when $u_k \rightharpoonup 0$, in E^α, then $u_k \to 0$ strongly in $C([0,T],\mathbb{R}^N)$. So we conclude $\beta = 0$ by (6.111).

In view of (A2), there exist two positive constants M_2 and C_6 such that

$$F(t,x) \leq M_2|x|^\mu$$

uniformly for a.e. $t \in [0,T]$ and $|x| \geq C_6$. We have

$$\begin{aligned}
\varphi(u) &= -\int_0^T \frac{1}{2}({}_0^C D_t^\alpha u(t), {}_t^C D_T^\alpha u(t))dt - \int_0^T F(t,u(t))dt \\
&\geq \frac{|\cos(\pi\alpha)|}{2}\|u\|_\alpha^2 - \int_{\Omega_5} F(t,u(t))dt - \int_{\Omega_6} F(t,u(t))dt \\
&\geq \frac{|\cos(\pi\alpha)|}{2}\|u\|_\alpha^2 - M_2\int_{\Omega_5} |u(t)|^\mu dt - \int_{\Omega_6} F(t,u(t))dt \\
&= \frac{|\cos(\pi\alpha)|}{2}\|u\|_\alpha^2 - M_2\int_0^T |u(t)|^\mu dt + M_2\int_{\Omega_6} |u(t)|^\mu dt - \int_{\Omega_6} F(t,u(t))dt \\
&\geq \frac{|\cos(\pi\alpha)|}{2}\|u\|_\alpha^2 - M_2{\beta_k}^\mu\|u\|_\alpha^\mu - C_7,
\end{aligned}$$

where $\Omega_5 := \{t \in [0,T] : |u(t)| \geq C_6\}$, $\Omega_6 := [0,T]\backslash\Omega_5$ and C_7 is a positive constant.

Choosing $r_k = 1/\beta_k$, it is obvious that $r_k \to \infty$ as $k \to \infty$, then

$$b_k := \inf_{\substack{u\in Z_k\\ \|u\|_\alpha=\rho_k}} \varphi(u) \to \infty, \quad \text{as } k \to \infty,$$

that is, the condition (H3) in Theorem 1.17 is satisfied.

In view of (6.110), let $\rho_k := \max\{d_k, r_k + 1\}$, then

$$a_k := \max_{\substack{u\in Y_k\\ \|u\|_\alpha=\rho_k}} \varphi(u) \leq 0,$$

and this shows the condition of (H2) in Theorem 1.17 is satisfied.

We have proved the functional φ satisfies all the conditions of Theorem 1.17, then φ has an unbounded sequence of critical values $c_n = \varphi(u_n)$ by Theorem 1.17. We only need to show $\|u_n\| \to \infty$ as $n \to \infty$.

In fact, since u_n is a critical point of the functional φ, that is

$$\langle \varphi'(u_n), u_n \rangle = -\int_0^T \left[\left({}_0^C D_t^\alpha u_n(t), {}_t^C D_T^\alpha u_n(t) \right) + (\nabla F(t, u_n(t)), u_n(t)) \right] dt = 0.$$

Hence, we have

$$\begin{aligned} c_n = \varphi(u_n) &= -\int_0^T \frac{1}{2} \left({}_0^C D_t^\alpha u_n(t), {}_t^C D_T^\alpha u_n(t) \right) dt - \int_0^T F(t, u_n(t)) dt, \\ &= \frac{1}{2} \int_0^T (\nabla F(t, u_n(t)), u_n(t)) dt - \int_0^T F(t, u_n(t)) dt, \qquad (6.112) \\ &\le \frac{1}{2} \int_0^T |\nabla F(t, u_n(t))| |u_n(t)| dt + \int_0^T |F(t, u_n(t))| dt, \end{aligned}$$

since $c_n \to \infty$, we conclude

$$\|u_n\| \to \infty, \quad \text{as } n \to \infty$$

by (6.112). In fact, if not, going to a subsequence if necessary, we may assume that

$$\|u_n\| \le M_3,$$

for all $n \in \mathbb{N}$ and some positive constant M_3.

Combining condition (A) and (6.112), we have

$$\begin{aligned} c_n &\le \frac{1}{2} \int_0^T |\nabla F(t, u_n(t))| |u_n(t)| dt + \int_0^T |F(t, u_n(t))| dt \\ &\le \frac{1}{2} (M_3 + 1) \max_{0 \le s \le M_3} m_1(s) \int_0^T m_2(t) dt, \end{aligned}$$

which contradicts the unboundness of c_n. This completes the proof of Theorem 6.6. □

Example 6.6. In BVP (6.1), let $F(t,x) = |x|^4$, and choose

$$\kappa = 4, \quad r = 2, \quad \mu = \mu' = 4 \ \text{ and } Q = Q' = 1,$$

so it is easy to verify that all the conditions (A1)-(A4) are satisfied. Then by Theorem 6.9, BVP (6.1) has infinite solutions $\{u_k\}$ on E^α for every positive integer k such that $\|u_k\| \to \infty$, as $k \to \infty$.

Theorem 6.10. *Assume that $F(t,x)$ satisfies the following assumption:*

(A5) *$F(t,x) := a(t)|x|^\gamma$, where $a(t) \in L^\infty([0,T], \mathbb{R}^+)$ and $1 < \gamma < 2$ is a constant.*

Then BVP (6.1) has infinite solutions $\{u_n\}$ on E^α for every positive integer n with $\|u_n\|_\alpha$ bounded.

Proof. Let us show that φ satisfies conditions in Theorem 1.18 item by item. First, we show that φ satisfies the $(\mathrm{PS})_c^*$ condition for every $c \in \mathbb{R}$.

Suppose $n_j \to \infty$, $u_{n_j} \in Y_{n_j}$, $\varphi(u_n) \to c$ and $(\varphi|_{Y_{n_j}})'(u_{n_j}) \to 0$, then $\{u_{n_j}\}$ is a bounded sequence, otherwise, $\{u_{n_j}\}$ would be unbounded sequence, passing to a subsequence, still denoted by $\{u_{n_j}\}$ such that $\|u_{n_j}\|_\alpha \geq 1$ and $\|u_{n_j}\|_\alpha \to \infty$. Note that

$$\langle \varphi'(u_{n_j}), u_{n_j}\rangle - \gamma\varphi(u_{n_j}) = \left(-1 + \frac{\gamma}{2}\right)\int_0^T ({}_0^C D_t^\alpha u_{n_j}(t), {}_t^C D_T^\alpha u_{n_j}(t))dt. \qquad (6.113)$$

However, from (6.113), we have

$$-\gamma\varphi(u_{n_j}) \geq \left(1 - \frac{\gamma}{2}\right)|\cos(\pi\alpha)|\|u_{n_j}\|_\alpha^2 - \|(\varphi|_{Y_{n_j}})'(u_{n_j})\|\|u_{n_j}\|_\alpha,$$

thus $\|u_{n_j}\|_\alpha$ is a bounded sequence in E^α. Going, if necessary, to a subsequence, we can assume that $u_{n_j} \rightharpoonup u$ in E^α. As $E^\alpha = \overline{\bigcup_{n_j} Y_{n_j}}$, we can choose $v_{n_j} \in Y_{n_j}$ such that $v_{n_j} \to u$.

Hence

$$\begin{aligned}
&\lim_{n_j\to\infty}\langle \varphi'(u_{n_j}), u_{n_j} - u\rangle \\
&= \lim_{n_j\to\infty}\langle \varphi'(u_{n_j}), u_{n_j} - v_{n_j}\rangle + \lim_{n_j\to\infty}\langle \varphi'(u_{n_j}), v_{n_j} - u\rangle \\
&= \lim_{n_j\to\infty}\langle (\varphi|_{Y_{n_j}})'(u_{n_j}), u_{n_j} - v_{n_j}\rangle \\
&= 0.
\end{aligned}$$

So we have

$$\begin{aligned}
&\lim_{n_j\to\infty}\langle \varphi'(u_{n_j}) - \varphi'(u), u_{n_j} - u\rangle \\
&= \lim_{n_j\to\infty}\langle \varphi'(u_{n_j}), u_{n_j} - u\rangle - \lim_{n_j\to\infty}\langle \varphi'(u), u_{n_j} - u\rangle \\
&= 0,
\end{aligned}$$

and

$$\begin{aligned}
&\langle \varphi'(u_{n_j}) - \varphi'(u), u_{n_j} - u\rangle \\
&= -\int_0^T ({}_0^C D_t^\alpha(u_{n_j}(t) - u(t)), {}_t^C D_T^\alpha(u_{n_j}(t) - u(t)))dt \\
&\quad - \int_0^T (\nabla F(t, u_{n_j}(t)) - \nabla F(t, u(t)), u_{n_j}(t) - u(t))dt \\
&\geq |\cos(\pi\alpha)|\|u_{n_j}(t) - u(t)\|_\alpha^2 \\
&\quad - \left|\int_0^T (\nabla F(t, u_{n_j}(t)) - \nabla F(t, u(t)))dt\right|\|u_{n_j}(t) - u(t)\|,
\end{aligned}$$

we can conclude $u_{n_j} \to u$ in E^α, furthermore, we have $\varphi'(u_{n_j}) \to \varphi'(u)$.

Let us prove $\varphi'(u) = 0$ below. Taking arbitrarily $w_k \in Y_k$, notice when $n_j \geq k$, we have

$$\begin{aligned}
\langle \varphi'(u), w_k\rangle &= \langle \varphi'(u) - \varphi'(u_{n_j}), w_k\rangle + \langle \varphi'(u_{n_j}), w_k\rangle \\
&= \langle \varphi'(u) - \varphi'(u_{n_j}), w_k\rangle + \langle (\varphi|_{Y_{n_j}})'(u_{n_j}), w_k\rangle.
\end{aligned}$$

Let $n_j \to \infty$ in the right side of above equation. Then

$$\langle \varphi'(u), w_k \rangle = 0, \quad \forall\, w_k \in Y_k,$$

so $\varphi'(u) = 0$, this shows that φ satisfies the $(\mathrm{PS})_c^*$ for every $c \in \mathbb{R}$.

For any finite dimensional subspace $E \subset E^\alpha$, there exists $\varepsilon > 0$ such that

$$\alpha\{t \in [0,T] : a(t)|u(t)|^\gamma \geq \varepsilon \|u\|_\alpha^\gamma\} \geq \varepsilon, \quad \forall\, u \in E\backslash\{0\}. \tag{6.114}$$

Otherwise, for any positive integer n, there exists $u_n \in E\backslash\{0\}$ such that

$$\alpha\left\{t \in [0,T] : a(t)|u_n(t)|^\gamma \geq \frac{1}{n}\|u_n\|_\alpha^\gamma\right\} < \frac{1}{n}.$$

Set $v_n := \frac{u_n(t)}{\|u_n\|_\alpha} \in E\backslash\{0\}$, then $\|v_n\|_\alpha = 1$ for all $n \in \mathbb{N}$ and

$$\alpha\left\{t \in [0,T] : a(t)|v_n(t)|^\gamma \geq \frac{1}{n}\right\} < \frac{1}{n}. \tag{6.115}$$

Since $\dim E < \infty$, it follows from the compactness of the unit sphere of E that there exists a subsequence, denoted also by $\{v_n\}$, such that $\{v_n\}$ converges to some v_0 in E. It is obvious that $\|v_0\|_\alpha = 1$.

By the equivalence of the norms on the finite-dimensional space, we have $v_n \to v_0$ in $L^2([0,T], \mathbb{R}^N)$, i.e.,

$$\int_0^T |v_n - v_0|^2 dt \to 0, \quad \text{as } n \to \infty. \tag{6.116}$$

By (6.116) and Hölder inequality, we have

$$\begin{aligned}\int_0^T a(t)|v_n - v_0|^\gamma dt &\leq \left(\int_0^T a(t)^{\frac{2}{2-\gamma}} dt\right)^{\frac{2-\gamma}{2}} \left(\int_0^T |v_n - v_0|^2 dt\right)^{\frac{\gamma}{2}} \\ &= \|a\|_{\frac{2}{2-\gamma}} \left(\int_0^T |v_n - v_0|^2 dt\right)^{\frac{\gamma}{2}} \\ &\to 0, \quad \text{as } n \to \infty.\end{aligned} \tag{6.117}$$

Thus, there exist $\xi_1, \xi_2 > 0$ such that

$$\alpha\{t \in [0,T] : a(t)|v_0(t)|^\gamma \geq \xi_1\} \geq \xi_2. \tag{6.118}$$

In fact, if not, we have

$$\alpha\left\{t \in [0,T] : a(t)|v_0(t)|^\gamma \geq \frac{1}{n}\right\} = 0$$

for all positive integer n.

It implies that

$$0 \leq \int_0^T a(t)|v_0|^{\gamma+2} dt < \frac{T}{n}\|v_0\|^2 \leq \frac{C_8^2 T}{n}\|v_0\|_\alpha^2 \to 0,$$

as $n \to \infty$, where

$$C_8 := \frac{T^{\alpha - \frac{1}{2}}}{\Gamma(\alpha)(2\alpha - 1)^{\frac{1}{2}}}$$

by (6.14). Hence $v_0 = 0$ which contradicts that $\|v_0\|_\alpha = 1$. Therefore, (6.118) holds.

Now let

$$\Omega_0 = \{t \in [0,T] : a(t)|v_0(t)|^\gamma \geq \xi_1\}, \quad \Omega_n = \{t \in [0,T] : a(t)|v_n(t)|^\gamma < \frac{1}{n}\},$$

and $\Omega_n^c = [0,T] \setminus \Omega_n = \{t \in [0,T] : a(t)|v_n(t)|^\gamma \geq \frac{1}{n}\}$.

By (6.115) and (6.118), we have

$$\begin{aligned}\alpha(\Omega_n \cap \Omega_0) &= \alpha(\Omega_0 \setminus (\Omega_n^c \cap \Omega_0)) \\ &\geq \alpha(\Omega_0) - \alpha(\Omega_n^c \cap \Omega_0) \\ &\geq \xi_2 - \frac{1}{n}\end{aligned}$$

for all positive integer n. Let n be large enough such that

$$\xi_2 - \frac{1}{n} \geq \frac{1}{2}\xi_2 \quad \text{and} \quad \frac{1}{2^{\gamma-1}}\xi_1 - \frac{1}{n} \geq \frac{1}{2^\gamma}\xi_1,$$

then we have

$$\begin{aligned}\int_0^T a(t)|v_n - v_0|^\gamma dt &\geq \int_{\Omega_n \cap \Omega_0} a(t)|v_n - v_0|^\gamma dt \\ &\geq \frac{1}{2^{\gamma-1}} \int_{\Omega_n \cap \Omega_0} a(t)|v_0|^\gamma dt - \int_{\Omega_n \cap \Omega_0} a(t)|v_n|^\gamma dt \\ &\geq \left(\frac{1}{2^{\gamma-1}}\xi_1 - \frac{1}{n}\right) \alpha(\Omega_n \cap \Omega_0) \\ &\geq \frac{\xi_1}{2^\gamma}\frac{\xi_2}{2} = \frac{\xi_1\xi_2}{2^{\gamma+1}} > 0\end{aligned}$$

for all large n, which is a contradiction to (6.117). Therefore, (6.114) holds.

For any $u \in Z_k$, let

$$\|u\|_2 := \left(\int_0^T |u(t)|^2 dt\right)^{\frac{1}{2}} \quad \text{and} \quad \gamma_k := \sup_{\substack{u \in Z_k \\ \|u\|_\alpha = 1}} \|u\|_2,$$

then we conclude $\gamma_k \to 0$ as $k \to \infty$ in the same way as in the proof of Theorem 6.6.

$$\begin{aligned}\varphi(u) &= -\int_0^T \frac{1}{2}({}_0^C D_t^\alpha u(t), {}_t^C D_T^\alpha u(t))dt - \int_0^T F(t,u(t))dt \\ &\geq \frac{1}{2}|\cos(\pi\alpha)|\|u\|_\alpha^2 - \int_0^T a(t)|u(t)|^\gamma dt \\ &\geq \frac{1}{2}|\cos(\pi\alpha)|\|u\|_\alpha^2 - \left(\int_0^T a(t)^{\frac{2}{2-\gamma}} dt\right)^{\frac{2-\gamma}{2}} \|u\|_2^\gamma \\ &\geq \frac{1}{2}|\cos(\pi\alpha)|\|u\|_\alpha^2 - \left(\int_0^T a(t)^{\frac{2}{2-\gamma}} dt\right)^{\frac{2-\gamma}{2}} \gamma_k^\gamma \|u\|_\alpha^\gamma.\end{aligned} \tag{6.119}$$

Let $\rho_k := \left(\frac{4c\gamma_k^\gamma}{|\cos(\pi\alpha)|}\right)^{\frac{1}{2-\gamma}}$, where $c := \left(\int_0^T a(t)^{\frac{2}{2-\gamma}} dt\right)^{\frac{1}{2-\gamma}}$, it is obvious that $\rho_k \to 0$, as $k \to \infty$.

In view of (6.119), we conclude

$$\inf_{\substack{u\in Z_k\\ \|u\|_\alpha=\rho_k}} \varphi(u) \geq \frac{|\cos(\pi\alpha)|}{4}\rho_k^2 > 0,$$

so the condition (H7) in Theorem 1.18 is satisfied.

Furthermore, by (6.119), for any $u \in Z_k$ with $\|u\|_\alpha \leq \rho_k$, we have

$$\varphi(u) \geq -c\gamma_k^\alpha \|u\|_\alpha^\gamma.$$

Therefore,

$$-c\gamma_k^\gamma \rho_k^\gamma \leq \inf_{\substack{u\in Z_k\\ \|u\|_\alpha\leq\rho_k}} \varphi(u) \leq 0.$$

So we have

$$\inf_{\substack{u\in Z_k\\ \|u\|_\alpha\leq\rho_k}} \varphi(u) \to 0,$$

for $\rho_k, \gamma_k \to 0$, as $k \to \infty$. Hence (H5) in Theorem 1.18 is satisfied.

For any $u \in Y_k \setminus \{0\}$,

$$\begin{aligned}
\varphi(u) &= -\int_0^T \frac{1}{2}({}_0^CD_t^\alpha u(t), {}_t^CD_T^\alpha u(t))dt - \int_0^T F(t, u(t))dt \\
&\leq \frac{1}{2|\cos(\pi\alpha)|}\|u\|_\alpha^2 - \int_0^T a(t)|u(t)|^\gamma dt \\
&\leq \frac{1}{2|\cos(\pi\alpha)|}\|u\|_\alpha^2 - \varepsilon\|u\|_\alpha^\gamma \alpha(\Omega_u) \\
&\leq \frac{1}{2|\cos(\pi\alpha)|}\|u\|_\alpha^2 - \varepsilon^2\|u\|_\alpha^\gamma,
\end{aligned}$$

where ε is given in (6.114), and $\Omega_u := \{t \in [0, T] : a(t)|u(t)|^\gamma \geq \varepsilon\|u\|_\alpha^\gamma\}$.

Choosing $0 < r_k < \min\{\rho_k, (|\cos(\pi\alpha)|\varepsilon^2)^{\frac{1}{2-\gamma}}\}$, we conclude

$$i_k := \max_{\substack{u\in Y_k\\ \|u\|_\alpha=r_k}} \varphi(u) < -\frac{1}{2|\cos(\pi\alpha)|}r_k^2 < 0, \quad \forall\, k \in \mathbb{N},$$

that is, the condition (H6) in Theorem 1.18 is satisfied.

We have proved the functional φ satisfies all the conditions of Theorem 1.18, then φ has a bounded sequence of negative critical values $c_n = \varphi(u_n)$ converging to 0 by Theorem 1.18, we only need to show $\|u_n\|_\alpha$ is bounded as for every positive integer n. Since

$$\begin{aligned}
c_n = \varphi(u_n &= -\int_0^T \frac{1}{2}({}_0^CD_t^\alpha u_n(t), {}_t^CD_T^\alpha u_n(t))dt - \int_0^T F(t, u_n(t))dt \\
&= -\int_0^T \frac{1}{2}({}_0^CD_t^\alpha u_n(t), {}_t^CD_T^\alpha u_n(t))dt - \int_0^T a(t)|u_n(t)|^\gamma dt \\
&\geq \frac{|\cos(\pi\alpha)|}{2}\|u_n\|_\alpha^2 - a_0\|u_n\|^\gamma T \\
&\geq \frac{|\cos(\pi\alpha)|}{2}\|u_n\|_\alpha^2 - a_0 T C_8^\gamma \|u_n\|_\alpha^\gamma,
\end{aligned} \tag{6.120}$$

where $a_0 = \operatorname{ess\,sup}\{a(t) : t \in [0,T]\}$, by Theorem 1.18, $c_n \to 0$ as $n \to \infty$. If $\|u_n\|_\alpha$ has an unbounded sequence, then c_n is unbounded by (6.120), which is a contradiction. The proof is completed. □

Example 6.7. In BVP (6.1), let $F(t,x) = a(t)|x|^{\frac{3}{2}}$, where

$$a(t) = \begin{cases} T, & t = 0, \\ t, & 0 < t \le T. \end{cases}$$

By Theorem 6.10, BVP (6.1) has infinite solutions $\{u_k\}$ on E^α for every positive integer k with $\|u_k\|_\alpha$ bounded.

6.5 Solutions for BVP with Left and Right Fractional Derivatives

6.5.1 *Introduction*

In Section 6.5, we consider the BVP of the following form

$$\begin{cases} {}_tD_T^\alpha({}_0D_t^\alpha u(t)) = \nabla F(t, u(t)), & \text{a.e. } t \in [0,T], \\ u(0) = u(T) = 0, \end{cases} \tag{6.121}$$

where ${}_tD_T^\alpha$ and ${}_0D_t^\alpha$ are the right and left Riemann-Liouville fractional derivatives of order $0 < \alpha \le 1$ respectively, F: $[0,T] \times \mathbb{R}^N \to \mathbb{R}$ is a given function satisfying some assumptions and $\nabla F(t,x)$ is the gradient of F at x.

In particular, if $\alpha = 1$, BVP (6.121) reduces to the standard second order boundary value problem of the following form

$$\begin{cases} u''(t) + \nabla F(t, u(t)) = 0, & \text{a.e. } t \in [0,T], \\ u(0) = u(T) = 0, \end{cases}$$

where F: $[0,T] \times \mathbb{R}^N \to \mathbb{R}$ is a given function and $\nabla F(t,x)$ is the gradient of F at x. Although many excellent results have been worked out on the existence of solutions for second order BVP (e.g., Li, Liang and Zhang, 2005; Nieto and O'Regan, 2009; Rabinowitz, 1986), it seems that no similar results were obtained in the literature for fractional BVP.

According to Benson, Wheatcraft and Meerschaert, 2000a, the one-dimensional form of the fractional ADE can be written as

$$\frac{\partial \mathcal{C}}{\partial t} = -v\frac{\partial \mathcal{C}}{\partial x} + \mathcal{D}j\frac{\partial^\gamma \mathcal{C}}{\partial x^\gamma} + \mathcal{D}(1-j)\frac{\partial^\gamma \mathcal{C}}{\partial(-x)^\gamma}, \tag{6.122}$$

where $\mathcal{C}$ is the expected concentration, t is time, v is a constant mean velocity, x is distance in the direction of mean velocity, $\mathcal{D}$ is a constant dispersion coefficient, $0 \le j \le 1$ describes the skewness of the transport process, and γ is the order of left and right fractional differential operators. For discussions of this equation, see Benson, Wheatcraft and Meerschaert, 2000b; Fix and Roop, 2004, when $\gamma = 2$, the dispersion operators are identical and the classical ADE is recovered. Fundamental (Green function) solutions are Lévy's γ-stable densities.

A special case of the fractional ADE (equation (6.122)) describes symmetric transitions, where $j = \frac{1}{2}$. Defining the symmetric operator equivalent to the Riesz potential in Samko, Kilbas and Marichev, 1993,

$$2\nabla^{\gamma} \equiv D_{+}^{\gamma} + D_{-}^{\gamma}$$

gives the mass balance equation for advection and symmetric fractional dispersion

$$\frac{\partial \mathcal{C}}{\partial t} = -v\nabla \mathcal{C} + \mathcal{D}\nabla^{\gamma}\mathcal{C}.$$

In Subsection 6.5.2, we shall establish a variational structure for BVP (6.121). We show that under some suitable assumptions, the critical points of the variational functional defined on a suitable Hilbert space are the solutions of BVP (6.121). In Subsection 6.5.3, the existence of weak solutions for BVP (6.121) with $\frac{1}{2} < \alpha \leq 1$ will be established, where α is the order of fractional derivative in BVP (6.121). In Subsection 6.5.4, we will give some existence results of solutions for BVP (6.121).

6.5.2 *Variational Structure*

Proposition 6.5. *Let $0 < \alpha \leq 1$ and $1 < p < \infty$. For all $u \in E_0^{\alpha,p}$, if $\alpha > \frac{1}{p}$, we have ${}_0D_t^{-\alpha}({}_0D_t^{\alpha}u(t)) = u(t)$. Moreover, we can get that $E_0^{\alpha,p} \in C_0([0,T], \mathbb{R}^N)$.*

Proof. Let $\frac{1}{p} + \frac{1}{q} = 1$ and $0 \leq t_1 < t_2 \leq T$. $\forall\ f \in L^p([0,T], \mathbb{R}^N)$, by using Hölder inequality and noting that $\alpha > \frac{1}{p}$, we have

$$\begin{aligned}
&|{}_0D_{t_1}^{-\alpha}f(t_1) - {}_0D_{t_2}^{-\alpha}f(t_2)| \\
&= \frac{1}{\Gamma(\alpha)}\left|\int_0^{t_1}(t_1-s)^{\alpha-1}f(s)ds - \int_0^{t_2}(t_2-s)^{\alpha-1}f(s)ds\right| \\
&= \frac{1}{\Gamma(\alpha)}\left|\int_0^{t_1}(t_1-s)^{\alpha-1}f(s)ds - \int_0^{t_1}(t_2-s)^{\alpha-1}f(s)ds\right| \\
&\quad + \frac{1}{\Gamma(\alpha)}\left|\int_{t_1}^{t_2}(t_2-s)^{\alpha-1}f(s)ds\right| \\
&\leq \frac{1}{\Gamma(\alpha)}\int_0^{t_1}\left((t_1-s)^{\alpha-1} - (t_2-s)^{\alpha-1}\right)|f(s)|ds \\
&\quad + \frac{1}{\Gamma(\alpha)}\int_{t_1}^{t_2}(t_2-s)^{\alpha-1}|f(s)|ds \\
&\leq \frac{1}{\Gamma(\alpha)}\left(\int_0^{t_1}\left((t_1-s)^{\alpha-1} - (t_2-s)^{\alpha-1}\right)^q ds\right)^{\frac{1}{q}}\|f\|_{L^p[0,T]} \\
&\quad + \frac{1}{\Gamma(\alpha)}\left(\int_{t_1}^{t_2}(t_2-s)^{(\alpha-1)q}ds\right)^{\frac{1}{q}}\|f\|_{L^p[0,T]} \\
&\leq \frac{1}{\Gamma(\alpha)}\left(\int_0^{t_1}(t_1-s)^{(\alpha-1)q} - (t_2-s)^{(\alpha-1)q}ds\right)^{\frac{1}{q}}\|f\|_{L^p[0,T]}
\end{aligned} \tag{6.123}$$

$$
\begin{aligned}
&+\frac{1}{\Gamma(\alpha)}\left(\int_{t_1}^{t_2}(t_2-s)^{(\alpha-1)q}ds\right)^{\frac{1}{q}}\|f\|_{L^p[0,T]}\\
&=\frac{\|f\|_{L^p[0,T]}}{\Gamma(\alpha)(1+(\alpha-1)q)^{\frac{1}{q}}}\left(t_1^{(\alpha-1)q+1}-t_2^{(\alpha-1)q+1}+(t_2-t_1)^{(\alpha-1)q+1}\right)^{\frac{1}{q}}\\
&+\frac{\|f\|_{L^p[0,T]}}{\Gamma(\alpha)(1+(\alpha-1)q)^{\frac{1}{q}}}\left((t_2-t_1)^{(\alpha-1)q+1}\right)^{\frac{1}{q}}\\
&\le\frac{2\|f\|_{L^p[0,T]}}{\Gamma(\alpha)(1+(\alpha-1)q)^{\frac{1}{q}}}(t_2-t_1)^{\alpha-1+\frac{1}{q}}\\
&=\frac{2\|f\|_{L^p[0,T]}}{\Gamma(\alpha)(1+(\alpha-1)q)^{\frac{1}{q}}}(t_2-t_1)^{\alpha-\frac{1}{p}}.
\end{aligned}
$$

For any $u \in E_0^{\alpha,p}$, as ${}_0D_t^{\alpha}u(t) \in L^p([0,T],\mathbb{R}^N)$, we apply (6.123) to obtain the continuity of the function ${}_0D_t^{-\alpha}({}_0D_t^{\alpha}u(t))$ on $[0,T]$. We complete the argument by using Propositions 1.6-1.7, and we have

$$
{}_0D_t^{-\alpha}({}_0D_t^{\alpha}u(t)) = u(t) + Ct^{\alpha-1}, \quad t \in [0,T],
$$

where $C \in \mathbb{R}^N$.

Since $u(0) = 0$ and ${}_0D_t^{-\alpha}({}_0D_t^{\alpha}u(t))$ is continuous in $[0,T]$, we can get that $C = 0$, which means that ${}_0D_t^{-\alpha}({}_0D_t^{\alpha}u(t)) = u(t)$ and u is continuous in $[0,T]$. □

Remark 6.6. In the case that $1-\alpha \ge \frac{1}{p}$, for any $u \in E_0^{\alpha,p}$, we also have ${}_0D_t^{-\alpha}({}_0D_t^{\alpha}u(t)) = u(t)$. In fact, set $f(t) = {}_0D_t^{\alpha-1}u(t)$. According to Propositions 1.6-1.7, we only need to prove that $f(0) = [{}_0D_t^{\alpha-1}u(t)]_{t=0} = 0$. Noting that $1-\alpha \ge \frac{1}{p}$, by using Hölder inequality, Lemma 6.1 and the similar method in the proof of Lemma 7 in Fix and Roop, 2004, we can obtain the desired result, i.e. $f(0) = 0$. We skip the proof since it is similar to Lemma 7 in Fix and Roop, 2004.

If $\alpha > \frac{1}{p}$, the following theorem is useful for us to establish the variational structure on the space $E_0^{\alpha,p}$ for BVP (6.121).

Theorem 6.11. *Let $1 < p < \infty$, $\frac{1}{p} < \alpha \le 1$ and $L : [0,T] \times \mathbb{R}^N \times \mathbb{R}^N \to \mathbb{R}$, $(t,x,y) \to L(t,x,y)$ be measurable in t for each $[x,y] \in \mathbb{R}^N \times \mathbb{R}^N$ and continuously differentiable in $[x,y]$ for almost every $t \in [0,T]$. If there exist $m_1 \in C(\mathbb{R}^+,\mathbb{R}^+)$, $m_2 \in L^1([0,T],\mathbb{R}^+)$ and $m_3 \in L^q([0,T],\mathbb{R}^+)$, $1 < q < \infty$, such that, for a.e. $t \in [0,T]$ and every $[x,y] \in \mathbb{R}^N \times \mathbb{R}^N$, one has*

$$
\begin{aligned}
&|L(t,x,y)| \le m_1(|x|)(m_2(t) + |y|^p),\\
&|D_xL(t,x,y)| \le m_1(|x|)(m_2(t) + |y|^p),\\
&|D_yL(t,x,y)| \le m_1(|x|)(m_3(t) + |y|^{p-1}),
\end{aligned}
$$

where $\frac{1}{p} + \frac{1}{q} = 1$, then the functional φ defined by

$$
\varphi(u) = \int_0^T L(t, u(t), {}_0D_t^{\alpha}u(t))dt
$$

is continuously differentiable on $E_0^{\alpha,p}$, *and* $\forall u, v \in E_0^{\alpha,p}$, *we have*

$$\begin{aligned}\langle \varphi'(u), v\rangle = \int_0^T & [(D_x L(t, u(t), {}_0D_t^\alpha u(t)), v(t)) \\ & + (D_y L(t, u(t), {}_0D_t^\alpha u(t)), {}_0D_t^\alpha v(t))]dt.\end{aligned} \tag{6.124}$$

Proof. It suffices to prove that φ has at every point u a directional derivative $\varphi'(u) \in (E_0^{\alpha,p})^*$ given by (6.124) and that the mapping

$$\varphi' : E_0^{\alpha,p} \to (E_0^{\alpha,p})^*, \quad u \to \varphi'(u)$$

is continuous.

We omit the rather technical proof which is similar to the proof of Theorem 1.4 in Mawhin and Willem, 1989. In fact, the only change we need is to replace the weak derivatives for u and v of Theorem 1.6 in Mawhin and Willem, 1989, by ${}_0D_t^\alpha u$ and ${}_0D_t^\alpha v$ respectively. The proof is completed. □

We are now in a position to give the definition for the solution of BVP (6.121).

Definition 6.3. A function $u : [0, T] \to \mathbb{R}^N$ is called a solution of BVP (6.121) if

(i) ${}_tD_T^{\alpha-1}({}_0D_t^\alpha u(t))$ and ${}_0D_t^{\alpha-1}u(t)$ are differentiable for almost every $t \in [0, T]$, and

(ii) u satisfies (6.121).

For a solution $u \in E^\alpha$ of BVP (6.121) such that $\nabla F(\cdot, u(\cdot)) \in L^1([0, T], \mathbb{R}^N)$, multiplying (6.121) by $v \in C_0^\infty([0, T], \mathbb{R}^N)$ yields

$$\begin{aligned}&\int_0^T [({}_tD_T^\alpha({}_0D_t^\alpha u(t)), v(t)) - (\nabla F(t, u(t)), v(t))]dt \\ &= \int_0^T [({}_0D_t^\alpha u(t), {}_0D_t^\alpha v(t)) - (\nabla F(t, u(t)), v(t))]dt \\ &= 0\end{aligned} \tag{6.125}$$

after applying (1.13) and Definition 6.3. Therefore, we can give the definition of weak solution for BVP (6.121) as follows.

Definition 6.4. By the weak solution of BVP (6.121), we mean that the function $u \in E^\alpha$ such that $\nabla F(\cdot, u(\cdot)) \in L^1([0, T], \mathbb{R}^N)$ and satisfies (6.125) for all $v \in C_0^\infty([0, T], \mathbb{R}^N)$.

Any solution $u \in E^\alpha$ of BVP (6.121) is a weak solution provided that $\nabla F(\cdot, u(\cdot)) \in L^1([0, T], \mathbb{R}^N)$. Our task is now to establish a variational structure on E^α with $\alpha \in (\frac{1}{2}, 1]$, which enables us to reduce the existence of weak solutions of BVP (6.121) to the one of finding critical points of corresponding functional.

Corollary 6.5. *Let* $L : [0, T] \times \mathbb{R}^N \times \mathbb{R}^N \to \mathbb{R}$ *be defined by*

$$L(t, x, y) = \frac{1}{2}|y|^2 - F(t, x),$$

where $F : [0,T] \times \mathbb{R}^N \to \mathbb{R}$ satisfies the condition (A) which is assumed as in Subsection 6.2.3.

If $\frac{1}{2} < \alpha \leq 1$ and $u \in E^\alpha$ is a solution of corresponding Euler equation $\varphi'(u) = 0$, where φ is defined as

$$\varphi(u) = \int_0^T \left(\frac{1}{2} |{}_0D_t^\alpha u(t)|^2 - F(t, u(t)) \right) dt, \quad \text{for } u \in E^\alpha, \tag{6.126}$$

then u is a weak solution of BVP (6.121) with $\frac{1}{2} < \alpha \leq 1$.

Proof. By Theorem 6.11, we have

$$0 = \langle \varphi'(u), v \rangle = \int_0^T [({}_0D_t^\alpha u(t), {}_0D_t^\alpha v(t)) - (\nabla F(t, u(t)), v(t))] dt$$

for all $u \in E^\alpha$ and hence for all $v \in C_0^\infty([0,T], \mathbb{R}^N)$. Thus, according to Definition 6.4, u is a weak solution of BVP (6.121). The proof is completed. □

Remark 6.7. Generally speaking, a critical point u of φ on E^α will be a weak solution of BVP (6.121). However, we shall show that every weak solution is also a solution of BVP (6.121).

6.5.3 *Existence of Weak Solutions*

According to Corollary 6.5, we know that in order to find weak solutions of BVP (6.121), it suffices to obtain the critical points of functional φ given by (6.126). We need to use some critical point theorems.

First, we use Theorem 1.14 to consider the existence of weak solutions for BVP (6.121). Assume that the condition (A) is satisfied. Recall that, in our setting in (6.126), the corresponding functional φ on E^α is continuously differentiable according to Corollary 6.5 and is also weakly lower semi-continuous functional on E^α as the sum of a convex continuous function (see Theorem 1.2 in Mawhin and Willem, 1989) and of a weakly continuous one (see Proposition 1.2 in Mawhin and Willem, 1989).

In fact, according to Proposition 6.3, if $u_k \rightharpoonup u$ in E^α, then $u_k \to u$ in $C([0,T], \mathbb{R}^N)$. Therefore, $F(t, u_k(t)) \to F(t, u(t))$ a.e. $t \in [0,T]$. By Lebesgue dominated convergence theorem, we have $\int_0^T F(t, u_k(t))dt \to \int_0^T F(t, u(t))dt$, which means that the functional $u \to \int_0^T F(t, u(t))dt$ is weakly continuous on E^α. Moreover, since fractional derivative operator is linear operator, the functional $u \to \int_0^T (|{}_0D_t^\alpha u(t)|^2/2)dt$ is convex and continuous on E^α.

If φ is coercive, by Theorem 1.14, φ has a minimum so that BVP (6.121) is solvable. It remains to find conditions under which φ is coercive on E^α, i.e. $\lim_{\|u\|_\alpha \to \infty} \varphi(u) = +\infty$, for $u \in E^\alpha$. We shall see that it suffices to require that $F(t,x)$ is bounded for a.e. $t \in [0,T]$ and all $x \in \mathbb{R}^N$.

Theorem 6.12. *Let $\alpha \in (\frac{1}{2}, 1]$ and assume that F satisfies (A). If*

$$|F(t,x)| \leq \bar{a}|x|^2 + \bar{b}(t)|x|^{2-\gamma} + \bar{c}(t), \quad \text{a.e. } t \in [0,T], \ x \in \mathbb{R}^N, \tag{6.127}$$

where $\bar{a} \in [0, \Gamma^2(\alpha+1)/2T^{2\alpha})$, $\gamma \in (0,2)$, $\bar{b} \in L^{2/\gamma}([0,T],\mathbb{R})$, *and* $\bar{c} \in L^1([0,T],\mathbb{R})$, *then BVP* (6.121) *has at least one weak solution which minimizes* φ *on* E^α.

Proof. According to the arguments above, our problem reduces to prove that φ is coercive on E^α. For $u \in E^\alpha$, it follows from (6.127) and (6.13) that

$$\begin{aligned}
\varphi(u) &= \frac{1}{2}\int_0^T |{}_0D_t^\alpha u(t)|^2 dt - \int_0^T F(t,u(t))dt \\
&\geq \frac{1}{2}\int_0^T |{}_0D_t^\alpha u(t)|^2 dt - \bar{a}\int_0^T |u(t)|^2 dt - \int_0^T \bar{b}(t)|u(t)|^{2-\gamma} dt - \int_0^T \bar{c}(t)dt \\
&= \frac{1}{2}\|u\|_\alpha^2 - \bar{a}\|u\|_{L^2}^2 - \int_0^T \bar{b}(t)|u(t)|^{2-\gamma} dt - \bar{c}_1 \\
&\geq \frac{1}{2}\|u\|_\alpha^2 - \bar{a}\|u\|_{L^2}^2 - \left(\int_0^T |\bar{b}(t)|^{2/\gamma} dt\right)^{\gamma/2}\left(\int_0^T |u(t)|^2 dt\right)^{1-\gamma/2} - \bar{c}_1 \\
&= \frac{1}{2}\|u\|_\alpha^2 - \bar{a}\|u\|_{L^2}^2 - \bar{b}_1\|u\|_{L^2}^{2-\gamma} - \bar{c}_1 \\
&\geq \frac{1}{2}\|u\|_\alpha^2 - \frac{\bar{a}T^{2\alpha}}{\Gamma^2(\alpha+1)}\|u\|_\alpha^2 - \bar{b}_1\left(\frac{T^\alpha}{\Gamma(\alpha+1)}\right)^{2-\gamma}\|u\|_\alpha^{2-\gamma} - \bar{c}_1 \\
&= \left(\frac{1}{2} - \frac{\bar{a}T^{2\alpha}}{\Gamma^2(\alpha+1)}\right)\|u\|_\alpha^2 - \bar{b}_1\left(\frac{T^\alpha}{\Gamma(\alpha+1)}\right)^{2-\gamma}\|u\|_\alpha^{2-\gamma} - \bar{c}_1,
\end{aligned}$$

where $\bar{b}_1 = (\int_0^T |\bar{b}(t)|^{2/\gamma} dt)^{\gamma/2}$ and $\bar{c}_1 = \int_0^T \bar{c}(t)dt$. Noting that $\bar{a} \in [0, \Gamma^2(\alpha+1)/2T^{2\alpha})$ and $\gamma \in (0,2)$, we have $\varphi(u) = +\infty$ as $\|u\|_\alpha \to \infty$, and hence φ is coercive, which completes the proof. □

Let $a_0 = \min_{\lambda\in[\frac{1}{2},1]}\{\Gamma^2(\lambda+1)/2T^{2\lambda}\}$. The following result follows immediately from Theorem 6.12.

Corollary 6.6. $\forall\alpha \in (\frac{1}{2},1]$ *and if* F *satisfies the condition (A) and* (6.127) *with* $a \in [0,a_0)$, *then BVP* (6.121) *has at least one weak solution which minimizes* φ *on* E^α.

Our task is now to use Theorem 1.15 (Mountain pass theorem) to find a nonzero critical point of functional φ on E^α.

Theorem 6.13. *Let* $\alpha \in (\frac{1}{2},1]$ *and suppose that* F *satisfies the condition (A). If*

(A1) $F \in C([0,T]\times\mathbb{R}^N,\mathbb{R})$ *and there exists* $\mu \in [0,\frac{1}{2})$ *and* $M > 0$ *such that* $0 < F(t,x) \leq \mu(\nabla F(t,x),x)$ *for all* $x \in \mathbb{R}^N$ *with* $|x| \geq M$ *and* $t \in [0,T]$;
(A2) $\limsup_{|x|\to 0} F(t,x)/|x|^2 < \Gamma^2(\alpha+1)/2T^{2\alpha}$ *uniformly for* $t \in [0,T]$ *and* $x \in \mathbb{R}^N$

are satisfied, then BVP (6.121) *has at least one nonzero weak solution on* E^α.

Proof. We will verify that φ satisfies all the conditions of Theorem 1.15.

First, we will prove that φ satisfies (PS) condition. Since $F(t,x)-\mu(\nabla F(t,x),x)$ is continuous for $t\in[0,T]$ and $|x|\le M$, there exists $c\in\mathbb{R}^+$, such that

$$F(t,x)\le\mu(\nabla F(t,x),x)+c,\quad t\in[0,T],\ |x|\le M.$$

By assumption (A1), we obtain

$$F(t,x)\le\mu(\nabla F(t,x),x)+c,\quad t\in[0,T],\ x\in\mathbb{R}^N. \tag{6.128}$$

Let $\{u_k\}\subset E^\alpha$, $|\varphi(u_k)|\le K,\ k=1,2,...,\ \varphi'(u_k)\to 0$. Notice that

$$\begin{aligned}\langle\varphi'(u_k),u_k\rangle&=\int_0^T[({}_0D_t^\alpha u_k(t),{}_0D_t^\alpha u_k(t))-(\nabla F(t,u_k(t)),u_k(t))]dt\\&=\|u_k\|_\alpha^2-\int_0^T(\nabla F(t,u_k(t)),u_k(t))dt.\end{aligned}\tag{6.129}$$

It follows from (6.128) and (6.129) that

$$\begin{aligned}K\ge\varphi(u_k)&=\frac{1}{2}\|u_k\|_\alpha^2-\int_0^T F(t,u_k(t))dt\\&\ge\frac{1}{2}\|u_k\|_\alpha^2-\mu\int_0^T(\nabla F(t,u_k(t)),u_k(t))dt-cT\\&=\left(\frac{1}{2}-\mu\right)\|u_k\|_\alpha^2+\mu\langle\varphi'(u_k),u_k\rangle-cT\\&\ge\left(\frac{1}{2}-\mu\right)\|u_k\|_\alpha^2-\mu\|\varphi'(u_k)\|_\alpha\|u_k\|_\alpha-cT,\quad k=1,2,\dots\ .\end{aligned}$$

Since $\varphi'(u_k)\to 0$, there exists $N_0\in\mathbb{N}$ such that

$$K\ge\left(\frac{1}{2}-\mu\right)\|u_k\|_\alpha^2-\|u_k\|_\alpha-cT,\quad k>N_0,$$

and this implies that $\{u_k\}\subset E^\alpha$ is bounded. Since E^α is a reflexive space, going to a subsequence if necessary, we may assume that $u_k\rightharpoonup u$ weakly in E^α, thus we have

$$\begin{aligned}\langle\varphi'(u_k)-\varphi'(u),u_k-u\rangle&=\langle\varphi'(u_k),u_k-u\rangle-\langle\varphi'(u),u_k-u\rangle\\&\le\|\varphi'(u_k)\|_\alpha\|u_k-u\|_\alpha-\langle\varphi'(u),u_k-u\rangle\\&\to 0,\quad\text{as } k\to\infty.\end{aligned}\tag{6.130}$$

Moreover, according to (6.14) and Proposition 6.3, we get that u_k is bounded in $C([0,T],\mathbb{R}^N)$ and $\|u_k-u\|=0$ as $k\to\infty$. Hence, we have

$$\int_0^T\nabla F(t,u_k(t))dt\to\int_0^T\nabla F(t,u(t))dt,\quad\text{as } k\to\infty. \tag{6.131}$$

Noting that

$$\begin{aligned}&\langle\varphi'(u_k)-\varphi'(u),u_k-u\rangle\\&=\int_0^T({}_0D_t^\alpha u_k(t)-{}_0D_t^\alpha u(t))^2dt-\int_0^T(\nabla F(t,u_k(t))-\nabla F(t,u(t)))(u_k(t)-u(t))dt\end{aligned}$$

$$\geq \|u_k - u\|_\alpha^2 - \left|\int_0^T (\nabla F(t, u_k(t)) - \nabla F(t, u(t)))dt\right| \|u_k - u\|.$$

Combining (6.130) and (6.131), it is easy to verify that $\|u_k - u\|_\alpha^2 \to 0$ as $k \to \infty$, and hence $u_k \to u$ in E_α. Thus, we obtain the desired convergence property.

From $\limsup_{|x|\to 0} F(t,x)/|x|^2 < \Gamma^2(\alpha+1)/2T^{2\alpha}$ uniformly for $t \in [0,T]$, there exists $\epsilon \in (0,1)$ and $\delta > 0$ such that $F(t,x) \leq (1-\epsilon)(\Gamma^2(\alpha+1)/2T^{2\alpha})|x|^2$ for all $t \in [0,T]$ and $x \in \mathbb{R}^N$ with $|x| \leq \delta$.

Let $\rho = \frac{\Gamma(\alpha)((\alpha-1)/2+1)^{\frac{1}{2}}}{T^{\alpha-\frac{1}{2}}}\delta$ and $\sigma = \epsilon\rho^2/2 > 0$. Then it follows from (6.14) that

$$\|u\| \leq \frac{T^{\alpha-\frac{1}{2}}}{\Gamma(\alpha)((\alpha-1)/2+1)^{\frac{1}{2}}}\|u\|_\alpha = \delta$$

for all $u \in E^\alpha$ with $\|u\|_\alpha = \rho$. Therefore, we have

$$\begin{aligned}
\varphi(u) &= \frac{1}{2}\|u\|_\alpha^2 - \int_0^T F(t,u(t))dt \\
&\geq \frac{1}{2}\|u\|_\alpha^2 - (1-\epsilon)\frac{\Gamma^2(\alpha+1)}{2T^{2\alpha}}\int_0^T |u(t)|^2 dt \\
&\geq \frac{1}{2}\|u\|_\alpha^2 - \frac{1}{2}(1-\epsilon)\|u\|_\alpha^2 \\
&= \frac{1}{2}\epsilon\|u\|_\alpha^2 = \sigma
\end{aligned}$$

for all $u \in E^\alpha$ with $\|u\|_\alpha = \rho$. This implies (ii) in Theorem 1.15 is satisfied.

It is obvious from the definition of φ and (A2) that $\varphi(0) = 0$, and therefore, it suffices to show that φ satisfies (iii) in Theorem 1.15.

Since $0 < F(t,x) \leq \mu(\nabla F(t,x), x)$ for all $x \in \mathbb{R}^N$ and $|x| \geq M$, a simple regularity argument then shows that there exists $r_1, r_2 > 0$ such that

$$F(t,x) \geq r_1|x|^{1/\mu} - r_2, \quad x \in \mathbb{R}^N, \quad t \in [0,T].$$

For any $u \in E^\alpha$ with $u \neq 0$, $\kappa > 0$ and noting that $\mu \in [0, \frac{1}{2})$, we have

$$\begin{aligned}
\varphi(\kappa u) &= \frac{1}{2}\|\kappa u\|_\alpha^2 - \int_0^T F(t, \kappa u(t))dt \\
&\leq \frac{1}{2}\kappa^2\|u\|_\alpha^2 - r_1\int_0^T |\kappa u(t)|^{1/\mu}dt + r_2 T \\
&= \frac{1}{2}\kappa^2\|u\|_\alpha^2 - r_1\kappa^{1/\mu}\|u\|_{L^{1/\mu}}^{1/\mu} + r_2 T \\
&\to -\infty, \quad \text{as } \kappa \to \infty.
\end{aligned}$$

Then there exists a sufficiently large κ_0 such that $\varphi(\kappa_0 u) \leq 0$. Hence (iii) holds.

Lastly noting that $\varphi(0) = 0$ while for our critical point u, $\varphi(u) \geq \sigma > 0$. Hence u is a nontrivial weak solution of BVP (6.121), and this completes the proof. □

Theorem 6.14. $\forall \alpha \in (\frac{1}{2}, 1]$, *suppose that F satisfies conditions (A) and (A1). If*

(A2)′ $F(t,x) = o(|x|^2)$, *as* $|x| \to 0$ *uniformly for* $t \in [0,T]$ *and* $x \in \mathbb{R}^N$

is satisfied, then BVP (6.121) *has at least one nonzero weak solution on* E^α.

Remark 6.8. The assumptions in Theorem 6.12 and Theorem 6.13 are classical and the examples can be found in many papers which use critical point theory to discuss differential equations, see, e.g., Li, Liang and Zhang, 2005; Mawhin and Willem, 1989; Rabinowitz, 1986 and references therein.

6.5.4 *Existence of Solutions*

We firstly give the following lemma which is useful for our further discussion.

Lemma 6.7. *Let* $0 < \alpha \le 1$. *If* $u \in E^\alpha$ *is a weak solution of BVP* (6.121), *then there exists a constant* $C \in \mathbb{R}^N$ *such that*

$$ {}_0D_t^\alpha u(t) = {}_tD_T^{-\alpha}\nabla F(t,u(t)) + C(T-t)^{\alpha-1}, \quad a.e.\ t \in [0,T]. $$

Proof. Since $u \in E^\alpha$ is a weak solution of BVP (6.121), i.e. $\forall h \in C_0^\infty([0,T],\mathbb{R}^N)$, we have

$$ \int_0^T [({}_0D_t^\alpha u(t), {}_0D_t^\alpha h(t)) - (\nabla F(t,u(t)), h(t))]dt = 0. \tag{6.132} $$

Noting that $\nabla F(\cdot,u(\cdot)) \in L^1([0,T],\mathbb{R}^N)$, and applying a similar argument as that for (6.5) in the proof of Lemma 6.1, we get that ${}_tD_T^{-\alpha}\nabla F(\cdot,u(\cdot)) \in L^1([0,T],\mathbb{R}^N)$. Let us define $w \in L^1([0,T],\mathbb{R}^N)$ by

$$ w(t) = {}_tD_T^{-\alpha}\nabla F(t,u(t)), \quad t \in [0,T], $$

so that

$$ \begin{aligned} \int_0^T (w(t), {}_0D_t^\alpha h(t))dt &= \int_0^T ({}_tD_T^\alpha w(t), h(t))dt \\ &= \int_0^T ({}_tD_T^\alpha({}_tD_T^{-\alpha}\nabla F(t,u(t))), h(t))dt \\ &= \int_0^T (\nabla F(t,u(t)), h(t))dt, \end{aligned} $$

by applying (1.13) and Proposition 1.5.

Hence, by (6.132) we have, for every $h \in C_0^\infty([0,T],\mathbb{R}^N)$,

$$ \int_0^T ({}_0D_t^\alpha u(t) - w(t), {}_0D_t^\alpha h(t))dt = 0. \tag{6.133} $$

According to Proposition 1.1 and in view of $h \in C_0^\infty([0,T],\mathbb{R}^N)$, we have ${}_0D_t^\alpha h(t) = {}_0D_t^{\alpha-1}h'(t)$. Since ${}_0D_t^\alpha u \in L^2([0,T],\mathbb{R}^N)$ and $w \in L^1([0,T],\mathbb{R}^N)$, using (1.12) and (6.133), we get that

$$ \int_0^T \left({}_tD_T^{\alpha-1}({}_0D_t^\alpha u(t) - w(t)), h'(t)\right) dt = 0. $$

If (e_j) denotes the Canonical basis of $\mathbb{R}^N$, we can choose

$$h(t) = \sin\frac{2k\pi t}{T}e_j \quad \text{or} \quad h(t) = e_j - \cos\frac{2k\pi t}{T}e_j, \qquad k = 1, \ldots \text{ and } j = 1, \ldots, N.$$

In view of ${}_tD_T^{\alpha-1}({}_0D_t^\alpha u - w) \in L^1([0,T], \mathbb{R}^N)$, and the theory of Fourier series implies that

$$ {}_tD_T^{\alpha-1}({}_0D_t^\alpha u(t) - w(t)) = \tilde{C} \tag{6.134}$$

a.e. on $[0,T]$ for some $\tilde{C} \in \mathbb{R}^N$. Using Proposition 1.5 and Proposition 1.3, we can get that

$${}_0D_t^\alpha u(t) = w(t) + C(T-t)^{\alpha-1}, \quad \text{a.e. } t \in [0,T],$$

for some $C \in \mathbb{R}^N$ and this completes the proof. □

Remark 6.9.

(i) According to (6.134) and Proposition 1.4, we have

$${}_tD_T^{\alpha-1}({}_0D_t^\alpha u(t)) = {}_tD_T^{\alpha-1}({}_tD_T^{-\alpha}\nabla F(t,u(t))) + \tilde{C} = {}_tD_T^{-1}\nabla F(t,u(t)) + \tilde{C}$$

a.e. on $[0,T]$ for some $\tilde{C} \in \mathbb{R}^N$. In view of Definition 1.1 and $\nabla F(\cdot, u(\cdot)) \in L^1([0,T], \mathbb{R}^N)$, we shall identify the equivalence class ${}_tD_T^{\alpha-1}({}_0D_t^\alpha u)$ and its continuous representant

$${}_tD_T^{\alpha-1}({}_0D_t^\alpha u(t)) = \int_t^T \nabla F(s, u(s))ds + \tilde{C} \tag{6.135}$$

for $t \in [0,T]$.

(ii) It follows from (6.135) and a classical result of Lebesgue theory that $-\nabla F(\cdot, u(\cdot))$ is the classical derivative of ${}_tD_T^{\alpha-1}({}_0D_t^\alpha u)$ a.e. on $[0,T]$.

We are now in a position to show that every weak solution of BVP (6.121) is also a solution of BVP (6.121).

Theorem 6.15. *Let $0 < \alpha \leq 1$. If $u \in E^\alpha$ is a weak solution of BVP* (6.121), *then u is also a solution of BVP* (6.121).

Proof. Firstly, we notice that ${}_0D_t^{\alpha-1}u(t)$ is derivative for almost every $t \in [0,T]$ and $({}_0D_t^{\alpha-1}u(t))' = {}_0D_t^\alpha u(t) \in L^2([0,T], \mathbb{R}^N)$ as $u \in E^\alpha$. On the other hand, Remark 6.9 implies that ${}_tD_T^{\alpha-1}({}_0D_t^\alpha u(t))$ is derivative a.e. on $[0,T]$ and $({}_tD_T^{\alpha-1}({}_0D_t^\alpha u(t)))' \in L^1([0,T], \mathbb{R}^N)$. Therefore, (i) in Definition 6.3 is verified.

It remains to show that u satisfies (6.121). In fact, according to Definition 1.2 and (6.135), we can get that

$${}_tD_T^\alpha({}_0D_t^\alpha u(t)) = -({}_tD_T^{\alpha-1}({}_0D_t^\alpha u(t)))' = \nabla F(t, u(t)), \quad \text{a.e. } t \in [0,T].$$

Moreover, $u \in E^\alpha$ implies that $u(0) = u(T) = 0$, and therefore (6.121) is verified. The proof is completed. □

The conclusions in Subsection 6.5.3 and Theorem 6.15 imply that BVP (6.121) with $\alpha \in (\frac{1}{2}, 1]$ possesses at least one solution if F satisfies some hypotheses. However, we would like to consider the existence of solutions for BVP (6.121) with $\alpha = \frac{1}{2}$ under the same hypotheses.

For any given $\epsilon_0 \in (0, \frac{1}{2})$, let $\epsilon \in (0, \epsilon_0)$ and $\delta = \delta(\epsilon) = \frac{1}{2} + \epsilon$. According to Corollary 6.6 and Theorem 6.14, if (A) and (6.127) with $a \in [0, a_0)$, or (A), (A1) and (A2)$'$ are satisfied, then $\forall \epsilon \in (0, \epsilon_0)$, the following BVP

$$\begin{cases} {}_tD_T^{\delta}\big({}_0D_t^{\delta}u(t)\big) = \nabla F(t, u(t)), \quad \text{a.e. } t \in [0, T], \\ u(0) = u(T) = 0 \end{cases} \tag{6.136}$$

has at least a weak solution $u_\epsilon \in E^\delta$. Moreover, according to Theorem 6.15, u_ϵ is also the solution of BVP (6.136). Now, our idea is to obtain the solution of BVP (6.121) with $\delta = \frac{1}{2}$ by considering the approximation of u_ϵ as $\epsilon \to 0$.

Theorem 6.16. *Assume that there exists $\epsilon_0 \in (0, \frac{1}{2})$ such that $\forall \epsilon \in (0, \epsilon_0)$ and $\delta = \delta(\epsilon) = \frac{1}{2} + \epsilon$, BVP (6.136) possesses a weak solution $u_\epsilon \in E^\delta$. Moreover, if the following conditions are satisfied*

(A3) *there exist $\beta > 2$ and $m \in L^\beta([0, T], \mathbb{R}^+)$ such that $|\nabla F(t, u_\epsilon(t))| \le m(t)$;*
(A4) *there exists $\beta_1 > 1/(\frac{1}{2} - \epsilon_0)$ such that ${}_0D_t^\delta u_\epsilon \in L^{\beta_1}([0, T], \mathbb{R}^N)$.*

Then there exists a sequence $\{\epsilon_n\}$ such that $\epsilon_1 > \epsilon_2 > \cdots > \epsilon_n \to 0$ as $n \to \infty$, $u(t) = \lim_{n\to\infty} u_{\epsilon_n}(t)$ exists uniformly on $[0, T]$ and u is a solution of BVP (6.121) with $\alpha = \frac{1}{2}$.

Proof. According to Theorem 6.15, u_ϵ is also a solution of BVP (6.136). Thus, we have

$${}_tD_T^{\delta}\big({}_0D_t^{\delta}u_\epsilon(t)\big) = \nabla F(t, u_\epsilon(t)), \quad \text{a.e. } t \in [0, T]. \tag{6.137}$$

Propositions 1.6-1.7 imply that the equation (6.137) is equivalent to the integral equation

$${}_0D_t^{\delta}u_\epsilon(t) = {}_tD_T^{-\delta}(\nabla F(t, u_\epsilon(t))) + C(T-t)^{\delta-1}, \quad \text{a.e. } t \in [0, T], \tag{6.138}$$

where $C = (1/\Gamma(\delta))[{}_tD_T^{\delta-1}({}_0D_t^{\delta}u_\epsilon(t))]_{t=T}$. Noting that ${}_0D_t^\delta u_\epsilon \in L^{\beta_1}([0, T], \mathbb{R}^N)$ according to (A4), direct calculation gives that

$$\begin{aligned} \left|{}_tD_T^{\delta-1}\big({}_0D_t^{\delta}u_\epsilon(t)\big)\right| &\le \frac{1}{\Gamma(1-\delta)} \int_t^T (s-t)^{-\delta} \left|{}_0D_s^{\delta}u_\epsilon(s)\right| ds \\ &\le \frac{1}{\Gamma(1-\delta)} \left(\int_t^T (s-t)^{\frac{-\delta\beta_1}{\beta_1 - 1}} ds \right)^{1-1/\beta_1} \|{}_0D_t^\delta u_\epsilon\|_{L^{\beta_1}} \\ &\le c(T-t)^{1-\delta-1/\beta_1} \|{}_0D_t^\delta u_\epsilon\|_{L^{\beta_1}}, \end{aligned}$$

where $c \in \mathbb{R}^+$ is a constant. It is obvious that $1-\delta-1/\beta_1 > 0$ since $\beta_1 > 1/(\frac{1}{2}-\epsilon_0) > 1/(1-\delta)$, then we have $C = (1/\Gamma(\delta))[{}_tD_T^{\delta-1}({}_0D_t^{\delta}u_\epsilon(t))]_{t=T} = 0$. Therefore, (6.138) can be written as

$${}_0D_t^{\delta}u_\epsilon(t) = {}_tD_T^{-\delta}(\nabla F(t, u_\epsilon(t))), \quad \text{a.e. } t \in [0, T]. \tag{6.139}$$

According to Proposition 6.5 and in view of the continuity of $u_\epsilon \in E^\delta$, (6.139) is equivalent to the integral equation

$$u_\epsilon(t) = {}_0D_t^{-\delta}({}_tD_T^{-\delta}\nabla F(t, u_\epsilon(t))), \quad t \in [0, T]. \tag{6.140}$$

On the other hand, we observe that $m \in L^\beta([0, T], \mathbb{R}^+)$ and $\beta > 2$ in (A3) imply that

$$\begin{aligned}
\left|{}_tD_T^{-\delta}m(t)\right| &\leq \frac{1}{\Gamma(\delta)}\int_t^T (s-t)^{\delta-1}|m(s)|ds \\
&\leq \frac{1}{\Gamma(\delta)}\left(\int_t^T (s-t)^{\frac{(\delta-1)\beta}{\beta-1}}ds\right)^{1-1/\beta}\|m\|_{L^\beta} \\
&\leq c_1 T^{\delta-1/\beta}\|m\|_{L^\beta} \\
&\leq c_1\|m\|_{L^\beta}\max_{\lambda\in[\frac{1}{2},1]}\{T^{\lambda-1/\beta}\}, \quad t \in [0, T],
\end{aligned}$$

where $c_1 \in \mathbb{R}^+$ is a constant. Therefore, there exists a constant $M \in \mathbb{R}^+$ such that $\|{}_tD_T^{-\delta}m\| \leq M$, which means that $|{}_tD_T^{-\delta}\nabla F(t, u_\epsilon(t))| \leq M$ on $[0, T]$ since $|\nabla F(t, u_\epsilon(t))| \leq m(t)$.

Set $G(t, u_\epsilon(t)) = {}_tD_T^{-\delta}\nabla F(t, u_\epsilon(t))$, and we have by (6.140)

$$\begin{aligned}
|u_\epsilon(t)| &\leq \frac{1}{\Gamma(\delta)}\int_0^t (t-s)^{\delta-1}|G(s, u_\epsilon(s))|ds \\
&\leq \frac{M}{\Gamma(\delta)}\int_0^t (t-s)^{\delta-1}ds \\
&\leq \frac{M}{\Gamma(\delta+1)}T^\delta \\
&\leq M\max_{\lambda\in[\frac{1}{2},1]}\left\{\frac{T^\lambda}{\Gamma(\lambda+1)}\right\}, \quad t \in [0, T].
\end{aligned} \tag{6.141}$$

The last inequality follows from the continuity of $T^\lambda/\Gamma(\lambda+1)$ with respect to $\lambda > 0$ and the fact that $\Gamma(\lambda) > 0$ for $\lambda > 0$. Furthermore, letting $0 \leq t_1 < t_2 \leq T$, we see that

$$\begin{aligned}
&|u_\epsilon(t_1) - u_\epsilon(t_2)| \\
&= \frac{1}{\Gamma(\delta)}\left|\int_0^{t_1}(t_1-s)^{\delta-1}G(s, u_\epsilon(s))ds - \int_0^{t_2}(t_2-s)^{\delta-1}G(s, u_\epsilon(s))ds\right| \\
&= \frac{1}{\Gamma(\delta)}\left|\int_0^{t_1}\left((t_1-s)^{\delta-1} - (t_2-s)^{\delta-1}\right)G(s, u_\epsilon(s))ds\right. \\
&\quad \left.+ \int_{t_1}^{t_2}(t_2-s)^{\delta-1}G(s, u_\epsilon(s))ds\right| \\
&\leq \frac{M}{\Gamma(\delta)}\left|\int_0^{t_1}(t_1-s)^{\delta-1} - (t_2-s)^{\delta-1}ds + \int_{t_1}^{t_2}(t_2-s)^{\delta-1}ds\right| \\
&= \frac{M}{\Gamma(\delta+1)}\left(2(t_2-t_1)^\delta + t_1^\delta - t_2^\delta\right)
\end{aligned} \tag{6.142}$$

$$\leq \frac{2M}{\Gamma(\delta+1)}(t_2-t_1)^{\delta}$$
$$\leq 2M \max_{\lambda\in[\frac{1}{2},1]} \left\{\frac{(t_2-t_1)^{\lambda}}{\Gamma(\lambda+1)}\right\}.$$

It then follows from (6.141) and (6.142) that the family $\{u_\epsilon\}$ forms an equicontinuous and uniformly bounded functions. Application of Arzela-Ascoli theorem shows the existence of a sequence $\{\epsilon_n\}$ such that $\epsilon_1 > \epsilon_2 > \cdots > \epsilon_n \to 0$ as $n \to \infty$, and $u(t) = \lim_{n\to\infty} u_{\epsilon_n}(t)$ exists uniformly on $[0,T]$. Since the continuity and boundness of $\nabla F(t,\cdot)$ imply the continuity of ${}_tD_T^{-\delta}\nabla F(t,\cdot)$, we obtain that

$${}_tD_T^{-\delta}\nabla F(t,u_{\epsilon_n}(t)) \to {}_tD_T^{-\delta}\nabla F(t,u(t)), \quad \text{as } n\to\infty,$$

and combining (6.139) yields

$$u(t) = {}_0D_t^{-\delta}({}_tD_T^{-\delta}\nabla F(t,u(t))), \quad t\in[0,T].$$

This proves that u is a solution of BVP (6.121) by using the Proposition 1.5 and Lemma 6.1. The proof is completed. □

Example 6.8. Set $F(t,x) = m(t)\sin(|x|)$, where $m \in L^{\beta}([0,T],\mathbb{R}^+)$ and $x \in \mathbb{R}^N$. Then (A3) is verified since $|F(t,x)| \leq m(t)$ for $x \in \mathbb{R}^N$. If for any $\epsilon \in (0,\epsilon_0)$, we have $u_\epsilon \in AC([0,T],\mathbb{R}^N)$ and $u'_\epsilon \in L^{\beta_1}([0,T],\mathbb{R}^N)$, then ${}_0D_t^{\delta}u_\epsilon \in L^{\beta_1}([0,T],\mathbb{R}^N)$ by using Proposition 1.1 and (6.5). Thus, (A4) is satisfied.

6.6 Notes and Remarks

The results in Subsections 6.2.1-6.2.4 are taken from Jiao and Zhou, 2011. The material in Subsections 6.2.5-6.2.6 due to Chen and Tang, 2012. The results in Section 6.3 are adopted from Kong, 2013. The main results of Section 6.4 are from Chen and Tang, 2013. The material in Section 6.5 due to Jiao and Zhou, 2012.

Chapter 7

Fractional Hamiltonian Systems

7.1 Introduction

The existence of homoclinic solutions is one of the most important problems in the history of Hamiltonian systems. It has been intensively studied by many mathematicians (see Ambrosetti and Zelati, 1993; Ding and Jeanjean, 2007; Izydorek and Janczewska, 2005; Makita, 2012; Omana and Willem, 1992; Paturel, 2001; Rabinowitz, 1990; Séré, 1992; Szulkin and Zou, 2001; Zelati, Ekeland and Séré, 1990; Zelati and Rabinowitz, 1991; Zou and Li, 2002). Variational methods to prove the existence of homoclinic solutions for second order Hamiltonian systems were first used by Rabinowitz, 1990, while the first multiplicity result, later improved by Paturel, 2001, is due to Ambrosetti and Zelati, 1993. In recent years, the existence and multiplicity of homoclinic solutions for the second order Hamiltonian systems have been extensively studied via variational methods in many papers (e.g. Ambrosetti and Zelati, 1993; Ding and Jeanjean, 2007; Izydorek and Janczewska, 2005; Makita, 2012; Paturel, 2001; Rabinowitz, 1990; Zelati and Rabinowitz, 1991; Zou and Li, 2002). It is worth to mention that the fractional Hamiltonian systems are not uniquely defined and many researchers have explored this area giving new insight into this problem (e.g. Baleanu, Golmankaneh and Golmankaneh, 2009; Tarasov, 2010; Toress, 2013; Nyamoradi and Zhou, 2016, 2017; Nyamoradi, Alsaedi, Ahmad and Zhou, 2017; Nyamoradi, Zhou, Alsaedi and Ahmad, 2017).

In this chapter, we give some results on existence and multiplicity of homoclinic solutions for fractional Hamiltonian systems.

7.2 Existence and Multiplicity of Homoclinic Solutions (I)

7.2.1 *Fractional Derivative Space*

In this section, we consider the following fractional Hamiltonian systems

$$\begin{cases} {}_tD^\alpha_\infty({}_{-\infty}D^\alpha_t u(t)) + L(t)u(t) = \nabla W(t, u(t)), \quad t \in \mathbb{R}, \\ u \in H^\alpha(\mathbb{R}), \end{cases} \tag{7.1}$$

where ${}_{-\infty}D_t^\alpha$ and ${}_tD_\infty^\alpha$ are left and right Liouville-Weyl fractional derivatives of order $\alpha \in (\frac{1}{2}, 1)$ on the whole axis $\mathbb{R}$ respectively, $u \in \mathbb{R}^n$, $W(t,u)$ is of indefinite sign and subquadratic as $|u| \to +\infty$ and $L \in C(\mathbb{R}, \mathbb{R}^{n\times n})$ is a positive definite symmetric matrix for all $t \in \mathbb{R}$.

As usual, we say that a solution $u(t)$ of (7.1) is homoclinic (to 0) if $u(t) \to 0$ as $t \to \pm\infty$. In addition, if $u(t) \neq 0$ then $u(t)$ is called a nontrivial homoclinic solution.

In particular, if $\alpha = 1$, (7.1) reduces to the standard second order Hamiltonian system of the following form

$$u''(t) + L(t)u(t) - \nabla W(t, u(t)) = 0, \quad t \in \mathbb{R}. \tag{7.2}$$

We define the Fourier transform $\mathcal{F}(u)(\xi)$ of $u(x)$ as

$$\mathcal{F}(u)(\xi) = \int_{-\infty}^{\infty} e^{-ix\cdot\xi} u(x)dx.$$

For any $\alpha > 0$, we define the semi-norm and norm respectively as Torres, 2013.

$$|u|_{I_{-\infty}^\alpha} = ||{}_{-\infty}D_x^\alpha u||_{L^2},$$

$$||u||_{I_{-\infty}^\alpha} = \left(||u||_{L^2}^2 + |u|_{I_{-\infty}^\alpha}^2\right)^{\frac{1}{2}}, \tag{7.3}$$

and let the space $I_{-\infty}^\alpha(\mathbb{R})$ denote the completion of $C_0^\infty(\mathbb{R})$ with respect to the norm $||\cdot||_{I_{-\infty}^\alpha}$.

Next, for $0 < \alpha < 1$, we give the relationship between classical fractional Sobolev space $H^\alpha(\mathbb{R})$ and $I_{-\infty}^\alpha(\mathbb{R})$, where $H^\alpha(\mathbb{R})$ is defined by

$$H^\alpha(\mathbb{R}) = \overline{C_0^\infty(\mathbb{R})}^{||\cdot||_\alpha},$$

with the norm

$$||u||_\alpha = \left(||u||_{L^2}^2 + |u|_\alpha^2\right)^{\frac{1}{2}}, \tag{7.4}$$

and semi-norm

$$|u|_\alpha = |||\xi|^\alpha \mathcal{F}(u)||_{L^2}.$$

Observe that the spaces $H^\alpha(\mathbb{R})$ and $I_{-\infty}^\alpha(\mathbb{R})$ are isomorphic and have equivalent norms (see Torres, 2013).

Therefore, we define

$$H^\alpha(\mathbb{R}) = \left\{u \in L^2(\mathbb{R}) | \ |\xi|^\alpha \mathcal{F}(u) \in L^2(\mathbb{R})\right\}.$$

7.2.2 *Some Lemmas*

We suppose the following conditions for $L(t)$ and $W(t,u)$:

(L) There exists a $l \in C(\mathbb{R}, (0, +\infty))$ such that $l(t) \to +\infty$ as $|t| \to +\infty$ and

$$(L(t)u, u) \geq l(t)|u|^2, \quad \text{for all } t \in \mathbb{R},\ u \in \mathbb{R}^n.$$

($\mathbf{L}_\nu$) There exists a constant $\nu < 2$ such that

$$\liminf_{|t|\to+\infty}\left[|t|^{\nu-2}\inf_{|\xi|=1}(L(t)\xi,\xi)\right] > 0.$$

(W1) $W(t,0) = 0$ for all $t \in \mathbb{R}$ and there exist constants $\max\{1, 2/(3-\nu)\} < \gamma_i < 2$ and $a_i \geq 0$ $(i = 1, 2, \ldots, m)$ such that

$$|W(t,u)| \leq \sum_{i=1}^{m} a_i|u|^{\gamma_i}, \quad \forall\,(t,u) \in \mathbb{R}\times\mathbb{R}^n.$$

(W2) There exists a function $\varphi \in C([0,+\infty),[0,+\infty))$ such that

$$|\nabla W(t,u)| \leq \varphi(|u|), \quad \forall\,(t,u) \in \mathbb{R}\times\mathbb{R}^n,$$

where $\varphi(x) = O(x^{\gamma_{m+1}-1})$ as $x \to 0^+$, $\max\{1, 2/(3-\nu)\} < \gamma_{m+1} < 2$.

(W3) There exists a constant $\delta_0 > 0$ such that

$$W(t,u) \geq \sum_{k=1}^{l} b_k(t)|u|^{\nu_k}, \quad \forall\, t \in \Omega,\ u \in \mathbb{R}^n,\ |u| \leq \delta_0,$$

for some positive measure subset Ω of $\mathbb{R}$, where $\max\{1, 2/(3-\nu)\} < \nu_k < 2$ are constants, $b_k : \mathbb{R} \to \mathbb{R}^+$ are bounded continuous functions for $k = 1, 2, \ldots, l$.

(W4) There exist $t_0 \in \mathbb{R}$ and $\max\{1, 2/(3-\nu)\} < \vartheta < 2$ such that

$$\lim_{(t,u)\to(t_0,0)} \frac{W(t,u)}{|u|^\vartheta} > 0.$$

(W5) $W(t,-u) = W(t,u)$ for all $t \in \mathbb{R}$ and $u \in \mathbb{R}^n$.

In order to establish our results via variational methods and the critical point theory, we firstly describe some properties of the space on which the variational associated with (7.1) is defined. Let

$$X^\alpha = \left\{u \in H^\alpha(\mathbb{R})\,\Big|\, \int_{\mathbb{R}} \left(|{}_{-\infty}D_t^\alpha u(t)|^2 + (L(t)u(t), u(t))\right)dt < \infty\right\}.$$

The space X^α is a Hilbert space with the inner product

$$\langle u, v\rangle_{X^\alpha} = \int_{\mathbb{R}} \left(({}_{-\infty}D_t^\alpha u(t), {}_{-\infty}D_t^\alpha v(t)) + (L(t)u(t), v(t))\right)dt,$$

and the corresponding norm

$$\|u\|_{X^\alpha}^2 = \langle u, u\rangle_{X^\alpha}.$$

Lemma 7.1. *(Torres, 2013) Let $\alpha > \frac{1}{2}$, then $H^\alpha(\mathbb{R},\mathbb{R}^n) \subset C(\mathbb{R},\mathbb{R}^n)$ and there is a constant $C = C_\alpha$ such that*

$$\|u\|_\infty = \sup_{x\in\mathbb{R}} |u(x)| \leq C\|u\|_\alpha. \tag{7.5}$$

So by Lemma 7.1, there exists a constant $C_\alpha > 0$ such that

$$\|u\|_\infty \leq C_\alpha \|u\|_{X^\alpha}. \tag{7.6}$$

By (L_ν), there exist integers $T_0 > 0$ and $M_0 > 0$ such that

$$|t|^{\nu-2} \inf_{|\xi|=1} (L(t)\xi, \xi) \geq M_0, \quad |t| > T_0,$$

which implies that

$$|t|^{\nu-2}(L(t)\xi, \xi) \geq M_0|\xi|^2, \quad |t| > T_0, \ \xi \in \mathbb{R}^n. \tag{7.7}$$

Lemma 7.2. *Suppose that L satisfies (L_ν). Then, for $1 \leq q \in (2/(3-\nu), 2)$, X^α is compactly embedded in $L^q(\mathbb{R}, \mathbb{R}^n)$. Moreover*

$$\int_{|t|>T} |u(t)|^q dt \leq \frac{\rho(q)}{T^{\frac{(3-\nu)q-2}{2}}} \|u\|_{X^\alpha}^q, \quad \forall\, u \in X^\alpha, \ T \geq T_0, \tag{7.8}$$

and

$$\|u\|_{L^q}^q \leq \left[\left(\int_{|t|\leq T} [l(t)]^{\frac{-q}{2-q}} dt \right)^{1-\frac{q}{2}} + \frac{\rho(q)}{T^{\frac{(3-\nu)q-2}{2}}} \right] \|u\|_{X^\alpha}^q, \quad \forall\, u \in X^\alpha, \ T \geq T_0, \tag{7.9}$$

where

$$\rho(q) = \left[\frac{2(2-q)}{(3-\nu)q-2} \right]^{1-\frac{q}{2}} M_0^{-\frac{q}{2}}, \tag{7.10}$$

and

$$l(t) = \inf_{x\in\mathbb{R}^n,\ |x|=1} (L(t)x, x). \tag{7.11}$$

Proof. Let $\varsigma = \frac{(3-\nu)q-2}{2-q}$. Then $\varsigma > 0$. For $u \in X^\alpha$ and $T \geq T_0$, it follows from (7.7) and (7.10) together with the Hölder inequality that

$$\begin{aligned}
\int_{|t|>T} |u(t)|^q dt &\leq \left(\int_{|t|>T} |t|^{-\frac{(2-\nu)q}{2-q}} dt \right)^{1-\frac{q}{2}} \left(\int_{|t|>T} |t|^{2-\nu} |u(t)|^2 dt \right)^{\frac{q}{2}} \\
&\leq \left(\frac{2}{\varsigma T^\varsigma} \right)^{1-\frac{q}{2}} \left(\frac{1}{M_0} \int_{|t|>T} (L(t)u(t), u(t)) dt \right)^{\frac{q}{2}} \\
&\leq \frac{2^{\frac{2-q}{2}}}{M_0^{\frac{q}{2}} \varsigma^{\frac{2-q}{2}} T^{\frac{(3-\nu)q-2}{2}}} \|u\|_{X^\alpha}^q \\
&\leq \frac{\rho(q)}{T^{\frac{(3-\nu)q-2}{2}}} \|u\|_{X^\alpha}^q.
\end{aligned}$$

This shows that (7.8) holds. Hence, from (7.8) and (7.11) and the Hölder inequality, one can get

$$\|u\|_{L^q}^q = \int_{|t|\leq T} |u(t)|^q dt + \int_{|t|>T} |u(t)|^q dt$$

$$\le \left(\int_{|t|\le T}[l(t)]^{-\frac{q}{2-q}}dt\right)^{1-\frac{q}{2}}\left(\int_{|t|\le T}l(t)|u(t)|^2dt\right)^{\frac{q}{2}}+\frac{\rho(q)}{T^{\frac{(3-\nu)q-2}{2}}}\|u\|_{X^\alpha}^q$$
$$\le \left(\int_{|t|\le T}[l(t)]^{-\frac{q}{2-q}}dt\right)^{1-\frac{q}{2}}\|u\|_{X^\alpha}^q+\frac{\rho(q)}{T^{\frac{(3-\nu)q-2}{2}}}\|u\|_{X^\alpha}^q.$$

This shows that (7.9) holds.

Finally, we prove that X^α is compactly embedded in $L^q(\mathbb{R},\mathbb{R}^n)$. Let $\{u_k\}\subset X^\alpha$ be a bounded sequence. Then by (7.6), there exists a constant $\Lambda>0$ such that

$$\|u_k\|_\infty \le C_\alpha\|u_k\|_{X^\alpha}\le \Lambda,\quad k\in\mathbb{N}. \tag{7.12}$$

Since X^α is reflexive, $\{u_k\}$ possesses a weakly convergent subsequence in X^α. Passing to a subsequence if necessary, we may assume that $u_k \rightharpoonup u_0$ weakly in X^α. It is easy to verify that

$$\lim_{k\to\infty}u_k(t)=u_0(t),\quad \forall\, t\in\mathbb{R}. \tag{7.13}$$

For any given number $\varepsilon>0$, we can choose $T_\varepsilon>0$ such that

$$\frac{2^{q-1}\rho(q)}{T_\varepsilon^{\frac{(3-\nu)q-2}{2}}}\left[\left(\frac{\Lambda}{C_\alpha}\right)^q+\|u_0\|_{X^\alpha}^q\right]<\varepsilon. \tag{7.14}$$

It follows from (7.13) that there exists $k_0\in\mathbb{N}$ such that

$$\int_{|t|\le T_\varepsilon}|u_k(t)-u_0(t)|^q dt<\varepsilon,\quad \forall\, k\ge k_0. \tag{7.15}$$

On the other hand, it follows from (7.8), (7.12) and (7.14) that

$$\begin{aligned}\int_{|t|>T_\varepsilon}|u_k(t)-u_0(t)|^q dt &\le 2^{q-1}\int_{|t|>T_\varepsilon}\left(|u_k(t)|^q+|u_0(t)|^q\right)dt\\ &\le \frac{2^{q-1}\rho(q)}{T_\varepsilon^{\frac{(3-\nu)q-2}{2}}}\left(\|u_k\|_{X^\alpha}^q+\|u_0\|_{X^\alpha}^q\right)\\ &\le \frac{2^{q-1}\rho(q)}{T_\varepsilon^{\frac{(3-\nu)q-2}{2}}}\left[\left(\frac{\Lambda}{C_\alpha}\right)^q+\|u_0\|_{X^\alpha}^q\right]<\varepsilon,\quad k\in\mathbb{N}.\end{aligned} \tag{7.16}$$

Since $\varepsilon>0$ is arbitrary, we obtain by (7.15) and (7.16) that

$$\|u_k-u_0\|_{L^q}^q=\int_{\mathbb{R}}|u_k(t)-u_0(t)|^q dt\to 0,\quad \text{as } k\to+\infty.$$

This shows that $\{u_k\}$ possesses a convergent subsequence in $L^q(\mathbb{R},\mathbb{R}^n)$. Therefore, X^α is compactly embedded in $L^q(\mathbb{R},\mathbb{R}^n)$ for $1\le q\in(2/(3-\nu),2)$. Therefore, the proof is completed. □

Also, by (L), since $l\in C(\mathbb{R},(0,\infty))$ and l is coercive, then $l_{\min}=\min_{t\in\mathbb{R}}l(t)$ exists, then we have

$$(L(t)u(t),u(t))\ge l(t)|u(t)|^2\ge l_{\min}|u(t)|^2,\quad \forall\, t\in\mathbb{R}. \tag{7.17}$$

Lemma 7.3. *Suppose that L satisfies (L). Then for $2 \le q < \infty$, X^α is compactly embedded in $L^q(\mathbb{R}, \mathbb{R}^n)$; moreover*

$$\int_{|t|>T} |u(t)|^q dt \le \frac{C_\alpha^{q-2}}{\min_{|s|\ge T} l(s)} \|u\|_{X^\alpha}^q, \quad \forall\, u \in X^\alpha, \;\; T \ge 1, \tag{7.18}$$

and

$$\|u\|_{L^q}^q \le l_{\min}^{-1} C_\alpha^{q-2} \|u\|_{X^\alpha}^q, \quad \forall\, u \in X^\alpha. \tag{7.19}$$

Proof. From (7.6) and (7.17), one can get

$$\begin{aligned}\int_{|t|>T} |u(t)|^q dt &\le \|u\|_\infty^{q-2} \int_{|t|>T} |u(t)|^2 dt \\ &\le \|u\|_\infty^{q-2} \int_{|t|>T} [l(t)]^{-1} (L(t)u(t), u(t)) dt \\ &\le \frac{\|u\|_\infty^{q-2}}{\min_{|s|\ge T} l(s)} \|u\|_{X^\alpha}^2 \\ &\le \frac{C_\alpha^{q-2}}{\min_{|s|\ge T} l(s)} \|u\|_{X^\alpha}^q,\end{aligned} \tag{7.20}$$

and

$$\begin{aligned}\|u\|_{L^q}^q &\le \|u\|_\infty^{q-2} \int_{t\in\mathbb{R}} |u(t)|^2 dt \\ &\le l_{\min}^{-1} \|u\|_\infty^{q-2} \int_{t\in\mathbb{R}} (L(t)u(t), u(t)) dt \\ &\le l_{\min}^{-1} C_\alpha^{q-2} \|u\|_{X^\alpha}^q,\end{aligned}$$

which, together with (7.20), shows that (7.18) and (7.19) hold.

We now can prove that X^α is compactly embedded in $L^q(\mathbb{R}, \mathbb{R}^n)$ for $2 \le q < \infty$ by (L). By Lemma 2.2 in Torres, 2013, we know that the embedding of X^α in $L^2(\mathbb{R}, \mathbb{R}^n)$ is continuous and compact. On the other hand, from Lemma 7.1, we know that if $u \in H^\alpha$ with $\frac{1}{2} < \alpha < 1$, then $u \in L^q(\mathbb{R}, \mathbb{R}^n)$ for all $q \in [2, +\infty)$, because

$$\int_{\mathbb{R}} |u(x)|^q dx \le \|u\|_\infty^{q-2} \|u\|_{L^2}^2.$$

So, it is easy to verify that the embedding of X^α in $L^q(\mathbb{R}, \mathbb{R}^n)$ is also continuous and compact for $2 \le q < \infty$. Therefore, combining this with Lemma 2.2 in Torres, 2013, we have the desired conclusion for $2 \le q < \infty$. Therefore, the proof is completed. □

Now, we establish the corresponding variational framework to obtain solutions of (7.1). To this end, define the functional $I : X^\alpha \to \mathbb{R}$ by

$$\begin{aligned}I(u) &= \frac{1}{2} \int_{\mathbb{R}} \Big(|{}_{-\infty}D_t^\alpha u(t)|^2 + (L(t)u(t), u(t)) \Big) dt - \int_{\mathbb{R}} W(t, u(t)) dt \\ &= \frac{1}{2} \|u\|_{X^\alpha}^2 - \int_{\mathbb{R}} W(t, u(t)) dt.\end{aligned} \tag{7.21}$$

Lemma 7.4. *Assume that the conditions* (L_ν), *(W1) and (W2) hold. Then the functional* I *is well defined and of class* $C^1(X^\alpha, \mathbb{R})$ *with*

$$\begin{aligned} I'(u)v = \int_{\mathbb{R}} \Big(& ({}_{-\infty}D_t^\alpha u(t), {}_{-\infty}D_t^\alpha v(t)) \\ & + (L(t)u(t), v(t)) - (\nabla W(t, u(t)), v(t)) \Big) dt. \end{aligned} \tag{7.22}$$

Furthermore, the critical points of I *in* X^α *are solutions of* (7.1) *with* $u(t) \to 0$ *as* $t \to \pm\infty$.

Proof. We firstly show that $I : X^\alpha \to \mathbb{R}$. For $T \geq T_0$, in view of (W1) and (7.19), we have

$$\begin{aligned} \left| \int_{\mathbb{R}} W(t, u(t)) dt \right| & \leq \sum_{i=1}^m a_i \int_{\mathbb{R}} |u(t)|^{\gamma_i} dt \\ & \leq \sum_{i=1}^m a_i \left[\left(\int_{|t| \leq T} [l(t)]^{\frac{-\gamma_i}{2-\gamma_i}} dt \right)^{1-\frac{\gamma_i}{2}} + \frac{\rho(\gamma_i)}{T^{\frac{(3-\nu)\gamma_i - 2}{2}}} \right] \|u\|_{X^\alpha}^{\gamma_i} \\ & \leq \sum_{i=1}^m \phi_i(T) \|u\|_{X^\alpha}^{\gamma_i}, \end{aligned} \tag{7.23}$$

where

$$\phi_i(T) := a_i \left[\left(\int_{|t| \leq T} [l(t)]^{\frac{-\gamma_i}{2-\gamma_i}} dt \right)^{1-\frac{\gamma_i}{2}} + \frac{\rho(\gamma_i)}{T^{\frac{(3-\nu)\gamma_i - 2}{2}}} \right].$$

Combining this with (7.21), it follows that $I : X^\alpha \to \mathbb{R}$.

Next, we prove that $I \in C^1(X^\alpha, \mathbb{R})$. Rewrite I as $I = I_1 - I_2$, where

$$\begin{aligned} I_1(u) & := \frac{1}{2} \int_{\mathbb{R}} \Big(|{}_{-\infty}D_t^\alpha u(t)|^2 + (L(t)u(t), u(t)) \Big) dt, \\ I_2(u) & := \int_{\mathbb{R}} W(t, u(t)) dt. \end{aligned} \tag{7.24}$$

It is easy to check that $I_1 \in C^1(X^\alpha, \mathbb{R})$, and that

$$I_1'(u)v = \int_{\mathbb{R}} \Big(({}_{-\infty}D_t^\alpha u(t), {}_{-\infty}D_t^\alpha v(t)) + (L(t)u(t), v(t)) \Big) dt.$$

Then, it is sufficient to show that $I_2 \in C^1(X^\alpha, \mathbb{R})$. So, we have

$$I_2'(u)v = \int_{\mathbb{R}} (\nabla W(t, u(t)), v(t)) dt, \quad \forall\, u, v \in X^\alpha. \tag{7.25}$$

By (W2), one can choose a constant $\varphi_0 > 0$ such that

$$\varphi(|u|) \leq \varphi_0 |u|^{\gamma_{m+1} - 1}, \quad \forall\, u \in \mathbb{R}^n,\ |u| \leq 1. \tag{7.26}$$

For any $u, v \in X^\alpha$, there exists $T_1 > 0$ such that $|u(t)| + |v(t)| < 1$ as $|t| > T_1$. Then for any function $\theta : \mathbb{R} \to (0,1)$ and any number $h \in (0,1)$, by (W2), (7.26) and Lemma 7.2, we have

$$\begin{aligned}
&\int_{\mathbb{R}} |(\nabla W(t, u(t) + \theta(t)hv(t)), v(t))|dt \\
&\leq \int_{|t|\leq T_1} |(\nabla W(t, u(t) + \theta(t)hv(t))||v(t))|dt \\
&\quad + \int_{|t|>T_1} |(\nabla W(t, u(t) + \theta(t)hv(t))||v(t))|dt \\
&\leq \int_{|t|\leq T_1} \max_{|x|\leq\|u\|_\infty+\|v\|_\infty} |\nabla W(t,x)||v(t))|dt \\
&\quad + \varphi_0 \int_{|t|>T_1} (|u(t)| + |v(t)|)^{\gamma_{m+1}-1}|v(t))|dt \\
&\leq \int_{|t|\leq T_1} \max_{|x|\leq\|u\|_\infty+\|v\|_\infty} |\nabla W(t,x)||v(t))|dt + \varphi_0 \int_{|t|>T_1} |v(t)|^{\gamma_{m+1}}dt \\
&\quad + \varphi_0 \left(\int_{|t|>T_1} |u(t)|^{\gamma_{m+1}}dt\right)^{1-\frac{1}{\gamma_{m+1}}} \left(\int_{|t|>T_1} |v(t)|^{\gamma_{m+1}}dt\right)^{\frac{1}{\gamma_{m+1}}} \\
&\leq \int_{|t|\leq T_1} \max_{|x|\leq\|u\|_\infty+\|v\|_\infty} |\nabla W(t,x)||v(t))|dt \\
&\quad + \varphi_0 \frac{\rho(\gamma_{m+1})}{T^{\frac{(3-\nu)\gamma_{m+1}-2}{2}}} (\|u\|_{X^\alpha}^{\gamma_{m+1}-1} + \|v\|_{X^\alpha}^{\gamma_{m+1}-1})\|v\|_{X^\alpha} < +\infty.
\end{aligned} \tag{7.27}$$

Then by (7.21) and (7.27), the mean value theorem and Lebesgue dominated convergence theorem, we get

$$\begin{aligned}
I_2'(u)v &= \lim_{h\to 0^+} \frac{I_2(u+hv) - I_2(u)}{h} \\
&= \lim_{h\to 0^+} \left[\int_{\mathbb{R}} \frac{W(t, u(t) + hv(t)) - W(t, u(t))}{h} dt\right] \\
&= \lim_{h\to 0^+} \left[\int_{\mathbb{R}} (\nabla W(t, u(t) + \theta(t)hv(t)), v(t))dt\right] \\
&= \int_{\mathbb{R}} (\nabla W(t, u(t)), v(t))dt.
\end{aligned}$$

This shows that (7.25) holds.

It remains to prove that I_2' is continuous. Suppose that $u_k \to u_0$ in X^α as $k \to \infty$, then, by the Banach-Steinhaus theorem, there exists a constant $\varrho > 0$ such that

$$\|u_0\|_{X^\alpha} \leq \frac{1}{C_\alpha}\varrho, \quad \sup_{k\in\mathbb{N}} \|u_k\|_{X^\alpha} \leq \frac{1}{C_\alpha}\varrho. \tag{7.28}$$

In view of (7.6), we have

$$\|u_0\|_\infty \leq \varrho, \quad \sup_{k\in\mathbb{N}} \|u_k\|_\infty \leq \varrho. \tag{7.29}$$

Now, by (W2), we can choose a constant $\varphi_1 > 0$ such that

$$\varphi(|u|) \le \varphi_1 |u|^{\gamma_{m+1}-1}, \quad \forall\, u \in \mathbb{R}^n,\ |u| \le \varrho. \tag{7.30}$$

Thus by (7.8), (7.22), (7.28)–(7.30), (W2) and the Hölder inequality, we obtain

$$\begin{aligned}
&|I_2'(u_k)v - I_2'(u_0)v| \\
&= \int_{\mathbb{R}} |(\nabla W(t,u_k(t)) - \nabla W(t,u_0(t)), v(t))|dt \\
&\le \int_{|t|\le T} |(\nabla W(t,u_k(t)) - \nabla W(t,u_0(t))||v(t))|dt \\
&\quad + \int_{|t|>T} |(\nabla W(t,u_k(t)) - \nabla W(t,u_0(t))||v(t))|dt \\
&\le o(1) + \varphi_1 \int_{|t|>T_1} \left(|u_k(t)|^{\gamma_{m+1}-1} + |u_0(t)|^{\gamma_{m+1}-1}\right) |v(t))|dt \\
&\le o(1) + \varphi_1 \left(\int_{|t|>T} |u_k(t)|^{\gamma_{m+1}} dt\right)^{1-\frac{1}{\gamma_{m+1}}} \left(\int_{|t|>T_1} |v(t)|^{\gamma_{m+1}} dt\right)^{\frac{1}{\gamma_{m+1}}} \\
&\quad + \varphi_1 \left(\int_{|t|>T} |u_0(t)|^{\gamma_{m+1}} dt\right)^{1-\frac{1}{\gamma_{m+1}}} \left(\int_{|t|>T_1} |v(t)|^{\gamma_{m+1}} dt\right)^{\frac{1}{\gamma_{m+1}}} \\
&\le o(1) + \varphi_1 \frac{\rho(\gamma_{m+1})}{T^{\frac{(3-\nu)\gamma_{m+1}-2}{2}}} (\|u_k\|_{X^\alpha}^{\gamma_{m+1}-1} + \|u_0\|_{X^\alpha}^{\gamma_{m+1}-1})\|v\|_{X^\alpha} \\
&= o(1), \quad \text{as } k \to +\infty,\ T \to +\infty,\ \forall\, v \in X^\alpha,
\end{aligned}$$

which shows the continuity of I_2'.

Finally, by a standard argument, it is easy to show that the critical points of I in X^α are solutions of (7.1) with $u(\pm\infty) = 0$. Therefore, the proof is completed. □

7.2.3 *Existence of Homoclinic Solutions*

Now, we can state existence results.

Theorem 7.1. *Suppose that L and W satisfy (L_ν) and (W1)–(W3). Then,* (7.1) *has at least one nontrivial homoclinic solution.*

Proof. In view of Lemma 7.4, $I \in C^1(X^\alpha, \mathbb{R})$. We show that I satisfies the hypotheses of Theorem 1.12.

Claim I. We first show that I is bounded from below. Selecting $T_2 > T_0$, it follows from (7.23) that

$$\int_{\mathbb{R}} W(t,u(t))dt \le \sum_{i=1}^{m} \phi_i(T_2)\|u\|_{X^\alpha}^{\gamma_i}, \quad \forall\, u \in X^\alpha. \tag{7.31}$$

From (7.21) and (7.31), we get

$$I(u) = \frac{1}{2}\|u\|_{X^\alpha}^2 - \int_{\mathbb{R}} W(t,u(t))dt \ge \frac{1}{2}\|u\|_{X^\alpha}^2 - \sum_{i=1}^{m} \phi_i(T_2)\|u\|_{X^\alpha}^{\gamma_i}. \tag{7.32}$$

Since $\max\{1, 2/(3-\nu)\} < \gamma_i < 2$, (7.32) implies that $I(u) \to +\infty$ as $\|u\|_{X^\alpha} \to +\infty$. Therefore, I is bounded from below.

Claim II. We show that I satisfies the Palais-Smale condition. Assume that $\{u_k\}_{k\in\mathbb{N}} \subset X^\alpha$ is a sequence such that $\{I(u_k)\}_{k\in\mathbb{N}}$ is bounded and $I'(u_k) \to 0$ as $k \to +\infty$. So by (7.6) and (7.32), there exists a constant $\Lambda_1 > 0$ such that

$$\|u_k\|_\infty \le C_\alpha \|u_k\|_{X^\alpha} \le \Lambda_1, \quad k \in \mathbb{N}. \tag{7.33}$$

Hence, passing to a subsequence if necessary, one may assume that $u_k \rightharpoonup u$ weakly in X^α. It is easy to verify that

$$\lim_{k\to\infty} u_k(t) = u(t), \quad \forall\, t \in \mathbb{R}. \tag{7.34}$$

So,

$$(I'(u_k) - I'(u))(u_k - u) \to 0, \quad \text{as } k \to \infty, \tag{7.35}$$

it follows from (7.33) and (7.34) that

$$\|u\|_{X^\alpha} \le \Lambda_1. \tag{7.36}$$

By (W2), we can choose $\varphi_2 > 0$ such that

$$\varphi(|u|) \le \varphi_2 |u|^{\gamma_{m+1}-1}, \quad \forall\, u \in \mathbb{R}^n,\ |u| \le \Lambda_1. \tag{7.37}$$

For any given number $\varepsilon > 0$, we can choose $T_3 > 0$ such that

$$\frac{\rho(\gamma_{m+1})}{T_3^{\frac{(3-\nu)\gamma_{m+1}-2}{2}}} \left[\left(\frac{\Lambda_1}{C_\alpha}\right)^{\gamma_{m+1}} + \|u\|_{X^\alpha}^{\gamma_{m+1}} \right] < \varepsilon. \tag{7.38}$$

It follows from (7.34) and the continuity of $\nabla W(t,x)$ on x that there exists $k_1 \in \mathbb{N}$ such that

$$\int_{|t|\le T_3} |\nabla W(t, u_k(t)) - \nabla W(t, u(t))||u_k(t) - u(t)| dt < \varepsilon, \quad \forall\, k \ge k_1. \tag{7.39}$$

Therefore, in view of (7.8), (7.33), (7.36)-(7.38) and (W2), we obtain

$$\begin{aligned}
&\int_{|t|>T_3} |\nabla W(t, u_k(t)) - \nabla W(t, u(t))||u_k(t) - u(t)| dt \\
&\le \varphi_2 \int_{|t|>T_3} \left(|u_k(t)|^{\gamma_{m+1}-1} + |u(t)|^{\gamma_{m+1}-1}\right) \Big(|u_k(t)| + |u(t)|\Big) dt \\
&\le 2\varphi_2 \int_{|t|>T_3} \left(|u_k(t)|^{\gamma_{m+1}} + |u(t)|^{\gamma_{m+1}}\right) dt \\
&\le 2\varphi_2 \frac{\rho(\gamma_{m+1})}{T_3^{\frac{(3-\nu)\gamma_{m+1}-2}{2}}} \left[\|u_k\|_{X^\alpha}^{\gamma_{m+1}} + \|u\|_{X^\alpha}^{\gamma_{m+1}} \right] \\
&\le 2\varphi_2 \frac{\rho(\gamma_{m+1})}{T_3^{\frac{(3-\nu)\gamma_{m+1}-2}{2}}} \left[\left(\frac{\Lambda_1}{C_\alpha}\right)^{\gamma_{m+1}} + \|u\|_{X^\alpha}^{\gamma_{m+1}} \right] \\
&< 2\varphi_2 \varepsilon, \quad k \in \mathbb{N}.
\end{aligned} \tag{7.40}$$

Since $\varepsilon > 0$ is arbitrary, so by (7.39) and (7.40), we get

$$\int_{\mathbb{R}} (\nabla W(t, u_k(t)) - \nabla W(t, u(t)), u_k(t) - u(t))dt \to 0, \quad \text{as } k \to +\infty. \tag{7.41}$$

On the other hand, we have

$$\begin{aligned} &(I'(u_k) - I'(u))(u_k - u) \\ &= \|u_k - u\|_{X^\alpha}^2 - \int_{\mathbb{R}} (\nabla W(t, u_k(t)) - \nabla W(t, u(t)), u_k(t) - u(t))dt. \end{aligned} \tag{7.42}$$

Hence, it follows from (7.35), (7.41) and (7.42) that $\|u_k - u\|_{X^\alpha} \to 0$ as $k \to +\infty$. Therefore, I satisfies Palais-Smale condition.

Then, by Theorem 1.12, $c = \inf_{X^\alpha} I(u)$ is a critical value of I, that is, there exists a critical point e such that $I(e) = c$.

Finally, we show that $e \neq 0$. Take some $u \in X^\alpha$ such that $\|u\|_{X^\alpha} = 1$. Then there exists a subset Ω of positive measure $|\Omega| < \infty$ of $\mathbb{R}$ such that $u(t) \neq 0$ for $t \in \Omega$. Take $\sigma > 0$ small enough so that $\sigma|u(t)| \leq \delta_0$ for $t \in \Omega$. By (W3), there exists a constant $\eta > 0$ such that

$$W(t, u) \geq \eta \sum_{k=1}^{l} |u|^{\nu_k}, \quad \forall\, t \in \Omega,\ u \in \mathbb{R}^n,\ |u| \leq \delta_0. \tag{7.43}$$

Then by (7.43), one can get

$$\begin{aligned} I(\sigma u) &= \frac{\sigma^2}{2}\|u\|_{X^\alpha}^2 - \int_{\mathbb{R}} W(t, \sigma u(t))dt \\ &\leq \frac{\sigma^2}{2} - \eta \sum_{k=1}^{l} \sigma^{\nu_k} \int_{\Omega} |u(t)|^{\nu_k} dt. \end{aligned} \tag{7.44}$$

Since $\max\{1, 2/(3-\nu)\} < \nu_k < 2$ ($k = 1, 2, \ldots, l$) and $\int_\Omega |u(t)|^\mu dt > 0$, (7.44) implies that $I(\sigma u) < 0$ for some $\sigma > 0$ with $\sigma|u(t)| \leq \delta_0$ for $t \in \Omega$. Thus, $I(e) = c = \inf_{X^\alpha} I(u) < 0$, therefore e is a nontrivial critical point of I, and hence $e = e(t)$ is a nontrivial homoclinic solution of system (7.1). The proof is completed. □

Theorem 7.2. *Suppose that L and W satisfy (L_ν), (W1), (W2), (W4) and (W5). Then, (7.1) has at least d ($\in \mathbb{N}$) distinct pairs of nontrivial homoclinic solutions.*

Proof. In view of Lemma 7.4 and the Proof of Theorem 7.1, $I \in C^1(X^\alpha, \mathbb{R})$ is bounded from below and satisfies the Palais-Smale condition. It is obvious that I is even and $I(0) = 0$. In order to apply Theorem 1.13, we show that there is a set $K \subset X^\alpha$ such that K is homeomorphic to S^{d-1} by an odd map, and $\sup_K I < 0$.

By (W4), there exist an open set $D \subset \mathbb{R}$ with $t_0 \in D$, $\sigma_1 > 0$ and $\eta > 0$ such that

$$W(t, u) \geq \eta |u|^\vartheta, \quad \forall\, (t, u) \in D \times \mathbb{R}^n,\ |u| < \sigma_1. \tag{7.45}$$

For any $d \in \mathbb{N}$, we take d disjoint open sets D_i such that $\bigcup_{i=1}^{d} D_i \subset D$. For $i = 1, 2, \ldots, d$, let $u_i \in (H_0^\alpha(D_i) \bigcap X^\alpha) \setminus \{0\}$ (for detail of $H_0^\alpha(D_i)$, see Ervin, 2006) and $\|u_i\|_{X^\alpha} = 1$, and

$$X_d = \operatorname{span}\{u_1, \ldots, u_d\}, \qquad S_d = \{u \in X_d : \|u\|_{X^\alpha} = 1\}. \tag{7.46}$$

For $u \in X_d$, there exist $\lambda_i \in \mathbb{R}$, $i = 1, 2, \ldots, d$ such that

$$u(t) = \sum_{i=1}^{d} \lambda_i u_i(t), \quad \text{for } t \in \mathbb{R}. \tag{7.47}$$

So

$$\|u\|_{L^\vartheta} = \left(\int_{\mathbb{R}} |u(t)|^\vartheta dt \right)^{\frac{1}{\vartheta}} = \left(\sum_{i=1}^{d} |\lambda_i|^\vartheta \int_{D_i} |u_i(t)|^\vartheta dt \right)^{\frac{1}{\vartheta}}, \tag{7.48}$$

and

$$\begin{aligned} \|u\|_{X^\alpha}^2 &= \int_{\mathbb{R}} \Big(|{}_{-\infty}D_t^\alpha u(t)|^2 + (L(t)u(t), u(t)) \Big) dt \\ &= \sum_{i=1}^{d} \lambda_i^2 \int_{D_i} \Big(|{}_{-\infty}D_t^\alpha u_i(t)|^2 + (L(t)u_i(t), u_i(t)) \Big) dt \\ &= \sum_{i=1}^{d} \lambda_i^2 \int_{\mathbb{R}} \Big(|{}_{-\infty}D_t^\alpha u_i(t)|^2 + (L(t)u_i(t), u_i(t)) \Big) dt \\ &= \sum_{i=1}^{d} \lambda_i^2 \|u_i\|_{X^\alpha}^2 = \sum_{i=1}^{d} \lambda_i^2. \end{aligned} \tag{7.49}$$

As all norms of a finite-dimensional normed space are equivalent, there is a constant $C' > 0$ such that

$$C' \|u\|_{X^\alpha} \le \|u\|_{L^\vartheta}, \quad \text{for } u \in X_d. \tag{7.50}$$

Note that $W(t, 0) = 0$, and so according to (7.45), (7.47)-(7.50), one can get

$$\begin{aligned} I(su) &= \frac{s^2}{2} \|u\|_{X^\alpha}^2 - \int_{\mathbb{R}} W(t, su(t)) dt \\ &= \frac{s^2}{2} \|u\|^2 - \sum_{i=1}^{d} \int_{D_i} W(t, s\lambda_i u_i(t)) \\ &\le \frac{s^2}{2} \|u\|_{X^\alpha}^2 - \eta s^\vartheta \sum_{i=1}^{d} |\lambda_i|^\vartheta \int_{D_i} |u_i(t)|^\vartheta dt \\ &\le \frac{s^2}{2} \|u\|_{X^\alpha}^2 - \eta s^\vartheta \|u\|_{L^\vartheta}^\vartheta \\ &\le \frac{s^2}{2} \|u\|_{X^\alpha}^2 - \eta (C's)^\vartheta \|u\|_{X^\alpha}^\vartheta, \quad \forall\, u \in S_d, \end{aligned} \tag{7.51}$$

and sufficiently small $s > 0$. In this case (7.45) is applicable, since u is continuous on $\overline{D}$ and so $|s\lambda_i u_i(t)| \leq \sigma_1$ for any $t \in D$, $i = 1, 2, \ldots, d$ can be true for sufficiently small s. Hence, it follows from (7.51) that there exist $\varepsilon > 0$ and $\sigma_2 > 0$ such that

$$I(\sigma_2 u) < -\varepsilon, \quad \forall\, u \in S_d. \tag{7.52}$$

Let

$$S_d^{\sigma_2} = \{\sigma_2 u : \ u \in S_d\}, \quad S^{d-1} = \left\{ \left(\frac{\lambda_1}{\sigma_2}, \frac{\lambda_2}{\sigma_2}, \ldots, \frac{\lambda_d}{\sigma_2} \right)^T \in \mathbb{R}^d : \sum_{i=1}^{d} \frac{\lambda_i^2}{\sigma_2^2} = 1 \right\}.$$

Then it follows from (7.49) that

$$S_d^{\sigma_2} = \left\{ \sum_{i=1}^{d} \lambda_i u_i : \ \sum_{i=1}^{d} \lambda_i^2 = \sigma_2^2 \right\}.$$

By (7.45), we define a map $\Psi : S_d^{\sigma_2} \to S^{d-1}$ as follows

$$\Psi(u) = \sigma_2^{-1} \left(\frac{\lambda_1}{\sigma_2}, \frac{\lambda_2}{\sigma_2}, \ldots, \frac{\lambda_d}{\sigma_2} \right)^T, \quad \forall\, u \in S_d^{\sigma}.$$

It is easy to verify that $\Psi : S_d^{\sigma_2} \to S^{d-1}$ is an odd homeomorphic map. On the other hand, by (7.52), we have

$$I(u) < -\varepsilon, \quad \forall\, u \in S_d^{\sigma_2},$$

and thus $\sup_{S_d^{\sigma_2}} I < -\varepsilon < 0$. By Theorem 1.13, I has at least d distinct pairs of critical points, and so system (7.1) possesses at least d distinct pairs of nontrivial homoclinic solutions. The proof is completed. □

Next, we replace the conditions (W1)–(W4) with the following conditions:

(W6) $W(t, 0) = 0$ for all $t \in \mathbb{R}$, there exist constants $\varpi_i \in [0, 2 - \nu)$, $g_i \geq 0$ and $\max\{1, 2(1 + \varpi_i)/(3 - \nu)\} < \tau_i < 2$ $(i = 1, 2, \ldots, r)$ such that

$$|W(t, u)| \leq \sum_{i=1}^{r} g_i (1 + |t|^{\varpi_i}) |u|^{\tau_i}, \quad \forall\, (t, u) \in \mathbb{R} \times \mathbb{R}^n.$$

(W7) There exist r functions $\chi_i \in C([0, +\infty), [0, +\infty))$ such that

$$|\nabla W(t, u)| \leq \sum_{i=1}^{r} (1 + |t|^{\varpi_i}) \chi_i(|u|), \quad \forall\, (t, u) \in \mathbb{R} \times \mathbb{R}^n,$$

where $\chi(x) = O(x^{\tau_{r+i} - 1})$ as $x \to 0^+$, $\max\{1, 2(1 + \varpi_i)/(3 - \nu)\} < \tau_{r+i} < 2$ $(i = 1, 2, \ldots, r)$.

(W8) There exists a constant $\delta_0^1 > 0$ such that

$$W(t, u) \geq \sum_{k=1}^{l} b_k^1(t) |u|^{\nu_k^1}, \quad \forall\, t \in \Omega, \ u \in \mathbb{R}^n, \ |u| \leq \delta_0^1,$$

for some positive measure subset Ω of $\mathbb{R}$, and where $\max\{1, 2(1 + \varpi_i)/(3 - \nu)\} < \nu_k^1 < 2$ are constants, $b_k^1 : \mathbb{R} \to \mathbb{R}^+$ are bounded continuous functions for $k = 1, 2, \ldots, l$.

(W9) There exist $t_0 \in \mathbb{R}$ and $\max\{1, 2(1+\varpi_i)/(3-\nu)\} < \vartheta < 2$ such that

$$\lim_{(t,u)\to(t_0,0)} \frac{W(t,u)}{|u|^\vartheta} > 0.$$

Then, we have the following results.

Theorem 7.3. *Suppose that L and W satisfy (L_ν) and (W6)–(W8). Then,* (7.1) *has at least one nontrivial homoclinic solution.*

Theorem 7.4. *Suppose that L and W satisfy (L_ν), (W5), (W6), (W7) and (W9). Then,* (7.1) *has at least d $(\in \mathbb{N})$ distinct pairs of nontrivial homoclinic solutions.*

Lemma 7.5. *Suppose that L satisfies (L_ν). Then for $\varpi \in [0,\nu)$ and $1 \le q \in (2(1+\varpi)/(3-\nu), 2)$, X^α is compactly embedded in $L^q(\mathbb{R},\mathbb{R}^n)$; moreover*

$$\int_{|t|>T} (1+|t|^\varpi)|u(t)|^q dt \le \frac{\rho(\varpi,q)}{T^{\frac{(3-\nu)q-2(1+\varpi)}{2}}} \|u\|^q_{X^\alpha}, \quad \forall\, u \in X^\alpha, \ T \ge T_0, \tag{7.53}$$

and

$$\begin{aligned}\int_{\mathbb{R}} (1+|t|^\varpi)|u(t)|^q dt \le & \left[\left(\int_{|t|\le T} (1+|t|^\varpi)^{\frac{2}{2-q}} [l(t)]^{\frac{-q}{2-q}} dt\right)^{1-\frac{q}{2}}\right. \\ & \left. + \frac{\rho(\varpi,q)}{T^{\frac{(3-\nu)q-2(1+\varpi)}{2}}}\right] \|u\|^q_{X^\alpha}, \quad \forall\, u \in X^\alpha, \ T \ge T_0,\end{aligned} \tag{7.54}$$

where

$$\rho(\varpi,q) = 2\left[\frac{2(2-q)}{(3-\nu)q-2(1+\varpi)}\right]^{1-\frac{q}{2}} M_0^{-\frac{q}{2}}, \tag{7.55}$$

and $l(t)$ is defined in (7.11).

Proof. Let $\zeta = \frac{(3-\nu)q-2(1+\varpi)}{2-q}$. Then $\zeta > 0$. For $u \in X^\alpha$ and $T \ge T_0$, it follows from (7.7) and (7.55) and the Hölder inequality that

$$\begin{aligned}\int_{|t|>T} (1+|t|^\varpi)|u(t)|^q dt &\le 2\left(\int_{|t|>T} |t|^{-\frac{(2-\nu)q-2\varpi}{2-q}} dt\right)^{1-\frac{q}{2}} \left(\int_{|t|>T} |t|^{2-\nu}|u(t)|^2 dt\right)^{\frac{q}{2}} \\ &= 2\left(\int_{|t|>T} |t|^{-(\zeta+1)} dt\right)^{1-\frac{q}{2}} \left(\int_{|t|>T} |t|^{2-\nu}|u(t)|^2 dt\right)^{\frac{q}{2}} \\ &\le 2\left(\frac{2}{\zeta T^\zeta}\right)^{1-\frac{q}{2}} \left(\frac{1}{M_0}\int_{|t|>T} (L(t)u(t),u(t))dt\right)^{\frac{q}{2}} \\ &\le \frac{2^{1+\frac{2-q}{2}}}{M_0^{\frac{q}{2}} \zeta^{\frac{2-q}{2}} T^{\frac{(3-\nu)q-2(1+\varpi)}{2}}} \|u\|^q_{X^\alpha} \\ &= \frac{\rho(\varpi,q)}{T^{\frac{(3-\nu)q-2(1+\varpi)}{2}}} \|u\|^q_{X^\alpha}.\end{aligned}$$

This shows that (7.53) holds. Hence, from (7.53) and (7.11) and the Hölder inequality, one can get

$$\begin{aligned}
&\int_{\mathbb{R}}(1+|t|^{\varpi})|u(t)|^q dt \\
&= \int_{|t|\le T}(1+|t|^{\varpi})|u(t)|^q dt + \int_{|t|>T}(1+|t|^{\varpi})|u(t)|^q dt \\
&\le \left(\int_{|t|\le T}(1+|t|^{\varpi})^{\frac{2}{2-q}}[l(t)]^{-\frac{q}{2-q}}dt\right)^{1-\frac{q}{2}}\left(\int_{|t|\le T} l(t)|u(t)|^2 dt\right)^{\frac{q}{2}} \\
&\quad + \frac{\rho(\varpi,q)}{T^{\frac{(3-\nu)q-2(1+\varpi)}{2}}}\|u\|_{X^\alpha}^q \\
&\le \left(\int_{|t|\le T}(1+|t|^{\varpi})^{\frac{2}{2-q}}[l(t)]^{-\frac{q}{2-q}}dt\right)^{1-\frac{q}{2}}\|u\|_{X^\alpha}^q \\
&\quad + \frac{\rho(\varpi,q)}{T^{\frac{(3-\nu)q-2(1+\varpi)}{2}}}\|u\|_{X^\alpha}^q.
\end{aligned}$$

This shows that (7.54) holds.

Finally, by similar argument in the proof of Lemma 7.2, it is easy to show that X^α is compactly embedded in $L^q(\mathbb{R},\mathbb{R}^n)$. Therefore, the proof is completed. □

In this case Lemma 7.5 holds again with replacing (W1) and (W2) by (W6) and (W7), and in view of (W6) and (7.54), we have

$$\begin{aligned}
\left|\int_{\mathbb{R}} W(t,u(t))dt\right| &\le \sum_{i=1}^{r} g_i\int_{\mathbb{R}}(1+|t|^{\varpi_i})|u(t)|^{\tau_i}dt \\
&\le \sum_{i=1}^{r} g_i\Bigg[\left(\int_{|t|\le T}(1+|t|^{\varpi_i})^{\frac{2}{2-\tau_i}}[l(t)]^{\frac{-\tau_i}{2-\tau_i}}dt\right)^{1-\frac{\tau_i}{2}} \\
&\quad + \frac{\rho(\varpi_i,\tau_i)}{T^{\frac{(3-\nu)\gamma_i-2(1+\varpi_i)}{2}}}\Bigg]\|u\|_{X^\alpha}^{\tau_i} \\
&\le \sum_{i=1}^{r}\Pi_i(T)\|u\|_{X^\alpha}^{\tau_i},
\end{aligned} \tag{7.56}$$

where

$$\Pi_i(T) := g_i\left[\left(\int_{|t|\le T}(1+|t|^{\varpi_i})^{\frac{2}{2-\tau_i}}[l(t)]^{\frac{-\tau_i}{2-\tau_i}}dt\right)^{1-\frac{\tau_i}{2}} + \frac{\rho(\varpi_i,\tau_i)}{T^{\frac{(3-\nu)\gamma_i-2(1+\varpi_i)}{2}}}\right].$$

Therefore, the proof of Theorems 7.3 and 7.4 are similar to Theorems 7.1 and 7.2, respectively, and are omitted.

We will use the following conditions on $W(t,u)$ to find infinitely many homoclinic solutions:

(W10) $\lim_{|u|\to\infty}\frac{W(t,u)}{|u|^2} = +\infty$ uniformly for all $t\in\mathbb{R}$.

(W11) There exists $\varrho > 0$ such that $W(t,u) \geq -\varrho|u|^2$ for all $(t,u) \in \mathbb{R} \times \mathbb{R}^n$.
(W12) $W(t,0) = 0$ and there exist $D > 0$ and $\gamma_j > 2$ $(j = 1, \ldots, l)$ such that

$$|\nabla W(t,u)| \leq D\Big(|u| + \sum_{j=1}^{l} |u|^{\gamma_j - 1}\Big), \quad \forall\, (t,u) \in \mathbb{R} \times \mathbb{R}^n.$$

(W13) There exist $\rho > 0$, p_j, $q_j > 0$, $0 \leq \sum_{j=1}^{l} q_j < \frac{\rho-2}{2}$ and $0 < \theta_j < 2$ $(j = 1, \ldots, l)$ such that for $\forall (t,u) \in \mathbb{R} \times \mathbb{R}^n$,

$$(\nabla W(t,u), u) - \rho W(t,u) \geq -\sum_{j=1}^{l} \Big[p_j|u|^2 + q_j(L(t)u,u) + M_j(t)|u|^{\theta_j}\Big],$$

where $M_j \in L^{\frac{2}{2-\theta_j}}(\mathbb{R}, \mathbb{R}^+)$ $(j = 1, \ldots, l)$.
(W14) There exist $\vartheta \geq \gamma_j - 1$ $(j = 1, \ldots, l), c > 0$ and $R_1 > 0$ such that

$$(\nabla W(t,u), u) - 2W(t,u) \geq c|u|^{\vartheta}, \quad \forall\, t \in \mathbb{R}, \ \forall\, |u| \geq R_1,$$
$$(\nabla W(t,u), u) \geq 2W(t,u), \quad \forall\, t \in \mathbb{R}, \ \forall\, |u| \leq R_1.$$

Remark 7.1. In view of (W12), we have

$$W(t,u) = \int_0^1 (\nabla W(t,su), u)ds \leq D\Big(\frac{1}{2}|u|^2 + \sum_{j=1}^{l} \frac{1}{\gamma_j}|u|^{\gamma_j}\Big), \quad \forall\, (t,u) \in \mathbb{R} \times \mathbb{R}^n.$$

Now, we can state our main results.

Theorem 7.5. *Suppose that L and W satisfy (L), (W5) and (W10)–(W13). Then, system* (7.1) *possesses an unbounded sequence of homoclinic solutions.*

Proof. Let $\{e_j\}_{j=1}^{\infty}$ be the standard orthogonal basis of X^α and define $X_j := \mathbb{R}e_j$, then Z_k and Y_k can be defined as that in Theorem 1.17. From (7.22) and (W5), we can obtain that $\Phi \in C^1(X^\alpha, \mathbb{R})$ is even. Let us prove that the functional Φ satisfies the required conditions in Theorem 1.17.

We firstly verify condition (H2) in Theorem 1.17. Let

$$\lambda_k = \sup_{u \in Z_k, \|u\|_{X^\alpha}=1} \|u\|_{L^2},$$
$$\beta_k^j = \sup_{u \in Z_k, \|u\|_{X^\alpha}=1} \|u\|_{L^{\gamma_j}}, \quad \text{for any } j = 1, \ldots, l,$$

then $\lambda_k \to 0$ and $\beta_k^j \to 0$ as $k \to +\infty$ for any $j = 1, \ldots, l$. Clearly the sequence $\{\lambda_k\}$ is nonnegative and nonincreasing, so we assume that $\lambda_k \to \overline{\lambda} \geq 0$, $k \to +\infty$. For every $k \geq 0$, there exists $u_k \in Z_k$ such that $\|u_k\|_{X^\alpha} = 1$ and $\|u_k\|_{L^2} > \frac{\lambda_k}{2}$. Then, up to a subsequence, we may assume that $u_k \rightharpoonup u$ weakly in X^α. Noticing that Z_k is a closed subspace of X^α, by Mazur theorem, we have $u \in Z_k$, for all $k > \widetilde{n}$. Consequently, we get $u \in \bigcap_{k=\widetilde{n}}^{\infty} Z_k = \{0\}$, which implies $u_k \rightharpoonup 0$ weakly in X^α. By Lemma 7.3, we have $u_k \to 0$ in $L^2(\mathbb{R}, \mathbb{R}^n)$. Thus we have proved that

$\overline{\lambda} = 0$. Similarly, we can prove that $\beta_k^j \to 0$ as $k \to +\infty$ for any $j = 1, \ldots, l$. In view of (7.21) and (W3), one can get

$$\begin{aligned}\Phi(u) &= \frac{1}{2}\|u\|_{X^\alpha}^2 - \int_{\mathbb{R}} W(t,u(t))dt \\ &\geq \frac{1}{2}\|u\|_{X^\alpha}^2 - D\left(\frac{1}{2}\|u\|_{L^2}^2 + \sum_{j=1}^{l}\frac{1}{\gamma_j}\|u\|_{L^{\gamma_j}}^{\gamma_j}\right) \\ &\geq \frac{1}{2}\|u\|_{X^\alpha}^2 - \frac{1}{2}D\lambda_k^2\|u\|_{X^\alpha}^2 - D\sum_{j=1}^{l}\frac{1}{\gamma_j}\beta_k^{\gamma_j}\|u\|_{X^\alpha}^{\gamma_j}.\end{aligned} \tag{7.57}$$

Since $\lambda_k \to 0$ as $k \to +\infty$, there exists a positive constant N_0 such that

$$D\lambda_k^2 \leq \frac{1}{2}, \quad \forall\, k \geq N_0. \tag{7.58}$$

By (7.57) and (7.58), we have

$$\Phi(u) \geq \frac{1}{4}\|u\|_{X^\alpha}^2 - D\sum_{j=1}^{l}\frac{1}{\gamma_j}\beta_k^{\gamma_j}\|u\|_{X^\alpha}^{\gamma_j}, \quad \forall\, k \geq N_0. \tag{7.59}$$

If we choose $r_k = \frac{1}{l}\max\left\{\left(8\frac{D}{\gamma_1}\beta_k^{\gamma_1}\right)^{\frac{1}{2-\gamma_1}}, \ldots, \left(8\frac{D}{\gamma_l}\beta_k^{\gamma_l}\right)^{\frac{1}{2-\gamma_l}}\right\}$, then

$$b_k = \inf_{u\in Z_k, \|u\|_{X^\alpha}=r_k} \Phi(u) \geq \frac{1}{8}r_k^2, \quad \forall\, k \geq N_0. \tag{7.60}$$

Since $\beta_k \to 0$ as $k \to \infty$ and $\gamma_j > 2$ for any $j = 1, \ldots, l$, we can obtain

$$b_k \to +\infty, \quad \text{as } k \to \infty.$$

We now verify condition (H1) in Theorem 1.17. Since $\dim Y_k < \infty$ and all norms of a finite-dimensional normed space are equivalent, there exists a constant $M_0 > 0$ such that

$$\|u\|_{X^\alpha} \leq M_0\|u\|_{L^2}, \quad \forall\, u \in Y_k. \tag{7.61}$$

By (W1), for $\varepsilon_1 = \left(1 + \varrho l_{\min}^{-1}\right)M_0^2$ where ϱ is given in (W2), there exists $\delta = \delta(\varepsilon_1) > 0$ such that

$$W(t,u) \geq \varepsilon_1|u|^2, \quad \forall\, |u| \geq \delta, \quad \forall\, t \in \mathbb{R}. \tag{7.62}$$

Then, for any $u \in Y_k$, in view of (7.19), (7.21) and (7.62), one has

$$\begin{aligned}\Phi(u) &= \frac{1}{2}\|u\|_{X^\alpha}^2 - \int_{\mathbb{R}} W(t,u(t))dt \\ &= \frac{1}{2}\|u\|_{X^\alpha}^2 - \int_{\{t\in\mathbb{R};\ |u(t)\geq\delta\}} W(t,u(t))dt - \int_{\{t\in\mathbb{R};\ |u(t)<\delta\}} W(t,u(t))dt \\ &\leq \frac{1}{2}\|u\|_{X^\alpha}^2 - \varepsilon_1\|u\|_{L^2}^2 + \varrho\|u\|_{L^2}^2 \\ &\leq \frac{1}{2}\|u\|_{X^\alpha}^2 - \frac{\varepsilon_1}{M_0^2}\|u\|_{X^\alpha}^2 + \varrho l_{\min}^{-1}\|u\|_{X^\alpha}^2\end{aligned}$$

$$= \left(\frac{1}{2} - \frac{\varepsilon_1}{M_0^2} + \varrho l_{\min}^{-1}\right) \|u\|_{X^\alpha}^2 = -\frac{1}{2}\|u\|_{X^\alpha}^2.$$

Hence, we can choose $\|u\|_{X^\alpha} = \varrho_k$ large enough ($\varrho_k > r_k > 0$) such that

$$a_k = \max_{u \in Y_k, \|u\| = \varrho_k} \Phi(u) \leq 0.$$

Finally, we prove that Φ satisfies the Palais-Smale condition. Let $\{u_i\}_{i\in\mathbb{N}} \subset X^\alpha$ be a Palais-Smale sequence, that is, $\{\Phi(u_i)\}_{n\in\mathbb{N}}$ is bounded and $\Phi'(u_i) \to 0$ as $i \to +\infty$. Then there exists a constant $M_1 > 0$ such that

$$|\Phi(u_i)| \leq M_1, \quad \|\Phi'(u_i)\|_{(X^\alpha)^*} \leq M_1 \tag{7.63}$$

for every $i \in \mathbb{N}$, where $(X^\alpha)^*$ is the dual space of X^α.

We now prove that $\{u_i\}$ is bounded in X^α. In fact, if not, we may assume that by contradiction that $\|u_i\|_{X^\alpha} \to \infty$ as $i \to +\infty$. Set $v_i = \frac{u_i}{\|u_i\|_{X^\alpha}}$. Clearly, $\|v_i\|_{X^\alpha} = 1$ and there is $v_0 \in X^\alpha$ such that, up to a subsequence

$$\begin{cases} v_i \rightharpoonup v_0, & \text{weakly in } X^\alpha, \\ v_i \to v_0, & \text{strongly in } L^q(\mathbb{R}, \mathbb{R}^n), \quad 2 \leq q < +\infty, \end{cases} \tag{7.64}$$

as $i \to +\infty$. Since $v_i \rightharpoonup v_0$ in X^α, it is easy to verify that

$$\lim_{i\to+\infty} v_i(t) = v_0(t), \quad \forall\, t \in \mathbb{R}. \tag{7.65}$$

Now, we consider the following two cases:

Case I. $v_0 = 0$. From (7.19), (7.63), (W13) and the Hölder inequality, we can obtain

$$\begin{aligned}
&\rho M_1 + M_1\|u_i\|_{X^\alpha} \geq \rho\Phi(u_i) - \Phi'(u_i)u_i \\
&= \left(\frac{\rho}{2} - 1\right)\|u_i\|_{X^\alpha}^2 + \int_{\mathbb{R}} [(\nabla W(t, u_i(t)), u_i(t)) - \rho W(t, u_i(t))]dt \\
&\geq \left(\frac{\rho}{2} - 1\right)\|u_i\|_{X^\alpha}^2 \\
&\quad - \sum_{j=1}^{l} \int_{\mathbb{R}} \left[p_j|u_i(t)|^2 + q_j(L(t)u_i(t), u_i(t)) + M_j(t)|u_i(t)|^{\theta_j}\right]dt \\
&\geq \left(\frac{\rho-2}{2} - \sum_{j=1}^{l} q_j\right)\|u_i\|_{X^\alpha}^2 - \sum_{j=1}^{l} p_j\|u_i\|_{L^2}^2 - \sum_{j=1}^{l} \|M_j\|_{L^{\frac{2}{2-\theta_j}}}\|u_i\|_{L^2}^{\theta_j} \\
&\geq \left(\frac{\rho-2}{2} - \sum_{j=1}^{l} q_j\right)\|u_i\|_{X^\alpha}^2 - \sum_{j=1}^{l} p_j\|u_i\|_{L^2}^2 \\
&\quad - \sum_{j=1}^{l} \|M_j\|_{L^{\frac{2}{2-\theta_j}}}(l_{\min}^{-1})^{\frac{\theta_j}{2}}\|u_i\|_{X^\alpha}^{\theta_j}.
\end{aligned} \tag{7.66}$$

Divided by $\|u_i\|_{X^\alpha}^2$ on both sides of (7.66), noting that $0 \leq \sum_{j=1}^{l} q_j < \frac{\rho-2}{2}$ and $0 < \theta_j < 2$ $(j = 1, \ldots, l)$, one has

$$\|v_i\|_{L^2}^2 \geq \frac{\frac{\rho-2}{2} - \sum_{j=1}^{l} q_j}{\sum_{j=1}^{l} p_j} > 0, \quad \text{as} \quad i \to \infty. \tag{7.67}$$

It follows from (7.64) and (7.67) that $v_0 \neq 0$. This is a contradiction.

Case II. $v_0 \neq 0$. Since $\{\Phi(u_i)\}_{n\in\mathbb{N}}$ is bounded, then by (7.63), we have

$$\Phi(u_i) = \frac{1}{2}\|u_i\|^2_{X^\alpha} - \int_{\mathbb{R}} W(t, u_i(t))dt \geq -M_1. \tag{7.68}$$

Divided by $\|u_i\|^2_{X^\alpha}$ on both sides of (7.68), we have

$$\int_{\mathbb{R}} \frac{W(t, u_i(t))}{\|u_i\|^2_{X^\alpha}} dt \leq \frac{1}{2} + \frac{M_1}{\|u_i\|^2_{X^\alpha}} < +\infty. \tag{7.69}$$

Let $\Lambda := \{t \in \mathbb{R} : \ v_0(t) \neq 0\}$, then $\Lambda \neq \emptyset$. Hence, by (7.65), we can obtain

$$\lim_{i\to+\infty} u_i(t) = \lim_{i\to+\infty} v_i(t)\|u_i\|_{X^\alpha} = +\infty, \quad \forall\, t \in \Lambda.$$

Combining (W10) and (W11), one has

$$\lim_{i\to+\infty} \left(\frac{W(t, u_i(t))}{|u_i(t)|^2} + \varrho\right) |v_i(t)|^2 = +\infty, \quad \forall\, t \in \Lambda. \tag{7.70}$$

So, by (W11), (7.64), (7.70) and Fatou lemma, one can get

$$\begin{aligned}
\int_{\mathbb{R}} \frac{W(t, u_i(t))}{\|u_i\|^2_{X^\alpha}} dt &= \int_{t\in\Lambda} \frac{W(t, u_i(t))}{\|u_i\|^2_{X^\alpha}} dt + \int_{t\in\mathbb{R}\setminus\Lambda} \frac{W(t, u_i(t))}{\|u_i\|^2_{X^\alpha}} dt \\
&\geq \int_{t\in\Lambda} \frac{W(t, u_i(t))}{\|u_i\|^2_{X^\alpha}} dt - \varrho \int_{t\in\mathbb{R}\setminus\Lambda} |v_i(t)|^2 dt \\
&= \int_{t\in\Lambda} \frac{W(t, u_i(t)) + \varrho|u_i(t)|^2}{|u_i(t)|^2} |v_i(t)|^2 dt - \varrho \int_{\mathbb{R}} |v_i(t)|^2 dt \\
&\to +\infty, \quad \text{as } i \to +\infty.
\end{aligned}$$

This contradicts (7.69). Therefore, $\{u_i\}$ is bounded in X^α, that is, there exists $\xi_1 > 0$ such that

$$\|u_i\|_{X^\alpha} \leq \xi_1. \tag{7.71}$$

Then the sequence $\{u_i\}$ has a subsequence, again denoted by $\{u_i\}$, and there exists $u \in X^\alpha$ such that $u_i \rightharpoonup u$ in X^α. Hence we will prove that $u_i \to u$ in X^α.

By (W13), (7.19) and (7.71), we have

$$
\begin{aligned}
&\int_{\mathbb{R}} (\nabla W(t,u_i(t)) - \nabla W(t,u(t)), u_i(t) - u(t))dt \\
&\le \int_{\mathbb{R}} (|\nabla W(t,u_i(t))| + |\nabla W(t,u(t))|)|u_i(t) - u(t)|dt \\
&\le D\int_{\mathbb{R}} \Big(|u_i(t)| + \sum_{j=1}^{l} |u_i(t)|^{\gamma_j - 1}\Big)|u_i(t) - u(t)|dt \\
&\quad + D\int_{\mathbb{R}} \Big((|u(t)| + \sum_{j=1}^{l} |u(t)|^{\gamma_j - 1}\Big)|u_i(t) - u(t)|dt \\
&\le D\Big(\|u_i\|_{L^2} + \sum_{j=1}^{l} \|u_i\|_{L^{2(\gamma_j-1)}}^{\gamma_j - 1}\Big)\|u_i - u\|_{L^2} \\
&\quad + D\Big(\|u\|_{L^2} + \sum_{j=1}^{l} \|u\|_{L^{2(\gamma_j-1)}}^{\gamma_j - 1}\Big)\|u_i - u\|_{L^2} \\
&\le D\Big(\sqrt{l_{\min}^{-1}}\|u_i\|_{X^\alpha} + \sum_{j=1}^{l} \sqrt{l_{\min}^{-1}} C_\alpha^{\gamma_j - 2}\|u_i\|_{X^\alpha}^{\gamma_j - 1}\Big)\|u_i - u\|_{L^2} \\
&\quad + D\Big(\|u\|_{L^2} + \sum_{j=1}^{l} \|u\|_{L^{2(\gamma_j-1)}}^{\gamma_j - 1}\Big)\|u_i - u\|_{L^2} \\
&\le D\Big(\sqrt{l_{\min}^{-1}}\xi_1 + \sum_{j=1}^{l} \sqrt{l_{\min}^{-1}} C_\alpha^{\gamma_j - 2}\xi_1^{\gamma_j - 1}\Big)\|u_i - u\|_{L^2} \to 0, \quad \text{as } i \to +\infty.
\end{aligned}
\tag{7.72}
$$

It follows from $u_i \rightharpoonup u$ weakly in X^α and (7.72) that

$$
(\Phi'(u_i) - \Phi'(u), u_i - u) = \|u_i - u\|_{X^\alpha}^2 - \int_{\mathbb{R}} (\nabla W(t,u_i(t)) - \nabla W(t,u(t)), u_i(t) - u(t))dt.
$$

It is easy to deduce that $\|u_i - u\|_{X^\alpha} \to 0$ as $i = +\infty$. Therefore, Φ satisfies the Palais-Smale condition.

Therefore, it follows from Theorem 1.17 that Φ possesses an unbounded sequence $\{d_i\}$ of critical values with $d_i = \Phi(u_i)$, where u_i is such that $\Phi'(u_i) = 0$ for $i = 1, 2, \ldots$. If $\|u_i\|_{X^\alpha}$ is bounded, then there exists $R > 0$ such that

$$
\|u_i\|_{X^\alpha} \le R, \quad \text{for } i \in \mathbb{N}. \tag{7.73}
$$

Hence, by virtue of (7.19) and (W12), we have

$$
\begin{aligned}
\frac{1}{2}\|u_i\|_{X^\alpha}^2 &= d_i + \int_{\mathbb{R}} W(t,u_i(t))dt \\
&\ge d_i - D\int_{\mathbb{R}} \Big(\frac{1}{2}|u_i(t)|^2 + \sum_{j=1}^{l} \frac{1}{\gamma_j}|u_i(t)|^{\gamma_j}\Big)dt
\end{aligned}
$$

$$\geq d_i - D\Big(\frac{1}{2}l_{\min}^{-1}\|u_i\|_{X^\alpha}^2 + \sum_{j=1}^{l}\frac{1}{\gamma_j}l_{\min}^{-1}C_\alpha^{\gamma_j-2}\|u_i\|_{X^\alpha}^{\gamma_j}\Big).$$

Thus, this follows that

$$d_i \leq \frac{1}{2}\|u_i\|_{X^\alpha}^2 + D\Big(\frac{1}{2}l_{\min}^{-1}\|u_i\|_{X^\alpha}^2 + \sum_{j=1}^{l}\frac{1}{\gamma_j}l_{\min}^{-1}C_\alpha^{\gamma_j-2}\|u_i\|_{X^\alpha}^{\gamma_j}\Big) < +\infty.$$

This contradicts the fact that $\{d_i\}$ is unbounded, and so $\|u_i\|_{X^\alpha}$ is unbounded. The proof is completed. □

Theorem 7.6. *Suppose that L and W satisfy (L), (W5), (W10)–(W12) and (W14). Then, system* (7.1) *possesses an unbounded sequence of homoclinic solutions.*

Proof. By a similar argument as that in Theorem 7.2, we can prove Theorem 7.6. In fact, we only need to prove that Φ satisfies the Palais-Smale condition. Let $\{u_i\}_{i\in\mathbb{N}} \subset X^\alpha$ be a Palais-Smale sequence, that is, $\{\Phi(u_i)\}_{n\in\mathbb{N}}$ is bounded and $\Phi'(u_i) \to 0$ as $i \to +\infty$. We now prove that $\{u_i\}$ is bounded in X^α. In fact, if not, we may assume that by contradiction that $\|u_i\|_{X^\alpha} \to \infty$ as $i \to +\infty$. We take v_i as in the proof of Theorem 7.2.

Case I. $v_0 = 0$. From (W14), one has

$$\begin{aligned} 2\Phi(u_i) - \Phi'(u_i)u_i &= \int_{\mathbb{R}}[(\nabla W(t,u_i(t)), u_i(t)) - 2W(t,u_i(t))]dt \\ &\geq \int_{\{t\in\mathbb{R}, |u_i(t)|\geq R_1\}}[(\nabla W(t,u_i(t)), u_i(t)) - 2W(t,u_i(t))]dt \\ &\geq c\int_{\{t\in\mathbb{R}, |u_i(t)|\geq R_1\}}|u_i(t)|^\vartheta dt, \end{aligned} \tag{7.74}$$

which implies that

$$\frac{\int_{t\in\mathbb{R}, |u_i(t)|\geq R_1}|u_i(t)|^\vartheta dt}{\|u_i\|_{X^\alpha}} \to 0, \quad \text{as} \quad i \to \infty. \tag{7.75}$$

It follows from (7.6), (W12), (W14) and Remark 7.1 that

$$\begin{aligned}
M_2 &\geq \Phi(u_i)\\
&= \frac{1}{2}\|u_i\|_{X^\alpha}^2 - \int_{\mathbb{R}} W(t,u_i(t))dt\\
&\geq \frac{1}{2}\|u_i\|_{X^\alpha}^2 - D\int_{\mathbb{R}} \left(\frac{1}{2}|u_i(t)|^2 + \sum_{j=1}^{l}\frac{1}{\gamma_j}|u_i(t)|^{\gamma_j}\right)dt\\
&\geq \frac{1}{2}\|u_i\|_{X^\alpha}^2 - \frac{1}{2}D\|u_i\|_{L^2}^2 - D\sum_{j=1}^{l}\frac{1}{\gamma_j}\int_{\{t\in\mathbb{R},|u_i(t)|\geq R_1\}}|u_i(t)|^{\gamma_j}dt\\
&\quad - D\sum_{j=1}^{l}\frac{1}{\gamma_j}\int_{\{t\in\mathbb{R},|u_i(t)|<R_1\}}|u_i(t)|^{\gamma_j}dt\\
&\geq \frac{1}{2}\|u_i\|_{X^\alpha}^2 - \frac{1}{2}D\|u_i\|_{L^2}^2 - D\|u_i\|_\infty\sum_{j=1}^{l}\frac{1}{\gamma_j}\int_{\{t\in\mathbb{R},|u_i(t)|\geq R_1\}}|u_i(t)|^{\gamma_j-1}dt\\
&\quad - D\sum_{j=1}^{l}\frac{1}{\gamma_j}R_1^{\gamma_j-2}\int_{\{t\in\mathbb{R},|u_i(t)|<R_1\}}|u_i(t)|^2dt\\
&\geq \frac{1}{2}\|u_i\|_{X^\alpha}^2 - \frac{1}{2}D\|u_i\|_{L^2}^2 - D\|u_i\|_\infty\sum_{j=1}^{l}\frac{1}{\gamma_j}R_1^{\gamma_j-\vartheta-1}\int_{\{t\in\mathbb{R},|u_i(t)|\geq R_1\}}|u_i(t)|^{\vartheta}dt\\
&\quad - D\sum_{j=1}^{l}\frac{1}{\gamma_j}R_1^{\gamma_j-2}\int_{\{t\in\mathbb{R},|u_i(t)|<R_1\}}|u_i(t)|^2dt\\
&\geq \frac{1}{2}\|u_i\|_{X^\alpha}^2 - D\left(\frac{1}{2}+\sum_{j=1}^{l}\frac{1}{\gamma_j}R_1^{\gamma_j-2}\right)\|u_i\|_{L^2}^2\\
&\quad - DC_\alpha\|u_i\|_{X^\alpha}\sum_{j=1}^{l}\frac{1}{\gamma_j}R_1^{\gamma_j-\vartheta-1}\int_{\{t\in\mathbb{R},|u_i(t)|\geq R_1\}}|u_i(t)|^{\vartheta}dt, \qquad (7.76)
\end{aligned}$$

for some $M_2 > 0$. Divided by $\|u_i\|_{X^\alpha}^2$ on both sides of (7.76), noting that (7.75), we have

$$\|v_i\|_{L^2}^2 \geq \frac{1}{2D\left(\frac{1}{2}+\sum_{j=1}^{l}\frac{1}{\gamma_j}R_1^{\gamma_j-2}\right)} > 0, \quad \text{as} \quad i\to\infty. \qquad (7.77)$$

It follows from (7.64) and (7.77) that $v_0 \neq 0$. This is a contradiction.

Case II. $v_0 \neq 0$. The proof is the same as that in Theorem 7.2, and we omit it here. Hence, $\{u_i\}$ is bounded in X^α. Similar to the proof of Theorem 7.2, we can prove that $\{u_i\}$ has a convergent subsequence in X^α. Hence, Φ satisfies the Palais-Smale condition. The proof is completed. □

7.3 Existence and Multiplicity of Homoclinic Solutions (II)

7.3.1 *Introduction*

In this section, we investigate the existence and multiplicity of the solutions of the following fractional Hamiltonian systems

$$\begin{cases} {}_tD_\infty^\alpha({}_{-\infty}D_t^\alpha u(t)) + L(t)u(t) = \nabla W(t, u(t)), & t \in \mathbb{R}, \\ u \in H^\alpha(\mathbb{R}), \end{cases} \tag{7.78}$$

where $\alpha \in (\frac{1}{2}, 1]$, ${}_{-\infty}D_t^\alpha$ and ${}_tD_\infty^\alpha$ are left and right Liouville-Weyl fractional derivatives of order $0 < \alpha < 1$ on the whole axis $\mathbb{R}$ respectively, $u \in \mathbb{R}^n$, $L \in C(\mathbb{R}, \mathbb{R}^{n\times n})$ is a positive definite symmetric matrix for all $t \in \mathbb{R}$ and $W : \mathbb{R} \times \mathbb{R}^n \to \mathbb{R}$ is a function that satisfy conditions which will be stated later and $\nabla W(t, u)$ is the gradient of W at u.

Throughout this section, we need the following assumptions:

[L]: There exists an $M > 0$ such that

$$(L(t)x, x) \geq M|x|^2, \quad \text{for all } t \in \mathbb{R},\ x \in \mathbb{R}^n;$$

and $W(t, x) = b(t)\pi(x)$ with the following assumptions:

[W1]: $b : \mathbb{R} \to (0, +\infty)$ is a continuous function such that $b(t) \to 0$ as $|t| \to \infty$;
[W2]: $\pi \in C^1(\mathbb{R}^n, \mathbb{R})$ and there is a constant $\mu > 2$ such that

$$0 < \mu\pi(x) \leq (\nabla\pi(x), x), \quad \forall\, x \in \mathbb{R}^n \setminus \{0\};$$

[W3]: $\nabla\pi(x) = o(|x|)$ as $|x| \to 0$;
[W4]: $\pi(-x) = \pi(x)$, for all $x \in \mathbb{R}^n$;
[W5]: for any $r > 0$, there exits $\alpha_0, \beta_0 > 0$ and $\varrho < 2$ such that

$$0 \leq \left(2 + \frac{1}{\alpha_0 + \beta_0|x|^\varrho}\right) W(t, x) \leq (\nabla W(t, x), x),$$

for all $(t, x) \in \mathbb{R} \times \{x \in \mathbb{R}^n : |x| \geq r\}$.

In this section, we present a new approach via variational methods and critical point theory to obtain sufficient conditions under which the Hamiltonian system has at least one homoclinic solution or multiple homoclinic solutions.

7.3.2 *Some Lemmas*

In order to establish our results via variational methods and the critical point theory, we firstly describe some properties of the space on which the variational associated with (7.1) is defined. Let

$$X^\alpha = \left\{ u \in H^\alpha(\mathbb{R}) \middle| \int_{\mathbb{R}} \left(|{}_{-\infty}D_t^\alpha u(t)|^2 + (b(t)u(t), u(t)) \right) dt < \infty \right\}.$$

The space X^α is a Hilbert space with the inner product

$$\langle u, v\rangle_{X^\alpha} = \int_{\mathbb{R}} \Big(({}_{-\infty}D_t^\alpha u(t), {}_{-\infty}D_t^\alpha v(t)) + (L(t)u(t), v(t)) \Big) dt,$$

and the corresponding norm

$$\|u\|_{X^\alpha}^2 = \langle u, u\rangle_{X^\alpha}.$$

Similar to the proofs of Lemmas 2.1 and 2.2 in Torres, 2013, we can get the following lemmas.

Lemma 7.6. *Suppose that $L(t)$ satisfies [L]. Then the space X^α is continuously embedded in $H^\alpha(\mathbb{R})$.*

Let $L_b^p(\mathbb{R}, \mathbb{R}^n)$ denote the weighted space of measurable functions $u : \mathbb{R} \to \mathbb{R}^n$ with the norm

$$\|u\|_{p,b} := \left(\int_{\mathbb{R}} b(t)|u(t)|^p dt \right)^{\frac{1}{p}}.$$

Lemma 7.7. *Suppose that [L] and [W1] hold. Then the imbedding of X^α in $L_b^2(\mathbb{R}, \mathbb{R}^n)$ is continuous and compact.*

Proof. It is easy to check that the embedding of $X^\alpha \hookrightarrow L_b^2(\mathbb{R}, \mathbb{R}^n)$ is continuous. Next, we prove that the embedding is compact. Let $\{u_n\}_{n\in\mathrm{N}} \subset X^\alpha$ be a sequence such that $u_n \rightharpoonup u$ in X^α, we show that $u_n \to u$ in $L_b^2(\mathbb{R}, \mathbb{R}^n)$. Suppose, without loss of generality, that $u_n \rightharpoonup 0$ in X^α. The Banach-Steinhaus theorem implies that

$$A = \sup_{n\in\mathrm{N}} \|u_n\| < \infty.$$

For any $\varepsilon > 0$, there is $T_0 < 0$ such that $b(t) \le \varepsilon$ for all $t \le T_0$. Similarly, there is $T_1 > 0$ such that $b(t) \le \varepsilon$ for all $t \ge T_1$. Sobolev theorem (see Stuart, 1995) implies that $u_n \to u$ uniformly on $\overline{\Omega} = [T_0, T_1]$, so there is k_0 such that

$$\int_\Omega b(t)|u_n(t)|^2 dt < \varepsilon, \quad \forall\, k \ge k_0. \tag{7.79}$$

Since $b(t) \le \varepsilon$ on $(-\infty, T_0]$, we have

$$\int_{-\infty}^{T_0} b(t)|u_n(t)|^2 dt \le \frac{\varepsilon}{M} \int_{-\infty}^{T_0} M|u_n(t)|^2 dt < \frac{\varepsilon}{M} A^2. \tag{7.80}$$

Similarly, since $b(t) \le \varepsilon$ on $(T_1, +\infty)$, one can get

$$\int_{T_1}^{+\infty} b(t)|u_n(t)|^2 dt \le \frac{\varepsilon}{M} \int_{T_1}^{+\infty} M|u_n(t)|^2 dt < \frac{\varepsilon}{M} A^2. \tag{7.81}$$

Combining (7.79)-(7.81), we get $u_n \to 0$ in $L_b^2(\mathbb{R}, \mathbb{R}^n)$. □

Remark 7.2. From Lemma 7.7, there is a constant C_b such that

$$\|u\|_{2,b} \le C_b \|u\|_{X^\alpha}, \quad \forall\, u \in X^\alpha.$$

Lemma 7.8. *(Torres, 2013) Let $\alpha > \frac{1}{2}$, then $H^\alpha(\mathbb{R}) \subset C(\mathbb{R})$ and there is a constant $C = C_\alpha$ such that*

$$\sup_{x\in\mathbb{R}} |u(x)| \leq C\|u\|_\alpha.$$

Also by Lemma 7.8, there is a constant $C_\alpha > 0$ such that

$$\|u\|_\infty \leq C_\alpha \|u\|_{X^\alpha}. \tag{7.82}$$

Let $I : X^\alpha \to \mathbb{R}$ be defined by

$$\begin{aligned} I(u) &:= \frac{1}{2}\int_{\mathbb{R}} \left(|{}_{-\infty}D_t^\alpha u(t)|^2 + (L(t)u(t), u(t))\right)dt - \int_{\mathbb{R}} W(t, u(t))dt \\ &:= \frac{1}{2}\|u\|_{X^\alpha}^2 - \int_{\mathbb{R}} W(t, u(t))dt. \end{aligned} \tag{7.83}$$

For convenience, we denote

$$\begin{aligned} J(u) &:= \frac{1}{2}\int_{\mathbb{R}} \left(|{}_{-\infty}D_t^\alpha u(t)|^2 + (L(t)u(t), u(t))\right)dt, \\ W(u) &:= \int_{\mathbb{R}} W(t, u(t))dt. \end{aligned}$$

Under the assumptions [L] and [W1]-[W5], with Lemma 3.1 in Riewe, 1996, we have

$$I'(u)v = \int_{\mathbb{R}} \left(({}_{-\infty}D_t^\alpha u(t), {}_{-\infty}D_t^\alpha v(t)) + (L(t)u(t), v(t)) - (\nabla W(t, u(t)), v(t))\right)dt, \tag{7.84}$$

for all $u, v \in X^\alpha$. Moreover, I is a continuously Fréchet-differentiable functional defined on X^α, i.e., $I \in C^1(X^\alpha, \mathbb{R})$.

Lemma 7.9. *Suppose that [L], [W1] and [W3] are satisfied. If $u_n \rightharpoonup u$ in X^α, then $\nabla\pi(u_n) \to \nabla\pi(u)$ in $L_b^2(\mathbb{R}, \mathbb{R}^n)$ as $n \to \infty$.*

Proof. The proof is similar to Lemma 2.4 in Torres, 2013 and is omitted. □

Now, from Lemma 7.8, it is well known that $X^\alpha \subset H^\alpha(\mathbb{R}, \mathbb{R}^n) \subset C(\mathbb{R}, \mathbb{R}^n)$, the space of continuous functions u on $\mathbb{R}$ such that $u(t) \to 0$ as $|t| \to +\infty$.

Lemma 7.10. *We have*

$$\begin{aligned} \pi(u) &\leq \pi\left(\frac{u}{|u|}\right)|u|^\mu, \quad |u| \leq 1, \\ \pi(u) &\geq \pi\left(\frac{u}{|u|}\right)|u|^\mu, \quad |u| \geq 1. \end{aligned} \tag{7.85}$$

The proof is similar to Lemma 2.3 in Riewe, 1996 and is omitted.

7.3.3 *Existence and Multiplicity*

The main results are the following theorems.

Theorem 7.7. *Let $\alpha > \frac{1}{2}$. Assume that L satisfies assumptions [L] and [W1]-[W3]. Then system (7.78) possesses a nontrivial homoclinic solution.*

Proof. It is clear that $I(0) = 0$. We show that I satisfies the hypotheses of the Theorem 1.15.

Step I. We show that I satisfies the Palais-Smale condition. Assume that $\{u_n\}_{n\in\mathrm{N}} \subset X^\alpha$ is a sequence such that $\{I(u_n)\}_{n\in\mathrm{N}}$ is bounded and $I'(u_n) \to 0$ as $n \to +\infty$. Then there exists a constant $c > 0$ such that

$$|I(u_n)| \le c, \qquad \|I'(u_n)\|_{X^\alpha} \le c, \qquad \text{for every } n \in \mathrm{N}. \tag{7.86}$$

We firstly prove that $\{u_n\}_{n\in\mathrm{N}}$ is bounded in X^α. By (7.83), (7.84), (7.86) and [W2], one can get

$$\begin{aligned}
c + \frac{c}{\mu}\|u_n\|_{X^\alpha} &\ge I(u_n) - \frac{1}{\mu}I'(u_n)u_n\\
&= \left(\frac{1}{2} - \frac{1}{\mu}\right)\int_{\mathbb{R}}\Big(|{}_{-\infty}D_t^\alpha u_n(t)|^2 + (L(t)u_n(t), u_n(t))\Big)dt\\
&\quad - \int_{\mathbb{R}} W(t, u_n(t))dt + \frac{1}{\mu}\int_{\mathbb{R}}(\nabla W(t, u_n(t)), u_n(t))dt\\
&= \left(\frac{1}{2} - \frac{1}{\mu}\right)\|u_n\|_{X^\alpha}^2 + \frac{1}{\mu}\int_{\mathbb{R}}((\nabla W(t, u_n(t)), u_n(t)) - \mu W(t, u_n(t)))dt\\
&\ge \left(\frac{1}{2} - \frac{1}{\mu}\right)\|u_n\|_{X^\alpha}^2, \qquad n \in \mathrm{N}.
\end{aligned}$$

Since $\mu > 2$, the above inequality shows that $\{u_n\}_{n\in\mathrm{N}}$ is bounded in X^α, i.e., that there exists a constant $\theta > 0$ such that

$$\|u_n\|_{X^\alpha} \le \theta, \qquad \text{for every } n \in \mathrm{N}.$$

Since X^α is a reflexive space (X^α is a Hilbert space), thus passing to a subsequence if necessary, by Lemma 7.7, we may assume that

$$u_n \rightharpoonup u, \quad \text{weakly in } X^\alpha, \quad u_n \to u, \quad \text{a.e. in } L_b^2(\mathbb{R}, \mathbb{R}^n).$$

Thus,

$$(I'(u_n) - I'(u))(u_n - u) \to 0, \qquad n \to \infty$$

and by Lemma 7.9 and the Hölder inequality, one can get

$$\int_{\mathbb{R}}(\nabla W(t, u_n(t)) - \nabla W(t, u(t)), u_n(t) - u(t))dt \to 0,$$

as $n \to +\infty$. On the other hand, we have

$$(I'(u_n) - I'(u))(u_n - u)$$

$$= \|u_n - u\|_{X^\alpha}^2 - \int_{\mathbb{R}} (\nabla W(t, u_n(t)) - \nabla W(t, u(t)), u_n(t) - u(t))dt.$$

Hence, $\|u_n - u\|_{X^\alpha} \to 0$ as $n \to +\infty$. Therefore, I satisfies Palais-Smale condition.

Step II. We now show that there exist constants $\rho > 0$ and $\alpha > 0$ such that I satisfies assumption (ii) of Theorem 1.15. For any given number $\varepsilon > 0$, from [W3], we can choose $\delta > 0$ such that

$$\pi(x) \leq \varepsilon |x|^2, \quad \text{for every } |x| \leq \delta.$$

By (7.82), if $\|u\|_{X^\alpha} = \frac{\delta}{C_\alpha} =: \rho$, then $|u(t)| \leq C_\alpha \cdot \rho = \delta$, so $\pi(u(t)) \leq \varepsilon |u(t)|^2$ for all $t \in \mathbb{R}$. Integrating on $\mathbb{R}$ and by Remark 7.2, we have

$$\int_{\mathbb{R}} W(t, u(t))dt \leq \varepsilon \|u\|_{2,b}^2 \leq \varepsilon C_b^2 \|u\|_{X^\alpha}^2. \tag{7.87}$$

Let

$$\beta = \frac{1}{4}\left(\frac{\delta}{C_\alpha}\right)^2.$$

For $\|u\|_{X^\alpha} = \rho \leq 1$, from (7.83) and (7.87), we obtain

$$\begin{aligned} I(u) &= \frac{1}{2}\int_{\mathbb{R}} \left(|{}_{-\infty}D_t^\alpha u(t)|^2 + (L(t)u(t), u(t))\right)dt - \int_{\mathbb{R}} W(t, u(t))dt \\ &\geq \frac{1}{2}\|u\|_{X^\alpha}^2 - \varepsilon C_b^2 \|u\|_{X^\alpha}^2 \\ &= \left(\frac{1}{2} - \varepsilon C_b^2\right)\|u\|_{X^\alpha}^2. \end{aligned} \tag{7.88}$$

Setting $\varepsilon = \frac{1}{4C_b^2}$, the inequality (7.88) implies that

$$I|_{\partial B_\rho} \geq \frac{1}{4}\left(\frac{\delta}{C_\alpha}\right)^2 = \beta. \tag{7.89}$$

Clearly, (7.89) shows that $\|u\|_{X^\alpha} = \rho$ implies $I(u) \geq \beta$, i.e., I satisfies assumption (ii) of Theorem 1.15.

Step III. We prove (iii) of Theorem 1.15, i.e., there exists $e \in X^\alpha$ such that $\|e\|_{X^\alpha} > \rho$ and $I(e) \leq 0$, where ρ is defined in Step II. By (7.85), there is $c_1 > 0$ such that

$$\pi(u(t)) \geq c_1 |u(t)|^\mu, \quad \text{for all } |u(t)| \geq 1. \tag{7.90}$$

Take some $u \in X^\alpha$ such that $\|u\|_{X^\alpha} = 1$. Then there exists a subset Ω of positive measure $|\Omega| < \infty$ of $\mathbb{R}$ such that $u(t) \neq 0$ for $t \in \Omega$. Take $\sigma > 0$ such that $\sigma|u(t)| \geq 1$ for $t \in \Omega$. Then by (7.83) and (7.90), one can get

$$I(\sigma u) \leq \frac{\sigma^2}{2} - c_1 \sigma^\mu \int_\Omega b(t)|u(t)|^\mu dt. \tag{7.91}$$

Since $\mu > 2$, $b(t) > 0$ and $\int_\Omega b(t)|u(t)|^\mu dt > 0$, (7.91) implies that $I(\sigma u) < 0$ for some $\sigma > 0$ with $\sigma|u(t)| \geq 1$ for $t \in \Omega$ and $\|\sigma u\|_{X^\alpha} > \rho$. Therefore, I possesses a critical value $c \geq \beta$ given by

$$c = \inf_{g \in \Gamma} \max_{s \in [0,1]} I(g(s)),$$

where $B_\rho(0)$ is an open ball in E of radius ρ centered at 0, and

$$\Gamma = \{g \in C([0,1], E) : \ g(0) = 0, \ g(1) = e\}.$$

Here, there is $u^* \in X^\alpha$ such that

$$I(u^*) = c, \quad I'(u^*) = 0.$$

Since $c > 0$, u^* is a nontrivial homoclinic solution. The proof is completed. □

Theorem 7.8. *Let $\alpha > \frac{1}{2}$. Assume that L and W satisfy [L] and [W1]-[W4]. Then there exists an unbounded sequence of homoclinic solutions for system* (7.78).

Proof. The conditions [W1] and [W4] imply that I is even. In view of the proof of Theorem 7.7, we see that $I \in C^1(X^\alpha, \mathbb{R})$, and I satisfies the Palais-Smale condition and assumptions (i) and (ii) of Theorem 1.16. To apply Theorem 1.16, it suffices to prove that I satisfies the condition (iv) of Theorem 1.16. Let $E' \subset X^\alpha$ be a finite-dimensional subspace. From Step III of Theorem 7.7, we know that, for any $u_0 \in E' \subset X^\alpha$ such that $\|u_0\|_{X^\alpha} = 1$, there is $m_{u_0} > 0$ such that

$$I(m_{u_0}) < 0, \quad \text{for } |m| \geq m_{u_0} > 0.$$

Since $E' \subset X^\alpha$ is a finite-dimensional subspace, we can choose an $R = r(E') > 0$ such that

$$I(\omega) < 0, \quad \forall\, \omega \in E' \setminus B_R(0).$$

Therefore, by Theorem 1.16, I possesses an unbounded sequence of critical values $\{c_j\}_{j\in \mathrm{N}}$ with $c_j \to +\infty$. Let u_j be the critical point of I corresponding to c_j, then (7.78) has infinitely many distinct homoclinic solutions. □

Theorem 7.9. *Let $\alpha > \frac{1}{2}$. Assume that L and W satisfy [L] and [W1], [W3]-[W5]. Then there exists an unbounded sequence of homoclinic solutions for system* (7.78).

Proof. In view of the proof of Theorem 7.8, we see that $I \in C^1(X^\alpha, \mathbb{R})$, and I satisfies assumptions (i), (ii) and (iv) of Theorem 1.16. To apply Theorem 1.16, it suffices to prove that I satisfies the condition (iv). Suppose that $\{u_n\}_{n\in\mathrm{N}} \subset X^\alpha$ is a (iv) sequence of I, that is, $\{I(u_n)\}$ is bounded and $(1 + \|u_n\|_{X^\alpha})\|I'(u_n)\|_{(X^\alpha)^*} \to 0$ as $n \to \infty$. Then, in view of (7.83) and (7.84), for a constant $C_0 > 0$, we have

$$\begin{aligned} C_0 &\geq 2I(u_n) - I'(u_n)u_n \\ &= \int_{\mathbb{R}} [(\nabla W(t, u_n(t)), u_n(t)) - 2W(t, u_n(t))]\, dt. \end{aligned} \tag{7.92}$$

Since $\pi(0) = 0$, then from [W3] that there exists $\eta \in]0, 1[$ such that

$$|W(t,x)| \leq \frac{1}{4} b(t)|x|^2, \quad \text{for every } t \in \mathbb{R}, \quad |x| \leq \eta. \tag{7.93}$$

By [W5], we have

$$(\nabla W(t,x), x) \geq 2W(t,x) \geq 0, \quad \forall (t,x) \in \mathbb{R} \times \mathbb{R}^n, \tag{7.94}$$

$$W(t,x) \leq (\alpha_0 + \beta_0|x|^{\varrho})\left[(\nabla W(t,x),x) - 2W(t,x)\right], \tag{7.95}$$

for all $(t,x) \in \mathbb{R} \times \{x \in \mathbb{R}^n : \ |x| > \eta\}$.

Now from (7.83), (7.82), (7.92)-(7.95) and Remark 7.2, we get

$$\begin{aligned}
&\frac{1}{2}\|u_n\|_{X^\alpha}^2 \\
&= I(u_n) + \int_{\mathbb{R}} W(t,u_n(t))dt \\
&= I(u_n) + \int_{\{t\in\mathbb{R},\ |u_n(t)|\leq\eta\}} W(t,u_n(t))dt + \int_{\{t\in\mathbb{R},\ |u_n(t)|>\eta\}} W(t,u_n(t))dt \\
&= I(u_n) + \frac{1}{4}\int_{\{t\in\mathbb{R},\ |u_n(t)|\leq\eta\}} b(t)|u_n(t)|^2 dt \\
&\quad + \int_{\{t\in\mathbb{R},\ |u_n(t)|>\eta\}} (\alpha_0 + \beta_0|u_n(t)|^{\varrho})\left[(\nabla W(t,u_n(t)),u_n(t)) - 2W(t,u_n(t))\right] dt \\
&\leq C_1 + \frac{1}{4}\|u_n\|_{2,b}^2 + \int_{\mathbb{R}} (\alpha_0 + \beta_0|u_n(t)|^{\varrho})\left[(\nabla W(t,u_n(t)),u_n(t)) - 2W(t,u_n(t))\right] dt \\
&\leq C_1 + \frac{1}{4}C_b^2\|u_n\|_{X^\alpha}^2 + C_0(\alpha_0 + \beta_0\|u_n\|_\infty^{\varrho}) \\
&\leq C_1 + \frac{1}{4}C_b^2\|u_n\|_{X^\alpha}^2 + C_0(\alpha_0 + \beta_0 C_\alpha^{\varrho}\|u_n\|_{X^\alpha}^{\varrho}).
\end{aligned}$$

Since $\varrho < 2$, it follows that $\{\|u_n\|\}$ is bounded. Next, similar to the proof of Theorem 7.7, we can also prove that $\{u_n\}$ has a convergent subsequence in X^α. Thus, I satisfies condition (iv). Therefore, the proof is completed. □

7.4 Notes and Remarks

The material in Section 7.2 are adopted Nyamoradi, Alsaedi, Ahmad and Zhou, 2017. The results in Section 7.3 are taken from Nyamoradi and Zhou, 2017.

Chapter 8

Fractional Partial Differential Equations

8.1 Introduction

Fractional calculus has become important topics thanks to its effective characterization of the ubiquitous power-law phenomena as well as its widespread applications in many areas of science and engineering such as porous media, turbulence, bioscience, geoscience, and viscoelastic material and so on. The most important mathematical equations among such models are fractional partial differential equations, which can be more relevant for describing the underlying anomalous features, non-local interactions, manifesting in memory-effects, sharp peaks, power law distributions, and self-similar structures. For such a kind of equations there are a large and rapidly growing number of publications. Although some results of qualitative analysis for fractional partial differential equations (FPDEs) can be similarly obtained, many classical PDEs' methods are hardly applicable directly to FPDEs. New theories and methods are thus required to be specifically developed for FPDEs, whose investigation becomes more challenging. Comparing with PDEs' classical theory, the researches on FPDEs are only on their initial stage of development.

The main objective of this chapter is to investigate the existence and regularity for a variety of time-fractional partial differential equations with applications. Section 8.2 is devoted to study of global and local existence, regularity of mild solutions for Navier-Stokes equations. In Section 8.3, an initial-boundary value problem for the nonlinear fractional Rayleigh-Stokes equation is studied in two cases, namely when the source term is globally Lipschitz or locally Lipschitz. In Section 8.4, we investigate the existence of a weak solution for Euler-Lagrange equations. In Section 8.5, we investigate the regularity and unique existence of the solution for initial-boundary value problems of diffusion equation with multiple time-fractional derivatives. Section 8.6 discusses the well-posedness and regularity results of the weak solution for a fractional wave equation.

8.2 Fractional Navier-Stokes Equations

8.2.1 *Introduction*

The Navier-Stokes equations describe the motion of the incompressible Newtonian fluid flows ranging from large scale atmospheric motions to the lubrication of ball bearings, and express the conservation of mass and momentum. For more details we refer to the monographs of Cannone, 1995 and Varnhorn, 1994. We find this system which is so rich in phenomena that the whole power of mathematical theory is needed to discuss the existence, regularity and boundary conditions; see, e.g., Lemarié-Rieusset, 2002 and von Wahl, 2013.

It is worth mentioning that Leray carried out an initial study that a boundary-value problem for the time-dependent Navier-Stokes equations possesses a unique smooth solution on some intervals of time provided the data are sufficiently smooth. Since then many results on the existence for weak, mild and strong solutions for the Navier-Stokes equations have been investigated intensively by many authors; see, e.g., de Almeida and Ferreira, 2013; Heck, Kim and Kozono, 2013; Iwabuchi and Takada, 2013; Koch *et al.*, 2009; Masmoudi and Wong, 2015 and Weissler, 1980. Moreover, one can find results on regularity of weak and strong solution from Amrouche and Rejaiba, 2014; Chemin and Gallagher, 2010; Chemin, Gallagher and Paicu, 2011; Choe, 2015; Danchin, 2000; Giga and Yoshikazu, 1991; Kozono, 1998; Raugel and Sell, 1993 and the references therein.

Theoretical analysis and experimental data have shown that classical diffusion equation fails to describe diffusion phenomenon in heterogeneous porous media that exhibits fractal characteristics. How is the classical diffusion equation modified to make it appropriate to depict anomalous diffusion phenomena? This problem is interesting for researchers. Fractional calculus have been found effective in modelling anomalous diffusion processes since it has been recognized as one of the best tools to characterize the long memory processes. Consequently, it is reasonable and significative to propose the generalized Navier-Stokes equations with Caputo fractional derivative operator, which can be used to simulate anomalous diffusion in fractal media. Its evolutions behave in a much more complex way than in classical inter-order case and the corresponding investigation becomes more challenging.

The main effort on time-fractional Navier-Stokes equations has been put into attempts to derive numerical solutions and analytical solutions; see Ganji *et al.*, 2010; El-Shahed and Salem, 2004 and Momani and Zaid, 2006. However, to the best of our knowledge, there are very few results on the existence and regularity of mild solutions for time-fractional Navier-Stokes equations. Recently, De Carvalho-Neto and Gabriela, 2015 dealt with the existence and uniqueness of global and local mild solutions for the time-fractional Navier-Stokes equations.

Motivated by above discussion, in this section we study the following time-fractional Navier-Stokes equations in an open set $\Omega \subset \mathbb{R}^n$ $(n \geq 3)$:

$$\begin{cases} \partial_t^\alpha u - \nu \Delta u + (u \cdot \nabla)u = -\nabla p + f, & t > 0, \\ \nabla \cdot u = 0, \\ u|_{\partial\Omega} = 0, \\ u(0, x) = a, \end{cases} \tag{8.1}$$

where ∂_t^α is Caputo fractional derivative of order $\alpha \in (0, 1)$, $u = (u_1(t, x), u_2(t, x), \ldots, u_n(t, x))$ represents the velocity field at a point $x \in \Omega$ and time $t > 0$, $p = p(t, x)$ is the pressure, ν the viscosity, $f = f(t, x)$ is the external force and $a = a(x)$ is the initial velocity. From now on, we assume that Ω has a smooth boundary.

Firstly, we get rid of the pressure term by applying Helmholtz projector P to equation (8.1), which converts equation (8.1) to

$$\begin{cases} \partial_t^\alpha u - \nu P \Delta u + P(u \cdot \nabla)u = Pf, & t > 0, \\ \nabla \cdot u = 0, \\ u|_{\partial\Omega} = 0, \\ u(0, x) = a. \end{cases}$$

The operator $-\nu P \Delta$ with Dirichlet type boundary conditions is, basically, the Stokes operator A in the divergence-free function space under consideration. Then we rewrite (8.1) as the following abstract form

$$\begin{cases} {}_0^C D_t^\alpha u = -Au + F(u, u) + Pf, & t > 0, \\ u(0) = a, \end{cases} \tag{8.2}$$

where $F(u, v) = -P(u \cdot \nabla)v$. If one can give sense to the Helmholtz projector P and the Stokes operator A, then the solution of equation (8.2) is also the solution of equation (8.1).

The objective of this section is to establish the existence and uniqueness of global and local mild solutions of problem (8.2) in $H^{\beta,q}$. Further, we prove the regularity results which state essentially that if Pf is Hölder continuous then there is a unique classical solution $u(t)$ such that Au and ${}_0^C D_t^\alpha u(t)$ are Hölder continuous in J_q.

In Subsection 8.2.2, we recall some notations, definitions, and preliminary facts. Subsection 8.2.3 is devoted to the existence and uniqueness of global mild solution in $H^{\beta,q}$ of problem (8.2). In Subsection 8.2.4, we proceed to study the local mild solution in $H^{\beta,q}$ and use the iteration method to obtain the existence and uniqueness of local mild solution in J_q of problem (8.2). Finally, Subsection 8.2.5 is concerned with the existence and regularity of classical solution in J_q of problem (8.2).

8.2.2 *Preliminaries*

In this subsection, we introduce notations, definitions, and preliminary facts which are used throughout this section.

Let $\Omega = \{(x_1, \ldots, x_n) : x_n > 0\}$ be open subset of $\mathbb{R}^n$, where $n \geq 3$. Let $1 < q < \infty$. Then there is a bounded projection P called on $(L^q(\Omega))^n$, whose range is the closure of

$$C_\sigma^\infty(\Omega) := \{u \in (C^\infty(\Omega))^n : \nabla \cdot u = 0, \ u \text{ has compact support in } \Omega\},$$

and whose null space is the closure of

$$\{u \in (C^\infty(\Omega))^n : u = \nabla\phi, \ \phi \in C^\infty(\Omega)\}.$$

For notational convenience, let $J_q := \overline{C_\sigma^\infty(\Omega)}^{|\cdot|_q}$, which is a closed subspace of $(L^q(\Omega))^n$. $(W^{m,q}(\Omega))^n$ is a Sobolev space with the norm $|\cdot|_{m,q}$.

$A = -\nu P\Delta$ denotes the Stokes operator in J_q whose domain is $D(A) = D(\Delta) \cap J_q$; here,

$$D(\Delta) = \{u \in (W^{2,q}(\Omega))^n : u|_{\partial\Omega} = 0\}.$$

It is known that $-A$ is a closed linear operator and generates the bounded analytic semigroup $\{e^{-tA}\}$ on J_q.

So as to state our results, we need to introduce the definitions of the fractional power spaces associated with $-A$. For $\beta > 0$ and $u \in J_q$, define

$$A^{-\beta}u = \frac{1}{\Gamma(\beta)} \int_0^\infty t^{\beta-1} e^{-tA} u dt.$$

Then $A^{-\beta}$ is a bounded, one-to-one operator on J_q. Let A^β be the inverse of $A^{-\beta}$. For $\beta > 0$, we denote the space $H^{\beta,q}$ by the range of $A^{-\beta}$ with the norm

$$|u|_{H^{\beta,q}} = |A^\beta u|_q.$$

It is easy to check that e^{-tA} extends (or restricts) to a bounded analytic semigroup on $H^{\beta,q}$. For more details, we refer to von Wahl, 2013.

Let X be a Banach space and J be an interval of $\mathbb{R}$. $C(J, X)$ denotes the set of all continuous X-valued functions. For $0 < \vartheta < 1$, $C^\vartheta(J, X)$ stands for the set of all functions which are Hölder continuous with the exponent ϑ.

Let $\alpha \in (0, 1]$ and $u : [0, \infty) \times \mathbb{R}^n \to \mathbb{R}^n$, Caputo fractional derivative with respect to time of the function u can be written as

$$\partial_t^\alpha u(t, x) = \partial_t \left(\int_0^t g_{1-\alpha}(t-s)\big(u(t, x) - u(0, x)\big) ds \right), \quad t > 0,$$

where $g_\alpha(t) = \frac{t^{\alpha-1}}{\Gamma(\alpha)}$.

Let us introduce the generalized Mittag-Leffler functions:

$$E_\alpha(-t^\alpha A) = \int_0^\infty M_\alpha(s) e^{-st^\alpha A} ds, \quad e_\alpha(-t^\alpha A) = \int_0^\infty \alpha s M_\alpha(s) e^{-st^\alpha A} ds,$$

where M_α is the Wright function (see Definition 1.9).

In the following, we give some properties of the generalized Mittag-Leffler functions:

Proposition 8.1.

(i) $e_\alpha(-t^\alpha A) = \frac{1}{2\pi i}\int_{\Gamma_\theta} e_\alpha(-\mu t^\alpha)(\mu I + A)^{-1}d\mu$;

(ii) $A^\gamma e_\alpha(-t^\alpha A) = \frac{1}{2\pi i}\int_{\Gamma_\theta} \mu^\gamma e_\alpha(-\mu t^\alpha)(\mu I + A)^{-1}d\mu$.

Proof. (i) In view of $\int_0^\infty \alpha s M_\alpha(s)e^{-st}ds = e_\alpha(-t)$ and Fubini theorem, we get

$$\begin{aligned} e_\alpha(-t^\alpha A) &= \int_0^\infty \alpha s M_\alpha(s) e^{-st^\alpha A}ds \\ &= \frac{1}{2\pi i}\int_0^\infty \alpha s M_\alpha(s)\int_{\Gamma_\theta} e^{-\mu s t^\alpha}(\mu I + A)^{-1}d\mu ds \\ &= \frac{1}{2\pi i}\int_{\Gamma_\theta} e_\alpha(-\mu t^\alpha)(\mu I + A)^{-1}d\mu, \end{aligned}$$

where Γ_θ is a suitable integral path.

(ii) A similar argument shows that

$$\begin{aligned} A^\gamma e_\alpha(-t^\alpha A) &= \int_0^\infty \alpha s M_\alpha(s) A^\gamma e^{-st^\alpha A}ds \\ &= \frac{1}{2\pi i}\int_0^\infty \alpha s M_\alpha(s)\int_{\Gamma_\theta} \mu^\gamma e^{-\mu s t^\alpha}(\mu I + A)^{-1}d\mu ds \\ &= \frac{1}{2\pi i}\int_{\Gamma_\theta} \mu^\gamma e_\alpha(-\mu t^\alpha)(\mu I + A)^{-1}d\mu. \end{aligned}$$

□

Moreover, we have the following results.

Lemma 8.1. *(Wang, Chen and Xiao, 2012) For $t > 0$, $E_\alpha(-t^\alpha A)$ and $e_\alpha(-t^\alpha A)$ are continuous in the uniform operator topology. Moreover, for every $r > 0$, the continuity is uniform on $[r, \infty)$.*

Lemma 8.2. *(Wang, Chen and Xiao, 2012) Let $0 < \alpha < 1$. Then*

(i) *for all $u \in X$, $\lim_{t\to 0^+} E_\alpha(-t^\alpha A)u = u$;*

(ii) *for all $u \in D(A)$ and $t > 0$, ${}_0^C D_t^\alpha E_\alpha(-t^\alpha A)u = -AE_\alpha(-t^\alpha A)u$;*

(iii) *for all $u \in X$, $E'_\alpha(-t^\alpha A)u = -t^{\alpha-1}Ae_\alpha(-t^\alpha A)u$;*

(iv) *for $t > 0$, $E_\alpha(-t^\alpha A)u = I_t^{1-\alpha}\big(t^{\alpha-1}e_\alpha(-t^\alpha A)u\big)$.*

Before presenting the definition of mild solution of problem (8.2), we give the following lemma for a given function $h : [0, \infty) \to X$. For more details we refer to Zhou, 2014, 2016.

Lemma 8.3. *If*

$$u(t) = a + \frac{1}{\Gamma(\alpha)}\int_0^t (t-s)^{\alpha-1}\big(-Au(s) + h(s)\big)ds, \ \text{for } t \geq 0 \tag{8.3}$$

holds, then we have

$$u(t) = E_\alpha(-t^\alpha A)a + \int_0^t (t-s)^{\alpha-1}e_\alpha(-(t-s)^\alpha A)h(s)ds.$$

We rewrite (8.2) as

$$u(t) = a + \frac{1}{\Gamma(\alpha)} \int_0^t (t-s)^{\alpha-1}\big(- Au(s) + F(u(s), u(s)) + Pf(s)\big)ds, \text{ for } t \geq 0.$$

Inspired by above discussion, we adopt the following concepts of mild solution to problem (8.2).

Definition 8.1. A function $u : [0, \infty) \to H^{\beta,q}$ is called a global mild solution of problem (8.2) in $H^{\beta,q}$, if $u \in C([0, \infty), H^{\beta,q})$ and for $t \in [0, \infty)$

$$\begin{aligned} u(t) = E_\alpha(-t^\alpha A)a &+ \int_0^t (t-s)^{\alpha-1} e_\alpha(-(t-s)^\alpha A) F(u(s), u(s))ds \\ &+ \int_0^t (t-s)^{\alpha-1} e_\alpha(-(t-s)^\alpha A) Pf(s)ds. \end{aligned} \tag{8.4}$$

Definition 8.2. Let $0 < T < \infty$. A function $u : [0, T] \to H^{\beta,q}$ (or J_q) is called a local mild solution of problem (8.2) in $H^{\beta,q}$ (or J_q), if $u \in C([0, T], H^{\beta,q})$ (or $C([0, T], J_q)$) and u satisfies (8.4) for $t \in [0, T]$.

For convenience, we define two operators Φ and $\mathcal{G}$ as follows:

$$\Phi(t) = \int_0^t (t-s)^{\alpha-1} e_\alpha(-(t-s)^\alpha A) Pf(s)ds,$$

$$\mathcal{G}(u, v)(t) = \int_0^t (t-s)^{\alpha-1} e_\alpha(-(t-s)^\alpha A) F(u(s), v(s))ds.$$

In subsequent proof, we use the following fixed point result.

Lemma 8.4. *(Cannone, 1995) Let $(X, |\cdot|_X)$ be a Banach space, $G : X \times X \to X$ a bilinear operator and L a positive real number such that*

$$|G(u, v)|_X \leq L|u|_X|v|_X, \ \forall \ u, \ v \in X.$$

Then for any $u_0 \in X$ with $|u_0|_X < \frac{1}{4L}$, the equation $u = u_0 + G(u, u)$ has a unique solution $u \in X$.

8.2.3 *Global Existence*

Our main purpose in this subsection is to establish sufficient conditions for existence and uniqueness of mild solution to problem (8.2) in $H^{\beta,q}$. To this end we assume that:

(f) Pf is continuous for $t > 0$ and $|Pf(t)|_q = o(t^{-\alpha(1-\beta)})$ as $t \to 0$ for $0 < \beta < 1$.

Lemma 8.5. *(Galdi, 1998; Weissler, 1980) Let $1 < q < \infty$ and $\beta_1 \leq \beta_2$. Then there is a constant $C = C(\beta_1, \beta_2)$ such that*

$$|e^{-tA}v|_{H^{\beta_2,q}} \leq Ct^{-(\beta_2-\beta_1)}|v|_{H^{\beta_1,q}}, \quad t > 0$$

for $v \in H^{\beta_1,q}$. Furthermore, $\lim_{t\to 0} t^{(\beta_2-\beta_1)}|e^{-tA}v|_{H^{\beta_2,q}} = 0$.

Now, we study an important technical lemma, that helps us to prove the main theorems of this subsection.

Lemma 8.6. *Let $1 < q < \infty$ and $\beta_1 \le \beta_2$. Then for any $T > 0$, there exists a constant $C_1 = C_1(\alpha, \beta_1, \beta_2) > 0$ such that*

$$|E_\alpha(-t^\alpha A)v|_{H^{\beta_2,q}} \le C_1 t^{-\alpha(\beta_2-\beta_1)}|v|_{H^{\beta_1,q}}$$

and

$$|e_\alpha(-t^\alpha A)v|_{H^{\beta_2,q}} \le C_1 t^{-\alpha(\beta_2-\beta_1)}|v|_{H^{\beta_1,q}}$$

for all $v \in H^{\beta_1,q}$ and $t \in (0,T]$. Furthermore,

$$\lim_{t\to 0} t^{\alpha(\beta_2-\beta_1)}|E_\alpha(-t^\alpha A)v|_{H^{\beta_2,q}} = 0.$$

Proof. Let $v \in H^{\beta_1,q}$. By Lemma 8.5, we estimate

$$\begin{aligned}|E_\alpha(-t^\alpha A)v|_{H^{\beta_2,q}} &\le \int_0^\infty M_\alpha(s)|e^{-st^\alpha A}v|_{H^{\beta_2,q}}ds\\ &\le \left(C\int_0^\infty M_\alpha(s)s^{-(\beta_2-\beta_1)}ds\right)t^{-\alpha(\beta_2-\beta_1)}|v|_{H^{\beta_1,q}}\\ &\le C_1 t^{-\alpha(\beta_2-\beta_1)}|v|_{H^{\beta_1,q}}.\end{aligned}$$

More precisely, Lebesgue dominated convergence theorem shows

$$\lim_{t\to 0} t^{\alpha(\beta_2-\beta_1)}|E_\alpha(-t^\alpha A)v|_{H^{\beta_2,q}} \le \int_0^\infty M_\alpha(s)\lim_{t\to 0} t^{\alpha(\beta_2-\beta_1)}|e^{-st^\alpha A}v|_{H^{\beta_2,q}}ds = 0.$$

Similarly,

$$\begin{aligned}|e_\alpha(-t^\alpha A)v|_{H^{\beta_2,q}} &\le \int_0^\infty \alpha s M_\alpha(s)|e^{-st^\alpha A}v|_{H^{\beta_2,q}}ds\\ &\le \left(\alpha C\int_0^\infty M_\alpha(s)s^{1-(\beta_2-\beta_1)}ds\right)t^{-\alpha(\beta_2-\beta_1)}|v|_{H^{\beta_1,q}}\\ &\le C_1 t^{-\alpha(\beta_2-\beta_1)}|v|_{H^{\beta_1,q}},\end{aligned}$$

where the constant $C_1 = C_1(\alpha, \beta_1, \beta_2)$ is such that

$$C_1 \ge C\max\left\{\frac{\Gamma(1-\beta_2+\beta_1)}{\Gamma(1+\alpha(\beta_1-\beta_2))}, \frac{\alpha\Gamma(2-\beta_2+\beta_1)}{\Gamma(1+\alpha(1+\beta_1-\beta_2))}\right\}.$$

□

For convenience, we denote

$$M(t) = \sup_{s\in(0,t]}\{s^{\alpha(1-\beta)}|Pf(s)|_q\},$$

$$B_1 = C_1\max\{B(\alpha(1-\beta), 1-\alpha(1-\beta)),\ B(\alpha(1-\gamma), 1-\alpha(1-\beta))\},$$

$$L \ge MC_1\max\left\{B(\alpha(1-\beta), 1-2\alpha(\gamma-\beta)),\ B(\alpha(1-\gamma), 1-2\alpha(\gamma-\beta))\right\},$$

where M will be given later.

Theorem 8.1. *Let $1 < q < \infty$, $0 < \beta < 1$ and (f) hold. For every $a \in H^{\beta,q}$, suppose that*

$$C_1|a|_{H^{\beta,q}} + B_1 M_\infty < \frac{1}{4L}, \tag{8.5}$$

where $M_\infty := \sup_{s\in(0,\infty)}\{s^{\alpha(1-\beta)}|Pf(s)|_q\}$. If $\frac{n}{2q} - \frac{1}{2} < \beta$, then there is a $\gamma > \max\{\beta, \frac{1}{2}\}$ and a unique function $u : [0,\infty) \to H^{\beta,q}$ satisfying:

(a) *$u : [0,\infty) \to H^{\beta,q}$ is continuous and $u(0) = a$;*
(b) *$u : (0,\infty) \to H^{\gamma,q}$ is continuous and $\lim_{t\to 0} t^{\alpha(\gamma-\beta)}|u(t)|_{H^{\gamma,q}} = 0$;*
(c) *u satisfies (8.4) for $t \in [0,\infty)$.*

Proof. Let $\gamma = \frac{(1+\beta)}{2}$. Define $X_\infty = X[\infty]$ as the space of all curves $u : (0,\infty) \to H^{\beta,q}$ such that:

(i) $u : [0,\infty) \to H^{\beta,q}$ is bounded and continuous;
(ii) $u : (0,\infty) \to H^{\gamma,q}$ is bounded and continuous, moreover,

$$\lim_{t\to 0} t^{\alpha(\gamma-\beta)}|u(t)|_{H^{\gamma,q}} = 0$$

with its natural norm

$$\|u\|_{X_\infty} = \max\left\{\sup_{t\geq 0}|u(t)|_{H^{\beta,q}},\ \sup_{t\geq 0} t^{\alpha(\gamma-\beta)}|u(t)|_{H^{\gamma,q}}\right\}.$$

It is obvious that X_∞ is a non-empty complete metric space.

From an argument of Weissler, 1980, we know that $F : H^{\gamma,q} \times H^{\gamma,q} \to J_q$ is a bounded bilinear map, then there exists M such that for $u, v \in H^{\gamma,q}$

$$\begin{aligned}|F(u,v)|_q &\leq M|u|_{H^{\gamma,q}}|v|_{H^{\gamma,q}},\\ |F(u,u) - F(v,v)|_q &\leq M(|u|_{H^{\gamma,q}} + |v|_{H^{\gamma,q}})|u-v|_{H^{\gamma,q}}.\end{aligned} \tag{8.6}$$

Claim I. The operator $\mathcal{G}(u(t),v(t))$ belongs to $C([0,\infty),H^{\beta,q})$ as well as $C((0,\infty),H^{\gamma,q})$ for $u,v \in X_\infty$. For arbitrary $t_0 \geq 0$ fixed and $\varepsilon > 0$ enough small, consider $t > t_0$ (the case $t < t_0$ follows analogously), we have

$$\begin{aligned}
&|\mathcal{G}(u(t),v(t)) - \mathcal{G}(u(t_0),v(t_0))|_{H^{\beta,q}}\\
&\leq \int_{t_0}^{t}(t-s)^{\alpha-1}|e_\alpha(-(t-s)^\alpha A)F(u(s),v(s))|_{H^{\beta,q}}ds\\
&\quad+\int_0^{t_0}\left|\left((t-s)^{\alpha-1} - (t_0-s)^{\alpha-1}\right)e_\alpha(-(t-s)^\alpha A)F(u(s),v(s))\right|_{H^{\beta,q}}ds\\
&\quad+\int_0^{t_0-\varepsilon}(t_0-s)^{\alpha-1}\left|\left(e_\alpha(-(t-s)^\alpha A) - e_\alpha(-(t_0-s)^\alpha A)\right)F(u(s),v(s))\right|_{H^{\beta,q}}ds\\
&\quad+\int_{t_0-\varepsilon}^{t_0}(t_0-s)^{\alpha-1}\left|\left(e_\alpha(-(t-s)^\alpha A) - e_\alpha(-(t_0-s)^\alpha A)\right)F(u(s),v(s))\right|_{H^{\beta,q}}ds\\
&=: I_{11}(t) + I_{12}(t) + I_{13}(t) + I_{14}(t).
\end{aligned}$$

We estimate each of the four terms separately. For $I_{11}(t)$, in view of Lemma 8.6, we obtain

$$\begin{aligned}I_{11}(t) &\le C_1\int_{t_0}^{t}(t-s)^{\alpha(1-\beta)-1}|F(u(s),v(s))|_q ds\\ &\le MC_1\int_{t_0}^{t}(t-s)^{\alpha(1-\beta)-1}|u(s)|_{H^{\gamma,q}}|v(s)|_{H^{\gamma,q}}ds\\ &\le MC_1\int_{t_0}^{t}(t-s)^{\alpha(1-\beta)-1}s^{-2\alpha(\gamma-\beta)}ds\sup_{s\in[0,t]}\left\{s^{2\alpha(\gamma-\beta)}|u(s)|_{H^{\gamma,q}}|v(s)|_{H^{\gamma,q}}\right\}\\ &= MC_1\int_{t_0/t}^{1}(1-s)^{\alpha(1-\beta)-1}s^{-2\alpha(\gamma-\beta)}ds\sup_{s\in[0,t]}\left\{s^{2\alpha(\gamma-\beta)}|u(s)|_{H^{\gamma,q}}|v(s)|_{H^{\gamma,q}}\right\}.\end{aligned}$$

By the properties of the Beta function, there exists $\delta > 0$ small enough such that for $0 < t - t_0 < \delta$,

$$\int_{t_0/t}^{1}(1-s)^{\alpha(1-\beta)-1}s^{-2\alpha(\gamma-\beta)}ds \to 0,$$

which follows that $I_{11}(t)$ tends to 0 as $t - t_0 \to 0$. For $I_{12}(t)$, since

$$\begin{aligned}I_{12}(t) &\le C_1\int_0^{t_0}\left((t_0-s)^{\alpha-1}-(t-s)^{\alpha-1}\right)(t-s)^{-\alpha\beta}|F(u(s),v(s))|_q ds\\ &\le MC_1\int_0^{t_0}\left((t_0-s)^{\alpha-1}-(t-s)^{\alpha-1}\right)(t-s)^{-\alpha\beta}s^{-2\alpha(\gamma-\beta)}ds\\ &\quad\times\sup_{s\in[0,t_0]}\left\{s^{2\alpha(\gamma-\beta)}|u(s)|_{H^{\gamma,q}}|v(s)|_{H^{\gamma,q}}\right\},\end{aligned}$$

noting that

$$\begin{aligned}&\int_0^{t_0}\left|(t_0-s)^{\alpha-1}-(t-s)^{\alpha-1}\right|(t-s)^{-\alpha\beta}s^{-2\alpha(\gamma-\beta)}ds\\ &\le\int_0^{t_0}(t-s)^{\alpha-1}(t-s)^{-\alpha\beta}s^{-2\alpha(\gamma-\beta)}ds+\int_0^{t_0}(t_0-s)^{\alpha-1}(t-s)^{-\alpha\beta}s^{-2\alpha(\gamma-\beta)}ds\\ &\le 2\int_0^{t_0}(t_0-s)^{\alpha(1-\beta)-1}s^{-2\alpha(\gamma-\beta)}ds\\ &= 2B(\alpha(1-\beta),1-2\alpha(\gamma-\beta)),\end{aligned}$$

then by Lebesgue dominated convergence theorem, we have

$$\int_0^{t_0}\left((t_0-s)^{\alpha-1}-(t-s)^{\alpha-1}\right)(t-s)^{-\alpha\beta}s^{-2\alpha(\gamma-\beta)}ds \to 0, \text{ as } t \to t_0,$$

one deduces that $\lim_{t\to t_0} I_{12}(t) = 0$. For $I_{13}(t)$, since

$$\begin{aligned}I_{13}(t) &\le\int_0^{t_0-\varepsilon}(t_0-s)^{\alpha-1}\left|\left(e_\alpha(-(t-s)^\alpha A)+e_\alpha(-(t_0-s)^\alpha A)\right)F(u(s),v(s))\right|_{H^{\beta,q}}ds\\ &\le C_1\int_0^{t_0-\varepsilon}(t_0-s)^{\alpha-1}\left((t-s)^{-\alpha\beta}+(t_0-s)^{-\alpha\beta}\right)|F(u(s),v(s))|_q ds\end{aligned}$$

$$
\begin{aligned}
&\leq 2MC_1 \int_0^{t_0-\varepsilon} (t_0-s)^{\alpha(1-\beta)-1} s^{-2\alpha(\gamma-\beta)} ds \\
&\quad \times \sup_{s\in[0,t_0]} \left\{ s^{2\alpha(\gamma-\beta)} |u(s)|_{H^{\gamma,q}} |v(s)|_{H^{\gamma,q}} \right\},
\end{aligned}
$$

using Lebesgue dominated convergence theorem again, the fact from the uniform continuity of the operator $e_\alpha(-t^\alpha A)$ due to Lemma 8.1 shows

$$
\begin{aligned}
\lim_{t\to t_0} I_{13}(t) &= \int_0^{t_0-\varepsilon} (t_0-s)^{\alpha-1} \\
&\quad \times \lim_{t\to t_0} \left| \left(e_\alpha(-(t-s)^\alpha A) - e_\alpha(-(t_0-s)^\alpha A)\right) F(u(s),v(s)) \right|_{H^{\beta,q}} ds \\
&= 0.
\end{aligned}
$$

For $I_{14}(t)$, by immediate calculation, we estimate

$$
\begin{aligned}
I_{14}(t) &\leq C_1 \int_{t_0-\varepsilon}^{t_0} (t_0-s)^{\alpha-1} \left((t-s)^{-\alpha\beta} + (t_0-s)^{-\alpha\beta}\right) |F(u(s),v(s))|_q ds \\
&\leq 2MC_1 \int_{t_0-\varepsilon}^{t_0} (t_0-s)^{\alpha(1-\beta)-1} s^{-2\alpha(\gamma-\beta)} ds \\
&\quad \times \sup_{s\in[t_0-\varepsilon,t_0]} \left\{ s^{2\alpha(\gamma-\beta)} |u(s)|_{H^{\gamma,q}} |v(s)|_{H^{\gamma,q}} \right\} \\
&\to 0, \text{ as } \varepsilon \to 0
\end{aligned}
$$

according to the properties of the Beta function. Thenceforth, it follows

$$
|\mathcal{G}(u(t),v(t)) - \mathcal{G}(u(t_0),v(t_0))|_{H^{\beta,q}} \to 0, \text{ as } t \to t_0.
$$

The continuity of the operator $\mathcal{G}(u,v)$ evaluated in $C((0,\infty),H^{\gamma,q})$ follows by the similar discussion as above. So, we omit the details.

Claim II. The operator $\mathcal{G} : X_\infty \times X_\infty \to X_\infty$ is a continuous bilinear operator. By Lemma 8.6, we have

$$
\begin{aligned}
|\mathcal{G}(u(t),v(t))|_{H^{\beta,q}} &\leq \left| \int_0^t (t-s)^{\alpha-1} e_\alpha(-(t-s)^\alpha A) F(u(s),v(s)) ds \right|_{H^{\beta,q}} \\
&\leq C_1 \int_0^t (t-s)^{\alpha(1-\beta)-1} |F(u(s),v(s))|_q ds \\
&\leq MC_1 \int_0^t (t-s)^{\alpha(1-\beta)-1} s^{-2\alpha(\gamma-\beta)} ds \\
&\quad \times \sup_{s\in[0,t]} \left\{ s^{2\alpha(\gamma-\beta)} |u(s)|_{H^{\gamma,q}} |v(s)|_{H^{\gamma,q}} \right\} \\
&\leq MC_1 B(\alpha(1-\beta), 1-2\alpha(\gamma-\beta)) \|u\|_{X_\infty} \|v\|_{X_\infty}
\end{aligned}
$$

and

$$
|\mathcal{G}(u(t),v(t))|_{H^{\gamma,q}} \leq \left| \int_0^t (t-s)^{\alpha-1} e_\alpha(-(t-s)^\alpha A) F(u(s),v(s)) ds \right|_{H^{\gamma,q}}
$$

$$\begin{aligned}
&\leq C_1 \int_0^t (t-s)^{\alpha(1-\gamma)-1} |F(u(s), v(s))|_q ds \\
&\leq MC_1 \int_0^t (t-s)^{\alpha(1-\gamma)-1} s^{-2\alpha(\gamma-\beta)} ds \\
&\quad \times \sup_{s\in[0,t]} \left\{ s^{2\alpha(\gamma-\beta)} |u(s)|_{H^{\gamma,q}} |v(s)|_{H^{\gamma,q}} \right\} \\
&\leq MC_1 t^{-\alpha(\gamma-\beta)} B(\alpha(1-\gamma), 1-2\alpha(\gamma-\beta)) \|u\|_{X_\infty} \|v\|_{X_\infty},
\end{aligned}$$

it follows that

$$\sup_{t\in[0,\infty)} t^{\alpha(\gamma-\beta)} |\mathcal{G}(u(t), v(t))|_{H^{\gamma,q}} \leq MC_1 B(\alpha(1-\gamma), 1-2\alpha(\gamma-\beta)) \|u\|_{X_\infty} \|v\|_{X_\infty}.$$

More precisely,

$$\lim_{t\to 0} t^{\alpha(\gamma-\beta)} |\mathcal{G}(u(t), v(t))|_{H^{\gamma,q}} = 0.$$

Hence, $\mathcal{G}(u,v) \in X_\infty$ and $\|\mathcal{G}(u(t), v(t))\|_{X_\infty} \leq L\|u\|_{X_\infty}\|v\|_{X_\infty}$.

Claim III. (c) holds. Let $0 < t_0 < t$. Since

$$\begin{aligned}
&|\Phi(t) - \Phi(t_0)|_{H^{\beta,q}} \\
&\leq \int_{t_0}^t (t-s)^{\alpha-1} |e_\alpha(-(t-s)^\alpha A) Pf(s)|_{H^{\beta,q}} ds \\
&\quad + \int_0^{t_0} \left((t_0-s)^{\alpha-1} - (t-s)^{\alpha-1}\right) |e_\alpha(-(t-s)^\alpha A) Pf(s)|_{H^{\beta,q}} ds \\
&\quad + \int_0^{t_0-\varepsilon} (t_0-s)^{\alpha-1} \left| \left(e_\alpha(-(t-s)^\alpha A) - e_\alpha(-(t_0-s)^\alpha A)\right) Pf(s) \right|_{H^{\beta,q}} ds \\
&\quad + \int_{t_0-\varepsilon}^{t_0} (t_0-s)^{\alpha-1} \left| \left(e_\alpha(-(t-s)^\alpha A) - e_\alpha(-(t_0-s)^\alpha A)\right) Pf(s) \right|_{H^{\beta,q}} ds \\
&\leq C_1 \int_{t_0}^t (t-s)^{\alpha(1-\beta)-1} |Pf(s)|_q ds \\
&\quad + C_1 \int_0^{t_0} \left((t_0-s)^{\alpha-1} - (t-s)^{\alpha-1}\right) (t-s)^{-\alpha\beta} |Pf(s)|_q ds \\
&\quad + C_1 \int_0^{t_0-\varepsilon} (t_0-s)^{\alpha-1} \left| \left(e_\alpha(-(t-s)^\alpha A) - e_\alpha(-(t_0-s)^\alpha A)\right) Pf(s) \right|_{H^{\beta,q}} ds \\
&\quad + 2C_1 \int_{t_0-\varepsilon}^{t_0} (t_0-s)^{\alpha(1-\beta)-1} |Pf(s)|_q ds \\
&\leq C_1 M(t) \int_{t_0}^t (t-s)^{\alpha(1-\beta)-1} s^{-\alpha(1-\beta)} ds \\
&\quad + C_1 M(t) \int_0^{t_0} \left((t_0-s)^{\alpha-1} - (t-s)^{\alpha-1}\right) (t-s)^{-\alpha\beta} s^{-\alpha(1-\beta)} ds \\
&\quad + C_1 M(t) \int_0^{t_0-\varepsilon} (t_0-s)^{\alpha-1} \left| \left(e_\alpha(-(t-s)^\alpha A) - e_\alpha(-(t_0-s)^\alpha A)\right) Pf(s) \right|_{H^{\beta,q}} ds
\end{aligned}$$

$$+ 2C_1 M(t) \int_{t_0-\varepsilon}^{t_0} (t_0 - s)^{\alpha(1-\beta)-1} s^{-\alpha(1-\beta)} ds.$$

By the properties of the Beta function, the first two integrals and the last integral tend to 0 as $t \to t_0$ as well as $\varepsilon \to 0$. In view of Lemma 8.1, the third integral also goes to 0 as $t \to t_0$, which implies

$$|\Phi(t) - \Phi(t_0)|_{H^{\beta,q}} \to 0, \text{ as } t \to t_0.$$

The continuity of $\Phi(t)$ evaluated in $H^{\gamma,q}$ follows by the similar argument as above.

On the other hand, we have

$$\begin{aligned}
|\Phi(t)|_{H^{\beta,q}} &\le \left| \int_0^t (t-s)^{\alpha-1} e_\alpha(-(t-s)^\alpha A) P f(s) ds \right|_{H^{\beta,q}} \\
&\le C_1 \int_0^t (t-s)^{\alpha(1-\beta)-1} |Pf(s)|_q ds \\
&\le C_1 M(t) \int_0^t (t-s)^{\alpha(1-\beta)-1} s^{-\alpha(1-\beta)} ds \\
&= C_1 M(t) B(\alpha(1-\beta), 1-\alpha(1-\beta)),
\end{aligned} \tag{8.7}$$

and

$$\begin{aligned}
|\Phi(t)|_{H^{\gamma,q}} &\le \left| \int_0^t (t-s)^{\alpha-1} e_\alpha(-(t-s)^\alpha A) P f(s) ds \right|_{H^{\gamma,q}} \\
&\le C_1 \int_0^t (t-s)^{\alpha(1-\gamma)-1} |Pf(s)|_q ds \\
&\le C_1 M(t) \int_0^t (t-s)^{\alpha(1-\gamma)-1} s^{-\alpha(1-\beta)} ds \\
&= t^{-\alpha(\gamma-\beta)} C_1 M(t) B(\alpha(1-\gamma), 1-\alpha(1-\beta)).
\end{aligned}$$

More precisely,

$$t^{\alpha(\gamma-\beta)} |\Phi(t)|_{H^{\gamma,q}} \le C_1 M(t) B(\alpha(1-\gamma), 1-\alpha(1-\beta)) \to 0, \text{ as } t \to 0,$$

since $M(t) \to 0$ as $t \to 0$ due to assumption (f). This ensures that $\Phi(t) \in X_\infty$ and $\|\Phi(t)\|_{X_\infty} \le B_1 M_\infty$.

For $a \in H^{\beta,q}$. By Lemma 8.1, it is easy to see that

$$E_\alpha(-t^\alpha A)a \in C([0,\infty), H^{\beta,q}) \text{ and } E_\alpha(-t^\alpha A)a \in C((0,\infty), H^{\gamma,q}).$$

This, together with Lemma 8.6, implies that for all $t \in (0,T]$,

$$\begin{aligned}
&E_\alpha(-t^\alpha A)a \in X_\infty, \\
&t^{\alpha(\gamma-\beta)} E_\alpha(-t^\alpha A)a \in C([0,\infty), H^{\gamma,q}), \\
&\|E_\alpha(-t^\alpha A)a\|_{X_\infty} \le C_1 |a|_{H^{\beta,q}}.
\end{aligned}$$

According to (8.5), the inequality

$$\|E_\alpha(-t^\alpha A)a + \Phi(t)\|_{X_\infty} \le \|E_\alpha(-t^\alpha A)a\|_{X_\infty} + \|\Phi(t)\|_{X_\infty} < \frac{1}{4L}$$

holds, which yields that (8.4) has a unique solution.

Claim IV. $u(t) \to a$ in $H^{\beta,q}$ as $t \to 0$. We need to verify

$$\lim_{t\to 0}\int_0^t (t-s)^{\alpha-1}e_\alpha(-(t-s)^\alpha A)Pf(s)ds = 0,$$
$$\lim_{t\to 0}\int_0^t (t-s)^{\alpha-1}e_\alpha(-(t-s)^\alpha A)F(u(s),u(s))ds = 0$$

in $H^{\beta,q}$. In fact, it is obvious that $\lim_{t\to 0}\Phi(t) = 0$ ($\lim_{t\to 0} M(t) = 0$) owing to (8.7). In addition,

$$\begin{aligned}
&\left|\int_0^t (t-s)^{\alpha-1}e_\alpha(-(t-s)^\alpha A)F(u(s),u(s))ds\right|_{H^{\beta,q}}\\
&\le C_1\int_0^t (t-s)^{\alpha(1-\beta)-1}|F(u(s),u(s))|_q ds\\
&\le MC_1\int_0^t (t-s)^{\alpha(1-\beta)-1}|u(s)|^2_{H^{\gamma,q}}ds\\
&\le MC_1\int_0^t (t-s)^{\alpha(1-\beta)-1}s^{-2\alpha(\gamma-\beta)}ds \sup_{s\in[0,t]}\left\{s^{2\alpha(\gamma-\beta)}|u(s)|^2_{H^{\gamma,q}}\right\}\\
&= MC_1B(\alpha(1-\beta),1-2\alpha(\gamma-\beta)) \sup_{s\in[0,t]}\left\{s^{2\alpha(\gamma-\beta)}|u(s)|^2_{H^{\gamma,q}}\right\} \to 0, \text{ as } t\to 0.
\end{aligned}$$

□

8.2.4 *Local Existence*

In this subsection, we study the local mild solution of problem (8.2) in $H^{\beta,q}$ and J_q.

Theorem 8.2. *Let $1 < q < \infty$, $0 < \beta < 1$ and (f) hold. Suppose*

$$\frac{n}{2q} - \frac{1}{2} < \beta. \tag{8.8}$$

Then there is a $\gamma > \max\{\beta, \frac{1}{2}\}$ such that for every $a \in H^{\beta,q}$ there exist $T_* > 0$ and a unique function $u : [0,T_*] \to H^{\beta,q}$ satisfying:

(a) *$u : [0,T_*] \to H^{\beta,q}$ is continuous and $u(0) = a$;*
(b) *$u : (0,T_*] \to H^{\gamma,q}$ is continuous and $\lim_{t\to 0} t^{\alpha(\gamma-\beta)}|u(t)|_{H^{\gamma,q}} = 0$;*
(c) *u satisfies (8.4) for $t \in [0,T_*]$.*

Proof. Let $\gamma = \frac{(1+\beta)}{2}$. Fix $a \in H^{\beta,q}$. Let $X = X[T]$ be the space of all curves $u : (0,T] \to H^{\beta,q}$ such that:

(i) $u : [0,T] \to H^{\beta,q}$ is continuous;
(ii) $u : (0,T] \to H^{\gamma,q}$ is continuous and $\lim_{t\to 0} t^{\alpha(\gamma-\beta)}|u(t)|_{H^{\gamma,q}} = 0$;

with its natural norm

$$\|u\|_X = \sup_{t\in[0,T]} \{t^{\alpha(\gamma-\beta)}|u(t)|_{H^{\gamma,q}}\}.$$

Similar to the proof of Theorem 8.1, it is easy to claim that $\mathcal{G} : X \times X \to X$ is continuous linear map and $\Phi(t) \in X$.

By Lemma 8.1, it is easy to see that for all $t \in (0, T]$,

$$E_\alpha(-t^\alpha A)a \in C([0,T], H^{\beta,q}),$$
$$E_\alpha(-t^\alpha A)a \in C((0,T], H^{\gamma,q}).$$

From Lemma 8.6, it follows that

$$E_\alpha(-t^\alpha A)a \in X,$$
$$t^{\alpha(\gamma-\beta)}E_\alpha(-t^\alpha A)a \in C([0,T], H^{\gamma,q}).$$

Hence, let $T_* > 0$ be sufficiently small such that

$$\|E_\alpha(-t^\alpha A)a + \Phi(t)\|_{X[T_*]} \le \|E_\alpha(-t^\alpha A)a\|_{X[T_*]} + \|\Phi(t)\|_{X[T_*]} < \frac{1}{4L},$$

which implies that (8.4) has a unique solution due to Lemma 8.4. □

In the following, we let $\gamma = \frac{(1+\beta)}{2}$.

Theorem 8.3. *Let $1 < q < \infty$, $0 < \beta < 1$ and (f) hold. Suppose that*

$$a \in H^{\beta,q} \text{ with } \frac{n}{2q} - \frac{1}{2} < \beta.$$

Then problem (8.2) has a unique mild solution u in J_q for $a \in H^{\beta,q}$. Moreover, u is continuous on $[0,T]$, $A^\gamma u$ is continuous in $(0,T]$ and $t^{\alpha(\gamma-\beta)}A^\gamma u(t)$ is bounded as $t \to 0$.

Proof. Step I. Set

$$K(t) := \sup_{s\in(0,t]} s^{\alpha(\gamma-\beta)}|A^\gamma u(s)|_q$$

and

$$\Psi(t) := \mathcal{G}(u,u)(t) = \int_0^t (t-s)^{\alpha-1}e_\alpha(-(t-s)^\alpha A)F(u(s),u(s))ds.$$

As an immediate consequence of Claim II in Theorem 8.1, $\Psi(t)$ is continuous in $[0,T]$, $A^\gamma\Psi(t)$ exists and is continuous in $(0,T]$ with

$$|A^\gamma\Psi(t)|_q \le MC_1B(\alpha(1-\gamma), 1-2\alpha(\gamma-\beta))K^2(t)t^{-\alpha(\gamma-\beta)}. \tag{8.9}$$

We also consider the integral $\Phi(t)$. Since (f) holds, the inequality

$$|Pf(s)|_q \le M(t)s^{-\alpha(1-\beta)}$$

is satisfied with a continuous function $M(t)$. From Claim III in Theorem 8.1, we derive that $A^\gamma\Phi(t)$ is continuous in $(0,T]$ with

$$|A^\gamma\Phi(t)|_q \le C_1M(t)B(\alpha(1-\gamma), 1-\alpha(1-\beta))t^{-\alpha(\gamma-\beta)}. \tag{8.10}$$

For $|Pf(t)|_q = o(t^{-\alpha(1-\beta)})$ as $t \to 0$, we have $M(t) = 0$. Here (8.10) means $|A^\gamma \Phi(t)|_q = o(t^{-\alpha(\gamma-\beta)})$ as $t \to 0$.

We prove that Φ is continuous in J_q. In fact, take $0 \le t_0 < t < T$, we have

$$\begin{aligned}
&|\Phi(t) - \Phi(t_0)|_q \\
&\le C_1 \int_{t_0}^{t} (t-s)^{\alpha-1} |Pf(s)|_q ds + C_1 \int_0^{t_0} \left((t_0-s)^{\alpha-1} - (t-s)^{\alpha-1}\right) |Pf(s)|_q ds \\
&\quad + \int_0^{t_0-\varepsilon} (t_0-s)^{\alpha-1} \|e_\alpha(-(t-s)^\alpha A) - e_\alpha(-(t_0-s)^\alpha A)\| |Pf(s)|_q ds \\
&\quad + 2C_1 \int_{t_0-\varepsilon}^{t_0} (t_0-s)^{\alpha-1} |Pf(s)|_q ds \\
&\le C_1 M(t) \int_{t_0}^{t} (t-s)^{\alpha-1} s^{-\alpha(1-\beta)} ds \\
&\quad + C_1 M(t) \int_0^{t_0} \left((t-s)^{\alpha-1} - (t_0-s)^{\alpha-1}\right) s^{-\alpha(1-\beta)} ds \\
&\quad + C_1 M(t) \int_0^{t_0-\varepsilon} (t_0-s)^{\alpha-1} s^{-\alpha(1-\beta)} ds \\
&\quad \times \sup_{s\in[0,t-\varepsilon]} \|e_\alpha(-(t-s)^\alpha A) - e_\alpha(-(t_0-s)^\alpha A)\| \\
&\quad + 2C_1 M(t) \int_{t_0-\varepsilon}^{t_0} (t_0-s)^{\alpha-1} s^{-\alpha(1-\beta)} ds \to 0, \text{ as } t \to t_0
\end{aligned}$$

by previous discussion.

Further, we consider the function $E_\alpha(-t^\alpha A)a$. It is obvious by Lemma 8.6 that

$$\begin{aligned}
&|A^\gamma E_\alpha(-t^\alpha A)a|_q \le C_1 t^{-\alpha(\gamma-\beta)} |A^\beta a|_q = C_1 t^{-\alpha(\gamma-\beta)} |a|_{H^{\beta,q}}, \\
&\lim_{t\to 0} t^{\alpha(\gamma-\beta)} |A^\gamma E_\alpha(-t^\alpha A)a|_q = \lim_{t\to 0} t^{\alpha(\gamma-\beta)} |E_\alpha(-t^\alpha A)a|_{H^{\gamma,q}} = 0.
\end{aligned}$$

Step II. Now we construct the solution by the successive approximation:

$$\begin{aligned}
&u_0(t) = E_\alpha(-t^\alpha A)a + \Phi(t), \\
&u_{n+1}(t) = u_0(t) + \mathcal{G}(u_n, u_n)(t), \ n = 0, 1, 2\ldots .
\end{aligned} \tag{8.11}$$

Making use of above results, we know that

$$K_n(t) := \sup_{s\in(0,t]} s^{\alpha(\gamma-\beta)} |A^\gamma u_n(s)|_q$$

are continuous and increasing functions on $[0, T]$ with $K_n(0) = 0$. Furthermore, in virtue of (8.9) and (8.11), $K_n(t)$ fulfils the following inequality

$$K_{n+1}(t) \le K_0(t) + MC_1 B(\alpha(1-\gamma), 1 - 2\alpha(\gamma-\beta)) K_n^2(t). \tag{8.12}$$

For $K_0(0) = 0$, we choose a $T > 0$ such that

$$4MC_1 B(\alpha(1-\gamma), 1 - 2\alpha(\gamma-\beta)) K_0(T) < 1. \tag{8.13}$$

Then a fundamental consideration of (8.12) ensures that the sequence $\{K_n(T)\}$ is bounded, i.e.,

$$K_n(T) \le \rho(T), \quad n = 0, 1, 2, ...,$$

where

$$\rho(t) = \frac{1 - \sqrt{1 - 4MC_1 B(\alpha(1-\gamma), 1 - 2\alpha(\gamma-\beta))K_0(t)}}{2MC_1 B(\alpha(1-\gamma), 1 - 2\alpha(\gamma-\beta))}.$$

Analogously, for any $t \in (0, T]$, $K_n(t) \le \rho(t)$ holds. In the same way we note that $\rho(t) \le 2K_0(t)$.

Let us consider the equality

$$w_{n+1}(t) = \int_0^t (t-s)^{\alpha-1} e_\alpha(-(t-s)^\alpha A)[F(u_{n+1}(s), u_{n+1}(s)) - F(u_n(s), u_n(s))]ds,$$

where $w_n = u_{n+1} - u_n$, $n = 0, 1, ...,$ and $t \in (0, T]$. Writing

$$W_n(t) := \sup_{s \in (0,t]} s^{\alpha(\gamma-\beta)} |A^\gamma w_n(s)|_q.$$

On account of (8.6), we have

$$|F(u_{n+1}(s), u_{n+1}(s)) - F(u_n(s), u_n(s))|_q \le M(K_{n+1}(s) + K_n(s))W_n(s)s^{-2\alpha(\gamma-\beta)},$$

which follows from Claim II in Theorem 8.1 that

$$t^{\alpha(\gamma-\beta)} |A^\gamma w_{n+1}(t)|_q \le 2MC_1 B(\alpha(1-\gamma), 1 - \alpha(1-\beta))\rho(t)W_n(t).$$

This inequality gives

$$\begin{aligned} W_{n+1}(T) &\le 2MC_1 B(\alpha(1-\gamma), 1 - 2\alpha(\gamma-\beta))\rho(T)W_n(T) \\ &\le 4MC_1 B(\alpha(1-\gamma), 1 - 2\alpha(\gamma-\beta))K_0(T)W_n(T). \end{aligned} \tag{8.14}$$

According to (8.13) and (8.14), it is easy to see that

$$\lim_{n\to\infty} \frac{W_{n+1}(T)}{W_n(T)} < 4MC_1 K_0(T) B(\alpha(1-\gamma), 1 - 2\alpha(\gamma-\beta)) < 1,$$

thus the series $\sum_{n=0}^{\infty} W_n(T)$ converges. It shows that $\sum_{n=0}^{\infty} t^{\alpha(\gamma-\beta)} A^\gamma w_n(t)$ converges uniformly for $t \in (0, T]$, therefore, the sequence $\{t^{\alpha(\gamma-\beta)} A^\gamma u_n(t)\}$ converges uniformly in $(0, T]$. This implies that

$$\lim_{n\to\infty} u_n(t) = u(t) \in D(A^\gamma)$$

and

$$\lim_{n\to\infty} t^{\alpha(\gamma-\beta)} A^\gamma u_n(t) = t^{\alpha(\gamma-\beta)} A^\gamma u(t) \text{ uniformly},$$

since $A^{-\gamma}$ is bounded and A^γ is closed. Accordingly, the function

$$K(t) = \sup_{s \in (0,t]} s^{\alpha(\gamma-\beta)} |A^\gamma u(s)|_q$$

also satisfies

$$K(t) \le \rho(t) \le 2K_0(t), \; t \in (0, T] \tag{8.15}$$

and

$$\begin{aligned}\varsigma_n &:= \sup_{s\in(0,T]} s^{2\alpha(\gamma-\beta)}|F(u_n(s),u_n(s))-F(u(s),u(s))|_q\\ &\le M(K_n(T)+K(T))\sup_{s\in(0,T]} s^{\alpha(\gamma-\beta)}|A^\gamma(u_n(s)-u(s))|_q\\ &\to 0, \text{ as } n\to\infty.\end{aligned}$$

Finally, it remains to verify that u is a mild solution of problem (8.2) in $[0,T]$. Since

$$\begin{aligned}|\mathcal{G}(u_n,u_n)(t)-\mathcal{G}(u,u)(t)|_q &\le C_1\int_0^t (t-s)^{\alpha-1}\varsigma_n s^{-2\alpha(\gamma-\beta)}ds\\ &= C_1B(\alpha,1-2\alpha(\gamma-\beta))t^{\alpha\beta}\varsigma_n\to 0, \quad \text{as } n\to\infty,\end{aligned}$$

we have $\mathcal{G}(u_n,u_n)(t)\to\mathcal{G}(u,u)(t)$. Taking the limits on both sides of (8.11), we derive

$$u(t)=u_0(t)+\mathcal{G}(u,u)(t). \tag{8.16}$$

Let $u(0)=a$, we find that (8.16) holds for $t\in[0,T]$ and $u\in C([0,T],J_q)$. What is more, the uniform convergence of $t^{\alpha(\gamma-\beta)}A^\gamma u_n(t)$ to $t^{\alpha(\gamma-\beta)}A^\gamma u(t)$ derives the continuity of $A^\gamma u(t)$ on $(0,T]$. From (8.15) and $K_0(0)=0$, we get that $|A^\gamma u(t)|_q = o(t^{-\alpha(\gamma-\beta)})$ is obvious.

Step III. We prove that the mild solution is unique. Suppose that u and v are mild solutions of problem (8.2). Let $w=u-v$, we consider the equality

$$w(t)=\int_0^t (t-s)^{\alpha-1}e_\alpha(-(t-s)^\alpha A)[F(u(s),u(s))-F(v(s),v(s))]ds.$$

Introducing the functions

$$\widetilde{K}(t):=\max\{\sup_{s\in(0,t]} s^{\alpha(\gamma-\beta)}|A^\gamma u(s)|_q,\ \sup_{s\in(0,t]} s^{\alpha(\gamma-\beta)}|A^\gamma v(s)|_q\}.$$

By (8.6) and Lemma 8.6, we get

$$|A^\gamma w(t)|_q\le 2MC_1\widetilde{K}(t)\int_0^t (t-s)^{\alpha(1-\gamma)-1}s^{-\alpha(\gamma-\beta))}|A^\gamma w(s)|_q ds.$$

Gronwall inequality shows that $A^\gamma w(t)=0$ for $t\in(0,T]$. This implies that $w(t)=u(t)-v(t)\equiv 0$ for $t\in[0,T]$. Therefore the mild solution is unique. □

8.2.5 *Regularity*

In this subsection, we consider the regularity of a solution u which satisfies problem (8.2). Throughout this part we assume that:

($\mathbf{f_1}$) $Pf(t)$ is Hölder continuous with an exponent $\vartheta\in(0,\alpha(1-\gamma))$, that is, there exists $L>0$

$$|Pf(t)-Pf(s)|_q\le L|t-s|^\vartheta, \quad \text{for all } 0<t,\ s\le T.$$

Definition 8.3. A function $u : [0,T] \to J_q$ is called a classical solution of problem (8.2), if $u \in C([0,T], J_q)$ with ${}_0^C D_t^\alpha u(t) \in C((0,T], J_q)$, which takes values in $D(A)$ and satisfies (8.2) for all $t \in (0,T]$.

Lemma 8.7. *Let (f_1) be satisfied. If*

$$\Phi_1(t) := \int_0^t (t-s)^{\alpha-1} e_\alpha(-(t-s)^\alpha A)\big(Pf(s) - Pf(t)\big)ds, \ \textit{for } t \in (0,T],$$

then $\Phi_1(t) \in D(A)$ and $A\Phi_1(t) \in C^\vartheta([0,T], J_q)$.

Proof. For fixed $t \in (0,T]$, from Lemma 8.6 and (f_1), we have

$$\begin{aligned} &(t-s)^{\alpha-1} |Ae_\alpha(-(t-s)^\alpha A)\big(Pf(s) - Pf(t)\big)|_q \\ &\le C_1(t-s)^{-1}|Pf(s) - Pf(t)|_q \\ &\le C_1 L(t-s)^{\vartheta-1} \in L^1([0,T], J_q), \end{aligned} \tag{8.17}$$

then

$$\begin{aligned} |A\Phi_1(t)|_q &\le \int_0^t (t-s)^{\alpha-1}|Ae_\alpha(-(t-s)^\alpha A)\big(Pf(s) - Pf(t)\big)|_q ds \\ &\le C_1 L \int_0^t (t-s)^{\vartheta-1} ds \\ &= \frac{C_1 L}{\vartheta} t^\vartheta < \infty. \end{aligned}$$

By the closeness of A, we obtain $\Phi_1(t) \in D(A)$.

We need to show that $A\Phi_1(t)$ is Hölder continuous. Since

$$\frac{d}{dt}\big(t^{\alpha-1} e_\alpha(-\mu t^\alpha)\big) = t^{\alpha-2} E_{\alpha,\alpha-1}(-\mu t^\alpha),$$

then

$$\begin{aligned} &\frac{d}{dt}\big(t^{\alpha-1} Ae_\alpha(-t^\alpha A)\big) \\ &= \frac{1}{2\pi i}\int_{\Gamma_\theta} t^{\alpha-2} E_{\alpha,\alpha-1}(-\mu t^\alpha) A(\mu I + A)^{-1} d\mu \\ &= \frac{1}{2\pi i}\int_{\Gamma_\theta} t^{\alpha-2} E_{\alpha,\alpha-1}(-\mu t^\alpha) d\mu - \frac{1}{2\pi i}\int_{\Gamma_\theta} t^{\alpha-2} \mu E_{\alpha,\alpha-1}(-\mu t^\alpha)(\mu I + A)^{-1} d\mu \\ &= \frac{1}{2\pi i}\int_{\Gamma'_\theta} -t^{\alpha-2} E_{\alpha,\alpha-1}(\xi)\frac{1}{t^\alpha} d\xi - \frac{1}{2\pi i}\int_{\Gamma'_\theta} t^{\alpha-2} E_{\alpha,\alpha-1}(\xi)\frac{\xi}{t^\alpha}(-\frac{\xi}{t^\alpha} I + A)^{-1}\frac{1}{t^\alpha} d\xi. \end{aligned}$$

In view of $\|(\mu I + A)^{-1}\| \le \frac{C}{|\mu|}$, we derive that

$$\left\|\frac{d}{dt}\big(t^{\alpha-1} Ae_\alpha(-t^\alpha A)\big)\right\| \le C_\alpha t^{-2}, \ 0 < t \le T.$$

By the mean value theorem, for every $0 < s < t \leq T$, we have

$$\begin{aligned}
&\left\|t^{\alpha-1}Ae_\alpha(-t^\alpha A) - s^{\alpha-1}Ae_\alpha(-s^\alpha A)\right\| \\
&= \left\|\int_s^t \frac{d}{d\tau}\big(\tau^{\alpha-1}Ae_\alpha(-\tau^\alpha A)\big)d\tau\right\| \\
&\leq \int_s^t \left\|\frac{d}{d\tau}\big(\tau^{\alpha-1}Ae_\alpha(-\tau^\alpha A)\big)\right\|d\tau \\
&\leq C_\alpha \int_s^t \tau^{-2}d\tau \\
&= C_\alpha\big(s^{-1} - t^{-1}\big).
\end{aligned} \tag{8.18}$$

Let $h > 0$ be such that $0 < t < t+h \leq T$, then

$$\begin{aligned}
&A\Phi_1(t+h) - A\Phi_1(t) \\
&= \int_0^t \big((t+h-s)^{\alpha-1}Ae_\alpha(-(t+h-s)^\alpha A) \\
&\quad - (t-s)^{\alpha-1}Ae_\alpha(-(t-s)^\alpha A)\big)\big(Pf(s) - Pf(t)\big)ds \\
&\quad + \int_0^t (t+h-s)^{\alpha-1}Ae_\alpha(-(t+h-s)^\alpha A)\big(Pf(t) - Pf(t+h)\big)ds \\
&\quad + \int_t^{t+h} (t+h-s)^{\alpha-1}Ae_\alpha(-(t+h-s)^\alpha A)\big(Pf(s) - Pf(t+h)\big)ds \\
&=: h_1(t) + h_2(t) + h_3(t).
\end{aligned} \tag{8.19}$$

We estimate each of the three terms separately. For $h_1(t)$, from (8.18) and (f_1), we have

$$\begin{aligned}
|h_1(t)|_q &\leq \int_0^t \|(t+h-s)^{\alpha-1}Ae_\alpha(-(t+h-s)^\alpha A) \\
&\qquad - (t-s)^{\alpha-1}Ae_\alpha(-(t-s)^\alpha A)\||Pf(s) - Pf(t)|_q ds \\
&\leq C_\alpha Lh \int_0^t (t+h-s)^{-1}(t-s)^{\vartheta-1}ds \\
&= C_\alpha Lh \int_0^t (s+h)^{-1}(t-s)^{\vartheta-1}ds \\
&\leq C_\alpha L \int_0^h \frac{h}{s+h}s^{\vartheta-1}ds + C_\alpha Lh \int_h^\infty \frac{s}{s+h}s^{\vartheta-1}ds \\
&\leq C_\alpha Lh^\vartheta.
\end{aligned} \tag{8.20}$$

For $h_2(t)$, we use Lemma 8.6 and (f_1),

$$\begin{aligned}
|h_2(t)|_q &\leq \int_0^t (t+h-s)^{\alpha-1}|Ae_\alpha(-(t+h-s)^\alpha A)\big(Pf(t) - Pf(t+h)\big)|_q ds \\
&\leq C_1 \int_0^t (t+h-s)^{-1}|Pf(t) - Pf(t+h)|_q ds \\
&\leq C_1 Lh^\vartheta \int_0^t (t+h-s)^{-1}ds \\
&= C_1 L\big(\ln(t+h) - \ln h\big)h^\vartheta.
\end{aligned} \tag{8.21}$$

Furthermore, for $h_3(t)$, by Lemma 8.6 and (f_1), we have

$$\begin{aligned}|h_3(t)|_q &\le \int_t^{t+h}(t+h-s)^{\alpha-1}|Ae_\alpha(-(t+h-s)^\alpha A)\big(Pf(s)-Pf(t+h)\big)|_q ds\\ &\le C_1\int_t^{t+h}(t+h-s)^{-1}|Pf(s)-Pf(t+h)|_q ds\\ &\le C_1L\int_t^{t+h}(t+h-s)^{\vartheta-1}ds\\ &= C_1L\frac{h^\vartheta}{\vartheta}. \end{aligned} \tag{8.22}$$

Combining (8.20), (8.21) with (8.22), we get that $A\Phi_1(t)$ is Hölder continuous. □

Theorem 8.4. *Let the assumptions of Theorem 8.3 be satisfied. If (f_1) holds, then for every $a\in D(A)$, the mild solution of* (8.2) *is a classical one.*

Proof. For $a\in D(A)$. Then Lemma 8.2(ii) ensures that $u(t)=E_\alpha(-t^\alpha A)a$ $(t>0)$ is a classical solution to the following problem

$$\begin{cases} {}_0^CD_t^\alpha u=-Au, & t>0,\\ u(0)=a. \end{cases}$$

Step I. We verify that

$$\Phi(t)=\int_0^t(t-s)^{\alpha-1}e_\alpha(-(t-s)^\alpha A)Pf(s)ds$$

is a classical solution to the problem

$$\begin{cases} {}_0^CD_t^\alpha u=-Au+Pf(t), & t>0,\\ u(0)=0. \end{cases}$$

It follows from Theorem 8.3 that $\Phi\in C([0,T],J_q)$. We rewrite $\Phi(t)=\Phi_1(t)+\Phi_2(t)$, where

$$\Phi_1(t)=\int_0^t(t-s)^{\alpha-1}e_\alpha(-(t-s)^\alpha A)\big(Pf(s)-Pf(t)\big)ds,$$

$$\Phi_2(t)=\int_0^t(t-s)^{\alpha-1}e_\alpha(-(t-s)^\alpha A)Pf(t)ds.$$

According to Lemma 8.7, we know that $\Phi_1(t)\in D(A)$. To prove the same conclusion for $\Phi_2(t)$. By Lemma 8.2(iii), we notice that

$$A\Phi_2(t)=Pf(t)-E_\alpha(-t^\alpha A)Pf(t).$$

Since (f_1) holds, it follows that

$$|A\Phi_2(t)|_q\le(1+C_1)|Pf(t)|_q,$$

thus

$$\Phi_2(t)\in D(A)\text{ for }t\in(0,T]\text{ and }A\Phi_2(t)\in C^\vartheta((0,T],J_q). \tag{8.23}$$

Next, we prove ${}_0^C D_t^\alpha \Phi \in C((0,T], J_q)$. In view of Lemma 8.2(iv) and $\Phi(0) = 0$, we have

$$ {}_0^C D_t^\alpha \Phi(t) = \frac{d}{dt}\big({}_0 D_t^{\alpha-1}\Phi(t)\big) = \frac{d}{dt}\big(E_\alpha(-t^\alpha A) * Pf\big). $$

It remains to prove that $E_\alpha(t^\alpha A) * Pf$ is continuously differentiable in J_q. Let $0 < h \le T - t$, one derives the following:

$$ \begin{aligned} &\frac{1}{h}\big(E_\alpha(-(t+h)^\alpha A) * Pf - E_\alpha(-t^\alpha A) * Pf\big) \\ &= \int_0^t \frac{1}{h}\big(E_\alpha(-(t+h-s)^\alpha A)Pf(s) - E_\alpha(-(t-s)^\alpha A)Pf(s)\big)ds \\ &\quad + \frac{1}{h}\int_t^{t+h} E_\alpha(-(t+h-s)^\alpha A)Pf(s)ds. \end{aligned} $$

Notice that

$$ \begin{aligned} &\int_0^t \frac{1}{h}\big|E_\alpha(-(t+h-s)^\alpha A)Pf(s) - E_\alpha(-(t-s)^\alpha A)Pf(s)\big|_q ds \\ &\le \frac{1}{h}\int_0^t |E_\alpha(-(t+h-s)^\alpha A)Pf(s)|_q ds + \frac{1}{h}\int_0^t |E_\alpha(-(t-s)^\alpha A)Pf(s)|_q ds \\ &\le C_1 M(t)\frac{1}{h}\int_0^t (t+h-s)^{-\alpha} s^{-\alpha(1-\beta)} ds + C_1 M(t)\frac{1}{h}\int_0^t (t-s)^{-\alpha} s^{-\alpha(1-\beta)} ds \\ &\le C_1 M(t)\frac{1}{h}\big((t+h)^{1-\alpha} + t^{1-\alpha}\big)B(1-\alpha, 1-\alpha(1-\beta)), \end{aligned} $$

then using the dominated convergence theorem, we find

$$ \begin{aligned} &\lim_{h\to 0}\int_0^t \frac{1}{h}\big(E_\alpha(-(t+h-s)^\alpha A)Pf(s) - E_\alpha(-(t-s)^\alpha A)Pf(s)\big)ds \\ &= \int_0^t (t-s)^{\alpha-1} A e_\alpha(-(t-s)^\alpha A)Pf(s)ds \\ &= A\Phi(t). \end{aligned} $$

On the other hand,

$$ \begin{aligned} &\frac{1}{h}\int_t^{t+h} E_\alpha(-(t+h-s)^\alpha A)Pf(s)ds \\ &= \frac{1}{h}\int_0^h E_\alpha(-s^\alpha A)Pf(t+h-s)ds \\ &= \frac{1}{h}\int_0^h E_\alpha(-s^\alpha A)\big(Pf(t+h-s) - Pf(t-s)\big)ds \\ &\quad + \frac{1}{h}\int_0^h E_\alpha(-s^\alpha A)\big(Pf(t-s) - Pf(t)\big)ds + \frac{1}{h}\int_0^h E_\alpha(-s^\alpha A)Pf(t)ds. \end{aligned} $$

From Lemmas 8.1, 8.6 and (f_1), we have

$$ \left|\frac{1}{h}\int_0^h E_\alpha(-s^\alpha A)\big(Pf(t+h-s) - Pf(t-s)\big)ds\right|_q \le C_1 L h^\vartheta, $$

$$\left|\frac{1}{h}\int_0^h E_\alpha(-s^\alpha A)\big(Pf(t-s)-Pf(t)\big)ds\right|_q \le C_1 L\frac{h^\vartheta}{\vartheta+1}.$$

Also Lemma 8.2(i) gives that $\lim_{h\to 0}\frac{1}{h}\int_0^h E_\alpha(s^\alpha A)Pf(t)ds = Pf(t)$. Hence

$$\lim_{h\to 0}\frac{1}{h}\int_t^{t+h} E_\alpha((t+h-s)^\alpha A)Pf(s)ds = Pf(t).$$

We deduce that $E_\alpha(t^\alpha A)*Pf$ is differentiable at t_+ and $\frac{d}{dt}\big(E_\alpha(t^\alpha A)*Pf\big)_+ = A\Phi(t)+Pf(t)$. Similarly, $E_\alpha(t^\alpha A)*Pf$ is differentiable at t_- and $\frac{d}{dt}\big(E_\alpha(t^\alpha A)*Pf\big)_- = A\Phi(t)+Pf(t)$.

We show that $A\Phi = A\Phi_1 + A\Phi_2 \in C((0,T],J_q)$. In fact, it is clear that $\Phi_2(t) = Pf(t) - E_\alpha(t^\alpha A)Pf(t)$ due to Lemma 8.2(iii), which is continuous in view of Lemma 8.1. Furthermore, according to Lemma 8.7, we know that $A\Phi_1(t)$ is also continuous. Consequently, ${}_0^C D_t^\alpha \Phi \in C((0,T],J_q)$.

Step II. Let u be the mild solution of (8.2). To prove that $F(u,u) \in C^\vartheta((0,T],J_q)$, in view of (8.6), we have to verify that $A^\gamma u$ is Hölder continuous in J_q. Take $h>0$ such that $0<t<t+h$.

Denote $\varphi(t) := E_\alpha(-t^\alpha A)a$, by Lemmas 8.2(iv) and 8.6, then

$$\begin{aligned}
|A^\gamma\varphi(t+h) - A^\gamma\varphi(t)|_q &= \left|\int_t^{t+h} -s^{\alpha-1}A^\gamma e_\alpha(-s^\alpha A)a ds\right|_q \\
&\le \int_t^{t+h} s^{\alpha-1}|A^{\gamma-\beta}e_\alpha(-s^\alpha A)A^\beta a|_q ds \\
&\le C_1\int_t^{t+h} s^{\alpha(1+\beta-\gamma)-1}ds|A^\beta a|_q \\
&= \frac{C_1|a|_{H^{\beta,q}}}{\alpha(1+\beta-\gamma)}\Big((t+h)^{\alpha(1+\beta-\gamma)} - t^{\alpha(1+\beta-\gamma)}\Big) \\
&\le \frac{C_1|a|_{H^{\beta,q}}}{\alpha(1+\beta-\gamma)} h^{\alpha(1+\beta-\gamma)}.
\end{aligned}$$

Thus, $A^\gamma\varphi \in C^\vartheta((0,T],J_q)$.

For every small $\varepsilon>0$, take h such that $\varepsilon \le t < t+h \le T$, since

$$\begin{aligned}
&|A^\gamma\Phi(t+h) - A^\gamma\Phi(t)|_q \\
&\le \left|\int_t^{t+h}(t+h-s)^{\alpha-1}A^\gamma e_\alpha(-(t+h-s)^\alpha A)Pf(s)ds\right|_q \\
&\quad + \left|\int_0^t A^\gamma\big((t+h-s)^{\alpha-1}e_\alpha(-(t+h-s)^\alpha A)\right. \\
&\quad \left. - (t-s)^{\alpha-1}e_\alpha(-(t-s)^\alpha A)\big)Pf(s)ds\right|_q \\
&= \phi_1(t) + \phi_2(t).
\end{aligned}$$

Applying Lemma 8.6 and (f), we get

$$\begin{aligned}\phi_1(t) &\le C_1\int_t^{t+h}(t+h-s)^{\alpha(1-\gamma)-1}|Pf(s)|_q ds\\ &\le C_1 M(t)\int_t^{t+h}(t+h-s)^{\alpha(1-\gamma)-1}s^{-\alpha(1-\beta)}ds\\ &\le M(t)\frac{C_1}{\alpha(1-\gamma)}h^{\alpha(1-\gamma)}t^{-\alpha(1-\beta)}\\ &\le M(t)\frac{C_1}{\alpha(1-\gamma)}h^{\alpha(1-\gamma)}\varepsilon^{-\alpha(1-\beta)}.\end{aligned}$$

To estimate ϕ_2, we give the inequality

$$\begin{aligned}\frac{d}{dt}\big(t^{\alpha-1}A^\gamma e_\alpha(-t^\alpha A)\big) &= \frac{1}{2\pi i}\int_\Gamma \mu^\gamma t^{\alpha-2}E_{\alpha,\alpha-1}(-\mu t^\alpha)(\mu I+A)^{-1}d\mu\\ &= \frac{1}{2\pi i}\int_{\Gamma'} -(-\frac{\xi}{t^\alpha})^\gamma t^{\alpha-2}E_{\alpha,\alpha-1}(\xi)(-\frac{\xi}{t^\alpha}I+A)^{-1}\frac{1}{t^\alpha}d\xi,\end{aligned}$$

this yields that $\|\frac{d}{dt}\big(t^{\alpha-1}A^\gamma e_\alpha(-t^\alpha A)\big)\| \le C_\alpha t^{\alpha(1-\gamma)-2}$. The mean value theorem shows

$$\begin{aligned}\|t^{\alpha-1}A^\gamma e_\alpha(-t^\alpha A) - s^{\alpha-1}A^\gamma e_\alpha(-s^\alpha A)\| &\le \int_s^t \left\|\frac{d}{d\tau}\big(\tau^{\alpha-1}A^\gamma e_\alpha(-\tau^\alpha A)\big)\right\|d\tau\\ &\le C_\alpha\int_s^t \tau^{\alpha(1-\gamma)-2}d\tau\\ &= C_\alpha\big(s^{\alpha(1-\gamma)-1}-t^{\alpha(1-\gamma)-1}\big),\end{aligned}$$

thus

$$\begin{aligned}\phi_2(t) &\le \int_0^t |A^\gamma\big((t+h-s)^{\alpha-1}e_\alpha(-(t+h-s)^\alpha A)\\ &\quad -(t-s)^{\alpha-1}e_\alpha(-(t-s)^\alpha A)\big)Pf(s)|_q ds\\ &\le C_\alpha\int_0^t \big((t-s)^{\alpha(1-\gamma)-1}-(t+h-s)^{\alpha(1-\gamma)-1}\big)|Pf(s)|_q ds\\ &\le C_\alpha M(t)\int_0^t (t-s)^{\alpha(1-\gamma)-1}s^{-\alpha(1-\beta)}ds\\ &\quad - C_\alpha M(t)\int_0^{t+h}(t-s+h)^{\alpha(1-\gamma)-1}s^{-\alpha(1-\beta)}ds\\ &\quad + C_\alpha M(t)\int_t^{t+h}(t-s+h)^{\alpha(1-\gamma)-1}s^{-\alpha(1-\beta)}ds\\ &\le C_\alpha M(t)\big(t^{\alpha(\beta-\gamma)}-(t+h)^{\alpha(\beta-\gamma)}\big)B(\alpha(1-\gamma),1-\alpha(1-\beta))\\ &\quad + C_\alpha M(t)h^{\alpha(1-\gamma)}t^{-\alpha(1-\beta)}\frac{1}{\alpha(1-\gamma)}\\ &\le C_\alpha M(t)h^{\alpha(\gamma-\beta)}[\varepsilon(\varepsilon+h)]^{\alpha(\beta-\gamma)} + C_\alpha M(t)h^{\alpha(1-\gamma)}\varepsilon^{-\alpha(1-\beta)}\frac{1}{\alpha(1-\gamma)},\end{aligned}$$

which ensures that $A^\gamma\Phi \in C^\vartheta([\varepsilon, T], J_q)$. Therefore $A^\gamma\Phi \in C^\vartheta((0, T], J_q)$ due to arbitrary ε.

Recall

$$\Psi(t) = \int_0^t (t-s)^{\alpha-1} e_\alpha(-(t-s)^\alpha A) F(u(s), u(s)) ds.$$

Since $|F(u(s), u(s))|_q \leq MK^2(t)s^{-2\alpha(\gamma-\beta)}$, where

$$K(t) := \sup_{s\in[0,t]} s^{\alpha(\gamma-\beta)}|u(s)|_{H^{\gamma,q}}$$

is continuous and bounded in $(0, T]$. A similar argument enable us to give the Hölder continuity of $A^\gamma\Psi$ in $C^\vartheta((0, T], J_q)$. Therefore, we have $A^\gamma u(t) = A^\gamma\varphi(t) + A^\gamma\Phi(t) + A^\gamma\Psi(t) \in C^\vartheta((0, T], J_q)$.

Since $F(u, u) \in C^\vartheta((0, T], J_q)$ is proved, according to Step II, this yields that ${}_0^C D_t^\alpha \Psi \in C((0, T], J_q)$, $A\Psi \in C((0, T], J_q)$ and ${}_0^C D_t^\alpha \Psi = -A\Psi + F(u, u)$. In this way we obtain that ${}_0^C D_t^\alpha u \in C((0, T], J_q)$, $Au \in C((0, T], J_q)$ and ${}_0^C D_t^\alpha u = -Au + F(u, u) + Pf$, we conclude that u is a classical solution. □

Theorem 8.5. *Assume that* (f_1) *holds. If u is a classical solution of* (8.2), *then* $Au \in C^\vartheta((0, T], J_q)$ *and* ${}_0^C D_t^\alpha u \in C^\vartheta((0, T], J_q)$.

Proof. If u is a classical solution of (8.2), then $u(t) = \varphi(t) + \Phi(t) + \Psi(t)$. It remains to show that $A\varphi \in C^{\alpha(1-\beta)}((0, T], J_q)$, it suffices to prove that $A\varphi \in C^{\alpha(1-\beta)}([\varepsilon, T], J_q)$ for every $\varepsilon > 0$. In fact, take h such that $\varepsilon \leq t < t + h \leq T$, by Lemma 8.2(iii),

$$\begin{aligned}
|A\varphi(t+h) - A\varphi(t)|_q &= \left| \int_t^{t+h} -s^{\alpha-1} A^2 e_\alpha(-s^\alpha A) a ds \right|_q \\
&\leq C_1 \int_t^{t+h} s^{-\alpha(1-\beta)-1} ds |a|_{H^{\beta,q}} \\
&= \frac{C_1|a|_{H^{\beta,q}}}{\alpha} \left(t^{-\alpha(1-\beta)} - (t+h)^{-\alpha(1-\beta)} \right) \\
&\leq \frac{C_1|a|_{H^{\beta,q}}}{\alpha} \frac{h^{\alpha(1-\beta)}}{(\varepsilon(\varepsilon+h))^{\alpha(1-\beta)}}.
\end{aligned}$$

Similar to Lemma 8.7, we write $\Phi(t)$ as

$$\begin{aligned}
\Phi(t) = \Phi_1(t) + \Phi_2(t) &= \int_0^t (t-s)^{\alpha-1} e_\alpha(-(t-s)^\alpha A)\big(Pf(s) - Pf(t)\big) ds \\
&\quad + \int_0^t (t-s)^{\alpha-1} e_\alpha(-(t-s)^\alpha A) Pf(t) ds,
\end{aligned}$$

for $t \in (0, T]$. It follows from Lemma 8.7 and (8.23) that $A\Phi_1(t) \in C^\vartheta([0, T], J_q)$ and $A\Phi_2(t) \in C^\vartheta((0, T], J_q)$, respectively.

Since $F(u, u) \in C^\vartheta((0, T], J_q)$, the result related to the function $\Psi(t)$ is proved by similar argument, which means that $A\Psi \in C^\vartheta((0, T], J_q)$. Therefore $Au \in C^\vartheta((0, T], J_q)$ and ${}_0^C D_t^\alpha u = Au + F(u, u) + Pf \in C^\vartheta((0, T], J_q)$. The proof is completed. □

8.3 Fractional Rayleigh-Stokes Equations

8.3.1 *Introduction*

In real-world applications, many of fluids are treated as non-Newtonian fluids. For instance, magma, lava, honey in natural substances, as well as paint, glue, ink in industry, etc, and a lot of fluids in food products, biology, cosmetics are noted as such fluids. In addition, it is well known that the fractional Rayleigh-Stokes equation plays a significant role in describing the behavior of non-Newtonian fluids. Therefore, the study on such equations is important in science and applications and needs particular attention.

In this section, we consider the Rayleigh-Stokes problem with regard to the time-fractional derivative and a nonlinear source term as follows

$$\begin{cases} \partial_t u - \Delta u - m\partial_t^\alpha \Delta u = F(u), & (x,t) \in \Omega \times (0,T), \\ u(x,t) = 0, & x \in \partial\Omega,\ t \in (0,T), \\ u(x,0) = u_0(x), & x \in \Omega. \end{cases} \tag{8.24}$$

Here Δ is the Laplacian, $\Omega \subset \mathbb{R}^d (d \geq 1)$ is a bounded domain with smooth boundary $\partial\Omega$, and $T > 0$ is a given time. The real constant m is positive, u_0 is the initial data in $L^2(\Omega)$, the notation ∂_t^α is the Riemann-Liouville fractional derivative of order $\alpha \in (0,1)$ defined by

$$\partial_t^\alpha v(x,t) = \frac{1}{\Gamma(1-\alpha)} \frac{\partial}{\partial t} \left(\int_0^t (t-\tau)^{-\alpha} v(x,\tau) d\tau \right), \tag{8.25}$$

where $\Gamma(\cdot)$ is the Gamma function. The function $F : \mathbb{R} \to \mathbb{R}$ is defined later.

The fractional Rayleigh-Stokes equation (8.24) has applications in describing the non-Newtonian behavior of fluids. Numerical solutions for problem (8.24) were considered by many authors, for example, the authors in Bazhlekova *et al.*, 2015 considered the numerical approximation of the solution using the Galerkin finite element method. To the best of author's knowledge, the fractional Rayleigh-Stokes problem with a nonlinear source, i.e., Problem (8.24), has not yet been studied. In comparison with the linear problem, the nonlinear problem is considerable complicated. The goal of this section is to develop a theory on existence and regularity of mild solutions to Problem (8.24) with two main cases of the source term, the globally Lipschitz case and the locally Lipschitz case. The behavior of the solution is different in each case, and making this difference clear is one of the novelties of this section. It is also worth emphasizing that in numerical analysis, the regularity of the mild solution to Problem (8.24) considerably contributes to obtaining a convergent scheme to approximate the solution.

In this section, the initial-boundary value problem (8.24) is studied in two cases, namely when the source term is globally Lipschitz or locally Lipschitz. The time-fractional derivative used in this work is the classical Riemann-Liouville derivative. Thanks to the spectral decomposition, a fixed point argument, and some useful function spaces, we establish global well-posed results for our problem. Furthermore, we demonstrate that the mild solution exists globally or blows up in finite time.

8.3.2 *Preliminaries*

8.3.2.1 *Space Settings*

We first set up some function spaces needed for our work. Let $L^2(\Omega)$ be the usual Lebesgue space with the norm $\|.\|_{L^2(\Omega)}$. Note that this space is generated by the inner product

$$\langle v, w\rangle = \int_\Omega v(x)w(x)dx, \quad v, w \in L^2(\Omega).$$

The notations $L^p(\Omega)$, $W^{k,p}(\Omega)$ stand for the Lebesgue spaces with $p \geq 1$ and the usual Sobolev spaces with a non-negative number k, respectively. We let $[s]$ denote the integer part of a non-negative real number s, and $\{s\} := s - [s]$ the decimal part of s. If $s = [s]$, we define $\mathcal{H}_p^s(\Omega) = W^{s,p}(\Omega)$. If $0 < s < 1$, we define the set of all functions $v \in L^p(\Omega)$ by $\mathcal{H}_p^s(\Omega)$ such that

$$\|v\|_{\mathcal{H}_p^s(\Omega)} := \|v\|_{L^p(\Omega)} + \left(\iint_{\Omega\times\Omega} \frac{|v(z) - v(x)|^p}{|z - x|^{n+ps}} dzdx\right)^{\frac{1}{p}} < \infty.$$

If $s > 1$ and $s \neq [s]$, we define the set of all functions $v \in L^p(\Omega)$ by $\mathcal{H}_p^s(\Omega)$ as follows

$$\|v\|_{\mathcal{H}_p^s(\Omega)} := \|v\|_{\mathcal{H}_p^{[s]}(\Omega)} + \sum_{|\alpha|=[s]} \left\|\frac{\partial^\alpha v}{\partial x^\alpha}\right\|_{\mathcal{H}_p^{\{s\}}(\Omega)}.$$

The space $\mathcal{H}_p^s(\Omega)$, for $s \in \mathbb{R}$ and $s \geq 0$, is called a Sobolev-Slobodeckij space. In addition, we set $\overset{\circ}{\mathcal{H}}{}_p^s(\Omega) = \overline{C_c^\infty(\Omega)}^{\mathcal{H}_p^s(\Omega)}$, and the duality $\mathcal{H}_p^{-s}(\Omega) = \left[\overset{\circ}{\mathcal{H}}{}_{p^*}^s(\Omega)\right]^*$, where p, p^* are the dual numbers of each other, i.e., $\frac{1}{p} + \frac{1}{p^*} = 1$. In this section, we consider an operator $\mathcal{A}$ on $\mathbb{V} := W^{2,2}(\Omega) \cap W_0^{1,2}(\Omega)$, and $\mathcal{A}$ has eigenvalues $\{\lambda_n\}_{n\in\mathbb{N}}$ that satisfy $0 \leq \lambda_1 \leq \lambda_2 \leq \cdots \leq \lambda_n \leq \ldots$, approach ∞ as n goes to ∞, and $\lambda_n \geq Cn^{\frac{2}{d}}$ for all $n \geq 1$. The corresponding eigenfunctions are denoted by $\varphi_n \in \mathbb{V}$. The most popular example of $\mathcal{A}$ is the negative Laplacian operator $-\Delta$ on $\mathbb{V}$. More generally $\mathcal{A}$ can be taken as the symmetric and uniformly elliptic operator.

For all $s \geq 0$, we define the following operator

$$\mathcal{A}^s h := \sum_{n=1}^\infty \langle h, \varphi_n\rangle \lambda_n^s \varphi_n, \quad h \in D(\mathcal{A}^s) := \left\{h \in L^2(\Omega) : \sum_{n=1}^\infty |\langle h, \varphi_n\rangle|^2 \lambda_n^{2s} < \infty\right\}.$$

The domain $D(\mathcal{A}^s)$ is a Banach space equipped with the norm

$$\|h\|_{D(\mathcal{A}^s)} := \left(\sum_{n=1}^\infty |\langle h, \varphi_n\rangle|^2 \lambda_n^{2s}\right)^{\frac{1}{2}}, \quad h \in D(\mathcal{A}^s).$$

The definition of the negative fractional power $\mathcal{A}^{-s}$ can be found in [7]. Its domain $D(\mathcal{A}^{-s})$ is a Hilbert space endowed with the dual inner product $\langle .,. \rangle_{-s,s}$ taken between $D(\mathcal{A}^{-s})$ and $D(\mathcal{A}^{s})$. This inner product generates the norm

$$\|h\|_{D(\mathcal{A}^{-s})} := \left(\sum_{n=1}^{\infty} \left| \langle h, \varphi_n \rangle_{-s,s} \right|^2 \lambda_n^{-2s} \right)^{\frac{1}{2}}.$$

Next, we recall some of the Sobolev embeddings necessary for the presentation of the main evaluations of the section.

For convenience, we let $\mathcal{H}^s(\Omega) := \mathcal{H}_2^s(\Omega)$. It is well-known that

$$\begin{aligned} &\mathcal{H}_p^0(\Omega) := L^p(\Omega), \quad \text{if } p \geq 1, \qquad D(\mathcal{A}^s) \hookrightarrow \mathcal{H}^{2s}(\Omega), \ \text{if } s \geq 0, \\ &\mathcal{H}^\sigma(\Omega) \hookrightarrow \mathcal{H}^\gamma(\Omega), \quad \text{if } 0 \leq \gamma < \sigma, \ \mathcal{H}^\gamma(\Omega) \hookrightarrow \mathcal{H}^\sigma(\Omega), \ \text{if } \sigma < \gamma \leq 0, \end{aligned}$$

and more generally, we have

$$\mathcal{H}_p^\sigma(\Omega) \hookrightarrow \mathcal{H}_q^\gamma(\Omega), \qquad \text{if} \quad \begin{cases} 1 < p \leq q < \infty, \\ -\infty < \gamma \leq \sigma < \infty, \\ \sigma - \gamma \geq \dfrac{d}{p} - \dfrac{d}{q}. \end{cases}$$

Throughout this section, for each $s \geq 0$, we denote the positive constant $\mathscr{C}_{s \to 2s}$ as follows

$$\|v\|_{\mathcal{H}^{2s}(\Omega)} \leq \mathscr{C}_{s \to 2s} \|v\|_{\mathcal{D}(\mathcal{A}^s)},$$

for all $v \in \mathcal{D}(\mathcal{A}^s)$. It is worth noting that $\langle h_1, h_2 \rangle_{-s,s} = \langle h_1, h_2 \rangle$, if $h_1 \in L^2(\Omega), h_2 \in D(\mathcal{A}^s)$. Hence, if we have the spectral decomposition $v(x) = \sum_{n=1}^{\infty} v_n \varphi_n(x)$, then

$$\|v\|_{D(\mathcal{A}^{-s})} = \left(\sum_{n=1}^{\infty} v_n \lambda_n^{-2s} \right)^{\frac{1}{2}}.$$

This equality is used as a basic computation. In the next part, we provide the formula for the mild solution to Problem (8.24).

8.3.2.2 *Solution Representation*

By using the Laplace transform, we find the formula for the mild solution to Problem (8.24) in the form of the Fourier series. Suppose that the mild solution u is described by the Fourier series $u(x,t) = \sum_{n=1}^{\infty} \Big\langle u(\cdot,t), \varphi_n(\cdot) \Big\rangle \varphi_n(x)$. Using Bazhlekova *et al.*, 2015, we get that

$$\Big\langle u(\cdot,t), \varphi_n(\cdot) \Big\rangle = \mathbb{P}_\alpha(n,t) u_{0,n} + \int_0^t \mathbb{P}_\alpha(n, t-\tau) \Big\langle F(u(\cdot,\tau)), \varphi_n(\cdot) \Big\rangle d\tau,$$

where $u_{0,n} = \Big\langle u_0(\cdot), \varphi_n(\cdot) \Big\rangle$ and $\mathbb{P}_\alpha(n,t)$ has the Laplace transform $(\mathcal{L})$ given by

$$\mathcal{L}\Big(\mathbb{P}_\alpha(n,t)\Big)(\zeta) = \frac{1}{\zeta + m\lambda_n \zeta^\alpha + \lambda_n}.$$

Therefore, we find that

$$u(x,t) = \sum_{n=1}^{\infty} \left[\mathbb{P}_\alpha(n,t) u_{0,n} + \int_0^t \mathbb{P}_\alpha(n,t-\tau) \Big\langle F(u(\cdot,\tau)), \varphi_n(\cdot) \Big\rangle d\tau \right] \varphi_n(x). \quad (8.26)$$

We can rewrite the above equation (8.26) in the following way

$$u(t) = \mathbf{S}(\alpha,t)u_0 + \int_0^t \mathbf{S}(\alpha,t-\tau) F(u(\cdot,\tau)) d\tau,$$

where $\mathbf{S}(\alpha,t)v = \sum_{n=1}^{\infty} \mathbb{P}_\alpha(n,t) \langle v, \varphi_n \rangle \varphi_n$ for any $v \in L^2(\Omega)$.

Before introducing the main results of this section, we demonstrate the following useful lemma.

Lemma 8.8. *Let $\alpha \in (0,1)$. Then, we get the following result*

$$\mathbb{P}_\alpha(n,t) \le R(m,\alpha) \left(1 + \lambda_n t^{1-\alpha}\right)^{-1}, \quad (8.27)$$

where

$$R(m,\alpha) = \frac{\Gamma(1-\alpha)}{\pi m \sin(\pi\alpha)} + 1.$$

Proof. First, thanks to Bazhlekova *et al.*, 2015, we obtain the equality

$$\mathbb{P}_\alpha(n,t) = \int_0^\infty e^{-rt} \mathbb{K}_\alpha(n,r) dr,$$

where

$$\mathbb{K}_\alpha(n,r) = \frac{m\lambda_n \sin(\alpha\pi) r^\alpha}{\pi \left[\Big(-r + \lambda_n m r^\alpha \cos(\alpha\pi) + \lambda_n \Big)^2 + \Big(\lambda_n m r^\alpha \sin(\alpha\pi) \Big)^2 \right]}.$$

It is easy to see that

$$\mathbb{K}_\alpha(n,r) \le \frac{m\lambda_n \sin(\alpha\pi) r^\alpha}{\pi \Big(m\lambda_n r^\alpha \sin(\alpha\pi) \Big)^2} = \frac{r^{-\alpha}}{\pi m \sin(\alpha\pi) \lambda_n}.$$

The preceding observations imply that

$$\begin{aligned} \mathbb{P}_\alpha(n,t) &\le \frac{\int_0^\infty e^{-rt} r^{-\alpha} dr}{\pi m \sin(\alpha\pi)\lambda_n} = \frac{\int_0^\infty e^{-rt} (rt)^{-\alpha} d(rt)}{\pi m \sin(\alpha\pi) t^{1-\alpha} \lambda_n} \\ &\le \frac{\int_0^\infty e^{-\nu} \nu^{-\alpha} d\nu}{\pi m \sin(\alpha\pi) t^{1-\alpha} \lambda_n} = \frac{\Gamma(1-\alpha)}{\pi m \sin(\alpha\pi) t^{1-\alpha} \lambda_n}. \end{aligned}$$

It should be noted that

$$\int_0^\infty e^{-\eta} \eta^{-\alpha} d\eta = \Gamma(1-\alpha).$$

By virtue of $0 < \mathbb{P}_\alpha(n,t) \le 1$, we obtain the desired result (8.27). □

8.3.3 *Globally Lispchitz Source Term*

In this subsection, we state the existence and regularity of the mild solution of Problem (8.24) under a globally Lipschitz condition for the source function F.

Theorem 8.6. *Suppose that* $F(0) = 0$ *and*

$$|F(z_1) - F(z_2)| \leq K_F|z_1 - z_2|, \; z_1, z_2 \in \mathbb{R} \tag{8.28}$$

with constant $K_F > 0$.

(a) *If the initial data* u_0 *belongs to* $L^2(\Omega)$, *for any* $T \in (0, \infty)$, *there exists a unique mild solution* u *on* $[0, T]$ *of Problem* (8.24) *and the regularity result below holds*

$$\sup_{0 \leq t \leq T} \|u(\cdot, t)\|_{L^2(\Omega)} + \left\| \frac{\partial u}{\partial t} \right\|_{L^q(0,T;D(\mathcal{A}^{-1}))} + \leq \overline{R}_T \|u_0\|_{L^2(\Omega)},$$

where $\overline{R}_T$ *is independent on* x, t, *and* q *satisfies* $1 < q < \frac{1}{\alpha}$.

(b) *If* $u_0 \in \mathbb{V}$, *we have* $u \in C([0, T]; L^2(\Omega))$. *More generally, for any* $0 \leq t < t + h \leq T$,

$$\|u(\cdot, t + h) - u(\cdot, t)\|_{L^2(\Omega)} \leq \mathbf{C} \max\{h^\alpha, h^{1-\alpha}\},$$

where $\mathbf{C}$ *is independent on* h.

(c) *Let* $u_0 \in L^p(\Omega)$ *and* p, ν, μ *satisfy*

$$\begin{cases} p \geq 1, \\ \max\{-1, -\frac{d}{4}\} < \nu \leq \min\left\{\frac{(p-2)d}{4p}, 0\right\}, \\ 0 \leq \mu < \min\left\{1 + \nu, \frac{d}{4}\right\}. \end{cases}$$

Then, there exists a mild solution $u \in L^\infty(0, T; L^{\frac{2d}{d-4\mu}}(\Omega))$ *to Problem* (8.24).

Proof. *Proof of the statement (a).*

Step I. Existence and uniqueness.

We define a mapping $\mathbf{J} : \mathbb{E} \to \mathbb{E}$ by

$$\mathbf{J}w(t) = \mathbf{S}(\alpha, t)u_0 + \int_0^t \mathbf{S}(\alpha, t - \tau)F(w(\tau))d\tau,$$

where

$$\mathbb{E} := \left\{u : [0, T] \to L^2(\Omega); \sup_{0 \leq t \leq T} \|\exp(-pt)u(\cdot, t)\|_{L^2(\Omega)} < \infty\right\}.$$

From the definition of $\mathbf{J}$ and Lemma 8.8, for any $w_1, w_2 \in \mathbb{E}$, we get

$$\begin{aligned}
&\left\| \exp(-pt)\left(\mathbf{J}w_1(\cdot,t) - \mathbf{J}w_2(\cdot,t)\right) \right\|_{L^2(\Omega)} \\
&= \left\| \exp(-pt) \int_0^t \mathbf{S}(\alpha, t-\tau)\left(F(w_1(\cdot,\tau)) - F(w_2(\cdot,\tau))\right) d\tau \right\|_{L^2(\Omega)} \\
&\le K_{\mathrm{F}} R(m,\alpha) {\lambda_1}^{-1} \exp(-pt) \int_0^t (t-\tau)^{\alpha-1} \|w_1(\cdot,\tau) - w_2(\cdot,\tau)\|_{L^2(\Omega)} d\tau \\
&\le K_{\mathrm{F}} R(m,\alpha) {\lambda_1}^{-1} \int_0^t (t-\tau)^{\alpha-1} \exp(-p(t-\tau)) \\
&\quad \times \sup_{\tau\in[0,T]} \left\| \exp(-p\tau)(w_1(\cdot,\tau) - w_2(\cdot,\tau)) \right\|_{L^2(\Omega)} d\tau \\
&= K_{\mathrm{F}} R(m,\alpha) {\lambda_1}^{-1} \|w_1 - w_2\|_{\mathbb{E}} \int_0^t (t-\tau)^{\alpha-1} \exp(-p(t-\tau)) d\tau.
\end{aligned} \tag{8.29}$$

With the aim of handling the integral $I_1 = \int_0^t (t-\tau)^{\alpha-1} \exp(-p(t-\tau)) d\tau$, we set $\tau = t\theta$, and immediately get

$$\begin{aligned}
I_1 &= \int_0^1 (t - t\theta)^{\alpha-1} \exp(-pt(1-\theta)) t d\theta \\
&= \int_0^1 t^\alpha (1-\theta)^{\alpha-1} \exp(-pt(1-\theta)) d\theta \\
&= \int_0^1 [pt(1-\theta)]^{\frac{\alpha}{2}} \exp(-pt(1-\theta)) \left(\frac{t}{p}\right)^{\frac{\alpha}{2}} (1-\theta)^{\frac{\alpha}{2}-1} d\theta.
\end{aligned}$$

Applying the inequality $z \le e^z$, we can find that

$$[pt(1-\theta)]^{\frac{\alpha}{2}} \le \exp\left(pt(1-\theta)\frac{\alpha}{2}\right) \le \exp\left(pt(1-\theta)\right),$$

provided that $\alpha < 1$. Furthermore, it is worth mentioning that

$$\int_0^1 (1-\theta)^{\frac{\alpha}{2}-1} d\theta = \frac{2}{\alpha}.$$

Then, we obtain

$$I_1 \le \frac{2}{\alpha}\left(\frac{T}{p}\right)^{\frac{\alpha}{2}}. \tag{8.30}$$

Combining (8.29) and (8.30) allows us to arrive at

$$\begin{aligned}
&\left\| \exp(-pt)\left(\mathbf{J}w_1(\cdot,t) - \mathbf{J}w_2(\cdot,t)\right) \right\|_{L^2(\Omega)} \\
&\le \frac{K_{\mathrm{F}} R(m,\alpha)}{\lambda_1} \frac{2}{\alpha} \left(\frac{T}{p}\right)^{\frac{\alpha}{2}} \|w_1 - w_2\|_{\mathbb{E}}.
\end{aligned} \tag{8.31}$$

Because the right-hand side of (8.31) is independent of t, we deduce that

$$\begin{aligned}
\|\mathbf{J}w_1 - \mathbf{J}w_2\|_{\mathbb{E}} &= \sup_{0\le t\le T} \left\| \exp(-pt)\left(\mathbf{J}w_1(\cdot,t) - \mathbf{J}w_2(\cdot,t)\right) \right\|_{L^2(\Omega)} \\
&\le \frac{2K_F R(m,\alpha)}{\alpha\lambda_1} \left(\frac{T}{p}\right)^{\frac{\alpha}{2}} \|w_1 - w_2\|_{\mathbb{E}}.
\end{aligned}$$

With a suitable choice of p, the last inequality implies

$$\|\mathbf{J}w_1 - \mathbf{J}w_2\|_{\mathbb{E}} \leq \tfrac{1}{2}\|w_1 - w_2\|_{\mathbb{E}}, \quad \forall\, w_1, w_2 \in \mathbb{E}.$$

Thus, $\mathbf{J}$ is a contraction mapping in $\mathbb{E}$. Applying the Banach fixed point theorem, we find that $\mathbf{J}$ has a unique fixed point u in $\mathbb{E}$. As a result, the function u is also the unique solution of Problem (8.24) in $\mathbb{E}$.

Step II. Regularity of the mild solution.

Firstly, we have

$$\begin{aligned}\|\mathbf{S}(\alpha,t)u_0\|^2_{L^2(\Omega)} &= \sum_{n=1}^{\infty}\Big|\mathbb{P}_\alpha(n,t)\Big|^2|u_{0,n}|^2\\ &\leq |R(m,\alpha)|^2\sum_{n=1}^{\infty}|u_{0,n}|^2\\ &\leq |R(m,\alpha)|^2\|u_0\|^2_{L^2(\Omega)}.\end{aligned}$$

This implies that

$$\|\mathbf{S}(\alpha,t)u_0\|_{L^2(\Omega)} \leq R(m,\alpha)\|u_0\|_{L^2(\Omega)}.$$

Using the triangle inequality, we obtain

$$\|u(\cdot,t)\|_{L^2(\Omega)} \leq \|\mathbf{S}(\alpha,t)u_0\|_{L^2(\Omega)} + \int_0^t \|\mathbf{S}(\alpha,t-\tau)F(u(\cdot,\tau))\|_{L^2(\Omega)}d\tau.$$

Because $\|F(u(\cdot,\tau))\|_{L^2(\Omega)} = \|F(u(\cdot,\tau)) - F(0)\|_{L^2(\Omega)} \leq K_{\mathrm{F}}\|u(\cdot,\tau)\|_{L^2(\Omega)}$, we see that

$$\begin{aligned}\|u(\cdot,t)\|_{L^2(\Omega)} &\leq R(m,\alpha)\|u_0\|_{L^2(\Omega)}\\ &\quad + K_{\mathrm{F}}R(m,\alpha)\int_0^t (t-\tau)^{\alpha-1}\|u(\cdot,\tau)\|_{L^2(\Omega)}d\tau.\end{aligned}$$

Grönwall inequality (see Ye, Gao and Ding, 2007) guarantees that

$$\begin{aligned}\|u(\cdot,t)\|_{L^2(\Omega)} &\leq R(m,\alpha)E_{\alpha,1}\Big(K_{\mathrm{F}}R(m,\alpha)\Gamma(\alpha)t^\alpha\Big)\|u_0\|_{L^2(\Omega)}\\ &:= \overline{R}(m,\alpha,T)\|u_0\|_{L^2(\Omega)}.\end{aligned} \tag{8.32}$$

Step III. Estimate of the term $\left\|\frac{\partial u}{\partial t}\right\|_{L^q(0,T;D(\mathcal{A}^{-1}))}$.

In view of the identity $\frac{d}{dt}\int_0^t G(t,s)ds = \int_0^t G_t(t,s)ds + G(t,t)$, we obtain

$$\begin{aligned}\frac{\partial u}{\partial t}(x,t) &= \underbrace{\sum_{n=1}^{\infty}\mathbb{Q}_\alpha(n,t)u_{0,n}\varphi_n(x)}_{\mathcal{I}_1(x,t)}\\ &\quad + \underbrace{\sum_{n=1}^{\infty}\left[\int_0^t \mathbb{Q}_\alpha(n,t-\tau)\Big\langle F(u(\cdot,\tau)),\varphi_n(\cdot)\Big\rangle d\tau\right]\varphi_n(x)}_{\mathcal{I}_2(x,t)}\\ &\quad + F(u(x,t)),\end{aligned}$$

where

$$\mathbb{Q}_\alpha(n,t) = \frac{d\mathbb{P}_\alpha(n,t)}{dt}.$$

Applying Theorem 2.1 of Bazhlekova *et al.*, 2015, we see that if $w \in \mathbb{V}$ then

$$\sum_{n=1}^{\infty} |\mathbb{Q}_\alpha(n,t)|^2 |w_n|^2 \le \mathcal{D}(\Omega, d, T, a)^2 t^{-2\alpha} \|w\|_{\mathbb{V}}^2,$$

where $\mathcal{D}(\Omega, d, T, a)$ depends only on Ω, d, T, a. Hence, if $w \in L^2(\Omega)$, we have

$$\sum_{n=1}^{\infty} \lambda_n^{-2} |\mathbb{Q}_\alpha(n,t)|^2 \Big\langle w, \varphi_n(x) \Big\rangle^2 \le \mathcal{D}^2(\Omega, N, T, a) t^{-2\alpha} \|w\|_{L^2(\Omega)}^2. \tag{8.33}$$

Consequently, we can estimate $\mathcal{I}_1$ in the following way

$$\begin{aligned}\|\mathcal{I}_1(.,t)\|_{D(\mathcal{A}^{-1})}^2 &= \sum_{n=1}^{\infty} \lambda_n^{-2} |\mathbb{Q}_\alpha(n,t)|^2 |u_{0,n}|^2 \\ &\le \mathcal{D}^2(\Omega, N, T, a) t^{-2\alpha} \|u_0\|_{L^2(\Omega)}^2.\end{aligned}$$

By means of $q \in (1, \frac{1}{\alpha})$, by applying Hölder inequality, we immediately get that

$$\begin{aligned}\|\mathcal{I}_1\|_{L^q(0,T;D(\mathcal{A}^{-1}))} &= \left(\int_0^T \|I_1(.,t)\|_{D(\mathcal{A}^{-1})}^q dt \right)^{\frac{1}{q}} \\ &\le \frac{\mathcal{D}(\Omega, N, T, a) T^{1-\alpha q}}{1-\alpha q} \|u_0\|_{L^2(\Omega)}.\end{aligned} \tag{8.34}$$

The term $\|\mathcal{I}_2(\cdot,t)\|_{D(\mathcal{A}^{-1})}$ is estimated as follows:

$$\begin{aligned}\|\mathcal{I}_2(\cdot,t)\|_{D(\mathcal{A}^{-1})} &= \left\| \int_0^t \left(\sum_{n=1}^{\infty} \mathbb{Q}_\alpha(n,t-\tau) \Big\langle F(u(\cdot,\tau)), \varphi_n(\cdot) \Big\rangle \varphi_n(x) \right) d\tau \right\|_{D(\mathcal{A}^{-1})} \\ &\le \int_0^t \left\| \sum_{n=1}^{\infty} \mathbb{Q}_\alpha(n,t-\tau) \Big\langle F(u(\cdot,\tau)) - F(0), \varphi_n(\cdot) \Big\rangle \varphi_n(x) \right\|_{D(\mathcal{A}^{-1})} d\tau \\ &= \int_0^t \left(\sum_{n=1}^{\infty} \lambda_n^{-2} \Big|\mathbb{Q}_\alpha(n,t-\tau)\Big|^2 \Big\langle F(u(\cdot,\tau)) - F(0), \varphi_n(\cdot) \Big\rangle^2 \right)^{\frac{1}{2}} d\tau.\end{aligned} \tag{8.35}$$

In view of (8.33), we find that

$$\begin{aligned}&\left(\sum_{n=1}^{\infty} \lambda_n^{-2} \Big|\mathbb{Q}_\alpha(n,t-\tau)\Big|^2 \Big\langle F(u(\cdot,\tau)) - F(0), \varphi_n(\cdot) \Big\rangle^2 \right)^{\frac{1}{2}} \\ &\le \mathcal{D}(\Omega, N, T, a)(t-\tau)^{-\alpha} \|F(u(\cdot,\tau)) - F(0)\|_{L^2(\Omega)} \\ &\le K_F \mathcal{D}(\Omega, N, T, a)(t-\tau)^{-\alpha} \|u(\cdot,\tau)\|_{L^2(\Omega)}.\end{aligned} \tag{8.36}$$

Combining (8.35) and (8.36) and (8.32) allows us to come to the conclusion

$$\begin{aligned}\|\mathcal{I}_2(\cdot,t)\|_{D(\mathcal{A}^{-1})} &\le K_{\mathrm{F}}\mathcal{D}(\Omega,d,T,a)\int_0^t (t-\tau)^{-\alpha}\|u(\cdot,\tau)\|_{L^2(\Omega)}d\tau \\ &\le K_{\mathrm{F}}\mathcal{D}(\Omega,N,T,a)\left(\int_0^t (t-\tau)^{-\alpha}d\tau\right)\sup_{0\le t\le T}\left\|u(\cdot,t)\right\|_{L^2(\Omega)} \\ &\le K_{\mathrm{F}}\mathcal{D}(\Omega,N,T,a)\overline{R}(m,\alpha,T)\frac{t^{1-\alpha}}{1-\alpha}\|u_0\|_{L^2(\Omega)}.\end{aligned}$$

This implies that

$$\begin{aligned}\|\mathcal{I}_2\|_{L^q(0,T;D(\mathcal{A}^{-1}))} &= \left(\int_0^T \|\mathcal{I}_2(.,t)\|^q_{D(\mathcal{A}^{-1})}dt\right)^{\frac{1}{q}} \\ &\le \frac{K_{\mathrm{F}}\mathcal{D}(\Omega,d,T,a)\overline{R}(m,\alpha,T)T^{1-\alpha+\frac{1}{q}}}{1-\alpha}\|u_0\|_{L^2(\Omega)}.\end{aligned} \tag{8.37}$$

In addition, using the Lipschitz property of F and (8.32), we also have

$$\|F(u)(\cdot,t)\|_{L^2(\Omega)} \le K_F\|u(\cdot,t)\|_{L^2(\Omega)} \le K_F\overline{R}(m,\alpha,T)\|u_0\|_{L^2(\Omega)}. \tag{8.38}$$

Combining (8.32), (8.34), (8.37), and (8.38), we get the desired results.

Proof of the statement (b).

By simple calculations, we get

$$\begin{aligned}u(x,\widetilde{t})-u(x,t) &= \sum_{n=1}^{\infty}\left[\mathbb{P}_\alpha(n,\widetilde{t})-\mathbb{P}_\alpha(n,t)\right]u_{0,n}\varphi_n(x) \\ &\quad+\sum_{n=1}^{\infty}\left[\int_t^{\widetilde{t}}\mathbb{P}_\alpha(n,\tau)\Big\langle F(u(\cdot,\widetilde{t}-\tau)),\varphi_n(\cdot)\Big\rangle d\tau\right]\varphi_n(x)+\sum_{n=1}^{\infty} \\ &\quad\times\left(\int_0^t \mathbb{P}_\alpha(n,\tau)\Big\langle F(u(\cdot,\widetilde{t}-\tau))-F(u(\cdot,t-\tau)),\varphi_n(\cdot)\Big\rangle d\tau\right)\varphi_n(x) \\ &=: \sum_{i=1}^{3}\mathcal{I}_i(x,\widetilde{t},t).\end{aligned}$$

Thanks to Lemma 8.8, we get

$$\begin{aligned}\left\|\mathcal{I}_2(\cdot,\widetilde{t},t)\right\|_{L^2(\Omega)} &\le \int_t^{\widetilde{t}}\left\|\mathbb{P}_\alpha(n,\tau)\Big\langle F(u(\cdot,\widetilde{t}-\tau)),\varphi_n(\cdot)\Big\rangle\right\|_{L^2(\Omega)}d\tau \\ &= \int_t^{\widetilde{t}}\sqrt{\sum_{n=1}^{\infty}\left|\mathbb{P}_\alpha(n,\tau)\right|^2\Big\langle F(u(\cdot,\widetilde{t}-\tau)),\varphi_n(\cdot)\Big\rangle^2}d\tau \\ &\le R(m,\alpha)\lambda_1^{-1}\int_t^{\widetilde{t}}\tau^{\alpha-1}\|F(u(\cdot,\widetilde{t}-\tau))\|_{L^2(\Omega)}d\tau \\ &\le K_{\mathrm{F}}R(m,\alpha)\overline{R}(m,\alpha,T)\lambda_1^{-1}\|u_0\|_{L^2(\Omega)}\frac{(\widetilde{t})^\alpha-t^\alpha}{\alpha} \\ &\le K_{\mathrm{F}}R(m,\alpha)\overline{R}(m,\alpha,T)\lambda_1^{-1}\|u_0\|_{L^2(\Omega)}\frac{(\widetilde{t}-t)^\alpha}{\alpha},\end{aligned}$$

it should be noted that

$$\|F(u(\cdot,\tau))\|_{L^2(\Omega)} \le K_F\|u(\cdot,\tau)\|_{L^2(\Omega)} \le K_{\mathrm{F}}\overline{R}(m,\alpha,T)\|u_0\|_{L^2(\Omega)},$$

and

$$(\widetilde{t})^\alpha - t^\alpha \le (\widetilde{t}-t)^\alpha.$$

With the view to considering the term $\mathcal{I}_1(x,\widetilde{t},t)$, we recall that $\mathbb{Q}_\alpha(n,t) = \frac{d\mathbb{P}_\alpha(n,t)}{dt}$. This implies that

$$\mathcal{I}_1(x,\widetilde{t},t) = \int_t^{\widetilde{t}} \Big(\sum_{n=1}^{\infty} \mathbb{Q}_\alpha(n,\tau)u_{0,n}\varphi_n(x)\Big)d\tau.$$

It follows from the previous result that

$$\begin{aligned}\left\|\mathcal{I}_1(\cdot,\widetilde{t},t)\right\|_{L^2(\Omega)} &\le \int_t^{\widetilde{t}} \sqrt{\sum_{n=1}^{\infty} |\mathbb{Q}_\alpha(n,\tau)|^2|u_{0,n}|^2 d\tau} \\ &\le \mathbf{M}_T\|u_0\|_{\mathbb{V}} \int_t^{\widetilde{t}} \tau^{-\alpha}d\tau \\ &\le \frac{\mathbf{M}_T(\widetilde{t}-t)^{1-\alpha}}{1-\alpha}\|u_0\|_{\mathbb{V}},\end{aligned}$$

where $\mathbf{M}_T$ is a constant and it is worth noting that $(\widetilde{t})^{1-\alpha} - t^{1-\alpha} \le (\widetilde{t}-t)^{1-\alpha}$.

In terms of the term $\mathcal{I}_3(\cdot,\widetilde{t},t)$, by using the argument similar to dealing with the term $\mathcal{I}_2(x,\widetilde{t},t)$, we get the following estimate

$$\begin{aligned}&\left\|\mathcal{I}_3(\cdot,\widetilde{t},t)\right\|_{L^2(\Omega)} \\ &\le K_{\mathrm{F}}R(m,\alpha)\lambda_1{}^{-1}\int_0^t (t-\tau)^{\alpha-1}\|u(\cdot,\widetilde{t}-\tau)-u(\cdot,t-\tau)\|_{L^2(\Omega)}d\tau \\ &= e^{pt}K_{\mathrm{F}}R(m,\alpha)\lambda_1{}^{-1}\int_0^t e^{-p(t-\tau)}(t-\tau)^{\alpha-1}e^{-p\tau}\|u(\cdot,\widetilde{t}-\tau) \\ &\quad - u(\cdot,t-\tau)\|_{L^2(\Omega)}d\tau.\end{aligned}$$

We now set

$$\mathcal{Z}(t) = e^{-pt}\|u(\cdot,\widetilde{t}) - u(\cdot,t)\|_{L^2(\Omega)}.$$

From the above observation, we deduce that

$$\begin{aligned}e^{pt}\mathcal{Z}(t) \le\ & \frac{K_F R(m,\alpha)\overline{R}(m,\alpha,T)}{\lambda_1}\frac{(\widetilde{t}-t)^\alpha}{\alpha}\|u_0\|_{L^2(\Omega)} \\ &+ \frac{\mathbf{M}_T(\widetilde{t}-t)^{1-\alpha}}{1-\alpha}\|u_0\|_{\mathbb{V}} \\ &+ \max_{\tau\in[0,T]} \mathcal{Z}(\tau)K_{\mathrm{F}}R(m,\alpha)\lambda_1{}^{-1}e^{pt}\int_0^t (t-\tau)^{\alpha-1}\exp(-p(t-\tau))d\tau.\end{aligned}$$

Thanks to (8.30), we get

$$\mathcal{Z}(t) \le C_\nu \max\left\{(\widetilde{t}-t)^\alpha, (\widetilde{t}-t)^{1-\alpha}\right\} + \frac{2}{\alpha}\left(\frac{T}{p}\right)^{\frac{\alpha}{2}} K_{\mathrm{F}} R(m,\alpha){\lambda_1}^{-1} \max_{\tau\in[0,T]} \mathcal{Z}(\tau),$$

where

$$C_\nu = K_{\mathrm{F}} R(m,\alpha)\overline{R}(m,\alpha,T){\lambda_1}^{-1}\|u_0\|_{L^2(\Omega)} + \frac{\mathbf{M}_T}{1-\alpha}\|u_0\|_{\mathbb{V}}.$$

Choosing p such that $\frac{2}{\alpha}\left(\frac{T}{p}\right)^{\frac{\alpha}{2}} K_{\mathrm{F}} R(m,\alpha){\lambda_1}^{-1} < \frac{1}{2}$, we have

$$\|u(\cdot,\widetilde{t}) - u(\cdot,t)\|_{L^2(\Omega)} \le 2C_\nu e^{pt}\max\{(\widetilde{t}-t)^\alpha, (\widetilde{t}-t)^{1-\alpha}\}.$$

This helps us to obtain the desired assertion.

Proof of the statement (c) To begin with, we note the following conclusion. For $d \ge 1$, we have

$$\left.\begin{aligned} L^p(\Omega) &\hookrightarrow D(\mathcal{A}^b), \quad \text{if} \quad -\frac{d}{4} < b \le 0, \quad p \ge \frac{2d}{d-4b}, \\ D(\mathcal{A}^b) &\hookrightarrow L^p(\Omega), \quad \text{if} \quad 0 \le b < \frac{d}{4}, \quad p \le \frac{2d}{d-4b}. \end{aligned}\right\}$$

Next, for $a, \vartheta > 0$ we consider the Banach space

$$L^\infty_{a,\vartheta}(0,T;L^{\frac{2d}{d-4\mu}}(\Omega))$$
$$:= \left\{ w \in L^\infty_{a,\vartheta}(0,T;L^{\frac{2d}{d-4\mu}}(\Omega)) \;\middle|\; t^\vartheta e^{-at}\big\|w(\cdot,t)\big\|_{L^{\frac{2d}{d-4\mu}}(\Omega)} < \infty,\ 0 \le t \le T \right\},$$

associated with the following norm

$$\|w\|_{L^\infty_{a,\vartheta}(0,T;L^q(\Omega))} := \operatorname*{esssup}_{0<t<T} t^\vartheta \Big\|\exp(-at)w(\cdot,t)\Big\|_{L^q(\Omega)}.$$

On account of the assumptions $u_0 \in L^p(\Omega)$ and $\max\{-1,-\frac{d}{4}\} < \nu \le \min\left\{\frac{(p-2)d}{4p}, 0\right\}$, we note that $u_0 \in D(\mathcal{A}^\nu)$. Furthermore, from $\mu - \nu < 1$, we find that

$$\begin{aligned} \Big\|\mathbf{S}(\alpha,t)u_0\Big\|^2_{D(\mathcal{A}^\mu)} &= \sum_{n=1}^\infty \lambda_n^{2\mu}\Big|\mathbb{P}_\alpha(n,t)\Big|^2 |u_{0,n}|^2 \\ &\le \sum_{n=1}^\infty \lambda_n^{2\mu}|R(m,\alpha)|^2\Big(\frac{1}{1+\lambda_n t^{1-\alpha}}\Big)^{2\mu-2\nu} \\ &\quad \times \Big(\frac{1}{1+\lambda_n t^{1-\alpha}}\Big)^{2-2\mu+2\nu} |u_{0,n}|^2 \\ &\le |R(m,\alpha)|^2 t^{2(1-\alpha)(\nu-\mu)} \sum_{n=1}^\infty \lambda_n^{2\nu}|u_{0,n}|^2 \\ &= |R(m,\alpha)|^2 t^{2(1-\alpha)(\nu-\mu)}\|u_0\|^2_{D(\mathcal{A}^\nu)}. \end{aligned}$$

The latter estimation and the embedding $L^p(\Omega) \hookrightarrow D(\mathcal{A}^\nu)$ gives

$$t^\vartheta e^{-at}\Big\|\mathbf{S}(\alpha,t)u_0\Big\|_{D(\mathcal{A}^\mu))} \le R(m,\alpha)t^{(1-\alpha)(\nu-\mu)+\vartheta}e^{-at}\|u_0\|_{L^p(\Omega)}.$$

Since $(1-\alpha)(\mu-\nu)<1$, we can choose $(1-\alpha)(\mu-\nu)\le\vartheta<1$. Then, by using the Sobolev embedding $D(\mathcal{A}^{\mu}))\hookrightarrow L^{\frac{2d}{d-4\mu}}(\Omega)$, we find that

$$\begin{aligned}t^{\vartheta}e^{-at}\Big\|\mathbf{S}(\alpha,t)u_0\Big\|_{L^{\frac{2d}{d-4\mu}}(\Omega)} &\lesssim t^{\vartheta}e^{-at}\Big\|\mathbf{S}(\alpha,t)u_0\Big\|_{D(\mathcal{A}^{\mu}))}\\ &\le R(m,\alpha)e^{-at}t^{(1-\alpha)(\nu-\mu)+\vartheta}\|u_0\|_{D(\mathcal{A}^{\nu})}\\ &\le C_{\nu,p}e^{-at}R(m,\alpha)T^{(1-\alpha)(\nu-\mu)+\vartheta}\|u_0\|_{L^p(\Omega)}.\end{aligned}$$

Therefore, we deduce that $\mathbf{S}(\alpha,t)u_0\in L^{\infty}_{a,\vartheta}(0,T;L^{\frac{2d}{d-4\mu}}(\Omega))$. By the same argument as in the previous step, since $\mu<1$, for $v_1,v_2\in L^{\frac{2d}{d-4\mu}}(\Omega)$ we have

$$\begin{aligned}&t^{\vartheta}e^{-at}\Big\|\int_0^t\mathbf{S}(\alpha,t-\tau)\Big(F(v_1(s))-F(v_2)(\tau)\Big)d\tau\Big\|_{D(\mathcal{A}^{\mu}))}\\ &\le t^{\vartheta}\int_0^t e^{-at}\Big\|\mathbf{S}(\alpha,t-\tau)\Big(F(v_1(s))-F(v_2)(\tau)\Big)\Big\|_{D(\mathcal{A}^{\mu}))}d\tau\\ &\le R(m,\alpha)\lambda_1^{\mu-1}t^{\vartheta}\int_0^t e^{-a(t-\tau)}(t-\tau)^{\alpha-1}\tau^{-\vartheta}\tau^{\vartheta}e^{-a\tau}\|F(v_1(\tau))\\ &\quad-F(v_2)(\tau)\|_{L^2(\Omega)}d\tau\\ &\le K_FR(m,\alpha)\lambda_1^{\mu-1}\Big(t^{\vartheta}\int_0^t(t-\tau)^{\alpha-1}e^{-a(t-\tau)}\tau^{-\vartheta}d\tau\Big)\\ &\quad\times\Big(\operatorname*{esssup}_{0<t<T}\tau^{\vartheta}e^{-a\tau}\|v_1(\tau)-v_2(\tau)\|_{L^2(\Omega)}\Big).\end{aligned}\tag{8.39}$$

Regarding the term $\int_0^t e^{-a(t-\tau)}(t-\tau)^{\alpha-1}\tau^{-\vartheta}d\tau$, using the inequality $e^{-a\tau}\le C_{\epsilon}(a\tau)^{-\epsilon}$, for any $0<\epsilon<\alpha$, we obtain

$$\begin{aligned}\int_0^t e^{-a(t-\tau)}(t-\tau)^{\alpha-1}\tau^{-\vartheta}d\tau &\le C_{\epsilon}a^{-\epsilon}\int_0^t(t-\tau)^{\alpha-\epsilon-1}\tau^{-\vartheta}d\tau\\ &=C_{\epsilon}a^{-\epsilon}t^{\alpha-\epsilon-\vartheta}\int_0^1(1-\tau)^{\alpha-\epsilon-1}\tau^{-\vartheta}d\tau\\ &=C_{\epsilon}a^{-\epsilon}B(\alpha-\epsilon,1-\vartheta)t^{\alpha-\epsilon-\vartheta}.\end{aligned}\tag{8.40}$$

It follows from (8.39) and (8.40) that

$$\begin{aligned}&t^{\vartheta}e^{-at}\Big\|\int_0^t\mathbf{S}(\alpha,t-\tau)\Big(F(v_1(\tau))-F(v_2)(\tau)\Big)d\tau\Big\|_{L^{\frac{2d}{d-4\mu}}(\Omega)}\\ &\le t^{\vartheta}e^{-at}\Big\|\int_0^t\mathbf{S}(\alpha,t-\tau)\Big(F(v_1(\tau))-F(v_2)(\tau)\Big)d\tau\Big\|_{D(\mathcal{A}^{\mu}))}\\ &\le K_F\lambda_1^{\mu-1}R(m,\alpha)C_{\epsilon}a^{-\epsilon}B(\alpha-\epsilon,1-\vartheta)a^{-\epsilon}t^{\alpha-\epsilon}\\ &\quad\times\Big(\operatorname*{esssup}_{0<t<T}\tau^{\vartheta}e^{-a\tau}\|v_1(\tau)-v_2(\tau)\|_{L^2(\Omega)}\Big)\\ &\le K_F\lambda_1^{\mu-1}R(m,\alpha)C_{\epsilon}a^{-\epsilon}B(\alpha-\epsilon,1-\vartheta)a^{-\epsilon}t^{\alpha-\epsilon}\\ &\quad\times\Big(\operatorname*{esssup}_{0<t<T}\tau^{\vartheta}e^{-a\tau}\|v_1(\tau)-v_2(\tau)\|_{L^{\frac{2d}{d-4\mu}}(\Omega)}\Big),\end{aligned}$$

where we have used the Sobolev embeddings $D(\mathcal{A}^{\mu})) \hookrightarrow L^{\frac{2d}{d-4\mu}}(\Omega) \hookrightarrow L^2(\Omega)$. Thus, we deduce that

$$\begin{aligned}&\|\mathbf{J}v_1 - \mathbf{J}v_2\|_{L^{\infty}_{a,\vartheta}(0,T;L^{\frac{2d}{d-4\mu}}(\Omega))}\\&\le K_F \lambda_1^{\mu-1} R(m,\alpha) C_\epsilon B(\alpha-\epsilon, 1-\vartheta) a^{-\epsilon} T^{\alpha-\epsilon} \|v_1 - v_2\|_{L^{\infty}_{a,\vartheta}(0,T;L^{\frac{2d}{d-4\mu}}(\Omega))}.\end{aligned}$$

Choosing a such that $K_F R(m,\alpha) C_\epsilon B(\alpha-\epsilon,1-\vartheta) a^{-\epsilon} T^{\alpha-\epsilon} < 1$, we come to the conclusion that $\mathbf{J}$ is a strict contraction on the space $L^{\infty}_{a,\vartheta}(0,T;L^{\frac{2d}{d-4\mu}}(\Omega))$. □

8.3.4 *Locally Lispchitz Source Term*

In this subsection, instead of (8.28) we consider the following assumption for the source term F.

Assumption (**H**): For every $z_0 \in \mathbb{R}$, there exist an open neighborhood of z_0 and a positive constant $\mathcal{Q}$ depending on this neighborhood with

$$\begin{aligned}\|F(v) - F(w)\|_{L^2(\Omega)} &\le \mathcal{Q}\|v - w\|_{L^2(\Omega)},\\ \|F(v)\|_{L^2(\Omega)} &\le \mathcal{Q}\|v\|_{L^2(\Omega)}\end{aligned}$$

for all v, w belonging to this neighborhood of z_0.

8.3.4.1 *Existence of the Mild Solution*

Theorem 8.7. *Suppose that F satisfies (H) and $u_0 \in D(\mathcal{A}^{\nu})$ for $0 < \nu < 1$. Then we can find a suitable real number $T^* > 0$ depending only on u_0 such that Problem* (8.24) *has a unique mild solution on* $(0, T^*)$.

Proof. Because of $u_0 \in D(\mathcal{A}^{\nu})$, by Lemma 8.8, we can deduce that

$$\widetilde{u}_0 := \mathbf{S}(\alpha,t) u_0 \in D(\mathcal{A}^{\nu}).$$

From (H), we can find a positive constant $R' > 0$ and a Lipschitz constant $\mathcal{Q} = \mathcal{Q}(\widetilde{u}_0, R')$ such that

$$\|F(v) - F(w)\|_{L^2(\Omega)} \le \mathcal{Q}\|v - w\|_{L^2(\Omega)}, \quad \forall v, w \in B(\widetilde{u}_0, R'). \tag{8.41}$$

We continue the proof by considering the following space

$$\begin{aligned}\mathbb{X}_{\nu,T^*} := \Bigg\{ & u \in C([0,T^*]; D(\mathcal{A}^{\nu})) \;\Big|\; u(t,0) = u_0,\\ & \text{and } \;\|u(\cdot,t) - \widetilde{u}_0\|_{D(\mathcal{A}^{\nu})} \le R^\star < R', \;\; \forall\, t \in [0,T^*] \Bigg\},\end{aligned}$$

and defining the mapping $\mathscr{B}$ on $\mathbb{X}_{\nu,T^*}$ by

$$\begin{aligned}\mathscr{B}(u)(t) = \sum_{n=1}^{\infty} \Bigg[& \mathbb{P}_\alpha(n,t) \Big\langle u_0(\cdot), \varphi_n(\cdot) \Big\rangle \\ & + \int_0^t \mathbb{P}_\alpha(n,t-\tau) \Big\langle F(u(\cdot,\tau)), \varphi_n(\cdot) \Big\rangle d\tau \Bigg] \varphi_n(x).\end{aligned} \tag{8.42}$$

First, we can see that $\|u\|_{\mathbb{X}_{\nu,T^*}} := \sup_{t\in[0,T^*]} \|u(\cdot,t)\|_{D(\mathcal{A}^\nu)}$ defines a norm on $\mathbb{X}_{\nu,T^*}$. Furthermore the Sobolev embedding $D(\mathcal{A}^\nu) \hookrightarrow L^2(\Omega)$ yields $\|w\|_{L^2(\Omega)} \leq C_\nu \|w\|_{D(\mathcal{A}^\nu)}$, for all $w \in D(\mathcal{A}^\nu)$, for some $C_\nu > 0$. With the purpose of using the Banach fixed point theorem, we take $u \in \mathbb{X}_{\nu,T^*}$ and set

$$(I) := \sum_{n=1}^{\infty} \left[\mathbb{P}_\alpha(n,t)\Big\langle u_0(\cdot), \varphi_n(\cdot)\Big\rangle\right] \varphi_n(x)$$

$$(II) := \sum_{n=1}^{\infty} \left[\int_0^t \mathbb{P}_\alpha(n,t-\tau)\Big\langle F(u(\cdot,\tau)), \varphi_n(\cdot)\Big\rangle d\tau\right] \varphi_n(x).$$

Applying Lemma 8.8, we get

$$\begin{aligned}\|(II)\|_{D(\mathcal{A}^\nu)} &= \left(\sum_{n=1}^{\infty} \lambda_n^{2\nu} \left[\int_0^t \mathbb{P}_\alpha(n,t-\tau)\Big\langle F(u(\cdot,\tau)), \varphi_n(\cdot)\Big\rangle d\tau\right]^2\right)^{\frac{1}{2}} \\ &\leq R(m,\alpha)\lambda_1^{\nu-1} \left(\sum_{n=1}^{\infty} \left[\int_0^t (t-\tau)^{\alpha-1}\Big\langle F(u(\cdot,\tau)), \varphi_n(\cdot)\Big\rangle d\tau\right]^2\right)^{\frac{1}{2}}.\end{aligned}$$

Using the Hölder inequality, we have

$$\begin{aligned}&\sum_{n=1}^{\infty} \left[\int_0^t (t-\tau)^{\alpha-1}\Big\langle F(u(\cdot,\tau)), \varphi_n(\cdot)\Big\rangle d\tau\right]^2 \\ &= \sum_{n=1}^{\infty} \left[\int_0^t (t-\tau)^{\frac{\alpha-1}{2}}(t-\tau)^{\frac{\alpha-1}{2}}\Big\langle F(u(\cdot,\tau)), \varphi_n(\cdot)\Big\rangle d\tau\right]^2 \\ &\leq \left[\int_0^t (t-\tau)^{\alpha-1} d\tau\right]\left[\int_0^t (t-\tau)^{\alpha-1}\sum_{n=1}^{\infty}\Big\langle F(u(\cdot,\tau)), \varphi_n(\cdot)\Big\rangle^2 d\tau\right] \\ &\leq \frac{t^{2\alpha}}{\alpha^2}\left(\sup_{t\in[0,T]} \|F(u(\cdot,t))\|_{L^2(\Omega)}\right)^2.\end{aligned}$$

This implies that

$$\|(II)\|_{D(\mathcal{A}^\nu)} \leq R_1(m,\alpha)t^\alpha \sup_{t\in[0,T]} \|F(u(\cdot,t))\|_{L^2(\Omega)}, \tag{8.43}$$

where $R_1(m,\alpha) = \frac{R(m,\alpha)\lambda_1^{\nu-1}}{\alpha}$. From the assumption (H), we obtain

$$\begin{aligned}\|F(u(\cdot,t))\|_{L^2(\Omega)} &\leq \mathcal{Q}\|u(\cdot,t) - \widetilde{u}_0\|_{L^2(\Omega)} + \|F(\widetilde{u}_0)\|_{L^2(\Omega)} \\ &\leq C_\nu \mathcal{Q}\left(R^\star + \|\widetilde{u}_0\|_{D(\mathcal{A}^\nu)}\right),\end{aligned}$$

for some $C_\nu > 0$. Hence, for $t \in [0,T^*]$, we get the following estimate

$$\begin{aligned}\|\mathscr{B}(u(\cdot,t)) - (I)\|_{D(\mathcal{A}^\nu)} &\leq R_1(m,\alpha)t^\alpha \|F(u)\|_{L^\infty((0,T);L^2(\Omega))} \\ &\leq R_1(m,\alpha)(T^*)^\alpha C_\nu \mathcal{Q}\left(R^\star + \|\widetilde{u}_0\|_{D(\mathcal{A}^\nu)}\right).\end{aligned}$$

Then, there exists a small time $T^* > 0$ such that

$$R_1(m,\alpha)(T^*)^\alpha C_\nu \mathcal{Q}\left(R^\star + \|\widetilde{u}_0\|_{D(\mathcal{A}^\nu)}\right) \leq R^\star.$$

Therefore, $\mathscr{B}(u) \in \mathbb{X}_{\nu,T^*}$ for any $u \in \mathbb{X}_{\nu,T^*}$. Next, we prove that if $T^* > 0$ is small enough, $\mathscr{B} : \mathbb{X}_{\nu,T^*} \to \mathbb{X}_{\nu,T^*}$ is a contraction on $C([0,T^*]; D(\mathcal{A}^\nu))$. Suppose that $u, v \in \mathbb{X}_{\nu,T^*}$, and by the same argument in considering (8.43), we obtain

$$\begin{aligned}
&\|\mathscr{B}(u)(\cdot,t) - \mathscr{B}(v)(\cdot,t)\|_{D(\mathcal{A}^\nu)} \\
&= \left(\sum_{n=1}^{\infty} \lambda_n^{2\nu} \left[\int_0^t \mathbb{P}_n(\alpha, t-\tau)\Big\langle F(u(\cdot,\tau)) - F(v(\cdot,\tau)), \varphi_n(\cdot)\Big\rangle d\tau\right]^2\right)^{\frac{1}{2}} \\
&\le R(m,\alpha)\lambda_1^{-1}\left(\sum_{n=1}^{\infty} \left[\int_0^t (t-\tau)^{\alpha-1}\Big\langle F(u(\cdot,\tau)) - F(v(\cdot,\tau)), \varphi_n(\cdot)\Big\rangle d\tau\right]^2\right)^{\frac{1}{2}} \\
&\le R_1(m,\alpha)t^\alpha \sup_{t\in[0,T]} \|F(u) - F(v)\|_{L^2(\Omega)}.
\end{aligned} \tag{8.44}$$

Thus, there exists a constant $C > 0$ such that

$$\|\mathscr{B}(u) - \mathscr{B}(v)\|_{\mathbb{X}_{\nu,T^*}} \le C(T^\star)^\alpha \|u - v\|_{\mathbb{X}_{\nu,T^*}}. \tag{8.45}$$

As a result, if we choose a sufficiently small time T^*, the mapping $\mathscr{B}$ will be a contraction on $\mathbb{X}_{\nu,T^*}$. In other words, $\mathscr{B}$ has a unique fixed point u in $\mathbb{X}_{\nu,T^*}$. □

8.3.4.2 *Continuation and Blow-Up Alternative*

In this subsection, we demonstrate a continuation result and a blow-up alternative for the mild solution which is proposed in Theorem 8.7. In particular, we also obtain the existence of a maximal time to this solution.

Theorem 8.8. *Suppose that F satisfies (H) and $u_0 \in D(\mathcal{A}^\nu)$ for any $0 < \nu < 1$. Then the solution on $(0, T^*)$ of Problem (8.24) is extended to $[0, T^* + \epsilon]$, for some $\epsilon > 0$.*

Proof. Letting T^* as in subsection 8.3.4.1, by virtue of (H), we can find a open ball $B\left(u(T^*), R''\right)$ and a constant $\mathcal{Q}_1 = \mathcal{Q}\left(u(T^*), R''\right)$ such that F is Lipschitz continuous when restricted to this ball. For $\epsilon > 0$, we introduce the following space

$$\mathbb{K} := \left\{ v \in C([0, T^* + \epsilon]; D(\mathcal{A}^\nu)) \,\middle|\, \begin{array}{ll} v(\cdot,t) = u(\cdot,t), & t \in [0,T^*], \\ \|v(\cdot,t) - u(\cdot,T^*)\|_{D(\mathcal{A}^\nu)} \le \mathfrak{R} < R'', & t \in [T^*, T^* + \epsilon] \end{array} \right\}$$

and define a mapping $\mathscr{P} : \mathbb{K} \to \mathbb{K}$ by

$$\mathscr{P}(v)(t) = \sum_{n=1}^{\infty} \left[\mathbb{P}_\alpha(n,t)u_{0,n} + \int_0^t \mathbb{P}_\alpha(n, t-\tau)\Big\langle F(u(\cdot,\tau)), \varphi_n(\cdot)\Big\rangle d\tau\right] \varphi_n(x).$$

To show that $\mathscr{P}$ has a fixed point in $\mathbb{K}$, we consider some steps.

Step I. It is easy to see that $\mathscr{P}(v)$ is continuously differentiable on $[0, T^* + \epsilon]$ provided that ε is sufficiently small. First, we verify that $\mathscr{P}$ is a mapping from $\mathbb{K}$

to $\mathbb{K}$. Let $v \in \mathbb{K}$. If $t \in [0, T^*]$, then $v(\cdot, t) = u(\cdot, t)$. Note $\mathscr{P}(v(\cdot, t)) = \mathscr{P}(u(\cdot, t)) = u(\cdot, t)$. If $t \in [T^*, T^* + \epsilon]$, then we get

$$\begin{aligned}
&\|\mathscr{P}(v)(t) - u(\cdot, T^*)\|_{D(\mathcal{A}^\nu)} \\
&\leq \left\| \mathbf{S}_\alpha(t) u_0 - \mathbf{S}_\alpha(T^*) u_0 \right\|_{D(\mathcal{A}^\nu)} \\
&\quad + \int_{T^*}^t \left\| \sum_{n=1}^\infty \mathbb{P}_n(\alpha, t-\tau) \Big\langle F(v(\cdot, \tau)), \varphi_n(\cdot) \Big\rangle \varphi_n(x) \right\|_{D(\mathcal{A}^\nu)} d\tau \\
&\quad + \int_0^{T^*} \left\| \sum_{n=1}^\infty [\mathbb{P}_n(\alpha, t-\tau) - \mathbb{P}_n(\alpha, T^* - \tau)] \right. \\
&\quad \times \left. \Big\langle F(v(\cdot, \tau)), \varphi_n(\cdot) \Big\rangle \varphi_n(x) \right\|_{D(\mathcal{A}^\nu)} d\tau.
\end{aligned}$$

From the fact that $\mathbf{S}_\alpha(t) u_0 \in C([0, \infty), D(\mathcal{A}^\nu))$, we suppose that $\epsilon > 0$ is small enough to get the following inequality

$$\mathcal{M}_1 := \left\| \mathbf{S}_\alpha(t) u_0 - \mathbf{S}_\alpha(T^*) u_0 \right\|_{D(\mathcal{A}^\nu)} \leq \frac{\mathfrak{R}}{3}. \tag{8.46}$$

Similar to the latter step, we have the following estimate

$$\begin{aligned}
\mathcal{M}_2 &:= \int_{T^*}^t \left\| \sum_{n=1}^\infty \mathbb{P}_n(\alpha, t-\tau) \Big\langle F(v(\cdot, \tau)), \varphi_n(\cdot) \Big\rangle \varphi_n(x) \right\|_{D(\mathcal{A}^\nu)} d\tau \\
&= \int_{T^*}^t \sqrt{\sum_{n=1}^\infty \lambda_n^{2\nu} \Big| \mathbb{P}_n(\alpha, t-\tau) \Big|^2 \Big\langle F(v(\cdot, \tau)), \varphi_n(\cdot) \Big\rangle^2} d\tau \\
&\leq \frac{R(m, \alpha)}{\lambda_1^{1-\nu}} \Big\| F(v) \Big\|_{L^\infty((0,T); L^2(\Omega))} \left(\int_{T^*}^t (t-\tau)^{\alpha-1} \right) d\tau \\
&= \frac{R(m, \alpha)}{\lambda_1^{1-\nu}} \big(\|F(v) - F(v(T^*))\|_{L^\infty((0,T); L^2(\Omega))} + \|F(u(T^*))\|_{L^2(\Omega)} \big) \\
&\quad \times \frac{(t - T^*)^\alpha}{\alpha},
\end{aligned}$$

which implies that we can choose a number $\epsilon > 0$ such that

$$\mathcal{M}_2 \leq \frac{\mathfrak{R}}{3}, \quad \text{for all } t \in [T^*, T^* + \epsilon]. \tag{8.47}$$

We consider the upper bound of $\mathbb{P}_n(\alpha, t-\tau) - \mathbb{P}_n(\alpha, T^* - \tau)$ for $0 \leq \tau \leq T^* \leq t$ for the following term

$$\begin{aligned}
\mathcal{M}_3 := \int_0^{T^*} &\left\| \sum_{n=1}^\infty \Big[\mathbb{P}_n(\alpha, t-\tau) - \mathbb{P}_n(\alpha, T^* - \tau) \Big] \right. \\
&\times \left. \Big\langle F(v(\cdot, \tau)), \varphi_n(\cdot) \Big\rangle \varphi_n(x) \right\|_{D(\mathcal{A}^\nu)} d\tau.
\end{aligned}$$

It should be noted that

$$\Big|\mathbb{P}_n(\alpha, t-\tau) - \mathbb{P}_n(\alpha, T^*-\tau)\Big| = \int_0^\infty \Big|e^{-r(t-\tau)} - e^{-r(T^*-\tau)}\Big| \mathbb{K}_\alpha(n,r)dr$$
$$\leq \frac{1}{m\pi\sin(\alpha\pi)\lambda_n}\int_0^\infty \Big|e^{-r(t-\tau)} - e^{-r(T^*-\tau)}\Big| \times r^{-\alpha}dr.$$

For $0<\beta<\alpha<1$, thanks to the fact that $1-e^{-w}\leq w^\beta, \forall w\geq 0$, we obtain

$$\int_0^\infty \Big|e^{-r(t-\tau)} - e^{-r(T^*-\tau)}\Big| r^{-\alpha}dr \leq (t-T^*)\int_0^\infty e^{-r(t-\tau)}r^{\beta-\alpha}dr$$
$$= \frac{(t-T^*)^\beta}{(t-\tau)^{1+\beta-\alpha}}\int_0^\infty e^{-y}y^{\beta-\alpha}dy \qquad (8.48)$$
$$= \frac{\Gamma(1+\beta-\alpha)(t-T^*)^\beta}{(t-\tau)^{1+\beta-\alpha}}$$

where $y=r(t-\tau)$. This together with (8.43) and (8.48) yield

$$\Big|\mathbb{P}_n(\alpha, t-\tau) - \mathbb{P}_n(\alpha, T^*-\tau)\Big| \leq \frac{\Gamma(1+\beta-\alpha)(t-T^*)^\beta}{m\pi\sin(\alpha\pi)\lambda_n(t-\tau)^{1+\beta-\alpha}}$$
$$\leq \frac{\Gamma(1+\beta-\alpha)}{m\pi\sin(\alpha\pi)\lambda_n}\frac{(t-T^*)^\beta}{(t-\tau)^{1+\beta-\alpha}}.$$

Then we find that

$$\Bigg\|\sum_{n=1}^\infty\Big[\mathbb{P}_n(\alpha, t-\tau) - \mathbb{P}_n(\gamma, T^*-\tau)\Big]\Big\langle F(v(\cdot,\tau)), \varphi_n(\cdot)\Big\rangle\varphi_n(x)\Bigg\|_{D(\mathcal{A}^\nu)}$$
$$= \left(\sum_{n=1}^\infty \lambda_n^{2\nu}\Big[\mathbb{P}_n(\alpha, t-\tau) - \mathbb{P}_n(\gamma, T^*-\tau)\Big]^2 \times \Big\langle F(v(\cdot,\tau)), \varphi_n(\cdot)\Big\rangle^2\right)^{\frac{1}{2}} \qquad (8.49)$$
$$\leq \frac{\lambda_1^{\nu-1}\Gamma(1+\beta-\alpha)}{m\pi\sin(\alpha\pi)}(t-T^*)^\beta(t-\tau)^{\alpha-\beta-1}\|F(v(\cdot,\tau))\|_{L^2(\Omega)},$$

where we note that $0<\alpha<1$. This implies that

$$\mathcal{M}_3 = \int_0^{T^*}\Bigg\|\sum_{n=1}^\infty\Big[\mathbb{P}_n(\alpha, t-\tau) - \mathbb{P}_n(\gamma, T^*-\tau)\Big] \times \Big\langle F(v(\cdot,\tau)), \varphi_n(\cdot)\Big\rangle\varphi_n(x)\Bigg\|_{D(\mathcal{A}^\nu)} d\tau$$
$$\leq \frac{\lambda_1^{\nu-1}\Gamma(1+\beta-\alpha)}{m\pi\sin(\alpha\pi)}(t-T^*)^\beta\Big\|F(v)\Big\|_{L^\infty((0,T);L^2(\Omega))} \times \left(\int_0^{T^*}(t-\tau)^{\alpha-\beta-1}d\tau\right)$$

$$= \frac{\lambda_1^{\nu-1}\Gamma(1+\beta-\alpha)}{m\pi\sin(\alpha\pi)}(t-T^*)^\beta \frac{t^{\alpha-\beta}-(t-T^*)^{\alpha-\beta}}{\alpha-\beta}\Big\|F(v)\Big\|_{L^\infty((0,T);L^2(\Omega))}.$$

By taking $\beta = \frac{\alpha}{2}$ and using the inequality $t^{\alpha-\beta}-(t-T^*)^{\alpha-\beta} \le (T^*)^{\alpha-\beta}$, we deduce that

$$\mathcal{M}_3 \le \frac{2\lambda_1^{\nu-1}\Gamma(1-\beta)}{m\alpha\pi\sin(\alpha\pi)}(T^*)^\beta(t-T^*)^\beta\Big\|F(v)\Big\|_{L^\infty((0,T);L^2(\Omega))}.$$

Therefore, there exists a small $\epsilon > 0$ such that

$$\mathcal{M}_3 \le \frac{\mathfrak{R}}{3}, \qquad \text{for } t \in [T^*, T^*+\epsilon] \tag{8.50}$$

In view of (8.46), (8.47) and (8.50), we have

$$\|\mathscr{P}(v)(t) - u(\cdot, T^*)\|_{D(\mathcal{A}^\nu)} \le \mathfrak{R}, \qquad \text{for } t \in [T^*, T^*+\epsilon],$$

provided that ϵ is suitably small.

Step II. We prove that $\mathscr{P}$ is a contraction on $\mathbb{K}$. Let $v, w \in \mathbb{K}$ and we have

$$\mathscr{P}(v)(t) - \mathscr{P}(w)(t) = \sum_{n=1}^{\infty}\left[\int_0^t \mathbb{P}_\alpha(n, t-\tau)\Big\langle F(v(\cdot,\tau)) - F(w(\cdot,\tau)), \varphi_n(\cdot)\Big\rangle d\tau\right]\varphi_n.$$

If $t \in [0, T^*]$, then in view of Theorem 8.7 (see (8.44) and (8.45)), we can find a constant $C > 0$ such that

$$\|\mathscr{P}(v)(t) - \mathscr{P}(w)(t)\|_{D(\mathcal{A}^\nu)} \le C(T^*)^\alpha \|v-w\|_{C([0,T^*];D(\mathcal{A}^\nu))}.$$

If $t \in [T^*, T^*+\epsilon]$, from Lemma 8.8, there exists $C > 0$ such that

$$\begin{aligned}
&\Big\|\mathscr{P}(v)(t) - \mathscr{P}(w)(t)\Big\|_{D(\mathcal{A}^\nu)} \\
&= \left\|\sum_{n=1}^{\infty}\left[\int_{T^*}^t \mathbb{P}_\alpha(n, t-\tau)\Big\langle F(v(\cdot,\tau)) - F(w(\cdot,\tau)), \varphi_n(\cdot)\Big\rangle d\tau\right]\varphi_n\right\|_{D(\mathcal{A}^\nu)} \\
&= \int_{T^*}^t \sqrt{\sum_{n=1}^{\infty}\lambda_n^{2\nu}\Big|\mathbb{P}_\alpha(n,t-\tau)\Big|^2\Big\langle F(v(\cdot,\tau)) - F(w(\cdot,\tau)), \varphi_n(\cdot)\Big\rangle^2}\, d\tau \\
&\le \frac{R(m,\alpha)}{\lambda_1^{1-\nu}}\Big\|F(v) - F(w)\Big\|_{L^\infty((T^*,t);L^2(\Omega))}\left(\int_{T^*}^t (t-s)^{\alpha-1}\right) ds \\
&= \frac{R(m,\alpha)}{\lambda_1^{1-\nu}}\Big\|F(v) - F(w)\Big\|_{L^\infty((T^*,t);L^2(\Omega))}\frac{(t-T^*)^\alpha}{\alpha} \\
&\le C\epsilon^\alpha \|F(v) - F(w)\|_{L^\infty((T^*,t);L^2(\Omega))} \\
&\le CQ\epsilon^\alpha \|v-w\|_{C([T^*,T^*+\epsilon];D(\mathcal{A}^\nu))}.
\end{aligned} \tag{8.51}$$

Consequently, we can find a number $C > 0$ such that

$$\|\mathscr{P}(v) - \mathscr{P}(w)\|_{\mathbb{K}} \le C\epsilon^\alpha \|v-w\|_{\mathbb{K}}.$$

Then, it is possible to find a positive number ϵ small enough to ensure that $\mathscr{P}$ is a contraction on $\mathbb{K}$. Therefore $\mathscr{P}$ has a unique fixed point v on $\mathbb{K}$. The proof is completed. □

Lemma 8.9. *Let $T > 0$ and $u : \Omega \times [0, T) \to L^2(\Omega)$ satisfy*

$$\sup_{t \in [0,T)} \Big(\|u(\cdot, t)\|_{D(\mathcal{A}^\nu)} \Big) < \infty.$$

Let $t_n \in [0, T)$ be a sequence such that $\lim_{n\to\infty} t_n = T$. Then

$$\lim_{n\to\infty} \int_0^{t_n} \left\| \sum_{k=1}^{\infty} \Big[\mathbb{P}_\alpha(k, t_n - \tau) - \mathbb{P}_\alpha(k, T - \tau) \Big] F(u(\cdot, \tau))\varphi_k \right\|_{D(\mathcal{A}^\nu)} d\tau = 0. \quad (8.52)$$

Proof. Let $\pounds := \sup_{s\in[0,T)} \|F(u(\cdot, s))\|_{L^2(\Omega)} < \infty$. Given $\epsilon > 0$, fix $\delta \in (0, T)$ such that

$$4\overline{B}_1 (T - \delta)^{\frac{\gamma}{2}} T^{\frac{\gamma}{2}} \pounds \leq \epsilon, \quad \overline{B}_1 = \frac{\lambda_1^{\nu-1}\Gamma(1 - \frac{\alpha}{2})}{m\alpha\pi \sin(\gamma\pi)}. \quad (8.53)$$

Similar to the argument in (8.49), we immediately infer that

$$\begin{aligned} &\left\| \sum_{k=1}^{\infty} \Big[\mathbb{P}_\alpha(k, t_n - \tau) - \mathbb{P}_\alpha(k, T - \tau) \Big] F(u(\cdot, \tau))\varphi_k \right\|_{D(\mathcal{A}^\nu)} \\ &\leq \overline{B}_1 \frac{(T - t_n)^{\frac{\alpha}{2}}}{(t_n - \tau)^{1-\frac{\alpha}{2}}} \Big\| F(u) \Big\|_{L^\infty((0,T);L^2(\Omega))}. \end{aligned}$$

By means of $0 < \delta < T$ and $\lim_{n\to\infty} t_n = T$, we can choose $\mathbf{N}_{1,\delta} \in \mathbb{N}$ such that $t_n > \delta$ for $n \geq \mathbf{N}_{1,\delta}$. This implies

$$\begin{aligned} &\int_\delta^{t_n} \left\| \sum_{k=1}^{\infty} \Big[\mathbb{P}_\alpha(k, t_n - \tau) - \mathbb{P}_\alpha(k, T - \tau) \Big] F(u(\cdot, \tau))\varphi_k \right\|_{D(\mathcal{A}^\nu)} d\tau \\ &\leq \overline{B}_1 (T - t_n)^{\frac{\alpha}{2}} \Big\| F(u) \Big\|_{L^\infty((0,T);L^2(\Omega))} \int_\delta^{t_n} (t_n - \tau)^{\frac{\alpha}{2}-1} d\tau \\ &\leq \overline{B}_1 T^{\frac{\alpha}{2}} \pounds (t_n - \delta)^{\frac{\alpha}{2}} \\ &\leq \overline{B}_1 T^{\frac{\gamma}{2}} (T - \delta)^{\frac{\alpha}{2}} \pounds. \end{aligned}$$

Also

$$\begin{aligned} &\int_0^{\delta} \left\| \sum_{k=1}^{\infty} \Big[\mathbb{P}_\alpha(k, t_n - \tau) - \mathbb{P}_\alpha(k, T - \tau) \Big] F(u(\cdot, \tau))\varphi_k \right\|_{D(\mathcal{A}^\nu)} d\tau \\ &\leq \overline{B}_1 (T - t_n)^{\frac{\gamma}{2}} \Big\| F(u) \Big\|_{L^\infty((0,T);L^2(\Omega))} \int_0^{\delta} (t_n - \tau)^{\frac{\gamma}{2}-1} d\tau \quad (8.54) \\ &\leq \overline{B}_1 (T - \delta)^{\frac{\gamma}{2}} \big[(t_n)^{\frac{\alpha}{2}} - (t_n - \delta)^{\frac{\gamma}{2}} \big] \pounds \\ &\leq \overline{B}_1 (T - \delta)^{\frac{\alpha}{2}} T^{\frac{\alpha}{2}} \pounds, \end{aligned}$$

provided that $t_n \geq \delta$. Combining the latter, the former, and (8.53) allows us to deduce that for $n \geq \mathbf{N}_{1,\delta}$

$$\int_0^{t_n} \left\| \sum_{k=1}^{\infty} \Big[\mathbb{P}_\alpha(k, t_n - \tau) - \mathbb{P}_\alpha(k, T - \tau) \Big] F(u(\cdot, \tau))\varphi_k \right\|_{D(\mathcal{A}^\nu)} d\tau \leq \epsilon.$$

This implies that (8.52) holds. $\square$

The next theorem is a result on global existence or noncontinuation by blowup. Namely, we have the following:

Theorem 8.9. *Assume that F satisfies (H) and $u_0 \in D(\mathcal{A}^\nu)$ for any $0 < \nu < 1$. Let u be the mild solution of Problem (8.24), u is defined on $[0, T_{\max})$, where $T_{\max}$ is the maximal time of existence of u. Then we have $T_{\max} = +\infty$ or*

$$\lim_{t \to T_{\max}^-} \|u(\cdot, t)\|_{D(\mathcal{A}^\nu)} = \infty. \tag{8.55}$$

Proof. Let

$$\mathcal{T} := \left\{ T \in [0, \infty) : \exists\, u : \Omega \times [0, T] \to L^2(\Omega) \ \text{ a solution to (8.24) in } \ (0, T) \right\},$$

and set $T_{\max} := \sup \mathcal{T}$. Assume that $T_{\max} < \infty$ and we demonstrate that (8.55) is true. Suppose that there exists $\overline{K}_0 < \infty$ such that $\|u(\cdot, t)\|_{D(\mathcal{A}^\nu)} \le \overline{K}_0, \ \ \forall\, t \in [0, T_{\max})$. Let a sequence $(t_n)_{n \in \mathbb{N}} \subset [0, T_{\max})$ satisfy $\lim_{n \to +\infty} t_n = T_{\max}$. Let $t_n > t_m$ and $\widetilde{K} := \sup_{t \in [0, T_{\max})} \|F(u(\cdot, t))\|_{L^2(\Omega)} < \infty$. Then using (8.54), we get from Lemma 8.9 that

$$\begin{aligned}
&\left\| \int_{t_m}^{t_n} \sum_{k=1}^{\infty} \mathbb{P}_\alpha(k, T_{\max} - \tau) F(u(\cdot, \tau)) \varphi_k \, d\tau \right\|_{D(\mathcal{A}^\nu)} \\
&\le \int_{t_m}^{t_n} \sum_{k=1}^{\infty} \left\| \mathbb{P}_\alpha(k, T_{\max} - \tau) F(u(\cdot, \tau)) \varphi_k \right\|_{D(\mathcal{A}^\nu)} d\tau \\
&\le \frac{R(m, \alpha)}{\lambda_1^{1-\nu}} \int_{t_m}^{t_n} (T_{\max} - s)^{\alpha - 1} \|F(u(\cdot, \tau))\|_{L^2(\Omega)} d\tau \\
&\le \frac{\widetilde{K_F} R(m, \alpha)}{\lambda_1^{1-\nu}} \int_{t_m}^{t_n} (T_{\max} - \tau)^{\alpha - 1} d\tau \\
&= \widetilde{K_F} R(m, \alpha) \lambda_1^{\nu - 1} \alpha^{-1} \Big[(T_{\max} - t_n)^\alpha - (T_{\max} - t_m)^\alpha \Big] \to 0, \quad \text{as } n, m \to \infty.
\end{aligned}$$

Then, we find that

$$\begin{aligned}
\|u(\cdot, t_n) - u(\cdot, t_m)\|_{D(\mathcal{A}^\nu)} \le\ & \left\| \mathbf{S}_\alpha(t_n) u_0 - \mathbf{S}_\alpha(t_m) u_0 \right\|_{D(\mathcal{A}^\nu)} \\
& + \left\| \int_0^{t_n} \sum_{k=1}^{\infty} \Big[\mathbb{P}_\alpha(k, t_n - \tau) - \mathbb{P}_\alpha(k, T_{\max} - \tau) \Big] \right. \\
& \left. \times F(u(\cdot, \tau)) \varphi_k d\tau \right\|_{D(\mathcal{A}^\nu)} \\
& + \left\| \int_0^{t_m} \sum_{k=1}^{\infty} \Big[\mathbb{P}_\alpha(k, t_n - \tau) - \mathbb{P}_\alpha(k, T_{\max} - \tau) \Big] \right. \\
& \left. \times F(u(\cdot, \tau)) \varphi_k d\tau \right\|_{D(\mathcal{A}^\nu)}
\end{aligned}$$

$$+ \left\| \int_{t_m}^{t_n} \sum_{k=1}^{\infty} \mathbb{P}_\alpha(k, T_{\max} - \tau) F(u(\cdot, \tau)) \varphi_k \, d\tau \right\|_{D(\mathcal{A}^\nu)}$$

$$\to 0, \quad \text{as } n, m \to \infty,$$

where we have used Lemma 8.9. It follows that $(u(\cdot, t_n))_{n \in \mathbb{N}}$ is a Cauchy sequence. This sequence converges to a function $u_{T_{\max}}(\cdot, t) \in D(\mathcal{A}^\nu)$. Therefore, we may extend u over $[0, T_{\max}]$ and we have the following equality

$$u(\cdot, t) = \mathbf{S}_\alpha(t) u_0 + \int_0^t \mathbf{S}_\alpha(t - \tau) F(u(\tau)) d\tau$$

for all $t \in [0, T_{\max}]$. Applying Theorem 8.8, the solution is extended to some larger interval, which leads to a contradiction with $T_{\max} > 0$. The proof is completed. □

8.4 Fractional Euler-Lagrange Equations

8.4.1 *Introduction*

In this section, we consider $a < b$ two reals, $d \in \mathbb{N}$ and the following Lagrangian functional

$$\mathfrak{L}(u) = \int_a^b L(u, {}_aD_t^\alpha u, t) dt,$$

where L is a Lagrangian, i.e. a map of the form:

$$L : \mathbb{R}^d \times \mathbb{R}^d \times [a, b] \to \mathbb{R},$$
$$(x, y, t) \to L(x, y, t),$$

where ${}_aD_t^\alpha$ is the left fractional derivative of Riemann-Liouville of order $0 < \alpha < 1$ and where the variable u is a function defined almost everywhere on (a, b) with values in $\mathbb{R}^d$. It is well-known that critical points of the functional L are characterized by the solutions of the fractional Euler-Lagrange equation:

$$\frac{\partial L}{\partial x}(u, {}_aD_t^\alpha u, t) + {}_tD_b^\alpha \left(\frac{\partial L}{\partial y}(u, {}_aD_t^\alpha u, t) \right) = 0, \tag{8.56}$$

where ${}_tD_b^\alpha$ is the right fractional derivative of Riemann-Liouville, see detailed proofs in Agrawal, 2002; Baleanu and Muslih, 2005 for example.

For any $p \geq 1, L^p := L^p((a, b), \mathbb{R}^d)$ denotes the classical Lebesgue space of p-integrable functions endowed with its usual norm $\| \cdot \|_{L^p}$. We denote by $| \cdot |$ the Euclidean norm of $\mathbb{R}^d$ and $C := C([a, b], \mathbb{R}^d)$ the space of continuous functions endowed with its usual norm $\| \cdot \|$. We remind that a function f is an element of AC if and only if $f' \in L^1$ and the following equality holds

$$f(t) = f(a) + \int_a^t f'(\xi) d\xi, \quad \forall \, t \in [a, b],$$

where f' denotes the derivative of f. We refer to Kolmogorov, Fomine and Tihomirov, 1974 for more details concerning the absolutely continuous functions. In addition, we denote by C_a (resp. AC_a or C_a^∞) the space of functions $f \in C$ (resp. AC or C^∞) such that $f(a) = 0$. In particular, $C_c^\infty \subset C_a^\infty \subset AC_a$.

Remark 8.1. In the whole section, an equality between functions must be understood as an equality holding for almost all $t \in (a, b)$. When it is not the case, the interval on which the equality is valid will be specified.

Definition 8.4. A function u is said to be a weak solution of (8.56) if $u \in C$ and if u satisfies (8.56) a.e. on $[a, b]$.

In the following, we will provide some properties concerning the left fractional integral operators of Riemann-Liouville. One can easily derive the analogous versions for the right ones. Proposition 8.2 is well-known and one can find their proofs in the classical literature on the subject (see Lemma 2.1 in Kilbas, Srivastava and Trujillo, 2006).

Proposition 8.2. *For any $\alpha > 0$ and any $p \geq 1$, ${}_aD_t^{-\alpha}$ is linear and continuous from L^p to L^p. Precisely, the following inequality holds*

$$\|{}_aD_t^{-\alpha} f\|_{L^p} \leq \frac{(b-a)^\alpha}{\Gamma(1+\alpha)}\|f\|_{L^p}, \quad \textit{for } f \in L^p.$$

The following classical property concerns the integration of fractional integrals. It is occasionally called fractional integration by parts:

Proposition 8.3. *Let $0 < \frac{1}{p} < \alpha < 1$ and $q = \frac{p}{p-1}$. Then, for any $f \in L^p$, we have*

(i) *${}_aD_t^{-\alpha}$ is Hölder continuous on $[a, b]$ with exponent $\alpha - \frac{1}{p} > 0$;*
(ii) $\lim_{t\to a} {}_aD_t^{-\alpha} f(t) = 0$.

Consequently, ${}_aD_t^{-\alpha} f(t)$ can be continuously extended by 0 in $t = a$. Finally, for any $f \in L^p$, we have ${}_aD_t^{-\alpha} f \in C_a$. Moreover, the following inequality holds

$$\|{}_aD_t^{-\alpha} f\| \leq \frac{(b-a)^{\alpha-\frac{1}{p}}}{\Gamma(\alpha)((\alpha-1)q+1)^{\frac{1}{q}}}\|f\|_{L^p}, \quad \textit{for } f \in L^p.$$

Proof. Let us note that this result is mainly proved in Section 6.2. Let $f \in L^p$. We first remind the following inequality

$$(\xi_1 - \xi_2)^q \leq \xi_1^q - \xi_2^q, \quad \text{for } \xi_1 \geq \xi_2 \geq 0.$$

Let us prove that ${}_aD_t^{-\alpha} f(t)$ is Hölder continuous on $[a, b]$. For any $a < t_1 < t_2 \leq b$, using Hölder inequality, we have

$$
\begin{aligned}
|{}_aD_t^{-\alpha}f(t_2) - {}_aD_t^{-\alpha}f(t_1)| &= \frac{1}{\Gamma(\alpha)}\left|\int_a^{t_2}(t_2-\xi)^{\alpha-1}f(\xi)d\xi - \int_a^{t_1}(t_1-\xi)^{\alpha-1}f(\xi)d\xi\right| \\
&\leq \frac{1}{\Gamma(\alpha)}\left|\int_{t_1}^{t_2}(t_2-\xi)^{\alpha-1}f(\xi)d\xi\right| \\
&\quad + \frac{1}{\Gamma(\alpha)}\left|\int_a^{t_1}((t_2-\xi)^{\alpha-1} - (t_1-\xi)^{\alpha-1})f(\xi)d\xi\right| \\
&\leq \frac{\|f\|_{L^p}}{\Gamma(\alpha)}\left(\int_{t_1}^{t_2}(t_2-\xi)^{(\alpha-1)q}d\xi\right)^{\frac{1}{q}} \\
&\quad + \frac{\|f\|_{L^p}}{\Gamma(\alpha)}\left(\int_a^{t_1}((t_1-\xi)^{\alpha-1} - (t_2-\xi)^{\alpha-1})^q d\xi\right)^{\frac{1}{q}} \\
&\leq \frac{\|f\|_{L^p}}{\Gamma(\alpha)}\left(\int_{t_1}^{t_2}(t_2-\xi)^{(\alpha-1)q}d\xi\right)^{\frac{1}{q}} \\
&\quad + \frac{\|f\|_{L^p}}{\Gamma(\alpha)}\left(\int_a^{t_1}(t_1-\xi)^{(\alpha-1)q} - (t_2-\xi)^{(\alpha-1)q}d\xi\right)^{\frac{1}{q}} \\
&\leq \frac{2\|f\|_{L^p}}{\Gamma(\alpha)((\alpha-1)q+1)^{\frac{1}{q}}}(t_2-t_1)^{\alpha-\frac{1}{p}}.
\end{aligned}
$$

The proof of the first point is completed. Let us consider the second point. For any $t \in [a,b]$, we can prove in the same manner that

$$
|{}_aD_t^{-\alpha}f(t)| \leq \frac{\|f\|_{L^p}}{\Gamma(\alpha)((\alpha-1)q+1)^{\frac{1}{q}}}(t-a)^{\alpha-\frac{1}{p}}, \quad \text{as } t \to 0.
$$

The proof is now completed. □

In Subsection 8.4.2, we introduce an appropriate space of functions. Subsection 8.4.3 is concerned with variational structure. Subsection 8.4.4 is devoted to the existence theorem of weak solution for (8.56).

8.4.2 *Functional Spaces*

In order to prove the existence of a weak solution of (8.56) using a variational method, we need the introduction of an appropriate space of functions. This space has to present some properties like reflexivity, see Dacorogna, 2008.

For any $0 < \alpha < 1$ and any $p \geq 1$, we define the following space of functions

$$
E_{\alpha,p} := \{u \in L^p |\ {}_aD_t^{\alpha}u \in L^p \text{ and } {}_aD_t^{-\alpha}({}_aD_t^{\alpha}u) = u \text{ a.e.}\}.
$$

We endow $E_{\alpha,p}$ with the following norm

$$
\begin{aligned}
&\|\cdot\|_{\alpha,p} : E_{\alpha,p} \to \mathbb{R}^+, \\
&u \mapsto \left(\|u\|_{L^p}^p + \|{}_aD_t^{\alpha}u\|_{L^p}^p\right)^{\frac{1}{p}}.
\end{aligned}
$$

Let us note that

$$\begin{aligned} |\cdot|_{\alpha,p} &: E_{\alpha,p} \to \mathbb{R}^+, \\ u &\mapsto \|{}_aD_t^\alpha u\|_{L^p} \end{aligned}$$

is an equivalent norm to $\|\cdot\|_{\alpha,p}$ for $E_{\alpha,p}$. Indeed, Proposition 8.2 leads to

$$\|u\|_{L^p} = \|{}_aD_t^{-\alpha}({}_aD_t^\alpha u)\|_{L^p} \leq \frac{(b-a)^\alpha}{\Gamma(1+\alpha)}\|{}_aD_t^\alpha u\|_{L^p}, \quad \text{for } u \in E_{\alpha,p}. \tag{8.57}$$

The goal of this section is to prove the following proposition:

Proposition 8.4. *Assuming* $0 < \frac{1}{p} < \alpha < 1$, $E_{\alpha,p}$ *is a reflexive separable Banach space and the compact embedding* $E_{\alpha,p} \hookrightarrow C_a$ *holds.*

Proof. Consider that

$$0 < \frac{1}{p} < \alpha < 1 \quad \text{and } q = \frac{p}{p-1}.$$

Now, we divide the proof into several steps.

Claim I. $E_{\alpha,p}$ is a reflexive separable Banach space.

Let us consider $(L^p)^2$ the set $L^p \times L^p$ endowed with the norm $\|(u,v)\|_{(L_p)^2} = (\|u\|_{L^p}^p + \|v\|_{L^p}^p)^{\frac{1}{p}}$. Since $p > 1, (L^p, \|\cdot\|_{L^p})$ is a reflexive separable Banach space and therefore, $((L^p)^2, \|\cdot\|_{(L^p)^2})$ is also a reflexive separable Banach space. We define $\Omega := \{(u, {}_aD_t^\alpha u) : \; u \in E_{\alpha,p}\}$. Let us prove that Ω is a closed subspace of $((L^p)^2, \|\cdot\|_{(L^p)^2})$. Let $(u_n, v_n)_{n\in\mathbb{N}} \subset \Omega$ such that

$$(u_n, v_n) \xrightarrow{(L^p)^2} (u, v).$$

Then, we prove that $(u,v) \in \Omega$. For any $n \in \mathbb{N}, (u_n, v_n) \in \Omega$. Thus, $u_n \in E_{\alpha,p}$ and $v_n = {}_aD_t^\alpha u_n$. Consequently, we have

$$u_n \xrightarrow{L^p} u \quad \text{and} \quad {}_aD_t^\alpha u_n \xrightarrow{L^p} v.$$

For any $n \in \mathbb{N}$, since $u_n \in E_{\alpha,p}$ and ${}_aD_t^{-\alpha}$ is continuous from L^p to L^p, we have

$$u_n = {}_aD_t^{-\alpha}({}_aD_t^\alpha u_n) \xrightarrow{L^p} {}_aD_t^{-\alpha} v.$$

Thus, $u = {}_aD_t^{-\alpha} v$, ${}_aD_t^\alpha u = {}_aD_t^\alpha({}_aD_t^{-\alpha} v) = v \in L^p$ and ${}_aD_t^{-\alpha}({}_aD_t^\alpha u) = {}_aD_t^{-\alpha} v = u$. Hence, $u \in E_{\alpha,p}$ and $(u,v) = (u, {}_aD_t^\alpha u) \in \Omega$. In conclusion, Ω is a closed subspace of $((L^p)^2, \; \|\cdot\|_{(L^p)^2})$ and then Ω is a reflexive separable Banach space. Finally, defining the following operator

$$\begin{aligned} A &: E_{\alpha,p} \to \Omega, \\ u &\mapsto (u, {}_aD_t^\alpha u), \end{aligned}$$

we prove that $E_{\alpha,p}$ is isometric isomorphic to Ω. This completes the proof of Claim I.

Claim II. The continuous embedding $E_{\alpha,p} \hookrightarrow C_a$.

Let $u \in E_{\alpha,p}$ and then ${}_aD_t^{\alpha}u \in L^p$. Since $0 < \frac{1}{p} < \alpha < 1$, Proposition 8.3 leads to ${}_aD_t^{-\alpha}({}_aD_t^{\alpha}u) \in C_a$. Furthermore, $u = {}_aD_t^{-\alpha}({}_aD_t^{\alpha}u)$ and consequently, u can be identified to its continuous representative. Finally, Proposition 8.3 also gives

$$\|u\| = \|{}_aD_t^{-\alpha}({}_aD_t^{\alpha}u)\| \leq \frac{(b-a)^{\alpha-\frac{1}{p}}}{\Gamma(\alpha)((\alpha-1)q+1)^{\frac{1}{q}}}|u|_{\alpha,p}, \quad \text{for } u \in E_{\alpha,p}.$$

Since $\|\cdot\|_{\alpha,p}$ and $|\cdot|_{\alpha,p}$ are equivalent norms, the proof of Claim II is completed.

Claim III. The compact embedding $E_{\alpha,p} \hookrightarrow C_a$.

Since $E_{\alpha,p}$ is a reflexive Banach space, we only have to prove that

$$\forall\ (u_n)_{n\in\mathbb{N}} \subset E_{\alpha,p} \quad \text{such that } u_n \overset{E_{\alpha,p}}{\rightharpoonup} u, \quad \text{then } u_n \overset{\mathcal{C}}{\to} u.$$

Let $(u_n)_{n\in\mathbb{N}} \subset E_{\alpha,p}$ such that

$$u_n \overset{E_{\alpha,p}}{\rightharpoonup} u.$$

Since $E_{\alpha,p} \hookrightarrow C_a$, we have

$$u_n \overset{\mathcal{C}}{\rightharpoonup} u.$$

Since $(u_n)_{n\in\mathbb{N}}$ converges weakly in $E_{\alpha,p}$, $(u_n)_{n\in\mathbb{N}}$ is bounded in $E_{\alpha,p}$. Consequently, $({}_aD_t^{\alpha}u_n)_{n\in\mathbb{N}}$ is bounded in L_p by a constant $M \geq 0$. Let us prove that $(u_n)_{n\in\mathbb{N}} \subset C_a$ is uniformly Lipschitzian on $[a,b]$. According to the proof of Proposition 8.3, for $\forall\ n \in \mathbb{N}, \forall\ a \leq t_1 < t_2 \leq b$, we have,

$$\begin{aligned}
|u_n(t_2) - u_n(t_1)| &\leq |{}_aD_t^{-\alpha}({}_aD_t^{\alpha}u_n(t_2)) - {}_aD_t^{-\alpha}({}_aD_t^{\alpha}u_n(t_1))| \\
&\leq \frac{2\|{}_aD_t^{\alpha}u_n\|_{L^p}}{\Gamma(\alpha)((\alpha-1)q+1)^{\frac{1}{p}}}(t_2-t_1)^{\alpha-\frac{1}{p}} \\
&\leq \frac{2M}{\Gamma(\alpha)((\alpha-1)q+1)^{\frac{1}{p}}}(t_2-t_1)^{\alpha-\frac{1}{p}}.
\end{aligned}$$

Hence, from Arzela-Ascoli theorem, $(u_n)_{n\in\mathbb{N}}$ is relatively compact in C. Consequently, there exists a subsequence of $(u_n)_{n\in\mathbb{N}}$ converging strongly in C and the limit is u by uniqueness of the weak limit.

Now, let us prove by contradiction that the whole sequence $(u_n)_{n\in\mathbb{N}}$ converges strongly to u in C. If not, there exist $\varepsilon > 0$ and a subsequence $(u_{n_k})_{k\in\mathbb{N}}$ such that

$$\|u_{n_k} - u\| > \varepsilon > 0, \quad \text{for } k \in \mathbb{N}. \tag{8.58}$$

Nevertheless, since $(u_{n_k})_{k\in\mathbb{N}}$ is a subsequence of $(u_n)_{n\in\mathbb{N}}$, then it satisfies

$$u_{n_k} \overset{E_{\alpha,p}}{\rightharpoonup} u.$$

In the same way (using Arzela-Ascoli theorem), we can construct a subsequence of $(u_{n_k})_{k\in\mathbb{N}}$ converging strongly to u in C which is a contradiction to (8.58). The proof of Claim III is now completed. □

Let us remind the following property

$$ {}_aD_t^{-\alpha}\varphi \in C_a^\infty, \quad \text{for } \varphi \in C_c^\infty. $$

From this result, we get the following results.

Proposition 8.5. *C_a^∞ is dense in $E_{\alpha,p}$.*

Proof. Indeed, let us first prove that $C_a^\infty \subset E_{\alpha,p}$. Let $u \in C_a^\infty \subset L^p$. Since $u \in AC_a$ and $u' \in L^p$, we have ${}_aD_t^\alpha u = {}_aD_t^{\alpha-1}u' \in L^p$. Since $u \in AC$, we also have ${}_aD_t^{-\alpha}({}_aD_t^\alpha u) = u$. Finally, $u \in E_{\alpha,p}$. Now, let us prove that C_a^∞ is dense in $E_{\alpha,p}$. Let $u \in E_{\alpha,p}$, then ${}_aD_t^\alpha u \in L^p$. Consequently, there exists $(v_n)_{n\in\mathbb{N}} \subset C_c^\infty$ such that

$$ v_n \xrightarrow{L^p} {}_aD_t^\alpha u \quad \text{and then} \quad {}_aD_t^{-\alpha}v_n \xrightarrow{L^p} {}_aD_t^{-\alpha}({}_aD_t^\alpha u) = u, $$

since ${}_aD_t^{-\alpha}$ is continuous from L^p to L^p. Defining $u_n := {}_aD_t^{-\alpha}v_n \in C_a^\infty$ for any $n \in \mathbb{N}$, we obtain

$$ u_n \xrightarrow{L^p} u \quad \text{and} \quad {}_aD_t^\alpha u_n = {}_aD_t^\alpha({}_aD_t^{-\alpha}v_n) = v_n \xrightarrow{L^p} {}_aD_t^\alpha u. $$

Finally, $(u_n)_{n\in\mathbb{N}} \subset C_a^\infty$ and converges to u in $E_{\alpha,p}$. The proof is completed. □

Proposition 8.6. *If $\frac{1}{p} < \min\{\alpha, 1-\alpha\}$, then $E_{\alpha,p} = \{u \in L^p \ : \ {}_aD_t^\alpha u \in L^p\}$.*

Proof. Indeed, let $u \in L^p$ satisfying ${}_aD_t^\alpha u \in L^p$ and let us prove that ${}_aD_t^{-\alpha}({}_aD_t^\alpha u) = u$. Let $\varphi \in C_c^\infty \subset L^1$. Since ${}_aD_t^\alpha u \in L^p$, Proposition 1.10 leads to

$$ \int_a^b {}_aD_t^{-\alpha}({}_aD_t^\alpha u)\cdot\varphi dt = \int_a^b {}_aD_t^\alpha u \cdot {}_tD_b^{-\alpha}\varphi dt = \int_a^b \frac{d}{dt}({}_aD_t^{\alpha-1}u)\cdot {}_tD_b^{-\alpha}\varphi dt. $$

Then, an integration by parts gives

$$ \int_a^b {}_aD_t^{-\alpha}({}_aD_t^\alpha u)\cdot\varphi dt = \int_a^b {}_aD_t^{\alpha-1}u \cdot {}_tD_b^{1-\alpha}u dt. $$

Indeed, ${}_tD_b^{-\alpha}\varphi(b) = 0$ since $\varphi \in C_c^\infty$ and ${}_aD_t^{\alpha-1}u(a) = 0$ since $u \in L^p$ and $\frac{1}{p} < 1-\alpha$. Finally, using Proposition 1.10 again, we obtain

$$ \int_a^b {}_aD_t^{-\alpha}({}_aD_t^\alpha u)\cdot\varphi dt = \int_a^b u \cdot {}_tD_b^{\alpha-1}({}_tD_b^{1-\alpha}\varphi)dt = \int_a^b u\cdot\varphi dt, $$

this completes the proof. □

Remark 8.2. In the Proposition 8.6, let us note that such a definition of $E_{\alpha,p}$ could lead us to name it fractional Sobolev space and to denote it by $W^{\alpha,p}$. Nevertheless, these notions and notations are already used, see Brezis, 2011.

8.4.3 Variational Structure

In this subsection, we assume that Lagrangian L is of class C^1 and we define the Lagrangian functional $\mathfrak{L}$ on $E_{\alpha,p}$ (with $0 < \frac{1}{p} < \alpha < 1$). Precisely, we define

$$\begin{aligned} \mathfrak{L} : E_{\alpha,p} &\to \mathbb{R}, \\ u &\mapsto \int_a^b L(u, {}_aD_t^\alpha u, t)dt. \end{aligned}$$

$\mathfrak{L}$ is said to be Gâteaux differentiable in $u \in E_{\alpha,p}$ if the map

$$\begin{aligned} D\mathfrak{L}(u) : E_{\alpha,p} &\to \mathbb{R}, \\ v &\mapsto D\mathfrak{L}(u)(v) := \lim_{h\to 0} \frac{\mathfrak{L}(u+hv) - \mathfrak{L}(u)}{h} \end{aligned}$$

is well-defined for any $v \in E_{\alpha,p}$ and if it is linear and continuous. A critical point $u \in E_{\alpha,p}$ of $\mathfrak{L}$ is defined by $D\mathfrak{L}(u) = 0$.

We introduce the following hypotheses:

(H1) there exist $0 \le d_1 \le p$ and $r_1, s_1 \in C(\mathbb{R}^d \times [a,b], \mathbb{R}^+)$ such that

$$|L(x,y,t) - L(x,0,t)| \le r_1(x,t)\|y\|^{d_1} + s_1(x,t), \quad \text{for } (x,y,t) \in \mathbb{R}^d \times \mathbb{R}^d \times [a,b];$$

(H2) there exist $0 \le d_2 \le p$ and $r_2, s_2 \in C(\mathbb{R}^d \times [a,b], \mathbb{R}^+)$ such that

$$\left\| \frac{\partial L}{\partial x}(x,y,t) \right\| \le r_2(x,t)\|y\|^{d_2} + s_2(x,t), \quad \text{for } (x,y,t) \in \mathbb{R}^d \times \mathbb{R}^d \times [a,b];$$

(H3) there exist $0 \le d_3 \le p-1$ and $r_3, s_3 \in C(\mathbb{R}^d \times [a,b], \mathbb{R}^+)$ such that

$$\left\| \frac{\partial L}{\partial y}(x,y,t) \right\| \le r_3(x,t)\|y\|^{d_3} + s_3(x,t), \quad \text{for } (x,y,t) \in \mathbb{R}^d \times \mathbb{R}^d \times [a,b];$$

(H4) *coercivity condition:* there exist $\gamma > 0, 1 \le d_4 < p, c_1 \in C(\mathbb{R}^d \times [a,b], [\gamma, \infty)), c_2, c_3 \in C([a,b], \mathbb{R})$ such that

$$\forall\ (x,y,t) \in \mathbb{R}^d \times \mathbb{R}^d \times [a,b], \quad L(x,y,t) \ge c_1(x,t)\|y\|^p + c_2(t)\|x\|^{d_4} + c_3(t);$$

(H5) *convexity condition:*

$$\forall\ t \in [a,b], \quad L(\cdot,\cdot,t) \text{ is convex.}$$

Hypotheses denoted by (H1)-(H3) are usually called regularity hypotheses (see Cesari, 1983; Dacorogna, 2008).

Let us prove the following results.

Lemma 8.10. *The following implications hold*

(i) *L satisfies (H1) $\Rightarrow$ for any $u \in E_{\alpha,p}$, $L(u, {}_aD_t^\alpha u, t) \in L^1$ and then $\mathfrak{L}(u)$ exists in $\mathbb{R}$;*

(ii) *L satisfies (H2) $\Rightarrow$ for any $u \in E_{\alpha,p}$, $\partial L/\partial x(u, {}_aD_t^\alpha u, t) \in L^1$;*

(iii) *L satisfies (H3) $\Rightarrow$ for any $u \in E_{\alpha,p}$, $\partial L/\partial y(u, {}_aD_t^\alpha u, t) \in L^q$, where $q = \frac{p}{p-1}$.*

Proof. Let us assume that $\mathfrak{L}$ satisfies (H1) and let $u \in E_{\alpha,p} \subset C_a$. Then, $\|{}_aD_t^\alpha u\|^{d_1} \in L^{p/d_1} \subset L^1$ and the three maps $t \to r_1(u(t),t)$, $s_1(u(t),t)$, $|L(u(t),0,t)| \in C([a,b],\mathbb{R}^+) \subset L^\infty \subset L^1$. Hypothesis (H1) implies for almost all $t \in [a,b]$

$$|L(u(t), {}_aD_t^\alpha u(t), t)| \leq r_1(u(t),t)\|{}_aD_t^\alpha u(t)\|^{d_1} + s_1(u(t),t) + |L(u(t),0,t)|.$$

Hence, $L(u, {}_aD_t^\alpha u(t), t) \in L^1$ and then $L(u)$ exists in $\mathbb{R}$. We proceed in the same manner in order to prove the second point of Lemma 8.10. Now, assuming that L satisfies (H3), we have $\|{}_aD_t^\alpha u\|^{d_3} \in L^{p/d_3} \subset L^q$ for any $u \in E_{\alpha,p}$. An analogous argument gives the third point of Lemma 8.10. This completes the proof. □

Lemma 8.11. *Assuming that L satisfies hypotheses (H1)-(H3), $\mathfrak{L}$ is Gâteaux differentiable in any $u \in E_{\alpha,p}$ and*

$$D\mathfrak{L}(u)(v) = \int_a^b \left(\frac{\partial L}{\partial x}(u, {}_aD_t^\alpha u, t) \cdot v + \frac{\partial L}{\partial y}(u, {}_aD_t^\alpha u, t) \cdot {}_aD_t^\alpha v\right) dt, \quad \textit{for } u, v \in E_{\alpha,p}.$$

Proof. Let $u, v \in E_{\alpha,p} \subset C_a$. Let $\psi_{u,v}$ defined for any $h \in [-1,1]$ and for almost all $t \in [a,b]$ by

$$\psi_{u,v}(t,h) := L\Big(u(t) + hv(t), {}_aD_t^\alpha u(t) + h\, {}_aD_t^\alpha v(t), t\Big).$$

Then, we define the following mapping

$$\begin{aligned}
&\phi_{u,v} : [-1,1] \to \mathbb{R}, \\
&h \mapsto \int_a^b L\Big(u + hv(t), {}_aD_t^\alpha u + h\, {}_aD_t^\alpha v, t\Big) dt = \int_a^b \psi_{u,v}(t,h) dt.
\end{aligned}$$

Our aim is to prove that the following term

$$D\mathfrak{L}(u)(v) = \lim_{h \to 0} \frac{\mathfrak{L}(u + hv) - \mathfrak{L}(u)}{h} = \lim_{h \to 0} \frac{\phi_{u,v}(h) - \phi_{u,v}(0)}{h} = \phi_{u,v}{}'(0)$$

exists in $\mathbb{R}$. In order to differentiate $\phi_{u,v}$, we use the theorem of differentiation under the integral sign. Indeed, we have for almost all $t \in [a,b]$, $\psi_{u,v}(t,\cdot)$ is differentiable on $[-1,1]$ with

$$\begin{aligned}
\frac{\partial \psi_{u,v}}{\partial h}(t,h) = {} & \frac{\partial L}{\partial x}\Big(u(t) + hv(t), {}_aD_t^\alpha u(t) + h\, {}_aD_t^\alpha v(t), t\Big) \cdot v(t) \\
& + \frac{\partial L}{\partial y}\Big(u(t) + hv(t), {}_aD_t^\alpha u(t) + h\, {}_aD_t^\alpha v(t), t\Big) \cdot {}_aD_t^\alpha v(t).
\end{aligned}$$

Then, from hypotheses (H2) and (H3), we have for any $h \in [-1,1]$ and for almost all $t \in [a,b]$

$$\begin{aligned}
&\left|\frac{\partial \psi_{u,v}}{\partial h}(t,h)\right| \\
&\leq \Big[r_2(u(t) + hv(t), t)\|{}_aD_t^\alpha u(t) + h\, {}_aD_t^\alpha v(t)\|^{d_2} + s_2(u(t) + hv(t), t)\Big] \|v(t)\| \\
&\quad + \Big[r_3(u(t) + hv(t), t)\|{}_aD_t^\alpha u(t) + h\, {}_aD_t^\alpha v(t)\|^{d_3} + s_3(u(t) + hv(t), t)\Big] \|{}_aD_t^\alpha v(t)\|.
\end{aligned}$$

We define

$$r_{2,0} := \max_{(t,h)\in[a,b]\times[-1,1]} r_2(u(t)+hv(t),t)$$

and we define similarly $s_{2,0}, r_{3,0}, s_{3,0}$. Finally, it holds

$$\begin{aligned}\left|\frac{\partial\psi_{u,v}}{\partial h}(t,h)\right| &\le 2^{d_2} r_{2,0} \underbrace{(\|{}_aD_t^\alpha u(t)\|^{d_2} + \|{}_aD_t^\alpha v(t)\|^{d_2})}_{\in L^{p/d_2}\subset L^1} \underbrace{\|v(t)\|}_{\in C_a\subset L^\infty} + s_{2,0} \underbrace{\|v(t)\|}_{\in C_a\subset L^1} \\ &\quad + 2^{d_3} r_{3,0} \underbrace{(\|{}_aD_t^\alpha u(t)\|^{d_3} + \|{}_aD_t^\alpha v(t)\|^{d_3})}_{\in L^{p/d_3}\subset L^q} \underbrace{\|{}_aD_t^\alpha v(t)\|}_{\in L^p} + s_{3,0} \underbrace{\|{}_aD_t^\alpha v(t)\|}_{\in L^p\subset L^1}.\end{aligned}$$

The right term is then a L^1 function independent of h. Consequently, applying the theorem of differentiation under the integral sign, $\phi_{u,v}$ is differentiable with

$$\phi_{u,v}'(h) = \int_a^b \frac{\partial\psi_{u,v}}{h}(t,h)dt, \quad \text{for } h\in[-1,1].$$

Hence

$$\begin{aligned}D\mathfrak{L}(u)(v) = \phi_{u,v}'(0) &= \int_a^b \frac{\partial\psi_{u,v}}{h}(t,0)dt \\ &= \int_a^b \left(\frac{\partial L}{\partial x}(u, {}_aD_t^\alpha, t)\ v + \frac{\partial L}{\partial y}(u, {}_aD_t^\alpha u, t)\ {}_aD_t^\alpha v\right) dt.\end{aligned}$$

From Lemma 8.10, it holds

$$\frac{\partial L}{\partial x}(u, {}_aD_t^\alpha u, t) \in L^1 \quad \text{and} \quad \frac{\partial L}{\partial y}(u, {}_aD_t^\alpha u, t) \in L^q.$$

Since $v\in C_a\subset L^\infty$ and ${}_aD_t^\alpha \in L^p$, $D\mathfrak{L}(u)(v)$ exists in $\mathbb{R}$. Moreover, we have

$$\begin{aligned}|D\mathfrak{L}(u)(v)| &\le \left\|\frac{\partial L}{\partial x}(u, {}_aD_t^\alpha u, t)\right\|_{L^1} \|v\| + \left\|\frac{\partial L}{\partial y}(u, {}_aD_t^\alpha u, t)\right\|_{L^q} \|{}_aD_t^\alpha v\|_{L^p} \\ &\le \left(\frac{(b-a)^{\alpha-\frac{1}{p}}}{\Gamma(\alpha)((\alpha-1)q+1)^{\frac{1}{q}}}\left\|\frac{\partial L}{\partial x}(u, {}_aD_t^\alpha u, t)\right\|_{L^1} + \left\|\frac{\partial L}{\partial y}(u, {}_aD_t^\alpha u, t)\right\|_{L^q}\right)|v|_{\alpha,p}.\end{aligned}$$

Consequently, $D\mathfrak{L}(u)$ is linear and continuous from $E_{\alpha,p}$ to $\mathbb{R}$. The proof is completed. □

8.4.4 *Existence of Weak Solution*

In this subsection, we will present the existence theorem of weak solution for (8.56). We firstly give two preliminary theorems.

Theorem 8.10. *Assume that L satisfies hypotheses (H1)-(H3). If u is a critical point of $\mathfrak{L}$, u is a weak solution of* (8.56).

Proof. Let u be a critical point of $\mathfrak{L}$. Then, we have in particular

$$D\mathfrak{L}(u)(v) = \int_a^b \left(\frac{\partial L}{\partial x}(u, {}_aD_t^\alpha u, t) \cdot v + \frac{\partial L}{\partial y}(u, {}_aD_t^\alpha u, t) \cdot {}_aD_t^\alpha v \right) dt = 0, \quad \text{for } v \in C_c^\infty.$$

For any $v \in C_c^\infty \subset AC_a$, ${}_aD_t^\alpha v = {}_aD_t^{\alpha-1} v' \in C_a^\infty$. Since $\partial L/\partial y(u, {}_aD_t^\alpha u, t) \in L^q$, Proposition 1.10 gives

$$\int_a^b \left[\frac{\partial L}{\partial x}(u, {}_aD_t^\alpha u, t) \cdot v + {}_tD_b^{\alpha-1}\left(\frac{\partial L}{\partial y}(u, {}_aD_t^\alpha u, t) \right) \cdot v' \right] dt = 0, \quad \text{for } v \in C_c^\infty.$$

Finally, we define

$$w_u(t) = \int_a^t \frac{\partial L}{\partial x}(u, {}_aD_t^\alpha u, t) dt, \quad \text{for } t \in [a, b].$$

Since $\partial L/\partial x(u, {}_aD_t^\alpha, t) \in L^1, w_u \in AC_a$ and $w_u' = \partial L/\partial x(u, {}_aD_t^\alpha u, t)$. Then, an integration by parts leads to

$$\int_a^b \left({}_tD_b^{\alpha-1}\left(\frac{\partial L}{\partial y}(u, {}_aD_t^\alpha u, t) \right) - w_u \right) v' dt = 0, \quad \text{for } v \in C_c^\infty.$$

Consequently, there exists a constant $C \in \mathbb{R}^d$ such that

$${}_tD_b^{\alpha-1}\left(\frac{\partial L}{\partial y}(u, {}_aD_t^\alpha u, t) \right) = C + w_u \in AC.$$

By differentiation, we obtain

$$-{}_tD_b^{\alpha}\left(\frac{\partial L}{\partial y}(u, {}_aD_t^\alpha u, t) \right) = \frac{\partial L}{\partial x}(u, {}_aD_t^\alpha u, t),$$

and then $u \in E_{\alpha,p} \subset C$ satisfies (8.56) a.e. on $[a, b]$. The proof is completed. □

As usual in a variational method, in order to prove the existence of a global minimizer of a functional, coercivity and convexity hypotheses need to be added on the Lagrangian. We have already define hypotheses (H4) (coercivity) and (H5) (convexity). Next, we introduce two different convexity hypotheses (H5)′ and (H5)″:

(H5)′ $\forall\ (x, t) \in \mathbb{R}^d \times [a, b]$, $L(x, \cdot, t)$ is convex and $(L(\cdot, y, t))_{(y,t) \in \mathbb{R}^d \times [a,b]}$ is uniformly equicontinuous on $\mathbb{R}^d$, i.e.,

$$\forall\ \varepsilon > 0,\ \exists\ \delta > 0,\ \forall\ (x_1, x_2) \in (\mathbb{R}^d)^2,\ \|x_2 - x_1\| < \delta$$
$$\Rightarrow \forall\ (y, t) \in \mathbb{R}^d \times [a, b],\ |L(x_2, y, t) - L(x_1, y, t)| < \varepsilon.$$

(H5)″ $\forall\ (x, t) \in \mathbb{R}^d \times [a, b]$, $L(x, \cdot, t)$ is convex.

Let us note that hypotheses (H5) and (H5)′ are independent, hypothesis (H5)″ is the weakest. Nevertheless, in this case, the detailed proof of Theorem 8.11 is more complicated. Consequently, in the case of hypothesis (H5)″, we do not develop the proof and we use a strong result proved in Dacorogna, 2008. Let us prove the following preliminary result.

Lemma 8.12. *Assume that L satisfies hypothesis (H4). Then, $\mathfrak{L}$ is coercive in the sense that*

$$\lim_{\|u\|_{\alpha,p} \to +\infty} \mathfrak{L}(u) = +\infty.$$

Proof. Let $u \in E_{\alpha,p}$, we have

$$\mathfrak{L}(u) = \int_a^b L(u, {}_aD_t^\alpha u, t)dt \geq \int_a^b c_1(u,t)\|{}_aD_t^\alpha u\|^p + c_2(t)\|u\|^{d_4} + c_3(t)dt.$$

Equation (8.57) implies that

$$\|u\|_{L^{d_4}}^{d_4} \leq (b-a)^{1-\frac{d_4}{p}}\|u\|_{L^p}^{d_4} \leq \frac{(b-a)^{\alpha+1-\frac{d_4}{p}}}{\Gamma(\alpha+1)}\|{}_aD_t^\alpha u\|_{L^p}^{d_4} = \frac{(b-a)^{\alpha+1-\frac{d_4}{p}}}{\Gamma(\alpha+1)}|u|_{\alpha,p}^{d_4}.$$

Finally, we conclude that

$$\begin{aligned}\mathfrak{L}(u) &\geq \gamma\|{}_aD_t^\alpha u\|_{L^p}^p - \|c_2\|\|u\|_{L^{d_4}}^{d_4} - (b-a)\|c_3\| \\ &\geq \gamma|u|_{\alpha,p}^p - \frac{\|c_2\|(b-a)^{\alpha+1-\frac{d_4}{p}}}{\Gamma(\alpha+1)}|u|_{\alpha,p}^{d_4} - (b-a)\|c_3\|, \quad \text{for } u \in E_{\alpha,p}.\end{aligned}$$

Since $d_4 < p$ and the norms $|\cdot|_{\alpha,p}$ and $\|\cdot\|_{\alpha,p}$ are equivalent, the proof is completed. □

Theorem 8.11. *Assume that L satisfies hypotheses (H1)-(H4) and one of hypotheses (H5), (H5)′ or (H5)″. Then, $\mathfrak{L}$ admits a global minimizer.*

Proof. Let $(u_n)_{n\in\mathbb{N}}$ be a sequence in $E_{\alpha,p}$ satisfying

$$\mathfrak{L}(u_n) \to \inf_{v\in E_{\alpha,p}} \mathfrak{L}(v) =: K.$$

Since L satisfies hypothesis (H1), $\mathfrak{L}(u) \in \mathbb{R}$ for any $u \in E_{\alpha,p}$. Hence, $K < +\infty$. Let us prove by contradiction that $(u_n)_{n\in\mathbb{N}}$ is bounded in $E_{\alpha,p}$. In the negative case, we can construct a subsequence $(u_{n_k})_{k\in\mathbb{N}}$ satisfying $\|u_{n_k}\|_{\alpha,p} \to \infty$. Since L satisfies hypothesis (H4), Lemma 8.12 gives:

$$K = \lim_{k\in\mathbb{N}} \mathfrak{L}(u_{n_k}) = +\infty,$$

which is a contradiction. Hence, $(u_n)_{n\in\mathbb{N}}$ is bounded in $E_{\alpha,p}$. Since $E_{\alpha,p}$ is reflexive, there exists a subsequence still denoted by $(u_n)_{n\in\mathbb{N}}$ converging weakly in $E_{\alpha,p}$ to an element denoted by $u \in E_{\alpha,p}$. Let us prove that u is a global minimizer of $\mathfrak{L}$. Since

$$u_n \overset{E_{\alpha,p}}{\rightharpoonup} u \text{ and } E_{\alpha,p} \hookrightarrow C_a,$$

we have

$$u_n \xrightarrow{C} u \text{ and } {}_aD_t^\alpha u_n \overset{L^p}{\rightharpoonup} {}_aD_t^\alpha u. \tag{8.59}$$

Case L satisfies (H5): by convexity, it holds for any $n \in \mathbb{N}$

$$\begin{aligned}\mathfrak{L}(u_n) &= \int_a^b L(u_n, {}_aD_t^\alpha u_n, t)dt \\ &\geq \int_a^b L(u, {}_aD_t^\alpha u, t)dt + \int_a^b \frac{\partial L}{\partial x}(u, {}_aD_t^\alpha u, t)\ (u_n - u)dt \\ &\quad + \int_a^b \frac{\partial L}{\partial y}(u, {}_aD_t^\alpha u, t)\ ({}_aD_t^\alpha u_n - {}_aD_t^\alpha u)dt.\end{aligned}$$

Since L satisfies hypotheses (H2) and (H3), $\partial L/\partial x(u, {}_aD_t^\alpha u, t) \in L^1$ and $\partial L/\partial y(u, {}_aD_t^\alpha u, t) \in L^q$. Consequently, using (8.59) and making n tend to $+\infty$, we obtain

$$K = \inf_{v \in E_{\alpha,p}} \mathfrak{L}(v) \geq \int_a^b L(u, {}_aD_t^\alpha u, t)dt = \mathfrak{L}(u).$$

Consequently, u is a global minimizer of $\mathfrak{L}$.

Case L satisfies (H5)$'$: let $\varepsilon > 0$. Since $(u_n)_{n\in\mathbb{N}}$ converges strongly in C to u, we have

$$\exists\, N \in \mathbb{N},\ \forall\, n \geq N, \quad \|u_n - u\| < \delta,$$

where δ is given in the definition of (H5)$'$. In consequence, it holds a.e. on $[a, b]$

$$|L(u_n(t), {}_aD_t^\alpha u_n(t), t) - L(u(t), {}_aD_t^\alpha u_n(t), t)| < \varepsilon, \quad \text{for } n \geq N. \tag{8.60}$$

Moreover, for any $n \geq N$, we have

$$\mathfrak{L}(u_n) = \int_a^b L(u, {}_aD_t^\alpha u, t)dt + \int_a^b \big(L(u_n, {}_aD_t^\alpha u_n, t) - L(u, {}_aD_t^\alpha u_n, t)\big)dt$$
$$+ \int_a^b \big(L(u, {}_aD_t^\alpha u_n, t) - L(u, {}_aD_t^\alpha u, t)\big)dt.$$

Then, for any $n \geq N$, it holds by convexity

$$\mathfrak{L}(u_n) \geq \int_a^b L(u, {}_aD_t^\alpha u, t)dt - \int_a^b |L(u_n, {}_aD_t^\alpha u_n, t) - L(u, {}_aD_t^\alpha u_n, t)|dt$$
$$+ \int_a^b \frac{\partial L}{\partial y}(u, {}_aD_t^\alpha u, t)\ \big({}_aD_t^\alpha u_n - {}_aD_t^\alpha u\big)dt.$$

And, using equation (8.60), we obtain for any $n \geq N$

$$\mathfrak{L}(u_n) \geq \int_a^b L(u, {}_aD_t^\alpha u, t)dt - \varepsilon(b-a) + \int_a^b \frac{\partial L}{\partial y}(u, {}_aD_t^\alpha u, t)\ \big({}_aD_t^\alpha u_n - {}_aD_t^\alpha u\big)dt.$$

We remind that $\partial L/\partial y(u, {}_aD_t^\alpha u, t) \in L^q$ since L satisfies (H3). Since $({}_aD_t^\alpha u_n)_{n\in\mathbb{N}}$ converges weakly in L^p to ${}_aD_t^\alpha u$ we obtain by making n tend to $+\infty$ and then by making ε tend to 0

$$K = \inf_{v \in E_{\alpha,p}} \mathfrak{L}(v) \geq \int_a^b L(u, {}_aD_t^\alpha u, t)dt = \mathfrak{L}(u).$$

Consequently, u is a global minimizer of $\mathfrak{L}$.

Case L satisfies (H5)$''$: we refer to Theorem 3.23 in Bacorogna, 2008. □

Finally, we give the existence theorem of weak solution for (8.56).

Theorem 8.12. *Let L be a Lagrangian of class C^1 and $0 < \frac{1}{p} < \alpha < 1$. If L satisfies the hypotheses denoted by (H1)-(H5). Then* (8.56) *admits a weak solution.*

Combining Theorem 8.10 and Theorem 8.11, the proof of Theorem 8.12 is obvious.

Let us consider some examples of Lagrangian L satisfying hypotheses of Theorem 8.12. Consequently, the fractional Euler-Lagrange equation (8.56) associated admits a weak solution $u \in E_{\alpha,p}$.

Example 8.1. The most classical example is the Dirichlet integral, i.e. the Lagrangian functional associated to the Lagrangian L given by

$$L(x, y, t) = \frac{1}{2}\|y\|^2.$$

In this case, L satisfies hypotheses (H1)-(H5) for $p = 2$. Hence, the fractional Euler-Lagrange equation (8.56) associated admits a weak solution in $E_{\alpha,p}$ for $\frac{1}{2} < \alpha < 1$.

In a more general case, the following Lagrangian L

$$L(x, y, t) = \frac{1}{p}\|y\|^p + a(x, t),$$

where $p > 1$ and $a \in C^1(\mathbb{R}^d \times [a, b], \mathbb{R}^+)$, satisfies hypotheses (H1)-(H4) and (H5)$''$. Consequently, the fractional Euler-Lagrange equation (8.56) associated to L admits a weak solution in $E_{\alpha,p}$ for any $\frac{1}{p} < \alpha < 1$. Let us note that if for any $t \in [a, b]$, $a(\cdot, t)$ is convex, then L satisfies hypothesis (H5).

In the unidimensional case $d = 1$, let us take a Lagrangian with a second term linear in its first variable, i.e.

$$L(x, y, t) = \frac{1}{p}|y|^p + f(t)x,$$

where $p > 1$ and $f \in C^1([a, b], \mathbb{R})$. Then, L satisfies hypotheses (H1)-(H5). Then, the fractional Euler-Lagrange equation (8.56) associated admits a weak solution in $E_{\alpha,p}$ for any $\frac{1}{p} < \alpha < 1$.

Theorem 8.12 is a result based on strong conditions on Lagrangian L. Consequently, some Lagrangian do not satisfy all hypotheses of Theorem 8.12. We can cite Bolza's example in dimension $d = 1$ given by

$$L(x, y, t) = (y^2 - 1)^2 + x^4.$$

L does not satisfy hypothesis (H4) neither hypothesis (H5)$''$. Nevertheless, as usual with variational methods, the conditions of regularity, coercivity and/or convexity can often be replaced by weaker assumptions specific to the studied problem. As an example, we can cite Ammi and Torres, 2008 and references therein about higher-order integrals of the calculus of variations. Indeed, in this subsection, it is proved that calculus of variations is still valid with weaker regularity assumptions.

8.5 Fractional Diffusion Equations

8.5.1 *Introduction*

We assume Ω to be a bounded domain in $\mathbb{R}^d$ with sufficiently smooth boundary $\partial\Omega$. We consider an initial-boundary value problem for a diffusion equation with two fractional time derivatives

$$\begin{cases} \partial_t^{\alpha_1} u(x,t) + q(x)\partial_t^{\alpha_2} u(x,t) = (-\mathcal{A}u)(x,t), & x \in \Omega, \quad t \in (0,T), \\ u(x,t) = 0, & x \in \partial\Omega, \quad t \in (0,T), \\ u(x,0) = a(x), & x \in \Omega. \end{cases} \tag{8.61}$$

Here $0 < \alpha_2 < \alpha_1 < 1$. For $\alpha \in (0,1)$, by ∂_t^α we denote the Caputo fractional derivative with respect to t

$$\partial_t^\alpha g(t) = \frac{1}{\Gamma(1-\alpha)} \int_0^t (t-\tau)^{-\alpha} \frac{d}{dt} g(\tau) d\tau$$

and Γ is the Gamma function and $q \in W^{2,\infty}(\Omega)$. The space $W^{2,\infty}(\Omega)$ is the usual Sobolev space (see, e.g., Adams, 1999).

The operator $\mathcal{A}$ denotes a second-order partial differential operator in the following form

$$(-\mathcal{A}u)(x) = \sum_{i,j=1}^{d} \frac{\partial}{\partial x_i}\left(a_{ij}(x) \frac{\partial}{\partial x_j} u(x) \right) + b(x)u(x), \quad x \in \Omega,$$

for $u \in H^2(\Omega) \bigcap H_0^1(\Omega)$, and we assume that $a_{ij} = a_{ji} \in C^1(\bar{\Omega})$, $1 \le i,j \le d, b \in C(\bar{\Omega})$, $b(x) \le 0$ for $x \in \bar{\Omega}$ and that there exists a constant $\nu > 0$ such that

$$\nu \sum_{j=1}^{d} \xi_j^2 \le \sum_{j,\ k=1}^{d} a_{jk}(x)\xi_j\xi_k, \quad x \in \bar{\Omega}, \quad \xi \in \mathbb{R}^d.$$

The classic diffusion models (diffusion equation with integer-order derivative) have played important roles in modelling contaminants diffusion processes. However, in recent two decades, more and more experimental data (e.g., the diffusion process in the highly heterogeneous media) showed that the classical model are inadequate to explain the phenomenon described by the experimental data, e.g., Adams and Gelhar pointed out that the field data in the saturated zone of a highly heterogeneous aquifer indicated the long-tailed profile in the spatial distribution of densities as the time passes, which is very different from the classical one (see, Adams and Gelhar, 1992). The above phenomenon of long-tailed profile has been investigated by many researchers, see Berkowitz, Scher and Silliman, 2000; Giona, Gerbelli and Roman, 1992; Hatano and Hatano, 1980, and the references therein. In the many researches, there is an effective one that being used to explain the long-tailed profile phenomenon, that is to replace the first-order time derivative with a fractional derivative of order $\alpha \in (0,1)$ since the fractional derivative possesses the memory effect which leads to the not too fast diffusion. This modified model is presented as a useful approach for the description of transport dynamics in complex system that are governed by anomalous diffusion and non-exponential relaxation patterns, and attracted great attention from different areas. For numerical calculation, see Benson, Schumer, Meerschaert *et al.*, 2001; Meerschaert and Tadjeran *et al.*, 2004; Diethelm and Luchko, 2004 and the references therein. For the theoretics, see Gorenflo, Luchko and Zabrejko, 1999; Hanyaga, 2002; Luchko, 2009a,b, 2010; Luchko and Gorenflo, 1999; Sakamoto and Yamamoto, 2011; Xu, Cheng and Yamamoto, 2011, etc. For the stochastic analysis, one can regard the time-fractional diffusion equation as a macroscopic model derived from the continuous-time random walk. Metzler and Klafter, 2000b demonstrated that a fractional diffusion equation describes a non-Markovian diffusion process with a memory. Roman and Alemany, 1994 investigated continuous-time random walks on fractals and showed that the average probability density of random walks on fractals obeys a diffusion equation with a fractional time derivative asymptotically. As for diffusion equations with multiple fractional time derivatives, see Jiang, Liu, Turner *et al.*, 2012; Daftardar-Gejji and Bhalekar, 2008; Luchko, 2011 and the references therein.

In this section, we consider the case of multiple fractional time derivatives. Such equations can be considered as more feasible model equations than equations with a single fractional time derivative in modeling diffusion in porous media. We apply the perturbation method and the theory of evolution equations to prove regularity as well as unique existence of solution to (8.61).

8.5.2 *Regularity and Unique Existence*

Let $L^2(\Omega)$ be a usual L^2-space with the scalar product $(\cdot,\cdot)$, and $H^l(\Omega)$, $H_0^m(\Omega)$ denote the usual Sobolev spaces (e.g., Adams, 1999). We set $\|a\|_{L^2(\Omega)} = (a,a)^{\frac{1}{2}}$.

We define the operator A in $L^2(\Omega)$ by

$$(Au)(x) = (\mathcal{A}u)(x), \quad x \in \Omega, \quad D(A) = H^2(\Omega) \cap H_0^1(\Omega).$$

Then the fractional power A^γ is defined for $\gamma \in \mathbb{R}$ (see, e.g., Pazy, 1983), and $D(A^\gamma) \subset H^{2\gamma}(\Omega)$, $D(A^{\frac{1}{2}}) = H_0^1(\Omega)$ for example. We note that $\|u\|_{D(A^\gamma)} := \|A^\gamma u\|_{L^2(\Omega)}$ is stronger than $\|u\|_{L^2(\Omega)}$ for $\gamma > 0$.

Since $-A$ is a symmetric uniformly elliptic operator, the spectrum of A is entirely composed of eigenvalues and counting according to the multiplicities, we can set $0 < \lambda_1 \le \lambda_2 \le \ldots$. By $\phi_n \in D(A)$, we denote the orthonormal eigenfunction corresponding to $\lambda_n : A\phi_n = \lambda_n\phi_n$. Then the sequence $\{\phi_n\}_{n\in\mathbb{N}}$ is orthonormal basis in $L^2(\Omega)$. Then we see that

$$D(A^\gamma) = \left\{\psi \in L^2(\Omega) : \sum_{n=1}^{\infty} \lambda_n^{2\gamma} |(\psi,\phi_n)|^2 < \infty\right\}$$

and that $D(A^\gamma)$ is a Hilbert space with the norm

$$\|\psi\|_{D(A^\gamma)} = \left(\sum_{n=1}^{\infty} \lambda_n^{2\gamma} |(\psi,\phi_n)|^2\right)^{\frac{1}{2}}.$$

Henceforth we associate with $u(x,t)$, provided that it is well-defined, a map $u(\cdot):(0,T)\to L^2(\Omega)$ by $u(t)(x)=u(x,t)$, $0<t<T$, $x\in\Omega$. Then we can write (8.61) as

$$\begin{cases} \partial_t^{\alpha_1}u(t)+q\partial_t^{\alpha_2}u(t)=-Au(t), & t>0 \text{ in } L^2(\Omega), \\ u(0)=a\in L^2(\Omega). \end{cases} \tag{8.62}$$

Remark 8.3. The interpretation of the initial condition should be made in a suitable function space. In our case, as Theorem 8.13 asserts, we have $\lim_{t\to 0}\|u(t)-a\|_{L^2(\Omega)}=0$.

Now we define the operator $S(t):L^2(\Omega)\to L^2(\Omega)$, $t\geq 0$, by

$$S(t)a:=\sum_{n=1}^{\infty}(a,\phi_n)E_{\alpha_1,1}(-\lambda_n t^{\alpha_1})\phi_n \quad \text{in} \quad L^2(\Omega)$$

for $a\in L^2(\Omega)$. Then we can prove that $S(t):L^2(\Omega)\to L^2(\Omega)$ is a bounded linear operator for $t\geq 0$ (see, e.g., Sakamoto and Yamamoto, 2011). Moreover the term-wise differentiations are possible and give

$$S'(t)a:=-\sum_{n=1}^{\infty}\lambda_n(a,\phi_n)t^{\alpha_1-1}E_{\alpha_1,\alpha_1}(-\lambda_n t^{\alpha_1})\phi_n \quad \text{in} \quad L^2(\Omega)$$

and

$$S''(t)a:=-\sum_{n=1}^{\infty}\lambda_n(a,\phi_n)t^{\alpha_1-2}E_{\alpha_1,\alpha_1-1}(-\lambda_n t^{\alpha_1})\phi_n \quad \text{in} \quad L^2(\Omega)$$

for $a\in L^2(\Omega)$.

For $F\in L^2(\Omega\times(0,T))$ and $a\in L^2(\Omega)$, there exists a unique solution in a suitable class (see, e.g., Sakamoto and Yamamoto, 2011) to the problem

$$\begin{cases} \partial_t^{\alpha_1}u(t)=-Au(t)+F, & 0<t<T, \\ u(0)=a. \end{cases}$$

This solution is given by

$$u(t)=\int_0^t A^{-1}S'(t-\tau)F(\tau)d\tau+S(t)a, \quad t>0. \tag{8.63}$$

In view of (8.63), we mainly discuss the equation

$$u(t)=S(t)a-\int_0^t A^{-1}S'(t-\tau)q\partial_t^{\alpha_2}u(\tau)d\tau, \quad 0<t<T, \tag{8.64}$$

in order to establish unique existence of solutions to (8.62). Henceforth, C denotes generic positive constants which are independent of a in (8.61), but may depend on T,α_1,α_2 and the coefficients of the operator A and q.

Now we are ready to state first main result in this section.

Theorem 8.13. *We assume that* $u \in C((0,T], L^2(\Omega))$ *satisfy* (8.64) *and*

$$\alpha_1 + \alpha_2 > 1.$$

Then

$$\|u(t)\|_{H^{2\gamma}(\Omega)} \le Ct^{-\alpha_1\gamma}\|a\|_{L^2(\Omega)}, \quad 0 < t \le T$$

for any $\gamma \in (0,1)$.

Proof. First we have

$$A^{\gamma-1}S'(t)a = -t^{\alpha_1-1}\sum_{n=1}^{\infty}\lambda_n^{\gamma}(a,\phi_n)E_{\alpha_1,\alpha_1}(-\lambda_n t^{\alpha_1})\phi_n \quad \text{in} \quad L^2(\Omega)$$

for $a \in L^2(\Omega)$ and $\gamma \ge 0$. Moreover, since

$$|E_{\alpha_1,\alpha_1}(-\eta)| \le \frac{C}{1+\eta}, \qquad \eta > 0$$

(see, e.g., Theorem 1.6 in Podlubny, 1999), we can prove

$$\|A^{\gamma-1}S'(t)\| \le Ct^{\alpha_1-1-\alpha_1\gamma}, \ \ t > 0 \tag{8.65}$$

and

$$\|A^{-1}S''(t)\| \le Ct^{\alpha_1-2}, \ \ t > 0. \tag{8.66}$$

Now we proceed to the proof of Theorem 8.13. We set

$$v(t) := \int_0^t A^{\gamma-1}S'(t-\eta)q\partial_t^{\alpha_2}u(\eta)d\eta, \quad 0 < t < T.$$

By (8.64), we have

$$A^{\gamma}u(t) = A^{\gamma}S(t)a - v(t), \quad 0 < t < T.$$

Therefore, using

$$\|u(t)\|_{H^{2\gamma}(\Omega)} \le C\|A^{\gamma}u(t)\|_{L^2(\Omega)},$$

it is sufficient to estimate $\|A^{\gamma}S(t)a\|_{L^2(\Omega)} + \|v(t)\|_{L^2(\Omega)}$. First we will estimate $\|v(t)\|_{L^2(\Omega)}$. Substituting the definition of $\partial_t^{\alpha_2}u$ and changing the order of integration, we have

$$\begin{aligned} v(t) &= \int_0^t A^{\gamma-1}S'(t-\eta)\frac{1}{\Gamma(1-\alpha_2)}\left(\int_0^{\eta}(\eta-\tau)^{-\alpha_2}qu'(\tau)d\tau\right)d\eta \\ &= \frac{1}{\Gamma(1-\alpha_2)}\int_0^t H(t,\tau)qu'(\tau)d\tau, \quad 0 < t < T. \end{aligned} \tag{8.67}$$

Here we have set

$$H(t,\tau) = \int_{\tau}^t A^{\gamma-1}S'(t-\eta)(\eta-\tau)^{-\alpha_2}d\eta.$$

Decomposing the integrand and introducing the change of variables $\eta - \tau \to \eta$ we obtain

$$\begin{aligned}
H(t,\tau) &= \int_\tau^t A^{\gamma-1}S'(t-\eta)(\eta-\tau)^{-\alpha_2}d\eta,\\
&= \int_\tau^t A^{\gamma-1}S'(t-\eta)[(\eta-\tau)^{-\alpha_2}-(t-\tau)^{-\alpha_2}]d\eta\\
&\quad + \int_\tau^t A^{\gamma-1}S'(t-\eta)d\eta(t-\tau)^{-\alpha_2}\\
&= \int_0^{t-\tau} A^{\gamma-1}S'(t-\eta-\tau)[\eta^{-\alpha_2}-(t-\tau)^{-\alpha_2}]d\eta \\
&\quad + \int_\tau^t A^{\gamma-1}S'(t-\eta)d\eta(t-\tau)^{-\alpha_2}\\
&= \int_0^{t-\tau} A^{\gamma-1}S'(t-\eta-\tau)[\eta^{-\alpha_2}-(t-\tau)^{-\alpha_2}]d\eta\\
&\quad - A^{\gamma-1}S(0)(t-\tau)^{-\alpha_2} + A^{\gamma-1}S(t-\tau)(t-\tau)^{-\alpha_2}\\
&=: I_1(t,\tau)+I_2(t,\tau).
\end{aligned} \tag{8.68}$$

On the other hand, we have

$$\begin{aligned}
\partial_\tau I_1(t,\tau) = &-\int_0^{t-\tau} A^{\gamma-1}S''(t-\eta-\tau)(\eta^{-\alpha_2}-(t-\tau)^{-\alpha_2})d\eta\\
&-\alpha_2\int_0^{t-\tau} A^{\gamma-1}S'(t-\eta-\tau)(t-\tau)^{-\alpha_2-1}d\eta\\
&-\lim_{\eta\to t-\tau} A^{\gamma-1}S'(t-\tau-\eta)[\eta^{-\alpha_2}-(t-\tau)^{-\alpha_2}].
\end{aligned}$$

By the estimate (8.65) we obtain

$$\begin{aligned}
&\|A^{\gamma-1}S'(t-\tau-\eta)(\eta^{-\alpha_2}-(t-\tau)^{-\alpha_2})\|_{L^2(\Omega)}\\
&\qquad\le C(t-\tau-\eta)^{\alpha_1-1-\alpha_1\gamma}\frac{|(t-\tau)^{\alpha_2}-\eta^{\alpha_2}|}{\eta^{\alpha_2}(t-\tau)^{\alpha_2}}.
\end{aligned}$$

According to the mean value theorem, we can choose $\theta\in(\eta,t-\tau)$ such that

$$|(t-\tau)^{\alpha_2}-\eta^{\alpha_2}| = |\alpha_2\theta^{\alpha_2-1}(t-\tau-\eta)| \le \alpha_2\eta^{\alpha_2-1}(t-\tau-\eta).$$

Hence we obtain

$$\begin{aligned}
&\|A^{\gamma-1}S'(t-\tau-\eta)(\eta^{-\alpha_2}-(t-\tau)^{-\alpha_2})\|_{L^2(\Omega)}\\
&\le C\alpha_2\eta^{-1}(t-\tau)^{-\alpha_2}(t-\tau-\eta)^{\alpha_1-\alpha_1\gamma}\to 0 \quad \text{as } \eta\to t-\tau
\end{aligned}$$

by $\alpha_1-\alpha_1\gamma>0$. This implies

$$\begin{aligned}
\partial_\tau I_1(t,\tau) = &-\int_0^{t-\tau} A^{\gamma-1}S''(t-\eta-\tau)(\eta^{-\alpha_2}-(t-\tau)^{-\alpha_2})d\eta\\
&-\alpha_2\int_0^{t-\tau} A^{\gamma-1}S'(t-\eta-\tau)(t-\tau)^{-\alpha_2-1}d\eta, \quad 0<t<T.
\end{aligned} \tag{8.69}$$

On the other hand, we have

$$\begin{aligned}\partial_\tau I_2(t,\tau) &= -\alpha_2 A^{\gamma-1}S(t-\tau)(t-\tau)^{-\alpha_2-1} + \alpha_2 A^{\gamma-1}S(0)(t-\tau)^{-\alpha_2-1}\\ &\quad + A^{\gamma-1}S'(t-\tau)(t-\tau)^{-\alpha_2}\\ &= \alpha_2\int_0^{t-\tau} A^{\gamma-1}S'(t-\eta-\tau)(t-\tau)^{-\alpha_2-1}d\eta\\ &\quad + A^{\gamma-1}S'(t-\tau)(t-\tau)^{-\alpha_2}.\end{aligned}$$

Adding this and (8.69) we obtain

$$\begin{aligned}\partial_\tau H(t,\tau) = &-\int_0^{t-\tau} A^{\gamma-1}S''(t-\eta-\tau)(\eta^{-\alpha_2}-(t-\tau)^{-\alpha_2})d\eta\\ &+ A^{\gamma-1}S'(t-\tau)(t-\tau)^{-\alpha_2}.\end{aligned} \tag{8.70}$$

Using (8.70) in (8.67), integrating by parts and using $H(t,t)=0$ we obtain

$$\begin{aligned}&(\Gamma(1-\alpha_2))v(t)\\ &= \int_0^t H(t,\tau)qu'(\tau)d\tau\\ &= -H(t,0)qa + \int_0^t\left[\int_0^{t-\tau} A^{\gamma-1}S''(t-\eta-\tau)(\eta^{-\alpha_2}-(t-\tau)^{-\alpha_2})d\eta\right.\\ &\quad\left. - A^{\gamma-1}S'(t-\tau)(t-\tau)^{-\alpha_2}\right]qu(\tau)d\tau\\ &=: I_3(t)+I_4(t).\end{aligned}$$

We set

$$B(\alpha,\beta) = \frac{\Gamma(\alpha)\Gamma(\beta)}{\Gamma(\alpha+\beta)}, \quad \alpha,\beta>0.$$

First, by (8.65) and $q\in W^{2,\infty}(\Omega)$ we have

$$\begin{aligned}\|I_3(t)\|_{L^2(\Omega)} &= \|-H(t,0)qa\|_{L^2(\Omega)}\\ &= \left\|-\int_0^t A^{\gamma-1}S'(t-\eta)\eta^{-\alpha_2}d\eta qa\right\|_{L^2(\Omega)}\\ &\le C\|a\|_{L^2(\Omega)}\int_0^t (t-\eta)^{\alpha_1-\alpha_1\gamma-1}\eta^{-\alpha_2}d\eta\\ &= C\|a\|_{L^2(\Omega)}B(1-\alpha_2,\alpha_1-\alpha_1\gamma)t^{\alpha_1-\alpha_1\gamma-\alpha_2},\end{aligned}$$

since $1-\alpha_2>0$ and $\alpha_1-\alpha_1\gamma>0$.

On the other hand, by $q\in W^{2,\infty}(\Omega)$ and $u|_{\partial\Omega}=0$, we have

$$\|A(qu(\tau))\|_{L^2(\Omega)} \le C\|qu(\tau)\|_{H^2(\Omega)} \le C\|u(\tau)\|_{H^2(\Omega)} \le C\|Au(\tau)\|_{L^2(\Omega)}$$

and $\|qu(\tau)\|_{L^2(\Omega)} \le C\|u(\tau)\|_{L^2(\Omega)}$, that is,

$$\|A^0(qu(\tau))\|_{L^2(\Omega)} \le C\|A^0u(\tau)\|_{L^2(\Omega)}.$$

Hence the interpolation theorem (see, e.g., Theorem 5.1 in Lions and Magenes, 1972) we obtain

$$\|A^\gamma(qu(\tau))\|_{L^2(\Omega)} \le C\|A^\gamma u(\tau)\|_{L^2(\Omega)}.$$

Therefore by (8.65) and (8.66), the second term of $I_4(t)$ can be estimated as follows:

$$\begin{aligned}
\|I_4(t)\|_{L^2(\Omega)} &\le C\int_0^t \bigg[\int_0^{t-\tau}(t-\eta-\tau)^{\alpha_1-2}(\eta^{-\alpha_2}-(t-\tau)^{-\alpha_2})d\eta \\
&\quad + (t-\tau)^{\alpha_1-1-\alpha_2}\bigg]\|A^\gamma(qu(\tau))\|_{L^2(\Omega)}d\tau \\
&\le C\int_0^t \bigg[\int_0^{t-\tau}(t-\eta-\tau)^{\alpha_1-2}\frac{(t-\tau-\eta)^{\alpha_2}}{\eta^{\alpha_2}(t-\tau)^{\alpha_2}}d\eta \\
&\quad + (t-\tau)^{\alpha_1-1-\alpha_2}\bigg]\|A^\gamma u(\tau)\|_{L^2(\Omega)}d\tau \\
&\le C\int_0^t \bigg[\int_0^{t-\tau}(t-\eta-\tau)^{\alpha_1+\alpha_2-2}\eta^{-\alpha_2}d\eta \\
&\quad + (t-\tau)^{\alpha_1-1-\alpha_2}\bigg]\|A^\gamma u(\tau)\|_{L^2(\Omega)}d\tau \\
&= C\int_0^t (B(1-\alpha_2,\alpha_1+\alpha_2-1)(t-\tau)^{\alpha_1-1} \\
&\quad + (t-\tau)^{\alpha_1-1-\alpha_2})\|A^\gamma u(\tau)\|_{L^2(\Omega)}d\tau.
\end{aligned}$$

For the last equality, we used $\alpha_1+\alpha_2>1$. Therefore we have

$$\begin{aligned}
\|\Gamma(1-\alpha_2)v(t)\|_{L^2(\Omega)} &\le C\|a\|^2_{L^2(\Omega)}B(1-\alpha_2,\alpha_1-\alpha_1\gamma)t^{\alpha_1-\alpha_1\gamma-\alpha_2} \\
&\quad + C\int_0^t (t-\tau)^{\alpha_1-1-\alpha_2}\|A^\gamma u(\tau)\|_{L^2(\Omega)}d\tau.
\end{aligned}$$

Thus the estimate of $\|v(t)\|_{L^2(\Omega)}$ is completed.

Next we estimate $\|A^\gamma S(t)a\|_{L^2(\Omega)}$. By Theorem 1.6 in Podlubny, 1999, we obtain

$$\begin{aligned}
\|A^\gamma S(t)a\|^2_{L^2(\Omega)} &= \left\|\sum_{n=1}^\infty (a,\phi_n)\lambda_n^\gamma E_{\alpha_1,1}(-\lambda_n t^{\alpha_1})\phi_n\right\|^2_{L^2(\Omega)} \\
&\le C\sum_{n=1}^\infty (a,\phi_n)^2 t^{-2\alpha_1\gamma}\left(\frac{(\lambda_n t^{\alpha_1})^\gamma}{1+\lambda_n t^{\alpha_1}}\right)^2 \\
&\le Ct^{-2\alpha_1\gamma}\|a\|^2_{L^2(\Omega)},
\end{aligned}$$

and hence

$$\begin{aligned}
\|A^\gamma u(t)\ \|_{L^2(\Omega)} &\le C\|a\|_{L^2(\Omega)}(t^{-\alpha_1\gamma}+t^{\alpha_1-\alpha_1\gamma-\alpha_2}) \\
&\quad + C\int_0^t (t-\tau)^{\alpha_1-1-\alpha_2}\|A^\gamma u(\tau)\|_{L^2(\Omega)}d\tau \\
&\le C\|a\|_{L^2(\Omega)}t^{-\alpha_1\gamma} + C\int_0^t (t-\tau)^{\alpha_1-1-\alpha_2}\|A^\gamma u(\tau)\|_{L^2(\Omega)}d\tau, \quad 0<t<T.
\end{aligned}$$

Therefore by an inequality of Gronwall type (see, Exercise 3 (p. 190) in Henry, 1981), we obtain

$$\|A^\gamma u(t)\|_{L^2(\Omega)} \le C\|a\|_{L^2(\Omega)}t^{-\alpha_1\gamma}, \quad 0<t\le T.$$

Thus the proof is completed. □

Remark 8.4. We may be able to remove the condition $\alpha_1+\alpha_2>1$. On the other hand, Prüss established regularity in case $\gamma=1$ for general $\alpha_1,\alpha_2\in(0,1)$ under a strong condition on $a\in\mathcal{D}(A)$ (see, Prüss, 1993).

On the basis of Theorem 8.13, a standard argument (see, Henry, 1981) yields:

Theorem 8.14. *For any* $\gamma \in (0,1)$ *there exists a mild solution to* (8.64) *in the space* $u \in C((0,T], \mathcal{D}(A^\gamma)) \bigcap C((0,T], L^2(\Omega))$.

The above results shall now be extended to the solution of linear diffusion equation with multiple fractional time derivatives

$$\partial_t^{\alpha_1} u(t) + \sum_{j=2}^{l} q_j \partial_t^{\alpha_j} u(t) = -Au(t), \quad t > 0$$

and

$$u(0) = a \in L^2(\Omega),$$

where $0 < \alpha_l < \cdots < \alpha_2 < \alpha_1 < 1$ and $q_j \in W^{2,\infty}(\Omega)$, $2 \le j \le l$.

As before the lower-order derivatives are regarded as source terms and we consider

$$u(t) = S(t)a - \int_0^t A^{-1}S'(t-\tau) \sum_{j=2}^{l} q_j \partial_t^{\alpha_j} u(\tau) d\tau, \quad 0 < t < T. \tag{8.71}$$

Similarly to Theorems 8.13 and 8.14, we can prove the following theorem.

Theorem 8.15. *Assume that* $u \in C((0,T], L^2(\Omega))$ *satisfies* (8.71) *and*

$$0 < \alpha_l < \cdots < \alpha_1, \quad \alpha_1 + \alpha_l > 1.$$

Then

$$\|u(t)\|_{H^{2\gamma}(\Omega)} \le Ct^{-\alpha_1 \gamma} \|a\|_{L^2(\Omega)}, \quad 0 < t \le T$$

for any $\gamma \in (0,1)$. *Moreover there exists a mild solution to* (8.71) *in the space* $C((0,T], \mathcal{D}(A^\gamma)) \cap C([0,T], L^2(\Omega))$ *with* $\gamma \in (0,1)$.

8.6 Fractional Wave Equations

8.6.1 *Introduction*

Fractional wave equations with time-dependent coefficients are natural generations of classical wave equations, which can be used to characterize propagation of wave in inhomogeneous media with frequency-dependent power-law behavior. This section discusses the well-posedness and regularity results of the weak solution for a fractional wave equation allowing that the coefficients may have low regularity. Our analysis relies on mollification arguments, Galerkin methods and energy arguments.

Consider the following fractional wave equation in a bound domain $\Omega \subset \mathbb{R}^N (N > 2)$ with smooth boundary $\partial\Omega$:

$$\begin{cases} \partial_t^\alpha u(t,x) - Au(t,x) = f(t,x), & (t,x) \in (0,T] \times \Omega, \\ u(t,x) = 0, & (t,x) \in [0,T] \times \partial\Omega, \\ u(0,x) = u_0, \ \partial_t u(0,x) = u_1, & x \in \Omega, \end{cases} \tag{8.72}$$

where ∂_t^α is a fractional derivative of order $\alpha \in (1, 2)$, which will be defined in the following contexts, and

$$\mathcal{A}u(t,x) = \sum_{i,j=1}^{N} \partial_i\big(a_{i,j}(t,x)\partial_j u(t,x)\big) + \sum_{j=1}^{N} b_j(t,x)\partial_j u(t,x) + c(t,x)u(t,x),$$

$\partial_i = \frac{\partial}{\partial x_i}$ for $i = 1, ..., N$ and $b_j \in L^\infty((0,T) \times \Omega)$, $c \in L^\infty(0, T, L^{\frac{2q}{q-2}}(\Omega))$ with $q \in [2, \frac{2N}{N-2})$, $a_{i,j} \in W^{1,\infty}(0, T; L^\infty(\Omega))$ and $a_{i,j} = a_{j,i}$. What is more, we assume that $\mathcal{A}$ is uniformly elliptic, i.e., there exist positive constants μ, ν such that

$$\mu|\zeta|^2 \leq \sum_{i,j=1}^{N} a_{i,j}(t,x)\zeta_i\zeta_j \leq \nu|\zeta|^2 \tag{8.73}$$

for a.a. $(t,x) \in [0,T] \times \Omega$, $\forall\ \zeta \in \mathbb{R}^N$.

Recall that the initial boundary value problem (8.72) would resolve itself into fractional diffusion equations when $\alpha \in (0,1)$. It has attracted a growing interest due to its widespread applications in sub-diffusive processes, the authors in Eidelman, 2004, constructed fundamental solutions to the problem using Fox's H-functions and Levi method, then the parametrix estimates were established. Zacher, 2009 studied the well-posedness of weak solutions of abstract evolutionary integro-differential equations based on the Galerkin method and energy estimates. Later, Kubica and Yamamoto, 2018 used the same method to obtain well-posedness of weak solutions of fractional diffusion equations with time-dependent coefficients. In Kim, Kim and Lim, 2017, the authors considered the problem with Caputo derivative on $\mathbb{R}^N$ in L^q-framework and then the uniqueness, existence, and $L^q(L^p)$-estimates of solutions are obtained. In Kian and Yamamoto, 2021, the authors investigated the well-posedness for this problem with time independent elliptic operators but general non-homogenous boundary conditions by mean of an eigenfunction representation involving the Mittag-Leffter functions.

Recently, the problem (8.72) has been the focus of many studies due to its significant application in super-diffusive model of anomalous diffusion such as diffusion in heterogeneous media and viscoelastic problems such as propagation of stress waves in viscoelastic solids. More specifically, significant development has been made in well-posedness as well as regularity results of the weak solution to fractional wave equations. For example, in Kian, 2017, the authors used Laplace transform to define weak solutions and used the Strichartz estimate to derive its well-posedness. Later, Otárola and Salgado, 2018 also gave the definition of weak solutions similar to that of inter-order case and established the well-posedness together with regularity estimates. In Alvarez, Gal, Keyantuo and Warma, 2019; Djida, Fernandez and Area, 2020, the authors obtained the results on existence and regularity of local and global weak solutions of semi-linear case. In Keyantuo, Lizama and Warma, 2017, the authors used integrated cosine family to give the representation of solutions and then provided the existence and regularity results of mild solutions. For other results for fractional wave equations, we refer to Bao, Caraballo, Tuan and Zhou,

2021, for existence and regularity, Bazhlekova, 2018, for subordination principle, D'Abbicco, Ebert, and Picon, 2017, for global existence of small data solutions, He and Peng, 2019 for approximate controllability, Kim and Lim, 2016, for asymptotic behavior, Zhou and He, 2021 for well-posedness and regularity, and the references therein.

In the literature mentioned on fractional wave equations, the main technique to construct solutions for deriving such existence and regularity results is based on Fourier series, cosine family or resolvent operators, and solutions are expressed by the Mittag-Leffler functions. In fact, as it is well known, the smoothness of solutions is followed by the properties of Mittag-Leffler functions. The main novelties of the present section lie in two aspects. Compared with the works of existing literatures on fractional wave equations, our analysis is rather general and relies on Galerkin methods and energy arguments, which can be applied to the general problem that Fourier expansive of solutions can't be used and it can't be converted to ordinary differential equations. On the other hand, in contrast with that on classical integer-order case, the main technical difficulty in the rigorous analysis on well-posedness and regularity of fractional wave equations stems from establishing the energy estimates of the problem. This is mainly due to the fact that integration by parts formula for integer-order derivatives can not be generalized directly to fractional-order case and properties of composition and conjugation on the fractional Caputo derivative ∂_t^α $(\alpha \in (1,2))$ do not exist. Therefore, we found it more challenging in dealing with the well-posedness and regularity of fractional wave equations.

The section is organized as follows. In Subsection 8.6.2 we recall some notations, definitions, and preliminary facts used throughout this work. In Subsection 8.6.3 we discuss approximation equations and show the existence of its solutions by means of mollification arguments and the Galerkin methods, which reduces the regularity of the coefficients $a_{i,j}, b_j, c, f$. The energy estimates of approximation solutions are established in Subsection 8.6.4. Finally, we derive the well-posedness and regularity results of fractional wave equations using the weak compactness arguments.

8.6.2 *Preliminaries*

Here we recall some notations, definitions, and preliminary facts which are used throughout this section.

Let X be a Banach space and $v : [0, \infty) \to X$. The left Riemann-Liouville fractional integral of order $\alpha > 0$ for the function v is defined as

$$ {}_0D_t^{-\alpha} v(t) = (g_\alpha * v)(t), \quad t > 0, $$

where $g_\alpha(t) = \frac{t^{\alpha-1}}{\Gamma(\alpha)}$ and $*$ denotes the convolution.

Further, ${}^L\partial_t^\alpha v$ and $\partial_t^\alpha v$ represent the left Riemann-Liouville fractional derivative and Caputo fractional derivative of order $\alpha > 0$ for the function v, respectively, which are defined by

$$^L\partial_t^\alpha v(t) = \frac{d^n}{dt^n}[{}_0D_t^{-(n-\alpha)}v(t)] \text{ and } \partial_t^\alpha v(t) = {}^L\partial_t^\alpha\left[v(t) - \sum_{k=0}^{n-1}\frac{v^{(k)}(0)}{k!}t^k\right], \quad t > 0,$$

where $n = [\alpha] + 1$, $[\alpha]$ denotes the integer part of α.

Here we denote by $AC^n([0,T], X)$ the space of functions v that $v \in C^{n-1}([0,T], X)$ and $v^{(n-1)} \in AC([0,T], X)$. In particular, $AC^1([0,T], X) = AC([0,T], X)$. It is worth mentioning that if $v \in AC^n([0,T], X)$, then the Caputo fractional derivative $\partial_t^\alpha v(t)$ exists almost everywhere on $[0,T]$, which is represented by

$$\partial_t^\alpha v(t) = [g_{n-\alpha} * v^{(n)}](t) \quad \text{for } t \in [0,T].$$

For more insight into the topic, see Kilbas, Srivastava and Trujillo, 2006 and Zhou, 2014.

Lemma 8.13. *(Kilbas, Srivastava and Trujillo, 2006) If $v \in AC^2([0,T], X)$ and $\alpha \in (1,2]$, then ${}_0D_t^{-\alpha}\partial_t^\alpha v(t) = v(t) - v(0) - v'(0)t$ and $\partial_t^\alpha {}_0D_t^{-\alpha}v(t) = v(t)$.*

Lemma 8.14. *Let $\alpha \in (1,2)$. If $v \in AC^2([0,T], X)$, then we have*

$$\partial_t^\alpha v(t) = \frac{d}{dt}\partial_t^{\alpha-1}v(t) - v'(0)g_{2-\alpha}(t) = \partial_t^{\alpha-1}v'(t)$$

for a.e. $t \in (0,T)$.

Proof. If $v \in AC^2([0,T], X)$, then $v'(t)$ exists for a.e. $t \in (0,T)$. From the definition of ∂_t^α we know that

$$\begin{aligned}
\partial_t^\alpha v(t) &= \frac{d^2}{dt^2}\int_0^t g_{2-\alpha}(s)[v(t-s) - v(0) - v'(0)(t-s)]ds \\
&= \frac{d}{dt}\int_0^t g_{2-\alpha}(s)[v'(t-s) - v'(0)]ds \\
&= \frac{d}{dt}\int_0^t g_{2-\alpha}(t-s)[v'(s) - v'(0)]ds \\
&= \frac{d}{dt}\partial_t^{\alpha-1}v(t) - v'(0)g_{2-\alpha}(t).
\end{aligned}$$

On the other hand, since $\alpha - 1 \in (0,1)$, we see that $\frac{d}{dt}\int_0^t g_{2-\alpha}(t-s)[v'(s) - v'(0)]ds = {}^L\partial_t^{\alpha-1}[v'(t) - v'(0)] = \partial_t^{\alpha-1}v'(t)$. Thus the proof is completed. □

Before proceeding further, we state an important lemma, which is a direct consequence of an estimate borrowed from Zacher, 2009.

Lemma 8.15. *Let $T > 0$ and H be a real Hilbert space with a scalar product $(\cdot,\cdot)$. Assume $k \in L^1(0,T)$, $k' \in L^{1,loc}(0,T)$, $k \geq 0$, $k' \leq 0$. Then for any $v \in H^1(0,T,H)$, there holds*

$$\int_0^t \left(\frac{d}{ds}(k * v)(s), v(s)\right)ds \geq \frac{1}{2}(k * \|v\|^2)(t) + \frac{1}{2}\int_0^t k(s)\|v(s)\|^2 ds$$

for any $t \in [0,T]$.

Next, a very significant example is given, which will pay the crucial role in the proof of energy estimates.

Example 8.2. For $\alpha \in (1, 2)$, we choose $k(t) = g_{2-\alpha}(t)$. Then for any $v \in H^2(0, T, H)$ and $t \in [0, T]$, there holds

$$\int_0^t \left(\frac{d}{ds}[\partial_s^{\alpha-1} v(s)], v'(s)\right) ds \geq \frac{1}{2}(g_{2-\alpha} * \|v'\|^2)(t) + \int_0^t \frac{g_{2-\alpha}(s)}{2}\|v'(s)\|^2 ds.$$

The following property presents the lower bound of the uniformly elliptic operator if the function has enough regularity, which was proved by Ladyzhenskaya, 1958 (see also Kubica and Yamamoto, 2018).

Lemma 8.16. *Assume that $\Omega \subset \mathbb{R}^N$ is a bounded domain with the boundary of C^2 class and (8.73) holds. If $u \in H^3(\Omega)$ and $u|_{\partial\Omega} = 0$ and $\Delta u|_{\partial\Omega} = 0$, then*

$$\frac{\mu}{4}\|\nabla^2 u\|^2 - C\|\nabla u\|^2 \leq \sum_{i,j=1}^{N} \int_\Omega \partial_i(a_{i,j}(t,x)\partial_j u)\Delta u dx,$$

where C depends continuously on $\max_{i,j} \|\nabla a_{i,j}(t,x)\|_{L^\infty}$ and the C^2-norm of $\partial\Omega$, and $\nabla^2 u = \{u_{x_i x_j}\}_{i,j=1}^N$.

We consider the space

$${}_0H^2(0,T) = \{v \in H^2(0,T) : \ v(0) = 0, v'(0) = 0\}.$$

Next, we introduce the definition of the weak solution of the equation (8.72).

Definition 8.5. Let $T \in (0, \infty)$ and $f \in L^2(0, T, L^2(\Omega))$. For given functions u_0 and u_1, we say a function

$$u \in L^2(0,T,H_0^1(\Omega)) \ \ \textit{with} \ \ {}_0D_t^{\alpha-2}(u - u_0 - u_1 t) \in {}_0H^2(0,T,H^{-1}(\Omega))$$

is a weak solution of the equation (8.72) provided

$$\frac{\partial^2}{\partial t^2}\int_\Omega {}_0D_t^{\alpha-2}(u(t,x) - u_0 - u_1 t)\omega(x)dx + \int_\Omega \sum_{i,j=1}^{N} a_{i,j}(t,x)\partial_j u(t,x)\cdot \partial_i\omega(x)dx$$

$$= \int_\Omega \sum_{j=1}^{N} b_j(t,x)\partial_j u(t,x)\omega(x)dx + \int_\Omega c(t,x)u(t,x)\omega(x)dx + \int_\Omega f(t)\omega(x)dx$$

for each $\omega \in H_0^1(\Omega)$ and a.e. $t \in [0, T]$.

The vector u_0 and u_1 can be regarded as initial data for $u(t)$ and $u'(t)$ at least in a weak sense, respectively. If e.g. $u \in AC^2([0,T], H^{-1}(\Omega))$, then the condition ${}_0D_t^{\alpha-2}(u - u_0 - u_1 t) \in {}_0H^2(0,T,H^{-1}(\Omega))$ implies $u(0) = u_0$ and $\partial_t u(0) = u_1$.

Remark 8.5. In view of Definition 8.5, we know $u' \in C([0,T], H^{-1}(\Omega))$ for $\alpha > \frac{3}{2}$.

Proof. Indeed, for $t_1, t_2 \in [0,T]$ with $t_1 < t_2$, it follows from Lemma 8.14 and Hölder inequality that

$$\begin{aligned}
&\|u'(t_2) - u'(t_1)\|_{H^{-1}} \\
&= \left\| \int_0^{t_2} g_{\alpha-1}(t_2 - s)\partial_s^\alpha u(s)ds - \int_0^{t_1} g_{\alpha-1}(t_1 - s)\partial_s^\alpha u(s)ds \right\|_{H^{-1}} \\
&\le \int_{t_1}^{t_2} g_{\alpha-1}(t_2 - s)\|\partial_s^\alpha u(s)\|_{H^{-1}}ds \\
&\quad + \int_0^{t_1} [g_{\alpha-1}(t_1 - s) - g_{\alpha-1}(t_2 - s)]\|\partial_s^\alpha u(s)\|_{H^{-1}}ds \\
&\le \frac{(t_2 - t_1)^{\alpha-\frac{3}{2}}}{(2\alpha-3)^{\frac{1}{2}}\Gamma(\alpha-1)}\|\partial_t^\alpha u\|_{L^2(0,T,H^{-1})} \\
&\quad + \|\partial_t^\alpha u\|_{L^2(0,T,H^{-1})}\left(\int_0^{t_1} [g_{\alpha-1}(t_1 - s) - g_{\alpha-1}(t_2 - s)]^2 ds\right)^{\frac{1}{2}}.
\end{aligned}$$

In view of the inequality $\xi_1^\sigma - \xi_2^\sigma \le (\xi_1 - \xi_2)^\sigma$ for $\xi_1, \xi_2 > 0$ and $0 \le \sigma \le 1$, we calculate the integral

$$\begin{aligned}
&\int_0^{t_1} [g_{\alpha-1}(t_1 - s) - g_{\alpha-1}(t_2 - s)]^2 ds \\
&\le \frac{1}{\Gamma^2(\alpha-1)}\int_0^{t_1} (t_1 - s)^{2(\alpha-2)} - (t_2 - s)^{2(\alpha-2)}ds \\
&\le \frac{(t_2 - t_1)^{2\alpha-3}}{(2\alpha-3)\Gamma^2(\alpha-1)}.
\end{aligned}$$

The second term is bounded by $\|\partial_t^\alpha u\|_{L^2(0,T,H^{-1})}\frac{(t_2-t_1)^{\alpha-\frac{3}{2}}}{(2\alpha-3)^{\frac{1}{2}}\Gamma(\alpha-1)}$. This ensures

$$\|u'(t_2) - u'(t_1)\|_{H^{-1}} \to 0, \quad \text{as } t_1 \to t_2.$$

The proof is completed. □

8.6.3 *Approximation Solution*

In this subsection we provide the Galerkin approximate scheme and derive the corresponding existence results. We will suppose initially that

$$\begin{aligned}
&A \in (W^{1,\infty}(0,T;L^\infty(\Omega)))^{N\times N}, \ b_j \in L^\infty((0,T)\times\Omega), \\
&c \in L^\infty(0,T,L^{\frac{2q}{q-2}}(\Omega)), \quad f \in L^2(0,T,L^2(\Omega))
\end{aligned} \tag{8.74}$$

for $q \in [2, \frac{2N}{N-2})$, where $A(t,x) = \{a_{i,j}(t,x)\}_{i,j=1}^N$ and $\mathbf{b} = (b_1, b_2, ..., b_N)$.

Let ϱ_ε be the standard mollifier satisfying

$$\varrho_\varepsilon \in C^\infty(\mathbb{R}), \quad \text{supp } \varrho_\varepsilon = \{t : |t| < \frac{\varepsilon}{T}\}, \quad \int_{\mathbb{R}} \varrho_\varepsilon(t)dt = 1.$$

Then we introduce the mollification v_ε of the function $v \in L^{1,loc}(\mathbb{R})$ as

$$v_\varepsilon(t) = (\varrho_\varepsilon * v)(t).$$

We note first that $v_\varepsilon \in C^\infty(\mathbb{R})$ and if $v \in L^p(\mathbb{R})$ for $p \geq 1$, then $v_\varepsilon \to v$ in $L^p(\mathbb{R})$ as $\varepsilon \to 0$.

Moreover, we denote $a^n_{i,j}, b^n_j, c^n, f_{\frac{1}{n}}$ by the mollification of $a_{i,j}$, b_j, c, f, which are defined by

$$a^n_{i,j}(t,x) = \big(\varrho_{\frac{1}{n}} * a_{i,j}(\cdot,x)\big)(t), \qquad b^n_j(t,x) = \big(\varrho_{\frac{1}{n}} * b_j(\cdot,x)\big)(t),$$
$$c^n(t,x) = \big(\varrho_{\frac{1}{n}} * c(\cdot,x)\big)(t), \qquad f_{\frac{1}{n}}(t) = \big(\varrho_{\frac{1}{n}} * f(\cdot,x)\big)(t),$$

where $a_{i,j}$ is the continuation by even reflection to $(-T,T)$ and zero elsewhere, b_j and c are the continuation by zero for $t \notin (0,T)$, and f is the continuation by odd reflection to $(-T,T)$ and zero elsewhere. Then $\lim_{n\to\infty} a^n_{i,j}(t) = a_{i,j}$ in $L^2((0,T)\times\Omega)$ for $a_{i,j} \in L^\infty((0,T)\times\Omega)$ (due to (8.73)).

Next, we seek approximate solutions $u_n(t,x)$ for the equation (8.72) in the form:

$$u_n(t,x) = \sum_{k=1}^{n} d_{n,k}(t) e_k(x), \quad \text{for } n \in \mathbb{N}, \tag{8.75}$$

where $\{e_k\}$ denotes the complete orthonormal system of eigenfunctions which forms an orthogonal basis of $L^2(\Omega) \cap H^1_0(\Omega)$ such that

$$-\Delta e_k = \lambda_k e_k \ \text{ in } \Omega, \qquad e_k|_{\partial\Omega} = 0, \ k = 1, 2, \ldots.$$

For the sake of selecting $d_{n,k}(t)$, one considers the following approximate equation:

$$\begin{cases} \partial^\alpha_t u_n(t,x) - \mathcal{A}^n u_n(t,x) = f^n(t), & (t,x) \in (0,T] \times \Omega, \\ u_n(0,x) = u_{n0}, \ \partial_t u_n(0,x) = u_{n1}, \end{cases} \tag{8.76}$$

where

$$\mathcal{A}^n u_n(t,x) = \sum_{i,j=1}^{N} \partial_i\big(a^n_{i,j}(t,x)\partial_j u_n(t,x)\big) + \sum_{j=1}^{N} b^n_j(t,x)\partial_j u_n(t,x) + c^n(t,x) u_n(t,x),$$

$$f^n(t,x) = \sum_{k=1}^{n} \big(f_{\frac{1}{n}}(t,\cdot), e_k(\cdot)\big) e_k(x),$$

$$u_{n0}(t,x) = \sum_{k=1}^{n} \big(u_0(\cdot), e_k(\cdot)\big) e_k(x), \quad u_{n1}(t,x) = \sum_{k=1}^{n} \big(u_1(\cdot), e_k(\cdot)\big) e_k(x).$$

Let us introduce the time-dependent bilinear form

$$\mathcal{B}^n[u,v;t] := \int_\Omega \sum_{i,j=1}^{N} a^n_{i,j}(t,x)\partial_j u \cdot \partial_i v - \sum_{j=1}^{N} b^n_j(t,x)\partial_j u v - c^n(t,x) u v dx.$$

Taking the scalar product of (8.76) with e_l for $l = 1, \ldots, n$, we obtain

$$\begin{cases} \big(\partial^\alpha_t u_n(t,\cdot), e_l\big) + \mathcal{B}^n[u_n, e_l; t] = \big(f_{\frac{1}{n}}(t), e_l\big), \\ \big(u_n(0,\cdot), e_l\big) = (u_{n0}, e_l), \ \big(\partial_t u_n(0,\cdot), e_l\big) = (u_{n1}, e_l). \end{cases} \tag{8.77}$$

More precisely, write

$$d_n(t) = \big(d_{n,1}(t), \ldots, d_{n,n}(t)\big),$$

$$\mathcal{L}^n(t) = \{L^n_{k,l}(t)\}^n_{k,l=1}, \qquad L^n_{k,l}(t) = \mathcal{B}^n[e_l, e_k; t],$$
$$F^n(t) = \big(f_{\frac{1}{n}}(t), e_l\big)^n_{l=1},$$
$$d_{n0} = \big(u_0, e_l\big)^n_{l=1}, \qquad d_{n1} = \big(u_1, e_l\big)^n_{l=1}.$$

Then (8.77) can be reduced to the following linear differential system for the functions d_n:

$$\begin{cases} \partial^\alpha_t d_n(t) + \mathcal{L}^n(t) d_n(t) = F^n(t), & \text{for } t \in (0,T], \\ d_n(0) = d_{n0}, \ d'_n(0) = d_{n1}. \end{cases} \tag{8.78}$$

Now we consider the nonlinear integral system for the functions

$$d_n(t) = d_{n0} + d_{n1}t + \big[g_\alpha * \big(\mathcal{L}^n(\cdot)d_n(\cdot)\big)\big](t) + \big[g_\alpha * F^n\big](t), \quad \text{for } t \in [0,T]. \tag{8.79}$$

We shall show that the system (8.79) has a unique solution d_n which belongs to $AC^2[0,T]$. By Lemma 8.13, then the solution d_n of the equation (8.79) is also the solution of the equation (8.78). To do this, we introduce the space

$$E_T = \big\{d \in C^1([0,T],\mathbb{R}^n) : \ d(0) = d_{n0}, \ d'(0) = d_{n1}, \ t^{2-\alpha}d''(t) \in C([0,T],\mathbb{R}^n)\big\},$$

and define a metric on E_T as

$$\|d\|_{E_T} = \|d\|_{C[0,T]} + \|d'\|_{C[0,T]} + \|t^{2-\alpha}d''\|_{C[0,T]}.$$

It is easy to show that $(E_T, \|\cdot\|_{E_T})$ is a complete metric space. We notice that $E_T \subset AC^2([0,T],\mathbb{R}^n)$.

Theorem 8.16. *Let $T \in (0,\infty)$ and (8.74) hold. For every $n \in \mathbb{N}$, the equation (8.79) has a unique solution in E_T.*

Proof. Consider the operator $\mathcal{T} : E_T \to E_T$ given by

$$\mathcal{T}d(t) = d_{n0} + d_{n1}t + \big[g_\alpha * \big(\mathcal{L}^n(\cdot)d(\cdot)\big)\big](t) + \big(g_\alpha * F^n\big)(t), \quad \text{for } t \in [0,T].$$

Then it is well-defined. Indeed, let $d \in E_T$, then $\mathcal{T}d(0) = d_{n0}$. Further, we immediately take the first and second derivatives of $\mathcal{T}d$ with respect to t to obtain

$$(\mathcal{T}d)'(t) = d_{n1} + g_\alpha(t)\big[\mathcal{L}^n(0)d(0) + F^n(0)\big] + \big[g_\alpha * \big(\mathcal{L}^n(\cdot)d(\cdot) + F^n(\cdot)\big)'\big](t), \text{ for } t \in [0,T],$$

and

$$\begin{aligned}(\mathcal{T}d)''(t) = {} & g_{\alpha-1}(t)\big[\mathcal{L}^n(0)d(0) + F^n(0)\big] \\ & + g_\alpha(t)\big[(\mathcal{L}^n)'(0)d(0) + \mathcal{L}^n(0)d'(0) + (F^n)'(0)\big] \\ & + g_\alpha * \big[\mathcal{L}^n(t)d(t) + F^n(t)\big]'', \quad \text{for } t \in (0,T].\end{aligned}$$

For convenience we let $G_d(t) = \mathcal{L}^n(t)d(t) + F^n(t)$. Then $G'_d(t) = (\mathcal{L}^n)'(t)d(t) + \mathcal{L}^n(t)d'(t) + (F^n)'(t)$ and $G_d, G'_d \in C([0,T],\mathbb{R}^n)$. We can easily check that $\mathcal{T}d$ and $(\mathcal{T}d)'$ are continuous on $C([0,T],\mathbb{R}^n)$, which also ensures that $(\mathcal{T}d)'(0) = d_{n1}$. Therefore it remains to consider the continuity of $t^{2-\alpha}(\mathcal{T}d)''(t)$. It is easy to verify the continuity of the first two component. To deal with the third one we estimate for $0 \le t_1 < t_2 \le T$

$$\begin{aligned}
&\left| t_2^{2-\alpha} \int_0^{t_2} g_\alpha(s) G_d''(t_2 - s) ds - t_1^{2-\alpha} \int_0^{t_1} g_\alpha(s) G_d''(t_1 - s) ds \right| \\
&\le |t_2^{2-\alpha} - t_1^{2-\alpha}| \int_0^{t_2} g_\alpha(s) |G_d''(t_2 - s)| ds + t_1^{2-\alpha} \int_{t_1}^{t_2} g_\alpha(s) |G_d''(t_2 - s)| ds \\
&\quad + t_1^{2-\alpha} \int_0^{t_1} g_\alpha(s) |G_d''(t_2 - s) - G_d''(t_1 - s)| ds \\
&=: I_1(t_1, t_2) + I_2(t_1, t_2) + I_3(t_1, t_2).
\end{aligned}$$

On the other hand, from the definition of G_d and $d \in E_T$, it follows that

$$G_d''(t) = (\mathcal{L}^n)''(t) d(t) + 2(\mathcal{L}^n)'(t) d'(t) + \mathcal{L}^n(t) d''(t) + (F^n)''(t).$$

From the representation of $\mathcal{L}^n(t)$ and F^n, we know that $\mathcal{L}^n$ and F^n belong to the space E_T, which yields that $G_d \in E_T$ and

$$|G_d''(t)| \le \|G_d\|_{E_T} t^{\alpha - 2}. \tag{8.80}$$

Thus one can immediately calculate $I_1(t_1, t_2)$ and $I_2(t_1, t_2)$ as follows

$$\begin{aligned}
I_1(t_1, t_2) &\le \|G_d\|_{E_T} |t_2^{2-\alpha} - t_1^{2-\alpha}| \int_0^{t_2} g_\alpha(s)(t_2 - s)^{\alpha - 2} ds \\
&= \frac{\|G_d\|_{E_T}}{\Gamma(\alpha)} B(\alpha, \alpha - 1) t_2^\alpha \left[1 - \left(\frac{t_1}{t_2} \right)^{2-\alpha} \right] \to 0, \quad \text{as } t_2 \to t_1,
\end{aligned}$$

and

$$\begin{aligned}
I_2(t_1, t_2) &\le \|G_d\|_{E_T} t_1^{2-\alpha} \int_{t_1}^{t_2} g_\alpha(s)(t_2 - s)^{\alpha - 2} ds \\
&\le \frac{\|G_d\|_{E_T}}{\Gamma(\alpha)} t_2^\alpha \left(\frac{t_1}{t_2} \right)^{2-\alpha} \int_{t_1/t_2}^1 s^{\alpha - 1}(1 - s)^{\alpha - 2} ds \to 0, \quad \text{as } t_2 \to t_1.
\end{aligned}$$

Finally, for $I_3(t_1, t_2)$, choosing $\delta \in (0, t_1)$ sufficient small for $t_1 > 0$, one can derive from the increasing property of g_α and (8.80) that

$$\begin{aligned}
I_3(t_1, t_2) &= t_1^{2-\alpha} \int_0^{t_1 - \delta} g_\alpha(s) |G_d''(t_2 - s) - G_d''(t_1 - s)| ds \\
&\quad + t_1^{2-\alpha} \int_{t_1 - \delta}^{t_1} g_\alpha(s) |G_d''(t_2 - s) - G_d''(t_1 - s)| ds \\
&\le t_1^{2-\alpha} g_\alpha(t_1 - \delta) \int_0^{t_1 - \delta} |G_d''(t_2 - s) - G_d''(t_1 - s)| ds \\
&\quad + 2\|G_d\|_{E_T} t_1^{2-\alpha} \int_{t_1 - \delta}^{t_1} g_\alpha(s)(t_1 - s)^{\alpha - 2} ds \\
&\le t_1^{2-\alpha} g_\alpha(t_1) \int_\delta^{t_1} |G_d''(t_2 - t_1 + s) - G_d''(s)| ds
\end{aligned}$$

$$+\frac{2\|G_d\|_{E_T}}{\Gamma(\alpha)}t_1^\alpha\int_{1-\delta/t_1}^1 s^{\alpha-1}(1-s)^{\alpha-2}ds.$$

It is clear that the second term tends to zero for some sufficient small δ. Then we choose one of such δ, it follows from the uniform continuity of G'' (due to the continuity of G''_d on $[\delta,T]$) that for any $\varepsilon>0$, there exists $\delta'<\delta$ with $|t_2-t_1|<\delta'$ such that $|G''_d(t_2-t_1+s)-G''_d(s)|<\varepsilon$. Thus this yields that the first term can be bounded by $\varepsilon t_1^{2-\alpha}g_\alpha(t_1)(t_1-\delta)$, which together with $I_3(0,t_2)=0$ shows that $I_3(t_1,t_2)\to 0$ as $t_2\to t_1$ for $0\le t_1<t_2\le T$.

Therefore, we have $\mathcal{T}d\in E_T$ for $d\in E_T$.

Moreover, for $d_1,d_2\in E_T$, we have

$$\begin{aligned}|\mathcal{T}d_1(t)-\mathcal{T}d_2(t)| &\le \Big[g_\alpha * \big|G_{d_1}(\cdot)-G_{d_2}(\cdot)\big|\Big](t)\\ &\le \frac{\|\mathcal{L}^n\|}{\Gamma(\alpha)}\|d_1-d_2\|_{E_T}\int_0^t(t-s)^{\alpha-1}ds \\ &\le \frac{\|\mathcal{L}^n\|t^\alpha}{\Gamma(1+\alpha)}\|d_1-d_2\|_{E_T},\end{aligned} \tag{8.81}$$

where we have used

$$\begin{aligned}\big|G_{d_1}(s)-G_{d_2}(s)\big| &\le |\mathcal{L}^n(s)||d_1(s)-d_2(s)|\\ &\le \|\mathcal{L}^n\|\|d_1-d_2\| \\ &\le \|\mathcal{L}^n\|\|d_1-d_2\|_{E_T},\quad \text{for } s\in[0,t].\end{aligned} \tag{8.82}$$

Similarly, in view of

$$\begin{aligned}\big|G'_{d_1}(s)-G'_{d_2}(s)\big| &\le |(\mathcal{L}^n)'(s)||d_1(s)-d_2(s)|+|\mathcal{L}^n(s)||d'_1(s)-d'_2(s)|\\ &\le \|(\mathcal{L}^n)'\|\|d_1-d_2\|+\|\mathcal{L}^n\|\|d'_1-d'_2\| \\ &\le 2\|\mathcal{L}^n\|_{C^1[0,T]}\|d_1-d_2\|_{E_T},\quad \text{for } s\in[0,t],\end{aligned} \tag{8.83}$$

we proceed to estimate $(\mathcal{T}d_1)'-(\mathcal{T}d_2)'$ as follows:

$$\begin{aligned}|(\mathcal{T}d_1)'(t)-(\mathcal{T}d_2)'(t)| &\le g_\alpha(t)\big|G_{d_1}(0)-G_{d_2}(0)\big|+\Big[g_\alpha*\big|G'_{d_1}(\cdot)-G'_{d_2}(\cdot)\big|\Big](t)\\ &\le \frac{2\|\mathcal{L}^n\|_{C^1[0,T]}}{\Gamma(\alpha)}\|d_1-d_2\|_{E_T}\int_0^t(t-s)^{\alpha-1}ds\\ &\le \frac{2\|\mathcal{L}^n\|_{C^1[0,T]}}{\Gamma(1+\alpha)}t^\alpha\|d_1-d_2\|_{E_T},\end{aligned} \tag{8.84}$$

where it is easy to show that $G_{d_1}(0)-G_{d_2}(0)=0$ due to $d_1(0)-d_2(0)=0$ and (8.82).

Finally, we will estimate $t^{2-\alpha}(\mathcal{T}d_1)''-t^{2-\alpha}(\mathcal{T}d_2)''$. Taking account of the following inequality

$$\begin{aligned}\big|G''_{d_1}(s)-G''_{d_2}(s)\big| &\le |(\mathcal{L}^n)''(s)||d_1(s)-d_2(s)|+2|(\mathcal{L}^n)'(s)||d'_1(s)-d'_2(s)|\\ &\quad+|\mathcal{L}^n(s)||d''_1(s)-d''_2(s)|\\ &\le \|(\mathcal{L}^n)''\|\|d_1-d_2\|+2\|(\mathcal{L}^n)'\|\|d'_1-d'_2\|\end{aligned}$$

$$+ s^{\alpha-2}\|\mathcal{L}^n\|\|s^{2-\alpha}d_1'' - s^{2-\alpha}d_2''\|$$
$$\leq 3\|\mathcal{L}^n\|_{C^2[0,T]}(1 + s^{\alpha-2})\|d_1 - d_2\|_{E_T}, \quad \text{for } s \in (0, t],$$

it holds that

$$\begin{aligned}
&|t^{2-\alpha}(\mathcal{T}d_1)''(t) - t^{2-\alpha}(\mathcal{T}d_2)''(t)| \\
&\leq t^{2-\alpha}g_{\alpha-1}(t)\big|G_{d_1}(0) - G_{d_2}(0)\big| + t^{2-\alpha}g_\alpha(t)\big|G_{d_1}'(0) - G_{d_2}'(0)\big| \\
&\quad + t^{2-\alpha}\big[g_\alpha * \big|G_{d_1}''(\cdot) - G_{d_2}''(\cdot)\big|\big](t) \\
&\leq \frac{3\|\mathcal{L}^n\|_{C^2[0,T]}}{\Gamma(\alpha)}\|d_1 - d_2\|_{E_T}t^{2-\alpha}\int_0^t (t-s)^{\alpha-1}(1 + s^{\alpha-2})ds \\
&\leq 3\|\mathcal{L}^n\|_{C^2[0,T]}\left(\frac{t^2}{\Gamma(1+\alpha)} + \frac{\Gamma(\alpha-1)}{\Gamma(2\alpha-1)}t^\alpha\right)\|d_1 - d_2\|_{E_T},
\end{aligned} \tag{8.85}$$

where we know from (8.83) that $G_{d_1}'(0) - G_{d_2}'(0) = 0$.

For the sake of convenience, we let

$$M(t) = 3t^\alpha\|\mathcal{L}^n\|_{C^2[0,T]}\left(\frac{1}{\Gamma(1+\alpha)} + \frac{\Gamma(\alpha-1)}{\Gamma(2\alpha-1)}\right) + \|\mathcal{L}^n\|_{C^2[0,T]}\frac{3t^2}{\Gamma(1+\alpha)}.$$

Then one can choose a $T_1 \in (0, T)$ small enough which ensures that $M(T_1) < 1$. Therefore combining (8.81), (8.84) with (8.85), we deduce that

$$\|\mathcal{T}d_1 - \mathcal{T}d_2\|_{E_{T_1}} \leq M(T_1)\|d_1 - d_2\|_{E_T}.$$

This also shows that the operator $\mathcal{T}$ is a strict contraction on $E(T_1)$. It follows that $\mathcal{T}$ has a fixed point, thus the equation (8.79) has a unique solution in E_{T_1}.

Now, we will deal with the continuation of the solution to the interval $[0, T]$. Let us make the assumption that we have obtained the solution $\bar{d}$ of the equation (8.79) on the interval $[0, T_l]$ for $T_l > 0$. We shall define the solution for $t \in [T_l, T_{l+1}]$ with $T_{l+1} > T_l$. To do this, we introduce the complete space

$$\bar{E}_{T_{l+1}} = \big\{d \in C^2((0, T_{l+1}], \mathbb{R}^n) : \ d(t) = \bar{d}(t) \ \text{ for } t \in [0, T_l]\big\},$$

with the distance $\|d\|_{\bar{E}_{T_{l+1}}} = \|d\|_{C^2[T_l, T_{l+1}]}$. Let $d \in \bar{E}_{T_{l+1}}$, then $d \in E_{T_{l+1}}$. According to the previous proof, we have that $\mathcal{T}d \in E_{T_{l+1}}$, which implies that $\mathcal{T}d \in C^1([0, T_{l+1}], \mathbb{R}^n)$ and $t^{2-\alpha}(\mathcal{T}d)'' \in C([0, T_{l+1}], \mathbb{R}^n)$. It holds that $(\mathcal{T}d)'' \in C((0, T_{l+1}], \mathbb{R}^n)$ and then $\mathcal{T}d \in \bar{E}_{T_{l+1}}$.

Next, we will show that the operator $\mathcal{T}$ is also a strict contraction on $\bar{E}_{T_{l+1}}$ when $T_{l+1} - T_l$ is sufficient small. We shall rewrite $\mathcal{T}$ in the following form:

$$\mathcal{T}d(t) = d_{n0} + d_{n1}t + \frac{1}{\Gamma(\alpha)}\int_0^{T_l} g_\alpha(t-s)G_d(s)ds + \frac{1}{\Gamma(\alpha)}\int_{T_l}^t g_\alpha(t-s)G_d(s)ds.$$

For $d_1, d_2 \in \bar{E}_{T_{l+1}}$, we have $d_1(t) - d_2(t) = 0$ and $G_{d_1}(t) - G_{d_2}(t) = 0$ for $t \in [0, T_l]$. Then

$$\mathcal{T}d_1(t) - \mathcal{T}d_2(t) = \frac{1}{\Gamma(\alpha)}\int_{T_l}^t g_\alpha(t-s)[G_{d_1}(s) - G_{d_2}(s)]ds.$$

This follows from (8.82) that

$$\|\mathcal{T}d_1 - \mathcal{T}d_2\|_{C[T_l,T_{l+1}]} \le \frac{\|\mathcal{L}^n\|}{\Gamma(1+\alpha)}\|d_1 - d_2\|_{C[T_l,T_{l+1}]}(T_{l+1} - T_l)^\alpha.$$

Similarly, we get

$$\begin{aligned}&\|(\mathcal{T}d_1)' - (\mathcal{T}d_2)'\|_{C[T_l,T_{l+1}]}\\ &\le \frac{1}{\Gamma(1+\alpha)}\big(\|(\mathcal{L}^n)'\|\|d_1 - d_2\|_{C[T_l,T_{l+1}]} + \|\mathcal{L}^n\|\|d_1' - d_2'\|_{C[T_l,T_{l+1}]}\big)(T_{l+1} - T_l)^\alpha,\end{aligned}$$

and

$$\begin{aligned}&\|(\mathcal{T}d_1)'' - (\mathcal{T}d_2)''\|_{C[T_l,T_{l+1}]}\\ &\le \frac{(T_{l+1} - T_l)^\alpha}{\Gamma(1+\alpha)}\Big[\|(\mathcal{L}^n)''\|\|d_1 - d_2\|_{C[T_l,T_{l+1}]}\\ &\quad + 2\|(\mathcal{L}^n)'\|\|d_1' - d_2'\|_{C[T_l,T_{l+1}]} + \|\mathcal{L}^n\|\|d_1'' - d_2''\|_{C[T_l,T_{l+1}]}\Big].\end{aligned}$$

Therefore

$$\|\mathcal{T}d_1 - \mathcal{T}d_2\|_{\bar{E}_{T_{l+1}}} \le \frac{4\|\mathcal{L}^n\|_{C^2[0,T]}}{\Gamma(1+\alpha)}\|d_1 - d_2\|_{\bar{E}_{T_{l+1}}}(T_{l+1} - T_l)^\alpha.$$

Moreover, we can choose one $T_{l+1} \in \left(T_l, T_l + \left(\frac{\Gamma(1+\alpha)}{4\|\mathcal{L}^n\|_{C^2[0,T]}}\right)^{\frac{1}{\alpha}}\right)$ such that $T_{l+1} - T_l$ is small enough, it also ensures that

$$0 < \frac{4\|\mathcal{L}^n\|_{C^2[0,T]}}{\Gamma(1+\alpha)}(T_{l+1} - T_l)^\alpha < 1.$$

Hence, the operator $\mathcal{T}$ is a strict contraction on $\bar{E}_{T_{l+1}}$, this also shows that the equation (8.79) has a unique solution on the interval $[T_l, T_{l+1}]$. We proceed to repeat the process on the intervals $[T_{l+1}, T_{l+2}], \ldots,$ until the equation (8.79) has a unique solution on the interval $[0, T]$. The claim then follows. □

8.6.4 *Energy Estimates*

The purpose of this subsection is to establish some priori estimates of approximation solutions through a mathematic analysis, which plays an important role in getting the main results, we can make it with the following lemma.

Lemma 8.17. *Assume that $u_0 \in H_0^1(\Omega)$, $u_1 \in L^2(\Omega)$ and recall the condition imposed to the parameters $a_{i,j}$, b_j, c and f. Then, for every $n \in \mathbb{N}$ and $t \in (0, T]$ the approximate solution u_n given by (8.75) and (8.79) satisfies the inequality*

$$\begin{aligned}&{}_0D_t^{\alpha-2}\|\partial_t u_n(t,\cdot)\|^2 + \int_0^t \|\partial_s u_n(s,\cdot)\|^2 ds + \|\nabla u_n(t,\cdot)\|^2\\ &\le \widetilde{M}_1\big(\|u_{n0}\|_{H_0^1}^2 + \|u_{n1}\|^2 t^{2-\alpha}\big) + \widetilde{M}_2 \int_0^t \|f_{\frac{1}{n}}(s,\cdot)\|^2 ds,\end{aligned}$$

where $\widetilde{M}_1$ and $\widetilde{M}_2$ are positive constants.

Proof. Multiply the equation (8.77) by $d'_{n,l}(t)$, sum it up from 1 to n and recall (8.75) to discover

$$\begin{cases} \left(\partial_t^\alpha u_n(t,\cdot), \partial_t u_n(t,\cdot)\right) + \mathcal{B}^n[u_n, \partial_t u_n; t] \\ = (f_{\frac{1}{n}}(t,\cdot), \partial_t u_n(t,\cdot)), & (t,x) \in (0,T] \times \Omega, \\ u_n(0,\cdot) = u_{n0},\ \partial_t u_n(0,\cdot) = u_{n1}. \end{cases} \tag{8.86}$$

Taking into account Lemma 8.14, we have

$$\partial_t^\alpha u_n(t,\cdot) = \frac{\partial}{\partial t}[\partial_t^{\alpha-1} u_n(t,\cdot)] - \frac{u_{n1}}{\Gamma(2-\alpha)} t^{1-\alpha}.$$

Using Example 8.2, it follows that

$$\begin{aligned} &\int_0^t \left(\partial_s^\alpha u_n(s,\cdot), \partial_s u_n(s,\cdot)\right) ds \\ &= \int_0^t \Big(\frac{\partial}{\partial s}[\partial_s^{\alpha-1} u_n(s,\cdot)], \partial_s u_n(s,\cdot)\Big) ds - \int_0^t \Big(\frac{u_{n1}}{\Gamma(2-\alpha)} s^{1-\alpha}, \partial_s u_n(s,\cdot)\Big) ds \\ &\geq \frac{1}{2}\, {}_0D_t^{\alpha-2} \|\partial_t u_n(t,\cdot)\|^2 + \frac{1}{2}\int_0^t g_{2-\alpha}(s) \|\partial_s u_n(s,\cdot)\|^2 ds \\ &\quad - \frac{1}{\Gamma(2-\alpha)} \int_0^t s^{1-\alpha} \left(u_{n1}, \partial_s u_n(s,\cdot)\right) ds. \end{aligned}$$

Therefore, we integrate the first equality of the equation (8.86) with respect to the time variable from 0 to t to obtain that

$$\begin{aligned} &\frac{1}{2}\, {}_0D_t^{\alpha-2} \|\partial_t u_n(t,\cdot)\|^2 + \frac{1}{2}\int_0^t g_{2-\alpha}(s) \|\partial_s u_n(s,\cdot)\|^2 ds \\ &+ \sum_{i,j=1}^N \int_0^t \int_\Omega a_{i,j}^n(s,x) \partial_j u_n(s,x) \partial_i \partial_s u_n(s,x) dx ds \\ &\leq \sum_{j=1}^N \int_0^t \int_\Omega b_j^n(s,x) \partial_j u_n(s,x) \partial_s u_n(s,x) dx ds \\ &+ \int_0^t \int_\Omega c^n(s,x) u_n(s,x) \partial_s u_n(s,x) dx ds \\ &+ \int_0^t \left(f_{\frac{1}{n}}(s,\cdot), \partial_s u_n(s,\cdot)\right) ds + \frac{1}{\Gamma(2-\alpha)} \int_0^t s^{1-\alpha} \left(u_{n1}, \partial_s u_n(s,\cdot)\right) ds \\ &=: J_1(t) + J_2(t). \end{aligned} \tag{8.87}$$

First we estimate the third term of the left-hand side of the above inequality. Using the integration by parts with respect to s, we derive that

$$\begin{aligned} &\sum_{i,j=1}^N \int_0^t \int_\Omega a_{i,j}^n(s,x) \partial_j u_n(s,x) \partial_i \partial_s u_n(s,x) dx ds \\ &= \sum_{i,j=1}^N \int_\Omega a_{i,j}^n(s,x) \partial_j u_n(s,x) \partial_i u_n(s,x) dx \Big|_0^t \end{aligned}$$

$$-\sum_{i,j=1}^{N}\int_0^t\int_\Omega[\partial_s a_{i,j}^n(s,x)\partial_j u_n(s,x)+a_{i,j}^n(s,x)\partial_j\partial_s u_n(s,x)]\partial_i u_n(s,x)dxds.$$

It follows from $a_{i,j}^n = a_{j,i}^n$ that

$$\sum_{i,j=1}^{N}\int_0^t\int_\Omega a_{i,j}^n(s,x)\partial_j u_n(s,x)\partial_i\partial_s u_n(s,x)dxds$$
$$=\sum_{i,j=1}^{N}\int_0^t\int_\Omega a_{i,j}^n(s,x)\partial_j\partial_s u_n(s,x)\partial_i u_n(s,x)dxds.$$

In addition, in view of the definition of $a_{i,j}^n$, we know that

$$\partial_s a_{i,j}^n(s,x) = \big(\varrho_{\frac{1}{n}} * \partial_t a_{i,j}(\cdot,x)\big)(s),$$

this yields that

$$|\partial_s a_{i,j}^n(s,x)| \le \|A\|_{W^{1,\infty}}.$$

Therefore, using (8.73) again, one can obtain that

$$\sum_{i,j=1}^{N}\int_0^t\int_\Omega a_{i,j}^n(s,x)\partial_j u_n(s,x)\partial_i\partial_s u_n(s,x)dxds$$
$$=\frac{1}{2}\sum_{i,j=1}^{N}\int_\Omega a_{i,j}^n(s,x)\partial_j u_n(s,x)\partial_i u_n(s,x)dx\Big|_0^t$$
$$-\frac{1}{2}\sum_{i,j=1}^{N}\int_0^t\int_\Omega \partial_s a_{i,j}^n(s,x)\partial_j u_n(s,x)\partial_i u_n(s,x)dxds$$
$$\ge\frac{\mu}{2}\int_\Omega|\nabla u_n(t,x)|^2dx-\frac{\nu}{2}\int_\Omega|\nabla u_n(0,x)|^2dx$$
$$-\frac{1}{2}\|A\|_{W^{1,\infty}}\int_0^t\int_\Omega\sum_{i,j=1}^{N}|\partial_j u_n(s,x)\partial_i u_n(s,x)|dsdx$$
$$\ge\frac{\mu}{2}\|\nabla u_n(t,\cdot)\|^2-\frac{\nu}{2}\|\nabla u_n(0,\cdot)\|^2-\frac{1}{2}\|A\|_{W^{1,\infty}}\int_0^t\|\nabla u_n(s,\cdot)\|^2ds.$$

Next we will estimate the upper bound of the right-handed side of (8.87). For $J_1(t)$, we use the Hölder inequality and Young inequality to obtain

$$J_1(t)=\sum_{j=1}^{N}\int_0^t\int_\Omega b_j^n(s,x)\partial_j u_n(s,x)\partial_s u_n(s,x)dxds$$
$$+\int_0^t\int_\Omega c^n(s,x)u_n(s,x)\partial_s u_n(s,x)dxds$$
$$\le\int_0^t\|\nabla u_n(s,\cdot)\|\|\partial_s u_n(s,\cdot)\|\|\mathbf{b}^n(s,\cdot)\|_{L^\infty}ds$$

$$
\begin{aligned}
&+\int_0^t \|u_n(s,\cdot)\|_{L^q}\|\partial_s u_n(s,\cdot)\|\|c^n(s,\cdot)\|_{L^{\frac{2q}{q-2}}}ds\\
&\leq \frac{C_\varepsilon}{2}\int_0^t \|\nabla u_n(s,\cdot)\|^2 ds + \frac{\varepsilon}{2}\int_0^t \|\partial_s u_n(s,\cdot)\|^2\|\mathbf{b}^n(s,\cdot)\|^2_{L^\infty}ds\\
&\quad+\frac{\varepsilon}{2}\int_0^t \|\partial_s u_n(s,\cdot)\|^2 ds + \frac{C_\varepsilon}{2}\int_0^t \|u_n(s,\cdot)\|^2_{L^q}\|c^n(s,\cdot)\|^2_{L^{\frac{2q}{q-2}}}ds\\
&\leq \frac{C_\varepsilon}{2}(1+C^2(q,N,\partial\Omega)\|c^n\|^2_{L^\infty(0,T,L^{\frac{2q}{q-2}})})\int_0^t \|\nabla u_n(s,\cdot)\|^2 ds\\
&\quad+\frac{\varepsilon}{2}\left(\|\mathbf{b}^n\|^2_{L^\infty((0,T)\times\Omega)}+1\right)\int_0^t \|\partial_s u_n(s,\cdot)\|^2 ds,
\end{aligned}
$$

where we have used $\|u_n(s,\cdot)\|_{L^q}\leq C(q,N,\Omega)\|\nabla u_n(s,\cdot)\|$ for $q\in[2,\frac{2N}{N-2})$ obtained by Evans, 2010. Moreover, $J_2(t)$ can be estimated by the Young's inequality

$$
\begin{aligned}
J_2(t) &= \int_0^t (f_{\frac{1}{n}}(s),\partial_s u_n(s,\cdot))ds + \frac{1}{\Gamma(2-\alpha)}\int_0^t s^{1-\alpha}(u_{n1},\partial_s u_n(s,\cdot))ds\\
&\leq \int_0^t \|f_{\frac{1}{n}}(s,\cdot)\|\|\partial_s u_n(s,\cdot)\|ds + \frac{1}{\Gamma(2-\alpha)}\int_0^t s^{1-\alpha}\|u_{n1}\|\|\partial_s u_n(s,\cdot)\|ds\\
&\leq \int_0^t (\frac{C_\varepsilon}{2}\|f_{\frac{1}{n}}(s,\cdot)\|^2+\frac{\varepsilon}{2}\|\partial_s u_n(s,\cdot)\|^2)ds\\
&\quad+\frac{1}{\Gamma(2-\alpha)}\int_0^t s^{1-\alpha}(\frac{C_\varepsilon}{2}\|u_{n1}\|^2+\frac{\varepsilon}{2}\|\partial_s u_n(s,\cdot)\|^2)ds\\
&\leq \frac{C_\varepsilon}{2}\frac{\|u_{n1}\|^2}{\Gamma(3-\alpha)}t^{2-\alpha}+\frac{C_\varepsilon}{2}\int_0^t \|f_{\frac{1}{n}}(s,\cdot)\|^2 ds\\
&\quad+\frac{\varepsilon}{2}\int_0^t (g_{2-\alpha}(s)+1)\|\partial_s u_n(s,\cdot)\|^2 ds.
\end{aligned}
$$

Let

$$
Q_n = \|\mathbf{b}^n\|^2_{L^\infty((0,T)\times\Omega)}+2 \text{ and } \tilde{Q}_n = 1+C^2(q,N,\Omega)\|c^n\|^2_{L^\infty(0,T,L^{\frac{2q}{q-2}})}.
$$

Then

$$
Q_n \leq C\left(\|\mathbf{b}\|^2_{L^\infty((0,T)\times\Omega)}+2\right) := Q,
$$

and

$$
\tilde{Q}_n \leq C\left(1+C^2(q,N,\Omega)\|c\|^2_{L^\infty(0,T,L^{\frac{2q}{q-2}})}\right) := \tilde{Q},
$$

for each n. We use the above inequalities in (8.87) and the decreasing property of $g_{2-\alpha}$ to obtain that

$$
\begin{aligned}
&\frac{1}{2}{}_0D_t^{\alpha-2}\|\partial_t u_n(t,\cdot)\|^2+\frac{(1-\varepsilon)g_{2-\alpha}(T)}{2}\int_0^t \|\partial_s u_n(s,\cdot)\|^2 ds+\frac{\mu}{2}\|\nabla u_n(t,\cdot)\|^2\\
&\leq \frac{\nu}{2}\|u_{n0}\|^2_{H_0^1}+\frac{C_\varepsilon}{2}\frac{\|u_{n1}\|^2}{\Gamma(3-\alpha)}t^{2-\alpha}+\frac{1}{2}(C_\varepsilon\tilde{Q}+\|A\|_{W^{1,\infty}})\int_0^t \|\nabla u_n(s,\cdot)\|^2 ds
\end{aligned}
$$

$$+\frac{\varepsilon Q}{2}\int_0^t \|\partial_s u_n(s,\cdot)\|^2 ds + \frac{C_\varepsilon}{2}\int_0^t \|f_{\frac{1}{n}}(s,\cdot)\|^2 ds.$$

For fixed $0 < \varepsilon < \frac{g_{2-\alpha}(T)}{g_{2-\alpha}(T)+Q}$, it follows that

$$\begin{aligned}\|\nabla u_n(t,\cdot)\|^2 \le{}& M_1\big(\|u_{n0}\|^2_{H_0^1} + \|u_{n1}\|^2 t^{2-\alpha}\big)\\ &+ M_1^* \int_0^t \|\nabla u_n(s,\cdot)\|^2 ds + \frac{C_\varepsilon}{\mu}\int_0^t \|f_{\frac{1}{n}}(s,\cdot)\|^2 ds,\end{aligned}$$

where $M_1 = \max\left\{\frac{\nu}{\mu}, \frac{C_\varepsilon}{\mu\Gamma(3-\alpha)}\right\}$ and $M_1^* = \frac{1}{\mu}(C_\varepsilon \tilde{Q} + \|A\|_{W^{1,\infty}})$, it yields from using the Gronwall inequality that

$$\|\nabla u_n(t,\cdot)\|^2 \le M_2\Big(\|u_{n0}\|^2_{H_0^1} + \|u_{n1}\|^2 t^{2-\alpha} + \int_0^t \|f_{\frac{1}{n}}(s,\cdot)\|^2 ds\Big), \quad \text{for } t \in [0,T],$$

where M_2 is a positive constant depending on M_1, M_1^* and T. Therefore, we have for $t \in [0,T]$

$$\begin{aligned}&{}_0D_t^{\alpha-2}\|\partial_t u_n(t,\cdot)\|^2 + [(1-\varepsilon)g_{2-\alpha}(T) - \varepsilon Q]\int_0^t \|\partial_s u_n(s,\cdot)\|^2 ds + \mu\|\nabla u_n(t,\cdot)\|^2\\ &\le M_1\big(\|u_{n0}\|^2_{H_0^1} + \|u_{n1}\|^2 t^{2-\alpha}\big) + C_\varepsilon \int_0^t \|f_{\frac{1}{n}}(s,\cdot)\|^2 ds\\ &\quad + M_2 t\big(C_\varepsilon \tilde{Q} + \|A\|_{W^{1,\infty}}\big)\Big(\|u_{n0}\|^2_{H_0^1} + \|u_{n1}\|^2 t^{2-\alpha} + \int_0^t \|f_{\frac{1}{n}}(s,\cdot)\|^2 ds\Big).\end{aligned}$$

The claim then follows. □

Lemma 8.18. *Assume that $u_0 \in H_0^1(\Omega)$, $u_1 \in L^2(\Omega)$ and recall the condition imposed to the parameters $a_{i,j}$, b_j, c and f. Then, for every $n \in \mathbb{N}$ and for every $t \in (0,T]$ the approximate solution u_n given by by (8.75) and (8.79) satisfies the inequality*

$$\begin{aligned}\int_0^t \|\partial_s^\alpha u_n(s,\cdot)\|^2_{H^{-1}} ds \le{}& 2\widetilde{M_1} M_3^2 t\big(\|u_{n0}\|^2_{H_0^1} + \|u_{n1}\|^2 t^{2-\alpha}\big)\\ &+ 2(\widetilde{M_2} M_3^2 t + 1)\int_0^t \|f_{\frac{1}{n}}(s,\cdot)\|^2 ds\end{aligned}$$

for $t \in (0,T]$.

Proof. For fixed $v \in H_0^1(\Omega)$, $\|v\|_{H_0^1} \le 1$, rewrite $v = v_1 + v_2$, where $v_1 \in \operatorname{span}\{e_k\}_{k=1}^n$ and $(v_2, e_k) = 0$ $(k = 1, ..., n)$. Observe $\|v_1\|_{H_0^1} \le 1$. Then (8.75) and (8.77) imply

$$\langle \partial_t^\alpha u_n(t,\cdot), v\rangle = (\partial_t^\alpha u_n(t,\cdot), v) = (\partial_t^\alpha u_n(t,\cdot), v_1) = (f_{\frac{1}{n}}(t,\cdot), v_1) - \mathcal{B}^n[u_n, v_1; t].$$

On the other hand, from the definition of $\mathcal{B}^n$ and Sobolev imbedding, we know that

$$\begin{aligned}
&\left|\mathcal{B}^n[u_n, v_1; t]\right| \\
&\leq \|A\|_{W^{1,\infty}}\|\nabla u_n(t,\cdot)\|\|\nabla v_1(\cdot)\| + \|\nabla u_n(t,\cdot)\|\|v_1(\cdot)\|\|\mathbf{b}^n(t,\cdot)\|_{L^\infty} \\
&\quad + \|u_n(t,\cdot)\|\|v_1(\cdot)\|_q\|c^n(t,\cdot)\|_{L^{\frac{2q}{q-2}}} \\
&\leq \|A\|_{W^{1,\infty}}\|\nabla u_n(t,\cdot)\|\|v_1(\cdot)\|_{H_0^1} + \|\nabla u_n(t,\cdot)\|\|v_1(\cdot)\|_{H_0^1}\|\mathbf{b}^n\|_{L^\infty((0,T)\times\Omega)} \\
&\quad + \|u_n(t,\cdot)\|\|v_1(\cdot)\|_{H_0^1}\|c^n\|_{L^\infty(0,T,L^{\frac{q}{q-2}})} \\
&\leq M_3\|\nabla u_n(t,\cdot)\|\|v_1(\cdot)\|_{H_0^1},
\end{aligned} \tag{8.88}$$

where $M_3 = C\big(\|A\|_{W^{1,\infty}} + \|\mathbf{b}\|_{L^\infty((0,T)\times\Omega)} + \|c\|_{L^\infty(0,T,L^{\frac{q}{q-2}})}\big)$. Moreover, we have $|(f_{\frac{1}{n}}(t,x), v_1)| \leq \|f_{\frac{1}{n}}(t,\cdot)\|\|v_1\| \leq \|f_{\frac{1}{n}}(t,\cdot)\|\|v_1\|_{H_0^1}$. Thus

$$\left|\left\langle \partial_t^\alpha u_n(t,\cdot), v\right\rangle\right| \leq \|f_{\frac{1}{n}}(t,\cdot)\| + M_3\|\nabla u_n(t,\cdot)\|$$

for $\|v_1\|_{H_0^1} \leq 1$. Consequently, $\|\partial_t^\alpha u_n(t,\cdot)\|_{H^{-1}} \leq \|f_{\frac{1}{n}}(t)\| + M_3\|\nabla u_n(t,\cdot)\|$, from Lemma 8.17 we can show

$$\begin{aligned}
\int_0^t \|\partial_s^\alpha u_n(s,\cdot)\|_{H^{-1}}^2 ds &\leq 2\int_0^t \|f_{\frac{1}{n}}(s,\cdot)\|^2 ds + 2M_3^2 t\big[\widetilde{M}_1\big(\|u_{n0}\|_{H_0^1}^2 + \|u_{n1}\|^2 t^{2-\alpha}\big) \\
&\quad + \widetilde{M}_2\int_0^t \|f_{\frac{1}{n}}(s)\|^2 ds\big].
\end{aligned}$$

The claim then follows. □

8.6.5 *Well-Posedness and Regularity*

In this subsection, we take the limit in approximate sequences and present the existence and uniqueness of weak solutions, and then we show the regularity results.

Theorem 8.17. *Suppose that $T > 0, u_0 \in H_0^1(\Omega), u_1 \in L^2(\Omega)$ and let $a_{i,j}$, b_j, c and f satisfy (3.1). Then there exists a weak solution $u \in C([0,T], L^2(\Omega)) \cap L^\infty(0,T,H_0^1(\Omega))$ of the equation (8.72) satisfying $u' \in L^2(0,T,L^2(\Omega))$, $\partial_t^\alpha u \in L^2(0,T,H^{-1}(\Omega))$. Moreover, u also satisfies the following estimate*

$$\begin{aligned}
&\max_{t\in[0,T]}\|u(t)\|_{H_0^1} + \|\partial_t u\|_{L^2(0,T,L^2)} + \|\partial_t^\alpha u\|_{L^2(0,T,H^{-1})} \\
&\leq \widetilde{M}\big(\|u_0\|_{H_0^1}^2 + \|u_1\|^2 + \|f\|_{L^2(0,T,L^2)}\big),
\end{aligned} \tag{8.89}$$

where $\widetilde{M}$ is a positive constant.

Proof. Step I. According to the energy estimate in Lemma 8.17, we see that the sequence $\{u_n(t)\}$is bounded in $H_0^1(\Omega)$ for $t \in [0,T]$, $\{\partial_t u_n\}$ is bounded in $L^2(0,T,L^2(\Omega))$, and Lemma 8.18 implies that the sequence $\partial_t^\alpha u_n$ is bounded in $L^2(0,T,H^{-1}(\Omega))$. This also implies that ${}_0D_t^{\alpha-2}(u - u_0 - u_1 t)$ is uniformly bounded in ${}_0H^2(0,T,H^{-1}(\Omega))$. As a consequence there exist $u \in C([0,T], L^2(\Omega)) \cap$

$L^\infty(0,T,H_0^1(\Omega))$ with $u'\in L^2(0,T,L^2(\Omega))$, $v\in L^2(0,T,H^{-1}(\Omega))$ and a subsequence of $\{u_n\}$, still denoted by $\{u_n\}$, such that

$$\begin{aligned} &u_n\to u \ \text{ in } C([0,T],L^2(\Omega)), \qquad u_n(t)\rightharpoonup u(t) \ \text{ in } L^2(0,T,H_0^1(\Omega)),\\ &\partial_t u_n\rightharpoonup \partial_t u \ \text{ in } L^2(0,T,L^2(\Omega)), \qquad \partial_t^\alpha u_n\rightharpoonup v \ \text{ in } L^2(0,T,H^{-1}(\Omega)). \end{aligned} \tag{8.90}$$

Since the continuity of ${}_0D_t^{\alpha-2}$ in $L^2(0,T)$ implies the weak continuity, it follows that

$${}_0D_t^{\alpha-2}\partial_t u_n\rightharpoonup {}_0D_t^{\alpha-2}\partial_t u \ \text{ in } L^2(0,T,L^2(\Omega)). \tag{8.91}$$

Next we would like to prove that $\partial_t^\alpha u=v$ in a weak sense. We take $\varphi\in C_0^\infty(0,T)$ and $\psi\in H_0^1(\Omega)$. Then

$$\begin{aligned} &\int_0^T \varphi(t)\langle v,\psi\rangle_{H^{-1}\times H_0^1}dt\\ &=\lim_{n\to\infty}\int_0^T \varphi(t)\langle \partial_t^\alpha u_n,\psi\rangle_{H^{-1}\times H_0^1}dt\\ &=\lim_{n\to\infty}\int_0^T \varphi(t)\Big\langle \partial_{tt}[{}_0D_t^{\alpha-2}(u_n-u_{n0}-u_{n1}t)],\psi\Big\rangle_{H^{-1}\times H_0^1}dt\\ &=\lim_{n\to\infty}\int_\Omega \psi(x)dx\int_0^T \varphi(t)\partial_{tt}[{}_0D_t^{\alpha-2}(u_n-u_{n0}-u_{n1}t)]dt\\ &=-\lim_{n\to\infty}\int_\Omega \psi(x)dx\int_0^T \varphi'(t)\partial_{t}[{}_0D_t^{\alpha-2}(u_n-u_{n0}-u_{n1}t)]dt\\ &=-\lim_{n\to\infty}\int_\Omega \psi(x)dx\int_0^T \varphi'(t){}_0D_t^{\alpha-2}(\partial_t u_n-u_{n1})dt\\ &=-\int_0^T \varphi'(t)\langle {}_0D_t^{\alpha-2}(\partial_t u-u_1),\psi\rangle dt\\ &=\int_0^T \varphi(t)\langle \partial_t^\alpha u,\psi\rangle_{H^{-1}\times H_0^1}dt, \end{aligned}$$

where we have used Lemma 8.14. So $\partial_t^\alpha u=v$ in a weak sense.

Step II. Fix an integer Λ and choose a function $w\in H_0^1(\Omega)$ of the form

$$\omega(x)=\sum_{k=1}^{\Lambda}\gamma_k e_k(x), \tag{8.92}$$

where $\{\gamma_k\}$ are arbitrary numbers. We select $n\ge\Lambda$, multiply (8.77) by γ_k and sum it up from 1 to Λ. Then we proceed to multiply the equation by $\varrho_\varepsilon(t+\tau)$ for fixed $\tau\in(0,T)$ and integrate with respect to t to discover

$$\begin{aligned} &\int_0^T \varrho_\varepsilon(t+\tau)\int_\Omega \partial_t^\alpha u_n(t,x)\omega(x)dxdt+\int_0^T \varrho_\varepsilon(t+\tau)\mathcal{B}^n[u_n,\omega;t]dt\\ &=\int_0^T \varrho_\varepsilon(t+\tau)\int_\Omega f_{\frac{1}{n}}(t,x)\omega(x)dxdt. \end{aligned} \tag{8.93}$$

For $\varepsilon < T - \tau$, we recall (8.91) to find that for a.e. $\tau \in (0, T)$,

$$\begin{aligned}
&\lim_{\varepsilon\to 0}\lim_{n\to\infty}\int_0^T \varrho_\varepsilon(t+\tau)\int_\Omega \partial_t^\alpha u_n(t,x)\omega(x)dxdt \\
&= -\lim_{\varepsilon\to 0}\lim_{n\to\infty}\int_0^T \varrho_\varepsilon{}'(t+\tau)\int_\Omega \partial_t[{}_0D_t^{\alpha-2}(u_n - u_{n0} - u_{n1}t)]\omega(x)dxdt \\
&= -\lim_{\varepsilon\to 0}\lim_{n\to\infty}\int_0^T \varrho_\varepsilon{}'(t+\tau)\int_\Omega {}_0D_t^{\alpha-2}(\partial_t u_n - u_{n1})\omega(x)dxdt \\
&= -\lim_{\varepsilon\to 0}\int_0^T \varrho_\varepsilon{}'(t+\tau)\int_\Omega {}_0D_t^{\alpha-2}(\partial_t u - u_1)\omega(x)dxdt \\
&= \lim_{\varepsilon\to 0}\int_0^T \varrho_\varepsilon(t+\tau)\int_\Omega \partial_t^\alpha u(t,x)\omega(x)dxdt \\
&= \int_\Omega \partial_t^\alpha u(\tau,x)\omega(x)dx.
\end{aligned}$$

We proceed similarly with remaining terms. We see that $\varrho_\varepsilon(t+\tau)\partial_i\omega(x)$ is smooth in $(0,T)\times\Omega$, From assumptions $a_{i,j}^n \in L^\infty((0,T)\times\Omega)$ thus $a_{i,j}^n \to a_{i,j}$ in $L^2((0,T)\times\Omega)$, and $\partial_j u_n(x,t) \rightharpoonup \partial_j u(x,t)$ in $L^2((0,T)\times\Omega)$ when $n\to\infty$, we get

$$\begin{aligned}
&\lim_{\varepsilon\to 0}\lim_{n\to\infty}\int_0^T \varrho_\varepsilon(t+\tau)\int_\Omega a_{i,j}^n(t,x)\partial_j u_n(t,x)\cdot\partial_i\omega(x)dxdt \\
&= \lim_{\varepsilon\to 0}\int_0^T \varrho_\varepsilon(t+\tau)\int_\Omega a_{i,j}(t,x)\partial_j u(t,x)\cdot\partial_i\omega(x)dxdt \\
&= \int_\Omega a_{i,j}(\tau,x)\partial_j u(\tau,x)\cdot\partial_i\omega(x)dx.
\end{aligned}$$

Similarly, since $b_j^n(t) \to b_j$ in $L^\infty((0,T)\times\Omega)$ and $c^n(t)\to c$ in $L^\infty(0,T,L^{\frac{2q}{q-2}}(\Omega))$, thus $b_j^n(t)\to b_j$ and $c^n(t)\to c$ in $L^2((0,T)\times\Omega)$, it together with (8.90) follows that

$$\lim_{\varepsilon\to 0}\lim_{n\to\infty}\int_0^T \varrho_\varepsilon(t+\tau)\int_\Omega b_j^n(t,x)\partial_j u_n(t,x)\cdot\omega(x)dxdt = \int_\Omega b_j(\tau,x)\partial_j u(\tau,x)\omega(x)dx,$$
$$\lim_{\varepsilon\to 0}\lim_{n\to\infty}\int_0^T \varrho_\varepsilon(t+\tau)\int_\Omega c^n(t,x)u_n(t,x)\cdot\omega(x)dxdt = \int_\Omega c(\tau,x)u(\tau,x)\omega(x)dx.$$

Therefore, one can find that

$$\lim_{\varepsilon\to 0}\lim_{n\to\infty}\int_0^T \varrho_\varepsilon(t+\tau)\mathcal{B}^n[u_n,\omega;t]dt = \mathcal{B}[u,\omega;\tau].$$

Moreover, we can derive that for a.e. $\tau\in(0,T)$,

$$\begin{aligned}
\lim_{\varepsilon\to 0}\lim_{n\to\infty}\int_0^T \varrho_\varepsilon(t+\tau)\int_\Omega f_{\frac{1}{n}}(t,x)\omega(x)dxdt &= \lim_{\varepsilon\to 0}\int_0^T \varrho_\varepsilon(t+\tau)\int_\Omega f(t,x)\omega(x)dxdt \\
&= (f(\tau,\cdot),\omega).
\end{aligned}$$

Therefore the following equality holds

$$\langle \partial_t^\alpha u(t,\cdot),\omega\rangle + \mathcal{B}[u,\omega;t] = (f(t,\cdot),\omega) \tag{8.94}$$

for $\omega = \sum_{k=1}^{\Lambda} \gamma_k e_k(x)$ and a.e. $t \in (0, T)$, since functions of the form (8.92) are dense in $H_0^1(\Omega)$, then the above equality also holds for all $\omega \in H_0^1(\Omega)$ and a.e. $t \in (0, T)$.

Finally, we note that

$$\int_0^t \|f_{\frac{1}{n}}(s, \cdot)\|^2 ds \leq \int_0^t \|f(s, \cdot)\|^2 ds + \int_t^{t+\frac{1}{n}} \|f(s, \cdot)\|^2 ds,$$

by the assumption of f, we know that $\sup_{t \in (0, T-\frac{1}{n})} \int_t^{t+\frac{1}{n}} \|f(s, \cdot)\|^2 ds \to 0$ uniformly with respect to t as $n \to \infty$. Therefore, Lemmas 8.17 and 8.18 give the estimate (8.89). □

Remark 8.6. If we give more smooth assumptions that the coefficients $a_{i,j}, b_j, c \in C^2((0, T], L^\infty(\Omega))$ and $f \in C^2((0, T], L^2(\Omega))$, then the mollification arguments imposed to the coefficients a, b, c and f can be avoided. Similar to the proof as we derived in Theorem 8.17, the existence result is obtained.

Theorem 8.18. *Under the assumptions of Theorem 8.17, we suppose that* $b_j \in W^{1,\infty}(0, T, W^{1,\infty}(\Omega))$, $c \in W^{1,\infty}(0, T, L^{\frac{2q}{q-2}}(\Omega))$. *Then a weak solution* u *of the equation* (8.72) *is unique.*

Proof. It suffices to show that the only weak solution of (8.72) with $f \equiv u_0 \equiv u_1 \equiv 0$ is $u \equiv 0$. To verify this, fix $0 \leq t \leq T$ and set $\omega(\tau) = \int_\tau^t u(s, \cdot) ds$ if $0 \leq \tau \leq t$ and $\omega(t) = 0$ if $0 \leq t \leq \tau \leq T$. Then $\omega(\tau) \in H_0^1(\Omega)$ for each $\tau \in [0, T]$ and we have

$$\int_0^t \langle \partial_\tau^\alpha u(\tau, \cdot), \omega(\tau) \rangle + \mathcal{B}[u, \omega; \tau] d\tau = 0.$$

Since $\partial_t u(0, \cdot) = \omega(t) = 0$, then $\partial_t^\alpha u = \partial_t \partial_t^{\alpha-1} u$, and so we obtain after integrating by parts in the first term above

$$\int_0^t -(\partial_\tau^{\alpha-1} u(\tau, \cdot), \omega'(\tau)) + \mathcal{B}[u, \omega; \tau] d\tau = 0.$$

Now $\omega' = -u$ for $0 \leq \tau \leq t \leq T$, and then

$$\int_0^t (\partial_\tau^{\alpha-1} u(\tau, \cdot), u(\tau, \cdot)) - \mathcal{B}[\omega', \omega; \tau] d\tau = 0.$$

From Lemma 8.15 and the decreasing property of $g_{2-\alpha}$ we know

$$\begin{aligned}\int_0^t (\partial_\tau^{\alpha-1} u(\tau, \cdot), u(\tau, \cdot)) d\tau &\geq \frac{1}{2}[g_{2-\alpha} * \|u(\tau, \cdot)\|^2](t) + \frac{g_{2-\alpha}(t)}{2} \int_0^t \|u(\tau, \cdot)\|^2 d\tau \\ &\geq \frac{g_{2-\alpha}(t)}{2} \int_0^t \|u(\tau, \cdot)\|^2 d\tau.\end{aligned}$$

Thus

$$\frac{g_{2-\alpha}(t)}{2} \int_0^t \|u(\tau, \cdot)\|^2 d\tau - \frac{1}{2} \int_0^t \partial_\tau \mathcal{B}[\omega, \omega; \tau] d\tau \leq \frac{1}{2} \int_0^t -\mathcal{C}[\omega, \omega; \tau] + \mathcal{D}[u, \omega; \tau] d\tau,$$

due to $2\mathcal{B}[\omega', \omega; \tau] = \partial_\tau \mathcal{B}[\omega, \omega; \tau] - \mathcal{C}[\omega, \omega; \tau] - \mathcal{D}[u, \omega; \tau]$, where

$$\mathcal{C}[u, v; \tau] = \int_\Omega \sum_{i,j=1}^N \partial_\tau a_{i,j}(\tau, x)\partial_j u \cdot \partial_i v - \sum_{j=1}^N \partial_\tau b_j(\tau, x)\partial_j uv - \partial_\tau c(\tau, x)uvdx,$$

$$\mathcal{D}[u, v; \tau] = \int_\Omega \sum_{j=1}^N \partial_j b_j(\tau, x)uv + 2\sum_{j=1}^N b_j(\tau, x)\partial_j vudx$$

for $u, v \in H_0^1(\Omega)$. Since

$$\begin{aligned}|\mathcal{C}[\omega, \omega; \tau]| &\le \|\partial_\tau a_{i,j}(\tau)\|_{L^\infty}\|\nabla\omega\|^2 + \|\nabla\omega\|\|\omega\|\|\partial_\tau b_j(\tau)\|_{L^\infty} \\ &\quad + \|\omega\|\|\omega\|_{L^q}\|\partial_\tau c(\tau)\|_{L^{\frac{2q}{q-2}}} \\ &\le \|\partial_\tau a_{i,j}(\tau)\|_{L^\infty(\Omega)}\|\nabla\omega\|^2 + \|\nabla\omega\|^2 + \|\omega\|^2\|\partial_\tau b_j(\tau)\|_{L^\infty}^2 \\ &\quad + \|\omega\|^2 + C^2(q, N, \Omega)\|\nabla\omega\|^2\|\partial_\tau c(\tau)\|^2_{L^{\frac{2q}{q-2}}} \\ &\le C\|\omega\|^2_{H_0^1},\end{aligned}$$

and

$$\begin{aligned}|\mathcal{D}[u, \omega; \tau]| &\le \|\partial_j b_j(\tau)\|_{L^\infty}\|u(\tau, \cdot)\|\|\omega\| + 2\|b_j(\tau)\|_{L^\infty}\|D\omega\|\|u(\tau, \cdot)\| \\ &\le \|u(\tau, \cdot)\|^2 + \|\partial_j b_j(\tau)\|^2_{L^\infty}\|\omega\|^2 + 2\|D\omega\|^2 + 2\|b_j(\tau)\|^2_{L^\infty}\|u(\tau, \cdot)\|^2 \\ &\le C(\|\omega\|^2_{H_0^1} + \|u(\tau, \cdot)\|^2).\end{aligned}$$

Hence

$$\frac{g_{2-\alpha}(t)}{2}\int_0^t \|u(\tau, \cdot)\|^2 d\tau + \frac{1}{2}\mathcal{B}[\omega(0), \omega(0); t] \le C\int_0^t \|\omega(\tau)\|^2_{H_0^1} + \|u(\tau, \cdot)\|^2 d\tau,$$

which together with

$$\begin{aligned}\mathcal{B}[\omega(0), \omega(0); t] &\ge \mu\|\nabla\omega(0)\|^2 - \|\nabla\omega(0)\|\|\omega(0)\|\|b_j(t)\|_{L^\infty} \\ &\quad - \|\omega(0)\|\|\omega(0)\|_{L^q}\|c(t)\|_{L^{\frac{2q}{q-2}}} \\ &\ge \mu\|\nabla\omega(0)\|^2 - (\frac{\mu}{4}\|\nabla\omega(0)\|^2 + \frac{1}{\mu}\|\omega(0)\|^2\|b_j(t)\|^2_{L^\infty}) \\ &\quad - (\frac{1}{4\varepsilon}\|\omega(0)\|^2 + \varepsilon C^2(q, N, \Omega)\|c\|^2_{L^\infty(0,T,L^{\frac{2q}{q-2}})}\|\nabla\omega(0)\|^2) \\ &\ge \frac{\mu}{2}\|\omega(0)\|^2_{H_0^1} - C\|\omega(0)\|^2\end{aligned}$$

for

$$\varepsilon = \frac{\mu}{4}\frac{1}{C^2(q, N, \partial\Omega)\|c\|^2_{L^\infty(0,T,L^{\frac{2q}{q-2}})}},$$

shows that

$$\begin{aligned}&g_{2-\alpha}(t)\int_0^t \|u(\tau, \cdot)\|^2 d\tau + \|\omega(0)\|^2_{H_0^1} \\ &\le C\Big(\int_0^t (\|\omega(\tau)\|^2_{H_0^1} + \|u(\tau, \cdot)\|^2)d\tau + \|\omega(0)\|^2\Big).\end{aligned} \tag{8.95}$$

Let us write

$$W(t) := \int_0^t u(\tau,\cdot)d\tau, \quad t \in [0,T],$$

whereupon (8.95) becomes

$$g_{2-\alpha}(t)\int_0^t \|u(\tau,\cdot)\|^2 d\tau + \|W(t)\|^2_{H_0^1}$$
$$\leq C\Big(\int_0^t (\|W(t)-W(\tau)\|^2_{H_0^1} + \|u(\tau,\cdot)\|^2)d\tau + \|W(t)\|^2\Big).$$

Since $\|W(t)-W(\tau)\|^2_{H_0^1} \leq 2\|W(\tau)\|^2_{H_0^1} + 2\|W(t)\|^2_{H_0^1}$, and $\|W(t)\| \leq \int_0^t \|u(\tau,\cdot)\|d\tau$, we can derive

$$g_{2-\alpha}(t)\int_0^t \|u(\tau,\cdot)\|^2 d\tau + (1-2tC_1)\|W(t)\|^2_{H_0^1} \leq C_1\int_0^t (\|W(\tau)\|^2_{H_0^1} + \|u(\tau,\cdot)\|^2)d\tau.$$

Choose T_1 so mall that

$$\frac{g_{2-\alpha}(T_1)}{2} \geq C_1 \text{ and } 1-2T_1C_1 \geq \frac{1}{2}.$$

Then if $0 < t \leq T_1$, we have

$$[g_{2-\alpha}(t) - C_1]\int_0^t \|u(\tau,\cdot)\|^2 d\tau + \frac{1}{2}\|W(t)\|^2_{H_0^1} \leq C_1\int_0^t \|W(\tau)\|^2_{H_0^1} d\tau.$$

Consequently the Gronwall inequality implies $W(t) \equiv 0$ on $[0,T_1]$. Then $\int_0^t \|u(\tau)\|^2 d\tau \equiv 0$ on $[0,T_1]$. This together with the continuity of u shows $u(t) \equiv 0$ on $[0,T_1]$.

We use the same argument on $[T_1,T_2]$, $[T_2,T_3]$,..., and then we can deduce $u \equiv 0$. □

8.7 Notes and Remarks

The results in Section 8.2 are adopted from Zhou and Peng, 2017b. Section 8.3 due to Luc, Lan, O'Regan, Tuan and Zhou, 2021. The results in Section 8.4 are taken from Bourdin, 2013. Section 8.5 due to Beckers and Yamamoto, 2013. The material in Section 8.6 are adopted from Peng and Zhou, 2022a.

Bibliography

N. Abada, M. Benchohra and H. Hammouche, Existence and controllability results for non-densely defined impulsive semilinear functional differential inclusions, J. Differential Equations, 246(2009), 3834-3863.

S. Abbas, M. Benchohra and Y. Zhou, Darboux problem for fractional order neutral functional partial hyperbolic differential equations, Int. J. Dyn. Syst. Diff. Equs., 2(2009), 301-312.

S. Abbas, M. Benchohra and Y. Zhou, Fractional order partial functional differential inclusions with infnite delay, Proc. A. Razmadze Math. Inst., 154(2010), 1-19.

S. Abbas, M. Benchohra and Y. Zhou, Fractional order partial hyperbolic functional differential equations with state-dependent delay, Int. J. Dyn. Syst. Diff. Equs., 3(2011), 459-490.

M. D'Abbicco, M.R. Ebert and T. Picon, Global existence of small data solutions to the semilinear fractional wave equation, New Trends in Analysis and Interdisciplinary Applications, Birkhäuser, Cham, (2017), 465-471.

R. A. Adams, Sobolev Spaces, Academic Press, New York, (1999).

E. E. Adams and L. W. Gelhar, Field study of dispersion in a heterogeneous aquifer 2: Spatial moments analysis, Water Resources Res., 28(1992), 3293-3307.

R. P. Agarwal and B. Ahmad, Existence of solutions for impulsive anti-periodic boundary value problems of fractional semilinear evolution equations, Dyn. Contin. Discrete Impuls. Syst. Ser. A: Math. Anal., 18(2011), 457-470.

R. P. Agarwal, M. Benchohra and S. Hamani, A survey on existence results for boundary value problems of nonlinear fractional differential equations, Acta Appl. Math., 109(2010), 973-1033.

R. P. Agarwal, Y. Zhou and Y. He, Existence of fractional neutral functional differential equations with bounded delay, Comput. Math. Appl., 59(2010), 1095-1100.

R. P. Agarwal, Y. Zhou, J. R. Wang *et al.*, Fractional functional differential equations with causal operators in Banach spaces, Math. Comput. Modelling, 54(2011), 1440-1452.

O. P. Agrawal, Formulation of Euler-Lagrange equations for fractional variational problems, J. Math. Anal. Appl., 272(2002), 368-379.

O. P. Agrawal, Generalized variational problems and Euler-Lagrange equations, Comput. Math. Appl., 59(2010), 1852-1864.

R. P. Agarwal, M. Belmekki and M. Benchohra, A survey on semilinear differential equations and inclusions involving Riemann-Liouville fractional derivative, Adv. Difference Equ., 2009(2009), Article ID 981728, 47 pages.

O. P. Agrawal, J. T. Machado and J. Sabatier, Fractional Derivatives and their Application: Nonlinear Dynamics, Springer-Verlag, Berlin, (2004).

B. Ahmad and P. Eloe, A nonlocal boundary value problem for a nonlinear fractional differential equation with two indices, Commun. Appl. Nonlinear Anal., 17(2010), 69-80.

B. Ahmad and J. J. Nieto, Existence results for a coupled system of nonlinear fractional differential equations with three-point boundary conditions, Comput. Math. Appl., 58(2009), 1838-1843.

B. Ahmad and J. J. Nieto, Existence of solutions for anti-periodic boundary value problems involving fractional differential equations via Leray-Schauder degree theory, Topol. Methods Nonlinear Anal., 35(2010), 295-304.

B. Ahmad, J. J. Nieto, A. Alsaedi *et al.*, A study of nonlinear langevin equation involving two fractional orders in different intervals, Nonlinear Anal.: RWA, 13(2012), 599-606.

B. Ahmad and S. Sivasundaram, Existence results for nonlinear impulsive hybrid boundary value problems involving fractional differential equations, Nonlinear Anal.: HS, 3(2009), 251-258.

B. Ahmad and S. Sivasundaram, Existence of solutions for impulsive integral boundary value problems of fractional order, Nonlinear Anal.: HS, 4(2010), 134-141.

B. Ahmad and G. Wang, Impulsive anti-periodic boundary value problem for nonlinear differential equations of fractional order, Comput. Math. Appl., 59(2010), 1341-1349.

N. U. Ahmed, Existence of optimal controls for a general class of impulsive systems on Banach space, SIAM J. Control Optimal, 42(2003), 669-685.

N. U. Ahmed, Optimal feedback control for impulsive systems on the space of finitely additive measures, Publ. Math. Debrecen, 70(2007), 371-393.

M. U. Akhmet, On the smoothness of solutions of impulsive autonomous systems, Nonlinear Anal.: TMA, 60(2005), 311-324.

R. Almeida, A. B. Malinowska and D. F. M. Torres, A fractional calculus of variations for multiple integrals with application to vibrating string, J. Math. Phys., 51(2010), 033503.

A. Alsaedi, B. Ahmad and M. Kirane, Maximum principle for certain generalized time and space fractional diffusion equations, Quart. Appl. Math., 73(2015), 163-175.

E. Alvarez, C. G. Gal, V. Keyantuo and M. Warma, Well-posedness results for a class of semi-linear super-diffusive equations, Nonlinear Anal., 181(2019), 24-61.

A. Ambrosetti and P. Rabinowitz, Dual variational methods in critical points theory and applications, J. Funct. Anal., 14(1973), 349-381.

A. Ambrosetti and V. C. Zelati, Multiple homoclinic orbits for a class of conservative systems, Rend. Sem. Mat. Univ. Padova, 89(1993), 177-194.

M. R. S. Ammi and D. F. M. Torres, Regularity of solutions to higher-order integrals of the calculus of variations, Internat. J. Systems Sci., 39(2008), 889-895.

C. Amrouche, A. Rejaiba, L^p-theory for Stokes and Navier-Stokes equations with Navier boundary condition, J. Differential Equations, 256(2014), 1515-1547.

V. V. Anh and N. N. Leonenko, Spectral analysis of fractional kinetic equations with randomdata, J. Statist. Phys., 104(2001), 1349-1387.

J. M. Arrieta, Rates of eigenvalues on a Dumbbell domain. Simple eigenvalue case, Tran. Amer. Math. Soc., 347(1995), 3503-3531.

J. M. Arrieta, A. Carvalho and G. Lozada-Cruz, Dynamics in dumbbell domains I. Continuity of the set of equilibria, J. Diff. Equs., 231(2006), 551-597.

J. M. Arrieta, A. Carvalho and G. Lozada-Cruz, Dynamics in dumbbell domains II. The limiting problem, J. Differential Equations, 247(2009a), 174-202.

J. M. Arrieta, A. Carvalho and G. Lozada-Cruz, Dynamics in dumbbell domains III. Continuity of attractors, J. Differential Equations, 247(2009b), 225-259.

A. Ashyralyev and F. Tetikoglu, A note on fractional spaces generated by the positive

operator with periodic conditions and applications, Bound. Value Probl., 1(2015), 1-17.
T. Atanackovic and B. Stankovic, On a class of differential equations with left and right fractional derivatives, ZAMM., 87(2007), 537-539.
C. Atkinson and A. Osseiran, Rational solutions for the time-fractional diffusion equation, SIAM J. Appl. Math., 71(2011), 92-106.
V. V. Au, Y. Zhou and D. O'Regan, On the well-posedness and blow-up for a semilinear biparabolic equation, Mediterr. J. Math., 19(2022), 35.
D. Averna and G. Bonanno, A mountain pass theorem for a suitable class of functions, Rocky Mountain J. Math., 39(2009), 707-727.
O. Baghani, On fractional Langevin equation involving two fractional orders, Commun. Nonlinear Sci. Numer. Simul., 42(2017), 675-681.
E. G. Bajlekova, Fractional evolution equations in Banach spaces, Thesis (Dr.) Technische Universiteit Eindhoven (The Netherlands), ProQuest LLC, Ann Arbor, MI, 2001, 113 pp.
Z. Bai, X. Dong and C. Yin, Existence results for impulsive nonlinear fractional differential equation with mixed boundary conditions, Bound. Value Probl., 2016(2016), 1-11.
Z. Bai and H. Lu, Positive solutions for boundary value problem of nonlinear fractional differential equation, J. Math. Anal. Appl., 311(2005), 495-505.
D. D. Bainov and P. S. Simeonov, Impulsive differential equations: periodic solutions and applications, New York, Longman Scientific and Technical Group Limited, (1993).
K. Balachandran and S. Kiruthika, Existence of solutions of abstract fractional impulsive semilinear evolution equations, E. J. Qualitative Theory of Diff. Equ., 4(2010), 1-12.
K. Balachandran, S. Kiruthika and J. J. Trujillo, Existence results for fractional impulsive integrodifferential equations in Banach spaces, Commun. Nonlinear Sci. Numer. Simul., 16(2011), 1970-1977.
K. Balachandran, Y. Zhou and J. Kokila, Relative controllability of fractional dynamical systems with delays in control, Commun. Nonlinear Sci. Numer. Simul., 17(2012a), 3508-3520.
K. Balachandran, Y. Zhou and J. Kokila, Relative controllability of fractional dynamical systems with distributed delays in control, Comput. Math. Appl., 64(2012b), 3201-3209.
D. Baleanu, J. A. T. Machado and A. C.-J. Luo, Fractional dynamics and control, Springer Science and Business Media, (2011).
D. Baleanu and S. I. Muslih, Lagrangian formulation of classical fields within Riemann-Liouville fractional derivatives, Phys. Scripta, 72(2-3)(2005), 119-121.
D. Baleanu and J. Trujillo, On exact solutions of a class of fractional Euler-Lagrange equations, Nonlinear Dynam., 52(2008), 331-335.
D. Baleanu, J. A. T. Machado and Z. B. Guvenc (Editors), New Trends in Nanotechnology and Fractional Calculus Applications, Springer, (2010).
Z. Bai, On positive solutions of a nonlocal fractional boundary value problem, Nonlinear Anal.: TMA, 72(2010), 916-924.
J. Banaš and K. Goebel, Measure of Noncompactness in Banach Spaces, Marcel Dekker, Inc., New York, (1980).
N. T. Bao, T. Caraballo, N. H. Tuan and Y. Zhou, Existence and regularity results for terminal value problem for nonlinear fractional wave equations, Nonlinearity, 34(2021), 1448-1502.
T. Bartsch, Infinitely many solutions of a symmetic Dirchlet problem, Nonlinear Anal.: TMA, 20(1993), 1205-1216.
T. Bartsch and M. Willem, On an elliptic equation with concave and convex nonlinearities, Proc. Amer. Math. Soc., 123(1995), 3555-3561.

F. Battelli and M. Fečkan, Chaos in singular impulsive O.D.E., Nonlinear Anal.: TMA, 28(1997), 655-671.

E. Bazhlekova and I. Bazhlekov, Subordination approach to multi-term time-fractional diffusion-wave equations, J. Comput. Appl. Math., 339(2018), 179-192.

E. Bazhlekova, B. Jin, R. Lazarov, and Z. Zhou, An analysis of the Rayleigh-Stokes problem for a generalized second-grade fluid, Numer. Math., 131(2015), 1-31.

A. Bazzani, G. Bassi and G. Turchetti, Diffusion and memory effects for stochastic processes and fractional langevin equations, Physica A: Stat. Mech. Appl., 324(2003), 530-550.

S. Beckers and M. Yamamoto, Regularity and unique existence of solution to linear diffusion equation with multiple time-fractional derivatives, Control and Optimization with PDE Constraints, K. Bredies, C. Clason, K. Kunisch and G. von Winckel (Editors), Basel: Birkhäuser, (2013).

M. Belmekki and M. Benchohra, Existence results for fractional order semilinear functional differential equations with nondense domain, Nonlinear Anal.: TMA, 72(2010), 925-932.

M. Benchohra, S. Hamani and S. K. Ntouyas, Boundary value problems for differential equations with fractional order and nonlocal conditions, Nonlinear Anal.: TMA, 71(2009), 2391-2396.

M. Benchohra, J. Henderson and S. K. Ntouyas, Impulsive differential equations and inclusions, Hindawi Publishing Corporation, vol. 2, New York, (2006).

M. Benchohra, J. Henderson, S. K. Ntouyas *et al.*, Existence results for fractional order functional differential equations with infinite delay, J. Math. Anal. Appl., 338(2008), 1340-1350.

M. Benchohra and D. Seba, Impulsive fractional differential equations in Banach spaces, E. J. Qualitative Theory of Diff. Equ., Spec. Ed. I, 8(2009), 1-14.

D. A. Benson, R. Schumer, M. M. Meerschaert *et al.*, Fractional dispersion, Lévy motion, and the MADE tracer test, Transp. Porous Media, 42(2001), 211-240.

D. A. Benson, S. W. Wheatcraft and M. M. Meerschaert, Application of a fractional advection-dispersion equation, Water Resour. Res., 36(2000a), 1403-1412.

D. A. Benson, S. W. Wheatcraft and M. M. Meerschaert, The fractional-order governing equation of Lévy motion, Water Resour. Res., 36(2000b), 1413-1423.

B. Berkowitz, H. Scher and S. E. Silliman, Anomalous transport in laboratory-scale heterogeneous porous media, Water Resource Res., 36(2000), 149-158.

G. Bonanno and P. Candito, Non-differentiable functionals and applications to elliptic problems with discontinuous nonlinearities, J. Differential Equations, 244(2008), 3031-3059.

B. Bonilla, M. Rivero and J. J. Trujillo, On systems of linear fractional differential equations with constant coefficients, Appl. Math. Comput., 187(2007), 68-78.

D. Bothe, Multivalued perturbantion of m-accretive differential inclusions, Israel J. Math., 108(1998), 109-138.

A. Boucherif and R. Precup, On the nonlocal initial value problem for first order differential equations, Fixed Point Theory, 4(2003), 205-212.

A. Boucherif and R. Precup, Semilinear evolution equations with nonlocal initial conditions, Dyn. Sys. Appl., 16(2007), 507-516.

L. Bourdin, Existence of a weak solution for fractional Euler-Lagrange equations, J. Math. Anal. Appl., 399(2013), 239-251.

L. Bourdin, J. Cresson and I. Greff, A continuous/discrete fractional Noether's theorem, Commun. Nonlinear Sci. Numer. Simul., 18(2013), 878-887.

H. Brezis, Functional Analysis, Sobolev Spaces and Partial Differential Equations, Springer, New York, (2011).

D. Bugajewska and P. Kasprzak, On the existence, uniqueness and topological structure of solution sets to a certain fractional differential equation, Comput. Math. Appl., 59(2009), 1108-1116.
S. Bushnaq, S. Momani and Y. Zhou, A reproducing kernel Hilbert space method for solving integro-differential equations of fractional order, J. Optim. Theory Appl., 156(2013), 96-105.
V. S. Buslaev and C. Sulem, On asymptotic stability of solitary waves for nonlinear Schrödinger equations, Ann. Inst. H. Poincaré Anal. Non Linéaire, 20(2003), 419-475.
L. Byszewski, Theorems about existence and uniqueness of solutions of a semi-linear evolution nonlocal Cauchy problem, J. Math. Anal. Appl., 162(1991), 494-505.
L. Byszewski and V. Lakshmikantham, Theorem about the existence and uniqueness of a solution of a nonlocal abstract Cauchy problem in a Banach space, Appl. Anal., 40(1991), 11-19.
L. Caffarelli and A. Vasseur, Drift diffusion equations with fractional diffusion and the quasi-geostrophic equation, Ann. of Math., 171(2010), 1903-1930.
M. Cannone, Ondelettes, Paraproduits et Navier-Stokes, Diderot Editeur, (1995).
J. Cao and H. Chen, Some results on impulsive boundary value problem for fractional differential inclusions, Electron. J. Qual. Theory Differ. Equ., 11(2010), 1-24.
J. Cao and H. Chen, Impulsive fractional differential equations with nonlinear boundary conditions, Math. Comput. Model., 55(2012), 303-311.
B. A. Carreras, V. E. Lynch and G. M. Zaslavsky, Anomalous diffusion and exit time distribution of particle tracers in plasma turbulence models, Phys. Plasmas, 8(2001), 5096-5103.
A. N. Carvalho, T. Dlotko and M. J. D. Nascimento, Nonautonomous semilinear evolution equations with almost sectorial operators, J. Evol. Equ., 8(2008), 631-659.
T. Cazenave, Stable solutions of the logarithmic Schrödinger equation, Nonlinear Anal.: TMA, 7(1983), 1127-1140.
T. Cazenave, Semilinear Schrödinger equations, Courant Lect. Notes Math., 10. New York University, Courant Institute of Mathematical Sciences, New York; American Mathematical Society, Providence, RI, (2003).
T. Cazenave and A. Haraux, An Introduction to Semilinear Evolution Equations, Clarendon Press, Oxford, (1998).
T. Cazenave and P.-L. Lions, Orbital stability of standing waves for some nonlinear Schrödinger equations, Comm. Math. Phys., 85(1982), 549-561.
G. Cerami, An existence criterion for the critical points on unbounded manifolds, Istit. Lombardo Accad. Sci. Lett. Rend. A, 112(1978), 332-336 (in Italian).
L. Cesari, Optimization-Theory and Applications, in: Applications of Mathematics (New York), vol. 17, Problems with ordinary differential equations, Springer-Verlag, New York, (1983).
Y. K. Chang and J. J. Nieto, Some new existence results for fractional differential inclusions with boundary conditions, Math. Comput. Modelling, 49(2009), 605-609.
A. S. Chaves, A fractional diffusion equation to decribe Lévy flights, Phys. Lett. A, 239(1998), 13-16.
A. V. Chechkin, R. Gorenflo and I. M. Sokolov, Retarding subdiffusion and accelerating superdiffusion governed by distributed order fractional diffusion equations, Phys. Rev. E (3), 66(2002), 1-7.
A. V. Chechkin, R. Gorenflo, I. M. Sokolov *et al.*, Distributed order time fractional diffusion equation, Fract. Calc. Appl. Anal., 6(2003), 259-279.
J. Y. Chemin and I. Gallagher, Large, global solutions to the Navier-Stokes equations, slowly varying in one direction, Trans. Amer. Math. Soc., 362(2010), 2859-2873.

J. Y. Chemin, I. Gallagher and M. Paicu, Global regularity for some classes of large solutions to the Navier-Stokes equations, Ann. of Math., 173(2011), 983-1012.

P. Chen, X. He and X. H. Tang, Infinitely many solutions for a class of fractional Hamiltonian systems via critical point theory, Math. Meth. Appl. Sci., (2015).

D. H. Chen, R. N. Wang and Y. Zhou, Nonlinear evolution inclusions: Topological characterizations of solution sets and applications, J. Funct. Anal., 265(2013), 2039-2073.

F. Chen, X. Luo and Y. Zhou, Existence results for nonlinear fractional difference equation, Adv. Difference Equ., vol. 2011, Article ID 713201, 12 pages.

F. Chen, J. J. Neito and Y. Zhou, Global attractivity for nonlinear fractional differential equations, Nonlinear Anal.: RWA, 13(2012), 287-298.

J. Chen and X. H. Tang, Existence and multiplicity of solutions for some fractional boundary value problem via critical point theory, Abstr. Appl. Anal., vol. 2012, Article ID 648635, 21 pages.

J. Chen and X. H. Tang, Infinitely many solutions for a class of fractional boundary value problem, Bull. Malays. Math. Sci. Soc., 36(2013), 1083-1097

F. Chen and Y. Zhou, Attractivity of fractional functional differential equation, Comput. Math. Appl., 62(2011), 1359-1369.

H. J. Choe, Boundary regularity of suitable weak solution for the Navier-Stokes equations, J. Funct. Anal., 268(2015), 2171-2187.

M. Choulli, Une Introduction aux Problems Inverses Elliptiques et Paraboliques. Springer-Verlag, (2009).

P. Clement and R. Zacher, Global smooth solutions to a fourth-order quasilinear fractional evolution equation. Functional analysis and evolution equations, 131-146, Birkhauser, Basel, (2008).

F. Comte, Opérateurs fractionnaires en économétrie et en finance, Prépublication MAP5, (2001).

C. Corduneanu, Functional Equations with Causal Operators, Taylor Francis, London, (2002).

J.-N. Corvellec, V. V. Motreanu and C. Saccon, Doubly resonant semilinear elliptic problems via nonsmooth critical point theory, J. Differential Equations, 248(2010), 2064-2091.

M. Cowling, I. Doust, A. McIntosh *et al.*, Banach space operators with a bounded H^∞ calculus, J. Austral. Math. Soc. Set. A 60 (1996), 51-89.

J. Cresson, Inverse problem of fractional calculus of variations for partial differential equations, Commun. Nonlinear Sci. Numer. Simul., 15(2010), 987-996.

J. Cresson, I. Greff and P. Inizan, Lagrangian for the convection-diffusion equation, Math. Meth. Appl. Sci., (2011).

J. Cresson and P. Inizan, Variational formulations of differential equations and asymmetric fractional embedding, J. Math. Anal. Appl., 385(2012), 975-997.

S. Cuccagna, Stabilization of solutions to nonlinear Schrödinger equations, Comm. Pure Appl. Math., 54(2001), 1110-1145.

B. Dacorogna, Direct Methods in the Calculus of Variations (2nd ed.), in: Applied Mathematical Sciences, vol. 78, Springer, New York, (2008).

V. Daftardar-Gejji and S. Bhalekar, Boundary value problems for multi-term fractional differential equations, J. Math. Anal. Appl., 345(2008), 754-765.

V. Daftardar-Gejji and H. Jafari, Analysis of a system of nonautonomous fractional differential equations involving caputo derivatives, J. Math. Anal. Appl., 328(2007), 1026-1033.

E. N. Dancer and D. Daners, Domain perturbation of elliptic equations subject to Robin boundary conditions, J. Diff. Equs., 74(1997), 86-132.

R. Danchin, Global existence in critical spaces for compressible Navier-Stokes equations, Invent. Math., 141(2000), 579-614.

M. A. Darwish, On monotonic solutions of an integral equation of Abel type, Mathematica Bohemica, 133(2008), 407-420.

M. A. Darwish, J. Henderson and S. K. Ntouyas, Fractional order semilinear mixed type functional differential equations and inclusions, Nonlinear Stud., 16(2009), 197-219.

M. F. de Almeida and L. C. F. Ferreira, On the Navier-Stokes equations in the half-space with initial and boundary rough data in Morrey spaces, J. Differential Equations, 254(2013), 1548-1570.

P. M. De Carvalho-Neto and P. Gabriela, Mild solutions to the time fractional Navier-Stokes equations in R^N, J. Differential Equations, 259(2015), 2948-2980.

K. Deimling, Nonlinear Functional Analysis, Springer-Verlag, (1985).

R. deLaubenfels, Existence families, functional calculi and evolution equations, Springer-Verlag, Berlin, (1994).

D. Delbosco and L. Rodino, Existence and uniqueness for a nonlinear fractional differential equation, J. Math. Anal. Appl., 204(1996), 609-625.

J. Deng and S. Wang, Existence of solutions of nonlocal Cauchy problem for some fractional abstract differential equation, Appl. Math. Lett., 55(2016), 42-48.

G. Desch and S. O. Londen, Evolutionary equations driven by fractional Brownian motion, Stoch. Partial Differ. Equ. Anal. Comput., 1(2013), 424-454.

J. V. Devi and V. Lakshmikantham, Nonsmooth analysis and fractional differential equations, Nonlinear Anal.: TMA, 70(2009), 4151-4157.

B. C. Dhage, Existence of extremal solutions for discontinuous functional integral equations, Applied Math. Letters, 19(2006), 881-886.

T. Diagana, G. Mophou and G. N'guérékata, On the existence of mild solutions to some semilinear fractional integro-differential equations, Electron. J. Qual. Theory Differ. Equ., 58(2010), 1-17.

J. Diblík and Z. Svoboda, Positive solutions of p-type retarded functional differential equations, Nonlinear Anal.: TMA, 63(2005), 813-821.

K. Diethelm, Analysis of fractional differential equations, J. Math. Anal. Appl., 265(2002), 229-248.

K. Diethelm, The Analysis of Fractional Differential Equations, Lecture Notes in Mathematics, Springer, (2010).

K. Diethelm and Y. Luchko, Numerical solution of linear multi-term initial value problems of fractional order. J. Comput. Anal. Appl., 6(2004), 243-263.

J. Dieudonne, Deux examples dequations differentielles, Acta. Sci. Math., 12B(1950), 38-40.

Y. Ding and L. Jeanjean, Homoclinic orbits for a nonperiodic Hamiltonian system, J. Differential Equations, 237(2007), 473-490.

J. D. Djida, A. Fernandez and I. Area, Well-posedness results for fractional semi-linear wave equations, Discrete Contin. Dyn. Syst. Ser. B, 25(2)(2020), 569-597.

X. W. Dong, J. R. Wang and Y. Zhou, On nonlocal problems for fractional differential equations in Banach spaces, Opuscula Mathematica, 31(2011), 341-357.

M. M. Dzhrbashyan and A. B. Nersessyan, Fractional derivatives and Cauchy problem for differential equations of fractional order, Izv. AN Arm. SSR. Mat., 3(1968), 3-29 (in Russian).

R. Eid, S. I. Muslih and D. Baleanu *et al.*, On fractional Schrödinger equation in α-dimensional fractional space, Nonlinear Anal.: RWA, 10(2009), 1299-1304.

S. D. Eidelman and A. N. Kochubei, Cauchy problem for fractional diffusion equations, J. Differential Equations, 199(2004), 211-255.

M. M. El-Borai, Some probability densities and fundamental solutions of fractional evolution equations, Chaos, Sol. Frac., 14(2002), 433-440.

M. M. El-Borai, The fundamental solutions for fractional evolution equations of parabolic type, J. Appl. Math. Stoch. Anal., 3(2004), 197-211.

A. M. A. El-Sayed, Nonlinear functional differential equations of arbitrary orders, Nonlinear Anal.: TMA, 33(1998), 181-186.

A. M. A. El-Sayed, W. G. El-Sayed and O. L. Moustafa, On some fractional functional equations, Pure Math. Appl., 6(1995), 321-332.

M. El-Shahed and A. Salem, On the generalized Navier-Stokes equations, Appl. Math. Comput., 156(1)(2004), 287-293.

V. J. Ervin and J. P. Roop, Variational formulation for the stationary fractional advection dispersion equation, Numer. Meth. Part. Diff. Equs., 22(2006), 558-576.

L. C. Evans, Partial Differential Equations, Second Edition, American Mathematical Society, Providence, Rhode Island, (2010).

K. S. Fa, Generalized langevin equation with fractional derivative and long-time correlation function, Phys. Rev., 73(2006), 061104.

K. S. Fa, Fractional langevin equation and Riemann-Liouville fractional derivative, The Eur. Phys. J., 24(2007), 139-143.

M. M. Fall and T. Weth, Nonexistence results for a class of fractional elliptic boundary value problems, J. Funct. Anal., 263(2012), 2205-2227.

Z. Fan, Impulsive problems for semilinear differential equations with nonlocal conditions, Nonlinear Anal.: TMA, 72(2010), 1104-1109.

Z. Fan and G. Li, Existence results for semilinear differential equations with nonlocal and impulsive conditions, J. Funct. Anal., 258(2010), 1709-1727.

M. Fečkan, Topological Degree Approach to Bifurcation Problems, Springer, (2008).

M. Fečkan, J. R. Wang and Y. Zhou, Controllability of fractional functional evolution equations of Sobolev type via characteristic solution operators, J. Optim. Theory Appl., 156(1)(2013), 79-95.

M. Fečkan, Y. Zhou and J. R. Wang, On the concept and existence of solutions for impulsive fractional differential equations, Commun. Nonlinear Sci. Numer. Simul., 17(2012), 3050-3060.

G. Fibich, Singular solutions of the subcritical nonlinear Schrödinger equation, Physica D: Nonlinear Phenomena, 240(2011), 1119-1122.

G. J. Fix and J. P. Roop, Least squares finite-element solution of a fractional order two-point boundary value problem, Comput. Math. Appl., 48(2004), 1017-1033.

A. Floer and A. Weinstein, Nonspreading wave packets for the cubic Schrödinger equation with a bounded potential, J. Funct. Anal., 69(1986), 397-408.

G. S. F. Frederico and D. F. M. Torres, A formulation of Noether's theorem for fractional problems of the calculus of variations, J. Math. Anal. Appl., 334(2007), 834-846.

R. R. Gadyl'shin, On the eigenvalues of a dumb-bell with a thin handle, Izv. Math., 69(2005), 265-329.

G. P. Galdi, An Introduction to the Mathematical Theory of the Navier-Stokes Equations: Nonlinear Steady Problems, Springer Tracts in Natural Philosophy, New York, (1998).

Z. Z. Ganji, D. D. Ganji, A. Ganji *et al.*, Analytical solution of time-fractional Navier-Stokes equation in polar coordinate by homotopy perturbation method, Numer. Methods Partial Diff. Equ., 26(2010), 117-124.

Y. Giga and S. Yoshikazu, Abstract L^p estimates for the Cauchy problem with applications to the Navier-Stokes equations in exterior domains, J. Funct. Anal., 102(1991), 72-94.

M. Giona, S. Gerbelli and H. E. Roman, Fractional diffusion equation and relaxation in complex viscoelastic materials, Physica A, 191(1992), 449-453.

C. Godbillon, Géométrie Différentielle et Mécanique Analytique, Hermann, Paris, (1969).

R. Gorenflo, Y. Luchko and P. P. Zabrejko, On solvability of linear fractional differential equations in Banach spaces, Frac. Calc. Appl. Anal., 2(1999), 163-176.

D. J. Guo, Existence of positive solutions for nth-order nonlinear impulsive singular integro-differential equations in Banach spaces, Nonlinear Anal.: TMA, 68(2008), 2727-2740.

D. J. Guo, V. Lakshmikantham and X. Z. Liu, Nonlinear Integral Equations in Abstract Spaces, Kluwer Academic, Dordrecht, (1996).

B. L. Guo and Y. P. Wu, Orbital stability of solitary waves for the nonlinear derivative Schrödinger equation, J. Differential Equations, 123(1995), 35-55.

I. Györi and G. Ladas, Oscillation Theory of Delay Differential Equations with Applications, Clarendon, Oxford, (1991).

J. K. Hale, Theory of Functional Differential Equations, Springer-Verlag, New York, (1977).

J. K. Hale and J. Kato, Phase space for retarded equations with infinite delay, Funkcial. Ekvac., 21(1978), 11-41.

B. Han, Refinable functions and cascade algorithms in weighted spaces with Hölder continuous masks, SIAM J. Math. Anal., 40(2008), 70-102.

A. Hanyga, Multidimensional solutions of time-fractional diffusion-wave equations, Proc. R. Soc. Lond. Ser. A: Math. Phys. Eng. Sci., 458(2002), 933-957.

D. Haroske and H. Triebel, Distributions, Sobolev spaces, elliptic equations, European Mathematical Society, Switzerland, (2008).

Y. Hatno and N. Hatano, Dispersive transport of ions in column experiments: an explanation of long-tailed profiles, Water Resources Res., 34(1980), 1027-1033.

N. Hayashi, E. I. Kaikina and P. I. Naumkin, Asymptotics for fractional nonlinear heat equations, J. London Math. Soc., 72(2005), 663-688.

J. W. He and L. Peng, Approximate controllability for a class of fractional stochastic wave equations, Comput. Math. Appl., 78(5)(2019), 1463-1476.

J. W. He and Y. Zhou, On a backward problem for nonlinear time fractional wave equations, Proc. Roy. Soc. Edinburgh Sect. A, 152(2022), 1589-1612.

J. W. He and Y. Zhou, Hölder regularity for non-autonomous fractional evolution equations, Frac. Calc. Appl. Anal., 25(2022), 378-407.

J. W. He, Y. Zhou, L. Peng and B. Ahmad, On well-posedness of semilinear Rayleigh-Stokes problem with fractional derivative on $\mathbb{R}^N$, Adv. Nonlinear Anal., 11(2022), 580-597.

H. Heck, H. Kim and H. Kozono, Weak solutions of the stationary Navier-Stokes equations for a viscous incompressible fluid past an obstacle, Math. Ann., 356(2013), 653-681.

S. Heikkilä and V. Lakshmikantham, Monotone Iterative for Discontinuous Nonlinear Differential Equations, Monographs and Texbooks in Pure and Applied Mathematics, 181. Marcel Dekker, Inc., New York, (1994).

H.-P. Heinz, On the behaviour of measure of noncompactness with respect to differentiation and integration of vector-valued functions, Nonlinear Anal.: TMA, 7(1983), 1351-1371.

D. Henry, Geometric Theory of Semilinear Parabolic Equations, LNM 840, Springer-Verlag, Berlin, Heidelberg, New York, (1981).

E. Hernández, D. O'Regan and K. Balachandran, On recent developments in the theory of abstract differential equations with fractional derivatives, Nonlinear Anal.: TMA, 73(2010), 3462-3471.

R. Herrmann, Fractional Calculus: An Introduction for Physicists, World Scientific, Singapore, (2011).

R. Hilfer, Applications of Fractional Calculus in Physics, World Scientific, Singapore, (2000).

Y. Hino, S. Murakami and T. Naito, Functional Differential Equations with Infinite Delay, Lecture Notes in Math., vol. 1473, Springer-Verlag, Berlin, (1991).

L. Hu and G. Kallianpur, Schröodinger equations with fractional Laplacians, Appl. Math. Optim., 42(2000), 281-290.

L. Hu, Y. Ren and R. Sakthivel, Existence and uniqueness of mild solutions for semilinear integro-differential equations of fractional order with nonlocal initial conditions and delays, Semigroup Forum, 79(2009), 507-514.

S. Hu and N. S. Papageorgiou, Handbook of Multivalued Analysis (Theory), Kluwer Academic Publishers, Dordrecht Boston, London, 1997.

S. Iqbal, J. Pečarić and Y. Zhou, Generalization of an inequality for integral transforms with kernel and related results, J. Inequalities and Applications, vol. 2010, Article ID 948430, 17 pages.

F. Isaia, On a nonlinear integral equation without compactness, Acta. Math. Univ. Comenianae., LXXV(2006), 233-240.

V. Isakov, Inverse Problems for Partial Differential Equations, Springer-Verlag, New York, (1998).

T. Iwabuchi and R. Takada, Global solutions for the Navier-Stokes equations in the rotational framework, Math. Ann., 357(2013), 727-741.

M. Izydorek and J. Janczewska, Homoclinic solutions for a class of the second order Hamiltonian systems, J. Differential Equations, 219(2005), 375-389.

A. Jaiwal and D. Bahuguna, Hilfer fractional differential equations with almost sectorial operators, Diff. Equs. Dyn. Syst., doi.org/10.1007/s12591-020-00514-y.

O. K. Jaradat, A. Al-Omari and S. Momani, Existence of the mild solution for fractional semilinear initial value problems, Nonlinear Anal.: TMA, 69(2008), 3153-3159.

W. Jiang, The controllability of fractional control systems with control delay, Comput. Math. Appl., 64(2012), 3153-3159.

H. Jiang, F. Liu, I. Turner *et al.*, Analytical solutions for the multi-term timespace Caputo-Riesz fractional advection-diffusion equations on a nite domain, J. Math. Anal. Appl., 389(2012), 1117-1127.

F. Jiao and Y. Zhou, Existence of solutions for a class of fractional boundary value problem via critical point theory, Comput. Math. Appl., 62(2011), 1181-1199.

F. Jiao and Y. Zhou, Existence results for fractional boundary value problem via critical point theory, Int. J. Bifurcation and Chaos, 22(2012), 1-17.

S. Jimbo, The singularly perturbed domain and the characterization for the eigenfunctions with Neumann boundary conditions, J. Differential Equations, 77(1989), 322-350.

F. Kappel and W. Schappacher, Some considerations to the fundamental theory of infinite delay equations, J. Differential Equations, 37(1980), 141-183.

V. Keyantuo, C. Lizama and M. Warma, Existence, Regularity and representation of solutions of time fractional wave equations, Electron. J. Differ. Equ., 2017(222)(2017), 1-42.

Y. Kian and M. Yamamoto, On existence and uniqueness of solutions for semilinear fractional wave equations, Fract. Calc. Appl. Anal., 20(2017), 117-138.

Y. Kian and M. Yamamoto, Well-posedness for weak and strong solutions of non-homogeneous initial boundary value problems for fractional diffusion equations, Frac. Calc. Appl. Anal., 24(2021), 168-201.

A. A. Kilbas, H. M. Srivastava and J. J. Trujillo, Theory and Applications of Fractional

Differential Equations, in: North-Holland Mathematics Studies, vol. 204, Elsevier Science B.V., Amsterdam, (2006).
K.H. Kim and S. Lim, Asymptotic behaviors of fundamental solution and its derivatives to fractional diffusion-wave equations, J. Korean Math. Soc., 53(2016), 929-967.
I. Kim, K.H. Kim and S. Lim, An $L^q(L^p)$-theory for the time fractional evolution equations with variable coefficients, Adv. Math., 306(2017), 123-176.
M. Kirane, Y. Laskri and N.-e. Tatar, Critical exponents of Fujita type for certain evolution equations and systems with spatiotemporal fractional derivatives, J. Math. Anal. Appl., 312(2005), 488-501.
V. Kiryakova, Generalized Fractional Calculus and Applications, in: Pitman Research Notes in Mathematics Series, vol. 301, Longman Scientific & Technical, Harlow, (1994).
J. Klafter, S. C. Lim and R. Metzler (Editors), Fractional Dynamics in Physics, World Scientific, Singapore, (2011).
M. Klimek, Lagrangian and Hamiltonian fractional sequential mechanics, Czechoslovak J. Phys., 52(2002), 1247-1253.
M. Klimek, Existence-uniqueness result for a certain equation of motion in fractional mechanics, Bull. Polish Acad. Sci., 58(4)(2010), 573-581.
V. Kobolev and E. Romanov, Fractional langevin equation to describe anomalous diffusion, Pro. Theor. Phys. Sup., 139(2000), 470-476.
G. Koch, N. Nadirashvili, G. A. Seregin *et al.*, Liouville theorems for the Navier-Stokes equations and applications, Acta Math., 203(2009), 83-105.
A. N. Kochubei, Distributed order calculus and equations of ultraslow diffusion, J. Math. Anal. Appl., 340(2008), 252-281.
A. Kolmogorov, S. Fomine and V. M. Tihomirov, Eléments de la théorie des fonctions et de l'analyse fonctionnelle, Éditions Mir, Moscow, (1974), Avec uncomplément sur les algèbres de Banach, par V.M. Tikhomirov, Traduit du russe par Michel Dragnev.
L. Kong, Existence of solutions to boundary value problems arising from the fractional advection dispersion equation, Electronic J. Differential Equations, 2013(106), 1-15.
P. Korman and A. C. Lazer, Homoclinic orbits for a class of symmetric Hamiltonian systems, Electronic J. Differential Equations, 01(1994), 1-10.
N. Kosmatov, Initial value problems of fractional order with fractional impulsive conditions, Results. Math., 63(2013), 1289-1310.
H. Kozono, L^1-solutions of the Navier-Stokes equations in exterior domains, Math. Ann., 312(1998), 319-340.
A. Kubica and M. Yamamoto, Initial-boundary value problems for fractional diffusion equations with time-dependent coefficients, Frac. Calc. Appl. Anal., 21(2018), 276-311.
S. Kumar and N. Sukavanam, Approximate controllability of fractional order semilinear systems with bounded delay, J. Differential Equations, 252(2012), 6163-6174.
O.A. Ladyzhenskaya, On integral estimates, convergence, approximate methods, and solution in functionals for elliptic operators (In Russian), Vestnik Leningrad. Univ., 13(1958), 60-69.
V. Lakshmikantham, Theory of fractional functional differential equations, Nonlinear Anal.: TMA, 69(2008), 3337-3343.
V. Lakshmikantham, D. D. Bainov and P. S. Simeonov, Theory of impulsive differential equations, World Scientific, Singapore-London, (1989).
V. Lakshmikantham and S. Leela, Nonlinear Differential Equations in Abstract Spaces, Pergamon Press, New York, (1969).
V. Lakshmikantham, S. Leela and J. V. Devi, Theory of Fractional Dynamic Systems, Cambridge Scientific Publishers, Cambridge, (2009).

V. Lakshmikantham, S. Leela, Z. Drici *et al.*, Theory of Causal Differential Equations, Atlantis Studies in Mathematics for Engineering and Science, vol. 5, World Scientific, (2010).
V. Lakshmikantham and A. S. Vatsala, General uniqueness and monotone iterative technique for fractional differential equations, Appl. Math. Lett., 21(2008a), 828-834.
V. Lakshmikantham and A. S. Vatsala, Basic theory of fractional differential equations, Nonlinear Anal.: TMA, 69(2008b), 2677-2682.
V. Lakshmikantham, L. Wen and B. Zhang, Theory of Differential Equations with Unbounded Delay, Kluwer Academic Publishers, Dordrecht, (1994).
Y.-H. Lan, H. X. Huang and Y. Zhou, Observer-based robust control of α $(1 < \alpha < 2)$ fractional-order uncertain systems: an LMI approach, IET Control Theory Appl., 6(2012), 229-234.
Y.-H. Lan and Y. Zhou, LMI-based robust control of fractional-order uncertain linear systems, Comput. Math. Appl., 62(2011), 1460-1471.
Y.-H. Lan and Y. Zhou, High-order D^{α}-type iterative learning control for fractional-order nonlinear time-delay systems, J. Optim. Theory Appl., 156(2013a), 153-166.
Y.-H. Lan and Y. Zhou, D^{α}-type iterative learning control for fractional-order linear time-delay systems, Asian J. Control, 15(2013b), 669-677.
P. G. Lemarié-Rieusset, Recent Developments in the Navier-Stokes Problem, Chapman CRC Press, (2002).
F. Li, Z. Liang and Q. Zhang, Existence of solutions to a class of nonlinear second order two-point boundary value problems, J. Math. Anal. Appl., 312(2005), 357-373.
C. Y. Li and Y. M. Long, Existence of nontrival homoclinic orbit in a class of Hamilton systems ü-L(t)u+W'_u(t,u)= 0, Acta Math. Sin., 45(2002), 349-354.
C. F. Li, X. N. Luo and Y. Zhou, Existence of positive solutions of boundary value problem for fractional differential equations, Comput. Math. Appl., 59(2010), 1363-1375.
B. Li, S. Sun and Y. Sun, Existence of solutions for fractional Langevin equation with infinite-point boundary conditions, J. Comput. Appl. Math., (2016), 1-10.
Z. Li and M. Yamamoto, Initial-boundary value problems for linear diffusion equation with multiple time-fractional derivatives, arXiv preprint arXiv:1306.2778, (2013).
K. Li, J. Peng and J. Jia, Cauchy problems for fractional differential equations with Riemann-Liouville fractional derivatives, J. Funct. Anal., 263(2012), 476-510.
J. Liang, J. H. Liu and T.-J. Xiao, Nonlocal impulsive problems for nonlinear differential equations in Banach spaces, Math. Comput. Model., 49(2009), 798-804.
S. C. Lim, M. Li and L. P. Teo, Langevin equation with two fractional orders, Phys. Lett., 372(2008), 6309-6320.
W. Lin, Global existence theory and chaos control of fractional differential equations, J. Math. Anal. Appl., 332(2007), 709-726.
J. L. Lions and E. Magenes, Non-homogeneous Boundary Value Problems and Applications, Springer-Verlag, (1972).
B. M. Liu and L. S. Liu, The decision theorems of relative compactness for two classes of abstract function groups in an infinite interval and its applications, J. Syst. Sci. Math. Sci., 30(2010), 1008-1019.
H. Liu, J. C. Chang, Existence for a class of partial differential equations with nonlocal conditions, Nonlinear Anal., 70(2009), 3076-3083.
J. Liu, Nonlinear impulsive evolution equations, Dyn. Contin. Discrete Impuls. Syst., 6(1999), 77-85.
L. Liu, F. Guo, C. Wu *et al.*, Existence theorems of global solutions for nonlinear Volterra type integral equations in Banach spaces, J. Math. Anal. Appl., 309(2005), 638-649.
Z. Liu and X. Li, On the controllability of impulsive fractional evolution inclusions in Banach spaces, J. Optim. Theory Appl., 156(2013), 167-182.

Z. Liu and X. Li, Approximate controllability of fractional evolution systems with Riemann-Liouville fractional derivatives, SIAM J. Control Optim., 53(2015), 1920-1933.

Z. B. Liu, L. S. Liu, and J. Zhao, The criterion of relative compactness for a class of abstract function groups in an infinite interval and its applications, J. Syst. Sci. Math. Sci., 28(2008), 370-378.

C. F. Lorenzo and T. T. Hartley, Variable order and distributed order fractional operators, Nonlinear Dynam., 29(2002), 57-98.

S. Lu, F. J. Molz and G. J. Fix, Possible problems of scale dependency in applications of the three-dimensional fractional advection-dispersion equation to natural porous media, Water Resources Res., 38(2002), 1165-1171.

Y. Luchko, Boundary value problems for the generalized time-fractional diffusion equation of distributed order, Fract. Calc. Appl. Anal., 12(2009a) 409-422.

Y. Luchko, Maximum principle for the generalized time-fractional diffusion equation. J. Math. Anal. Appl., 351(2009b), 218-223.

Y. Luchko, Some uniqueness and existence results for the initial-boundary-value problems for the generalized time-fractional diffusion equation, Comput. Math. Appl., 59(2010), 1766-1772.

Y. Luchko, Initial-boundary-value problems for the generalized multi-term time-fractional diffusion equation, J. Math. Anal. Appl., 374(2011), 538-548.

Y. Luchko and R. Gorenflo, An operational method for solving fractional differential equations with the Caputo derivarives, Acta Math. Vietnam., 24(1999), 207-233.

N. H. Luc, D. Lan, D. O'Regan, N. A. Tuan and Y. Zhou, On the initial value problem for the nonlinear fractional Rayleigh-Stokes equation, J. Fixed Point Theory Appl., 23(4)(2021), 1-28.

A. Lunardi, Analytic Semigroups and Optimal Regularity in Parabolic Problems, Birkhäuser, Basel, (1995).

A. C. J. Luo and V. S. Afraimovich (Editors), Long-range Interaction, Stochasticity and Fractional Dynamics, Springer, (2010).

V. Lupulescu, Causal functional differential equations in Banach spaces, Nonlinear Anal.: TMA, 69(2008), 4787-4795.

E. Lutz, Fractional langevin equation, Phys. Rev., 64(2001), 051106.

J. T. Machado, V. Kiryakova and F. Mainardi, Recent history of fractional calculus, Commun. Nonlinear Sci. Numer. Simul., 16(2011), 1140-1153.

M. Magdziarz and T. Zorawik, Stochastic representation of a fractional subdiffusion equation, The case of infinitely divisible waiting times, Lévy noise and space-time-dependent coefficients, Proc. Amer. Math. Soc., (2015).

R. L. Magin, Fractional calculus models of complex dynamics in biological tissues, Comput. Math. Appl., 59(5)(2010), 1586-1593.

F. Mainardi, P. Paraddisi and R. Gorenflo, Probability Distributions Generated by Fractional Diffusion Equations, in: J. Kertesz, I. Kondor (Eds.), Econophysics: An Emerging Science, Kluwer, Dordrecht, (2000).

F. Mainardi and P. Pironi, The Fractional Langevin equation: Brownian motion revisited, Extracta Math., 10(1996), 140-154.

P. D. Makita, Homoclinic orbits for second order Hamiltonian equations in R, J. Dyn. Diff. Equs., 24(2012), 857-871.

A. B. Malinowska and D. F. M. Torres, Introduction to the Fractional Calculus of Variations, Imp. Coll. Press, London, (2012).

H. Markus, The Functional Valculus for Sectorial Operators, Oper. Theory Adv. Appl., vol. 69, Birkhauser-Verlag, Basel, (2006).

N. Masmoudi and T. K. Wong, Local-in-time existence and uniqueness of solutions to the Prandtl equations by energy methods, Comm. Pure Appl. Math., 68(2015), 1683-1741.

J. Mawhin, Topological degree methods in nonlinear boundary value problems, CMBS Regional Conference Series in Mathematics, 40, AMS, Providence, R.I., (1979).

J. Mawhin and M. Willem, Critical Point Theorey and Hamiltonian Systems, Springer, New York, (1989).

A. McIntosh, Operators which have an H^{∞} functional calculus, Proc. Centre Math. Anal. Austral. Nat. Univ., 14(1986), 210-231.

M. M. Meerschaert, D. A. Benson, H. Scheffler *et al.*, Stochastic solution of space-time fractional diffusion equations, Phys. Rev. E, 65(2002), 1103-1106.

M. M. Meerschaert, E. Nane and H. P. Scheffler, Stochastic model for ultraslow diffusion, Stochastic Process. Appl., 116(2006), 1215-1235.

M. M. Meerschaert, E. Nane and P. Vellaisamy, Distributed-order fractional Cauchy problems on bounded domains, arXiv preprint arXiv:0912.2521v1, (2009).

M. M. Meerschaert and C. Tadjeran, Finite difference approximations for fractional advection-dispersion flow equations, J. Comput. Appl. Math., 172(2004), 65-77.

R. Metzler and J. Klafter, The Random Walk's Guide to Anomalous Diffusion: A Fractional Dynamics Approach, Phys. Rep., 339(2000a), 1-77.

R. Metzler and J. Klafter, Boundary value problems for fractional diffusion equations, Phys. A, 278(2000b), 107-125.

R. Metsler and J. Klafter, The restaurant at the end of the random walk: recent developments in the description of anomalous transport by fractional dynamics, J. Phys. A, 37(2004), 161-208.

M. W. Michalski, Derivatives of noninteger order and their applications, Inst. Math., Polish Acad. Sci., (1993).

R. Mikulevicius, H. Pragarauskas and N. Sonnadara, On the Cauchy-Dirichlet problem in the half space for parabolic SPDEs in weighted Hölder spaces, Acta Appl. Math., 97(2007), 129-149.

K. S. Miller and B. Ross, An Introduction to the Fractional Calculus and Differential Equations, John Wiley, New York, (1993).

S. Momani and O. Zaid, Analytical solution of a time-fractional Navier-Stokes equation by Adomian decomposition method, Appl. Math. Comput., 177(2006), 488-494.

H. Mönch, Boundary value problems for nonlinear ordinary differential equations of second order in Banach spaces, Nonlinear Anal.: TMA, 4(1980), 985-999.

G. M. Mophou, Existence and uniqueness of mild solution to impulsive fractional differential equations, Nonlinear Anal.: TMA, 72(2010), 1604-1615.

G. M. Mophou and G. M. N'Guérékata, Existence of mild solutions of some semilinear neutral fractional functional evolution equations with infinite delay, Appl. Math. Comput., 216(2010), 61-69.

V. Mureşan, Existence, uniqueness and data dependence for the solutions of some integro-differential equations of mixed type in Banach space, J. Anal. Appl., 23(2004), 205-216.

I. P. Natanson, Theory of functions of a real variable, vol. II, Frederick Ungar Publishing Co., New York, (1960).

J. M. A. M. van Neerven and B. Straub, On the existence and growth of mild solutions of the abstract Cauchy problem for operators with polynomially bounded resolvent, Houston J. Math., 24(1998),137-171.

G. M. N'Guerekata, A Cauchy problem for some fractional abstract differential equation with nonlocal conditions, Nonlinear Anal.: TMA, 70(2009), 1873-1876.

J. J. Nieto and D. O'Regan, Variational approach to impulsive differential eqquations, Nonlinear Anal.: RWA, 10(2009), 680-690.
N. Nyamoradi, A. Alsaedi, B. Ahmad and Y. Zhou, Multiplicity of homoclinic solutions for fractional Hamiltonian systems with subquadratic potential, Entropy, 19(2017), 50.
N. Nyamoradi and Y. Zhou, Infinitely many solutions for a perturbed nonlinear fractional boundary value problems depending on two parameters, Eur. Phys. J. Special Topics, 222(2013), 1997-2011.
N. Nyamoradi and Y. Zhou, Bifurcation results for a class of fractional Hamiltonian systems with Liouville-Weyl fractional derivatives, J. Vibration and Control, 22(2016), 1358-1368.
N. Nyamoradi and Y. Zhou, Homoclinic orbits for a class of fractional Hamiltonian systems via variational methods, J. Optim. Theory Appl., 174(2017), 210-222.
N. Nyamoradi, Y. Zhou, B. Ahmad and A. Alsaedi, Variational approach to Homoclinic solutions for fractional Hamiltonian systems, J. Optim. Theory Appl., 174(2017), 223-237.
T. Odzijewicz, A. B. Malinowska and D. F. M. Torres, Fractional variational calculus with classical and combined Caputo derivatives, Nonlinear Anal.: TMA, 75(2012a), 1507-1515.
T. Odzijewicz, A. B. Malinowska and D. F. M. Torres, Fractional calculus of variations in terms of a generalized fractional integral with applications to physics, Abstr. Appl. Anal., (2012b).
T. Odzijewicz, A. B. Malinowska and D. F. M. Torres, Generalized fractional calculus with applications to the calculus of variations, Comput. Math. Appl., (2012c).
W. Omana and M. Willem, Homoclinic orbits for a class of Hamiltonian systems, Differ. Integr. Equ., 5(5)(1992), 1115-1120.
D. O'Regan and R. Precup, Existence criteria for integral equations in Banach spaces, J. Inequal. & Appl., 6(2001), 77-97.
E. Otárola and A.J. Salgado, Regularity of solutions to space-time fractional wave equations, a PDE approach, Frac. Calc. Appl. Anal., 21(2018), 1262-1293.
D. N. Pandey, A. Ujlayan and D. Bahuguna, On a solution to fractional order integro-differential equations with analytic semigroups, Nonlinear Anal.: TMA, 71(2009), 3690-3698.
E. Paturel, Multiple homoclinic orbits for a class of Hamiltonian systems, Calc. Val. Partial Differ. Equ., 12(2001), 117-143.
A. Pazy, Semigroups of linear operators and applications to partial differential equations. In Applied Mathematical Sciences, vol. 44. Springer-Verlag, New York, (1983).
L. Peng and Y. Zhou, Well-posedness and regularity results for fractional wave equations with time-dependent coefficients, Fractal Fract., 6(2022a), 644.
L. Peng and Y. Zhou, The existence of mild and classical solutions for time fractional Fokker-Planck equations, Monatsh. Math., 199(2022b), 377-410.
L. Peng, Y. Zhou and J. W. He, The well-posedness analysis of distributed order fractional diffusion problems on $\mathbb{R}^N$, Monatsh. Math., 198(2022), 445-463.
F. Periago and B. Straub, A functional calculus for almost sectorial operators and applications to abstract evolution equations, J. Evol. Equs., 2(2002), 41-68.
S. Picozzi and B. West, Fractional Langevin model of memory in financial markets, Phys. Rev., 66(2002), 046118.
I. Podlubny, Fractional Differential Equations, Academic Press, San Diego, (1999).
R. Ponce, Hölder continuous solutions for fractional differential equations and maximal regularity, J. Differential Equations, 255(2013), 3284-3304.

T. R. Prabhakar, A singular integral equation with a generalized Mittag Leffler function in the kernel, Yokohama Math. J., 19(1971), 7-15.
E. M. Rabei, K. I. Nawafleh, R. S. Hijjawi *et al.*, The Hamilton formalism with fractional derivatives, J. Math. Anal. Appl., 327(2007), 891-897.
P. H. Rabinowitz, Minimax Methods in Critical Point Theory with Applications to Differential Equations, CBMS, American Mathematical Society, vol. 65, (1986).
P. H. Rabinowitz, Homoclinic orbits for a class of Hamiltonian systems, Proc. Roy. Soc. Edinburgh Sect., A: 114(1990), 33-38.
P. H. Rabinowitz and K. Tanaka, Some result on connecting orbits for a class of Hamiltonian systems, Math. Z., 206(1991), 473-499.
M. H. M. Rashid and Y. El-Qaderi, Semilinear fractional integro-differential equations with compact semigroup, Nonlinear Anal.: TMA, 71(2009), 6276-6282.
G. Raugel and G. R. Sell, Navier-Stokes equations on thin 3D domains, I, Global attractors and global regularity of solutions, J. Amer. Math. Soc., 6(1993), 503-568.
F. Riewe, Nonconservative Lagrangian and Hamiltonian mechanics, Phys. Rev. E, 53(1996), 1890-1899.
F. Riewe, Mechanics with fractional derivatives, Phys. Rev. E, 55(3, part B)(1997), 3581-3592.
H. E. Roman and P. A. Alemany, Continuous-time random walks and the fractional diffusion equation, J. Phys. A, 27(1994), 3407-3410.
A. Rontó and M. Rontó, Successive approximation method for some linear boundary value problems for differential equations with a special type of argument deviation, Miskolc Math. Notes, 10(2009), 69-95.
W. Rudin, Functional Analysis (2nd ed.), McGraw-Hill, (1991).
I. A. Rus, Metrical fixed point theorems, Univ. of Cluj-Napoca (Romania), (1979).
I. A. Rus, Picard mappings: results and problems, Babes-Bolyai Univ., Cluj-Napoca, Seminar on Fixed Point Theory, Preprint 6(1987), 55-64.
I. A. Rus, Weakly Picard mappings, Comment. Math. Univ. Carolin., 34(1993), 769-773.
I. A. Rus, Picard operators and applications, Sci. Math. Jpn., 58(2003), 191-219.
I. A. Rus and S. Mureşan, Data dependence of the fixed points set of some weakly Picard operators, Proc. Itinerant Seminar, Srima Publishing House, Cluj-Napoca (Romania), (2000), 201-207.
J. Sabatier, O. Agrawal and J. T. Machado, Advances in fractional calculus. Theoretical developments and applications in physics and engineering, Springer-Verlag, Berlin, (2007).
K. Sakamoto and M. Yamamoto, Initial value/boundary value problems for fractional diffusionwave equations and applications to some inverse problems, J. Math. Anal. Appl., 382(2011), 426-447.
H. A. H. Salem, On the nonlinear Hammerstein integral equations in Banach spaces and application to the boundary value problem of fractional order, Math. Comput. Modelling, 70(2009), 1873-1876.
S. G. Samko, A. A. Kilbas and O. I. Marichev, Fractional Integral and Derivatives: Theory and Applications, Gordon and Breach Science Publishers, Longhorne, PA, (1993).
T. Sandev, R. Metzler and Ž. Tomovski, Velocity and displacement correlation functions for fractional generalized Langevin equations, Fract. Calc. Appl. Anal., 15(2012), 426-450.
T. Sandev, Ž. Tomovski and J. L. A. Dubbeldam, Generalized Langevin equation with a three parameter Mittag-Leffler noise, Physica A: Stat. Mech. Appl., 390(2011), 3627-3636.
W. R. Schneider and W. Wayes, Fractional diffusion and wave equation, J. Math. Phys., 30(1989), 134-144.

J. B. Seaborn, Hypergeometric functions and their applications, Springer-Verlag, New York, (1991).

M. A. Şerban, I. A. Rus and A. Petruşl, A class of abstract Volterra equations, via weakly Picard operators technique, Math. Inequal. Appl., 13(2010), 255-269.

E. Séré, Existence of infinitely many homoclinic orbits in Hamiltonian systems, Math. Z., 209(1992), 27-42.

M. F. Shlesinger, B. J. West and J. Klafter, Lévy dynamics of enhanced diffusion: applications to turbulence, Phys. Rev. Lett., 58(1987), 1100-1103.

X. B. Shu, Y. Z. Lai and Y. M. Chen, The existence of mild solutions for impulsive fractional partial differential equations, Nonlinear Anal.: TMA, 74(2011), 2003-2011.

L. M. Sokolov, A. V. Chechkin and J. Klafter, Distributed-order fractional kinetics, Acta Phys. Polon. B, 35(2004), 1323-1341.

C. A. Stuart, Bifurcation into Spectral Gaps, Société Mathématique de Belgique, (1995).

P.-L. Sulem and C. Sulem, The Nonlinear Schrödinger Equation: Self-focusing and Wave Collapse, Hardback, (1999).

H.-R. Sun and Q.-G. Zhang, Existence of solutions for a fractional boundary value problem via the Mountain Pass method and an iterative technique, Comput. Math. Appl., 64(2012), 3436-3443.

W. Sudsutad, B. Ahmad, S. K. Ntouyas *et al.*, Impulsively hybrid fractional quantum Langevin equation with boundary conditions involving Caputo q_k-fractional derivatives, Chaos, Solitons & Fractals, 91(2016), 47-62.

W. Sudsutad, S. K. Ntouyas and J. Tariboon, Systems of fractional Langevin equations of Riemann-Liouville and Hadamard types, Adv. Differ. Equ., 2015(2015), 1-24.

A. Szulkin and W. Zou, Homoclinic orbits for asymptotically linear Hamiltonian systems, J. Funct. Anal., 187(2001), 25-41.

C. L. Tang and X. P. Wu, Periodic solutions for second order systems with not uniformly coercive potential, J. Math. Anal. Appl., 259(2001), 386-397.

C. L. Tang and X. P. Wu, Some critical point theorems and their applications to periodic solution for second order Hamiltonian systems, J. Differential Equations., 248(2010), 660-692.

V. E. Tarasov, Fractional generalization of gradient and Hamiltonian systems, J. Phys. A: Math. Gen., 38(2005), 5929-5943.

V. E. Tarasov, Fractional Dynamics: Application of Fractional Calculus to Dynamics of Particles, Fields and Media, Springer, (2010).

V. E. Tarasov, Fractional dynamics: Application of fractional calculus to dynamics of particles, fields and media, Springer Science and Business Media, (2011).

V. E. Tarasov, Theoretical physics models with integro-differentiation of fractional order, IKI, RCD, (2011). (in Russian)

A. Zain and A. M. Tazali, Local existence theorems for ordinary differential equations of fractional order, Ordinary and Partial Differential Equations, Lecture Notes in Math., Springer, Dundee, 964(1982), 652-665.

C. Thaiprayoon, S. K. Ntouyas and J. Tariboon, On the nonlocal Katugampola fractional integral conditions for fractional Langevin equation, Adv. Differ. Equ., 2015(2015), 1-16.

Y. Tian and Z. Bai, Existence results for the three-point impulsive boundary value problem involving fractional differential equations, Comput. Math. Appl., 59(2010), 2601-2609.

Y. Tian and Y. Zhou, Positive solutions for multipoint boundary value problem of fractional differential equations, J. Appl. Math. Comput., 38(2012), 417-427.

S. P. Toropova, Relation of semigroups with singularity to integrated semigroups, Russian Math., 47(2003), 66-74.

C. Torres, Existence of solution for a class of fractional Hamiltonian systems, Electronic J. Differential Equations, 259(2013), 1-12.
T.-P. Tsai, Asymptotic dynamics of nonlinear Schrödinger equations with many bound states, J. Differential Equations, 192(1995), 225-282.
S. Umarov and R. Gorenflo, Cauchy and nonlocal multi-point problems for distributed order pseudo-differential equations, Z. Anal. Anwend., 24(2005), 449-466.
V. S. Vladimirov, Equations of Mathematical Physics, Nauka, Moscow, (1981).
W. Varnhorn, The Stokes Equations, Akademie Verlag, Berlin, (1994).
W. von Wahl, The Equations of Navier-Stokes and Abstract Parabolic Equations, Springer-Verlag, (2013).
W. von Wahl, Gebrochene potenzen eines elliptischen operators und parabolische differentialgleichungen in Räumen hölderstetiger Funktionen, Nachr. Akad. Wiss. Göttingen, Math.-Phys. Klasse, 11(1972), 231-258.
Y. Wang, Global existence and blow up of solutions for the inhomogeneous nonlinear Schrödinger equation in $\mathbb{R}^2$, J. Math. Anal. Appl., 338(2008), 1008-1019.
X. Wang, Impulsive boundary value problem for nonlinear differential equations of fractional order, Comput. Math. Appl., 62(2011), 2383-2391.
G. Wang, B. Ahmad and L. Zhang, Some existence results for impulsive nonlinear fractional differential equations with mixed boundary conditions, Comput. Math. Appl., 59(2010), 1389-1397.
G. Wang, B. Ahmad and L. Zhang, Impulsive anti-periodic boundary value problem for nonlinear differential equations of fractional order, Nonlinear Anal.: TMA, 74(2011), 792-804.
G. Wang, L. Zhang and G. Song, Systems of first order impulsive functional differential equations with deviating arguments and nonlinear boundary conditions, Nonlinear Anal.: TMA, 74(2011), 974-982.
R. N. Wang, D. H. Chen and T. J. Xiao, Abstract fractional Cauchy problems with almost sectorial operators, J. Differential Equations, 252(2012), 202-235.
J. R. Wang, X. W. Dong and Y. Zhou, Existence, attractive and stability of solutions for quadratic urysohon fractional integral equations, Commun. Nonlinear Sci. Numer. Simul., 17(2012), 545-554.
J. R. Wang, X. W. Dong and Y. Zhou, Analysis of nonlinear integral equations with Erdelyi-Kober fractional operator, Commun. Nonlinear. Sci. Numer. Simul., 17(2012), 3129-3139.
J. R. Wang, Z. Fan and Y. Zhou, Nonlocal controllability of semilinear dynamic systems with fractional derivative in Banach spaces, J. Optim. Theory Appl., 154(2012), 292-302.
J. R. Wang, M. Fečkan and Y. Zhou, On the new concept of solutions and existence results for impulsive fractional evolution equations, Dyn. Partial Differ. Equ., 8(2011), 345-361.
J. R. Wang, M. Fečkan and Y. Zhou, Ulam's type stability of impulsive ordinary differential equations, J. Math. Anal. Appl., 395(2012), 258-264.
J. R. Wang, M. Fečkan and Y. Zhou, Nonexistence of periodic solutions and asymptotically periodic solutions for fractional differential equations, Commun. Nonlinear Sci. Numer. Simul., 18(2013a), 246-256.
J. R. Wang, M. Fečkan and Y. Zhou, Relaxed controls for nonlinear fractional impulsive evolution equations, J. Optim. Theory Appl., 156(1)(2013b), 13-32.
J. R. Wang, M. Fečkan and Y. Zhou, Fractional order iterative functional differential equations with parameter, Appl. Math. Modelling, 37(2013c), 6055-6067.
J. R. Wang, M. Fečkan and Y. Zhou, Presentation of solutions of impulsive fractional

Langevin equations and existence results, Eur. Phys. J. Special Topics, 222(2013d), 1855-1872.

J. R. Wang, L. Lv and Y. Zhou, Ulam stability and data dependence for fractional differential equations with Caputo derivative, Electron. J. Qual. Theory Differ. Equ., 63(2011), 1-10.

J. R. Wang, L. Lv and Y. Zhou, Boundary value problems for fractional differential equations involving Caputo derivative in Banach spaces, J. Appl. Math. Comput., 38(2012a), 209-224.

J. R. Wang, L. Lv and Y. Zhou, New Concepts and results in stability of fractional differential equations, Commun. Nonlinear Sci. Numer. Simul., 17(2012b), 2530-2538.

J. R. Wang and W. Wei, A class of nonlocal impulsive problems for integrodifferential equations in Banach spaces, Results Math., 58(2010), 379-397.

J. R. Wang, W. Wei and Y. Zhou, Fractional finite time delay evolution systems and optimal controls in infinite dimensional spaces, J. Dyn. Control Syst., 17(2011), 515-535.

J. R. Wang, X. Xiang and Y. Peng, Periodic solutions of semilinear impulsive periodic system on Banach space, Nonlinear Anal.: TMA, 71(2009), e1344-e1353.

J. R. Wang and Y. Zhou, A class of fractional evolution equations and optimal controls, Nonlinear Anal.: RWA, 12(2011a), 262-272.

J. R. Wang and Y. Zhou, Study of an approximation process of time optimal control for fractional evolution systems in Banach spaces, Adv. Difference Equ., vol. 2011b, Article ID 385324, 15 pages.

J. R. Wang and Y. Zhou, Analysis of nonlinear fractional control systems in Banach spaces, Nonlinear Anal.: TMA, 74(2011c), 5929-5942.

J. R. Wang and Y. Zhou, Time optimal controls problem of a class of fractional distributed system, Int. J. Dyn. Syst. Differ. Equs., 3(2011d), 363-382.

J. R. Wang and Y. Zhou, Existence of mild solutions for fractional delay evolution systems, Appl. Math. Comput., 218(2011e), 357-367.

J. R. Wang and Y. Zhou, Existence and controllability results for fractional semilinear differential inclusions, Nonlinear Anal.: RWA, 12(2011f), 3642-3653.

J. R. Wang and Y. Zhou, Mittag-Leffer-Ulam stabilities of fractional evolution equations, Appl. Math. Lett., 25(2012a), 723-728.

J. R. Wang and Y. Zhou, Complete controllability of fractional evolution systems, Commun. Nonlinear Sci. Numer. Simul., 17(2012b), 4346-4355.

J. R. Wang, Y. Zhou and M. Fečkan, Alternative results and robustness for fractional evolution equations with periodic boundary conditions, Electron. J. Qual. Theory Differ. Equ., 97(2011), 1-15.

J. R. Wang, Y. Zhou and M. Fečkan, On recent developments in the theory of boundary value problems for impulsive fractional differential equations, Comput. Math. Appl., 64(2012a), 3008-3020.

J. R. Wang, Y. Zhou and M. Fečkan, Nonlinear impulsive problems for fractional differential equations and Ulam stability, Comput. Math. Appl., 64(2012b), 3389-3405.

J. R. Wang, Y. Zhou and M. Fečkan, Abstract Cauchy problem for fractional differential equations, Nonlinear Dynam., 71(2013), 685-700.

J. R. Wang, Y. Zhou and M. Fečkan, On the nonlocal Cauchy problem for semilinear fractional order evolution equations, Cent. Eur. J. Math., online, (2014).

J. R. Wang, Y. Zhou and M. Medved, On the solvability and optimal controls of fractional integrodifferential evolution systems with infinite delay, J. Optim. Theory Appl., 152(2012a), 31-50.

J. R. Wang, Y. Zhou and M. Medved, Picard and weakly Picard operators technique for nonlinear differential equations in Banach spaces, J. Math. Anal. Appl., 389(2012b), 261-274.

J. R. Wang, Y. Zhou and M. Medved, Qualitative analysis for nonlinear fractional differential equations via topological degree method, Topol. Methods Nonlinear Anal., 40(2012c), 245-271.

J. R. Wang, Y. Zhou and M. Medved, Existence and stability of fractional differential equations with Hadamard derivative, Topol. Methods Nonlinear Anal., 41(2013), 113-133.

J. R. Wang, Y. Zhou and W. Wei, A class of fractional delay nonlinear integrodifferential controlled systems in Banach spaces, Commun. Nonlinear Sci. Numer. Simul., 16(2011a), 4049-4059.

J. R. Wang, Y. Zhou and W. Wei, Impulsive problems fractional evolution equations and optimal controls in infinite dimensional spaces, Topol. Methods Nonlinear Anal., 38(2011b), 17-43.

J. R. Wang, Y. Zhou and W. Wei, Study in fractional differential equations by means of topological degree method, Numer. Funct. Anal. Optim., 33(2012a), 216-238.

J. R. Wang, Y. Zhou and W. Wei, Optimal feedback control for semilinear fractional evolution equations in Banach spaces, Systems Control Lett., 61(2012b), 472-476.

J. R. Wang, Y. Zhou and W. Wei, Stabilization of solutions to nonlinear impulsive evolution equations, Kybernetika, 48(2012c), 1211-1228.

J. R. Wang, Y. Zhou and W. Wei, Fractional Schrödinger equations with potential and optimal controls, Nonlinear Anal.: RWA, 13(2012d), 2755-2766.

J. R. Wang, Y. Zhou and W. Wei, Cauchy problems for fractional differential equations via Picard and weakly Picard operators technique, Fixed Point Theory, 14(2013), 219-234.

J. R. Wang, Y. Zhou and W. Wei *et al.*, Nonlocal problems for fractional integrodifferential equations via fractional operators and optimal controls, Comput. Math. Appl., 62(2011), 1427-1441.

W. Wei, S. H. Hou and K. L. Teo, On a class of strongly nonlinear impulsive differential equation with time delay, Nonlinear Dyn. Syst. Theory, 6(2006), 281-293.

W. Wei, X. Xiang and Y. Peng, Nonlinear impulsive integro-differential equation of mixed type and optimal controls, Optimization, 55(2006), 141-156.

F. B. Weissler, The Navier-Stokes initial value problem in L^p, Arch. Rational Mech. Anal., 74(1980), 219-230.

B. West, M. Bologna and P. Grigolini, Physics of Fractal Operators, Springer-Verlag, Berlin, (2003).

S. Westerlund, Dead matter has memory! Causal consulting, Kalmar, Sweden, (2002).

J. H. Wu, Unified treatment of local theory of NFDEs with infinite delay, Tamkang J. Math., 22(1991), 51-72.

T. J. Xiao and J. Liang, The Cauchy Problem for Higher Order Abstract Differential Equations, Lecture Notes in Math., vol. 1701, Springer, Berlin, New York, (1998).

X. Xu, J. Cheng and M. Yamamoto, Carleman estimate for a fractional diffusion equation with half order and application, Appl. Anal., 90(2011), 1355-1371.

A. Yagi, Abstract Parabolic Evolution Equations and their Applications, Springer-Verlag, Berlin Heidelberg, (2010).

T. Yang, Impulsive control theory, Springer, Berlin, (2001).

L. Yang and H. Chen, Nonlocal boundary value problem for impulsive differential equations of fractional order, Adv. Differ. Equ., 2011(2011), Article ID 404917, 16 pages.

H. Ye, J. Gao and Y. Ding, A generalized Gronwall inequality and its application to a fractional differential equation, J. Math. Anal. Appl., 328(2007), 1075-1081.

R. Yuan and Z. Zhang, Homoclinic solutions for a class of second order Hamiltonian systems, Results Math., 61(2012), 195-208.

R. Zacher, A De Giorgi-Nash type theorem for time fractional diffusion equations, Math. Ann., 356(2013), 99-146.

R. Zacher, Weak solutions of abstract evolutionary integro-differential equations in Hilbert spaces, Funkcial. Ekvac., 52(2009), 1-18.

G. M. Zaslavsky, Fractional kinetic equation for hamiltonian chaos, chaotic advection, tracer dynamics and turbulent dispersion, Phys. D, 76(1994), 110-122.

G. M. Zaslavsky, Chaos, fractional kinetics, and anomalous transport, Phys. Rep., 371(2002), 461-580.

G. M. Zaslavsky, D. Stevens and H. Weitzner, Self-similar transport in incomplete chaos, Phys. Rev. E., 48(1993), 1683-1694.

E. Zeidler, Nonlinear Functional Analysis and its Applications, vol. II A and B, Springer-Verlag, New York, (1990).

V. C. Zelati, I. Ekeland and E. Séré, A variational approach to homoclinic orbits in Hamiltonian systems, Math. Ann., 288(1990), 133-160.

V. C. Zelati and P. H. Rabinowitz, Homoclinic orbits for second order Hamiltonian systems possessing superquadratic potentials, J. Amer. Math. Soc., 4(1991), 693-727.

S. Zhang, Existence of positive solution for some class of nonlinear fractional differential equations, J. Math. Anal. Appl., 278(2003), 136-148.

S. Zhang, Existence of solution for a boundary value problem of fractional order, Acta Mathematica Scientia, 26(2006), 220-228.

S. Zhang, Positive solutions to singular boundary value problem for nonlinear fractional differential equation, Comput. Math. Appl., 59(2010), 1300-1309.

M. Zheng, F. Liu, I. Turner *et al.*, A novel high order space-time spectral method for the time fractional Fokker-Planck equation, SIAM J. Sci. Comput., 37(2015), A701-A724.

M. Zhou, B. Ahmad and Y. Zhou, Existence of attractive solutions for Hilfer fractional evolution equations with almost sectorial operators, Symmetry, 14(2022), 392.

M. Zhou, C.F. Li and Y. Zhou, Existence of mild solutions for Hilfer fractional evolution equations with almost sectorial operators, Axioms, 11(2022), 144.

Y. Zhou, Existence and uniqueness of fractional functional differential equations with unbounded delay, Int. J. Dyn. Syst. Diff. Equs., 1(4)(2008), 239-244.

Y. Zhou, Existence and uniqueness of solutions for a system of fractional differential equations, Fract. Calc. Appl. Anal., 12(2009), 195-204.

Y. Zhou, Basic Theory of Fractional Differential Equations, World Scientific (1st ed.), Singapore, (2014).

Y. Zhou, Fractional Evolution Equations and Inclusions: Analysis and Control, Academic Press, London, (2016).

Y. Zhou, Attractivity for fractional evolution equations with almost sectorial operators, Frac. Calc. Appl. Anal., 21(3)(2018a), 786-800.

Y. Zhou, Attractivity for fractional differential equations, Applied Math. Lett., 75(2018b), 1-6.

Y. Zhou, Infinite interval problems for fractional evolution equations, Mathematics, 10(2022), 900.

Y. Zhou, B. Ahmad, F. Chen and A. Alsaedi, Oscillation for fractional partial differential equations, Bull. Malays. Math. Sci. Soc., 42(2)(2019), 449-465.

Y. Zhou, B. Ahmad and A. Alsaedi, Existence of nonoscillatory solutions for fractional functional differential equations, Bull. Malays. Math. Sci. Soc., 42(2)(2019a), 751-766.

Y. Zhou, B. Ahmad and A. Alsaedi, Necessary and sufficient conditions for oscillation of second-order dynamic equations on time scales, Math. Meth. Appl. Sci., 42(2019b), 4488-4497.

Y. Zhou, A. Alsaedi, B. Ahmad, Oscillation for Fractional Neutral Functional Differential Systems, J. Comput. Anal. Appl., 25(2018), 965-974.

Y. Zhou, B. Ahmad and A. Alsaedi, Existence of nonoscillatory solutions for fractional neutral differential equations, Applied Math. Lett., 72(2017), 70-74.

Y. Zhou and J. W. He, Well-posedness and regularity for fractional damped wave equations, Monatsh. Math., 194(2) (2021), 425-458.

Y. Zhou and J. W. He, New results on controllability of fractional evolution systems with order $\alpha \in (1,2)$, Evol. Equ. Control Theory, 10(2021), 491-509.

Y. Zhou and J. W. He, A Cauchy problem for fractional evolution equations with Hilfer's fractional derivative on semi-infinite interval, Frac. Calc. Appl. Anal., 25(2022), 924-961.

Y. Zhou, J. W. He, B. Ahmad and N. H. Tuan, Existence and regularity results of a backward problem for fractional diffusion equations, Math. Meth. Appl. Sci., 42(2019), 6775-6790.

Y. Zhou and F. Jiao, Existence of extremal solutions for discontinuous fractional functional differential equations, Int. J. Dyn. Syst. Diff. Equs., 2(2009), 237-252.

Y. Zhou and F. Jiao, Nonlocal Cauchy problem for fractional evolution equations, Nonlinear Anal.: RWA, 11(2010a), 4465-4475.

Y. Zhou and F. Jiao, Existence of mild solutions for fractional neutral evolution equations, Comput. Math. Appl., 59(2010b), 1063-1077.

Y. Zhou, F. Jiao and J. Li, Existence and uniqueness for p-type fractional neutral differential equations, Nonlinear Anal.: TMA, 71(2009a), 2724-2733.

Y. Zhou, F. Jiao and J. Li, Existence and uniqueness for fractional neutral differential equations with infinite delay, Nonlinear Anal.: TMA, 71(2009b), 3249-3256.

Y. Zhou, F. Jiao and J. Pečarić, Abstract Cauchy problem for fractional functional differential equations in Banach spaces, Topol. Methods Nonlinear Anal., 42(2013), 119-136.

Y. Zhou, J. Manimaran, L. Shangerganesh and A. Debbouche, Weakness and Mittag-Leffler Stability of Solutions for Time-Fractional Keller-Segel Models, Int. J. Nonlinear Sci. Numer. Simul., 19(7-8)(2018), 753-761.

Y. Zhou and L. Peng, Topological Structure of Solution Sets for Semilinear Evolution Inclusions, Z. Anal. Anwend., 37(2018), 189-207.

Y. Zhou and L. Peng, Topological properties of solutions sets for partial functional evolution inclusions, C. R. Math. Acad. Sci. Paris, 355(2017a), 45-64.

Y. Zhou and L. Peng, On the time-fractional Navier-Stokes equations, Comput. Math. Appl., 73(6)(2017b), 874-891.

Y. Zhou and L. Peng, Weak solutions of the time-fractional Navier-Stokes equations and optimal control, Comput. Math. Appl., 73(6)(2017c), 1016-1027.

Y. Zhou, L. Peng and B. Ahmad, Topological properties of solution sets for stochastic evolution inclusions, Stoch. Anal. Appl., 36(2018), 114-137.

Y. Zhou, L. Peng, B. Ahmad and A. Alsaedi, Energy Methods for the Time-Fractional Navier-Stokes Equations, Chaos, Solitons Fractals, 102(2017), 78-85.

Y. Zhou L. Peng and Y. Q. Huang, Duhamel's formula for time-fractional Schrödinger equations, Math. Meth. Appl. Sci., 41(2018a), 8345-8349.

Y. Zhou L. Peng and Y. Q. Huang, Existence and Hölder continuity of solutions for time-fractional Navier-Stokes equations, Math. Meth. Appl. Sci., 41(2018b), 7830-7838.

Y. Zhou, X. H. Shen and L. Zhang, Cauchy problem for fractional evolution equations with Caputo derivative, Eur. Phys. J. Special Topics, 222(2013), 1747-1764.

Y. Zhou, S. Suganya and M. M. Arjunan, Existence and controllability for impulsive evolution inclusions without compactness, J. Dyn. Control Syst., 24(2018), 297-311.

Y. Zhou, S. Suganya and B. Ahmad, Approximate controllability of impulsive fractional integro-differential equation with state-dependent delay in Hilbert spaces, IMA J. Math. Control Inform., 36(2019), 603-622.

Y. Zhou, L. Shangerganesh, J. Manimaran and A. Debbouche, A class of time-fractional reaction-diffusion equation with nonlocal boundary condition, Math. Meth. Appl. Sci., 41(2018), 2987-2999.

Y. Zhou, Y. Tian and Y.-Y. He, Floquet boundary value problem of fractional functional differential equations, Electron. J. Qual. Theory Differ. Equ., 50(2010), 1-13.

Y. Zhou and J. N. Wang, The nonlinear Rayleigh-Stokes problem with Riemann-Liouville fractional derivative, Math. Meth. Appl. Sci., 44(2021), 2431-2438.

Y. Zhou, V. Vijayakumar and R. Murugesu, Controllability for fractional evolution inclusions without compactness, Evol. Equ. Control The., 4(2015), 507-524.

Y. Zhou, V. Vijayakumar, C. Ravichandran and R. Murugesu, Controllability results for fractional order neutral functional differential inclusions with infinite delay, Fixed Point Theory, 18(2017), 773-798.

Y. Zhou, L. Zhang and X. H. Shen, Existence of mild solutions for fractional evolutions, J. Integr. Equs. Appl., 25(2013), 557-586.

L. P. Zhu and G. Li, Nonlocal differential equations with multivalued perturbations in Banach spaces, Nonlinear Anal.: TMA, 69(2008), 2843-2850.

A. Zoia, M.-C. Néel and A. Cortis, Continuous-time random-walk model of transport in variably saturated heterogeneous porous media, Phys. Rev. E, 81(2010), 031104.

A. Zoia, M.-C. Néel and M. Joelson, Mass transport subject to time-dependent flow with nonuniform sorption in porous media, Phys. Rev. E, 80(2009), 056301.

Index

p-function, 44, 45, 59

Arzela-Ascoli theorem, 3, 43, 53, 67, 72, 76, 135, 149, 161, 162, 175, 176, 195, 222, 237, 238, 284, 293, 303, 355, 435
  PC-type, 3, 260, 268
asymptotic expansion, 12, 205
atomic, 45, 46, 48, 56, 61–63, 67

Banach contraction mapping principle, 19, 53, 67, 68, 140, 217, 258, 267, 283, 290
Beta function, 217, 273–275, 395, 396, 398
blow-up alternative, 425
Bochner
  integrable, 3, 16, 124
  integral, 197
  theorem, 3, 124
bounded
  norm-bounded, 20
  totally, 21, 41, 43
  uniformly, 3, 34, 42, 53, 154, 162, 169, 174–176

Caputo
  fractional derivative, 4, 6, 7, 9, 30, 44, 60, 73, 121, 123, 132, 136, 149, 168, 187, 204, 224, 225, 254, 256, 262, 278, 285, 300, 324, 389, 390
  fractional derivative operator, 311, 388
Carathéodory, 36
  $L^{\frac{1}{\beta}}$-Carathéodory, 30, 36, 37
  condition, 121, 123, 181
  function, 36
Cauchy sequence, 1
Chandrabhan, 36
  $L^{\frac{1}{\gamma}}$-Chandrabhan, 30, 36, 37
  function, 36
compact
  mapping, 176
  operator, 87, 154, 170, 222, 292–294, 326
compactness, 134, 161, 162, 175, 176, 211, 219, 268, 292, 340
condition
  $L^{\frac{1}{\beta}}$-Carathéodory, 30
  $L^{\frac{1}{\delta}}$-Lipschitz, 30
  $L^{\frac{1}{\gamma}}$-Chandrabhan, 30
  $(PS)_c$, 21, 23
  $(PS)_c^*$, 23
  Ambrosetti-Rabinowitz, 310, 315
  boundary, 12, 297
  Carathéodory, 121, 123, 181
  Dirichlet type boundary, 323, 389
  initial, 150, 151, 180
  Neumann boundary, 201, 202, 225
  nonlocal, 150, 151, 179, 180
  Palais-Smale (or (PS)), 21, 22, 312, 315, 349
cone, 20, 21, 35, 43
continuation, 425, 457, 461
continuous
  absolutely, 36, 39, 40, 61, 123, 125, 126, 131, 152, 322, 431
  completely, 19–21, 31, 33–35, 41, 51, 53, 65, 67, 75, 76, 86, 87, 164, 183, 185, 260, 261, 284
  equi-uniformly, 302, 303
  Hölder, 74, 79, 200, 213, 223, 389, 390, 403, 404, 406, 408, 432

Lipschitz, 225, 226
strongly, 24, 154, 170, 212, 326
uniformly, 24, 70, 207
uniformly Lipschtiz, 261
critical point, 23, 24, 298
theorem, 21, 298, 347
theory, 295, 297, 298, 359, 379

degree
Brouwer, 132
coincidence, 132
Leray-Schauder, 132
topological, 121
delay
bounded, 29, 30, 45
infinite, 60
unbounded, 30
Dirichlet integral, 443
Dunford-Riesz integral, 26

embedding
compact, 434, 435
continuous, 434
energy estimate, 452, 453, 455, 462, 467
equation
differential, 132
Euler, 347
Euler-Lagrange, 297, 387, 431, 443
evolution, 149
functional differential, 29, 45, 73, 84, 88, 108
integral, 122, 151, 162, 168, 176, 211, 256, 264, 271, 286, 288, 297, 353, 354
Langevin, 270, 271
Navier-Stokes, 387–389
neutral functional differential, 45, 60
partial differential, 132, 150, 202, 295, 297
equicontinuity, 149, 292
equicontinuous, 3, 16, 31, 34, 35, 42, 53, 67, 72, 100, 127, 129, 134, 154–156, 158–161, 165, 166, 170–175, 177, 178, 194, 195, 198, 220, 228, 229, 231, 233–235, 240, 244, 245, 247, 248, 250, 260, 268, 293, 355
function, 3, 42, 43
set, 259, 284
uniformly, 440
existence, 17, 18, 21, 29, 30, 33, 46, 48, 53, 62, 63, 67, 73, 74, 76, 136, 149–151, 168, 179, 181, 183, 197, 204, 205, 211, 218, 236, 253, 254, 257, 262, 265, 267, 270, 271, 278, 279, 284–286, 288, 295, 297, 298, 303, 310, 317, 322, 323, 329, 331, 343, 344, 346, 347, 351, 353, 355, 357, 365, 379, 387–389, 392, 399, 411, 415, 423, 425, 430, 433, 439, 440, 442, 445, 446, 452, 453, 456, 467, 470

Fatou lemma, 321
Fick's law, 296
fixed point theorem
Darbo, 149
Darbo-Sadovskii, 20
Hybrid, 21, 30
Krasnoselskii, 20, 35, 48, 62, 67, 149
Leray-Schauder, 187
O'Regan, 20, 183
Schaefer, 19, 181, 259, 260, 294
Schauder, 19, 73, 76, 86, 149, 195, 196, 293
Fountain theorem, 22
Dual, 23
Fourier transform, 297, 305
Fréchet
derivative, 21, 46, 48
differentiable, 21
fractional
p-type neutral differential equation, 29
p-type neutral functional differential equation, 44
abstract differential equation, 121, 149
abstract evolution equation, 149
advection dispersion equation, 295, 296, 343, 344
boundary value problem, 21, 295
calculus of variation, 297, 298
Cauchy problem, 136–138, 140, 141, 143, 226, 254
derivative, 4, 150, 270, 297, 304, 343, 344, 431
derivative operator, 9, 10, 297, 347
differential equation, 29, 121, 136, 262, 271, 278, 295, 297
differential operator, 295, 297
diffusion equation, 150, 224, 273, 444, 445, 452
evolution equation, 149–151, 168, 179, 180, 226

functional differential equation, 29, 30, 43, 123
Hamiltonian system, 357, 379
impulsive boundary value equation, 253
impulsive differential equation, 253, 254, 262
impulsive evolution equation, 254, 285
impulsive initial value equation, 253
impulsive Langevin equation, 254, 270, 275
initial-boundary value problem, 225
integral equation, 132, 133, 137, 141–144, 254, 255, 271
integral operator, 9, 10, 297
iterative functional differential equation, 30, 74
neutral differential equation, 29, 30
neutral functional differential equation, 60
optimal control, 197
ordinary differential equation, 121, 273
partial differential equation, 119, 164, 251, 387
Rayleigh-Stokes equation, 411
Rayleigh-Stokes problem, 411
variational principle, 297
variational problem, 297
wave equation, 387, 451–453
Fredholm alternative theorem, 180
Fubini Theorem, 206

Gâteaux
derivative, 24
differentiable, 24, 437, 438
Galerkin approximation, 297
Green's formula, 115, 118
Gronwall inequality, 187, 196, 199, 223, 264, 403, 466, 472
singular, 253

Hölder
continuous, 74, 79, 200, 213, 223, 389, 390, 403, 404, 406, 408, 432
inequality, 2, 31, 37, 47, 50, 62, 64, 77, 78, 124, 127, 128, 138, 139, 163, 257, 261, 265, 267, 282, 283, 302, 340, 344, 345, 433, 464
Helmholtz projector, 389
Hilfer
fractional derivative, 227
Hilfer fractional derivative, 237
Hodge projection, 390
homoclinic solution, 357, 379, 382, 384

inequality
Gronwall, 223, 264, 403
Hölder, 2, 31, 37, 47, 50, 62, 64, 77, 78, 124, 127, 128, 138, 139, 163, 257, 261, 265, 267, 282, 283, 302, 340, 344, 345, 433
Poincaré, 15
singular Gronwall, 253
Young, 305, 464
infinite interval, 237, 238, 251
infinitesimal generator, 24, 150, 154, 168, 169, 288
iteration method, 389

Lagrange
mean value theorem, 274
Lagrangian functional, 437
Laplace
operator, 202, 203, 225
transform, 153, 154, 168, 169, 211, 212, 287
Laplacian operator, 201, 225, 412
Lebesgue
dominated convergence theorem, 3, 41, 73, 127, 134, 139, 158, 159, 172, 173, 208, 220, 242, 246, 248, 291, 310, 347, 364, 393, 395, 396
integrable, 3, 31, 37, 38, 40, 47, 62, 124, 126, 197
measurable, 31, 35, 36, 46, 47, 62, 242
space, 412
Liouville-Weyl
fractional derivative, 8, 305, 358, 379
fractional integral, 8
fractional integral operator, 296
Lipschitz, 387, 411, 415, 419, 423, 425
$L^{\frac{1}{\delta}}$-Lipschitz, 30, 37

mean value theorem, 405, 409
measurable function, 3, 37, 38, 47, 67, 124
measure of noncompactness, 15, 121, 123, 129
Hausdorff, 15, 16, 149, 151, 168
Kuratowski, 15, 16, 149, 235, 249
Mittag-Leffler function, 12, 204, 205, 254, 256, 270, 272, 273, 390, 453

multiplicity, 357, 379, 382

Nemitskiĭ operator, 225, 226
Neumann theorem, 180
nonlinear alternative of Leray-Schauder type, 20, 261
nonlocal term, 132
norm
  Bielecki, 136
  Chebyshev, 75, 136
  Euclidean, 431
  graph, 27, 218
  operator, 200
  supremum, 48

optimal control, 187
oscillation, 30, 88, 90, 93–95, 109, 112, 116

Peano theorem, 121
Picard operator, 18, 19, 121, 136, 140, 141
  weak, 18, 19, 136, 143, 144
Pompeiu-Hausdorff functional, 18, 19, 145–147
priori
  bound, 184, 260
  estimate, 18, 132
probability density function, 187
problem
  bifurcation, 132
  boundary value, 132, 253, 262, 295
  Cauchy, 121, 123, 132, 136, 149–151, 154, 160–162, 164, 165, 167–169, 174–179, 200, 204, 211, 216, 226, 227, 253, 254, 285, 286
  Dirichlet, 116
  impulsive, 255, 262
  initial value, 29, 30, 60, 121, 123
  initial-boundary value, 411
  nonlocal, 132, 133, 135, 149, 179, 253, 254, 288

regularity, 314, 350, 387–389, 403, 411, 415, 417, 437, 443, 445, 450–453, 455, 467
relative compactness, 149, 162, 176
relatively compact, 2, 3, 15, 19, 42, 43, 58, 59, 130, 131, 134, 135, 161, 162, 167, 174–176, 268, 303
resolvent operator, 28, 200, 204, 205
Riemann-Liouville
  fractional derivative, 4–6, 9, 12, 88–90, 108, 109, 114, 125, 149–151, 343, 431
  fractional integral, 5, 6, 150, 227, 295, 322, 334, 453
  fractional integral operator, 296, 298, 432
  fractional partial derivative, 164
  integral, 237
Riesz representation theorem, 46

sectorial operator, 27, 200, 202, 205
  almost, 25, 149, 200, 203, 204, 226, 227
semigroup, 27, 28
  C_0-semigroup, 24, 149–151, 154, 160, 164, 168, 169, 174, 179, 237, 285
  analytic, 27, 187, 201, 390
  integrated, 226
  operator, 187
  property, 10, 170
Sobolev embeddings, 413, 423
solution, 29–33, 36, 40, 45–48, 53, 55, 57, 58, 60–63, 69, 70, 72–77, 80, 82–85, 121–123, 126, 127, 131–133, 135–138, 141–144, 156, 162, 165, 171, 176, 179, 181, 183–186, 204, 205, 212, 213, 218, 220, 253–258, 260–262, 264, 265, 268–272, 275, 278, 279, 284, 286, 295–298, 303, 304, 306, 308, 310, 319, 322, 324, 325, 328–334, 338, 343, 344, 346, 347, 352, 353, 355, 389, 403, 431
  PC-mild, 254, 286, 288, 290, 293, 294
  α-mild, 187, 188
  approximation, 453, 462
  classical, 150, 204, 205, 211, 213, 216, 223, 226, 297, 389, 404, 406, 410
  eventually positive, 113, 114, 117, 118
  extremal, 29, 30, 35
  homoclinic, 357, 358, 365, 367, 369–372, 377, 379, 384
  lower, 36, 37, 44
  maximal, 40, 43, 44
  mild, 149–152, 154, 156, 160–162, 164, 167–169, 171, 174–181, 183, 186, 204, 205, 211–213, 216–219, 222, 223, 225, 226, 230, 236, 237, 243, 286, 287, 387–389, 391, 392, 399, 400, 403, 406, 408, 411
  minimal, 40, 43, 44
  nonoscillatory, 30, 88, 98, 108

operator, 73, 149
positive, 84, 264, 269
upper, 36, 37, 44
weak, 314, 344, 346–348, 350–353, 387, 432, 433, 439, 442, 443, 452, 467
space
Banach, 1–3, 7, 12, 15, 19–24, 27, 48, 61, 63, 76, 86, 87, 121, 123, 132, 136, 150, 154, 168, 169, 179, 196, 200, 203, 218, 227, 230, 285, 300, 301, 315, 324, 334, 390, 392, 434, 435
Cartesian product, 300
complete metric, 19, 76, 137, 217, 394
divergence-free function, 389
fractional derivative, 297, 298, 300
fractional power, 390
fractional Sobolev, 436
Hilbert, 15, 298, 304, 334, 344
Lebesgue, 200
metric, 18, 19, 216
of p-integrable functions, 431
of absolutely continuous functions, 2
of continuous and bounded functions, 60
of continuous functions, 2, 30, 44, 61, 123, 132, 136, 150, 200, 431
of distributions, 14
of linear bounded operators, 24
of measurable functions, 2, 61, 143
phase, 60
Sobolev, 297, 358, 390
Sobolev-Slobodeckij, 412
Stokes operator, 389, 390
successive approximation, 136, 138, 141
symmetric matrix, 358, 379

topological degree, 17, 121
transform
Fourier, 297, 305
Laplace, 153, 154, 168, 169, 211, 212, 238, 287, 413, 452

uniform operator topology, 155, 170, 206–208, 229, 240, 391
uniformly Lipschtiz continuous, 261
uniqueness, 29, 30, 46, 53, 55, 62, 67, 73, 79, 86, 136, 150, 204, 211, 212, 218, 236, 253, 254, 257, 261, 262, 265, 269, 270, 278, 279, 286, 288, 303, 388, 389, 392, 415, 435, 452, 467

variational
functional, 297, 298, 344
method, 295, 297, 334, 433, 440, 443
structure, 297, 298, 300, 303, 306, 344–346, 433, 437
variational methods, 357, 359, 379

weak solution, 387, 439, 442, 443, 451, 452, 455, 467, 470
well-posedness, 387, 451–453, 467
Wright function, 13, 390

Young
inequality, 305
theorem, 298

www.ingramcontent.com/pod-product-compliance
Ingram Content Group UK Ltd.
Pitfield, Milton Keynes, MK11 3LW, UK
UKHW051044150926
8624IPUK00006B/394